Operator Theory: Advances and Applications

Vol. 190

Modern Analysis and Applications

The Mark Krein Centenary Conference
Volume 1: Operator Theory and Related Topics

Vadim Adamyan
Yurij Berezansky
Israel Gohberg
Myroslav Gorbachuk
Valentyna Gorbachuk
Anatoly Kochubei
Heinz Langer
Gennadiy Popov
Editors

Birkhäuser
Basel · Boston · Berlin

Editors:

Vadim M. Adamyan
Department of Theoretical Physics
Odessa National I.I. Mechnikov University
Dvoryanska st. 2
65026 Odessa, Ukraine
e-mail: vadamyan@paco.net

Yurij Berezansky
Institute of Mathematics
Ukrainian National Academy of Sciences
Tereshchenkivska st.
01601 Kyiv-4, Ukraine
e-mail: berezan@mathber.carrier.kiev.ua

Israel Gohberg
Department of Mathematical Sciences
Raymond and Beverly Sackler
Faculty of Exact Sciences
Tel Aviv University
69978 Ramat Aviv, Israel
e-mail: gohberg@post.tau.ac.il

Myroslav Gorbachuk
Valentyna Gorbachuk
Insitute of Mathematics
Ukrainian National Academy of Sciences
Tereshchenkivska st.
01601 Kyiv-4, Ukraine
e-mail: imath@horbach.kiev.ua

Anatoly Kochubei
Insitute of Mathematics
Ukrainian National Academy of Sciences
Tereshchenkivska st.
01601 Kyiv-4, Ukraine
e-mail: kochubei@i.com.ua

Heinz Langer
Institute of Analysis and
Scientific Computing
Technical University of Vienna
Wiedner Hauptstrasse 8–10
1040 Vienna, Austria
e-mail: hlanger@mail.zserv.tuwien.ac.at

Gennadiy Popov
Department of Mathematical Physics
Odessa National I.I. Mechnikov University
Dvoryanska st. 2
65026 Odessa, Ukraine
e-mail: popov@onu.edu.ua

2000 Mathematical Subject Classification: 47; 22E, 30E, 32A, 33C, 35P, 35Q, 41A, 42A, 46C, 46F, 49K, 65N, 60G, 60H, 60J; 01-99

Library of Congress Control Number: 2009925156

Bibliographic information published by Die Deutsche Bibliothek.
Die Deutsche Bibliothek lists this publication in the Deutsche Nationalbibliografie; detailed bibliographic data is available in the Internet at http://dnb.ddb.de

ISBN 978-3-7643-9918-4 Birkhäuser Verlag AG, Basel - Boston - Berlin

Basel · Boston · Berlin
P.O. Box 133, CH-4010 Basel, Switzerland
Part of Springer Science+Business Media
Printed on acid-free paper produced from chlorine-free pulp. TCF∞
Printed in Germany
ISBN 978-3-7643-9918-4 (Vol. 1/OT 190) e-ISBN 978-3-7643-9919-1
ISBN 978-3-7643-9920-7 (Vol. 2/OT 191) e-ISBN 978-3-7643-9921-4
ISBN 978-3-7643-9924-5 (Set)

9 8 7 6 5 4 3 2 1

www.birkhauser.ch

Contents

Preface

This is the first of two volumes containing peer-reviewed research and survey papers based on invited talks at the International Conference on Modern Analysis and Applications. The conference, which was dedicated to the 100th anniversary of the birth of Mark Krein, one of the greatest mathematicians of the 20th century, was held in Odessa, Ukraine, on April 9–14, 2007. The conference focused on the main ideas, methods, results, and achievements of M.G. Krein.

This first volume is devoted to the operator theory and related topics. It opens with the biography papers about M.G. Krein and a number of survey papers about his work. The main part of the book consists of original research papers presenting the state of the art in operator theory and its application.

The second volume of these proceedings, entitled Differential Operators and Mechanics, concerns other aspects of the conference. The two volumes will be of interest to a wide-range of readership in pure and applied mathematics, physics and engineering sciences.

The editors are sincerely grateful to the persons who contributed to the preparation of these proceedings: Sergei Marchenko, Myroslav Sushko, Kostyantyn Yusenko and Vladimir Zavalnyuk.

Mark Grigorievich Krein, 1907–1989

Operator Theory:
Advances and Applications, Vol. 190, xi–xx

Mark Grigorievich Krein (on his 100th birthday anniversary)

V.M. Adamyan, D.Z. Arov, Yu.M. Berezansky, V.I. Gorbachuk, M.L. Gorbachuk, V.A. Mikhailets and A.M. Samoilenko

April 3, 2007, is the 100th anniversary of the birth of Mark Grigorievich Krein, one of the most celebrated mathematicians of the 20th century, whose whole life was closely connected with Ukraine. He was born in Kyïv to a family of modest income and with seven children. His father had a timber selling business which, after the 1917 Revolution, he had to leave.

Mark Grigorievich first exhibited his extraordinary mathematical talents during his teens. Beginning at the age of 14, he started to systematically attend lectures of D.O. Grave, scientific seminars of Grave and B.M. Delone in Kyïv University, and lectures of Delone at Kyïv Polytechnic Institute. At the age of 17, influenced by the autobiographical work of M. Gorky "My Universities", he decided that it was time to start his own "universities". Together with a friend, he went to Odessa to join one of its many circus troupes, for he had a dream of becoming an acrobat.

However, fate had its way and produced for the world, in the person of M.G. Krein, not an acrobat but a prominent mathematician whose influence on the development of mathematics can not be overestimated. The acrobat's job for which he applied had already been filled. While waiting for a new opening, he met N.G. Chebotarev, a famous algebraist and a wonderful person, to whom he had a recommendation letter from D.O. Grave. At that time, Chebotarev was conducting research at Odessa University. Sensing through their conversations that this young person had mathematical gifts, he convinced Krein to abandon his dreams of performing in a circus and proposed that he pursue a PHD in a school of mathematics. Together with a colleague, S.I. Shatunovski, he succeeded in getting the young man admitted to a Doctorate Program, even though Krein was then only 19 and did not have even a high school diploma, to say nothing about a university degree. In this way, he became a PhD student at Odessa University with Chebotarev's guidance. Since then, M.G. Krein always had a photograph of N.G.

Chebotarev over his desk. In his "Mathematical Autobiography", Chebotarev, talking about the 17 year old Krein, recalled that he "without having graduated from a high school, had brought to him a personal work with a very distinguished content". Chebotarev was very proud of his first student and regarded him as "one of the best mathematicians in Ukraine".

A well-known specialist in mechanics, G.K. Suslov, also became interested in working with Krein. Together with F.R. Gantmacher, Krein attended Suslov's seminar at the Odessa Polytechnic Institute. Krein's later interests were formed under the direct influence of Chebotarev and Suslov. Chebotarev instilled in him a love for algebraic techniques and for algebra in general, an interest in various problems in the theory of functions, in particular, the problem of distribution of zeros for some classes of functions, interpolation and extension theory. At the same time Krein's interest in Mechanics came from working with G.K. Suslov and this interest can be traced in many of his mathematical works.

In 1928, Chebotarev moved to Kazan' and became a Professor of Kazan' University. A year after that, Krein finished his Doctoral studies and started to teach at the Donetsk Mining Institute, where he had been working for two years. At the time, he was already married; in 1927 he had happily married Raisa L'vovna Romen, a faithful friend who was his assistant for sixty years. She specialized in ship architecture, having graduated from the Odessa Marine Institute. Their only child, a daughter Irma, obtained a PhD degree in Philosophy, became a specialist in Cybernetics, and founded a new branch of the discipline called "Humanitarian Cybernetics". Their only grandson Alexey, who graduated from the Department of Mathematics in Odessa University and worked in the theory of systems, died at a very early age from a blood illness, which greatly influenced the health of both Krein and his wife, with whom he lived all his short life. The only great grandson they had, also a mathematician, now lives outside of Ukraine.

In 1931, Krein returned to Odessa for an appointment as Professor at Odessa University. He worked together with B.Ya. Levin, with whom he had friendly relations starting with their first meeting and until his last days. In 1934, at the age of 31, Krein obtained a Doctoral degree in Physics and Mathematics from Moscow University without submitting a thesis, and shortly after that, in 1939, he was elected a corresponding member of the Academy of Sciences of the Ukrainian Soviet Socialist Republic.

The early flourishing of Krein's talent as a scientist was accompanied by an equally early recognition of his pedagogical talents. At the age of 25, he started at Odessa University, a scientific seminar that soon became one of the leading centers for research in functional analysis, a young branch of mathematics at the time that became the main area of his research. During this period, M.G. Krein's mathematical interests included oscillation matrices and kernels, geometry of Banach spaces, the Nevanlinna–Pick interpolation problem, extension of positive definite functions and their applications. His first students were A.V. Artemenko, M.S. Livshic, D.P. Mil'man, M.A. Naimark, M.A. Rutman, S.A. Orlov, V.L. Shmul'yan. Works of these mathematicians became an indispensable part of modern mathematics.

At the same time, Krein also worked in the Research Institute at Khar'kov University (1934–1940), and during the periods 1940–1941 and 1944–952, he was the Head of the Department of Algebra and Functional Analysis at the Institute of Mathematics of the Academy of Sciences of the Ukrainian SSR (one of the researchers working here during the period of 1940–1941 was the great S. Banach with whom Krein had maintained scientific contacts since his trip to L'vov in 1940). Many of the results he obtained at that time, as well as those obtained with his students, friends and colleagues, including N.I. Akhiezer and F.R. Gantmacher, have now become classic and can be found in major monographs and textbooks on functional analysis.

During World War II, Krein headed the Chair of Theoretical Mechanics at Kuibyshev Industrial Institute (Russia). He had preferred working there as opposed to the Chair of Mathematics, since for a technical school, it deals with more scientific directions and gives wider possibilities.

In 1944, he returned to Odessa and never left it again. He loved this city, knew the history of its streets, and was fond of the local "Odessa language" and Odessa jokes; he often visited the Odessa Operetta. However, very soon Krein was dismissed from Odessa University along with his closest friend B.Ya. Levin. These dismissals were a consequence of the anti-Semitic politics of the Stalin regime and the corruption of the University administration. A principal scientific attitude of these scientists, their opposition to pushing through illiterate doctoral theses, was regarded as an indication of Zionism. The official dismissal was given to him on the day of his fortieth year anniversary as a "present" from his director who favored "more reliable" staff, taking into account the political situation in the country and the "new personnel policy" in the 1940s, which was conducted under the thesis of fighting Zionism and Cosmopolitanism. This meant an end of the center of functional analysis in Odessa University, and an end of the official scientific career of M.G. Krein.

In 1944–1954, Krein worked in the Department of Theoretical Mechanics in the Odessa Marine Institute. Regardless of the difficulties pertaining to these times, he initiated a number of important new directions in mathematics and mechanics and became a world famous scientist. Together with the theoretical value of his results, their applied importance had also increased, especially those related to parametric resonance. V. Veksler, a renowned physicist, has remarked that "without works of M.G. Krein, we would not have a synchrophasotron". In a popular book of N. Wiener, the father of cybernetics, "I Am a Mathematician", the name of M.G. Krein is associated with the name of A.M. Kolmogorov, which was a way to acknowledge the value of their researches, published in the Proceedings of the Academy of Sciences of the USSR, in prediction and control theories during and shortly after the war. New official and unofficial students of M.G. Krein are the renowned mathematicians and physicists I.Ts. Gohberg, I.S. Iokhvidov, I.S. Kats, A.A. Kostyukov, G.Ya. Lyubarskii, A.A. Nudel'man, G.Ya. Popov, V.G. Sizov, and Yu.L. Shmul'yan.

In 1952, Krein again suffered dismissal, this time from the Institute of Mathematics of the Academy of Sciences of the Ukrainian SSR, where he had also founded a school of functional analysis, representatives of which are Yu.M. Berezansky, Yu.L. Daletsky, G.I. Kats, M.A. Krasnosel'skii, B.I. Korenblyum, and S.G. Krein. The official explanation was that he lived in Odessa but not in Kiev. However, as I.Ts. Gohberg recalls in his memoirs, "it is easy to twig: at this time there was a known tragedy with Jewish doctors".

Starting in 1954 and until his retirement, Mark Grigorievich held the Chair of Theoretical Mechanics at the Odessa Institute of Civil Engineering. At the end of his life, he worked as a consultant in the Institute of Physical Chemistry of the Academy of Sciences of the Ukrainian SSR. A younger generation of his students includes V.M. Adamyan, D.Z. Arov, H. Langer, F.E. Melik–Adamyan, I.E. Ovcharenko, Sh.N. Saakyan, I.M. Spitkovskii, V.A. Yavryan, and others.

M.G. Krein wrote or co-authored more than 300 papers and monographs all of which with no exception were published abroad in translation, some of them several times. These works are of an excellent analytic level and quality, broad in topic, and have opened a number of new directions in mathematics, while significantly enriching traditional directions. They initiated and continue to inspire many mathematicians, engineers, and physicists throughout the world. The following is an incomplete list of branches of mathematics where M.G. Krein's research became fundamental and, to an extent, have determined the direction for a later development: oscillation kernels and matrices, the moment problem, orthogonal polynomials and approximation theory, cones and convex sets in Banach spaces, operators on spaces with two norms, extension theory for Hermitian operators, the theory of extension of positive definite functions and spiral arcs, the theory of entire operators, integral operators, direct and inverse spectral problems for inhomogeneous strings and Sturm–Liouville equations, the trace formula and scattering theory, the method of directing functionals, stability theory for differential equations, Wiener–Hopf and Toeplitz integrals and singular integral operators, the theory of operators on spaces with an indefinite metric, indefinite extension problems, non-selfadjoint operators, characteristic operator-valued functions and triangular models, perturbation theory and the Fredholm theory, interpolation theory and factorization theory, prediction theory for stationary stochastic processes, problems in elasticity theory, the theory of vessel waves and wave resistance. A characteristic feature of his works is a deep inner unity, an interlacing of abstract and geometric ideas with concrete analytic results and their applications. Since the range of mathematical interests of Mark Grigorievich Krein is very broad, let us consider only the main, in our opinion, directions of his research.

An important role in the development of functional analysis and its application was played by Krein's papers on geometry of Banach spaces and linear topological spaces, and operators that act on them. We note first of all an introduction to and study of Banach spaces with a fixed cone of vectors and dual to them, spaces with two norms, convex sets and weak topologies in Banach spaces. Two theorems that became very famous are the Krein–Mil'man theorem on extreme points of convex sets, and the Krein–Kakutani theorem on an isomorphism of an abstract Banach space with identity and endowed with a vector structure to the space of continuous functions on a compact Hausdorff space.

A unification of algebraic and geometrical methods is clearly seen in Krein's studies in the theory of topological groups and homogeneous spaces. Harmonic analysis on a locally compact commutative group and finding a duality-like principle for compact noncommutative groups (for commutative groups, the dual object becomes the group of characters), in particular, the fact that the structure of a homogeneous compact space is completely determined by the algebra of harmonic functions on it, had a significant impact on the subsequent development of abstract harmonic analysis.

Krein completely described positive selfadjoint extensions of positive symmetric operators and developed their classification. An essential role here is played by two extreme extensions, the strict (Friedrichs extension) and the mild extensions, later called the Krein–von Neumann extension. These results were applied to a study of boundary-value problems for ordinary differential equations. Using and expanding the theory of analytic functions, he studied Hermitian operators with equal deficiency indices and found a new interesting class of operators, which he called entire, in the theory of which he found analogues of main constructions in the classical moment problem in the undetermined case. This theory permitted connection of the following problems, distinct at a first glance: the moment problem, the problem of continuation of positive definite functions and spiral arcs, a description of spectral functions of a string, and others, and, in some sense, this permitted solution of these problems completely. The theory also led to the discovery and solution of new original problems in the theory of analytic functions, and once more confirmed the foresight of M.G. Krein who said that "significant successes in functional analysis can be achieved by involving the broader modern machinery of the theory of analytic functions; in its turn functional analysis, as 'a customer', will stimulate the latter".

Krein developed a general method of directing functionals with which he obtained an eigenfunction decomposition of ordinary selfadjoint differential operators. This has extended the research over many years of J. Sturm, J. Liouville, V.A. Steklov, and H. Weyl on second-order equations to include differential equations of an arbitrary order. The theory also provided a basis for developing a theory of integral representations of positive definite kernels in terms of elementary ones. As particular cases, it led to well-known theorems of S. Bochner, S.N. Bernstein, etc. on integral representations of functions that are positive definite, exponentially convex, and other classes of functions. Here again, Mark Grigorievich has shown his extraordinary ability to see behind almost any concrete problem "an impressive thing, some selfadjoint unbounded operator" such that its spectral decomposition solves the problem.

For many years, Mark Grigorievich put much effort into working on problems connected with stability of solutions of differential equations, although he never regarded himself as a specialist in this field. He considered this as something like a hobby. The theory of stability zones, developed by A.M. Lyapunov for second-order equations, was finally generalized by M.G. Krein, after a 50 year intermission due to serious problems, to canonical systems with periodic coefficients using methods of functional analysis. A foundation of stability theory that he created for differential equations in a Banach space made it possible to achieve this goal in a simpler way and, sometimes, in a more complete form even in the case of a system with one degree of freedom.

Krein made a significant contribution to the theory of inverse problems for a Sturm–Liouville equation, a more general equation of a string, and canonical systems of differential equations. In particular, he solved the problem of recovering a Sturm–Liouville equation from two spectra, and a canonical system from its spectral function or from a scattering matrix. This work used the analytic machinery developed when studying entire operators and the theory of systems of Wiener–Hopf equations. The state of the theory at the time was not satisfactory for Krein, and he succeeded to go far ahead in constructing a general theory of such systems. This theory has achieved perfection and completeness in the series of his publications honored by the Krylov Prize in 1979. The theory was constructed on the basis of factorization of matrix-valued functions. Factorization problems for matrix- and operator-valued functions were always interesting to Krein in themselves. Let us also mention that, in the course of these studies, the theory of accelerants appeared, a theory that can be considered in the case of canonical systems with two unknown functions as a continuous analogue of orthogonal polynomials on the circle. Developing further the methods that he proposed for solving the inverse spectral problem for a string, Krein and his students solved the problem of recovering a string, possibly singular, with friction at one end, from the sequence of eigenfrequencies as well as function theory problems related to such a string, and considered the existence problem for a special representation of a polynomial that is positive on a system of closed intervals. This problem and also the extremal problems for polynomials, which he solved, generalize the corresponding problem of A.A. Markov who worked with only one interval.

Krein's ideas and methods have also deeply penetrated the theory of nonselfadjoint operators. They helped this theory, which was considered in his talk at the congress in Moscow in 1966 as one of the links of a "certain connected set of events taking place in the area of Hilbert spaces", to look nowadays like a real mountain chain that has its own architecture, its own analytic tools and, one can even say, a special calculus, all of which have unexpected applications in different areas of analysis. So, here again, M.G. Krein has achieved his "Cape of Good Hope", which we can also call his "Cape of Previsions".

Krein was one of the creators of the theory of operators on spaces with indefinite metric. His idea of a defining polynomial and the method of directing functionals created a foundation for the theory of integral representations and continuation of Hermitian-indefinite functions with a finite number of negative squares, the theory of spectral decompositions of selfadjoint and unitary operators on Π_κ-type Pontryagin spaces, which now has achieved a level of development similar to the one of the corresponding theory on a Hilbert space. The geometry of spaces named after Krein and operators on these spaces have attracted increasingly more attention of both theorists and practitioners. On the basis of the obtained results, generalized classes of Schur functions, Carathéodory functions, and Nevanlinna functions were investigated, which are generalized in the sense that the quadratic forms connected with them have a finite number of negative squares. In these classes, the corresponding generalizations of classic discrete and continuous problems were studied, such as the trigonometric and power moment problems, the Schur problem, the Nevanlinna–Pick problem, and others. For this case the theory of accelerants, continuous analogues of orthogonal polynomials, and spectral theory of canonical systems, constructed earlier in the definite situation, were developed. A next step is a continuous version of Nehari's problem for rectangular contracting matrix-valued functions on the real axis, and an application for solving matrix-continuous analogs of the Schur problem and the Carathéodory–Toeplitz problem.

In the problems described above, and in other problems of harmonic analysis, a description of solutions of the problem is given in terms of a fractional-linear transformation over a class of contracting analytic matrix-valued functions such that the coefficient matrix of the transformation has certain properties. This formula became a starting point for finding solutions with an extremal value of the so-called entropy functional that plays a special role in a number of applications.

Close ties between theoretical and applied aspects in the work of M.G. Krein were reflected in multiple applications of his results in numerous branches of science and technology. As we have already mentioned, his studies of the moment problem were connected with optimal control problems with distributed parameters, the theory of extensions of positive definite functions was related to linear predictions for stationary processes, the proposed method for determining critical frequencies in the parametric resonance phenomenon is used in the synchrophasotron theory. Let us also recall, in this connection, his results in the theory of vessel waves and wave resistance. His method of calculating the number of negative eigenvalues of an extension of a positive Hermitian operator is used for studying stability of structures. Contact problems in elasticity theory and the theory of molecular interactions also use results of Mark Grigorievich. His studies of topological groups were recently applied in graph theory, and the Krein algebras are used in modern combinatorial analysis. We also would like to recall a number of works that developed Krein's ideas, written together with his students, on infinite-dimensional Hankel matrices and the generalized Schur problem (Nehari's problem), which gave an impulse to a new direction in control theory, the H_∞-optimal control. Nowadays it is a subject of many papers, monographs and conferences.

M.G. Krein was not only a prominent scientist but an excellent pedagogue. He has taught many world famous scientists among whom there are 20 Doctors of Sciences and 50 Doctoral Candidates. He generously shared his knowledge and plans with them as well as with his other colleagues. For more than half a century, Krein has been conducting a City Mathematical Seminar that he created, which was held in the House of Scientists in Odessa for a long time, and then in the Institute of Civil Engineering, and still later in the Southern Science Center. Older and younger scientists, Krein's students and friends, participated in the work of the Seminar. They included V.M. Adamyan, D.Z. Arov, M.P. Brodskii, Yu.P. Ginzburg, I.Ts. Gohberg, G.M. Gubreyev, I. S. Iohvidov, I.S. Kats, K.R. Kovalenko, G. Langer, F.E. Melik–Adamyan, S.M. Mkhitaryan, A.A. Nudel'man, I.E. Ovcharenko, G.Ya. Popov, Sh.N. Saakyan, L.A. Sakhnovich,I.M. Spitkovskii, Yu.P. Shmul'yan, V.A. Yavryan. Giving a talk at this seminar was considered an honor for a mathematician in the former Soviet Union. Also Krein conducted smaller seminars at the institutes where he worked. In the Odessa Marine Institute, he organized a seminar on hydrodynamics. Among the participants there were Yu.L. Vorob'iev, A.A. Kostyukov and V.G. Sizov. In the Kuibyshev Industrial Institute, there was a seminar, which he headed, with G.Ya. Lyubarskii, O.V. Svirskii, and A.V. Shtraus participating in its work. Also, as was already mentioned, while working in the Institute of Mathematics of the Academy of Sciences of the Ukrainian SSR, he headed a seminar on functional analysis, where also worked Yu.M. Berezansky, Yu.L. Daletsky, G.I. Kats, B.I. Korenblyum, M.A. Krasnosel'skii, S.G. Krein. Almost every year, Mark Grigorievich gave a course of lectures to students and young researchers, based on his new results. Many of them remain unpublished. Only in 1997 were notes of his lectures on the theory of entire operators, given at Odessa Pedagogical Institute, made available by V.M. Adamyan and D.Z. Arov, prepared and enlarged by V.I. Gorbachuk and M.L. Gorbachuk, and published by Birkhäuser with the support of I.Ts. Gohberg. A similar thing happened with his course of lectures given at Moscow State University, where he discussed his results on the theory of forecasting for many-dimensional stochastic processes, where Yu.A. Rozanov was one of the participants. Later his notes became a part of his review published in "Uspekhi Matematicheskikh Nauk". The courses of lectures that he gave at USSR mathematical schools, namely, "On some new investigations in perturbation theory" (Kanev, 1963), "An introduction to geometry of indefinite J-spaces and a theory of operators on it" (Katsiveli, 1964) left an unforgettable impression with its depth and the number of posed problems. At the International Mathematical Congress (Moscow, 1966), his one-hour talk "Analytic problems and results of the theory of operators on a Hilbert space" was followed with loud applause from a large overcrowded auditorium, to which L.V. Kantorovich, the head of the session, responded that even the most famous actors do not always get such an ovation.

Mark Grigorievich was a benevolent, fair person, although demanding of others and himself. The level of his scientific ethics is shown by the following example. When studying entire operators with the deficiency index $(1, 1)$, it is important to consider the resolvent matrix, the use of which gives a description of all spectral functions of such operators. Krein had shown that this matrix is a monodromy matrix of a certain canonical system, and made a hypothesis that there is a unique Hamiltonian of this system, with a certain normalization, which he proved only for positive operators. In the general case, this was proved by Louis de Branges with a use of functional but not operator methods. During one of his talks at a meeting of the Moscow Mathematical Society, Mark Grigorievich gave the following account of the work of Louis de Branges: "I consider it brilliant. In a short time he (Louis de Branges) has succeeded in finishing the distance that took me so many years. Louis de Branges has repeated many of my propositions but the final result belongs to him. I was heading to it but never reached it". From the ethical point of view, these words can be compared with those of Euler about solving the isoperimetrical problem by Lagrange. In a letter to LaGrange, he wrote "Your analytic solution of the isoperimetrical problem contains everything that one can wish. I am quite happy that the theory I work on almost by myself is brought by you to the highest level of perfection".

Krein's scientific achievements were widely acknowledged by the international mathematical community. He was one of four Soviet mathematicians who was elected a foreign member of the American Academy of Arts and Sciences, a member of the US National Academy of Sciences, a member of many mathematical societies and editorial boards of leading mathematical journals. In 1982, Mark Grigorievich was awarded with the International Wolf Prize in Mathematics, which is an analogue of the Nobel Prize in mathematics. The foreword to the prize contains the words: "His work is the culmination of the noble line of research begun by Tchebycheff, Stieltjes, S. Bernstein, and Markov, and continued by F. Riesz, Banach, and Szegö. Krein brought the full force of mathematical analysis to bear on problems of function theory, operator theory, probability and mathematical physics. His contributions led to important developments in the applications of mathematics to different fields ranging from theoretical mechanics to electrical engineering. His style in mathematics and his personal leadership and integrity have set standards of excellence". One of the best books of well-known American mathematicians P. Lax and R. Phillips "Scattering theory for automorphic functions" (Princeton University Press and University of Tokyo Press, 1976) was dedicated to Mark Grigorievich Krein, "one of the giants in mathematics of the 20th century, as an homage to his extraordinarily broad and deep contribution to mathematics".

Regardless of all this, his scientific career in his own country, we have already said, was finished already in 1939. Accused of Jewish nationalism and cosmopolitanism, in that he cited foreign authors very often and conversely, of being cited by foreign mathematicians (Krein is one of the most cited authors in the world), he never became a member of the Academy of Sciences of the Soviet Union nor of the Academy of Sciences of the Ukrainian SSR. It may seem that the standards required by them were "too high" for Krein. Many of his students could not obtain, from the Highest Attestation Committee, a confirmation of their PhD degree. Multiple proposals by the Moscow Mathematical Society and other influential mathematical institutions and individual mathematicians such as P.S. Aleksandrov, A.N. Kolmogorov, and I. G. Petrovsky for awarding Mark Grigorievich the State Prize or any other prestigious prize always ended, despite all their arguments, by excluding the name of M.G. Krein from the list of candidates for the prize. No state or governmental institution supported him, so the presidents of neither the Academy of Sciences of the Soviet Union nor the Academy of Sciences of the Ukrainian SSR did anything for him, although they were well aware of the level of his research. The President of the Academy of Sciences of the Soviet Union, M.V. Keldysh, would ask the President of the Academy of Sciences of the Ukrainian SSR, B.E. Paton, why the most famous mathematician of Ukraine is still not a member of the Academy, although Paton could address the same question to Keldysh.

He was never permitted to leave the Soviet Union. During all his life he never crossed the border of the country. He was not even allowed to personally receive the Wolf Prize. One time he obtained a permission to take part in a conference in Hungary (near the Balaton lake), however he did not use the visa because there was a cholera epidemic in Odessa at this time and he could not leave the city for public health reasons. When I.Ts. Gohberg, who participated in the conference, informed B.Sz.-Nagy, the head of the Organizing Committee, about the reason for Krein not coming to the conference, knowing the attitude of the officials to the subject he said with a smile, “So now it is called cholera, is it?”. That was the only time when the reason of M.G. Krein’s absence was true. Also, no foreign scientist who wished to meet him could travel to Odessa. This happened, for example, to J. Helton and R. Phillips. All this was attributed to the technicality that there is no branch of the Academy of Science in Odessa.

During the Perestroika times, the situation changed for the better. Mark Grigorievich, together with N.N. Bogolyubov, was awarded the State Prize in the area of Science and Technology in 1987. During the years of independence, the Institute of Mathematics of the National Academy of Sciences of Ukraine published a three volume collection of his works that had appeared in not easily accessible journals. In 2006 the Presidium of the National Academy of Sciences of Ukraine decided to found the M.G. Krein Prize to be given for distinguished achievements in functional analysis. It is a pity that Mark Grigorievich did not see the day when a country called the USSR had disappeared from geographical maps and Ukraine became independent. He died on October 17, 1989, and did not see the Pomarancheva (Orange) Revolution. It is certain that, in his thoughts, he would be in the Square (Maidan), together with his daughter Irma, with his “scientific children and grand children”.

However, regardless of all the problems he experienced in his time, Mark Grigorievich was a happy person, since happiness is granted only to those who know a lot, and the more the person knows the stronger and distinctly he sees the poetry in the world where one with a poor knowledge may never find it. Looking into a puddle in the dusk some people see the water, others see the stars. Mark Grigorievich saw the stars. We are happy to be his contemporaries.

Operator Theory:
Advances and Applications, Vol. 190, xxi–xxii

Mark Grigorievich Krein (on his 100th birthday anniversary)

I. Gohberg

It is a great pleasure and a special honor to participate in this conference dedicated to the 100th birthday of a great mathematician of our time, Mark Grigorievich Krein. The conference was organized in the wonderful Southern town of Odessa, the town that he regarded with love and affection. The main organizer of this conference is Odessa University, the university where M.G. was a doctoral student, where he completed his studies under the instructorship of the famous N.G. Chebotaryov. In the thirties Krein created one of the strongest centers of functional analysis throughout the world at Odessa University. Many of his results of this period, as well as joint results with his friends, colleagues and outstanding students are now characterized as classical and appear in all textbooks on functional analysis.

During the years after World War Two the situation in Odessa and at the university changed to the worse. M.G. had to contend with hostile elements, which fought against him using the officially supported anti-Semitism which was very rife in Odessa. He was accused of Jewish nationalism. This accusation was certainly included in his classified file and was held against him all his life. M.G. was not allowed to have Jewish students and was deprived of a University base. He was dismissed from Odessa University and this university was not any more an international center of mathematics. Nevertheless the famous Krein center of functional analysis and its applications continued to live and work battling hostilities on his way. The group around M.G. (of which I was also a member) existed almost on a private basis, holding many of his meetings in his house or in the Odessa Scientists Club. M.G. and his group continued to have a great impact on the development of mathematics and its applications throughout the World. Even though he was never allowed to travel abroad his brilliant work knew no borders. M.G. Krein is the author and coauthor of hundreds papers and monographs of unsurpassed breadth and quality. His work opened up new areas of mathematics and greatly enriched the more traditional ones. He educated dozens of brilliant students at home and inspired the work of many mathematicians, engineers and physicists all over the world. This conference may serve as a witness of the popularity and influence of M.G. Krein's work. It also witnesses that the situation and political atmosphere changed considerably. The National Academy of Sciences of Ukraine recently announced the inauguration of a new prize: Krein's prize, and the Odessa university opened a memorial plaque. It is a great pity that Mark Grigorievich cannot see all of this and benefit of it.

The work of M.G. continues to be highly appreciated. Here is the citation for M.G. prestigious Wolf prize awarded to him in Jerusalem in 1982: "His work is the culmination of the noble line of research begun by Chebyshev, Stieltjes, Bernstein and Markov and continued by Riesz, Banach and Szegö. Krein brought the full force of mathematical analysis to bear on problems of function theory, operator theory, probability and mathematical physics. His contributions led to important developments in the applications of mathematics to different fields ranging from theoretical mechanics to electrical engineering. His style in mathematics and his personal leadership and integrity had set standards of excellence." Krein was a very fine pedagogue and lecturer. He was known for his scientific generosity and enthusiasm, as well as kindness and attention to young mathematicians. The author of these lines was very privileged to have during many years such a teacher, coauthor and friend. He will always remember M.G. Krein with gratitude, affection and admiration.

I would like to end this note by remembering the time when two of us were ending the work on the monographs on nonselfadjoint operators. This work took a very long time and our friends made fun of us. Some jokes were circulating in Odessa and I.S. Iohvidov wrote a little poem in Russian (translated into English by C. Davis)

Around the festive table, all our friends
Have come to mark our new book's publication
The fresh and shiny volume in their hands,
They offer Izya and me congratulations.
The long awaited hour is here at last.
The sourest skeptic sees he was mistaken.
And, smiling comes to cheer us like the rest,
And I am so delighted..., I awaken.

From M.G. Krein's dream, New Year's Eve, 1963

The interest to Krein's work and achievements is growing. This conference will certainly show how high is appreciated and successfully continued and developed Krein's mathematical inheritance today. The high number of participants from all the world gives already a clear indication for this. Professor Vadim Adamyan a former outstanding student of Mark Grigorievich successfully continues the work and activities of M.G. in Odessa university. Allow me to thank him for the excellent organization of this conference.

Part 1
Plenary Talks

Operator Theory:
Advances and Applications, Vol. 190, 3–26

Modified Krein Formula and Analytic Perturbation Procedure for Scattering on Arbitrary Junction

V. Adamyan, B. Pavlov and A. Yafyasov

To the memory of Mark Grigorjevich Krein

Abstract. The single-electron transport through a junction of quantum network is modeled as a specific scattering problem for the Schrödinger operator on the system of semi-infinite cylindrical domains (quantum wires) short-circuited by a compact domain (quantum well or dot). For calculation of the one-body scattering parameters of any junction having form of a compact domain with piece-wise smooth boundary and attached thin wires a semi-analytic perturbation procedure based on a specially selected intrinsic large parameter is suggested. The approximate scattering matrix of a thin junction obtained in this way is the scattering matrix of the corresponding solvable model.

Mathematics Subject Classification (2000). Primary 35P25, 35Q40, 47A40; Secondary 47A10, 47A55.

Keywords. Schrödinger operator, scattering matrix, quantum network, junction, Dirichle-to-Neumann.

1. Introduction

In this paper we denote by Ω_s the vertex domains (the quantum wells) and by ω^m the quantum wires of equal width δ connecting the wells to each other or extending to infinity. It is convenient to assume, that the domains and the leads are separated from each other by imaginable orthogonal bottom sections γ_m, $\cup_m \gamma_m = \Gamma$, see Fig. 1 below. The dynamic of single electron on the network $\Omega : \{\cup_s \Omega_s\} \cup \{\cup_m \omega^m\}$ is determined by the one-particle Schrödinger equation which is transformed, after

This work was partially supported by the US Civilian Research and Development Foundation (CRDF) and the Government of Ukraine under Grant UM2-2811-OD06.

separation of time and scaling of energy $E \to \lambda = 2m_0 E\hbar^{-2}$, to the spectral problem for the Schrödinger operator $\mathcal{L}$ on the network:

$$\mathcal{L}\psi = -\frac{1}{2\mu}\nabla^2\psi + \mathrm{V}\psi = \lambda\psi, \tag{1.1}$$

where m_0 is the bare electron mass, and $\mu = m^*/m_0$ is the ratio of the effective mass and the bare mass. The potential V is a real constant V_δ on the wires and it is a piecewise continuous function V_s on the quantum wells Ω_s. We consider hereafter a star shaped network – *a junction* – with a single well $\Omega_{\rm int}$ (the inner part of the network), and few quantum wires ω^n attached to it. Hereafter we denote by $\omega := \cup_n \omega^n = \Omega\backslash\Omega_{\rm int}$ the "exterior part" of the network.

Theoretical analysis of the electron transport through the junction is usually reduced to one-electron scattering problem, see [1–3], for the pair of Hamiltonians: the one electron Schrödinger operator $\mathcal{L}$ on the whole junction Ω and the splitting of it $\mathcal{L} \to L_{\rm int} \oplus l^\omega := \mathcal{L}_0$ which is an orthogonal sum of $L_{\rm int} = \mathcal{L}|_{L_2(\Omega_{\rm int})}$ and $l^\omega = \mathcal{L}|_{L_2(\omega)}$, obtained via imposing of additional zero boundary condition – a mathematical version of the "solid wall" – on Γ. The part

$$l^\omega := -2\mu^{-1} \triangle + V_\delta$$

of the split operator on the exterior part of the junction plays a role of a standard unperturbed Hamiltonian in the above scattering problem. This scattering problem is a sophisticated perturbation problem for the operator $\mathcal{L}_0$ which has embedded eigenvalues. Under the perturbation – removing the solid wall on Γ – the standing waves in the vertex domain $\Omega_{\rm int}$ are bred with the running waves in the wires, resulting in resonances which define the resonance character of the transmission across the junction. Analytical calculation of the scattering matrix of the two-dimensional junction is a difficult mathematical problem concerning perturbation of embedded eigenvalues. For practical needs of mathematical design of quantum networks with prescribed transport properties physicists substitute the networks by quantum graphs, with an appropriate boundary condition at the vertices, see [3, 4, 5]. Validity of such a solvable model was confirmed by smooth approximation of the graph by thin manifold shrinking to the graph, see for instance [6, 7]. This analysis showed that, in particular, for uniform shrinking, the eigenvalues, *at the lower spectral threshold,* $\lambda = 0$ of the two-dimensional Laplace equation on the manifold – the "fattened graph", with Neumann boundary conditions, – converge to the eigenvalues of the one-dimensional Schrödinger equation on the graph, with the boundary condition at the vertex $a : \sum_s \frac{d\psi_s}{dx}(a) = 0$, called conditionally "Kirchhoff". This mathematical result is proved in [7], and remains valid for various spectral problems on fattened graphs, in particular, for the spectral problem of diffusion, where the spectral properties of the relevant second-order partial differential operator near the threshold $\lambda = 0$ are important.[1]

[1]Note that in [8] a violation of some version of Kirchhoff boundary condition is noticed for electrons on a quantum network.

Contrary to diffusion, scattering of electrons in quantum networks for low temperatures is observed on the small interval of essential spectra centered at the Fermi level [9], which can be situated well above the lower threshold. In [10, 11] the *resonance mechanism* of conductance across the junction is considered. For thin junction the role of main detail of the transmitting mechanism is played by the resonance eigenfunction φ_0, which corresponds to the eigenvalue λ_0 of the Schrödinger operator on the vertex domain, which is the closest to the scaled Fermi-level Λ^F. The magnitude of the transmission coefficient is defined by the shape of the resonance eigenfunction of the Schrödinger operator on the vertex domain of the junction. The resonance mechanism permits to interpret the phenomenological parameter in the boundary condition suggested by Datta ([2]) for T-junction.

In this paper we suggest a modified analytic perturbation procedure for calculation of the scattering matrix of *arbitrary junction*, on a given interval of the essential spectrum centered at the Fermi level and containing no spectral thresholds. For thin junction the role of the first step – "jump-start" – in this analytic perturbation procedure is played by the solvable model of the junction which is completely based on spectral data of the Schrödinger operator on the vertex domain of the junction.

2. Scattering in quantum networks

Consider a junction Ω constructed of the vertex domain – a quantum well Ω_{int} – and the straight leads – quantum wires ω^m, of equal width δ connecting the well to the infinity, $\cup_m \omega^m := \omega$. It is convenient to assume, that the domain Ω_{int} and the wires ω^m are separated from each other by imaginable orthogonal bottom sections γ^m, $\cup_m \gamma^m = \Gamma$, see (2).

The dynamic of a single electron on the network is governed by the Schrödinger equation which becomes equivalent, after separation of time and scaling of energy $E \to \lambda = 2m_0 E \hbar^{-2}$, to the spectral problem for the Schrödinger operator $\mathcal{L}$ on Ω, see below (1.1).

We assume that the temperature is low and the Fermi level $\Lambda^F = 2m_0 E_F \hbar^{-2}$ lies deep enough below the potential on the complement $R_3 \backslash \Omega$, to assume that ψ vanishes on the boundary $\partial\Omega$ of the network. The above one-electron Hamiltonian $\mathcal{L}$ is selfadjoint in the Hilbert space $L_2(\Omega_{\text{int}} \cup \omega)$ of all square-integrable functions. The transport properties of the junction are determined by the structure of the scattered waves – the eigenfunctions of continuous spectrum of $\mathcal{L}$. We consider also the Schrödinger equation $L_{\text{int}}\psi = \lambda\psi$, on the quantum well Ω_{int} with L_{int} defined by the the potential and additional zero boundary condition on Γ. The one-body transport problem for the quantum network is solved on the spectral interval Δ if all scattered waves are constructed for the values of energy $\lambda \in \Delta$. For given temperature T an essential spectral interval $\Delta_T := \Delta$ is centered on the scaled Fermi level, see [9], $E^F \to \Lambda^F = 2m_0 E^F \hbar^{-2}$, as

$$\Delta = \left[\Lambda^F - \Theta, \Lambda^F + \Theta\right], \quad \Theta \simeq 2m_0 \kappa T \,\hbar^{-2}. \tag{2.1}$$

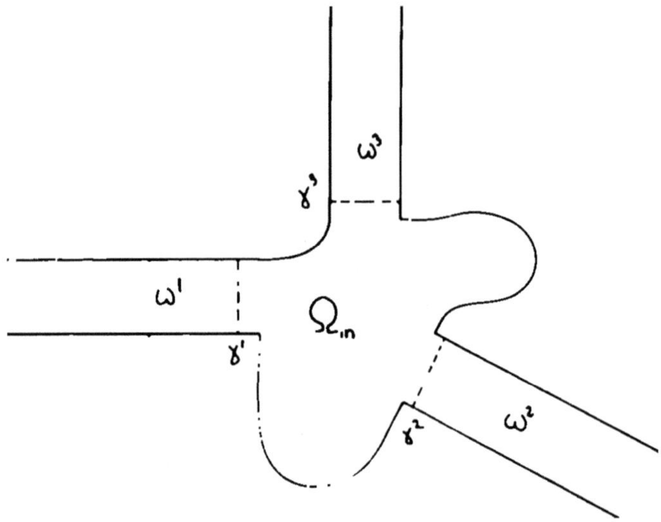

FIGURE 1. A junction.

Hereafter we assume that neither of spectral thresholds $\pi^2 l^2 \delta^{-2} + V_\delta$ is situated on the Δ. For given scaled Fermi level the spectral branches $\sigma_l := \left[\pi^2 l^2 \delta^{-2} + V_\delta, \infty\right)$ can be classified into two categories: open branches $[\pi^2 l^2 \delta^{-2}, \infty)$, corresponding to the lower group of thresholds

$$\pi^2 l^2 \delta^{-2} + V_\delta < \Lambda^F,$$

and closed branches, corresponding to the upper group of thresholds

$$\pi^2 l^2 \delta^{-2} + V_\delta > \Lambda^F.$$

These branches are characterized by the behavior of the corresponding exponential modes on the leads for $\lambda \in \Delta$.

1. Oscillating modes:

$$\chi^l_\pm(x) := \exp\left(\pm i\sqrt{\lambda - V_\delta - \pi^2 l^2 \delta^{-2}}\, x^{\|}\right) e_l(x^\perp),$$

in open branches of the wires, $\lambda - V_\delta - \pi^2 l^2 \delta^{-2} > 0$, with cross-section eigenfunctions $e_l(x^\perp) = \sqrt{2/\delta}\ \sin \pi l x^\perp/\delta\ l = 1, 2, \ldots$, $0 < x^{\|} < \infty$, $0 < x^\perp < \delta$ for $x = (x^\perp, x^{\|}) \in \omega^m$, and

2. Similar exponentially decreasing modes

$$\xi^l(x) := \exp\left(-\sqrt{\pi^2 l^2 \delta^{-2} + V_\delta - \lambda}\, x^{\|}\right) e_l(x^\perp),$$

in closed branches of the wires $\pi^2 l^2 \delta^{-2} + V_\delta - \lambda > 0$.

The above modes satisfy formally the differential equation:

$$\mathcal{L}\chi^l_\pm = \lambda \chi^l_\pm,\ \mathcal{L}\xi^l(x) = \lambda \xi^l(x),$$

and vanish on both shores of the leads. The scattered waves are obtained via matching on $\Gamma := \{x : x^{\|} = 0\}$ of the solution of the Schrödinger equation $L_{\rm int}\psi = \lambda\psi$ in $\Omega_{\rm int}$, $\psi\Big|_{\partial\Omega_{\rm int}\setminus\Gamma} = 0$, to the scattering Ansatz $\psi(x,\lambda) = \{\psi_l^m(x,\lambda)\}$ in the wires ω^m. The scattering Ansatz on the exterior part $\omega := \cup_m \omega^m$ of the network is combined of the above exponential modes as

$$\psi_l^m(x) = \begin{cases} \chi_+^l(x) + \sum_{\pi^2 r^2/\delta^2<\lambda} S_{l,r}^{m,m}\chi_-^r(x) + \sum_{\pi^2 r^2/\delta^2>\lambda} s_{l,r}^{m,m}\xi^r(x), & x\in\omega^m \\ \sum_{\pi^2 r^2/\delta^2<\lambda} S_{l,r}^{m,n}\chi_-^r(x) + \sum_{\pi^2 r^2/\delta^2>\lambda} s_{l,r}^{n,m}\xi^r(x), x\in\omega^n, & n\neq m. \end{cases} \tag{2.2}$$

The subspaces

$$\bigvee_{\pi^2 l^2\delta^{-2}+V_\delta-\Lambda^F<0} e_l := E_+ \subset L_2(\Gamma),$$

$$\bigvee_{\pi^2 l^2\delta^{-2}+V_\delta-\Lambda^F>0} e_l := E_- \subset L_2(\Gamma)$$

are called the subspaces of open and closed channels, respectively, $E_+\oplus E_- = L_2(\Gamma)$. The subspaces $\mathcal{H}_\pm := E_\pm \times L_2(0,\infty) \in L_2(\omega)$ are called *the channel spaces* of the open and closed channels. They may be interpreted as invariant subspaces of the unperturbed Schrödinger operator l^ω in $L_2(\omega)$, defined by the restriction of the differential expression $\mathcal{L}$ onto $L_2(\omega^{out})$ with zero boundary condition ("solid wall") on Γ and zero boundary conditions on both shores of the leads.

Matching on Γ the scattering Ansatz ψ to the solution of the Schrödinger equation inside the quantum well gives an infinite linear system for the coefficients S_{lr}^n, s_{lr}^n, see [12]. Formally this system can be solved, if the Green function $G_{\rm int}$ (resolvent kernel) of the Schrödinger operator $L_{\rm int}$ on $\Omega_{\rm int}$, with zero boundary condition, and Meixner conditions at the inner corners of the boundary, is known.

Really, according to general theory of the second-order linear partial equations, see [13], for regular points λ of $L_{\rm int}$ the solution u of the boundary problem with the data $u\Big|_\Gamma = u_\Gamma$ is represented by the Poisson map with the kernel:

$$u(x) = \int_\Gamma \mathcal{P}_{\rm int}(x,\gamma)u_\Gamma(\gamma)d\gamma = -\int_\Gamma \frac{\partial G_{\rm int}(x,\gamma)}{\partial n_\gamma}u_\Gamma(\gamma)d\gamma\Big|_\Gamma.$$

The corresponding boundary current is calculated formally as

$$\frac{\partial u}{\partial n}\Big|_\Gamma = -\int_\Gamma \frac{\partial^2 G_{\rm int}(x,\gamma)}{\partial n_x\partial n_\gamma}u_\Gamma(\gamma)d\gamma\Big|_\Gamma := \mathcal{DN}_{\rm int}u_\Gamma.$$

The operator $\mathcal{DN}_{\rm int}$ is called Dirichlet-to-Neumann map of $L_{\rm int}$. It is correctly defined on the appropriate Sobolev class on Γ, see [14, 15]. Besides, it is a Nevanlinna analytic operator function in λ on the set of regular points of $L_{\rm int}$. In particular, $\mathcal{DN}_{\rm int}$ takes self-adjoint values for real regular λ. More about modern DN-techniques and its applications in spectral analysis can be found in [16, 17, 18, 19, 20]. In this paper we study connection between the one-body scattering matrix and the DN-map on the two-dimensional junction. For "thin" junctions the connection is used in transport problems in [11, 21, 22].

Denoting by $K_\pm$ the exponents of the above Ansatz in the open and closed channels in the wires and by the tables S, s the coefficients in front of the corresponding exponentials, we represent the Ansatz for the scattered wave on the leads in form

$$\Psi(x,\nu) = e^{iK_+ x}\,\nu + e^{-iK_+ x}\,S\nu + e^{-K_- x}\,s\nu. \tag{2.3}$$

In fact only the oscillating component of the scattered wave in the open channels contains an essential information on details of the scattering process. The direct problem of scattering is: to find the coefficients S in front of the oscillating exponentials $e^{-iK_+ x}\,S\nu$ – the scattering matrix. In case when the solid wall is erected on the bottom sections $\Gamma = \cup_m \gamma^m$ of the wires, the scattering matrix is $-I$. The removal of the wall results in breeding of the standing waves on the quantum well with the running exponential waves in the wires. This breeding generates the exponentially decreasing "evanescent waves" $e^{-K_- x}\,s\nu$, which do not contribute to results of scattering at infinity, but affect the shape of the oscillating modes and add serious computational obstacles, see for instance [12].

The difficulty of the direct scattering problem is defined by the fact, that the above matching is a major perturbation of

$$L_{\Omega_{\rm int}} \oplus l^\omega := \mathcal{L}_0 \longrightarrow \mathcal{L},$$

caused by the removal of the "solid wall" on Γ via replacement of the zero boundary condition by the matching condition. It is a typical perturbation problem on continuous spectrum, for an operator $\mathcal{L}_0$ which has embedded eigenvalues. Breeding of the standing waves in the quantum well with the running waves in the wires gives non-square-integrable resonance states, which define resonance character of the transmission across the quantum well. Unfortunately this breeding can't be interpreted only in terms of the spectral theory of self-adjoint operators. Nevertheless we are able to suggest an algebraic version of the analysis of this breeding, based on the corresponding Krein formula.

3. Krein formula for the scattering matrix

Krein formula for the scattering matrix which corresponds to the generalized resolvent of a general symmetric operator was obtained first in [27]. In [28] this formula was used in analysis of zero-range solvable models. In [3] the Krein formula is used for analysis of the one-dimensional model of the quantum network in form of a quantum graph. In this paper we aim at the problem of fitting of the model suggested in [3].

We begin with derivation of the Krein formula for the scattering matrix of a realistic two-dimensional junction. The parameters of the fitted solvable model of the junction can be selected based on comparison of the special representation of the Krein formula of the junction, see next section, Theorem 3.1, with the corresponding formula of the model.

Consider the decomposition of the cross-section subspace $E := L_2(\Gamma)$ into an orthogonal sum of the entrance subspaces $E_\pm$ of the open and closed channels. Assuming that the Dirichlet-to-Neumann map $\mathcal{DN}_{\rm int}$ of the Schrödinger operator $L_{\rm int}$ is known, construct the matrix representation of $\mathcal{DN}_{\rm int}$ with respect to the orthogonal decomposition $E = E_+ \oplus E_-$, denoting by $P_\pm$ the corresponding orthogonal projections $I = P_+ \oplus P_-$:

$$\mathcal{DN}_{\rm int} = \begin{pmatrix} P_+\mathcal{DN}P_+ & P_+\mathcal{DN}P_+ \\ P_-\mathcal{DN}P_+ & P_-\mathcal{DN}P_- \end{pmatrix} := \begin{pmatrix} \mathcal{DN}_{++} & \mathcal{DN}+- \\ \mathcal{DN}_{-+} & \mathcal{DN}_{--} \end{pmatrix}. \tag{3.1}$$

The Cauchy data of the scattering Ansatz on the bottom sections Γ are:

$$\Psi(\nu)\Big|_\Gamma = (I+S)\nu + s\nu,$$

$$\Psi'(\nu)\Big|_\Gamma = iK_+(I-S)\nu - K_- s\nu. \tag{3.2}$$

Inserting the boundary values of the scattering Ansatz on the bottom sections into the DN-map, we obtain:

$$\mathcal{DN}\{[I+S]\nu + s\nu\} = iK_+[I-S]\nu - K_- s\nu.$$

The orthogonal components of the result in $E_\pm$ are equal to

$$\begin{aligned} \mathcal{DN}_{++}[I+S]\nu + \mathcal{DN}_{+-}s\nu &= iK_+(I-S)\nu, \\ \mathcal{DN}_{-+}[I+S]\nu + \mathcal{DN}_{--}s\nu &= -K_- s\nu, \end{aligned} \tag{3.3}$$

respectively.

Proposition 3.1. *The operator $\mathcal{DN}_{--} + K_-$ is invertible in E_- on a complex neighborhood of the essential spectral interval.*

Proof. The claim of proposition follows immediately from the Krein formula for the generalized resolvent of the concerned Schrödinger operator $\mathcal{L}$. Indeed, let $g_{{\rm ext},z}(x,y)$ be the Green function (resolvent kernel) of the self-adjoint operator $l_\omega - zI, {\rm Im} z \neq 0$, on wires and $\hat{g}_z$ be the operator from $L_2(\Gamma)$ into $L_2(\omega)$ defined the expression

$$(\hat{g}_z u)(x) := -\int_\Gamma \frac{\partial g_{{\rm ext},z}(x,\gamma)}{\partial n_\gamma} u_\Gamma(\gamma)d\gamma\Big|_\Gamma, \quad u_\Gamma \in L_2(\Gamma).$$

We write P_- for the orthogonal projector onto E_- in $L_2(\Gamma)$ and π_- for the orthogonal projector onto subspace $\mathcal{H}_-$ in $L_2(\Omega)$. Then the Krein formula for $\pi_-(\mathcal{L}-z)^{-1}|_{\mathcal{H}_-}$ can be written in the form

$$\pi_-(\mathcal{L}-z)^{-1}|_{\mathcal{H}_-} = (l_\omega - z)^{-1}|_{\mathcal{H}_-} - \hat{g}_z P_- \left(\mathcal{DN}_{--} + K_-\right)^{-1} P_- \hat{g}_{\bar z}^*|_{\mathcal{H}_-}.$$

Hence, $(\mathcal{DN}_{--} + K_-)^{-1}$ cannot have non-real singularities. The real singularities (poles) serve eigenvalues of the corresponding intermediate Hamiltonian, see [22]. □

Therefore

$$s\nu = -\frac{I}{\mathcal{DN}_{--} + K_-}\mathcal{DN}_{-+}[I + S]\nu,$$

which implies the following Krein formula for the scattering matrix:

$$S = \left[iK_+ - \mathcal{DN}^F\right]^{-1}\left[iK_+ + \mathcal{DN}^F\right], \tag{3.4}$$

where

$$\mathcal{DN}^F := \mathcal{DN}_{++} - \mathcal{DN}_{+-}\frac{I}{\mathcal{DN}_{--} + K_-}\mathcal{DN}_{-+}. \tag{3.5}$$

$\mathcal{DN}^F$ has a structure typical for the classical Krein formula and can be interpreted as a Dirichlet-to-Neumann map of an *intermediate Hamiltonian* and (3.4) has form of the scattering matrix for the one-dimensional scattering systems, see for instance [23]. Similar formula plays an important role in modern approach to the one-dimensional inverse spectral problem, see [24, 25]. It is worth mentioning that S is K_+-unitary, that is

$$S^* K_+ S = K_+. \tag{3.6}$$

To transform the Krein formula (3.4) of the two-dimensional scattering system to the convenient quasi-one-dimensional form, we have to analyze the expression (3.5) in details.

The intermediate Hamiltonian was introduced in [11, 21, 22] as a component L_F of the splitting

$$\mathcal{L} \longrightarrow l^F \oplus L_F$$

of $\mathcal{L}$ defined by an additional "partial" zero boundary condition imposed at the bottom sections $\cup_m \gamma^m := \Gamma$ onto the elements from the domain of $\mathcal{L}$:

$$P_+ u\Big|_\Gamma = 0. \tag{3.7}$$

Here l^F, L_F are selfadjoint operators in $\mathcal{H}_+$, $\mathcal{H}_- \oplus L_2(\Omega_{\text{int}})$ respectively, see [22]. The absolutely continuous spectra of l^F, L_F coincide with the union of all open and closed branches $\sigma_l = \left[\frac{\pi^2 l^2}{\delta^2} + V_\infty, \infty\right)$ respectively:

$$\sigma(l^F) = \cup_{\text{open}}\, \sigma_l, \; \sigma_{ac}(L_F) = \cup_{\text{closed}}\, \sigma_l.$$

It is proven in [22], that the restriction

$$P_{L_2(\Omega_{\text{int}})}[L_F - \lambda I]^{-1} P_{L_2(\Omega_{\text{int}})\oplus\mathcal{H}_-}$$

of the resolvent of L_F, acting as an operator from $L_2(\Omega_{\text{int}}) \oplus \mathcal{H}_-$ onto $L_2(\Omega_{\text{int}})$ can be represented by an integral operator with a kernel G^F. Then the statement (3.4) can be verified based on the Poisson formula for the solution of an *intermediate boundary problem* for the Schrödinger equation

$$\mathcal{L}u - \lambda u = 0, \; P_+ u(x) = 0 \text{ if } x > 0, \quad P_+ u\Big|_\Gamma = u_+ \in E_+.$$

$$u(x) = -\int_\Gamma \frac{\partial G^F}{\partial n_\gamma}(x, \gamma)\, u_+ d\Gamma, \quad x \in \Omega_{\text{int}}.$$

Then, denoting by $u_\pm$ the components of $u\Big|_\Gamma$ in $E_\pm$, and taking into account that $P_- \frac{\partial u}{\partial n_\gamma} = -K_- u_-$, we obtain:

$$\begin{pmatrix} \mathcal{DN}_{++} & \mathcal{DN}+- \\ \mathcal{DN}_{-+} & \mathcal{DN}_{--} \end{pmatrix} \begin{pmatrix} u_+ \\ u_- \end{pmatrix} = \begin{pmatrix} P_+ \frac{\partial u_+}{\partial n_\gamma} \\ -K_- u_- \end{pmatrix}.$$

One can see that the expression (3.5) is the Schur complement, see [26], of the matrix

$$\begin{pmatrix} \mathcal{DN}_{++} & \mathcal{DN}+- \\ \mathcal{DN}_{-+} & \mathcal{DN}_{--} + K_- \end{pmatrix}.$$

This implies the announced statement, once we define the DN-map of the intermediate Hamiltonian as

$$P_+ \frac{\partial u_+}{\partial n_\gamma}\Big|_\Gamma := \mathcal{DN}^F u_+. \tag{3.8}$$

The DN-map of the selfadjoint operator L_F has a negative imaginary part in the upper half-plane $\Im\lambda > 0$ and simple poles at the eigenvalues of L_F. Practically, for relatively thin wires, see [22], we are able to substitute the DN-map $\mathcal{DN}^F$ in the above Krein formula (3.4) for the scattering matrix by an appropriate rational approximation

$$\mathcal{DN}^F \longrightarrow \mathcal{DN}^F_\Delta \tag{3.9}$$

on the essential spectral interval Δ, *with the same poles and residues on* Δ. The corresponding approximate scattering matrix takes the form

$$S(\lambda) \to S_\Delta(\lambda) = \frac{iK_+ - \mathcal{DN}^F_\Delta}{iK_+ + \mathcal{DN}^F_\Delta}. \tag{3.10}$$

Rational expressions of the above form (3.10) are typical for one-dimensional scattering systems. Sometimes they can be interpreted as scattering matrices of zero-range solvable models with "Inner Hamiltonian", see [27, 28]. These zero-range models are automatically fitted on the essential spectral interval Δ, once $\mathcal{DN}^F_\Delta$ serves a rational approximation of $\mathcal{DN}^F$ on Δ.

The above expression (3.10) is scattering matrix for a solvable model if the rational approximation $\mathcal{DN}^F_\Delta$ does not contain a polynomial part. To develop the corresponding perturbation technique for *arbitrary junction* we need explicit expression for the poles and residues of the DN-map $\mathcal{DN}^F$ of the intermediate Hamiltonian. While the corresponding data for L_{int} can be obtained via straightforward computing with standard programs, the similar problem for the intermediate Hamiltonian appears to be more difficult. Fortunately, for "relatively thin" junctions, the spectral data can be obtained via special analytic perturbation procedure based on a certain "modified" representation of the Krein formula (3.5).

4. Analytic perturbation procedure for the Krein denominator and compensation of singularities

Both terms in the left side of (3.5) have singularities on the spectrum of the non-perturbed operator L_{int}. It is normally expected, that the singularities of the first and second term at the eigenvalues of L_{int} compensate each other, so that only the zeros of the denominator $\mathcal{DN}_{--}+K_-$ arise as singularities of $\mathcal{DN}^F$. In this section we produce analysis supporting this statement, see a one-dimensional version of the statement in [31]. But we obtain in course of the relevant calculation even more important "byproduct": we derive an algebraic equation for the eigenvalues of the intermediate Hamiltonian and calculate the residues at the corresponding poles of the intermediate DN-map. Then we are able to do the first step announced above, calculating, based on (3.4), the scattering matrix of a "relatively thin" junction.

For given temperature T we consider an *essential spectral interval* $\Delta_T := \Delta$, see (2.1). We assume that the temperature is *low*, so that Δ is situated inside the conductivity band Δ_F between the lower threshold $\lambda_{\min}$ of the closed channels and the upper threshold $\lambda_{\max}$ of the open channels

$$\Delta \subset (\lambda_{\max}, \lambda_{\min}) = \Delta_F.$$

Our aim is: to construct on Δ a convenient local "quasi-one-dimensional" representation of the intermediate DN-map and for the scattering matrix of the junction, (3.4), *with compensated singularities* inherited from the L_{int}. Later, in next section, we will use this construction as a basement for an analytic perturbation procedure, with an "intrinsic" large parameter, to calculate approximately the scattering matrix of *arbitrary* junction.

Let us present the DN-map $\mathcal{DN}$ of L_{int} on the essential spectral interval as a sum

$$\mathcal{DN}_{\text{int}} = \sum_{\lambda_s \in \Delta} \frac{\left|\frac{\partial \varphi_s}{\partial n}\right\rangle \left\langle \frac{\partial \varphi_s}{\partial n}\right|_\Gamma}{\lambda - \lambda_s} + \mathcal{K} := \mathcal{DN}^{\Delta} + \mathcal{K} \tag{4.1}$$

of the rational expression constituted by the polar terms with singularities at the eigenvalues $\lambda_s \in \Delta$ of the operator L_{int} and an analytic operator-function $\mathcal{K}$ on G_Δ. We will also use the operators obtained from $\mathcal{DN}_{\text{int}}$ via framing of it by the projections $P_\pm$, for instance:

$$P_+ \mathcal{DN}_{\text{int}} P_- = P_+ \mathcal{DN}^{\Delta} P_- + P_+ \mathcal{K} P_- = \mathcal{DN}^{\Delta}_{+-} + \mathcal{K}_{+-}.$$

We introduce also the linear hull $E_\Delta = \bigvee_s \{\varphi_s\}$ - an invariant subspace of L_{int} corresponding to the essential spectral interval Δ and the part

$$L^{\Delta} := \sum_{\lambda_s \in \Delta} \lambda_s \mid \varphi_s \rangle \langle \varphi_s \mid$$

of L_{int} in it.

To calculate the intermediate DN-map (3.5) in terms of the standard DN-map of L_{int} we have to solve the equation:

$$[\mathcal{DN}_{--} + K_{-}]u = \mathcal{DN}_{-+}g \tag{4.2}$$

on the essential spectral interval Δ. *It can be solved based on Banach principle if K_{-} can play a role of a large parameter, so that the operator*

$$[\mathcal{K}_{--} + K_{-}]^{-1} \tag{4.3}$$

exists on Δ. Then, due to continuity of $\mathcal{K}_{--}, K_{-}$ there exist also a complex neighborhood of Δ where the inverse exists. We assume that this complex neighborhood is G_{Δ}. The junction, for which the condition (4.3) is fulfilled, we call *relatively thin junction*, based on the following motivation. The DN-map of L_{int} is homogeneous degree -1. It acts from $W_2^{3/2}(\Gamma)$ to $W_2^{1/2}(\Gamma)$, see [14]. If Ω_{int} has a small diameter d then, the norm of the correcting term $\mathcal{K}$ is estimated as Const $1/d$. The same estimate remains true for $P_{-}\mathcal{K}P_{-} := \mathcal{K}_{--}$. The exponent K_{-} also acts from $W_2^{3/2}(\Gamma)$ to $W_2^{1/2}(\Gamma)$ and the norm of its inverse is estimated as Const δ. Then the $W_2^{3/2}-$ norm of $K_{-}^{-1}\,\mathcal{K}_{--}$ is estimated as Const δ/d. Hence, in particular, $K_{-} + \mathcal{K}_{--} = K_{-}\left[I + K_{-}^{-1}\,\mathcal{K}_{--}\right]$ is invertible if $\delta/d \ll 1$, see more comments in [22]. Notice, that for an arbitrary junction the *auxiliary* Fermi level $\Lambda_1^F := \Lambda_1$ can be selected such that the condition (4.3) is fulfilled. We will use this option in the following section, when calculating the scattering matrix. Now we proceed in this section assuming that (4.3) is fulfilled.

Denoting by T the map

$$T = \sum_{\lambda_s \in \Delta_T} |\varphi_s\rangle\left\langle \frac{\partial \varphi_s}{\partial n}\right|, \quad \text{and} \quad T\frac{I}{\mathcal{K}_{--} + K_{-}}T^{+} := Q(\lambda) : E_{\Delta} \to E_{\Delta}.$$

We also denote

$$\left(P_{+} - \mathcal{K}_{+-}\frac{I}{\mathcal{K}_{--} + K_{-}}P_{-}\right) := \mathcal{J}(\lambda).$$

Then we discover, after some cumbersome calculation, that all singularities in the Krein formula, arising from the eigenvalues λ_s of L_{int} are compensated.

Theorem 4.1. *The Krein formula* (3.5) *for the intermediate DN-map, can be rewritten, for a thin junction, on the essential spectral interval, as:*

$$\mathcal{DN}^F = \mathcal{K}_{++} - \mathcal{K}_{+-}\frac{I}{\mathcal{K}_{--} + K_{-}}\mathcal{K}_{-+} + \mathcal{J}T^{+}\rangle\frac{I}{\lambda I - L^{\Delta} + Q(\lambda)}\langle T\mathcal{J}^{+}. \tag{4.4}$$

The representation (4.4) *remains valid on a complex neighborhood* G_{Δ} *of the essential spectral interval.*

Remark 4.2. The announced representation (4.4) of the Krein formula (3.5) for the DN-map of the intermediate Hamiltonian, has on the essential spectral interval only non-compensated singularities, at the eigenvalues of the intermediate

Hamiltonian, calculated as zeros of the denominator $\lambda I^{\Delta} - L^{\Delta} + Q(\lambda) := \mathcal{D}(\lambda)$:

$$\mathcal{D}(\lambda_s^F)\, \nu_s^F = 0.$$

These singularities coincide with the eigenvalues of the intermediate Hamiltonian.

We call the above formula (4.4) for $\mathcal{DN}^F$ *the modified Krein formula.* Inserting (4.4) into the above formula (3.4) gives a convenient representation for the scattering matrix of the relatively thin junction, which permits, in particular, to calculate the resonances based on eigenvalues of the intermediate operator.

In case of one-dimensional zeros of the denominator $\mathcal{D}$ the corresponding residues are calculated as projections onto the subspaces

$$\mathcal{E}_s^F = \mathcal{J}(\lambda_s^F) T^+ \, \nu_s^F.$$

For multidimensional zeros of the denominator, $\mathcal{D}(\lambda_s^F) N_s^F = 0$, dim $N_s^F \ > 1$ the residues are projections onto the images of the corresponding null-spaces $N_s^F = \bigvee_s \nu_s^N$

$$\mathcal{E}_s^F = \mathcal{J}(\lambda_s^F) T^+ \, N_s^F.$$

Proof. We begin with the standard Krein formula (3.5) for the DN-map of the intermediate Hamiltonian L_F on the essential spectral interval Δ. Denote by $\mathcal{DN}^{\Delta}$ the component of the DN-map of L_{int} on Δ defined by the formula (4.1), and introduce similar notations for the matrix elements of $\mathcal{DN}$ with respect to the orthogonal decomposition $E = E_+ \oplus E_-$, for instance

$$\mathcal{DN}_{+-} = \mathcal{DN}^{\Delta}_{+-} + \mathcal{K}_{+-}.$$

To calculate explicitly the second addendum in (3.5), we re-write (4.2)as:

$$\left[\mathcal{DN}^{\Delta}_{--} u + (K_- + \mathcal{K}_{--})\right] u = \left[\mathcal{DN}^{\Delta}_{-+} + \mathcal{K}_{-+}\right] g.$$

If $(K_- + \mathcal{K}_{--})$ is invertible on Δ, then the above equation is equivalent to:

$$\frac{I}{K_- + \mathcal{K}_{--}} \mathcal{DN}^{\Delta}_{--} u + u = \frac{I}{K_- + \mathcal{K}_{--}} \left[\mathcal{DN}^{\Delta}_{-+} g + \mathcal{K}_{-+} g\right]. \qquad (4.5)$$

Denote

$$\frac{\langle \frac{\partial \varphi_s}{\partial n}, u\rangle}{\lambda - \lambda_s} := v_s, \ \sum_s \varphi_s\rangle v_s = \frac{I}{\lambda I^{\Delta} - L^{\Delta}} T u := \mathbf{v},$$

and take into account that

$$\mathcal{DN}^{\Delta}_{-+} = P_- T^+ \frac{I}{\lambda I^{\Delta} - L^{\Delta}} T P_+.$$

Then multiplying (4.5) by T we obtain an equation for $\mathbf{v}$:

$$\left[\lambda I^{\Delta} - L^{\Delta} + Q(\lambda)\right] \mathbf{v} = Q \frac{I}{\lambda I^{\Delta} - L^{\Delta}} T P_+ g + T \frac{I}{K_- + \mathcal{K}_{--}} \mathcal{K}_{-+} g.$$

This gives the following representation for $\mathbf{v}$

$$\mathbf{v} = \frac{I}{\lambda I^{\Delta} - L^{\Delta} + Q(\lambda)} \left[Q \frac{I}{\lambda I^{\Delta} - L^{\Delta}} T P_+ g + T \frac{I}{K_- + \mathcal{K}_{--}} \mathcal{K}_{-+} g\right].$$

and permits to calculate, based on (4.5)

$$\begin{aligned}
u &= -\frac{I}{K_- + \mathcal{K}_{--}} T^+ \mathbf{v} + \frac{I}{K_- + \mathcal{K}_{--}} T^+ \frac{I}{\lambda I^\Delta - L^\Delta} T P_+ g \\
&\quad + \frac{I}{K_- + \mathcal{K}_{--}} \mathcal{K}_{-+} g \\
&= -\frac{I}{K_- + \mathcal{K}_{--}} T^+ \frac{I}{\lambda I^\Delta - L^\Delta + Q(\lambda)} \\
&\quad \times \left[Q \frac{I}{\lambda I^\Delta - L^\Delta} T P_+ g + T \frac{I}{K_- + \mathcal{K}_{--}} \mathcal{K}_{-+} g \right] \\
&\quad + \frac{I}{K_- + \mathcal{K}_{--}} T^+ \frac{I}{\lambda I^\Delta - L^\Delta} T P_+ g + \frac{I}{K_- + \mathcal{K}_{--}} \mathcal{K}_{-+} g.
\end{aligned}$$

Now we substitute this expression into the formula (3.5):

$$\begin{aligned}
\mathcal{DN}^F g &= \mathcal{DN}^{\Delta_T}_{++} g + \mathcal{K}_{++} g - \mathcal{DN}^{\Delta_T}_{+-} u - \mathcal{K}_{+-} u \qquad (4.6)\\
&= P_+ T^+ \frac{I}{\lambda I^\Delta - L^\Delta} T P_+ g + \mathcal{K}_{++} g - P_+ T^+ \frac{I}{\lambda I^\Delta - L^\Delta} T P_- u - \mathcal{K}_{+-} u \\
&:= I_1 + I_2 + I_3 + I_4,
\end{aligned}$$

where

$$\begin{aligned}
I_3 &= -P_+ T^+ \frac{I}{\lambda I^\Delta - L^\Delta} T P_- u \\
&= -P_+ T^+ \frac{I}{\lambda I^\Delta - L^\Delta} Q(\lambda) \frac{I}{\lambda I^\Delta - L^\Delta} T P_+ g \\
&\quad - P_+ T^+ \frac{I}{\lambda I^\Delta - L^\Delta} T \frac{I}{K_- + \mathcal{K}_{--}} \mathcal{K}_{-+} g \\
&\quad + P_+ T^+ \frac{I}{\lambda I^\Delta - L^\Delta} Q \frac{I}{\lambda I^\Delta - L^\Delta + Q} Q \frac{I}{\lambda I^\Delta - L^\Delta} T P_+ g \\
&\quad + P_+ T^+ \frac{I}{\lambda I^\Delta - L^\Delta} Q \frac{I}{\lambda I^\Delta - L^\Delta + Q} T \frac{I}{K_- + \mathcal{K}_{--}} \mathcal{K}_{-+} g \\
&= I_{31} + I_{32} + I_{33} + I_{34},
\end{aligned}$$

and

$$\begin{aligned}
I_4 &= -\mathcal{K}_{+-} u \\
&= \mathcal{K}_{+-} \frac{I}{K_- + \mathcal{K}_{--}} T^+ \frac{I}{\lambda I^\Delta - L^\Delta + Q(\lambda)} Q \frac{I}{\lambda I^\Delta - L^\Delta} T P_+ g \\
&\quad + \mathcal{K}_{+-} \frac{I}{K_- + \mathcal{K}_{--}} T^+ \frac{I}{\lambda I^\Delta - L^\Delta + Q(\lambda)} T \frac{I}{K_- + \mathcal{K}_{--}} \mathcal{K}_{-+} g \\
&\quad + -\mathcal{K}_{+-} \frac{I}{K_- + \mathcal{K}_{--}} T^+ \frac{I}{\lambda I^\Delta - L^\Delta} T P_+ g - \mathcal{K}_{+-} \frac{I}{K_- + \mathcal{K}_{--}} \mathcal{K}_{-+} g \\
&= I_{41} + I_{42} + I_{43} + I_{44}.
\end{aligned}$$

Insert these results into the above formula (4.6) and collect the terms in the right side which contain the second power of $[\lambda I^{\Delta} - L^{\Delta}]^{-1}$:

$$\begin{aligned} I_{31} + I_{33} &= -P_+T^+\frac{I}{\lambda I^{\Delta} - L^{\Delta}}Q(\lambda)\frac{I}{\lambda I^{\Delta} - L^{\Delta}}TP_+g \\ &+ P_+T^+\frac{I}{\lambda I^{\Delta} - L^{\Delta}}Q\frac{I}{\lambda I^{\Delta} - L^{\Delta} + Q}Q\frac{I}{\lambda I^{\Delta} - L^{\Delta}}TP_+g \\ &= -P_+T^+\frac{I}{\lambda I^{\Delta} - L^{\Delta}}\,\frac{I}{\lambda I^{\Delta} - L^{\Delta} + Q}TP_+g. \end{aligned} \tag{4.7}$$

This result, combined with I_1 yields:

$$\begin{aligned} &P_+T^+\,\frac{I}{\lambda I^{\Delta} - L^{\Delta}}TP^+g - P_+T^+\frac{I}{\lambda I^{\Delta} - L^{\Delta}}\,\frac{I}{\lambda I^{\Delta} - L^{\Delta} + Q}TP_+g \\ &\quad = P_+T^+\frac{I}{\lambda I^{\Delta} - L^{\Delta} + Q}TP_+g := J_1g. \end{aligned} \tag{4.8}$$

Now we combine the terms $I_{32} + I_{34}$ and $I_{41} + I_{43}$ containing $[\lambda I^{\Delta} - L^{\Delta}]^{-1}$:

$$\begin{aligned} I_{32} + I_{34} &= -P_+T^+\frac{I}{\lambda I^{\Delta} - L^{\Delta}}\left[-I + Q\frac{I}{\lambda I^{\Delta} - L^{\Delta} + Q}\right]T\frac{I}{K_- + \mathcal{K}_{--}}\mathcal{K}_{-+}g \\ &= -P_+T^+\frac{I}{\lambda I^{\Delta} - L^{\Delta} + Q}T\frac{I}{K_- + \mathcal{K}_{--}}\mathcal{K}_{-+}g := J_2g, \end{aligned} \tag{4.9}$$

$$\begin{aligned} I_{41} + I_{43} &= \mathcal{K}_{+-}\frac{I}{K_- + \mathcal{K}_{--}}T^+\left[-I + \frac{I}{\lambda I^{\Delta} - L^{\Delta} + Q}Q\right]TP_+g \\ &= -\mathcal{K}_{+-}\frac{I}{K_- + \mathcal{K}_{--}}T^+\frac{I}{\lambda I^{\Delta} - L^{\Delta} + Q}TP_+g := J_3g. \end{aligned} \tag{4.10}$$

We see that no terms left in the right side of (4.6) with singularities $[\lambda I^{\Delta} - L^{\Delta}]^{-1}$ inherited from the unperturbed operator – all these singularities are compensated. Assembling separately the terms J_1g, J_2g, J_3g, I_{43} containing $[\lambda I^{\Delta} - L^{\Delta} + Q]^{-1}$ and regular terms I_2, I_{44}, we obtain the announced expression $\mathcal{DN}^F g$

$$\begin{aligned} \mathcal{DN}^F g &= P_+T^+\frac{I}{\lambda I^{\Delta} - L^{\Delta} + Q}TP_+g \\ &- P_+T^+\frac{I}{\lambda I^{\Delta} - L^{\Delta} + Q}T\frac{I}{K_- + \mathcal{K}_{--}}\mathcal{K}_{-+}g \\ &- \mathcal{K}_{+-}\frac{I}{K_- + \mathcal{K}_{--}}T^+\frac{I}{\lambda I^{\Delta} - L^{\Delta} + Q}TP_+g \\ &+ \mathcal{K}_{+-}\frac{I}{K_- + \mathcal{K}_{--}}T^+\frac{I}{\lambda I^{\Delta} - L^{\Delta} + Q(\lambda)}T\frac{I}{K_- + \mathcal{K}_{--}}\mathcal{K}_{-+}g \\ &+ \mathcal{K}_{++}g - \mathcal{K}_{+-}\frac{I}{K_- + \mathcal{K}_{--}}\mathcal{K}_{-+}g \end{aligned}$$

$$= P_+T^+ - \mathcal{K}_{+-}\frac{I}{K_- + \mathcal{K}_{--}}T^+\rangle\frac{I}{\lambda I^\Delta - L^\Delta + Q(\lambda)}\langle TP_+$$
$$- T\frac{I}{K_- + \mathcal{K}_{--}}\mathcal{K}_{-+}g + \mathcal{K}_{++}g - \mathcal{K}_{+-}\frac{I}{K_- + \mathcal{K}_{--}}\mathcal{K}_{-+}g. \qquad (4.11)$$

The announced expression (4.4) for $\mathcal{DN}^F$ is obtained from the above formula by introducing the notation $P_+ - \mathcal{K}_{+-}\frac{I}{K_-+\mathcal{K}_{--}} := \mathcal{J}$. The derived formula is extended onto the complex neighborhood G_Δ of the essential spectral interval due to analyticity. Further analytical continuation is possible as well, but the estimates of leading and subordinate terms are obviously lost. □

The scattering matrix of the original problem on the essential spectral interval may be obtained via replacement in (3.4) the intermediate DN-map by the expression (4.4) with compensated singularities. This substitution is possible for thin junctions, when the exponent K_- in closed channels can play a role of a large parameter, compared with the error $\mathcal{K}_{--}$ of the rational approximation $\mathcal{DN}^\Delta$ of $\mathcal{DN}$. This condition may be not satisfied for given Fermi level $\Lambda := \Lambda_0$.

Remark 4.3. If the rational approximation of the intermediate DN-map (4.4) does not contain a polynomial part, but only a real poles, then it can be interpreted as a DN-map of a solvable model, see [30]. In case when a polynomial part is present, one may hope to interpret the corresponding scattering matrix as one of the solvable model in a Pontryagin space.

5. Intrinsic large parameter and an analytic perturbation procedure for the scattering matrix of an arbitrary junction

In previous section the existence of the inverse $[\mathcal{K}_{--} + K_-]^{-1}$ was guaranteed by the presence of the large parameter $\lambda_{\min} - \lambda$, with $\lambda_{\min}$ selected as the nearest to Λ^F threshold $\pi^2 n^2 \delta^{-2} > \Lambda^F$. In fact the choice of the "technical" Fermi level $\Lambda^F := \Lambda_0$ is in our hands, so we are able to select another value $\Lambda_1 \gg \Lambda_0$, such that the condition (4.3) is fulfilled. The corresponding splitting of the original Hamiltonian would be defined by the orthogonal decomposition of the entrance space $E = \left[E_+^0 \oplus E_+^1\right] \oplus E_-^1$, such that few closed channels with thresholds $V_\infty + \frac{\pi^2 l^2}{\delta^2}$ situated between Λ_0 and Λ_1 are formally included into the lower group of channels, with an extended entrance subspace $E_+^0 \oplus E_+^1 := E_+$. We will use hereafter the intermediate DN-map $\mathcal{DN}^1$ of the operator L_1, defined by the semi-transparent boundary condition *hight* Λ_1 associated with the above decomposition of the entrance space $L_2(\Gamma) = \left[E_+^0 \oplus E_+^1\right] \oplus E_-^1$.

$$P_+ u\Big|_\Gamma = 0, \text{ with } P_+ := P_{E_+^0 \oplus E_+^1}.$$

Denote by $K_+^0, K_\pm^1$ the exponents of the oscillating and decreasing solutions of the Schrödinger equation in the channels associated with E_+^0, $E_\pm^1$, respectively.

Consider the orthogonal decomposition $E = E^0_+ \oplus E^1_+ \oplus E^1_-$ and represent the DN-map $\mathcal{DN}$ of L_{int} by the matrix

$$\mathcal{DN} = \begin{pmatrix} \mathcal{DN}^{00}_{++} & \mathcal{DN}^{01}_{++} & \mathcal{DN}^{01}_{+-} \\ \mathcal{DN}^{10}_{++} & \mathcal{DN}^{11}_{++} & \mathcal{DN}^{11}_{+-} \\ \mathcal{DN}^{10}_{-+} & \mathcal{DN}^{11}_{-+} & \mathcal{DN}^{11}_{--} \end{pmatrix} := \mathbf{DN}. \tag{5.1}$$

Hereafter we consider the Schrödinger operator (1.1) on an *arbitrary* junction $\Omega = \Omega_{\text{int}} \cup \omega$, assuming that the compact domain Ω_{int} has a piecewise smooth boundary and the Meixner conditions are imposed at the inner corners of the boundary of Ω_{int}. Consider the rational approximation of the DN-map of the Schrödinger operator L_{int} on the essential spectral interval Δ:

$$\mathcal{DN} = \mathcal{DN}(\Delta) + \mathcal{K},$$

including into $\mathcal{DN}(\Delta)$ the polar terms corresponding to the eigenvalues $\lambda_s \in \Delta$:

$$\sum_{\lambda_s \in \Delta} \frac{\left| \frac{\partial \varphi_s}{\partial n} \right\rangle \left\langle \frac{\partial \varphi_s}{\partial n} \right|}{\lambda - \lambda_s} =: \mathcal{DN}(\Delta),$$

with a properly selected self-adjoint operator C_Δ and denote appropriately the corresponding matrix elements, for instance:

$$\mathcal{DN}^{00}_{++} = \mathcal{DN}(\Delta)^{00}_{++} + \mathcal{K}^{00}_{++}.$$

Now we select, for the junction Ω, the technical Fermi-level Λ_1 from the condition, that the junction is thin, with respect to the new Fermi level Λ_1:

Definition 5.1. We say that the the quantum network is relatively thin on the level Λ_1 if the operator $\mathcal{K}^{11}_{--} + K^1_-$ is invertible on some complex neighborhood G_Δ of the essential spectral interval Δ.

This condition may be substituted by a stronger, but more convenient condition

$$\sup_{\lambda \in G_\Delta} \| \mathcal{K}^{00}_{+-}(\lambda) \| < \sqrt{\Lambda^F_1 - \Lambda^F - 2m_0 \kappa T \hbar^{-2}}. \tag{5.2}$$

If Λ_1 is defined from (5.2), we construct the corresponding decomposition $E = E_+ \oplus E_-$, with $E_+ = \left[E_+ \oplus E^1_+\right]$, $E_- = E^1_-$

$$E = \left[E^0_+ \oplus E^1_+\right] \oplus E^1_-.$$

and define the intermediate Hamiltonian L_1 as a non-trivial component of the corresponding splitting of $\mathcal{L}$:

$$\mathcal{L} = L^1_F \oplus l^F_1, \tag{5.3}$$

obtained by imposing on Γ the additional boundary condition

$$P_+ u \Big|_\Gamma = 0.$$

Note that the trivial part l_1^F of this splitting contains additional channels in the "lover" group of channels : $E_+ = E_+^0 \oplus E_+^1$, which correspond to exponentially decreasing modes $e^{-K_+^1 x}\nu$. The matrix (5.1) connects the boundary data $\Psi(0)$, $\Psi'(0)$ of the scattering Ansatz

$$\Psi(x,\lambda) = e^{iK_+ x}\nu + e^{-iK_+ x} S\nu + e^{-K_+^1 x} s_+^1 \nu + e^{-K_-^1 x} s_-^1 \nu, \tag{5.4}$$

$$\begin{pmatrix} iK_+(\nu - S\nu) \\ -K_+^1 s_+^1 \nu \\ -K_-^1 s_-^1 \nu \end{pmatrix} = \mathbf{DN} \begin{pmatrix} (\nu + S\nu) \\ s_+^1 \nu \\ s_-^1 \nu \end{pmatrix}.$$

Eliminating $s_-^1\nu$ from the last equation,

$$s_-^1 \nu = \frac{I}{\mathcal{DN}_{--}^{11} + K_-^1} \left[\mathcal{DN}_{-+}^{01}(\nu + S\nu) + \mathcal{DN}_{-+}^{11} s_+^1 \nu\right]$$

we obtain *a finite-dimensional equation for the components of the scattering Ansatz in* $E_+^0 \oplus E_+^1$

$$\begin{pmatrix} iK_+(\nu - S\nu) \\ -K_+^1 s_+^1 \nu \end{pmatrix} = \tilde{\mathbf{DN}} \begin{pmatrix} (\nu + S\nu) \\ s_+^1 \nu \end{pmatrix}.$$

Here

$$\tilde{\mathbf{DN}} := \begin{pmatrix} \tilde{\mathcal{DN}}_{++}^{00} & \tilde{\mathcal{DN}}_{++}^{01} \\ \tilde{\mathcal{DN}}_{++}^{10} & \tilde{\mathcal{DN}}_{++}^{11} \end{pmatrix},$$

where

$$\tilde{\mathcal{DN}}_{++}^{00} = \mathcal{DN}_{++}^{00} - \mathcal{DN}_{+-}^{01} \frac{I}{\mathcal{DN}_{--}^{11} + K_-^1} \mathcal{DN}_{-+}^{10},$$

$$\tilde{\mathcal{DN}}_{++}^{01} = \mathcal{DN}_{++}^{01} - \mathcal{DN}_{+-}^{01} \frac{I}{\mathcal{DN}_{--}^{11} + K_-^1} \mathcal{DN}_{-+}^{11},$$

$$\tilde{\mathcal{DN}}_{++}^{10} = \mathcal{DN}_{++}^{10} - \mathcal{DN}_{+-}^{11} \frac{I}{\mathcal{DN}_{--}^{11} + K_-^1} \mathcal{DN}_{-+}^{10},$$

$$\tilde{\mathcal{DN}}_{++}^{11} = \mathcal{DN}_{++}^{11} - \mathcal{DN}_{+-}^{11} \frac{I}{\mathcal{DN}_{--}^{11} + K_-^1} \mathcal{DN}_{-+}^{11}.$$

Eliminating $s_+^1\nu$ from the second equation we obtain a finite-dimensional expression for the Scattering matrix of the junction

$$S(\lambda) = \frac{iK_+ - \left[\tilde{\mathcal{DN}}_{++}^{00} - \tilde{\mathcal{DN}}_{++}^{01} \frac{I}{\tilde{\mathcal{DN}}_{++}^{11} + K_+^1} \tilde{\mathcal{DN}}_{++}^{01}\right]}{iK_+ + \left[\tilde{\mathcal{DN}}_{++}^{00} - \tilde{\mathcal{DN}}_{++}^{01} \frac{I}{\tilde{\mathcal{DN}}_{++}^{11} + K_+^1} \tilde{\mathcal{DN}}_{++}^{01}\right]}, \tag{5.5}$$

with the denominator preceding the numerator. The ultimate representation (5.5) of the scattering matrix is completely finite-dimensional, hence more convenient for the computational process. The large parameter Λ_1 permits to eliminate the infinite-dimensional part K_-^1 of K_- and obtain a completely finite-dimensional formula (5.5) for the scattering matrix, without any additional assumptions on geometrical or physical parameters of the network. Actually essential details of the analytic perturbation process which are still present in (5.5) are mostly reloaded by (5.5) on the direct computing with finite matrices. Hence the formula (5.5) opens, in particular, a semi-analytic way of calculating of transmission coefficients across any junction. Comparison of the formula (5.5) with (3.4) implies the equation

$$\tilde{\mathcal{DN}}^{00}_{++} - \tilde{\mathcal{DN}}^{01}_{++} \frac{I}{\tilde{\mathcal{DN}}^{11}_{++} + K^1_+} \tilde{\mathcal{DN}}^{01}_{++} = \mathcal{DN}^F. \tag{5.6}$$

The terms of (5.6) contain sophisticated singularities inherited from the operator L_{int}. Again, we are able to transform this expression to another form, with all singularities compensated. We observe first the compensation singularities in **DN**, representing it in Krein's form. Denote

$$\mathcal{T}_+ = \sum_{\lambda_s \in \Delta} |\varphi_s\rangle \left\langle P^0_+ \frac{\partial \varphi_s}{\partial n} + P^1_+ \frac{\partial \varphi_s}{\partial n} \right|,$$

$$\mathcal{T}_- = \sum_{\lambda_s \in \Delta} |\varphi_s\rangle \left\langle P^1_- \frac{\partial \varphi_s}{\partial n} \right|,$$

and consider the rational approximation of **DN**

$$\mathbf{DN} = \mathbf{DN}(\Delta) + \mathcal{K}:$$

$$\mathbf{DN}_{++} := \begin{pmatrix} \mathcal{DN}^{00}_{++} & \mathcal{DN}^{01}_{++} \\ \mathcal{DN}^{10}_{++} & \mathcal{DN}^{11}_{++} \end{pmatrix} = \mathcal{T}^+_+ \frac{I}{\lambda I^\Delta - L^\Delta} \mathcal{T}_+ + \begin{pmatrix} \mathcal{K}^{00}_{++} & \mathcal{K}^{01}_{++} \\ \mathcal{K}^{10}_{++} & \mathcal{K}^{11}_{++} \end{pmatrix},$$

$$\mathbf{DN}_{+-} := \begin{pmatrix} \mathcal{DN}^{00}_{+-} & \mathcal{DN}^{01}_{+-} \\ \mathcal{DN}^{10}_{+-} & \mathcal{DN}^{11}_{+-} \end{pmatrix} = \mathcal{T}^+_+ \frac{I}{\lambda I^\Delta - L^\Delta} \mathcal{T}_- + \begin{pmatrix} \mathcal{K}^{01}_{+-} \\ \mathcal{K}^{11}_{+-} \end{pmatrix},$$

$$\mathbf{DN}_{-+} := \begin{pmatrix} \mathcal{DN}^{10}_{-+} & \mathcal{DN}^{11}_{-+} \\ \mathcal{DN}^{10}_{-+} & \mathcal{DN}^{11}_{-+} \end{pmatrix} = \mathcal{T}^+_- \frac{I}{\lambda I^\Delta - L^\Delta} \mathcal{T}_+ + (\mathcal{K}^{10}_{-+}, \mathcal{K}^{11}_{-+}).$$

Consider the Krein formula for $\tilde{\mathbf{DN}}$

$$\tilde{\mathbf{DN}} = \mathbf{DN}_{++} - \mathbf{DN}_{+-} \frac{I}{\mathcal{DN}^{11}_{--}(\Delta) + \mathcal{K}^{11}_{--} + K^1_-} \mathbf{DN}_{-+}. \tag{5.7}$$

Compensation of singularities in (5.7) inherited from the spectrum of L_{int} can be observed in the same way as the compensation of singularities in (3.5). Introduce

$$\mathcal{T}_{+-} \frac{I}{\mathcal{K}_{--} + K_-} \mathcal{T}^+_{-+} := Q(\lambda) : E_\Delta \to E_\Delta,$$

and

$$P_+ - \begin{pmatrix} \mathcal{K}^{01}_{+-} \\ \mathcal{K}^{11}_{+-} \end{pmatrix} \frac{I}{\mathcal{K}^{11}_{--} + K^1_-} P_- := \mathcal{J}(\lambda).$$

Theorem 5.2. *The Krein formula* (4.11) *for the* $\tilde{\mathbf{DN}}$ *can be re-written on the essential spectral interval, as:*

$$\tilde{\mathbf{DN}} = \begin{pmatrix} \mathcal{K}^{00}_{++} & \mathcal{K}^{01}_{++} \\ \mathcal{K}^{10}_{++} & \mathcal{K}^{11}_{++} \end{pmatrix} - \begin{pmatrix} \mathcal{K}^{01}_{+-} \\ \mathcal{K}^{11}_{+-} \end{pmatrix} \frac{I}{\mathcal{K}^{11}_{--} + K^{1}_{-}} \left(\mathcal{K}^{01}_{-+}, \mathcal{K}^{11}_{-+}\right)$$
$$+ \mathcal{J}T^{+}\rangle \frac{I}{\lambda I - L^{\Delta} + Q(\lambda)} \langle T\mathcal{J}^{+} = \mathcal{K}^{F}_{\Delta} + \mathcal{DN}^{F}_{\Delta}, \tag{5.8}$$

with

$$\mathcal{K}^{F}_{\Delta} := \begin{pmatrix} \mathcal{K}^{00}_{++} & \mathcal{K}^{01}_{++} \\ \mathcal{K}^{10}_{++} & \mathcal{K}^{11}_{++} \end{pmatrix} - \begin{pmatrix} \mathcal{K}^{01}_{+-} \\ \mathcal{K}^{11}_{+-} \end{pmatrix} \frac{I}{\mathcal{K}^{11}_{--} + K^{1}_{-}} \left(\mathcal{K}^{01}_{-+}, \mathcal{K}^{11}_{-+}\right)$$

and

$$\mathcal{DN}^{F}_{\Delta} := \mathcal{J}T^{+}\rangle \frac{I}{\lambda I - L^{\Delta} + Q(\lambda)} \langle T\mathcal{J}^{+} = \sum_{\lambda^{F}_{s}} \frac{\phi^{F}_{s}\rangle\,\langle\phi^{F}_{s}}{\lambda - \lambda^{F}_{s}}. \tag{5.9}$$

Here λ^{F}_{s} *are the eigenvalues of the intermediate Hamiltonian which arose from the eigenvalues of* L_{int} *on the essential spectral interval, and* ϕ^{F}_{s} *are the projections of the boundary currents of the corresponding normalized eigenfunctions* φ^{F}_{s} *of the intermediate Hamiltonian* L^{F} *onto the entrance subspace* E_{+} *of the open channels,*

$$\phi^{F}_{s} = P_{+} \frac{\partial \varphi^{F}_{s}}{\partial n}\Big|_{\Gamma}.$$

The summation on s in the above formula (5.9) *is spread over all (vector-) zeros of* $L^{\Delta} - \lambda I^{\Delta} + Q(\lambda)$ *which arose from the eigenvalues of* L_{int} *on the essential spectral interval. The representation* (4.11) *remains valid on some complex neighborhood* G_{Δ} *of the essential spectral interval.*

Note that the expression (5.6) is the Schur complement, see [26], of the matrix

$$\tilde{\mathbf{DN}} + \begin{pmatrix} 0 & 0 \\ 0 & K^{1}_{-} \end{pmatrix} = \begin{pmatrix} \mathcal{DN}^{00}_{++} & \mathcal{DN}^{01}_{++} \\ \mathcal{DN}^{10}_{++} & \mathcal{DN}^{11}_{++} + K^{1}_{-} \end{pmatrix}.$$

Absence of singularities at the spectrum of L_{int} in (5.8) is inherited by the Schur complement. Inserting the Schur complement into (5.5) gives an explicit formula for the scattering matrix of the junction in form:

$$S(\lambda) = \left\{iK_{+} - \left[\mathcal{DN}^{F}_{\Delta} + \mathcal{K}^{F}_{\Delta}\right]\right\} \left\{iK_{+} + \left[\mathcal{DN}^{F}_{\Delta} + \mathcal{K}^{F}_{\Delta}\right]\right\}^{-1}, \tag{5.10}$$

with the denominator preceding the numerator. The details of this representation can be recovered, if needed, from the above theorem 5.2. We leave this calculation to the reader. Note that the above expression (5.10) for the scattering matrix can be simplified if some additional assumption is imposed on K_{+}, $\mathcal{K}^{F}_{\Delta}$.

Definition 5.3. *We call the junction* Ω *thin in open channels on the essential spectral interval if*

$$\| K_{+}^{-1/2} \| \, \| K_{+}^{-1/2} \mathcal{K}^{F}_{\Delta} \| < 1.$$

Theorem 5.4. *If the junction Ω is thin in open channels on the essential spectral interval, then it can be obtained by the analytic perturbation procedure from the essential scattering matrix*

$$S_{\mathrm{ess}}(\lambda) = [iK_+ - \mathcal{DN}^F_\Delta][iK_+ + \mathcal{DN}^F_\Delta]^{-1}, \tag{5.11}$$

where denominator precedes the numerator and the intermediate DN-map $\mathcal{DN}^F = \mathcal{DN}^F_\Delta + \mathcal{K}^F_\Delta$ is substituted by the essential polar part $\mathcal{DN}^F_\Delta$.

Proof. Represent the numerator and the denominator of the right side of (5.10) as:

$$iK_+ - \left[\mathcal{DN}^F_\Delta + \mathcal{K}^F_\Delta\right] = \left(iK_+ - \mathcal{DN}^F_\Delta\right)\left[I - \left(iK_+ - \mathcal{DN}^F_\Delta\right)^{-1}\mathcal{K}^F_\Delta\right]$$
$$iK_+ + \left[\mathcal{DN}^F_\Delta + \mathcal{K}^F_\Delta\right] = \left(iK_+ + \mathcal{DN}^F_\Delta\right)\left[I + \left(iK_+ + \mathcal{DN}^F_\Delta\right)^{-1}\mathcal{K}^F_\Delta\right].$$

Notice that

$$\sup_{\lambda\in\Delta} \| \left(iK_+ + \mathcal{DN}^F_\Delta\right)^{-1}\mathcal{K}^F_\Delta \| < 1, \tag{5.12}$$

if

$$\| K_+^{-1/2} \| \, \| K_+^{-1/2}\mathcal{K}^F_\Delta \| < 1$$

Indeed, denote $\mathcal{DN}^F_\Delta := A$, $\mathcal{K}^F_\Delta := B$. Then

$$\begin{aligned}\| \frac{I}{iK_+ + A} Bu \| &\le \| K_+^{-1/2} \frac{I}{iI + K_+^{-1/2}AK_+^{-1/2}} K_+^{-1/2}Bu \| \\ &\le \| K_+^{-1/2} \| \, \| \frac{I}{iI + K_+^{-1/2}AK_+^{-1/2}} K_+^{-1/2}Bu \| \\ &\le \| K_+^{-1/2} \| \, \| K_+^{-1/2}Bu \| \le \| K_+^{-1/2} \| \, \| K_+^{-1/2}\mathcal{K}^F_\Delta u \| .\end{aligned}$$

This result implies (5.12). Now we can represent the scattering matrix as a product of three factors:

$$S(\lambda) = \left[I + \left(iK_+ + \mathcal{DN}^F_\Delta\right)^{-1}\mathcal{K}^F_\Delta\right]^{-1} S_{\mathrm{ess}}(\lambda)\left[I - \left(iK_+ - \mathcal{DN}^F_\Delta\right)^{-1}\mathcal{K}^F_\Delta\right]. \tag{5.13}$$

The central factor coincides with the essential scattering matrix, and the left and right factors contain the small parameter $\left(iK_+ \pm \mathcal{DN}^F_\Delta\right)^{-1}\mathcal{K}^F_\Delta$. Hence the first factor can be decomposed into the geometrically convergent series. Thus the scattering matrix can be obtained from the essential scattering matrix via standard analytic perturbation procedure, with the above small parameter. □

Remark 5.5. Denote by λ_0 the vector zero of the numerator of the essential scattering matrix:

$$\left[iK_+ - \mathcal{DN}^F_\Delta\right] e_0 = 0.$$

For the network which is *sufficiently thin* on the open channels the estimate

$$\sup_{\lambda\in\Sigma_\epsilon} \| \left[iK_+ - \mathcal{DN}^F_\Delta\right]^{-1}\mathcal{K}^F_\Delta \| < 1$$

is valid on a small circle $\Sigma_\epsilon = \{|\lambda - \lambda_0| = \epsilon\}$ centered at λ_0. Then, due to the operator-valued Rouche theorem, [34] the numerators of the original and the essential scattering matrices have equal total multiplicity of vector zeros inside Σ_ϵ, because

$$\sup_{\Sigma_\epsilon} \| I - \left[iK_+ - \mathcal{DN}^F_\Delta\right]^{-1} \left[iK_+ - \mathcal{DN}^F_\Delta - \mathcal{K}^F_\Delta\right] \| = \sup_{\Sigma_\epsilon} \| \left[iK_+ - \mathcal{DN}^F_\Delta\right]^{-1} \mathcal{K}^F_\Delta \| < 1.$$

Thus the zeros of the original scattering matrix – resonances – of the thin junction are situated close to the zeros of the essential scattering matrix. A relevant perturbation procedure may be developed for calculation of the resonances. Localization of zeros is important for estimation of speed of transition processes in the junction, if it is used as a switch.

Remark 5.6. The above statement (5.11) and the formula (5.13) permits to substitute, on the essential spectral interval, the scattering matrix of a thin junction by the essential scattering matrix. According to [30], the essential scattering matrix can be interpreted as a scattering matrix of a solvable model.

6. Conclusion: role of solvable models in analytic perturbation procedure and a relevant realization problem

The solvable model of thin junction fitted on a certain essential spectral interval can serve as a first step – *a jump-start*, see [29] – of the modified analytic perturbation procedure which is applied to perturbation of embedded eigenvalues, see extended comments in [22]. The proposed jump-start procedure confirms the hypothesis of H. Poincaré, about the role of resonances in analytic perturbation procedure: elimination, due to the chain-rule for the scattering matrices, of resonances on the essential spectral interval Δ permits to construct a convergent analytic perturbation procedure. Unfortunately none finite degree of precision in our approximations for the scattering matrix allows to construct the solvable model with exactly the same resonances on Δ as in original scattering problem. Nevertheless one can say, that zero-range solvable models of the quantum system, see for instance [36, 37, 38, 39] could, after appropriate fitting, play a role of the jump start. Our jump-start solvable models, see also [29, 30, 40, 41, 22] are automatically fitted, because the corresponding scattering matrix serves an approximation of the whole scattering matrix of the original perturbed operator.

It may be interesting that Nobel Prize winner 1972 Iliya Prigogine, see [42], inspired by the above-mentioned idea of H. Poincaré, [35], about the role of resonances in analytic perturbation procedure, attempted to construct an "intermediate operator" – a version of our jump-start – as a tool of analytic perturbation procedure on continuous spectrum. His attempt was not successful, because he imposed, in advance, too strong conditions on the object of his search. In particular, he assumed that the intermediate operator should be a function of the non-perturbed Hamiltonian. Our jump-start is obtained based on local rational approximation of the corresponding DN-map, and it is constructed via finite-dimensional perturbation of the original Hamiltonian, with the same leading resonances on the essential spectral interval.

We were able, see Sections 4, 5 and references therein, to construct a solvable model of a *thin junction* in the Hilbert space with a standard positive metric. We conjecture, that a similar solvable model can be constructed for *arbitrary junction* when using operators in Pontryagin space based on the corresponding realization theorems, see for instance [43]. Note that solvable models in Pontryagin space are more flexible, but yet reduce to a standard selfadjoint operators in the positive invariant subspace of scattered waves, see for instance [40, 41].

References

[1] Yu.A. Bychkov and E.I. Rashba, *Oscillatory effects and the magnetic susceptibility of carriers in inversion layers.* J. Phys. C. **171** (1984), 6039–6045.

[2] S. Datta and B. Das Sarma, *Electronic analog of the electro-optic modulator.* Appl. Phys. Lett. **56** (1990), no. 7, 665–667.

[3] V. Adamyan, *Scattering matrices for microschemes.* Operator Theory: Adv. and Appl. **59** (1992), no. 1.

[4] J. Splettstoesser, M. Governale, and U. Zülicke, *Persistent current in ballistic mesoscopic rings with Rashba spin-orbit coupling.* Phys. Rev. B, 68:165341, (2003).

[5] I.A. Shelykh, N.G. Galkin, and N.T. Bagraev, *Quantum splitter controlled by Rashbe spin-orbit coupling.* Phys. Rev. B **72** (2005), 235–316.

[6] P. Kuchment and Zeng, *Convergence of spectra of mesoscopic systems collapsing onto graph.* Journal of Mathematical Analysis and Application **258** (2001), 671–700.

[7] P. Exner, O. Post, *Convergence of graph-like thin manifolds.* J. Geom. Phys. **541** (2005), 77–115.

[8] J. Gabelli, G. Feve, J.-M. Berroir, B. Placais, A. Cavanna, B. Etienne, Y. Jin, D.C. Glatti, *Violation of Kirchhoff's Laws for a coherent RC Cirquit.* Science **313** (2006), 499–502.

[9] O. Madelung. *Introduction to solid-state theory.* Translated from German by B.C. Taylor. Springer Series in Solid-State Sciences, 2. Springer-Verlag, Berlin, New York, 1978.

[10] M. Harmer, B. Pavlov, A. Yafyasov *Boundary conditions at the junction.* International Workshop on Computational Electronics (IWCE-11), Vienna, 25 May–29 May 2006, book of abstracts, 241–242.

[11] N. Bagraev, A. Mikhailova, B.S. Pavlov, L.V. Prokhorov, and A. Yafyasov. *Parameter regime of a resonance quantum switch.* Phys. Rev. B, 71:165308, 2005.

[12] R. Mittra, S. Lee *Analytical techniques in the theory of guided waves* The Macmillan Company, NY, Collier-Macmillan Limited, London, 1971.

[13] R. Courant, D. Hilbert, *Methods of mathematical physics. Vol. II. Partial differential equations.* Reprint of the 1962 original. Wiley Classics Library. A Wiley-Interscience Publication. John Wiley & Sons, Inc., New York, 1989. xxii+830 pp.

[14] J. Sylvester, G. Uhlmann *The Dirichlet to Neumann map and applications.* Proceedings of the Conference "Inverse problems in partial differential equations (Arcata, 1989)", SIAM, Philadelphia **101** (1990).

[15] B. Pavlov, *S-Matrix and Dirichlet-to-Neumann Operators.* Encyclopedia of Scattering, ed. R. Pike, P. Sabatier, Academic Press, Harcourt Science and Tech. Company (2001), 1678–1688.

[16] W.O. Amrein, D.B. Pearson, *M-operators: a generalization of Weyl-Titchmarsh Theory.* Journal of Computational and Applied Mathematics **171** (2004), 1-2, 1–26.

[17] M. Marletta, *Eigenvalue problems on exterior domains and Dirichlet-to-Neumann map.* J. Comp. Appl. Math. **171** (2004), 1-2 (2004), 367–391.

[18] B.M. Brown, M. Marletta, *Spectral inclusion and spectral exactness for PDEs on exterior domain.* IMA J. Numer. Anal. **24** (2004), 1, 21–43.

[19] B.N. Brown, M.S.P. Eeasham, W.D. Evans, *Laudatum* [Norrie Everitt]. J. Comput. Appl. Math. **171** (2004), 1–2.

[20] F. Gesztesy, Y. Latushkin, M. Mitrea, M. Zinchenko, *Non-selfadjoint operators, infinite determinants and some applications.*, Russian journal of Mathematical Physics **12** (2005), 443–471.

[21] Mikhailova, B. Pavlov, L. Prokhorov, *Modelling of quantum networks.* arXiv math-ph/031238, 2004, 69 p.

[22] A. Mikhailova, B. Pavlov, L. Prokhorov, *Intermediate Hamiltonian via Glazman splitting and analytic perturbation for meromorphic matrix-functions.* Mathematische Nachrichten **280** (2007), 12, 1376–1416.

[23] B. Pavlov *On one-dimensional scattering of plane waves on an arbitrary potential.* Teor. i Mat. Fiz. **16** (1973), no. 1, 105–119.

[24] F. Gesztesy, B. Simon, *Inverse spectral analysis with partial information on the potential. I. The case of an a.c. component in the spectrum.* Papers honouring the 60th birthday of Klaus Hepp and of Walter Hunziker, Part II (Zürich, 1995). Helv. Phys. Acta **70** (1997), no. 1-2, 66–71.

[25] F. Gesztesy, R. Nowell, W. Pötz, *One-dimensional scattering theory for quantum systems with nontrivial spatial asymptotics.* Differential Integral Equations **10** (3), (1997), 521–546.

[26] V. Adamyan, H. Langer, R. Mennicken, *Spectral decomposition of selfadjoint block operator matrices with unbounded entries.* Mathematische Nachrichten **178** (1996), 43–80.

[27] V. Adamyan, B. Pavlov, *Zero-radius potentials and M.G. Krein's formula for generalized resolvents.* Proc. LOMI **149** (1986), 7–23.

[28] B. Pavlov, *The theory of extensions and explicitly solvable models.* Russian Math. Surveys **42:6** (1987), 127–168.

[29] B. Pavlov, I. Antoniou, *Jump-start in analytic perturbation procedure for Friedrichs model.* J. Phys. A: Math. Gen. **38** (2005), 4811–4823.

[30] B. Pavlov, *A star-graph model via operator extension.* Mathematical Proceedings of the Cambridge Philosophical Society, **142**, Issue 02, March 2007, 365–384 doi: 10.1017/S0305004106009820, Published online by Cambridge University Press 10 Apr. 2007.

[31] V. Bogevolnov, A. Mikhailova, B. Pavlov, A. Yafyasov, *About Scattering on the Ring.* Operator Theory: Advances and Applications, (Israel Gohberg Anniversary Conference, Groningen), Ed. A. Dijksma, A.M. Kaashoek, A.C.M. Ran), Birkhäuser, Basel (2001), 155–187.

[32] N.I. Akhiezer, I.M. Glazman, *Theory of Linear Operators in Hilbert Space.*, (Frederick Ungar, Publ., New York, vol. 1, 1966) (Translated from Russian by M. Nestel).

[33] M. Harmer, *Hermitian symplectic geometry and extension theory.* Journal of Physics A: Mathematical and General **33** (2000), 9193–9203.

[34] I.S. Gohberg and E.I. Sigal, *Operator extension of the theorem about logarithmic residue and Rouché theorem.* Mat. Sbornik. **84** (1971), 607.

[35] H. Poincaré, *Méthodes nouvelles de la mécanique céleste.* **1**, 1892, Second edition: Dover, New York, 1957.

[36] E. Fermi, *Sul motto dei neutroni nelle sostance idrogenate.* (in Italian) Richerca Scientifica **7** (1936), 13.

[37] F.A. Berezin, L.D. Faddeev, *A remark on Schrödinger equation with a singular potential.* Soviet Math. Dokl. **2** (1961), 372–376.

[38] Yu.N. Demkov, V.N. Ostrovskij, *Zero-range potentials and their applications in Atomic Physics*, (Plenum Press, NY-London, 1988).

[39] S. Albeverio, P. Kurasov, *Singular perturbations of differential operators. Solvable Schrödinger type operators.* London Mathematical Society Lecture Note Series, **271**. Cambridge University Press, Cambridge, 2000. xiv+429 pp.

[40] B. Pavlov, V. Kruglov, *Operator Extension technique for resonance scattering of neutrons by nuclei.* Hadronic Journal **28** (2005), June, 259–268.

[41] B.S. Pavlov, V.I. Kruglov, *Symplectic operator-extension techniques and zero-range quantum models.* New Zealand J. Math. **34** (2005), no. 2, 125–142.

[42] I. Prigogine *Irreversibility as a Symmetry-breaking Process.* Nature, **246** (1973), 9.

[43] S. Belyi, S. Hassi, H. de Snoo, E. Tsekanovskii, *A general realization theorem for matrix-valued Herglotz-Nevanlinna functions.* Linear Algebra Appl. **419** (2006), no. 2-3, 331–358.

V. Adamyan
Department of Theorertical Physics, Odessa I.I. Mechnikov National University
65082 Odessa, Ukraine
e-mail: `vadamyan@paco.net`

B. Pavlov
NZ Institute for Advanced Studies, Institute of Information and Mathematical Sciences
Massey University, Albany Campus, Private Bag 102-904, North Shore Mail Centre
Auckland, New Zealand
e-mail: `B.Pavlov@massey.ac.nz`

A. Yafyasov
V. Fock Institute for Physics, St. Petersbourg University
198904 St. Petersburg, Russian Federation
e-mail: `yafyasov@desse.phys.spbu.ru`

Operator Theory:
Advances and Applications, Vol. 190, 27–63

The Schur Transformation for Nevanlinna Functions: Operator Representations, Resolvent Matrices, and Orthogonal Polynomials

D. Alpay, A. Dijksma and H. Langer

Dedicated to Mark Krein on the occasion of his 100th anniversary

Abstract. A Nevanlinna function is a function which is analytic in the open upper half-plane and has a non-negative imaginary part there. In this paper we study a fractional linear transformation for a Nevanlinna function n with a suitable asymptotic expansion at ∞, that is an analogue of the Schur transformation for contractive analytic functions in the unit disk. Applying the transformation p times we find a Nevanlinna function n_p which is a fractional linear transformation of the given function n. The main results concern the effect of this transformation to the realizations of n and n_p, by which we mean their representations through resolvents of self-adjoint operators in Hilbert space. Our tools are block operator matrix representations, u-resolvent matrices, and reproducing kernel Hilbert spaces.

Mathematics Subject Classification (2000). Primary 47A57, 47B32, Secondary 42C05, 30E05.

Keywords. Schur transformation, Nevanlinna function, realization, symmetric operator, self-adjoint operator, moment problem, reproducing kernel Hilbert space, orthogonal polynomials.

D. Alpay acknowledges with thanks the Earl Katz family for endowing the chair which supported this research. The research of A. Dijksma and H. Langer was supported in part by the Center for Advanced Studies in Mathematics (CASM) of the Department of Mathematics, Ben-Gurion University.

1. Introduction

In the papers [4] and [5] the Schur transformation for generalized Nevanlinna functions with a reference point z_1 in the open upper half-plane was considered. An analogous transformation for Nevanlinna functions (for the definition of a Nevanlinna function see Section 2) and for the reference point ∞ is defined in [1, Lemma 3.3.6], see [3]. This transformation or a simple modification of it we call here the *Schur transformation for Nevanlinna functions*, and it is the starting point for the present paper. To give more details, we consider a Nevanlinna function n which has for some integer $p \geq 1$ an asymptotic expansion of order $2p+1$ at ∞, for example

$$n(z) = -\frac{s_0}{z} - \frac{s_1}{z^2} - \cdots - \frac{s_{2p}}{z^{2p+1}} + \mathrm{o}\left(\frac{1}{z^{2p+1}}\right), \quad z = \mathrm{i}y,\ y \to \pm\infty. \tag{1.1}$$

The Schur transform $\widehat{n}$ of n is the function

$$\widehat{n}(z) := -\frac{s_0}{n(z)} - z + \frac{s_1}{s_0}; \tag{1.2}$$

the relation between n and $\widehat{n}$ can also be written as

$$n(z) = -\frac{s_0}{z - \dfrac{s_1}{s_0} + \widehat{n}(z)}.$$

The transformed function $\widehat{n} =: n_1$ is again a Nevanlinna function, but in general with an asymptotic expansion of the form (1.1) of lower order $2p-1$, and if $p > 1$ the Schur transformation can be again applied to n_1 etc. As a result we obtain a finite sequence of Nevanlinna functions $n_1 = \widehat{n}, n_2 = \widehat{n}_1, \ldots, n_p = \widehat{n}_{p-1}$; this is the sequence of functions that appears in the asymptotic expansion of n by continued fractions, see [1, Section 3.3.6].

The transformation (1.2) is closely related to the finite Hamburger moment problem. We recall that the Nevanlinna function n with an asymptotic expansion (1.1) admits an integral representation

$$n(z) = \int_{-\infty}^{\infty} \frac{d\sigma(t)}{t-z}, \quad z \neq z^*,$$

where σ is a bounded non-decreasing function on $\mathbb{R}$. The coefficients s_j in (1.1) are the *moments* of the function σ:

$$s_j = \int_{-\infty}^{\infty} t^j \, d\sigma(t), \quad j = 0, 1, \ldots, 2p. \tag{1.3}$$

The moment problem we have in mind is the problem to determine all Nevanlinna functions n with an expansion (1.1) and given coefficients s_j, $j = 0, 1, \ldots, p$, see [1, Theorem 3.2.1].

An essential feature in our studies are operator representations or so-called realizations of Nevanlinna functions, see [14], [11], and [10]. In fact, if the Nevanlinna function n admits an asymptotic expansion (1.1) its operator representation takes the simple form

$$n(z) = \big((A - z)^{-1}u, u\big), \quad z \neq z^*, \tag{1.4}$$

with some Hilbert space $\mathcal{H}$ with inner product $(\cdot, \cdot)$, $u \in \mathcal{H}$, and a self-adjoint operator A in $\mathcal{H}$. We study the corresponding operator representation of the Schur transform $\widehat{n}$, and also of the functions $n_2, \dots, n_p$. For example, the function $\widehat{n}$ admits an operator representation of the form (1.4) with a Hilbert space $\widehat{\mathcal{H}}$, an operator $\widehat{A}$, and an element $\widehat{u}$ which are the orthogonal complement of the element u in $\mathcal{H}$, the compression of A to $\widehat{\mathcal{H}}$, and a multiple of the projection of Au onto $\widehat{\mathcal{H}}$, respectively. After applying the Schur transformation p times, the resulting function n_p admits an operator representation of the form (1.4) with the space

$$\mathcal{H}_p' = \mathcal{H} \ominus \mathcal{H}_p, \quad \mathcal{H}_p := \operatorname{span}\{u, Au, \dots, A^{p-1}u\},$$

the operator that is the compression of A to this space, and an element which is a multiple of the projection of $A^p u$ onto $\mathcal{H}_p'$.

Since n_p is obtained by subsequent application of fractional linear transformations of the form (1.2), there is a fractional linear relation between the function n and the transformed function n_p. We derive an explicit form for the defining 2×2 matrix function V of this relation in three ways: By calculating the resolvent of the operator A in its 2×2 block matrix operator form corresponding to the decomposition $\mathcal{H} = \mathcal{H}_p \oplus \mathcal{H}_p'$, by means of the description of all generalized resolvents of a certain symmetric restriction of A with defect one in the space $\mathcal{H}_{p+1}$, and via reproducing kernel methods using the non-negative Nevanlinna kernel

$$L_n(z, w) = \frac{n(z) - n(w)^*}{z - w^*}, \quad z, w \in \mathbb{C} \setminus \mathbb{R}, \ z \neq w^*.$$

For the Nevanlinna function n with an asymptotic expansion (1.1), polynomials e_j and d_j, $j = 1, 2, \dots, p$, of first and second kind can be defined by the well-known formulas, see [1, Chapter I]. Recall that e_j is a polynomial of degree j, and that d_j is a polynomial of degree $j - 1$. We show that the polynomials $\widehat{e}_j$ of first kind of the transformed function $\widehat{n}$ coincide, up to constant factor, with the polynomials $d_{j+1}(z)$ of second kind for the given function n, whereas the polynomials $\widehat{d}_j$ of second kind for $\widehat{n}$ are linear combinations of e_{j+1} and d_{j+1}. As a consequence, the polynomials of second kind for n are orthogonal with respect to the measure generated by the non-decreasing function $\widehat{\sigma}$ in the representation of the form (1.4) of the Nevanlinna function $\widehat{n}$; in this statement $\widehat{n}$ can be replaced by the function $-1/n$. As in the classical moment problem, the 2×2 matrix function V, which determines the fractional linear relation between n and n_p, can be represented by the polynomials of first and second kind.

A short synopsis is as follows. The Schur transformation is defined in the next section. We start with weaker forms of the asymptotic expansion (1.1), for example

$$n(z) = -\frac{s_0}{z} - \frac{s_1}{z^2} + \mathrm{o}\left(\frac{1}{z^2}\right), \quad z = \mathrm{i}y, \ y \to \pm\infty,$$

and consider also a weaker form of the Schur transformation. In Section 3 we mention three concrete forms of the operator representation of n. The basic result of this section is Theorem 3.1 which describes the operator model for the transformed function. Higher-order approximations and the corresponding polynomials of first and second kind are introduced in Section 4. In the operator model an asymptotic expansion (1.1) can be characterized by the fact that $u \in \operatorname{dom} A^p$. The main result of this section is the relation between the polynomials of first and second kind of n and $\widehat{n}$ which was mentioned above. The reduction via a p-dimensional subspace, that corresponds to p subsequent applications of the Schur transformation, is given in Section 5 by means of a block operator matrix representation of A. In Section 6 the corresponding transformation matrix V is expressed in terms of the polynomials of first and second kind. Although the final formulas are well known (see for example [1]) this approach seems to be new.

In Section 7, applying the theory of u-resolvent matrices, we derive a representation of a transformation matrix in an explicit form by means of the given moments; it corresponds to Potapov's formula for the solution matrix of the Nevanlinna–Pick problem, compare also [2]. Finally, in Section 8 we explain the connection between n and n_p through some basic results from the theory of resolvent invariant reproducing kernel spaces, and give another proof for the representation of the transformation matrix by orthogonal polynomials.

2. The Schur transformation

1. A Nevanlinna function is a complex function n which is defined and analytic in the upper half-plane $\mathbb{C}^+$ and has the property

$$z \in \mathbb{C}^+ \quad \Longrightarrow \quad \operatorname{Im} \, n(z) \geq 0.$$

We always suppose that n is extended to the lower half-plane $\mathbb{C}^-$ by the relation

$$n(z) = n(z^*)^*, \quad z \in \mathbb{C}^-, \tag{2.1}$$

and to those points of the real axis into which it can be continued analytically. The set of all Nevanlinna functions is denoted by $\mathbf{N}_0$. Recall that $n \in \mathbf{N}_0$ if and only if n is analytic in $\mathbb{C} \setminus \mathbb{R}$ and the kernel

$$L_n(z,w) = \frac{n(z) - n(w)^*}{z - w^*}, \quad z, w \in \mathbb{C} \setminus \mathbb{R}, \; z \neq w^*,$$

is positive definite.

Let $n \in \mathbf{N}_0$ and consider the following properties of n:

$$(1_0)\ n(z) = -\frac{s_0}{z} + \mathrm{o}\left(\frac{1}{z}\right), \qquad (2_0)\ n(z) = -\frac{s_0}{z} + \mathrm{O}\left(\frac{1}{z^2}\right),$$

$$(3_0)\ n(z) = -\frac{s_0}{z} - \frac{s_1}{z^2} + \mathrm{o}\left(\frac{1}{z^2}\right),$$

where here and in the following, the limit relations are understood to hold for $z = \mathrm{i}y$, $y \to \pm\infty$. The assumption (2.1) implies that s_0 and s_1 are real numbers. Evidently, $(3_0) \Longrightarrow (2_0) \Longrightarrow (1_0)$. The function n satisfies the assumption (1_0) if and only if it belongs to the class (R_0) of [13], which means that it admits an integral representation

$$n(z) = \int_{-\infty}^{+\infty} \frac{1}{t-z}\, d\sigma(t), \quad z \in \mathbb{C} \setminus \mathbb{R}, \tag{2.2}$$

where σ is a bounded non-decreasing function on $\mathbb{R}$. Then

$$\int_{-\infty}^{+\infty} d\sigma(t) = s_0,$$

hence $s_0 \geq 0$, and if $s_0 = 0$ then $n(z) \equiv 0$. With the representation (2.2) of n the assumption

$$\int_{-\infty}^{\infty} |t|\, d\sigma(t) < \infty \tag{2.3}$$

implies that (3_0) is satisfied. Indeed, (2.3) implies that

$$s_1 = \int_{-\infty}^{\infty} t\, d\sigma(t)$$

exists and with $z = iy$

$$z^2\left(n(z) + \frac{s_0}{z} + \frac{s_1}{z^2}\right) = \int_{-\infty}^{\infty} \frac{t^2}{t-z}\, d\sigma(t) = \int_{-\infty}^{\infty} \frac{t^2 + iyt}{t^2+y^2}\, t\, d\sigma(t) = \mathrm{o}(1).$$

The assumptions (1_0), (2_0), and (3_0) are all different. To see that $(1_0) \not\Longrightarrow (2_0)$ we show that if $n \in \mathbf{N}_0$ has the representation (2.2) with $\operatorname{supp} \sigma \subset [0, \infty)$ and

$$\int_0^{\infty} d\sigma(t) < \infty, \qquad \int_0^{\infty} t\, d\sigma(t) = \infty,$$

then (2_0) does not hold: Let $c > 0$ be given arbitrarily (large) and choose $K > 0$ such that $\int_0^K t\, d\sigma(t) \geq c$. If y is chosen large enough then for $0 \leq t \leq K$ we have

$$\frac{y^2}{t^2+y^2} \geq \frac{1}{2},$$

and hence

$$\int_{-\infty}^{\infty} \frac{y^2 t}{t^2+y^2}\, d\sigma(t) \geq \frac{c}{2},$$

and therefore, with $z = iy$,

$$z^2\left(n(z) + \frac{s_0}{z}\right) = z\int_{-\infty}^{\infty} \frac{t}{t-z}\, d\sigma(t) = \int_{-\infty}^{\infty} \frac{-y^2 t + iyt^2}{t^2+y^2}\, d\sigma(t) \neq \mathrm{O}(1),$$

which implies that (2_0) does not hold. Thus, for example, the function

$$n(z) = \frac{-1}{z-\sqrt{-z}} = \int_0^\infty \frac{1}{t-z}\,\frac{dt}{\pi(t+1)\sqrt{t}}$$

satisfies (1_0) but not (2_0).

Let n be the Nevanlinna function, defined in the upper half-plane by

$$n(z) = \frac{-s_0}{z+\gamma+f(z)},\ z\in\mathbb{C}^+,$$

where s_0 is a positive real number, γ is a complex number with $\operatorname{Im}\gamma > 0$, and f is a Nevanlinna function such that $f(z) = \mathrm{o}(1)$. It has the properties

$$\lim_{z=iy,\,y\to\infty} z^2\left(n(z)+\frac{s_0}{z}\right) = \gamma s_0, \qquad \lim_{z=iy,\,y\to-\infty} z^2\left(n(z)+\frac{s_0}{z}\right) = \gamma^* s_0,$$

and hence n satisfies (2_0) but, since the two limits are different (and non-real), it does not satisfy (3_0).

Instead of the assumption (3_0) also the assumption

$$n(z) = -\frac{s_0}{z} - \frac{s_1}{z^2} + \mathrm{O}\left(\frac{1}{z^3}\right) \tag{2.4}$$

seems reasonable. However, according to [14, Bemerkung 1.11], (2.4) implies the existence of a real number s_2 such that

$$n(z) = -\frac{s_0}{z} - \frac{s_1}{z^2} - \frac{s_2}{z^3} + \mathrm{o}\left(\frac{1}{z^3}\right); \tag{2.5}$$

this relation will be considered in Section 4 as assumption (1_1). The implication (2.4) $\Longrightarrow$ (2.5) can also be seen from the integral representation (2.2) of n: (2.4) implies

$$z^3\left(\int_{-\infty}^\infty \frac{1}{t-z}+\frac{1}{z}\,d\sigma(t)+\frac{s_1}{z^2}\right) = \mathrm{O}(1),$$

and hence with $z = iy$

$$-\int_{-\infty}^\infty \frac{y^2t(t+iy)}{t^2+y^2}\,d\sigma(t) + iys_1 = \mathrm{O}(1), \quad y\to\infty.$$

Taking the imaginary part we see that

$$s_1 = \lim\nolimits_{y\to\infty}\int_{-\infty}^\infty \frac{y^2t}{t^2+y^2}\,d\sigma(t)$$

and taking the real part we see that there exist real numbers C and y_0 such that

$$\int_{-\infty}^\infty \frac{y^2t^2}{t^2+y^2}\,d\sigma(t) \le C, \quad y\ge y_0.$$

This implies that

$$s_2 := \int_{-\infty}^\infty t^2\,d\sigma(t) < \infty,$$

hence

$$\int_{-\infty}^{\infty} |t|\, d\sigma(t) < \infty$$

and

$$s_1 = \int_{-\infty}^{\infty} t\, d\sigma(t).$$

Now (2.5) easily follows from the integral representations of n and the expressions for the real numbers s_0, s_1, and s_2: With $z = iy$ we have

$$z^3\left(n(z) + \frac{s_0}{z} + \frac{s_1}{z^2} + \frac{s_2}{z^3}\right) = \int_{-\infty}^{\infty} \frac{t^3}{t-z}\, d\sigma(t) = \int_{-\infty}^{\infty} \frac{t^4 + it^3 y}{t^2 + y^2}\, d\sigma(t) = \mathrm{o}(1).$$

2. Now we define the basic transformations considered this paper.

Definition 2.1. If $n \in \mathbf{N}$ satisfies the assumption (1_0) or (2_0), the *Schur type transform* $\widetilde{n}$ *of* n is the function

$$\widetilde{n}(z) = \frac{-s_0}{n(z)} - z, \tag{2.6}$$

if $n \in \mathbf{N}$ satisfies the assumption (3_0) the *Schur transform* $\widehat{n}$ *of* n is the function

$$\widehat{n}(z) = \frac{-s_0}{n(z)} - z + \frac{s_1}{s_0}. \tag{2.7}$$

The difference between the formulas (2.6) and (2.7) is just in the additive real constant s_1/s_0: under the stronger assumption (3_0) this constant assures that the transform tends to zero if z tends to $\pm\infty$ along the imaginary axis, see (2.10) below.

The relations (2.6) and (2.7) can also be written as a first step of a continued fraction expansion

$$n(z) = -\frac{s_0}{z + \widetilde{n}(z)}, \qquad \text{or } n(z) = -\frac{s_0}{z - \frac{s_1}{s_0} + \widehat{n}(z)}.$$

Theorem 2.2. *The following equivalences hold:*

$$n \in \mathbf{N}_0 \text{ and satisfies } (1_0) \Longleftrightarrow \widetilde{n} \in \mathbf{N}_0,\ \widetilde{n}(z) = \mathrm{o}(z), \tag{2.8}$$

$$n \in \mathbf{N}_0 \text{ and satisfies } (2_0) \Longleftrightarrow \widetilde{n} \in \mathbf{N}_0,\ \widetilde{n}(z) = \mathrm{O}(1), \tag{2.9}$$

$$n \in \mathbf{N}_0 \text{ and satisfies } (3_0) \Longleftrightarrow \widehat{n} \in \mathbf{N}_0,\ \widehat{n}(z) = \mathrm{o}(1). \tag{2.10}$$

Proof. We have

$$\widetilde{n}(z) = \widehat{n}(z) - \frac{s_1}{s_0} = -\frac{s_0 + zn(z)}{n(z)}.$$

A straightforward calculation yields

$$\operatorname{Im}\widetilde{n}(z) = \operatorname{Im}\widehat{n}(z) = \frac{\operatorname{Im} z}{|n(z)|^2}\left(s_0 \frac{\operatorname{Im} n(z)}{\operatorname{Im} z} - |n(z)|^2\right),$$

and the estimate

$$|n(z)|^2 = \left|\int_{-\infty}^{+\infty} \frac{d\sigma(t)}{t-z}\right|^2 \leq \int_{-\infty}^{+\infty} \frac{d\sigma(t)}{|t-z|^2} \int_{-\infty}^{+\infty} d\sigma(t) = \frac{\operatorname{Im} n(z)}{\operatorname{Im} z} s_0$$

implies $\widetilde{n}$, $\widehat{n} \in \mathbf{N}_0$. The asymptotic properties of $\widetilde{n}$ follow in case (2.8) from the relation

$$\frac{\widetilde{n}(z)}{z} = -\frac{s_0}{zn(z)} - 1 = \mathrm{o}(1),$$

in case (2.9) from the relation

$$\widetilde{n}(z) = -\frac{s_0 + zn(z)}{n(z)} = \frac{z\,\mathrm{O}\left(\frac{1}{z}\right)}{zn(z)},$$

and for $\widehat{n}(z)$ in case (2.10) in a similar way or from [1, Lemma 3.3.6].

Conversely, starting from $\widetilde{n}(z)$ as in (2.8), the relation

$$z\left(n(z) + \frac{s_0}{z}\right) = s_0 \frac{\dfrac{\widetilde{n}(z)}{z}}{1 + \dfrac{\widetilde{n}(z)}{z}}$$

implies that from $\widetilde{n}(z) = \mathrm{o}(z)$ it follows that n satisfies (1_0). The corresponding proofs for (2.9) and (2.10) are similar. □

3. Self-adjoint operator representations

A function $n \in \mathbf{N}_0$ admits a *self-adjoint operator representation* or *realization* with a self-adjoint relation A in some Hilbert space $\mathcal{H}$ of the form

$$n(z) = n(z_0)^* + (z - z_0^*)\left(\left(I + (z - z_0)(A - z)^{-1}\right) v, v\right) \tag{3.1}$$

with z_0 an arbitrary non-real number z_0 and an element $v \in \mathcal{H}$, see [14], [11], and [10]. If v is chosen to be a *generating element for* A, which means that

$$\mathcal{H} = \overline{\mathrm{span}}\left(\{v\} \cup \{(A - z)^{-1}v \,|\, z \in \mathbb{C} \setminus \mathbb{R}\}\right)$$

and which is always possible, then the operator representation (3.2) is called *minimal* and then it is unique up to unitary equivalence. We have the following equivalences, see [17]:

$$A \text{ is an operator} \Longleftrightarrow n(z) = \mathrm{o}(z),$$
$$v \in \mathrm{dom} A \Longleftrightarrow \lim_{y \to \infty} y \,\mathrm{Im}\, n(iy) < \infty;$$

for $n \in \mathbf{N}_0$ the latter limit always exists: it is either a non-negative number or ∞.

If the Nevanlinna function n satisfies the assumption (1_0) (or any of the assumptions (2_0), (3_0)) the representation (3.1) can be simplified to

$$n(z) = \left((A - z)^{-1}u, u\right), \qquad z \in \mathbb{C} \setminus \mathbb{R}, \tag{3.2}$$

where A is a self-adjoint operator in some Hilbert space $\mathcal{H}$, $u \in \mathcal{H}$, $(u, u) = s_0$. If u is chosen to be a generating element for A, or equivalently,

$$\mathcal{H} = \overline{\mathrm{span}}\,\{(A - z)^{-1}u \,|\, z \in \mathbb{C} \setminus \mathbb{R}\}$$

which is always possible, then the operator representation (3.2) is also called *minimal* and then it is unique up to unitary equivalence. The representation (3.2) follows from (3.1) and the above-mentioned equivalences by taking $u = c(A - z_0)v$ with some unimodular complex number c.

Here are three examples for a more concrete choice of the triplet $\mathcal{H}$, A, u in (3.2) for the given function $n \in \mathbf{N}_0$ with integral representation (2.2).

(1) $\mathcal{H} = L^2(\sigma)$, A is the operator of multiplication with the independent variable, and $u(t) \equiv 1$, $t \in \mathbb{R}$.

(2) $\mathcal{H}$ is the completion of the linear span of the functions $\mathbf{r}_z$, $z \in \mathbb{C} \setminus \mathbb{R}$:

$$r_z(t) := \frac{1}{t - z}, \quad t \in \mathbb{R},$$

with inner product defined by

$$(r_z, r_\zeta) = \frac{n(z) - n(\zeta)^*}{z - \zeta^*}, \quad z, \zeta \in \mathbb{C} \setminus \mathbb{R}, \ z \neq \zeta^*,$$

A is the operator of multiplication by t, and $u(t) \equiv 1$, $t \in \mathbb{R}$.

(3) $\mathcal{H}$ is the reproducing kernel Hilbert space $\mathcal{L}(n)$ with reproducing kernel

$$L_n(z, w) = \frac{n(z) - n(w)^*}{z - w^*}, \quad z, w \in \mathbb{C} \setminus \mathbb{R}, \ z \neq w^*,$$

A is the self-adjoint operator whose resolvent $(A - z)^{-1}$ is the difference-quotient operator R_z:

$$(R_z f)(\zeta) = \frac{f(\zeta) - f(z)}{\zeta - z}, \quad f \in \mathcal{L}(n),$$

and take $u = n$; this function belongs to the space $\mathcal{L}(n)$, since n satisfies the condition (1_0). Recall that the reproducing property of the kernel L_n is reflected in the inner product of the space $\mathcal{L}(n)$:

$$\langle f, L_n(\,\cdot\,, z)\rangle_{\mathcal{L}(n)} = f(z), \quad f \in \mathcal{L}(n), \ z \in \mathbb{C} \setminus \mathbb{R}.$$

That (3.2) holds follows from

$$(R_z n)(\zeta) = L_n(\zeta, z^*)$$

and the reproducing property of the kernel L_n:

$$\begin{aligned} \langle (A - z)^{-1} u, u\rangle_{\mathcal{L}(n)} &= \langle R_z n, n\rangle_{\mathcal{L}(n)} \\ &= \langle n, L_n(\,\cdot\,, z^*)\rangle^*_{\mathcal{L}(n)} = n(z^*)^* = n(z). \end{aligned}$$

The unitary equivalence of the representations in (1) and (2) follows easily from the relation

$$(r_z, r_\zeta) = \int_{-\infty}^{\infty} \frac{d\sigma(t)}{(t - z)(t - \zeta^*)}, \quad z, \zeta \in \mathbb{C} \setminus \mathbb{R},$$

and the fact that the functions r_z, $z \neq z^*$, form a total set in $L^2(\sigma)$. The unitary equivalence between the two representations of n in (2) and (3) is given by the mapping U:

$$U(r_z) = L_n(\,\cdot\,, z^*);$$

in particular, we have $Uu = n$ where u is the function $u(t) \equiv 1$, $t \in \mathbb{R}$. The space $L^2(\sigma)$ (or the equivalent space in (2)) we denote also by $\mathcal{H}(n)$. We mention that the definition of the spaces in (2) and (3) can also be used for generalized Nevanlinna functions, whereas in this case the space $L^2(\sigma)$ need not be defined. In Sections 4–7 we will prove theorems using the representation of n in (1), in Section 8 we reprove some of these results using the representation in the reproducing kernel Hilbert space $\mathcal{L}(n)$.

Since, according to Theorem 2.2, the functions $\widetilde{n}$ and $\widehat{n}$ in Definition 2.1 belong to the class $\mathbf{N}_0$ and are $\mathrm{o}(z)$ for $z = iy$, $y \to \infty$, they admit again an operator representation of the form (3.1), for example,

$$\widehat{n}(z) = \widehat{n}(z_0)^* + (z - z_0^*)\left((\widehat{A} - z_0)(\widehat{A} - z)^{-1}\widehat{v}, \widehat{v}\right) \tag{3.3}$$

with a self-adjoint operator $\widehat{A}$ in some Hilbert space $\widehat{\mathcal{H}}$, z_0 an arbitrary non-real number, and an element $\widehat{v} \in \widehat{\mathcal{H}}$. Clearly, as the difference between the functions $\widetilde{n}$ and $\widehat{n}$ is just an additive real constant, the operator representation for $\widetilde{n}$ can be chosen the same, that is, in (3.3) $\widehat{n}$ can be replaced by $\widetilde{n}$.

Theorem 3.1. *Let $n \in \mathbf{N}_0$ satisfying the condition (1_0) and with operator representation (3.2) be given, and let*

$$\widehat{n}(z) = \frac{-s_0}{n(z)} - z + \frac{s_1}{s_0}$$

be the Schur transform of n from (2.7). Then in the operator representation (3.3) of $\widehat{n}$ we can choose $\widehat{\mathcal{H}} = \{u\}^\perp$, $\widehat{A}$ in $\widehat{\mathcal{H}}$ as the compression of A to $\widehat{\mathcal{H}} : \widehat{A} = \widehat{P}A|_{\widehat{\mathcal{H}}}$, where $\widehat{P}$ is the orthogonal projection in $\mathcal{H}$ onto $\widehat{\mathcal{H}}$, and the element $\widehat{v}$ as

$$\widehat{v} = \frac{\|u\|}{((A - z_0)^{-1}u, u)}\widehat{P}(A - z_0)^{-1}u.$$

If $\widehat{n}$ also satisfies the condition (1_0)[1]*, then*

$$\widehat{n}(z) = \left((\widehat{A} - z)^{-1}\widehat{u}, \widehat{u}\right), \quad \widehat{u} := \frac{\widehat{P}Au}{\|u\|}.$$

Remark 3.2. The resolvent of $\widehat{A}$ is given by

$$(\widehat{A} - z)^{-1} = (A - z)^{-1} - \frac{((A - z)^{-1}\,\cdot\,, u)}{((A - z)^{-1}u, u)}(A - z)^{-1}u,$$

[1]This is the case when n satisfies condition (1_1) defined in Section 4, see Lemma 4.1.

and

$$\widehat{v} = \frac{\|u\|(A-z_0)^{-1}u - ((A-z_0)^{-1}u,u)\dfrac{u}{\|u\|}}{|((A-z_0)^{-1}u,u)|}.$$

Note that $\big((A-z_0)^{-1}u,u\big) = n(z_0) \neq 0$, otherwise $n(z) \equiv 0$.

Proof of Theorem 3.1. (1) Suppose that n satisfies (1_0). Then we have $\|u\| = \sqrt{s_0}$ and

$$\widehat{n}(z) - \widehat{n}(z_0)^* = -\frac{\|u\|^2}{r(z)} - z + \frac{\|u\|^2}{r(z_0)^*} + z_0^*, \tag{3.4}$$

where we have put $r(z) := ((A-z)^{-1}u,u)$. It remains to show that the expression on the right-hand side of (3.4) equals

$$(z-z_0^*)\left((\widehat{A}-z_0)(\widehat{A}-z)^{-1})\widehat{v},\widehat{v}\right) = (z-z_0^*)\|\widehat{v}\|^2+(z-z_0^*)(z-z_0)\left((\widehat{A}-z)^{-1}\widehat{v},\widehat{v}\right).$$

This is a straightforward calculation, we only indicate some formulas:

$$(z-z_0)(\widehat{A}-z)^{-1}\widehat{v} = \frac{\|u\|}{|r(z_0)|}\left(\frac{r(z_0)}{r(z)}(A-z)^{-1}u - (A-z_0)^{-1}u\right),$$

$$\|\widehat{v}\|^2 = \|u\|^2\frac{\|(A-z_0)^{-1}u\|^2}{|r(z_0)|^2} - 1,$$

and

$$(z-z_0^*)(z-z_0)\left((\widehat{A}-z)^{-1}\widehat{v},\widehat{v}\right)$$
$$= \|u\|^2\left(\frac{1}{r(z_0)^*} - \frac{1}{r(z)} - \frac{(z-z_0^*)\|(A-z_0)^{-1}u\|^2}{|r(z_0)|^2}\right).$$

(2) Now assume that $\widehat{n}$ satisfies (1_0). Then $\widehat{v} \in \operatorname{dom}\widehat{A} \subset \operatorname{dom} A$ and the equality

$$\widehat{v} = \frac{\|u\|}{r(z_0)}(A-z_0)^{-1}u - \frac{u}{\|u\|}$$

shows that also $u \in \operatorname{dom} A$. If we take $\widehat{u} = -(\widehat{A}-z_0)\widehat{v}$, then (see after (3.2)) $\widehat{n}$ has the asserted representation. It remains to show that $\widehat{u} = \widehat{P}Au/\|u\|$. We have

$$\widehat{u} = -(A-z_0)\widehat{v} = \frac{1}{\|u\|}(A-z_0)u - \frac{\|u\|}{r(z_0)}u.$$

Taking the inner product of both sides with u and using $(\widehat{u},u) = 0$, we see that

$$\frac{\|u\|}{r(z_0)} = \frac{((A-z_0)u,u)}{\|u\|^3}$$

and hence

$$\widehat{u} = \frac{1}{\|u\|}(A-z_0)u - \frac{((A-z_0)u,u)}{\|u\|^3}u = \frac{1}{\|u\|}\left(Au - \frac{(Au,u)}{\|u\|^2}u\right) = \frac{1}{\|u\|}\widehat{P}Au. \qquad \square$$

4. Higher-order asymptotics. Orthogonal polynomials

1. For $n \in \mathbf{N}_0$ and some integer $p \geq 1$ we introduce the assumptions

$$(1_p)\ n(z) = -\frac{s_0}{z} - \frac{s_1}{z^2} - \cdots - \frac{s_{2p}}{z^{2p+1}} + \mathrm{o}\left(\frac{1}{z^{2p+1}}\right),$$

$$(2_p)\ n(z) = -\frac{s_0}{z} - \frac{s_1}{z^2} - \cdots - \frac{s_{2p}}{z^{2p+1}} + \mathrm{O}\left(\frac{1}{z^{2p+2}}\right),$$

$$(3_p)\ n(z) = -\frac{s_0}{z} - \frac{s_1}{z^2} - \cdots - \frac{s_{2p+1}}{z^{2p+2}} + \mathrm{o}\left(\frac{1}{z^{2p+2}}\right).$$

Again, $(3_p) \Longrightarrow (2_p) \Longrightarrow (1_p)$, and by [14, Satz 1.10] for the operator representation the assumption (1_p) is equivalent to $u \in \operatorname{dom} A^p$. That is, for the above representation with the space $\mathcal{H}(n)$ the functions

$$\mathbf{t}_k(t) := t^k, \quad k = 0, 1, \ldots, p,$$

belong to $\mathcal{H}(n)$ and the first p of these elements, $\mathbf{t}_0, \mathbf{t}_1, \ldots, \mathbf{t}_{p-1}$, belong to $\operatorname{dom} A$. Moreover, the formal relation

$$(A - z)^{-1} = -\sum_{j=0}^{2p} \frac{A^j}{z^{j+1}} + \frac{A^{2p+1}}{z^{2p+1}}(A - z)^{-1}$$

implies easily

$$\begin{aligned} n(z) &= ((A - z)^{-1}u, u) \\ &= -\sum_{j=0}^{p} \frac{(A^j u, u)}{z^{j+1}} - \sum_{j=p+1}^{2p} \frac{(A^n u, A^{j-p}u)}{z^{j+1}} + \frac{1}{z^{2p+1}}\left(A(A - z)^{-1}A^p u, A^p u\right). \end{aligned}$$

It follows that

$$s_j = \begin{cases} (A^j u, u) & \text{if} \quad j = 0, 1, \ldots, p, \\ (A^p u, A^{j-p}u) & \text{if} \quad j = p+1, p+2, \ldots, 2p. \end{cases}$$

Therefore the above assumptions are equivalent to the following relations for the operator A and the generating element u:

$$\begin{aligned} &(1_p) \Longleftrightarrow u \in \operatorname{dom} A^p, \\ &(2_p) \Longleftrightarrow u \in \operatorname{dom} A^p,\ z(A(A - z)^{-1}A^p u, A^p u) = \mathrm{O}(1), \\ &(3_p) \Longleftrightarrow u \in \operatorname{dom} A^p,\ z(A(A - z)^{-1}A^p u, A^p u) + \alpha = \mathrm{o}(1) \text{ with } \alpha \in \mathbb{R}; \end{aligned} \tag{4.1}$$

in fact, in the last equivalence we have $\alpha = s_{2p+1}$.

Now we consider a function $n \in \mathbf{N}_0$ with the property (1_p) for some $p > 1$. For $0 \leq k \leq p$, by S_k we denote the $(k+1) \times (k+1)$ Hankel matrix

$$S_k := \begin{pmatrix} s_0 & s_1 & \cdots & s_k \\ s_1 & s_2 & \cdots & s_{k+1} \\ \vdots & \vdots & & \vdots \\ s_k & s_{k+1} & \cdots & s_{2k} \end{pmatrix}; \tag{4.2}$$

it is the Gram matrix associated with the $k+1$ functions $\mathbf{t}_0, \mathbf{t}_1, \ldots, \mathbf{t}_k$, and we introduce the Gram determinants

$$D_k := \det S_k = \begin{vmatrix} s_0 & s_1 & \cdots & s_k \\ s_1 & s_2 & \cdots & s_{k+1} \\ \vdots & \vdots & & \vdots \\ s_k & s_{k+1} & \cdots & s_{2k} \end{vmatrix}, \quad k = 0, 1, \ldots p. \tag{4.3}$$

Further, for $k = 1, \ldots, p$, $\mathcal{H}_k$ denotes the k-dimensional subspace

$$\mathcal{H}_k := \operatorname{span}\{\mathbf{t}_0, \mathbf{t}_1, \ldots, \mathbf{t}_{k-1}\}$$

of $\mathcal{H}(n)$. Evidently, the subspace $\mathcal{H}_k$ is non-degenerated if and only if $D_{k-1} \neq 0$.

In the rest of this section we suppose that $D_{p-1} \neq 0$, that is, the subspace $\mathcal{H}_p$ is non-degenerated. If $D_p = 0$, then the function n with the given asymptotics (j_p) is uniquely determined and rational of Mac Millan degree p, in fact, see [1, pp. 22, 23]

$$n(z) = -\frac{d_p(z)}{e_p(z)},$$

where the polynomials e_p of degree p and d_p of degree $p-1$ are defined below. To exclude this (simple) case we often suppose that even $D_p \neq 0$; clearly, this implies $D_{p-1} \neq 0$.

As a basis in $\mathcal{H}_p$ we choose a system of elements $e_k \in \mathcal{H}(n) = L^2(\sigma)$, $k = 0, 1, \ldots, p-1$, which is obtained from the system $\mathbf{t}_0, \mathbf{t}_1, \ldots, \mathbf{t}_p$ by the Gram–Schmidt orthonormalization procedure. This so-called system of *orthogonal polynomials of first kind, associated with the function* n is defined by the following properties, $j, k = 0, 1, \ldots, p-1$:

1. $e_0(z) \equiv 1/\sqrt{s_0}$,
2. $e_k(z)$ is a real polynomial of degree k with positive leading coefficient,
3. $(e_j, e_k) = \delta_{jk}$.

Then, see [1, (1.4)],

$$e_k(z) = \frac{1}{\sqrt{D_{k-1} D_k}} \begin{vmatrix} s_0 & s_1 & \cdots & s_k \\ s_1 & s_2 & \cdots & s_{k+1} \\ \vdots & \vdots & \cdots & \vdots \\ s_{k-1} & s_k & \cdots & s_{2k-1} \\ 1 & z & \cdots & z^k \end{vmatrix}, \quad k = 1, 2, \ldots, p-1, \tag{4.4}$$

and by this formula with $k = p$ also a polynomial e_p can be defined. Evidently, $e_p \in \mathcal{H}_p^{[\perp]}$, and

$$\operatorname{span}\{\mathbf{t}_0, \mathbf{t}_1, \ldots, \mathbf{t}_k\} = \operatorname{span}\{e_0, e_1, \ldots, e_k\}, \quad k = 0, 1, \ldots, p.$$

The orthogonal polynomials e_j, $j = 0, 1, \ldots, p$, satisfy the difference equations

$$b_{k-1}e_{k-1}(z) + a_k e_k(z) + b_k e_{k+1}(z) = z e_k(z), \quad k = 0, 1, \ldots, p-1, \tag{4.5}$$

with real numbers a_k, $k = 0, 1, \ldots, p-1$, $b_{-1} = 0$, and positive numbers b_k, $k = 1, \ldots, p-1$, and the 'initial condition' $e_0(z) = 1/\sqrt{s_0}$. Explicit formulas for a_k, b_k can be given, see [1]; we note that

$$a_0 = \frac{s_1}{s_0}, \quad b_0 = \frac{\sqrt{s_2 s_0 - s_1^2}}{s_0}. \tag{4.6}$$

The relation (4.5) implies that with respect to the basis $e_0, e_1, \ldots, e_{p-1}$ of the space $\mathcal{H}_p$ the compression A_p of the operator A to $\mathcal{H}_p$ is given by the Jacobi matrix

$$\mathcal{A}_p := \begin{pmatrix} a_0 & b_0 & 0 & \cdots & 0 & 0 \\ b_0 & a_1 & b_1 & \cdots & 0 & 0 \\ \vdots & \vdots & \vdots & \vdots & \vdots & \vdots \\ 0 & 0 & 0 & \cdots & a_{p-2} & b_{p-2} \\ 0 & 0 & 0 & \cdots & b_{p-2} & a_{p-1} \end{pmatrix}, \tag{4.7}$$

and that

$$A e_{p-1} = b_{p-2} e_{p-2} + a_{p-1} e_{p-1} + b_{p-1} e_p.$$

The latter relation means for the orthogonal polynomials

$$b_{p-2} e_{p-2}(z) + a_{p-1} e_{p-1}(z) + b_{p-1} e_p(z) = z e_{p-1}(z),$$

therefore, the eigenvalues of $\mathcal{A}_p$ are the zeros of the polynomial e_p. For later use we write the last $p-1$ difference equations (4.5) explicitly in the form

$$\left\{ \begin{array}{lcl} b_0 e_0 + a_1 e_1 + b_1 e_2 & = & z e_1 \\ b_1 e_1 + a_2 e_2 + b_2 e_3 & = & z e_2 \\ b_2 e_2 + a_3 e_3 + b_3 e_4 & = & z e_3 \\ & \vdots & \\ b_{p-2} e_{p-2} + a_{p-1} e_{p-1} + b_{p-1} e_p & = & z e_{p-1}; \end{array} \right. \tag{4.8}$$

this system of homogeneous equations for $e_0, e_1, \ldots, e_p$ determines the orthogonal polynomials uniquely if we add the initial conditions

$$e_0(z) = \frac{1}{\sqrt{s_0}}, \quad e_1(z) = \frac{z - a_0}{b_0 \sqrt{s_0}}; \tag{4.9}$$

the second condition is just the first equation in (4.5).

The *polynomials of second kind, associated with the function* $n \in \mathbf{N}_0$, are the functions d_k, $k = 0, 1, \ldots, p$, defined as follows:

$$d_k(z) = \sqrt{s_0}\left(\frac{e_k(z) - e_k(\cdot)}{z - \cdot}, e_0\right) = \left(\frac{e_k(z) - e_k(\cdot)}{z - \cdot}, u\right), \quad k = 0, 1, \ldots, p. \tag{4.10}$$

Hence $d_0(z) = 0$ and d_k is a polynomial of degree $k - 1$, $k \geq 1$. The definition of d_k and the relation (4.5) imply that

$$b_{k-1}d_{k-1}(z) + a_k d_k(z) + b_k d_{k+1}(z) = z d_k(z), \quad k = 1, \ldots, p - 1. \tag{4.11}$$

Therefore the polynomials e_k and d_k satisfy for $k = 1, 2, \ldots, p-1$ the same difference equations but with different initial conditions:

$$d_0(z) = 0, \quad d_1(z) = \frac{\sqrt{s_0}}{b_0}. \tag{4.12}$$

For later use we write the difference equations (4.11) in the form

$$\left\{\begin{array}{lcl} a_1 d_1 + b_1 d_2 & = & z d_1 \\ b_1 d_1 + a_2 d_2 + b_2 d_3 & = & z d_2 \\ b_2 d_2 + a_3 d_3 + b_3 d_4 & = & z d_3 \\ & \vdots & \\ b_{p-2}d_{p-2} + a_{p-1}d_{p-1} + b_{p-1}d_p & = & z d_{p-1}. \end{array}\right. \tag{4.13}$$

For any two solutions $u_0, \ldots, u_p$ and $v_0, \ldots, v_p$ of the difference equations (4.5) with $b_{-1} = 0$:

$$\begin{array}{rcl} z u_k(z) & = & b_{k-1}u_{k-1}(z) + a_k u_k(z) + b_k u_{k+1}(z), \\ \zeta v_k(\zeta) & = & b_{k-1}v_{k-1}(\zeta) + a_k v_k(\zeta) + b_k v_{k+1}(\zeta), \end{array} \quad k = 0, 1, \ldots, p - 1,$$

the Christoffel–Darboux formulas hold:

$$\begin{array}{rcl} \displaystyle\sum_{k=m}^{p-1} (z - \zeta) u_k(z) v_k(\zeta) & = & b_{p-1}\big(u_p(z)v_{p-1}(\zeta) - u_{p-1}(z)v_p(\zeta)\big) \\ & & -b_{m-1}\big(u_m(z)v_{m-1}(\zeta) - u_{m-1}(z)v_m(\zeta)\big); \end{array} \tag{4.14}$$

in particular,

$$d_p(z)e_{p-1}(z) - e_p(z)d_{p-1}(z) = \frac{1}{b_{p-1}}. \tag{4.15}$$

2. In this subsection we assume that $n \in \mathbf{N}_0$ satisfies the assumption (1_p) for some $p \geq 1$, and we consider its Schur transform $\widehat{n}$ from (2.7). For the following lemma see [9, Lemma 2.1], we sketch the proof.

Lemma 4.1. *Suppose that* n *satisfies* (1_p) *for some* $p \geq 1$:

$$n(z) = -\frac{s_0}{z} - \frac{s_1}{z^2} - \cdots - \frac{s_{2p}}{z^{2p+1}} + \mathrm{o}\left(\frac{1}{z^{2p+1}}\right),$$

then its Schur transform $\widehat{n}$ *satisfies* (1_{p-1}) :

$$\widehat{n}(z) = -\frac{\widehat{s}_0}{z} - \frac{\widehat{s}_1}{z^2} - \cdots - \frac{\widehat{s}_{2p-2}}{z^{2p-1}} + \mathrm{o}\left(\frac{1}{z^{2p-1}}\right),$$

with for $j = 0, 1, \ldots, 2p-2$

$$\widehat{s}_j = \frac{(-1)^{j+1}}{s_0^{j+2}} \begin{vmatrix} s_1 & s_0 & 0 & \cdots & 0 & 0 \\ s_2 & s_1 & s_0 & \cdots & 0 & 0 \\ \vdots & \vdots & \ddots & \ddots & \vdots & \vdots \\ s_{j+1} & s_j & s_{j-1} & \cdots & s_1 & s_0 \\ s_{j+2} & s_{j+1} & s_j & \cdots & s_2 & s_1 \end{vmatrix}. \tag{4.16}$$

Proof. Write

$$\widehat{n}(z) = z\left(1 + \frac{s_1}{s_0 z} + \frac{s_2}{s_0 z^2} + \cdots + \frac{s_{2p}}{s_0 z^{2p}} + \mathrm{o}\left(\frac{1}{z^{2p}}\right)\right)^{-1} - z + \frac{s_1}{s_0}.$$

If we set

$$q(z) = \frac{s_1}{s_0 z} + \frac{s_2}{s_0 z^2} + \cdots + \frac{s_{2p}}{s_0 z^{2p}} + \mathrm{o}\left(\frac{1}{z^{2p}}\right),$$

then

$$\begin{aligned} \frac{1}{1+q(z)} &= 1 - q(z) + \cdots - q(z)^{2p-1} + \frac{q(z)^{2p}}{1+q(z)} \\ &= 1 - q(z) + \cdots - q(z)^{2p-1} + \mathrm{o}\left(\frac{1}{z^{2p-1}}\right) \end{aligned}$$

and

$$\begin{aligned} \widehat{n}(z) &= \frac{z}{1+q(z)} - \left(z - \frac{s_1}{s_0}\right)\frac{1+q(z)}{1+q(z)} \\ &= \left(-zq(z) + \frac{s_1}{s_0}(1+q(z))\right)\frac{1}{1+q(z)} \\ &= \left(-z\left(\frac{s_2}{s_0 z^2} + \cdots + \frac{s_{2p}}{s_0 z^{2p}} + \mathrm{o}\left(\frac{1}{z^{2p}}\right)\right) + \frac{s_1}{s_0}q(z)\right)\frac{1}{1+q(z)} \\ &= \left(-\frac{s_2}{s_0 z} - \cdots - \frac{s_{2p}}{s_0 z^{2p-1}} + \mathrm{o}\left(\frac{1}{z^{2p-1}}\right) + \frac{s_1}{s_0}q(z)\right)\frac{1}{1+q(z)}, \end{aligned}$$

which is of the needed form. Formula (4.16) for the coefficients $\widehat{s}_j$ can be obtained by equating powers of z from both sides of the equality

$$n(z)\left(z - \frac{s_1}{s_0} + \widehat{n}(z)\right) = -s_0.$$

□

Now we can formulate the main result of this subsection. See also [16, Corollary 6.4] for a similar formula in the continuous case and [8, Theorem 6.2.5] for a related result.

Theorem 4.2. *Let* $n \in \mathbf{N}_0$ *satisfy condition* (1_p) *for some* $p \geq 2$. *If* e_k *and* d_k, $k = 0, 1, \ldots, p$, *denote the polynomials of first and second kind associated with the function* n, *and* $\widehat{e}_k$ *and* $\widehat{d}_k$, $k = 0, 1, \ldots, p-1$, *denote the polynomi-*

als of first and second kind associated with the Schur transform $\widehat{n}$ *of* n*, then for* $k = 0, 1, \ldots, p-1$ *the following relations hold:*

$$\widehat{e}_k(z) = \frac{1}{\sqrt{s_0}} d_{k+1}(z), \tag{4.17}$$

$$\begin{aligned} \widehat{d}_k(z) &= -\sqrt{s_0}\, e_{k+1}(z) + \frac{1}{\sqrt{s_0}}\left(z - \frac{s_1}{s_0}\right) d_{k+1}(z) \\ &= b_0\left(\frac{e_{k+1}(z) - e_{k+1}(\cdot)}{z - \cdot}, e_1\right) = \frac{1}{\sqrt{s_0}}\left(\frac{e_{k+1}(z) - e_{k+1}(\cdot)}{z - \cdot}, \cdot - a_0\right). \end{aligned} \tag{4.18}$$

Remark 4.3. (i) Here we write $\widehat{e}$ and $\widehat{d}$ for the polynomials of first and second kind associated with the Schur transform $\widehat{n}$ of n, but the reader is reminded that these functions are *not* the Schur transforms of the polynomials e and d.
(ii) If $\check{e}_k$ and $\check{d}_k$ stand for the polynomials of first and second kind associated with the Nevanlinna function $-1/n$, then for $k = 0, 1, \ldots, p-1$

$$\check{e}_k(z) = d_{k+1}(z), \quad \check{d}_k(z) = -e_{k+1}(z) + \frac{1}{s_0}\left(z - \frac{s_1}{s_0}\right) d_{k+1}(z).$$

The first equality readily follows from the fact that the spectral functions of $\widehat{n}$ and $-1/n$ only differ by a factor s_0. The second equality can be obtained by tracing the proof below; the only difference lies in (4.19): with evident notation, it should be replaced by

$$\check{s}_0 = \frac{s_0 s_2 - s_1^2}{s_0^3}.$$

Proof of Theorem 4.2. For the function $\widehat{n}$, again with evident notation, we have

$$\widehat{s}_0 = \frac{s_0 s_2 - s_1^2}{s_0^2} \tag{4.19}$$

and, as a consequence of Theorem 3.1,

$$\widehat{\mathcal{A}}_{p-1} = \begin{pmatrix} \widehat{a}_0 & \widehat{b}_0 & 0 & \cdots & 0 & 0 \\ \widehat{b}_0 & \widehat{a}_1 & \widehat{b}_1 & \cdots & 0 & 0 \\ \vdots & \vdots & \vdots & \vdots & \vdots & \vdots \\ 0 & 0 & 0 & \cdots & \widehat{a}_{p-3} & \widehat{b}_{p-3} \\ 0 & 0 & 0 & \cdots & \widehat{b}_{p-3} & \widehat{a}_{p-2} \end{pmatrix} = \begin{pmatrix} a_1 & b_1 & 0 & \cdots & 0 & 0 \\ b_1 & a_2 & b_2 & \cdots & 0 & 0 \\ \vdots & \vdots & \vdots & \vdots & \vdots & \vdots \\ 0 & 0 & 0 & \cdots & a_{p-2} & b_{p-2} \\ 0 & 0 & 0 & \cdots & b_{p-2} & a_{p-1} \end{pmatrix}. \tag{4.20}$$

For the $\widehat{e}_j$, $j = 0, \ldots, p-1$, we find

$$\widehat{e}_0(z) = \frac{1}{\sqrt{\widehat{s}_0}} = \frac{s_0}{\sqrt{s_0 s_2 - s_1^2}} = \frac{1}{b_0} = \frac{1}{\sqrt{s_0}} d_1(z)$$

and $\widehat{e}_1, \widehat{e}_2, \ldots, \widehat{e}_{p-1}$ follow from the equations (see (4.5)):

$$\begin{cases} a_1\widehat{e}_0 + b_1\widehat{e}_1 & = & z\widehat{e}_0 \\ b_1\widehat{e}_0 + a_2\widehat{e}_1 + b_2\widehat{e}_2 & = & z\widehat{e}_1 \\ b_2\widehat{e}_1 + a_3\widehat{e}_2 + b_3\widehat{e}_3 & = & z\widehat{e}_2 \\ & \vdots & \\ b_{p-2}\widehat{e}_{p-3} + a_{p-1}\widehat{e}_{p-2} + b_{p-1}\widehat{e}_{p-1} & = & z\widehat{e}_{p-2}. \end{cases}$$

Since these equation coincide with (4.13) we obtain

$$\widehat{e}_j = c\, d_{j+1}, \quad j = 0, 1, \ldots, p-1.$$

The constant c can be determined from the initial condition $\widehat{e}_0 = c\, d_1$, which gives $c = 1/\sqrt{s_0}$. Therefore

$$\widehat{e}_j(z) = \frac{1}{\sqrt{s_0}}\, d_{j+1}(z), \quad j = 0, 1, \ldots, p-1,$$

and (4.17) is proved.

For the polynomials of second kind $\widehat{d}_j$ we obtain in a similar way

$$\widehat{d}_0(z) = 0, \quad \widehat{d}_1(z) = \frac{\sqrt{\widehat{s}_0}}{\widehat{b}_0},$$

$$\begin{cases} \widehat{a}_1\widehat{d}_1 + \widehat{b}_1\widehat{d}_2 & = & z\widehat{d}_1 \\ \widehat{b}_1\widehat{d}_1 + \widehat{a}_2\widehat{d}_2 + \widehat{b}_2\widehat{d}_3 & = & z\widehat{d}_2 \\ \widehat{b}_2\widehat{d}_2 + \widehat{a}_3\widehat{d}_3 + \widehat{b}_3\widehat{d}_4 & = & z\widehat{d}_3 \\ & \vdots & \\ \widehat{b}_{p-3}\widehat{d}_{p-3} + \widehat{a}_{p-2}\widehat{d}_{p-2} + \widehat{b}_{p-2}\widehat{d}_{p-1} & = & z\widehat{d}_{p-2} \end{cases}$$

(one equation less than in (4.13)). These equations can be written as

$$\begin{cases} a_2\widehat{d}_1 + b_2\widehat{d}_2 & = & z\widehat{d}_1 \\ b_2\widehat{d}_1 + a_3\widehat{d}_2 + b_3\widehat{d}_3 & = & z\widehat{d}_2 \\ b_3\widehat{d}_2 + a_4\widehat{d}_3 + b_4\widehat{d}_4 & = & z\widehat{d}_3 \\ & \vdots & \\ b_{p-2}\widehat{d}_{p-3} + a_{p-1}\widehat{d}_{p-2} + b_{p-1}\widehat{d}_{p-1} & = & z\widehat{d}_{p-2}. \end{cases} \tag{4.21}$$

The last $p-3$ equations of this system coincide with the last $p-3$ equations of (4.8) and (4.13). Therefore a solution vector $(\widehat{d}_j)_1^{p-1}$ of the last $p-3$ equations of (4.21) can be obtained as a linear combination of the solution vectors $(e_j)_2^p$ and $(d_j)_2^p$ of the last $p-3$ equations in (4.8) and (4.13):

$$\widehat{d}_j = \gamma e_{j+1} + \delta d_{j+1}, \quad j = 1, 2, \ldots, p-1.$$

Now γ, δ have to be found such that these relations hold also for $j = 0$ with $\widehat{d}_0(z) = 0$, and for $j = 1$ with $\widehat{d}_1(z) = \dfrac{\sqrt{\widehat{s}_0}}{\widehat{b}_0}$. Since $\widehat{d}_0(z) = 0$ it follows that

$$0 = \gamma e_1(z) + \delta d_1(z) = \gamma \frac{z - a_0}{b_0} \frac{1}{\sqrt{s_0}} + \delta \frac{\sqrt{s_0}}{b_0} = \gamma \frac{z - \dfrac{s_1}{s_0}}{b_0} \frac{1}{\sqrt{s_0}} + \delta \frac{\sqrt{s_0}}{b_0},$$

which is satisfied for

$$\gamma = -\varepsilon, \quad \delta = \varepsilon \left(\frac{z}{s_0} - \frac{s_1}{s_0^2} \right).$$

The relation $\widehat{d}_1(z) = \dfrac{\sqrt{\widehat{s}_0}}{\widehat{b}_0}$ implies

$$\begin{aligned}
\widehat{d}_1(z) &= \varepsilon \left(-e_2(z) + \left(\frac{z}{s_0} - \frac{a_0}{s_0} \right) d_2(z) \right) \\
&= \varepsilon \left(-\frac{(z - a_1)e_1 - b_0 e_0}{b_1} + \frac{z - a_0}{s_0} \frac{(z - a_1)d_1}{b_1} \right) \\
&= \varepsilon \left(-\frac{(z - a_1)\dfrac{z - a_0}{b_0}\dfrac{1}{\sqrt{s_0}} - b_0 \dfrac{1}{\sqrt{s_0}}}{b_1} + \frac{z - a_0}{s_0} \frac{(z - a_1)\dfrac{\sqrt{s_0}}{b_0}}{b_1} \right) \\
&= \varepsilon \frac{b_0}{b_1 \sqrt{s_0}} = \frac{\sqrt{\widehat{s}_0}}{\widehat{b}_0} = \frac{b_0}{\widehat{b}_0},
\end{aligned}$$

hence

$$\varepsilon = \frac{b_1}{\widehat{b}_0} \sqrt{s_0}.$$

According to (4.20) we find $\varepsilon = \sqrt{s_0}$. This proves the first equality in (4.18). The remaining equalities follow from (4.10) and the second equality in (4.9). □

3. In this subsection we give a second proof of Theorem 4.2 using asymptotic expansions, see [1, (1.34b)]. Assume that $n \in \mathbf{N}_0$ satisfies (1_p) for some $p \geq 2$, that is,

$$n(z) = -\frac{s_0}{z} - \frac{s_1}{z^2} - \cdots - \frac{s_{2p}}{z^{2p+1}} + \mathrm{o}\left(\frac{1}{z^{2p+1}} \right),$$

then

$$-\frac{d_p(z)}{e_p(z)} = -\frac{s_0}{z} - \frac{s_1}{z^2} - \cdots - \frac{s_{2p-1}}{z^{2p}} + \mathrm{O}\left(\frac{1}{z^{2p+1}} \right). \tag{4.22}$$

According to [1, the second to last formula on p. 22] the function $-d_p/e_p$ is a Nevanlinna function and by [14, Bemerkung 1.11] there is a real number t_{2p} such that $-d_p/e_p$ has the asymptotic expansion

$$-\frac{d_p(z)}{e_p(z)} = -\frac{s_0}{z} - \frac{s_1}{z^2} - \cdots - \frac{s_{2p-1}}{z^{2p}} - \frac{t_{2p}}{z^{2p+1}} + \mathrm{o}\left(\frac{1}{z^{2p+1}} \right). \tag{4.23}$$

By Lemma 4.1,

$$\widehat{n}(z) = -\frac{\widehat{s}_0}{z} - \frac{\widehat{s}_1}{z^2} - \cdots - \frac{\widehat{s}_{2p-2}}{z^{2p-1}} + \mathrm{o}\left(\frac{1}{z^{2p-1}}\right),$$

and hence by (4.22)

$$-\frac{\widehat{d}_{p-1}(z)}{\widehat{e}_{p-1}(z)} = -\frac{\widehat{s}_0}{z} - \frac{\widehat{s}_1}{z^2} - \cdots - \frac{\widehat{s}_{2p-3}}{z^{2p-2}} + \mathrm{O}\left(\frac{1}{z^{2p-1}}\right). \tag{4.24}$$

By Lemma 4.1, the Schur transform of the function $-d_p/e_p$ in (4.23) has the asymptotic expansion

$$\begin{aligned}\widehat{\left(-\frac{d_p}{e_p}\right)}(z) &= \frac{s_0 e_p(z)}{d_p(z)} - z + \frac{s_1}{s_0} =: \frac{r(z)}{d_p(z)} \\ &= -\frac{\widehat{s}_0}{z} - \frac{\widehat{s}_1}{z^2} - \cdots - \frac{\widehat{s}_{2p-3}}{z^{2p-2}} - \frac{\widehat{t}_{2p-2}}{z^{2p-1}} + \mathrm{o}\left(\frac{1}{z^{2p-1}}\right), \qquad (4.25)\\ &= -\frac{\widehat{s}_0}{z} - \frac{\widehat{s}_1}{z^2} - \cdots - \frac{\widehat{s}_{2p-3}}{z^{2p-2}} + \mathrm{O}\left(\frac{1}{z^{2p-1}}\right),\end{aligned}$$

where only the number $\widehat{t}_{2p-2}$ depends on t_{2p} according to formula (4.16). Here the polynomial r, defined via the second equality sign, is given by

$$r(z) = s_0 e_p(z) - \left(z - \frac{s_1}{s_0}\right) d_p(z)$$

and its degree is $\leq p$. Comparing (4.24) with (4.25), we find that

$$\frac{r(z)}{d_p(z)} - \frac{\widehat{d}_{p-1}(z)}{\widehat{e}_{p-1}(z)} = \mathrm{O}\left(\frac{1}{z^{2p-1}}\right).$$

The degree of the product $d_p\,\widehat{e}_{p-1}$ equals $2p-2$ and hence

$$\frac{r(z)}{d_p(z)} = \frac{\widehat{d}_{p-1}(z)}{\widehat{e}_{p-1}(z)},$$

which readily implies that for some number $k \neq 0$

$$\widehat{e}_{p-1}(z) = k d_p(z), \quad \widehat{d}_{p-1}(z) = k\left(s_0 e_p(z) - \left(z - \frac{s_1}{s_0}\right) d_p(z)\right). \tag{4.26}$$

We claim $k = 1/\sqrt{s_0}$. With the proof of the claim the proof of the theorem is complete.

To prove the claim we note that the leading coefficient of the polynomial e_k is equal to $\sqrt{D_{k-1}/D_k}$ and that, by (4.5),

$$\sqrt{\frac{D_k}{D_{k+1}}} = \frac{1}{b_k}\sqrt{\frac{D_{k-1}}{D_k}}.$$

Hence

$$\sqrt{\frac{D_{p-1}}{D_p}} = \frac{1}{b_{p-1}} \cdots \frac{1}{b_1}\sqrt{\frac{D_0}{D_1}} = \frac{1}{b_{p-1}} \cdots \frac{1}{b_1}\frac{1}{b_0}\frac{1}{\sqrt{s_0}},$$

and, similarly, because of (4.20) and (4.19),

$$\sqrt{\frac{\widehat{D}_{p-2}}{\widehat{D}_{p-1}}} = \frac{1}{\widehat{b}_{p-2}} \cdots \frac{1}{\widehat{b}_0} \frac{1}{\sqrt{\widehat{s}_0}} = \frac{1}{b_{p-1}} \cdots \frac{1}{b_1} \frac{1}{b_0}.$$

From (4.26) we obtain

$$\frac{1}{k} \sqrt{\frac{\widehat{D}_{p-2}}{\widehat{D}_{p-1}}} = \lim_{z\to\infty} \frac{d_p(z)}{z^{p-1}} = s_0 \lim_{z\to\infty} \frac{e_p(z)}{z^p} = s_0 \sqrt{\frac{D_{p-1}}{D_p}},$$

that is,

$$\sqrt{s_0}\, \frac{1}{b_{p-1}} \cdots \frac{1}{b_1} \frac{1}{b_0} = \frac{1}{k} \frac{1}{b_{p-1}} \cdots \frac{1}{b_1} \frac{1}{b_0}.$$

Therefore, $k = 1/\sqrt{s_0}$ and the claim holds.

5. Reduction via a p-dimensional subspace

Let again $n \in \mathbf{N}_0$ with the property (1_p) be given. We decompose the space $\mathcal{H}(n)$ with $\mathcal{H}_p = \operatorname{span}\{\mathbf{t}_0, \mathbf{t}_1, \dots, \mathbf{t}_{p-1}\}$ as follows:

$$\mathcal{H}(n) = \mathcal{H}_p \oplus \mathcal{H}'_p. \tag{5.1}$$

Then, evidently, $e_p \in \mathcal{H}'_p$. The corresponding matrix representation of the operator A is

$$A = \begin{pmatrix} A_0 & \widetilde{B} \\ B & D \end{pmatrix} \tag{5.2}$$

with A_0 given by the Jacobi matrix $\mathcal{A}_0$ from (4.7),

$$B = b_{p-1}(\,\cdot\,, e_{p-1})e_p, \quad \widetilde{B} = b_{p-1}(\,\cdot\,, e_p)e_{p-1}.$$

The operator A_0 is bounded and self-adjoint in $\mathcal{H}_p$, and D is self-adjoint (possibly unbounded) in $\mathcal{H}'_p$. In the following theorem we express the function n by means of the entries of the matrix in (5.2). We set

$$a_{00}(z) := ((A_0 - z)^{-1}u, u), \quad a_{11}(z) := ((A_0 - z)^{-1}e_{p-1}, e_{p-1}),$$

$$a(z) := ((A_0 - z)^{-1}u, e_{p-1}) = ((A_0 - z)^{-1}e_{p-1}, u), \tag{5.3}$$

and

$$\check{r}(z) := \begin{vmatrix} a_{00}(z) & a(z) \\ a(z) & a_{11}(z) \end{vmatrix}.$$

The last equality in (5.3) follows from $a(z^*)^* = a(z)$, in fact, by Cramer's rule, $a(z) = \sqrt{s_0}/(a_0 - z)$ if $p = 1$ and $a(z) = (-1)^{p-1}\sqrt{s_0}\, b_0 \dots b_{p-2}/\det(\mathcal{A}_0 - z)$ if $p \geq 2$.

Theorem 5.1. *Suppose that the function $n \in \mathbf{N}_0$ satisfies for some $p \geq 1$ one of the assumptions $(j_p), j = 1, 2, 3$, and that $D_p \neq 0$, see (4.3). Then*

$$n(z) = \frac{\check{r}(z)n_p(z) - a_{00}(z)}{a_{11}(z)n_p(z) - 1}, \tag{5.4}$$

where $n_p(z) := ((D - z)^{-1}u_p, u_p)$ with $u_p := b_{p-1}e_p$. The function n_p belongs to $\mathbf{N}_0$ and satisfies the assumption (j_0). Moreover, for $k \geq 1$ we have

$$u \in \operatorname{dom} A^{p+k} \iff u_p \in \operatorname{dom} D^k, \tag{5.5}$$

and n satisfies the assumption (j_{p+k}) if and only if n_p satisfies the assumption (j_k).

Remark 5.2. If the operator representation (3.2) of n is $n(z) = ((A-z)^{-1}u, u)$ with the space $\mathcal{H} = \mathcal{H}(n)$, then, according to Theorem 3.1, the operator representation of $n_1 = \widehat{n}$ is given by

$$\mathcal{H}_1' = \mathcal{H} \ominus \{u\}, \quad A_1 = P_{\mathcal{H}_1'} A|_{\mathcal{H}_1'}, \quad u_1 = \frac{P_{\mathcal{H}_1'} Au}{\|u\|},$$

the operator representation of $n_2 = \widehat{n}_1$ by

$$\mathcal{H}_2' = \mathcal{H} \ominus \{u, Au\}, \quad A_2 = P_{\mathcal{H}_2'} A|_{\mathcal{H}_2'}, \quad u_2 = \frac{P_{\mathcal{H}_2'} A^2 u}{\|u\| \|u_1\|},$$

and, via induction, the operator representation of $n_p = \widehat{n}_{p-1}$ by

$$\mathcal{H}_p' = \mathcal{H} \ominus \{u, Au, \ldots, A^{p-1}u\}, \quad A_p = P_{\mathcal{H}_p'} A|_{\mathcal{H}_p'}, \quad u_p = \frac{P_{\mathcal{H}_p'} A^p u}{\|u\| \|u_1\| \ldots \|u_{p-1}\|}.$$

Note that $\|u\| = \sqrt{s_0}$ and, by Theorem 5.1, $\|u_j\| = b_{j-1}$, $j = 1, \ldots, p$.

Proof of Theorem 5.1. With the matrix (5.2), the equation $(A - z)x = u$ becomes

$$\begin{aligned} (A_0 - z)x_1 \,+\, b_{p-1}(x_2, e_p)e_{p-1} &= u, \\ b_{p-1}(x_1, e_{p-1})e_p \,+\, (D - z)x_2 &= 0, \end{aligned} \tag{5.6}$$

where x is written as $x = \begin{pmatrix} x_1 & x_2 \end{pmatrix}^\top$ according to the decomposition (5.1). The second equation implies

$$x_2 = -b_{p-1}(x_1, e_{p-1})(D - z)^{-1}e_p.$$

We insert this into (5.6) and apply $(A_0 - z)^{-1}$ to get

$$x_1 = (A_0 - z)^{-1}u + b_{p-1}^2(x_1, e_{p-1}) \left((D - z)^{-1}e_p, e_p\right) (A_0 - z)^{-1}e_{p-1}. \tag{5.7}$$

Now take the inner product with e_{p-1} and solve the obtained equation for (x_1, e_{p-1}):

$$(x_1, e_{p-1}) = \frac{\left((A_0 - z)^{-1}u, e_{p-1}\right)}{1 - b_{p-1}^2 \left((D - z)^{-1}e_p, e_p\right) \left((A_0 - z)^{-1}e_{p-1}, e_{p-1}\right)}.$$

Observing that $n(z) = ((A - z)^{-1}u, u) = (x_1, u)$, the relation (5.7) yields (5.4). To prove (5.5), denote by P' the orthogonal projection onto $\mathcal{H}_p'$ in (5.1). If $k = 1$, then $u \in \operatorname{dom} A^{p+1}$ is equivalent to $v := A^p u \in \operatorname{dom} A$. Since D is the only entry in the matrix of (5.2) which is possibly unbounded (B and $\widetilde{B}$ are even one-dimensional), $v \in A$ is equivalent to the fact that the non-zero component $P'v$, which is a multiple of e_p, belongs to $\operatorname{dom} D$. If $k = 2$ we observe that

$$A^2 = \begin{pmatrix} A_0^2 + \widetilde{B}B & A_0\widetilde{B} + \widetilde{B}D \\ BA_0 + DB & B\widetilde{B} + D^2 \end{pmatrix}. \tag{5.8}$$

Since $e_p \in \operatorname{dom} D$ the operators DB and $\widetilde{B}D$ and hence all the entries in the matrix representation of A^2 except possibly D^2 are bounded. Now $u \in \operatorname{dom} A^{p+2}$ is equivalent to $v = A^p u \in \operatorname{dom} A^2$, and hence, by (5.8), $P'v \in \operatorname{dom} D^2$. The claim for arbitrary k follows by induction.

The last claim of the theorem for $j = 1$ follows immediately from (5.5) and the first equivalence in (4.1). For $j = 2, 3$ we also use the equivalences in (4.1). A simple calculation yields

$$(A - z)^{-1} = \begin{pmatrix} R_{11}(z) & R_{12}(z) \\ R_{12}(z^*)^* & R_{22}(z) \end{pmatrix} \tag{5.9}$$

with

$$R_{11}(z) = S_1(z)^{-1}, \quad R_{12}(z) = -b_{p-1}\left((D - z)^{-1}\cdot, e_p\right) S_1(z)^{-1}e_{p-1},$$

$$R_{22}(z) = (D - z)^{-1} + b_{p-1}^2 \left(S_1(z)^{-1}e_{p-1}, e_{p-1}\right)\left((D - z)^{-1}\cdot, e_p\right)(D - z)^{-1}e_p,$$

where $S_1(z) := A_0 - z - b_{p-1}^2\left((D - z)^{-1}e_p, e_p\right)(\cdot, e_{p-1})e_{p-1}$, the first Schur complement. It is easy to see that for $f, g \in \mathcal{H}_p$ we have

$$\lim_{y\to\infty} iy(S_1(iy)^{-1}f, g) = -(f, g).$$

Now we observe the relation

$$A(A - z)^{-1} = \begin{pmatrix} A_0R_{11}(z) + \widetilde{B}R_{12}(z^*)^* & A_0R_{12}(z) + \widetilde{B}R_{22}(z) \\ BR_{11}(z) + DR_{12}(z^*)^* & BR_{12}(z) + DR_{22}(z) \end{pmatrix}$$

and the fact that for $z = iy$, $y \to \infty$, for example $zA_0R_{11}(z) = zA_0S_1(z)$ has a limit,

$$\begin{aligned} zA_0R_{12}(z) &= -zb_{p-1}\big((D - z)^{-1}\cdot, e_p\big)A_0S_1(z)^{-1}e_{p-1} = o(1), \\ zDR_{22}(z) &= zD(D-z)^{-1} + zb_{p-1}^2\big(S_1(z)^{-1}e_p, e_p\big)\big((D-z)^{-1}\cdot, e_p\big)D(D-z)^{-1}e_p \\ &= zD(D - z)^{-1} + o(1), \end{aligned}$$

etc. These relations imply for example with $v = A^{p+k}u$

$$z\left(A(A - z)^{-1}v, v\right) = z\left(D(D - z)^{-1}P'v, P'v\right) + O(1).$$

Since $P'v \in \operatorname{span}\{e_p, e_{p+1}, \dots, e_{p+k}\}$, $e_p, e_{p+1}, \dots, e_{p+k-1} \in \operatorname{dom} D$ and hence

$$z\left(D(D - z)^{-1}x', x'\right) = O(1) \text{ for } x' \in \operatorname{span}\{e_p, e_{p+1}, \dots, e_{p+k}\},$$

and since $P'v$ has a non-zero component in the direction of e_{p+k} the claim follows from (4.1). □

6. Representation of the transformation matrix by orthogonal polynomials

The 2×2 matrix function which generates the fractional linear transformation (5.4) we denote in the following by V:

$$V(z) := \frac{1}{a(z)} \begin{pmatrix} \check{r}(z) & -a_{00}(z) \\ a_{11}(z) & -1 \end{pmatrix}. \tag{6.1}$$

In this section we express V by the polynomials of first and second kind. To this end, the elements of $\mathcal{H}_p$ are considered as column vectors with respect to the basis $e_0, e_1, \ldots, e_{p-1}$.

First we solve the equation $(A_0 - z)x = e_{p-1}$ in $\mathcal{H}_p$. With the Jacobi matrix $\mathcal{A}_0$ from (4.7) this equation becomes

$$\mathcal{A}_0 x - zx = e_{p-1},$$

or

$$(\mathcal{A}_0 - z) \begin{pmatrix} \xi_1 \\ \vdots \\ \xi_{p-1} \\ \xi_p \end{pmatrix} = \begin{pmatrix} 0 \\ \vdots \\ 0 \\ 1 \end{pmatrix}.$$

According to the definition of the orthogonal polynomials of first kind, the solution of the system with the 1 in the last component of the vector on the right-hand side replaced by $-b_{p-1}e_p(z)$ is the vector with components $e_0(z), e_1(z), \ldots, e_{p-1}(z)$. It follows that

$$\xi_j = -\frac{e_{j-1}(z)}{b_{p-1}e_p(z)}, \quad j = 1, 2, \ldots, p,$$

and hence

$$\begin{aligned} \left((A_0 - z)^{-1}e_{p-1}, e_{p-1}\right) &= (x, e_{p-1}) = -\frac{e_{p-1}(z)}{b_{p-1}e_p(z)}, \\ \left((A_0 - z)^{-1}e_{p-1}, e_0\right) &= (x, e_0) = -\frac{e_0(z)}{b_{p-1}e_p(z)}, \end{aligned}$$

that is,

$$a_{11}(z) = -\frac{e_{p-1}(z)}{b_{p-1}e_p(z)}, \quad a(z) = -\frac{1}{b_{p-1}e_p(z)}. \tag{6.2}$$

Next we solve the equation $(A_0 - z)x = u = \sqrt{s_0}\, e_0$. As above, in matrix form it becomes

$$(\mathcal{A}_0 - z) \begin{pmatrix} \xi_1 \\ \xi_2 \\ \vdots \\ \xi_p \end{pmatrix} = \begin{pmatrix} \sqrt{s_0} \\ 0 \\ \vdots \\ 0 \end{pmatrix}.$$

According to the definition of the polynomials of the second kind and because of $b_0\, d_1(z) = \sqrt{s_0}$ we have

$$(\mathcal{A}_0 - z)\begin{pmatrix} 0 \\ d_1(z) \\ \vdots \\ d_{p-2}(z) \\ d_{p-1}(z) \end{pmatrix} = \begin{pmatrix} \sqrt{s_0} \\ 0 \\ \vdots \\ 0 \\ -b_{p-1}d_p(z) \end{pmatrix} = \begin{pmatrix} \sqrt{s_0} \\ 0 \\ \vdots \\ 0 \\ 0 \end{pmatrix} - b_{p-1}d_p(z)\begin{pmatrix} 0 \\ 0 \\ \vdots \\ 0 \\ 1 \end{pmatrix}.$$

It follows that

$$\begin{pmatrix} \xi_1 \\ \xi_2 \\ \vdots \\ \xi_p \end{pmatrix} = (\mathcal{A}_0 - z)^{-1}\begin{pmatrix} \sqrt{s_0} \\ 0 \\ \vdots \\ 0 \end{pmatrix} = \begin{pmatrix} 0 \\ d_1(z) \\ \vdots \\ d_{p-1}(z) \end{pmatrix} - \frac{d_p(z)}{e_p(z)}\begin{pmatrix} e_0(z) \\ e_1(z) \\ \vdots \\ e_{p-1}(z) \end{pmatrix}$$

and hence

$$a_{00}(z) = ((A_0 - z)^{-1}u, u) = -\frac{d_p(z)}{e_p(z)}. \tag{6.3}$$

Inserting the expressions from (6.2) and (6.3) into (5.4) and observing the relation (4.15) we find that V can be written as

$$V(z) = \begin{pmatrix} -d_{p-1}(z) & -b_{p-1}d_p(z) \\ e_{p-1}(z) & b_{p-1}e_p(z) \end{pmatrix},$$

and hence we obtain the following theorem.

Theorem 6.1. *If, for some integer $p \geq 1$, the Nevanlinna function n satisfies one of the assumptions (j_p), $j \in \{1,2,3\}$, and $D_p \neq 0$ then the following relation holds:*

$$n(z) = ((A - z)^{-1}u, u) = -\frac{d_{p-1}(z)n_p(z) + b_{p-1}d_p(z)}{e_{p-1}(z)n_p(z) + b_{p-1}e_p(z)}, \tag{6.4}$$

where $n_p(z) = ((D - z)^{-1}u_p, u_p)$, $u_p = b_{p-1}e_p$.

Remark 6.2. (i) Using (4.12) and (4.6) we obtain from (6.4) with $p = 1$:

$$n(z) = -\frac{s_0}{z - \dfrac{s_1}{s_0} + n_1(z)}$$

and hence, because $n_1(z) = \mathrm{o}(1)$, n_1 is the Schur transform of n: $n_1 = \widehat{n}$. For $p \geq 2$ we obtain from (6.4) and (6.4) with p replaced by $p-1$ and with the help of (4.5) and (4.15) that

$$n_{p-1}(z) = -\frac{b_{p-2}^2}{z - a_{p-1} + n_p(z)},$$

hence

$$n_{p-1}(z) = -\frac{b_{p-2}^2}{z} - \frac{a_{p-1}b_{p-2}^2}{z^2} + \mathrm{o}\left(\frac{1}{z^2}\right)$$

and $n_p = \widehat{n}_{p-1}$.

(ii) From (6.4), (4.17), (4.18), and (4.20), we obtain

$$\widehat{n}(z) = -\frac{s_0}{n(z)} - (z - a_0) = -\frac{\widehat{d}_{p-2}(z)n_p(z) + \widehat{b}_{p-2}\widehat{d}_{p-1}(z)}{\widehat{e}_{p-2}(z)n_p(z) + \widehat{b}_{p-2}\widehat{e}_{p-1}(z)}.$$

Since, according to Theorem 4.2, $\widehat{d}_k$ and $\widehat{e}_k$ are the polynomials of first and second kind associated with $\widehat{n}$, this formula implies that the function n_p is the $p-1$-th Schur transform of $\widehat{n}$.

The 2×2 matrix polynomial V, which generates the fractional linear transformation (6.4), has the property

$$\det V(z) = b_{p-1}\left(d_p(z)e_{p-1}(z) - e_p(z)d_{p-1}(z)\right) = 1.$$

With

$$J = \begin{pmatrix} 0 & 1 \\ -1 & 0 \end{pmatrix},$$

V is J-*unitary* on the real line, that is,

$$V(z)JV(z)^* = J, \quad z \in \mathbb{R}.$$

Therefore $V(z)^{-1}$ exists for all $z \in \mathbb{C}$ and we can form the polynomial matrix function

$$V_0(z) = V(z)V(0)^{-1} = \begin{pmatrix} p_1^{(p)}(z) & p_0^{(p)}(z) \\ q_1^{(p)}(z) & q_0^{(p)}(z) \end{pmatrix}$$

with

$$\begin{aligned}
p_0^{(p)}(z) &= b_{p-1}\left(d_p(z)\,d_{p-1}(0) - d_{p-1}(z)\,d_p(0)\right), \\
p_1^{(p)}(z) &= b_{p-1}\left(d_p(z)\,e_{p-1}(0) - d_{p-1}(z)\,e_p(0)\right), \\
q_0^{(p)}(z) &= b_{p-1}\left(e_{p-1}(z)\,d_p(0) - e_p(z)\,d_{p-1}(0)\right), \\
q_1^{(p)}(z) &= b_{p-1}\left(e_{p-1}(z)\,e_p(0) - e_p(z)\,e_{p-1}(0)\right).
\end{aligned}$$

A straightforward calculation leads to the relation

$$n(z) \equiv \left((A - z)^{-1}u, u\right) = \frac{p_1^{(p)}(z)h_p(z) + p_0^{(p)}(z)}{q_1^{(p)}(z)h_p(z) + q_0^{(p)}(z)},$$

where

$$h_p(z) = -\frac{d_{p-1}(0)n_p(z) + b_{p-1}d_p(0)}{e_{p-1}(0)n_p(z) + b_{p-1}e_p(0)}. \tag{6.5}$$

With relation (4.14), the following formulas can be obtained, compare [1, I.2.4]:

$$p_0^{(p)}(z) = z\sum_{k=0}^{p-1} d_k(z)\, d_k(0),$$

$$p_1^{(p)}(z) = 1 + z\sum_{k=0}^{p-1} d_k(z)\, e_k(0),$$

$$q_0^{(p)}(z) = 1 - z\sum_{k=0}^{p-1} e_k(z)\, d_k(0),$$

$$q_1^{(p)}(z) = -z\sum_{k=0}^{p-1} e_k(z)\, e_k(0).$$

7. Transformation by means of a u-resolvent matrix

Given again a function $n \in \mathbf{N}_0$ with one of the properties (j_p), $j = 1, 2, 3$. Besides the decomposition (5.1) we consider the decomposition

$$\mathcal{H}(n) = \mathcal{H}_{p+1} \oplus \mathcal{H}'', \qquad \mathcal{H}_{p+1} = \operatorname{span}\{\mathcal{H}_p,\, e_p\} = \operatorname{span}\{e_0, e_1, \dots, e_p\},$$

and in the space $\mathcal{H}_{p+1}$ the restriction

$$S := A|_{\mathcal{H}_p} = \begin{pmatrix} A_0 \\ B \end{pmatrix}.$$

This restriction is a non-densely defined symmetric operator in $\mathcal{H}_{p+1}$ with defect index $(1, 1)$, and, evidently, the given function $n = \big((A - z)^{-1}u, u\big)$ is one of the u-resolvents of this operator S. Hence n can be represented as a fractional linear transformation of some function $g \in \mathbf{N}_0$ by means of the u-resolvent matrix $W = (w_{k\ell})_{k,\ell=1}^2$ of S:

$$n(z) = \frac{w_{11}(z)g(z) + w_{12}(z)}{w_{21}(z)g(z) + w_{22}(z)}. \tag{7.1}$$

Such a u-resolvent matrix W can easily be calculated. To this end we fix a self-adjoint extension of S in $\mathcal{H}_{p+1}$, which means that we fix some $\gamma \in \mathbb{R}$ in the right lower corner of the matrix representation of S with respect to the basis $e_0, e_1, \dots, e_p$ of $\mathcal{H}_{p+1}$. Denote this matrix or self-adjoint extension of S in $\mathcal{H}_{p+1}$ by $A_{0,\gamma}$:

$$A_{0,\gamma} = \begin{pmatrix} A_0 & \widetilde{B} \\ B & \gamma \end{pmatrix}.$$

According to [15] this u-resolvent matrix W is given by the formula

$$W(z) = \frac{1}{(u, \varphi(z^*))}\begin{pmatrix} (R_z^\gamma u, u) & (R_z^\gamma u, u)Q(z) - (u, \varphi(z^*))(\varphi(z), u) \\ 1 & Q(z) \end{pmatrix}, \tag{7.2}$$

where $R_z^\gamma = (A_{0,\gamma} - z)^{-1}$, $\varphi(z)$ is a defect function of S corresponding to the self-adjoint extension $A_{0,\gamma}$, and Q is the corresponding Q-function. An easy calculation yields

$$(R_z^\gamma u, u) = \big((A_{0,\gamma} - z)^{-1} u, u\big) = a_{00}(z) - \frac{b_{p-1}^2 a(z)^2}{\Delta(z)},$$

where

$$\Delta(z) = z - \gamma + b_{p-1}^2 a_{11}(z).$$

Since $S = A_{0,\gamma}|_{\mathcal{H}_p}$ and hence, in terms of linear relations,

$$S^* = \{\{x, A_{0,\gamma}x + \lambda e_p\} \,|\, x \in \mathcal{H}_{p+1}, \lambda \in \mathbb{C}\},$$

it is easy to check that for $\varphi(z)$ with $\{\varphi(z), z\varphi(z)\} \in S^*$ we can choose

$$\varphi(z) = (A_{0,\gamma} - z)^{-1} e_p = \frac{-1}{\Delta(z)} \begin{pmatrix} -b_{p-1}(A_0 - z)^{-1} e_{p-1} \\ 1 \end{pmatrix},$$

and then the Q-function, which is the solution (up to a real additive constant) of the equation

$$\frac{Q(z) - Q(\zeta)^*}{z - \zeta^*} = (\varphi(z), \varphi(\zeta)),$$

becomes

$$Q(z) = \frac{-1}{\Delta(z)}.$$

Inserting these expressions into W from (7.2) we find

$$W(z) = \frac{\Delta(z)}{b_{p-1}a(z)} \begin{pmatrix} a_{00}(z) - b_{p-1}^2 \dfrac{a(z)^2}{\Delta(z)} & -\dfrac{a_{00}(z)}{\Delta(z)} \\ 1 & \dfrac{-1}{\Delta(z)} \end{pmatrix}. \tag{7.3}$$

Observe that $W(z)$ is J-unitary on the real line. Next we establish the connection between the matrix functions V from (6.1) and W from (7.2), in fact we find a simple expression for $V^{-1}W$. We have

$$V(z)^{-1} = \frac{1}{a(z)} \begin{pmatrix} -1 & a_{00}(z) \\ -a_{11}(z) & a_{00}(z)a_{11}(z) - a(z)^2 \end{pmatrix}.$$

Multiplying this matrix from the right by $W(z)$ from (7.3) we obtain

$$V(z)^{-1}W(z) = -\begin{pmatrix} -b_{p-1} & 0 \\ \dfrac{z-\gamma}{b_{p-1}} & -\dfrac{1}{b_{p-1}} \end{pmatrix}. \tag{7.4}$$

Theorem 7.1. *If the function $n \in \mathbf{N}_0$ has one of the properties (j_p), $j \in \{1,2,3\}$, then the matrix functions V from (6.1) and W from (7.2) are connected by the* *relation* (7.4). *Therefore for the Nevanlinna functions* n_p *in* (5.4) *and* g *in* (7.1) *the following relation holds*:

$$n_p(z) = -\frac{b_{p-1}^2}{z - \gamma - \frac{1}{g(z)}}. \tag{7.5}$$

If $1/g(z) = \mathrm{o}(1)$, then formula (7.5) implies that n_p admits the asymptotic expansion

$$n_p(z) = -\frac{b_{p-1}^2}{z} - \frac{\gamma b_{p-1}^2}{z^2} + \mathrm{o}\left(\frac{1}{z^2}\right)$$

and $-1/g$ is its Schur transform: $-1/g = \widehat{n}_p$. Hence the number γ, which defines the self-adjoint extension of S, corresponds to the number a_p.

Remark 7.2. If instead of a self-adjoint *operator* extension $A_{0,\gamma}$ of S we choose the (multi-valued) self-adjoint *relation* extension of S:

$$A_{0,\infty} = S + \mathrm{span}\,\{0, e_p\} = A_0 + \mathrm{span}\,\{0, e_p\},$$

then we obtain

$$R_z^\infty = (A_{0,\infty} - z)^{-1} = (A_0 - z)^{-1}P,$$

where P is the orthogonal projection in $\mathcal{H}_{p+1}$ onto $\mathcal{H}_p$,

$$\varphi(z) = \begin{pmatrix} -b_{p-1}(A_0 - z)^{-1}e_{p-1} \\ 1 \end{pmatrix}, \quad Q(z) = z + b_{p-1}^2 a_{11}(z),$$

so that

$$W(z) = -\frac{1}{b_{p-1}^2 a(z)} \begin{pmatrix} a_{00}(z) & a_{00}(z + b_{p-1}^2 a_{11}(z)) - b_{p-1}^2 a(z)^2 \\ 1 & z + b_{p-1}^2 a_{11}(z) \end{pmatrix},$$

$$V(z)^{-1}W(z) = \begin{pmatrix} 0 & -b_{p-1} \\ \dfrac{1}{b_{p-1}} & \dfrac{z}{b_{p-1}} \end{pmatrix},$$

and instead of (7.5) we have

$$n_p(z) = -\frac{b_{p-1}^2}{z + g(z)}.$$

Thus if $g(z) = \mathrm{o}(1)$, then n_p has the asymptotic expansion

$$n_p(z) = -\frac{b_{p-1}^2}{z} + \mathrm{o}\left(\frac{1}{z^2}\right)$$

and g is the Schur transform of n_p: $g = \widehat{n}_p$.

A more explicit form of the resolvent matrix W from (7.2) can be obtained following [15] and [2]. To this end we decompose the space $\mathcal{H}_{p+1}$ as

$$\mathcal{H}_{p+1} = \mathrm{ran}(S - z) \dot{+} \mathrm{span}\, u, \quad z \in \mathbb{C},\ a(z) \neq 0,$$

($\dot{+}$ stands for direct sum) and denote for $y \in \mathcal{H}_{p+1}$ by $\mathsf{P}(z)y$ the coefficient of u in the corresponding decomposition of y:

$$y = (S - z)x + (\mathsf{P}(z)y)\,u \tag{7.6}$$

with some $x \in \operatorname{dom} S = \mathcal{H}_p$. Further, define $\mathsf{Q}(z)\,y = \big((S-z)^{-1}(y - (\mathsf{P}(z)y)u), u\big)$. Then, according to [15], the resolvent matrix can be chosen to be

$$W^0(z) = I_2 + z\begin{pmatrix} \mathsf{Q}(z) \\ -\mathsf{P}(z) \end{pmatrix} (\mathsf{Q}(0)^* \;\; -\mathsf{P}(0)^*)\, J, \quad J = \begin{pmatrix} 0 & 1 \\ -1 & 0 \end{pmatrix}. \tag{7.7}$$

We derive an explicit expression for $W^0(z) := W(z)W(0)^{-1}$, following [2]. To this end, for the vectors and operators we use matrix representations with respect to the basis $\mathbf{t}_0(=u), \mathbf{t}_1, \dots, \mathbf{t}_p$. Recall that S_p is the Gram matrix associated with this basis. We denote by $\mathfrak{S}$ the $(p+1)\times(p+1)$-matrix

$$\mathfrak{S} = \begin{pmatrix} 0 & 0 & \cdots & 0 & 0 \\ 1 & 0 & \cdots & 0 & 0 \\ 0 & 1 & \cdots & 0 & 0 \\ \vdots & \vdots & & \vdots & \vdots \\ 0 & 0 & \cdots & 1 & 0 \end{pmatrix}$$

and by e the first column in the $(p+1)\times(p+1)$ identity matrix. Then S and u correspond to $\mathfrak{S}\,|_{\mathbb{C}^p \dot{+} \{0\}}$ and e. First we apply the operator $(I - z\mathfrak{S}^*)^{-1}$ to (7.6) and observe the relation

$$e^*(I - z\mathfrak{S}^*)^{-1}(\mathfrak{S} - z)x = 0, \quad x \in \mathbb{C}^p \dot{+} \{0\}.$$

It follows that

$$e^*(I - z\mathfrak{S}^*)^{-1}y = (\mathsf{P}(z)y)\, e^*(I - z\mathfrak{S}^*)^{-1}u = \mathsf{P}(z)y. \tag{7.8}$$

Further, observing that $e^*S_p = \begin{pmatrix} s_0 & s_1 & \cdots & s_p \end{pmatrix}$ we obtain

$$\begin{aligned} \mathsf{Q}(z)y &= \big((S-z)^{-1}(y - (\mathsf{P}(z)y)u), u\big) \\ &= e^*S_p\left((\mathfrak{S}-z)^{-1}y - (\mathfrak{S}-z)^{-1}e\left(e^*(I - z\mathfrak{S}^*)^{-1}y\right)\right) \\ &= \begin{pmatrix} s_0 & s_1 & \cdots & s_p \end{pmatrix}\left((\mathfrak{S}-z)^{-1}(I - z\mathfrak{S}^*) - (\mathfrak{S}-z)^{-1}ee^*\right)(I - z\mathfrak{S}^*)^{-1}y \\ &= \begin{pmatrix} s_0 & s_1 & \cdots & s_p \end{pmatrix}\mathfrak{S}^*(I - z\mathfrak{S}^*)^{-1}y \\ &= \begin{pmatrix} 0 & s_0 & s_1 & \cdots & s_{p-1} \end{pmatrix}(I - z\mathfrak{S}^*)^{-1}y, \end{aligned}$$

where for the second last equality sign we have used that

$$(\mathfrak{S}-z)^{-1}(I - z\mathfrak{S}^*) - (\mathfrak{S}-z)^{-1}ee^* = \mathfrak{S}^*.$$

Together with (7.8) we find

$$\begin{pmatrix} \mathsf{Q}(z) \\ -\mathsf{P}(z) \end{pmatrix} = \begin{pmatrix} 0 & s_0 & s_1 & \cdots & s_{p-1} \\ -1 & 0 & 0 & \cdots & 0 \end{pmatrix}(I - z\mathfrak{S}^*)^{-1},$$

and (7.7) becomes

$$W^0(z) = I_2 + z \begin{pmatrix} 0 & s_0 & \cdots & s_{p-1} \\ -1 & 0 & \cdots & 0 \end{pmatrix} (I - z\mathfrak{S}^*)^{-1} S_p^{-1} \begin{pmatrix} 0 & -1 \\ s_0 & 0 \\ \vdots & \vdots \\ s_{p-1} & 0 \end{pmatrix} \begin{pmatrix} 0 & 1 \\ -1 & 0 \end{pmatrix}.$$

8. Reproducing kernel spaces: reduction via resolvent invariant subspaces

In this section we start from the operator representation of the Nevanlinna function n in the corresponding reproducing kernel space $\mathcal{L}(n)$ with kernel

$$L_n(z, w) = \frac{n(z) - n(w)^*}{z - w^*}, \quad z, \zeta \in \mathbb{C} \setminus \mathbb{R},$$

see Section 2, (3). The operator A is introduced via its resolvent $(A - z)^{-1}$ which is the difference-quotient operator R_z defined by

$$(R_z f)(\zeta) = \frac{f(\zeta) - f(z)}{\zeta - z}, \quad f \in \mathcal{L}(n). \tag{8.1}$$

If n satisfies one of the assumptions (j_p), then, by [3, Lemma 5.1], the functions

$$f_0(\zeta) = n(\zeta),\ f_1(\zeta) = \zeta n(\zeta) + s_0, \ldots,\ f_p(\zeta) = \zeta^p n(\zeta) + \zeta^{p-1} s_0 + \cdots + s_{p-1}$$

all belong to $\mathcal{L}(n)$ and

$$\langle f_k, f_j \rangle_{\mathcal{L}(n)} = s_{j+k}, \quad j, k = 0, 1, \ldots, p. \tag{8.2}$$

In particular, $u := n \in \mathcal{L}(n)$, and by the reproducing property of the kernel L_n we have

$$n(z) = ((A - z)^{-1} u, u)_{\mathcal{L}(n)}.$$

By $\mathcal{U}_J$ we denote the class of all 2×2 matrix polynomials Θ which are J-unitary on $\mathbb{R}$ and for which the kernel

$$K_\Theta(z, w) = \frac{J - \Theta(z) J \Theta(w)^*}{z - w^*}$$

is non-negative. The reproducing kernel Hilbert space with this kernel will be denoted by $\mathcal{H}(\Theta)$; its elements are 2-vector functions. The matrix polynomials V and W considered in the previous sections belong to $\mathcal{U}_J$: this follows from the Christoffel–Darboux formulas (4.14) for V and from (7.4) for W. Note that if Θ belongs to $\mathcal{U}_J$, then $\det \Theta(z) \equiv c$, where is c is a unimodular complex number, because the determinant $\det \Theta(z)$ is a non-vanishing polynomial in z.

The following theorem was proved in [4, Theorem 8.1], even in an indefinite setting.

Theorem 8.1. *Let $n \in \mathbf{N}_0$ and suppose that there exists a matrix polynomial*

$$\Theta = \begin{pmatrix} a & b \\ c & d \end{pmatrix} \in \mathcal{U}_\ell$$

such that the mapping

$$\mathbf{u} \longrightarrow \begin{pmatrix} 1 & -n \end{pmatrix} \mathbf{u}$$

is an isometry from $\mathcal{H}(\Theta)$ into $\mathcal{L}(n)$. Define the function $\check{n}$ by

$$n(z) = \frac{a(z)\check{n}(z) + b(z)}{c(z)\check{n}(z) + d(z)}.$$

Then the following statements hold.

(i) *$\check{n}$ is Nevanlinna function.*

(ii) *The mapping $g \mapsto f$:*

$$f(\zeta) = \big(a(\zeta) - n(\zeta)c(\zeta)\big)g(\zeta)$$

is an isometry from $\mathcal{L}(\check{n})$ into $\mathcal{L}(n)$.

(iii) *We have*

$$\mathcal{L}(n) = \begin{pmatrix} 1 & -n \end{pmatrix} \mathcal{H}(\Theta) \oplus (a - nc)\mathcal{L}(\check{n})$$

and the mapping

$$W : \mathcal{L}(n) \ni f \mapsto \begin{pmatrix} \mathbf{u} \\ g \end{pmatrix} \in \begin{pmatrix} \mathcal{H}(\Theta) \\ \mathcal{L}(\check{n}) \end{pmatrix},$$

where $f, \mathbf{u}$, and g are connected by the relation

$$f(\zeta) = \begin{pmatrix} 1 & -n(\zeta) \end{pmatrix} \mathbf{u}(\zeta) + \big(a(\zeta) - n(\zeta)c(\zeta)\big)g(\zeta),$$

is a unitary mapping from $\mathcal{L}(n)$ onto $\mathcal{H}(\Theta) \oplus \mathcal{L}(\check{n})$.

(iv) *The mapping WR_zW^* is of the form*

$$WR_zW^* = \begin{pmatrix} R_{11}(z) & R_{12}(z) \\ R_{21}(z) & R_{22}(z) \end{pmatrix} : \begin{pmatrix} \mathcal{P}(\Theta) \\ \mathcal{L}(\check{n}) \end{pmatrix} \to \begin{pmatrix} \mathcal{P}(\Theta) \\ \mathcal{L}(\check{n}) \end{pmatrix}, \tag{8.3}$$

with

$$\begin{aligned} R_{11}(z) &= R_z - \frac{1}{k(z)}(R_z\Theta)(\,\cdot\,) \begin{pmatrix} \check{n}(z) \\ 1 \end{pmatrix} \begin{pmatrix} 0 & 1 \end{pmatrix} E_z \\ &= R_z - K_\Theta(\,\cdot\,, z^*) \begin{pmatrix} 1 \\ -n(z) \end{pmatrix} \begin{pmatrix} 0 & 1 \end{pmatrix} E_z, \\ R_{12}(z) &= \frac{1}{k(z)}(R_z\Theta)(\,\cdot\,) \begin{pmatrix} d(z) \\ -c(z) \end{pmatrix} E_z \\ &= -(a(z) - n(z)c(z))K_\Theta(\,\cdot\,, z^*) \begin{pmatrix} 0 \\ 1 \end{pmatrix} E_z, \\ R_{21}(z) &= -\frac{1}{k(z)}(R_z\check{n})(\,\cdot\,) \begin{pmatrix} 0 & 1 \end{pmatrix} E_z \end{aligned}$$

$$= -\frac{1}{k(z)} L_{\check{n}}(\,\cdot\,, z^*) \begin{pmatrix} 0 & 1 \end{pmatrix} E_z,$$

$$R_{22}(z) = R_z - \frac{c(z)}{k(z)} (R_z \check{n})(\,\cdot\,) E_z$$

$$= R_z - \frac{c(z)}{k(z)} L_{\check{n}}(\,\cdot\,, z^*) E_z,$$

where R_z is the difference-quotient operator, E_z is the operator of evaluation at the point z on any reproducing kernel space, and

$$k(z) = c(z)\check{n}(z) + d(z) = \frac{\det \Theta(z)}{a(z) - n(z)c(z)}.$$

We mention that formula (8.3) corresponds to the relation (5.9) above.

A space of functions is called *resolvent-invariant* if it is invariant under the difference-quotient operator R_z as defined in (8.1). In the following lemma, with a resolvent-invariant non-degenerate subspace of a certain inner product space a 2×2 matrix function is associated.

Lemma 8.2. *Let $\mathcal{M}$ be a finite-dimensional resolvent-invariant space of 2-vector polynomials endowed with an inner product $\langle\,\cdot\,,\,\cdot\,\rangle$ such that*

$$\langle R_z f, g\rangle - \langle f, R_w g\rangle - (z - w^*)\langle R_z f, R_w g\rangle = g(w)^* J f(z), \tag{8.4}$$

and let $\mathcal{M}_1$ be a resolvent-invariant non-degenerate subspace of $\mathcal{M}$. Then there exists a $\Theta_1 \in \mathcal{U}_J$ such that

(i) $\mathcal{M}_1 = \mathcal{H}(\Theta_1)$,

(ii) $\mathcal{M} = \mathcal{H}(\Theta_1) \oplus \Theta_1 \mathcal{N}$ *where $\mathcal{N} = \Theta_1^{-1}\mathcal{M}_1^{\perp}$ is a resolvent-invariant space of 2-vector polynomials, for which the relation* (8.4) *holds if equipped with the inner product*

$$(\Theta_1^{-1} f, \Theta_1^{-1} g)_{\mathcal{N}} = \langle f, g\rangle, \quad f, g \in \mathcal{M}_1{}^{\perp}.$$

Relation (8.4) is often called *de Branges identity*, see [7] and, for further references, [12]. That $\mathcal{N}$ consists of 2-vector polynomials is due to fact that $\Theta_1^{-1}(z) = -J\Theta_1(z^*)^* J$ is a matrix polynomial. The other claims of the lemma follow from [6, Theorem 3.1].

Now we formulate and prove Theorem 6.1 again in the context of reproducing kernel spaces.

Theorem 8.3. *If, for some integer $p \geq 1$, the Nevanlinna function n satisfies one of the assumptions (j_p), $j \in \{1, 2, 3\}$, and $D_p \neq 0$ then the following relation holds:*

$$n(z) = -\frac{d_{p-1}(z)\check{n}_p(z) + b_{p-1}d_p(z)}{e_{p-1}(z)\check{n}_p(z) + b_{p-1}e_p(z)}. \tag{8.5}$$

where $\check{n}_p$ is a Nevanlinna function such that $\check{n}_p \in \mathcal{L}(\check{n}_p)$, and

$$\check{n}_p(z) = (R_z \check{n}_p, \check{n}_p)_{\mathcal{L}(\check{n}_p)}. \tag{8.6}$$

Comparing (6.4) and (8.5) we find that $\breve{n}_p(z) = n_p(z)$, the pth element in the sequence obtained by applying the Schur transformation p times starting with n.

Proof of Theorem 8.3. Let $\mathcal{M}$ be the linear space spanned by the $p+1$ 2-vector functions

$$\mathbf{f}_0(\zeta) = \begin{pmatrix} 0 \\ -1 \end{pmatrix}, \mathbf{f}_1(\zeta) = \begin{pmatrix} s_0 \\ -\zeta \end{pmatrix}, \dots, \mathbf{f}_p(\zeta) = \begin{pmatrix} s_0\zeta^{p-1} + \cdots + s_{p-1} \\ -\zeta^p \end{pmatrix} \tag{8.7}$$

and equipped with the inner product which makes the map $\mathbf{u} \mapsto \begin{pmatrix} 1 & -n \end{pmatrix} \mathbf{u}$ an isometry from $\mathcal{M}$ into $\mathcal{L}(n)$, see (8.2). Note that if n has the integral representation (2.2), the elements of $\mathcal{M}$ are of the form

$$\begin{pmatrix} \int_{-\infty}^{\infty} (R_\zeta f)(t)\, d\sigma(t) \\ -f(\zeta) \end{pmatrix},$$

where f is a polynomial of degree $\le p$. Indeed, it suffices to show this for the basis elements of $\mathcal{M}$: If $f(\zeta) = \zeta^j$, then

$$R_\zeta f(t) = \frac{\zeta^j - t^j}{\zeta - t} = \zeta^{j-1} + t\zeta^{j-2} + \cdots + t^{j-2}\zeta + t^{j-1},$$

and hence, on account of (1.3),

$$\int_{-\infty}^{\infty} (R_\zeta f)(t)\, d\sigma(t) = s_0\zeta^{j-1} + s_1\zeta^{j-2} + \cdots + s_{p-2}\zeta + s_{p-1}.$$

It follows that $\mathcal{M}$ is also spanned by the polynomial vectors

$$\begin{pmatrix} -d_j \\ e_j \end{pmatrix}, \quad j = 0, 1, \dots, p,$$

where e_j and d_j are the polynomials of first and second kind associated with n, see (4.4) and (4.10).

Let $\mathcal{M}_p$ be the space spanned by the first p of the 2-vector functions in (8.7). Since S_{p-1} from (4.2) is a positive matrix, the space $\mathcal{M}_p$ is non-degenerate and

$$\mathcal{M} = \mathcal{M}_p \oplus \operatorname{span} \begin{pmatrix} -d_p \\ e_p \end{pmatrix}.$$

As both $\mathcal{M}$ and $\mathcal{M}_p$ are resolvent-invariant spaces, by Lemma 8.2 we have that for some $\Theta_1 \in \mathcal{U}_J$, which is normalized by $\Theta_1(0) = I_2$ (and hence $\det \Theta(z) \equiv 1$),

$$\mathcal{M}_p = \mathcal{H}(\Theta_1), \quad \mathcal{M} = \mathcal{H}(\Theta_1) \oplus \Theta_1 \mathcal{N}.$$

Here $\mathcal{N}$ is a one-dimensional resolvent-invariant space, which, when equipped with the induced inner product, satisfies the de Branges identity and therefore is spanned by a constant J-neutral vector $\begin{pmatrix} \alpha & \beta \end{pmatrix}^\top$ such that

$$\Theta_1(z) \begin{pmatrix} \alpha \\ \beta \end{pmatrix} = b_{p-1} \begin{pmatrix} -d_p(z) \\ e_p(z) \end{pmatrix}.$$

For $\lambda \in \mathbb{R}$ denote by C_λ the constant J-unitary matrix

$$C_\lambda = \begin{cases} \begin{pmatrix} \lambda\alpha & \alpha \\ -\alpha^{-*} + \lambda\beta & \beta \end{pmatrix}, & \alpha \neq 0, \\ \begin{pmatrix} \beta^{-*} & 0 \\ \lambda\beta & \beta \end{pmatrix}, & \alpha = 0. \end{cases}$$

Then there exists a λ such that

$$\Theta(z) := \Theta_1(z)C_\lambda = \begin{pmatrix} a(z) & -b_{p-1}d_p(z) \\ c(z) & b_{p-1}e_p(z) \end{pmatrix},$$

where a and c are polynomials such that $\deg c < \deg e_p = p$. The inclusion

$$R_0 \begin{pmatrix} a \\ c \end{pmatrix} \in \mathcal{H}(\Theta) = \mathcal{H}(\Theta_1) = \mathcal{M}_p$$

implies that $\deg a < p - 1$. From $\det \Theta(z) \equiv 1$ it follows that $x = a$ and $y = c$ are polynomial solutions of the equation

$$x(z)e_p(z) + y(z)d_p(z) = \frac{1}{b_{p-1}}.$$

Since all polynomial solutions of this equation are given by

$$x(z) = a(z) - s(z)d_p(z), \quad y(z) = c(z) + s(z)e_p(z)$$

with some polynomial s, the solutions $x = a$ and $y = c$ have minimal degrees and because of that they are unique. Observing (4.15), we find

$$a(z) = -d_{p-1}(z), \quad c(z) = e_{p-1}(z).$$

Hence

$$\Theta(z) = \begin{pmatrix} -d_{p-1}(z) & -b_{p-1}d_p(z) \\ e_{p-1}(z) & b_{p-1}e_p(z) \end{pmatrix} = V(z)$$

and $C_\lambda = \Theta(0)$ is the coefficient matrix of the fractional linear transformation (6.5).

Define the function $\check{n}_p$ by (8.5). Then, according to Theorem 8.1, it is a Nevanlinna function. We show that $\check{n}_p \in \mathcal{L}(\check{n}_p)$. The function $f_p(\zeta) = \begin{pmatrix} 1 & -n(\zeta) \end{pmatrix} \mathbf{f}_p(\zeta)$ belongs to $\mathcal{L}(n)$ and, according to Theorem 8.1 (iii), it can be written as

$$\begin{pmatrix} 1 & -n(\zeta) \end{pmatrix} \mathbf{f}_p(\zeta) = \begin{pmatrix} 1 & -n(\zeta) \end{pmatrix} \mathbf{u}_p(\zeta) + (a(\zeta) - n(\zeta)c(\zeta))g_p(\zeta)$$

with $\mathbf{u}_p \in \mathcal{M}_p$, $g_p \in \mathcal{L}(\check{n}_p)$, and the two summands on the right-hand side are orthogonal. This orthogonality and the isometry of the multiplication by $\begin{pmatrix} 1 & -n \end{pmatrix}$ imply that $(0 \neq)\, \mathbf{f}_p - \mathbf{u}_p \in \mathcal{M}_p^\perp$ and hence there is a non-zero complex number γ such that

$$\mathbf{f}_p - \mathbf{u}_p = \gamma \begin{pmatrix} -d_p(\zeta) \\ e_p(\zeta) \end{pmatrix}.$$

Therefore

$$(a(\zeta) - n(\zeta)c(\zeta))g_p(\zeta) = -\gamma\,(d_p(\zeta) + e_p(\zeta)n(\zeta))$$

and

$$g_p(\zeta) = -\gamma \frac{e_p(\zeta)\, n(\zeta) + d_p(\zeta)}{-n(\zeta)\, c(\zeta) + a(\zeta)} = -\frac{\gamma}{b_{p-1}} \check{n}_p(\zeta).$$

Hence $\check{n}_p \in \mathcal{L}(\check{n}_p)$. Equality (8.6) follows from item (3) in Section 3. □

References

[1] N.I. Akhiezer, *The classical moment problem and some related topics in analysis.* Fizmatgiz, Moscow, 1961; English transl.: Hafner, New York, 1965.

[2] D. Alpay, R.W. Buursema, A. Dijksma, and H. Langer, *The combined moment and interpolation problem for Nevanlinna functions.* Operator Theory, Structured Matrices, and Dilations, Theta Ser. Adv. Math., 7, Theta, Bucharest, 2007, 1–28.

[3] D. Alpay, A. Dijksma, and H. Langer, *Factorization of J-unitary matrix polynomials on the line and a Schur algorithm for generalized Nevanlinna functions.* Linear Algebra Appl. **387** (2004), 313–342,

[4] D. Alpay, A. Dijksma, H. Langer, and Y. Shondin, *The Schur transform for generalized Nevanlinna functions: interpolation and self-adjoint operator realizations.* Complex Anal. Oper. Theory **1** (2007), 189–120.

[5] D. Alpay, A. Dijksma, and H. Langer, *The transformation of Issai Schur and related topics in an indefinite setting.* Oper. Theory Adv. Appl. **176**, Birkhäuser, Basel, 2007, 1–98.

[6] D. Alpay and H. Dym, *Structure invariant spaces of vector valued functions, Hermitian matrices and a generalization of the Iohvidov laws.* Linear Algebra Appl. **137/138** (1990) 137–181.

[7] L. de Branges, *Some Hilbert spaces of analytic functions*, I. Trans. Amer. Math. Soc. **106** (1963), 445–468.

[8] A. Bultheel, P. González-Vera, E. Hendriksen, and O. Njåstad, *Orthogonal rational functions*, Cambridge Monographs on Appl. and Comp. Math. **5**, Cambridge University Press, Cambridge, 1999.

[9] M. Derevyagin, *On the Schur algorithm for indefinite moment problem.* Methods Funct. Anal. Topology **9**(2) (2003), 133–145.

[10] A. Dijksma, H. Langer, A. Luger, and Y. Shondin, *Minimal realizations of scalar generalized Nevanlinna functions related to their basic factorization.* Oper. Theory Adv. Appl. **154**, Birkhäuser, Basel, 2004, 69–90.

[11] A. Dijksma, H. Langer, and H.S.V. de Snoo, *Eigenvalues and pole functions of Hamiltonian systems with eigenvalue depending boundary conditions.* Math. Nachr. **161** (1993), 107–154.

[12] H. Dym, *J-contractive matrix functions, reproducing kernel Hilbert spaces and interpolation.* CBMS Regional Conference Series in Mathematics **71**, Amer. Math. Soc., Providence, RI, 1989.

[13] I.S. Kac and M.G. Krein, *R-functions-analytic functions mapping the upper half-plane into itself.* Amer. Math. Soc. Transl. **103**(2) (1974), 1–18.

[14] M.G. Krein and H. Langer, *Über einige Fortsetzungsprobleme, die eng mit der Theorie hermitescher Operatoren in Raume* Π_κ *zusammenhängen.* I. *Einige Funktionenklassen und ihre Darstellungen.* Math. Nachr. **77** (1977), 187–236.

[15] M.G. Krein and H. Langer, *Über einige Fortsetzungsprobleme, die eng mit der Theorie hermitescher Operatoren in Raume* Π_κ *zusammenhängen. II. Verallgemeinerte Resolventen, u-Resolventen und ganze Operatoren.* J. Functional Analysis **30**(3) (1978), 390–447.

[16] M.G. Krein and H. Langer, *On some continuation problems which are closely related to the theory of Hermitian operators in spaces* Π_κ. *IV: Continuous analogues of orthogonal polynomials on the unit circle with respect to an indefinite weight and related continuation problems for some classes of functions*, J. Operator Theory **13** (1985), 299–417.

[17] H. Langer and B. Textorius, *On generalized resolvents and Q-functions of symmetric linear relations (subspaces)*. Pacific J. Math. **72** (1977), 135–165.

D. Alpay
Department of Mathematics
Ben-Gurion University of the Negev
P.O. Box 653
84105 Beer-Sheva, Israel
e-mail: dany@math.bgu.ac.il

A. Dijksma
Department of Mathematics
University of Groningen
P.O. Box 407
9700 AK Groningen, The Netherlands
e-mail: a.dijksma@math.rug.nl

H. Langer
Institute for Analysis and Scientific Computing
Vienna University of Technology
Wiedner Hauptstrasse 8–10
A-1040 Vienna, Austria
e-mail: hlanger@mail.zserv.tuwien.ac.at

Operator Theory:
Advances and Applications, Vol. 190, 65–112

M. Kreĭn's Research on Semi-Bounded Operators, its Contemporary Developments, and Applications

Yu. Arlinskiĭ and E. Tsekanovskiĭ

To the memory of Mark Grigor'evich Kreĭn

Abstract. We are going to consider the M. Kreĭn classical papers on the theory of semi-bounded operators and the theory of contractive self-adjoint extensions of Hermitian contractions, and discuss their impact and role in the solution of J. von Neumann's problem about parametrization in terms of his formulas of all nonnegative self-adjoint extensions of nonnegative symmetric operators, in the solution of the Phillips–Kato extension problems (in restricted sense) about existence and parametrization of all proper sectorial (accretive) extensions of nonnegative operators, in bi-extension theory of nonnegative operators with the exit into triplets of Hilbert spaces, in the theory of singular perturbations of nonnegative self-adjoint operators, in general realization problems (in system theory) of Stieltjes matrix-valued functions, in Nevanlinna-Pick system interpolation in the class of sectorial Stieltjes functions, in conservative systems theory with accretive main Schrödinger operator, in the theory of semi-bounded symmetric and self-adjoint operators invariant with respect to some groups of transformations. New developments and applications to the singular differential operators are discussed as well.

Mathematics Subject Classification (2000). Primary 47A63, 47B25; Secondary 47B65.

Keywords. Semi-bounded and nonnegative operators, accretive and sectorial extensions, the von Neumann and the Phillips–Kato extension problems, singular perturbations, system theory.

This paper is an extended version of a plenary talk given by one of the authors (E.T.) at the International Conference in Odessa, Ukraine, and dedicated to the centenary of Mark Kreĭn. The bitterness of Kreĭn's life was recalled by Yury Berezansky and Israel Gohberg in their comments at the Conference. Below are

This work was completed with the support of our TEX-pert.

some recollections about Kreĭn's life that were presented at the beginning of the talk. One day E.T. called M. Kreĭn asking for an appointment to discuss some problems. Mark Grigor'evich picked up the phone and after warm welcome said: "Why don't you congratulate me?" " I do not know what I should congratulate you with." " I was elected to the American Academy of Arts and Sciences together with C. Chaplin, A. Solzhenitsin and A. Sakharov", M. Kreĭn told. "I did not know about that. There was no information on radio, newspapers and television." "Yes", answered Mark Grigor'evich, and "there will be no information about my election in any media in the USSR as I was informed by local officials."

1. Introduction

The literature on extension theory of semi-bounded operators is too extensive to be discussed exhaustively in this paper. We use references [1]–[175] and cover in this survey only topics closely related directly or indirectly to our scientific interests as well as results and developments we were involved in personally.

In 1929 John von Neumann published a paper [144] where for the first time an Extension Theory and his well-known formulas appeared. These formulas describe the domains (in terms of operator-valued parameter) of all self-adjoint extensions of a given symmetric operator on some Hilbert space. The von Neumann formulas can be presented in the following way: If S is a symmetric operator acting on some Hilbert space $\mathfrak{H}$ with the dense in $\mathfrak{H}$ domain $D(S)$ and $\widetilde{S}$ is self-adjoint extension of S, then

$$\begin{aligned} D(S^*) &= D(S)\dot{+}\mathfrak{N}_i\dot{+}\mathfrak{N}_{-i} \\ D(\widetilde{S}) &= D(S)\dot{+}(I+U)\mathfrak{N}_i \end{aligned}$$

where $\mathfrak{N}_{\pm i}$ are defect subspaces of S, that is, $S^*x = \pm ix$, $x \in \mathfrak{N}_{\pm i}$, U is an isometry of $\mathfrak{N}_i$ onto $\mathfrak{N}_{-i}$, and

$$\begin{aligned} f_{\widetilde{S}} &= f_S + x_i + Ux_i \\ \widetilde{S} f_{\widetilde{S}} &= S f_S + i x_i - iUx_i, \\ x_i &\in \mathfrak{N}_i, Ux_i \in \mathfrak{N}_{-i}. \end{aligned}$$

These formulas establish a one-to-one correspondence between isometries $U : \mathfrak{N}_i \to \mathfrak{N}_{-i}$ and all self-adjoint extensions $\widetilde{S}$ of a given symmetric operator S.

Operator S is called semi-bounded from below if there exists a number m, called a lower bound, such that for all $x \in D(S)$

$$(Sx, x) \geq m(x, x).$$

We will assume further that lower bound m is the greatest lower bound of S. In the same paper J. von Neumann formulated a problem about existence and description of all self-adjoint extensions preserving the greatest lower bound of a given densely defined symmetric operator semi-bounded from below. In particular, the von Neumann problem is about the existence of an extension $\widetilde{S} = \widetilde{S}^*$ of the given symmetric operator S such that

$$(\widetilde{S}x, x) \geq m(x, x), \quad x \in D(\widetilde{S}).$$

For any small $\epsilon > 0$ J. von Neumann established only existence of a self-adjoint operator $\widetilde{S} = \widetilde{S}^*$ such that

$$(\widetilde{S}x, x) \geq (m - \epsilon)(x, x), \quad x \in D(\widetilde{S}).$$

Existence of self-adjoint extensions preserving the greatest lower bound has been established by K. Friedrichs, M. Stone and H. Freudental [82], [163], [81]. If S is semi-bounded symmetric operator with the lower bound m and $\widetilde{S}$ its self-adjoint extension with the same lower bound, then

$$S - mI \geq 0$$
$$\widetilde{S} - mI \geq 0$$

and the von Neumann problem can be reduced to the problem of existence and description of all nonnegative self-adjoint extensions of a given nonnegative densely defined symmetric operator.

In 1947 M. Kreĭn published a paper appeared in two parts [112], [113] where based on his theory of self-adjoint contractive extensions of a given non-densely defined Hermitian (symmetric) contraction he gave a description (in an implicit form) of all nonnegative self-adjoint extensions of a nonnegative symmetric operator by means of linear fractional transforms of nonnegative operators. He has discovered two extremal nonnegative self-adjoint extensions, the maximal and the minimal ones, such that the maximal one coincides with the nonnegative extension discovered earlier by Friedrichs and the minimal one in the case of positive lower bound can be obtained in the framework of von Neumann's approach. Following [12], [6] we call these two extremal extensions *the Friedrichs and the Kreĭn-von Neumann extensions respectively.*

2. Closed sectorial sesquilinear forms and associated maximal sectorial operators

Recall some definitions and results from [104].

Let $\tau[\cdot,\cdot]$ be a sesquilinear form in a Hilbert space H defined on a linear manifold $\mathcal{D}[\tau]$. The form τ is called symmetric if $\tau[u,v] = \overline{\tau[v,u]}$ for all $u, v \in \mathcal{D}[\tau]$ and nonnegative if $\tau[u] := \tau[u,u] \geq 0$ for all $u \in \mathcal{D}[\tau]$.

The form τ is called sectorial with the vertex at the point $\gamma \in \mathbb{C}$ and a semi-angle $\theta \in [0, \pi/2)$ if its numerical range

$$W(\tau) = \{\tau[u],\ u \in \mathcal{D}[\tau],\ ||u|| = 1\}$$

is contained in the sector

$$S_\gamma = \{z \in \mathbb{C} : |\arg(z - \gamma)| \leq \theta\},$$

i.e.,

$$\left|\mathrm{Im}\,(\tau[u]-\gamma\|u\|^2)\right| \le \tan\theta\,\mathrm{Re}\,(\tau[u]-\gamma\|u\|^2), \qquad u\in\mathcal{D}[\tau].$$

Thus, τ is sectorial with vertex at γ if and only if the form $\tau[u,v]-\gamma(u,v)$ has vertex at the origin.

Let τ be a sesquilinear form. The form $\tau^*[u,v]:=\overline{\tau[v,u]}$ is called the adjoint to τ, and the forms

$$\tau_{\mathrm{R}}[u,v]:=\frac{1}{2}\left(\tau[u,v]+\tau^*[u,v]\right),$$

$$\tau_{\mathrm{I}}[u,v]:=\frac{1}{2i}\left(\tau[u,v]-\tau^*[u,v]\right), \qquad u,v\in\mathcal{D}[\tau].$$

are called the real and the imaginary parts of τ, respectively.

A sequence $\{u_n\}$ is called τ-converging to the vector $u\in H$ if

$$\lim_{n\to\infty}u_n=u \quad\text{and}\quad \lim_{n,m\to\infty}\tau[u_n-u_m]=0.$$

The form τ is called closed if for every sequence $\{u_n\}$ τ- converging to a vector u it follows that $u\in\mathcal{D}[\tau]$ and $\lim\limits_{n\to\infty}\tau[u-u_n]=0$. A sectorial form τ with vertex at the origin is closed if and only if the linear manifold $\mathcal{D}[\tau]$ is a Hilbert space with the inner product $(u,v)_\tau=\tau_{\mathrm{R}}[u,v]+(u,v)$ [104]. The form τ is called closable if it has a closed extension; in this case the closure of τ is the smallest closed extension of τ.

A linear operator T in H is called sectorial with vertex at γ and a semi-angle $\theta\in[0,\pi/2)$ if the form $\tau(u,v):=(Tu,v)$ called sectorial with vertex at γ and a semi-angle θ. As proved in [104] for such T the corresponding form is closable. A sectorial operator T is called maximal sectorial (m-sectorial) if one of the equivalent conditions is fulfilled

(i) T does not admit sectorial extensions without exit from the given Hilbert space,
(ii) the resolvent set of T is nonempty,
(iii) the adjoint operator T^* is sectorial.

If τ is a closed, densely defined sectorial form, then according to the first representation theorem [112], [104] there exists a unique m-sectorial operator T in H, associated with τ in the following sense:

$$(Tu,v)=\tau[u,v] \quad\text{for all}\quad u\in\mathcal{D}(T) \quad\text{and}\quad \text{for all}\quad v\in\mathcal{D}[\tau].$$

If τ is closed and nonnegative form, then by the second representation theorem [112], [104] the identities hold:

$$\mathcal{D}[\tau]=\mathcal{D}(T^{\frac{1}{2}}),\ \tau[u,v]=(T^{\frac{1}{2}}u,T^{\frac{1}{2}}v), \qquad u,v\in\mathcal{D}[\tau].$$

If τ is a closed nonnegative but non-densely defined form, then we will associate with τ the nonnegative self-adjoint linear relation [155]

$$\mathbf{T}=\left\{\langle u,Tu+h\rangle,\ u\in\mathcal{D}(T),\ h\in H\ominus\overline{\mathcal{D}[\tau]}\right\},$$

where T is a nonnegative self-adjoint operator associated with τ in the subspace $\overline{\mathcal{D}[\tau]}$. Clearly, $\mathbf{T}(0) = H \ominus \overline{\mathcal{D}[\tau]}$. The inverse linear relation $\mathbf{T}^{-1}$ is associated with the form

$$\tau^{-1}[f + h_1, g + h_2] := (\widehat{T}^{-\frac{1}{2}} f, \widehat{T}^{-\frac{1}{2}} g), \ f, g \in \mathcal{R}(T^{\frac{1}{2}}), \qquad h_1, h_2 \in \mathbf{T}(0),$$

where $\widehat{T}^{-\frac{1}{2}} = \left(T^{\frac{1}{2}} | \overline{\mathcal{R}(T^{\frac{1}{2}})}\right)^{-1}$. We will denote by $\mathcal{R}[\mathbf{T}]$ the linear manifold $\mathcal{R}(T^{\frac{1}{2}}) \oplus \mathbf{T}(0)$.

If S is a nonnegative Hermitian operator ($(Su, u) \geq 0$ for all $u \in \mathcal{D}(S)$) then the form $\tau[u, v] := (Su, v)$ is a closable. Following the M. Kreĭn notations we denote by $S[\cdot, \cdot]$ the closure of the form τ and by $\mathcal{D}[S]$ its domain. By definition $S[u] = S[u, u]$ for all $u \in \mathcal{D}[S]$.

Let S be a closed densely defined sectorial operator in a Hilbert space $\mathfrak{H}$ and let S^* be its adjoint. The *Friedrichs extension* [104] S_F of S is defined as a maximal sectorial operator associated with the form $S[\cdot, \cdot]$. If S is symmetric and nonnegative then $\mathcal{D}(S_F) = \mathcal{D}[S] \cap \mathcal{D}(S^*)$, $S_F = S^* | \mathcal{D}(S_F)$. The Friedrichs extension S_F is a unique nonnegative self-adjoint extension having the domain in $\mathcal{D}[S]$. If $\mathfrak{N}_z = \operatorname{Ker}(S^* - zI_{\mathfrak{H}})$ are the defect subspaces, then $\mathcal{D}[S] \cap \mathfrak{N}_z = \{0\}$, $z \in \rho(S_F)$.

Note that if $\mathbf{S}$ is a nonnegative linear relation, then its Friedrichs extension $\mathbf{S}_F$ is a nonnegative self-adjoint linear relation associated with the closed form $\mathbf{S}[\cdot, \cdot]$. In particular, if S is a non-densely defined nonnegative operator, then $\mathbf{S_F}(0) = H \ominus \overline{\mathcal{D}(S)}$ [155].

3. The M. Kreĭn approach

Let S be a closed symmetric nonnegative operator in $\mathfrak{H}$ with the domain $\mathcal{D}(S)$. The fractional-linear transformation

$$A = (I_{\mathfrak{H}} - S)(I_{\mathfrak{H}} + S)^{-1}$$

is a Hermitian contraction ($||A|| \leq 1$) in $\mathfrak{H}$ defined on the subspace $\mathcal{D}(A) = (I_{\mathfrak{H}} + S)\mathcal{D}(S)$. The following M. Kreĭn's result plays essential role in the sequel.

Theorem 3.1 (M. Kreĭn [112]**).** *There exists at least one self-adjoint contractive* (*sc*) *extension $\widetilde{A}$ of A. Moreover the set of all self-adjoint contractive* (*sc*) *extensions of A forms an operator interval $[A_\mu, A_M]$, where the endpoints possess the properties*

$$\begin{aligned} &\inf_{\varphi \in D(A)} ((I_{\mathfrak{H}} + A_\mu)(f + \varphi), (f + \varphi)) = 0, \\ &\inf_{\varphi \in D(A)} ((I_{\mathfrak{H}} - A_M)(f + \varphi), (f + \varphi)) = 0 \end{aligned} \tag{3.1}$$

for all $f \in \mathfrak{H}$.

The operator A admits a unique sc-extension if and only if

$$\sup_{\varphi \in \mathcal{D}(A)} \frac{|(A\varphi, h)|^2}{||\varphi||^2 - ||A\varphi||^2} = \infty$$

for all nonzero vectors h from the orthogonal complement $\mathfrak{H} \ominus \mathcal{D}(A)$.

With the aim to establish Theorem 3.1 M. Kreĭn introduced and studied the following operator transformation. Let H be a bounded nonnegative self-adjoint operator in $\mathfrak{H}$ and let $\mathfrak{L}$ be a subspace in $\mathfrak{H}$. M. Kreĭn proved that the set of all bounded self-adjoint operators C in $\mathfrak{H}$ such that

$$0 \leq C \leq H, \quad \mathcal{R}(C) \subset \mathfrak{L}$$

has a maximal element which M. Kreĭn denoted by $H_{\mathfrak{L}}$. It is established by M. Kreĭn that

$$H_{\mathfrak{L}} = H^{\frac{1}{2}} P_{\Omega} H^{\frac{1}{2}}, \tag{3.2}$$

where P_Ω is the orthogonal projection in $\mathfrak{H}$ onto the subspace

$$\Omega = \{f \in \mathfrak{H} : H^{\frac{1}{2}} f \in \mathfrak{L}\}.$$

In addition

$$(H_{\mathfrak{L}} f, f) = \inf_{\varphi \in \mathfrak{H} \ominus \mathfrak{L}} ((H(f+\varphi), f+\varphi))$$

for all $f \in \mathfrak{H}$. From (3.2) it follows that proved in

$$\mathcal{R}\left((H_{\mathfrak{L}})^{\frac{1}{2}}\right) = \mathcal{R}(H^{\frac{1}{2}}) \cap \mathfrak{L}$$

and hence [112]

$$H_{\mathfrak{L}} = 0 \iff \mathcal{R}(H^{\frac{1}{2}}) \cap \mathfrak{L} = \{0\}.$$

The operator $H_{\mathfrak{L}}$ was called in [8] the *shorted operator*. Properties of the shorted operators were studied in [8], [9], [10], [80], [118], [145], [147], [159].

Let $\widetilde{A}$ be any *sc*-extension of Hermitian contraction A. Then the operators A_μ and A_M can be defined as follows [112]:

$$A_\mu = \widetilde{A} - (I_{\mathfrak{H}} + \widetilde{A})_{\mathfrak{N}}, \; A_M = \widetilde{A} + (I_{\mathfrak{H}} - \widetilde{A})_{\mathfrak{N}}.$$

Thus, for the extremal *sc*-extensions A_μ and A_M of A one has

$$(I_{\mathfrak{H}} + A_\mu)_{\mathfrak{N}} = (I_{\mathfrak{H}} - A_M)_{\mathfrak{N}} = 0.$$

The construction of some *sc* extension of A was given by M. Kreĭn in [112] (see also [2], [126]).

The general description of contractive extensions for nondensely defined contraction A has been established by M.G. Crandall [64]. Denote by $[\mathfrak{H}_1, \mathfrak{H}_2]$ the set of all bounded linear operators acting from Hilbert space $\mathfrak{H}_1$ into Hilbert space $\mathfrak{H}_2$. If A is a contraction defined on the subspace $\mathcal{D}(A)$ of the Hilbert space $\mathfrak{H}$, then obviously $A \in [\mathcal{D}(A), \mathfrak{H}]$. Let $A^* \in [\mathfrak{H}, \mathcal{D}(A)]$ be the adjoint to A. Put $\mathfrak{N} = \mathfrak{H} \ominus \mathcal{D}(A)$ and denote by P_A, $P_{\mathfrak{N}}$ the orthogonal projections onto $\mathcal{D}(A)$ and $\mathfrak{N}$, respectively. Let $D_T = (I - T^*T)^{1/2}$ be the defect operator for a contraction T and let $\mathfrak{D}_T = \overline{\mathcal{R}(D_A)}$.

Theorem 3.2 (Crandall [64]**).** *The formula*

$$\widetilde{A} = AP_A + D_{A^*} K P_{\mathfrak{N}} \tag{3.3}$$

establishes a one-to-one correspondence between all contractive extensions of A and all contractions $K \in [\mathfrak{N}, \mathfrak{D}_{A^*}]$.

Next, we would like to present the explicit formulas for *sc*-extensions A_μ and A_M of a Hermitian contraction A established in [32]. Let

$$G = \overline{D_{A^*}\mathcal{D}(A)}, \quad L = \mathfrak{D}_{A^*} \ominus G$$

Denote by P_L the orthogonal projections onto L and define the contraction $Z \in [G, \mathfrak{N}]$ by the relation

$$Z D_{A^*} x = P_{\mathfrak{N}} A x, \; x \in \mathcal{D}(A).$$

Then A_μ and A_M take the form

$$\begin{aligned} A_\mu &= AP_A + D_{A^*}(Z^* P_{\mathfrak{N}} - P_L D_{A^*}), \\ A_M &= AP_A + D_{A^*}(Z^* P_{\mathfrak{N}} + P_L D_{A^*}). \end{aligned}$$

Here $Z^* \in [\mathfrak{N}, G]$ is the adjoint to Z. Note that in fact M. Kreĭn constructed in [112] the *sc*-extension

$$\widetilde{A} = AP_A + D_{A^*} Z^* P_{\mathfrak{N}} = \frac{A_\mu + A_M}{2}.$$

The operator $A_0 := P_A A$ is a selfadjoint contraction in the Hilbert space $\mathcal{D}(A)$. Since A is a contraction in $\mathfrak{H}$, the following contractive operator $K_0 \in [\mathfrak{D}_{A_0}, \mathfrak{N}]$

$$K_0 D_{A_0} f = P_{\mathfrak{N}} A f, \qquad f \in \mathcal{D}(A).$$

is well defined. This gives the following matrix representation for A

$$A = A_0 + K_0 D_{A_0} = \begin{pmatrix} A_0 \\ K_0 D_{A_0} \end{pmatrix}.$$

Theorem 3.3. *Let A be a closed symmetric contraction A in $\mathfrak{H} = \mathcal{D}(A) \oplus \mathfrak{N}$. Then: the formula*

$$\widetilde{A} = \begin{pmatrix} A_0 & D_{A_0} K_0^* \\ K_0 D_{A_0} & -K_0 A_0 K_0^* + D_{K_0^*} X D_{K_0^*} \end{pmatrix} : \begin{pmatrix} \mathfrak{H}_0 \\ \mathfrak{N} \end{pmatrix} \to \begin{pmatrix} \mathfrak{H}_0 \\ \mathfrak{N} \end{pmatrix} \tag{3.4}$$

gives a one-to-one correspondence between all sc-extensions $\widetilde{A}$ of the symmetric contraction A and all contractions X in the subspace $\mathfrak{D}_{K_0^*}$. The operators $X = -I_{\mathfrak{D}_{K_0^*}}$ and $X = I_{\mathfrak{D}_{K_0^*}}$ correspond to A_μ and A_M, i.e.,

$$A_\mu = \begin{pmatrix} A_0 & D_{A_0} K_0^* \\ K_0 D_{A_0} & -K_0 A_0 K_0^* - D_{K_0^*}^2 \end{pmatrix}, \; A_M = \begin{pmatrix} A_0 & D_{A_0} K_0^* \\ K_0 D_{A_0} & -K_0 A_0 K_0^* + D_{K_0^*}^2 \end{pmatrix}.$$

Another matrix forms for A_μ and A_M is derived in [11]. Properties of *sc*-extensions were studied in [34] and [102]. Theorem 3.3 can be obtained from general parametrization of contractive block-operator matrices

$$\begin{pmatrix} A_{11} & A_{12} \\ A_{21} & A_{22} \end{pmatrix} : \begin{pmatrix} \mathfrak{H} \\ \mathfrak{N} \end{pmatrix} \to \begin{pmatrix} \mathfrak{K} \\ \mathfrak{M} \end{pmatrix}$$

considered by Arsene–Gheondea [44], Shmulyan–Yanovskaya [160], Davis–Kahan–Weinberger [65], Dritschel–Rovnyak [75] (see also [128], [108], [137], [102], [25],

[26]). This parametrization is given by the formula

$$T = \begin{pmatrix} A_{11} & D_{A_{11}^*}M \\ KD_{A_{11}} & -KA_{11}^*M + D_{K^*}XD_M \end{pmatrix}, \tag{3.5}$$

where $A_{11} \in [\mathfrak{H}, \mathfrak{K}]$, $M \in [\mathfrak{N}, \mathfrak{D}_{A_{11}^*}]$, $K \in [\mathfrak{D}_{A_{11}}, \mathfrak{M}]$, and $X \in [\mathfrak{D}_M, \mathfrak{D}_{K^*}]$ are contractions.

The investigation of *sc*-extensions of a nondensely defined Hermitian contraction has been continued in the papers of M. Kreĭn and I. Ovčarenko [117] and [118]. Here there were introduced so-called Q_μ- and Q_M-functions of the form

$$Q_\mu(z) = \left[(A_M - A_\mu)^{1/2}(A_\mu - zI)^{-1}(A_M - A_\mu)^{1/2} + I_{\mathfrak{H}}\right]\Big|\mathfrak{N}, \tag{3.6}$$

and

$$Q_M(z) = \left[(A_M - A_\mu)^{1/2}(A_M - zI)^{-1}(A_M - A_\mu)^{1/2} - I_{\mathfrak{H}}\right]\Big|\mathfrak{N}, \tag{3.7}$$

with $z \in \mathbb{C}\setminus[-1,1]$. These functions take values in the class $[\mathfrak{N}, \mathfrak{N}]$. Moreover, they belong to the class of Herglotz-Nevanlinna functions, i.e., they are holomorphic on $\mathbb{D} \setminus \mathbb{R}$ and satisfy the relations $Q(z)^* = Q(\bar{z})$ and $\operatorname{Im} z \operatorname{Im} Q(z) \geq 0$ for $z \in \mathbb{D} \setminus \mathbb{R}$. As indicated above, these functions admit also analytical continuations to $\operatorname{Ext}[-1,1] := \mathbb{C}\setminus[-1,1]$. In addition they are connected by $Q_\mu(z)Q_M(z) = Q_M(z)Q_\mu(z) = -I_{\mathfrak{N}}$. In these papers M. Kreĭn and I. Ovčarenko established the following formulas

$$\begin{aligned}
(\widetilde{A} - zI_{\mathfrak{H}})^{-1} &= (A_\mu - zI_{\mathfrak{H}})^{-1} \\
&\quad - (A_\mu - zI_{\mathfrak{H}})^{-1}C^{\frac{1}{2}}K\left(I_{\mathfrak{N}} + (Q_\mu(z) - I_{\mathfrak{N}})^{-1})K\right)^{-1}C^{\frac{1}{2}}(A_\mu - zI_{\mathfrak{H}})^{-1}, \\
(\widetilde{A} - zI_{\mathfrak{H}})^{-1} &= (A_M - zI_{\mathfrak{H}})^{-1} \\
&\quad + (A_M - zI_{\mathfrak{H}})^{-1}C^{\frac{1}{2}}\widetilde{K}\left(I_{\mathfrak{N}} - (Q_M(z) + I_{\mathfrak{N}})^{-1})\widetilde{K}\right)^{-1}C^{\frac{1}{2}}(A_M - zI_{\mathfrak{H}})^{-1}
\end{aligned}$$

which give a one-to-one correspondence between resolvents of all *sc*-extensions and all operators K and $\widetilde{K}$ from the operator interval $[0, I_{\mathfrak{N}}]$. Here $C := A_M - A_\mu$. Another resolvent formulas for *sc* extensions have been obtained in [166] and [26]. V. Adamyan [1] considered a nonnegative contraction B and gave a description (in block operator matrix form) of all nonnegative contractive extensions $\widetilde{B}$ of B and their canonical resolvents.

In [118] the inverse problem for Q_μ- and Q_M-functions was teated. It has been suggested some analytical characterizations for these functions among the class of Herglotz-Nevanlinna functions holomorphic on $\operatorname{Ext}[-1,1]$. Here the limiting behaviour of these functions at ∞ and at the points $z = 1$, $z = -1$ plays a fundamental role. If Q is the Q_μ-function of some nondensely defined Hermitian contraction A, then it follows from the operator representation (3.6) and the extremal properties of the *sc*-extensions A_μ and A_M that Q satisfies the following limit conditions:

1) $s-\lim_{z\to\infty} Q(z) = I_{\mathfrak{N}}$;
2) $\lim_{z\uparrow -1} (Q(z)h,h) = +\infty$, for all $h\in\mathfrak{N}\setminus\{0\}$;
3) $s-\lim_{z\downarrow 1} Q(z) = 0$.

One of the principal results in [118], cf. [118, Theorem 2.2], contains the following assertion: if a Nevanlinna function Q, holomorphic on $\operatorname{Ext}[-1,1]$ and with values in the class $[\mathfrak{N},\mathfrak{N}]$ has the properties 1)–3), then there is a Hilbert space $\mathfrak{H}$ extending $\mathfrak{N}$ and a Hermitian contraction A in $\mathfrak{H}$ defined on $\mathfrak{H}\ominus\mathfrak{N}$, such that Q is the Q_μ-function of A, i.e., it admits an operator representation of the form (3.6).

In the paper [27] it has been shown that this statement is true only when $\mathfrak{N}$ is finite-dimensional.

Theorem 3.4 (Arlinskiĭ–Hassi–de Snoo, [27]). *Let $\mathfrak{N}$ be a separable Hilbert space and let $Q(z)$ be an operator-valued Nevanlinna function, which is holomorphic on* $\operatorname{Ext}[-1,1]$ *and takes values in $L(\mathfrak{N})$. Assume that $Q(z)$ satisfies the conditions* $1)-3)$*. Then there exist a Hilbert space $\mathfrak{H}\supset\mathfrak{N}$, a Hermitian operator A in $\mathfrak{H}$ defined on $\mathcal{D}(A)=\mathfrak{H}\ominus\mathfrak{N}$, and sc-extensions $\widetilde{A}_1$ and $\widetilde{A}_2$ of A, such that*

(i) $\widetilde{A}_1\le\widetilde{A}_2$;
(ii) $\operatorname{Ker}(\widetilde{A}_2-\widetilde{A}_1)=\mathcal{D}(A)$;
(iii) $\mathcal{R}((\widetilde{A}_2-\widetilde{A}_1)^{1/2})\cap\mathcal{R}((\widetilde{A}_1-A_\mu)^{1/2})$
$=\mathcal{R}((\widetilde{A}_2-\widetilde{A}_1)^{1/2})\cap\mathcal{R}((A_M-\widetilde{A}_2)^{1/2})=\{0\}$;

and such that $Q(z)$ has the operator representation

$$Q(z)=\left[(\widetilde{A}_2-\widetilde{A}_1)^{1/2}(\widetilde{A}_1-zI)^{-1}(\widetilde{A}_2-\widetilde{A}_1)^{1/2}+I_{\mathfrak{H}}\right]\Big|\mathfrak{N}.$$

If in addition $\dim\mathfrak{N}<\infty$, then $\widetilde{A}_1=A_\mu$ and $\widetilde{A}_2=A_M$, and therefore in this case $Q(z)$ is the Q_μ-function of A.

In [27] for the case $\dim\mathfrak{N}=\infty$ were constructed pairs $\{\widetilde{A}_1,\widetilde{A}_2\}$ of *sc*-extensions of A which satisfy the properties (i)–(iii), but which in general differ from the pair $\{A_\mu,A_M\}$ of the endpoints of the corresponding operator interval. As a consequence it was obtained examples of operator-valued Q-functions of Hermitian contractions which are not Q_μ-functions (or Q_M-functions), but still satisfy all the limit conditions 1)–3) above. Also the precise characterizations for the Q_μ- and Q_M-functions of M. Kreĭn and I. Ovcharenko were given in [27] in the general case $\dim\mathfrak{N}\le\infty$.

Suppose that $\mathcal{D}(S)$ is dense in $\mathfrak{H}$. Then the operators

$$S_\mu=(I_{\mathfrak{H}}-A_\mu)(I_{\mathfrak{H}}+A_\mu)^{-1},\quad S_M=(I_{\mathfrak{H}}-A_M)(I_{\mathfrak{H}}+A_M)^{-1}$$

are well defined and are nonnegative self-adjoint extensions of S. M. Kreĭn called them the *hard* and the *soft* extensions of A, respectively. In the sequel the nonnegative self-adjoint extension S_M given by the relation $S_M=(I_{\mathfrak{H}}-A_M)(I_{\mathfrak{H}}+A_M)^{-1}$ we will call *the Kreĭn-von Neumann extension* of S and will denote by S_K. M. Kreĭn proved that the operator S_μ coincides with the Friedrichs extension S_F of S and established the following theorem.

Theorem 3.5 (M. Kreĭn [112]**).** *The following conditions are equivalent:*

(i) *the nonnegative self-adjoint operator $\widetilde{S}$ is the extension of S,*

(ii) $$(S_F + aI_{\mathfrak{H}})^{-1} \le (\widetilde{S} + aI_{\mathfrak{H}})^{-1} \le (S_K + aI_{\mathfrak{H}})^{-1}$$

for any (then for all) positive number a.

(iii) $S_F \le \widetilde{S} \le S_K$ *in the sense of quadratic forms, i.e.,*

$$\mathcal{D}[S] \subseteq \mathcal{D}[\widetilde{S}] \subseteq \mathcal{D}[S_K],$$
$$\widetilde{S}[u] \ge S_K[u] \quad \text{for all} \quad u \in \mathcal{D}[\widetilde{S}],$$
$$\widetilde{S}[v] = S[v] \quad \text{for all} \quad v \in \mathcal{D}[S].$$

The operator S admits a unique nonnegative self-adjoint extension if and only if

$$\inf_{v \in \mathcal{D}(S)} \frac{(Sv, v)}{|(v, \varphi_{-a})|^2} = 0$$

for all nonzero vectors φ_{-a} from the defect subspace $\mathfrak{N}_{-a}$, where $a > 0$.

If the lower bound of S is positive, then for the Kreĭn-von Neumann extension the following formulas hold [112]:

$$D(S_K) = D(S) \dot{+} \operatorname{Ker} S^*, \quad D[S_K] = D[S] \dot{+} \operatorname{Ker} S^*, \tag{3.8}$$

and moreover

$$D[\widetilde{S}] = D[S] \dot{+} (D[\widetilde{S}] \cap \operatorname{Ker} S^*)$$

for every nonnegative self-adjoint extension $\widetilde{S}$ of S.

Observe that the construction of the Friedrichs and the Kreĭn-von Neumann extensions via a factorization of S over a fixed auxiliary Hilbert space has been given by V. Prokaj, Z. Sebestyén and J. Stochel in [152] and [157].

In [112] M. Kreĭn also considered densely defined symmetric operators having the spectral gap (a, b), i.e., such operators H that

$$\left\| \left(H - \frac{a+b}{2} I \right) h \right\| \ge \frac{b-a}{2} ||h||, \quad h \in \mathcal{D}(H).$$

The latter condition is equivalent to the inequality

$$||H_1 h|| \ge ||h||, \quad h \in \mathcal{D}(H)$$

for

$$H_1 = \frac{2}{b-a} \left(H - \frac{a+b}{2} I \right),$$

and

$$||Hh||^2 - (a+b)(Hh, h) + ab||h|^2 \ge 0, \quad h \in \mathcal{D}(H).$$

Letting $a \to -\infty$ one has

$$(Hh, h) \ge b||h||^2, \quad h \in \mathcal{D}(H).$$

Thus, the case of a semi-bounded symmetric operator is a limit case of an operator with the finite gap.

Let $B = H_1^{-1}$. Then B is a nondensely defined Hermitian contraction with a dense range. Let $\widetilde{B}$ be a *sc*-extension of B. Then

$$\widetilde{H} = \frac{b-a}{2}\widetilde{B}^{-1} + \frac{a+b}{2}I$$

is a self-adjoint extension of H preserving the gap (a,b). So, we arrive to the Kreĭn theorem [112]:

If H has a gap (a,b) then there exists at least one self-adjoint extension of H preserving the gap (a,b).

Further developments in the theory of self-adjoint extensions of a symmetric operators with gaps can be found in [67], [57], [58], [59], [60], [5].

4. The Birman-Vishik approach

In [175] M. Vishik considered the following extension problem. Let L_0 and M_0 be two closed densely defined linear operators in the Hilbert space $\mathfrak{H}$. Suppose

$$(L_0 f, g) = (f, M_0 g) \quad \text{for all } f \in \mathcal{D}(L_0),\ g \in \mathcal{D}(M_0).$$

Such a pair of operators $\{L_0, M_0\}$ are often called a *dual pair*. Additionally suppose

L_0 and M_0 have bounded inverses.

Let the operator $\widetilde{A}$ satisfies the condition

$$L_0 \subset \widetilde{A} \subset M_0^*.$$

The latter condition is equivalent to that $\widetilde{A}$ is an extension of L_0 and $\widetilde{A}^*$ is an extension of M_0. Such operators $\widetilde{A}$ are called the extensions of a dual pair $\{L_0, M_0\}$.

M. Vishik proved that under the above condition there exists an extension $\widetilde{A}_0$ of the dual pair $\{L_0, M_0\}$ such that $\widetilde{A}_0$ and $\widetilde{A}_0^*$ have bounded inverses ($\Longleftrightarrow$ the point 0 belongs to the resolvent set of $\widetilde{A}_0$). Further M. Vishik described domains of extensions of the dual pair $\{L_0, M_0\}$ by means of $\widetilde{A}_0$ and subspaces $V \subset \operatorname{Ker} L_0^*$ and $U \subset \operatorname{Ker} M_0^*$, and operators $\widetilde{C} : V \to U$. The main objective of M. Vishik was an application of the abstract results to the elliptic boundary vales problems. In the particular case $L_0 = M_0 = S$, where S is a positive definite symmetric operator with dense domain $\mathcal{D}(S)$ the M. Vishik abstract formula takes the form

$$\mathcal{D}(\widetilde{S}) = \mathcal{D}(S) \dot{+} (S_F^{-1} + \widetilde{C})\mathcal{D}(\widetilde{C}) \dot{+} (\operatorname{Ker} S^* \ominus \mathcal{D}(\widetilde{C})) \tag{4.1}$$

and gives the parametrization of domains for all self-adjoint extensions of S. Here S_F is the Friedrichs extension of S and $\widetilde{C}$ is a self-adjoint operator with the domain $\mathcal{D}(\widetilde{C})$ in the Hilbert space $N = \overline{\mathcal{D}(\widetilde{C})} \subseteq \operatorname{Ker} S^*$.

M. Birman using Vishik's results gave a description of all nonnegative self-adjoint extensions of a positive definite symmetric operator.

Theorem 4.1 (Birman [55]**).** *Let S be a symmetric operator with a lower bound $m > 0$. A self-adjoint extension $\widetilde{S}$ given by (4.1) has a lower bound $\gamma < m$ if and only if the corresponding operator $\widetilde{C}$ satisfies the inequalities*

$$(\widetilde{C}^{-1}v, v) \geq \gamma||v||^2 + \gamma^2((S_F - \gamma I_{\mathfrak{H}})^{-1}v, v), \quad v \in \mathcal{D}(\widetilde{C}^{-1}),$$

where $\mathcal{D}(\widetilde{C}^{-1}) = \mathcal{R}(\widetilde{C}) \oplus (\operatorname{Ker} S^* \ominus N)$, $\widetilde{C}^{-1}(\operatorname{Ker} S^* \ominus N) = 0$. In particular, $\widetilde{S} \geq 0$ if and only if $\widetilde{C} \geq 0$. If that is the case then the corresponding associated closed form is given by

$$\mathcal{D}[\widetilde{S}] = \mathcal{D}[S] \dot{+} \mathcal{R}(\widetilde{C}^{\frac{1}{2}})) \dot{+} (\operatorname{Ker} S^* \ominus \overline{N}),$$
$$\widetilde{S}[v + h] = S[v] + ||\widetilde{C}^{-1/2}h||^2,$$

where $v \in \mathcal{D}[S]$, $h \in \mathcal{R}(\widetilde{C}^{\frac{1}{2}}) \oplus (\operatorname{Ker} S^* \ominus N)$.

As a consequence it is shown in [55] that if c is the lower bound of $\widetilde{C}^{-1}$ and $c > -m$ then $\widetilde{S}$ is semi-bounded and its lower bound γ admits the estimate

$$\gamma \geq \frac{mc}{m + c}.$$

A further development of the abstract Vishik and Birman approaches and their applications to the boundary value problems for partial elliptic differential operators and systems has been given by G. Grubb in [93], [94], [95], [96], [97], [98] (see also [99]). In particular it is proved in [97] that if S_F^{-1} is a compact operator and if $\widetilde{C}^{-1}$ is a semi-bounded (from below) then $\widetilde{S}$ is semi-bounded. The latter results also has been obtained by M. Gorbachuk and V. Mihaĭlets in [92]. The aspect of Birman–Kreĭn–Vishik (BKV) theory connected with closed forms associates with nonnegative self-adjoint extensions is emphasized by Alonso–Simon in [6]. The connections of BKV theory with singular perturbations of self-adjoint operators can be found in [109], [56], [110]. Note that the positivity of a lower bound is essential in Birman's approach. For the applications of Birman's formula to the operator with zero lower bound it is necessary to consider operator $S + aI$ with an arbitrary positive a. But in this case the domains of the Kreĭn-von Neumann extremal extensions of S and $S+aI$, generally speaking, are different and Birman's approach does not "catch" the Kreĭn-von Neumann extension S_K of S.

5. The Weyl-Titchmarsh functions approach

There are several approaches relating to the Weyl-Titchmarsh function $M(z)$ that is a unitary invariant for the symmetric operator S and its self-adjoint extension $\widetilde{S}$:

1. **Aronjain-Donohue** (scalar case), Kreĭn-Saakyan, Gesztesy-Tsekanovskiĭ (matrix-valued case), Gesztesy-Makarov-Tsekanovskiĭ, and Gesztesy-Kalton-Makarov-Tsekanovskiĭ [40], [73], [88], [84], [83].

If S is densely defined symmetric operator, $\mathfrak{N}_{\pm i}$ are defect subspaces of S and $\widetilde{S}$ is its self-adjoint extension, then

$$M_{\widetilde{S}}(z) = P_{\mathfrak{N}_i}(z\widetilde{S}+I)(\widetilde{S}-zI)^{-1}\Big|_{\mathfrak{N}_i}. \tag{5.1}$$

2. **The Kreĭn-Langer** [114], [115], [116].

Denote by $[\mathfrak{H}_1, \mathfrak{H}_2]$ the set of all bounded linear operators from Hilbert space $\mathfrak{H}_1$ into Hilbert space $\mathfrak{H}_2$ and let $\gamma(z_0) \in [E, \mathfrak{N}_{z_0}]$, $\gamma^{-1}(z_0) \in [\mathfrak{N}_{z_0}, E]$ where E is some Hilbert space, $\mathfrak{N}_{z_0}$ is the defect subspace of a given symmetric operator S, z_0 is a regular point of the self-adjoint extension $\widetilde{S}$ of S. Consider

$$\gamma(z) = (\widetilde{S}-z_0I)(\widetilde{S}-zI)^{-1}\gamma(z_0).$$

An operator-valued function $Q(z)$ that maps Hilbert space E into itself and satisfies the equation

$$Q(z) - Q^*(\zeta) = (z-\bar{\zeta})\gamma^*(\zeta)\gamma(z)$$

is called the Kreĭn-Langer Q-function.

3. **Derkach-Malamud** [66], [67].

Let S be a closed densely defined symmetric operator with equal defect numbers in $\mathfrak{H}$. Let E be some Hilbert space, Γ_1 and Γ_2 be linear mappings of $D(S^*)$ into E. A triplet $\{E, \Gamma_1, \Gamma_2\}$ is called a space of boundary values (s.b.v.) or a boundary triplet for S^* [62], [106] [90], [91] if

a) $(S^*x, y) - (x, S^*y) = (\Gamma_1 x, \Gamma_2 y)_E - (\Gamma_2 x, \Gamma_1 y)_E$ for all $x, y \in \mathcal{D}(S^*)$,

b) a mapping $\Gamma : x \to \{\Gamma_1 x, \Gamma_2 x\}$, $x \in D(S^*)$ is a surjection of $D(S^*)$ onto $E \times E$.

From this definition it follows that $\operatorname{Ker}\Gamma_k \supset \mathcal{D}(S)$, $k = 1, 2$, the operators

$$\widetilde{S}_1 = S^*|\operatorname{Ker}\Gamma_1,\ \widetilde{S}_2 = S^*|\operatorname{Ker}\Gamma_2$$

are self-adjoint extensions of S, and moreover, they are transversal in the sense

$$\mathcal{D}(S^*) = \mathcal{D}(\widetilde{S}_1) + \mathcal{D}(\widetilde{S}_2).$$

A function

$$M(z)(\Gamma_2 x_z) = \Gamma_1 x_z, \quad x_z \in \mathfrak{N}_z,$$

where $\mathfrak{N}_z$ is a defect subspace of S is called the Weyl-Titchmarsh (Weyl) function of the boundary triplet. All three approaches determine a Herglotz-Nevanlinna function $M(z)$ or $Q(z)$ and this function is a unitary invariant of the symmetric operator and its corresponding self-adjoint extension.

Theorem 5.1 (Derkach–Malamud–Tsekanovskiĭ [70], [71], [67], [68]**).** *Let S be a closed densely defined nonnegative symmetric operator and let $\{E, \Gamma_1, \Gamma_2\}$ be the boundary triplet such that $\widetilde{S}_2 = S^*|\operatorname{Ker}\Gamma_2$ is a nonnegative extension of S. Then S has a non-unique nonnegative self-adjoint extension if and only if*

$$\mathcal{D} = \left\{h \in E : \lim_{x\uparrow 0} (M(x)h, h)_E < \infty\right\} \neq \{0\},$$

and the quadratic form

$$\tau[h] = \lim_{x \uparrow 0} (M(x)h, h)_E, \ \mathcal{D}[\tau] = \mathcal{D}$$

is bounded from below. If $M(0)$ *is a self-adjoint linear relation in* E *associated with* τ*, then the Kreĭn-von Neumann extension* S_K *can be defined by the boundary condition*

$$\mathcal{D}(S_K) = \{u \in \mathcal{D}(S^*) : \langle \Gamma_2 u, \Gamma_1 u \rangle \in M(0)\}.$$

The relation $M(0)$ *is also the strong resolvent limit of* $M(x)$ *when* $x \to -0$*. Moreover,* $\widetilde{S}_2$ *and* S_K *are disjoint if and only if* $\overline{\mathcal{D}} = E$ *and transversal iff* $\mathcal{D} = E$*. In addition, if* $\widetilde{S}_2 = S_F$*, then there is a one-to-one correspondence given by*

$$\mathcal{D}(\widetilde{S}_{\widetilde{\mathbf{T}}}) = \left\{u \in \mathcal{D}(S^*) : \left\langle \Gamma_2 u, \Gamma_1 u \right\rangle \in \widetilde{\mathbf{T}}\right\}, \ \widetilde{S}_{\widetilde{\mathbf{T}}} = S^* | \mathcal{D}(\widetilde{S}_{\widetilde{\mathbf{T}}})$$

between nonnegative self-adjoint extensions $\widetilde{S}_{\widetilde{\mathbf{T}}}$ *and self-adjoint relations* $\widetilde{\mathbf{T}}$ *satisfying the condition*

$$\widetilde{\mathbf{T}} \geq M(0).$$

The operator version of this theorem ($M(0)$ is operator) was established in [70], [71], and this form of the theorem by Derkach–Malamud [67], [68].

The domain of the extremal Friedrichs extension can be described as well in external terms by means of the limit values of the Weyl-Titchmarsh functions at $-\infty$. This description in different approaches was obtained by Aronjain-Donohue [40], [73] (scalar case and Friedrichs extension only), Kreĭn-Ovcharenko [119], [120] in terms of the Kreĭn-Langer-Ovcharenko so-called Q_μ and Q_M functions [116], [118], Gesztesy-Tsekanovskiĭ (matrix case) [88], Gesztesy-Makarov-Tsekanovskiĭ [84], Derkach-Malamud [66], [67], Arlinskiĭ-Hassi-de Snoo [26]. The Weyl-Titchmarsh functions approach is an essential part in Kreĭn's famous resolvent formula describing canonical and generalized resolvents and was used by Kreĭn-Langer, Derkach-Malamud, Gesztesy-Makarov-Tsekanovskiĭ, Arlinskiĭ-Tsekanovskiĭ, Belyi-Menon-Tsekanovskiĭ in parametrization of all those resolvents [114], [67], [84], [31], [48]. The theory of boundary value spaces was developed by Straus, Rofe-Beketov, V. Gorbachuk, M. Gorbachuk, Lyantse–Storozh, Kochubeĭ, Bruk, Michailets, Vaĭnerman, Derkach–Malamud, Arlinskiĭ [161], [153], [154], [89], [90], [91], [131], [106], [107], [62], [140], [174], [66], [67], [13].

6. Non-densely defined nonnegative operators

Ando and Nishio [12] found a necessary and sufficient condition for nonnegative operator S with generally speaking non-dense domain to admit a nonnegative self-adjoint extension and it was shown that if any one of such extensions exists, then the Kreĭn-von Neumann (von Neumann in [12]) extension exists as well and the domain of its square root is explicitly determined.

A nonnegative symmetric operator S with, generally speaking, non-dense domain is called positively closable if the relations

$$\lim_{n\to\infty}(Sx_n, x_n) = 0, \quad \lim_{n\to\infty} Sx_n = y$$

implies $y = 0$. If a nonnegative operator S is densely defined, then it is positively closable.

Theorem 6.1 (Ando-Nishio [12]). *Let S be a closed nonnegative symmetric operator in a Hilbert space $\mathfrak{H}$. Define the functional*

$$\theta(h) := \sup_{x\in\mathcal{D}(S)} \frac{|(Sx,h)|^2}{(Sx,x)}.$$

Then the following conditions are equivalent

(i) *S admits a nonnegative self-adjoint extension;*
(ii) *S is positively closable;*
(iii) *the functional $\theta(h)$ is finite on a dense set.*

If one of the conditions (i)–(iii) *holds true then S admits a nonnegative self-adjoint extension S_K such that*

$$\mathcal{D}(S_K^{\frac{1}{2}}) = \{h \in \mathfrak{H} : \theta(h) < \infty\}, \quad ||S_K^{\frac{1}{2}}h||^2 = \theta(h),$$

and moreover if $\widetilde{S}$ is a nonnegative self-adjoint extension of S then $\widetilde{S} \geq S_K$ (in the sense of quadratic forms).

It is proved in [12] that the relations

$$\mathcal{D}(S_K) = \mathcal{D}(S)\dot{+}\mathfrak{N}_0,\ S_K(v+f_0) = Sv,\ v \in \mathcal{D}(S), \quad f_0 \in \mathfrak{N}_0 := \mathfrak{H} \ominus \operatorname{Ker} S^*$$

remain true for a positive definite symmetric operator S with nondense domain (cf. (3.8)). The following result is also established in [12].

Theorem 6.2 (Ando-Nishio [12]). *Let S be a nonnegative, closed, and positively closable operator in the Hilbert space $\mathfrak{H}$. Let $a > 0$ and let the self-adjoint extension $\widetilde{S}_a$ of S be defined as follows:*

$$\begin{aligned} &\mathcal{D}(S_a) = \mathcal{D}(S)\dot{+}\mathfrak{N}_{-a},\\ &\widetilde{S}_a(v + f_{-a}) = Sv - af_{-a},\ v \in \mathcal{D}(S),\ f_{-a} \in \mathfrak{N}_{-a}, \end{aligned}$$

where $\mathfrak{N}_{-a} = \mathfrak{H} \ominus (S + aI_{\mathfrak{H}})\mathcal{D}(S)$. Then

$$(S_K + I_{\mathfrak{H}})^{-1} = \lim_{a\to+0}(\widetilde{S}_a + I_{\mathfrak{H}})^{-1}. \tag{6.1}$$

If in addition S is densely defined then also

$$(S_F + I_{\mathfrak{H}})^{-1} = \lim_{a\to+\infty}(\widetilde{S}_a + I_{\mathfrak{H}})^{-1}. \tag{6.2}$$

Formulas (6.1) and (6.2) were established independently by A.V. Shtrauss in [161]. Also in [161] nonnegative extensions of bounded nondensely defined nonnegative Hermitian operators were studied. Nonnegative extensions of nonnegative subspaces have been considered by Coddington and Coddington-de Snoo [63]. In the recent paper of Sebestyén–Stochel [158] the following development of the Kreĭn Theorem 3.5 is obtained:

Let S be a nonnegative operator with not necessarily dense domain and let R and Q be two nonnegative self-adjoint extensions of S. If T is nonnegative self-adjoint operator such that $R \le T \le Q$ (in the sense of quadratic forms), then T is an extension of S.

Note that Kreĭn's Theorem 3.1 can be obtained from the Ando-Nishio Theorem 6.1 [18]. Indeed, let A be a Hermitian contraction defined on the space $\mathcal{D}(A)$ in $\mathfrak{H}$. Then the operators $I_{\mathfrak{H}} \pm A$ are bounded nondensely defined nonnegative operators. From the relations

$$||(I_{\mathfrak{H}} \pm A)x||^2 + ||D_A x||^2 = 2\left((I_{\mathfrak{H}} \pm A)x, x\right), \qquad x \in \mathcal{D}(A),$$

and Theorem 6.1 it follows that the Kreĭn-von Neumann extensions $(I_{\mathfrak{H}} \pm A)_K$ are bounded self-adjoint operators (with domain $\mathfrak{H}$) and moreover

$$((I_{\mathfrak{H}} \pm A)_K h, h) \le 2||h||^2, \qquad h \in \mathfrak{H}.$$

Hence the operators

$$(I_{\mathfrak{H}} + A)_K - I_{\mathfrak{H}} \text{ and } I_{\mathfrak{H}} - (I_{\mathfrak{H}} - A)_K$$

are *sc*-extensions of A. It can be proved that the equalities

$$A_\mu = (I_{\mathfrak{H}} + A)_K - I_{\mathfrak{H}}, \quad A_M = I_{\mathfrak{H}} - (I_{\mathfrak{H}} - A)_K$$

hold. Let $\widetilde{A}$ be a *sc*-extension of A Then $I_{\mathfrak{H}} + \widetilde{A}$ and $I_{\mathfrak{H}} - \widetilde{A}$ are nonnegative self-adjoint extensions of $I_{\mathfrak{H}} + A$ and $I_{\mathfrak{H}} - A$, respectively. Applying once again the Ando-Nishio Theorem 6.1 we get

$$I_{\mathfrak{H}} - \widetilde{A} \ge (I_{\mathfrak{H}} - A)_K, \quad I_{\mathfrak{H}} + \widetilde{A} \ge (I_{\mathfrak{H}} + A)_K.$$

Hence $A_\mu \le \widetilde{A} \le A_M$.

7. More about the Kreĭn-von Neumann extension

In this section we present results related to the properties of the Kreĭn-von Neumann extension (see [112], [113], [17], [18] [19] [28], [37]).

Apart from the relation $S_K = (I_{\mathfrak{H}} - A_M)(I_{\mathfrak{H}} + A_M)^{-1}$ (see Section 3) the operator S_K can be defined as follows [12], [63]:

$$S_K = \left((S^{-1})_F\right)^{-1},$$

where S^{-1} denotes in this context the inverse linear relation. From this definition for every $f \in \mathcal{D}(S_K)$ it follows that

$$\inf\left\{\|S_K f - S\varphi\|^2 + S_K[f - \varphi] : \varphi \in \mathcal{D}(S)\right\} = 0, \tag{7.1}$$

and

$$\mathcal{R}(S_K^{1/2}) \cap \mathfrak{N}_z = \{0\},\ z \in \rho(S_F),\ \overline{\mathcal{R}(S_K)} = \overline{\mathcal{R}(S)}.$$

It can be easily proved that for every nonnegative self-adjoint extension $\widetilde{S}$ of S hold the relations

$$\begin{aligned}\widetilde{S}[f,u] &= (f, S^*u),\ f \in \mathcal{D}[S], \qquad u \in \mathcal{D}[\widetilde{S}] \cap \mathcal{D}(S^*),\\ \mathcal{D}[\widetilde{S}] &= \mathcal{D}[S] \dot{+} \mathfrak{N}_z \cap \mathcal{D}[\widetilde{S}].\end{aligned}$$

This yields

$$u \in \mathcal{D}(S^*) \cap \mathcal{D}[S_K] \iff S^*u \in \mathcal{R}(S_F^{1/2}),$$

and the equality

$$S_K[u] = S_F^{-1}[S^*u], \qquad u \in \mathcal{D}(S^*) \cap \mathcal{D}[S_K].$$

In particular, $\mathfrak{N}_z \cap \mathcal{D}[S_K] = \mathfrak{N}_z \cap \mathcal{R}(S_F^{1/2})$ and $S_K[\varphi_z] = |z|^2 S_F^{-1}[\varphi_z]$ for $\varphi_z \in \mathfrak{N}_z \cap \mathcal{D}[S_K]$, and moreover

$$S_K[f+\varphi_z, g+\psi_z] = \left(S_F^{1/2} f + z\widehat{S}_F^{-1/2}\varphi_z, S_F^{1/2} g + z\widehat{S}_F^{-1/2}\psi_z\right)$$

for $f, g \in \mathcal{D}[S]$, $\varphi_z, \psi_z \in \mathfrak{N}_z \cap \mathcal{D}[S_K]$ and $z \in \rho(S_F)$.

Using the relation $(S_F - zI)(S_F - \lambda I)^{-1}\mathfrak{N}_z = \mathfrak{N}_\lambda$ and one can obtain for $\varphi_\lambda \in \mathfrak{N}_\lambda \cap \mathcal{R}(S_F^{1/2})$, $\varphi_z \in \mathfrak{N}_z \cap \mathcal{R}(S_F^{1/2})$

$$S_K[\varphi_\lambda, \varphi_z] = \lambda\bar{z}\, S_F^{-1}[\varphi_\lambda, \varphi_z] = \left(\lambda\widehat{S}_F^{-1/2}\varphi_\lambda, z\widehat{S}_F^{-1/2}\varphi_z\right).$$

A nonnegative self-adjoint extension $\widetilde{S}$ of S is called extremal if the relation

$$\inf\left\{\widetilde{S}[u-\varphi] : \varphi \in \mathcal{D}(S)\right\} = 0$$

holds for every $u \in \mathcal{D}[\widetilde{S}]$. A characterization of the Kreĭn-von Neumann extension S_K is obtained in [17] and [18]: *the Kreĭn-von Neumann extension S_K is the unique extremal nonnegative self-adjoint extension of S having maximal domain of its closed associated sesquilinear form.*

The next theorem gives a descriptions of all closed forms associated with nonnegative self-adjoint extensions of S.

Theorem 7.1 (Arlinskiĭ [17]**).** *If $\widetilde{S}$ is a nonnegative self-adjoint extension of a nonnegative symmetric operator S, then the form*

$$(\widetilde{S}u, v) - S_K[u,v], \qquad u, v \in \mathcal{D}(\widetilde{S})$$

is a nonnegative and closable in the Hilbert space $\mathcal{D}[S_K]$. Moreover, the formulas

$$\begin{aligned}\mathcal{D}[\widetilde{S}] &= \mathcal{D}[\tau],\\ \widetilde{S}[u,v] &= S_K[u,v] + \tau[u,v], \qquad u, v \in \mathcal{D}[\widetilde{S}]\end{aligned} \tag{7.2}$$

give a one-to-one correspondence between all closed forms $\widetilde{S}[\cdot,\cdot]$ associated with nonnegative self-adjoint extensions $\widetilde{S}$ of S and all nonnegative forms $\tau[\cdot,\cdot]$ closed in the Hilbert space $\mathcal{D}[S_K]$ and such that $\tau[\varphi] = 0$ for all $\varphi \in \mathcal{D}[S]$.

In addition, the closed form associated with an extremal extensions are closed restrictions of the form $S_K[\cdot,\cdot]$ on the linear manifolds $\mathcal{M}$ such that

$$\mathcal{D}[S] \subseteq \mathcal{M} \subseteq \mathcal{D}[S_K].$$

Investigations of all extremal extensions in more detail and their applications are presented in the paper of Arlinskiĭ–Hassi–Sebestyén–H. de Snoo [28].

Note that the formula (7.2) is a particular case of more general descriptions obtained by Yu. Arlinskiĭ in [17] (see also [18, 19]) for all closed forms associated with m-sectorial extensions with vertex at the origin of a given sectorial operator S.

8. The von Neumann problem

In this section we present the solution of von Neumann's problem in terms of his formulas. Let S be densely defined, closed nonnegative symmetric operator in Hilbert space $\mathfrak{H}$. Consider the linear manifold $\mathcal{D}(S^*)$ as a Hilbert space $\mathfrak{H}_+$ with the inner product

$$(f,g)_+ = (f,g) + (S^*f, S^*g).$$

Then $\mathcal{D}(S_F)$ is a subspace in $\mathfrak{H}_+$. Let $\mathfrak{N}_F$ be the orthogonal complement to $\mathcal{D}(S)$ in $\mathcal{D}(S_F)$ with respect to the inner product $(\cdot,\cdot)_+$. Then the orthogonal decomposition

$$\mathfrak{H}_+ = \mathcal{D}(S) \oplus \mathfrak{N}_F \oplus S_F\mathfrak{N}_F$$

is valid. Let

$$\mathfrak{N}_0 = \mathcal{R}(S_F^{\frac{1}{2}}) \cap \mathfrak{N}_F.$$

Clearly, $S_F^{-\frac{1}{2}}(\mathfrak{N}_0) \subset \mathcal{D}(S_F)$.

Theorem 8.1 (Arlinskiĭ-Tsekanovskiĭ [35], [36], [39]**).** *The condition $\mathfrak{N}_0 = \{0\}$ is necessary and sufficient for the uniqueness of nonnegative self-adjoint extension of S. Suppose $\mathfrak{N}_0 \neq \{0\}$. Then the formulas*

$$\mathcal{D}(\widetilde{S}) = \mathcal{D}(S) \oplus (I + S_F\widetilde{U})\mathcal{D}(\widetilde{U})$$

$$\widetilde{S}(x + h + S_F U h) = S_F(x+h) - \widetilde{U}h$$

$$x \in \mathcal{D}(S), \quad h \in \mathcal{D}(\widetilde{U})$$

give a one-to-one correspondence between all nonnegative self-adjoint extensions $\widetilde{S}$ of S and all $(+)$-self-adjoint operators $\widetilde{U}$ in $\mathfrak{N}_F$ satisfying the condition

$$0 \le \widetilde{U} \le W_0^{-1}$$

where W_0^{-1} determines the operator inverse with respect to the $(+)$-nonnegative self-adjoint relation $\mathbf{W_0}$ in $\mathfrak{N}_F$ associated with the form

$$\omega_0[x,y] = (\widehat{S}_F^{-\frac{1}{2}}x, \widehat{S}_F^{-\frac{1}{2}}y)_+ = (S_F^{\frac{1}{2}}x, S_F^{\frac{1}{2}}y) + (\widehat{S}_F^{-\frac{1}{2}}x, \widehat{S}_F^{-\frac{1}{2}}y), \quad x, y \in \mathfrak{N}_0.$$

Here $\widehat{S}_F^{-\frac{1}{2}}$ is the Moore-Penrose pseudo-inverse. Operator $\widetilde{S}$ coincides with the Kreĭn-von Neumann nonnegative self-adjoint extension S_K if and only if $\widetilde{U} = W_0^{-1}$.

As a result of this theorem we get

$$\mathcal{D}[\widetilde{S}] = \mathcal{D}[S]\dot{+}S_F\mathcal{R}(\widetilde{U}^{\frac{1}{2}})$$

where $\widetilde{S}$ is any nonnegative self-adjoint extension, $\widetilde{U}$ is the corresponding parameter (from theorem). For the Kreĭn-von Neumann nonnegative extension we get

$$\mathcal{D}[S_K] = \mathcal{D}[S]\dot{+}S_F\mathfrak{N}_0.$$

Let P_i^+ be an orthogonal projection of +-orthogonal decomposition of $\mathfrak{H} = \mathcal{D}(S) \oplus \mathfrak{N}_i \oplus \mathfrak{N}_{-i}$ onto $\mathfrak{N}_i$ and

$$\mathcal{D}(S_F) = \mathcal{D}(S)\dot{+}(I + V_F)\mathfrak{N}_i.$$

An operator

$$\widetilde{V}P_i^+h = -V_F P_i^+(\widetilde{U} + iI)(\widetilde{U} - iI)^{-1}h, \quad h \in \mathfrak{N}_F,$$

where $\widetilde{U}$ is a (+)-self-adjoint operator in $\mathfrak{N}_F$ satisfying $0 \leq \widetilde{U} \leq W_0^{-1}$, defines a nonnegative self-adjoint extension $\widetilde{S}$ by the von Neumann formula

$$\mathcal{D}(\widetilde{S}) = \mathcal{D}(S)\dot{+}(I + \widetilde{V})\mathfrak{N}_i.$$

9. Kreĭn's resolvent formula

In this section we consider only canonical resolvents formula (resolvents of two self-adjoint extensions without exit from Hilbert space). The generalized resolvents formula(with the exit from Hilbert space) is the subject of so many publications that deserves special attention.

We call two self-adjoint extensions $\widetilde{S}_1$ and $\widetilde{S}_2$ of S relatively prime if

$$\mathcal{D}(\widetilde{S}_1) \cap \mathcal{D}(\widetilde{S}_2) = \mathcal{D}(S).$$

Theorem 9.1 (Kreĭn–Saakyan [121], [156] **).** *Let $\widetilde{S}_1$ and $\widetilde{S}_2$ be relatively prime self-adjoint extensions of densely defined symmetric operator S. Then for $z \in \rho(\widetilde{S}_1) \cap \rho(\widetilde{S}_2)$*

$$\begin{aligned}(\widetilde{S}_2 - zI)^{-1} &= (\widetilde{S}_1 - zI) + (\widetilde{S}_1 - iI)(\widetilde{S}_1 - zI)^{-1}P_{1,2}(z)(\widetilde{S}_1 + iI)(\widetilde{S}_1 - zI)^{-1} \\ &= (\widetilde{S}_1 - zI)^{-1} + (\widetilde{S}_1 - iI)(\widetilde{S}_1 - zI)^{-1}P_i(\mathcal{T} - M_{\widetilde{S}_1}(z))^{-1}P_i(\widetilde{S}_1 + iI)(\widetilde{S}_1 - zI)^{-1}\end{aligned}$$

where P_i is the orthogonal projector onto $Ker(S^* - iI)$, $\mathcal{T} = \mathcal{T}^*$ in this defect subspace of S, and

$$P_{1,2}(z) = (\widetilde{S}_1 - zI)(\widetilde{S}_1 - iI)^{-1}((\widetilde{S}_2 - zI)^{-1} - (\widetilde{S}_1 - zI)^{-1})(\widetilde{S}_1 - zI)(\widetilde{S}_1 + iI)^{-1}.$$

Apparently, Kreĭn's formula was first derived independently by Kreĭn and Naimark in the special case of defect indices $(1, 1)$. The case of finite defect indices (n, n) is due to Kreĭn. Saakyan extended Kreĭn's formula to the general case of infinite defect indices but as it was before without any connections of Kreĭn's formula with von Neumann's parametrization. The part of Kreĭn's resolvent formula with connection of von Neumann's parametrization was obtained by Gesztesy-Makarov-Tsekanovskiĭ [84] as well as explicitly derived linear fractional transformation for the Weyl-Titchmarsh functions corresponding to the given two self-adjoint extensions.

Theorem 9.2 (Gesztesy-Makarov-Tsekanovskiĭ [84]**).** *Let $\widetilde{S}_1$, $\widetilde{S}_2$ be relatively prime self-adjoint extensions of the densely defined, closed symmetric operator S. Then for $z \in \rho(\widetilde{S}_1) \cap \rho(\widetilde{S}_2)$*

1. $(\widetilde{S}_2 - zI)^{-1} = (\widetilde{S}_1 - zI)^{-1}$
   $+(\widetilde{S}_1 - iI)(\widetilde{S}_1 - zI)^{-1}P_i(\tan\alpha_{1,2} - M_{\widetilde{S}_1}(z))^{-1}P_i(\widetilde{S}_1 + iI)(\widetilde{S}_1 - zI)^{-1}$,
   where P_i is the orthogonal projection onto defect subspace $\mathrm{Ker}\,(S^* - iI)$ *and* $\exp(-2i\alpha_{1,2}) = -\widetilde{U}_2^{-1}\widetilde{U}_1$, $\widetilde{U}_1, \widetilde{U}_2$ *are isometries in von Neumann's formulas, $M_{\widetilde{S}_1}(z)$ is the Weyl-Titchmarsh function of the form* (5.1).
2. *The Weyl-Titchmarsh functions of the form* (5.1) *that correspond to $\widetilde{S}_1$ and $\widetilde{S}_2$ satisfy the following relation*
   $$M_{\widetilde{S}_2}(z) = \exp(-i\alpha_{1,2})(\cos\alpha_{1,2} + \sin\alpha_{1,2}M_{\widetilde{S}_1}(z)) \times (\sin\alpha_{1,2} - \cos\alpha_{1,2}M_{\widetilde{S}_1}(z))^{-1}\exp(i\alpha_{1,2}).$$

For non-densely defined symmetric operator S the analog of this theorem was obtained by Belyi–Menon–Tsekanovskii [48]. Kreĭn's resolvent formula has been used in a large variety of problems in mathematical physics. A complete bibliography on Kreĭn's formula is beyond the scope of this paper and obviously resolvent formula deserves a special separate survey. We only would like to mention some publications on Kreĭn's resolvent formula and its applications by Kreĭn–Langer, Langer–Textorius, Nenciu, Straus, Derkach–Malamud, Hassi–Langer–de Snoo, Kurasov–Pavlov, Kurasov, Albeverio–Kurasov, Pavlov, Posilicano, Arlinskiĭ–Tsekanovskiĭ, Gesztesy–Mitrea, Exner [115], [128], [143], [161], [66], [67], [124], [123], [4], [146], [151], [37], [85], [86], [78], [79]. Many applications of Kreĭn's resolvent formula in mathematical physics were presented in the survey by K.A. Makarov [132].

For nonnegative, densely defined symmetric operators S Kreĭn's formula for canonical resolvents was obtained by Arlinskiĭ-Tsekanovskiĭ [33], [39] in terms of parametrization of nonnegative self-adjoint operators in von Neumann's problem and has the form

$$(\widetilde{S} - zI)^{-1} = (S_F - zI)^{-1} + ((S_F - zI)^{-1}(I + zS_F) + S_F)\widetilde{U}(I - M(z)\widetilde{U})^{-1}P^+_{\mathfrak{N}_F}(S_F - zI)^{-1}. \quad (9.1)$$

This formula establishes a one-to-one correspondence between resolvents of all nonnegative self-adjoint extensions $\widetilde{S}$ of S and all $(+)$-nonnegative self-adjoint operators $\widetilde{U}$ in $\mathfrak{N}_F$ from theorem (8.1). The function $M(z)$ is the Weyl–Titchmarsh function of the form

$$M(z) = P^+_{\mathfrak{N}_F}(S_F - zI)^{-1}(I + zS_F)\Big|_{\mathfrak{N}_F}.$$

10. Point-interaction model

Consider the following operator

$$S\varphi = -\Delta\varphi, \quad y_1, y_2, \ldots, y_m \in \mathbb{R}^3,$$
$$\mathcal{D}(S) = \{\varphi(x) \in H_2^2(\mathbb{R}^3) : \varphi(y_j) = 0, \quad j = 1, \ldots, m\},$$

where Δ is the Laplacian, $H_2^2(\mathbb{R}^3)$ is the Sobolev space, $x \in \mathbb{R}^3$. It is well known that S ia a nonnegative operator in $L^2(\mathbb{R}^3, dx)$ with defect numbers (m, m) and zero lower bound. This is the point-interaction model with m points of interaction. In the book by Albeverio–Gesztesy–Hoegh-Krohn–Holden [3] it is shown that the Friedrichs extension S_F of S has the form

$$\mathcal{D}(S_F) = H_2^2(\mathbb{R}^3), \quad S_F = -\Delta.$$

The approach developed by Arlinskiĭ-Tsekanovskiĭ in the solution of von Neumann's problem [36], [39] and discussed above allows to obtain description of all nonnegative self-adjoint extensions and parametrization of their domains including the domain of the Kreĭn-von Neumann extension S_K. For the simplicity we consider in this paper one point of interaction and refer to the paper [35], [39], where the general case of m points of interaction was considered. Consider symmetric , nonnegative operator in one point of interaction model.

$$\mathcal{D}(S) = \{\varphi(x) \in H_2^2(\mathbb{R}^3) : \varphi(y) = 0, \ y \in \mathbb{R}^3\}, \quad S\varphi = -\Delta\varphi.$$

This operator as we mentioned is nonnegative with the zero lower bound and has defect indices $(1, 1)$. The Kreĭn-von Neumann extension and its domain has the following description [36], [39]:

$$\mathcal{D}(S_K) = \left\{ \varphi(x) = \varphi_S(x) + \lambda \frac{\exp(-\frac{|x-y|}{\sqrt{2}})}{|x-y|} (\sin \frac{|x-y|}{\sqrt{2}} + \cos \frac{|x-y|}{\sqrt{2}}) \right\},$$
$$\varphi_S(x) \in \mathcal{D}(S), \quad \varphi_S(y) = 0, \quad \lambda \in \mathbb{C},$$
$$S_K\varphi(x) = -\Delta\varphi_S + \lambda \frac{\exp(-\frac{|x-y|}{\sqrt{2}})}{|x-y|} \left(\sin \frac{|x-y|}{\sqrt{2}} - \cos \frac{|x-y|}{\sqrt{2}} \right).$$

The description of all nonnegative self-adjoint extensions of the symmetric operator S in the point interaction model with one point of interaction can be presented by the formulas:

$$\mathcal{D}(\widetilde{S}_u) = \left\{ \varphi(x) = \varphi_S(x) + \lambda \frac{\exp(-\frac{|x-y|}{\sqrt{2}})}{|x-y|} \left(\sin \frac{|x-y|}{\sqrt{2}} + u \cos \frac{|x-y|}{\sqrt{2}} \right) \right\},$$
$$\varphi_S(x) \in \mathcal{D}(S), \quad \varphi_S(y) = 0, \quad \lambda \in \mathbb{C}, \quad 0 \le u \le 1,$$
$$\widetilde{S}_u\varphi(x) = -\Delta\varphi_S + \lambda \frac{\exp(-\frac{|x-y|}{\sqrt{2}})}{|x-y|} \left(\sin \frac{|x-y|}{\sqrt{2}} - u \cos \frac{|x-y|}{\sqrt{2}} \right).$$

These formulas establish a one-to-one correspondence between the set of all nonnegative self-adjoint extensions $\widetilde{S}_u$ of S and the set of all real numbers satisfying the inequality $0 \le u \le 1$.

It should be pointed out that in the case $\mathbb{R}^2$ the nonnegative symmetric operator

$$\mathcal{D}(S) = \{\varphi(x) \in H_2^2(\mathbb{R}^2) : \varphi(y) = 0,\ y \in \mathbb{R}^2\}, \quad S = -\Delta|\mathcal{D}(S)$$

has a unique nonnegative self-adjoint extension. This fact was established by Gesztesy–Kalton–Makarov–Tsekanovskii [83] and independently by Adamyan [1]. It follows from our Theorem 8.1 as well.

11. $\mathfrak{H}_2$-construction and nonnegative extensions

Let A be an unbounded selfadjoint operator acting on a separable Hilbert space $\mathfrak{H}$ and let $\mathfrak{H}_{+2} \subset \mathfrak{H}_{+1} \subset \mathfrak{H} \subset \mathfrak{H}_{-1} \subset \mathfrak{H}_{-2}$ be the chain of rigged Hilbert spaces constructed by means of A [54], i.e., $\mathfrak{H}_{+2} = \mathcal{D}(A)$, $\mathfrak{H}_{+1} = \mathcal{D}(|A|^{1/2})$ equipped by norms $||f||_{\mathfrak{H}_{+k}} = \left(\||A|^{k/2}\|^2 + \|f\|^2\right)^{1/2}$, $k = 1,2$. The operator A has the continuation $\mathbf{A}$ which continuously maps $\mathfrak{H}_p$, $p = 0,1$ into $\mathfrak{H}_{p-2}$ and $|\mathbf{A}|^{1/2}$ is the continuation of $|A|^{1/2}$ and maps $\mathfrak{H}_p$, $p = -1,0$ into $\mathfrak{H}_{p-1}$. Moreover, the resolvent $(\mathbf{A}-\lambda I)^{-1}$, $\lambda \in \rho(A)$ maps $\mathfrak{H}_{-p}$, $p = 0,1,2$ onto $\mathfrak{H}_{-p+2}$ and the resolvent identity holds:

$$(\mathbf{A}-\lambda I)^{-1} - (\mathbf{A}-zI)^{-1} = (\lambda - z)\,(A-\lambda I)^{-1}\,(\mathbf{A}-zI)^{-1}.$$

Note that $(Af,g) = (f,\mathbf{A}g)$, $\left(|A|^{1/2}h,g\right) = \left(h,|\mathbf{A}|^{1/2}g\right)$. Suppose, for the simplicity, that $\operatorname{Ker} A = \{0\}$.

Let Φ be a subspace in $\mathfrak{H}_{-2}$ such that

$$\Phi \cap \mathfrak{H} = \{0\}.$$

Then the operator defined as follows

$$\mathcal{D}(\mathring{A}) = \left\{f \in \mathfrak{H}_{+2} : (f,\varphi) = 0, \quad \text{for all} \quad \varphi \in \Phi\right\}, \quad \mathring{A} = A|\mathcal{D}(\mathring{A}) \tag{11.1}$$

is closed, densely defined and nonnegative with the defect numbers equal to $\dim \Phi$. For the defect subspace $\mathfrak{N}_z$ of $\mathring{A}$ holds the identity $\mathfrak{N}_z = (\mathbf{A}-zI)^{-1}\Phi$.

Operators $\mathbf{A} \pm iI$ are unitary from $\mathfrak{H}$ onto $\mathfrak{H}_{-2}$, the operator

$$\mathbf{J} := \left(\mathbf{A}^2 + I\right)^{-1} = (A-iI)^{-1}(\mathbf{A}+iI)^{-1} = \tfrac{1}{2i}\left((\mathbf{A}-iI)^{-1} - (\mathbf{A}+iI)^{-1}\right).$$

unitarily maps $\mathfrak{H}_{-2}$ onto $\mathfrak{H}_{+2}$ and for all $f \in \mathfrak{H}_{+2}$, $\varphi \in \mathfrak{H}_{-2}$ hold the relations

$$(f,\varphi) = (f,\mathbf{J}\varphi)_{\mathfrak{H}_{+2}} = \left(\mathbf{J}^{-1}f,\varphi\right)_{\mathfrak{H}_{-2}}.$$

It is evident the relation $A\mathbf{J} = \frac{1}{2}\left((\mathbf{A}-iI)^{-1} + (\mathbf{A}+iI)^{-1}\right)$. The Hilbert space $\mathfrak{H}_{+2}$ has the $(+2)$-orthogonal decomposition $\mathfrak{H}_{+2} = \mathcal{D}(\mathring{A}) \oplus \mathbf{J}\Phi$. Let $\mathring{A}^*$ be the adjoint operator and let $\mathfrak{H}_{+} = \mathcal{D}(\mathring{A}^*)$ be the corresponding Hilbert space. According to the von Neumann formula we have the $(+)$-orthogonal decomposition of H_+

$$\mathfrak{H}_+ = \mathfrak{H}_{+2} \oplus A\mathbf{J}\Phi = \mathcal{D}(\mathring{A}) \oplus \mathbf{J}\Phi \oplus A\mathbf{J}\Phi$$

and $\mathring{A}^* A\mathbf{J}\varphi = -\mathbf{J}\varphi$, $\varphi \in \Phi$. Let $\overset{0}{\mathfrak{H}}_{+1}$ be the closure in $\mathfrak{H}_{+1}$ of $\mathcal{D}(\mathring{A})$, then

$$\overset{0}{\mathfrak{H}}_{+1} = \mathcal{D}[\mathring{A}],$$

and $\overset{0}{\mathfrak{H}}_{+1} \oplus \mathfrak{N}_{-1} = \mathfrak{H}_{+1}$ (the orthogonal sum in $\mathfrak{H}_{+1}$.)

Theorem 11.1 (Arlinskiĭ–Tsekanovskiĭ [35]). *Let A be a nonnegative self-adjoint operator. Suppose $\Phi \cap \mathfrak{H}_{-1} = \{0\}$. Then the operator A is the Friedrichs extension of $\mathring{A}$. The operator A is a unique nonnegative self-adjoint extension of $\mathring{A}$ if and only if $\Phi \cap \mathbf{A}^{1/2}\mathfrak{H}_{-1} = \{0\}$. The Kreĭn-von Neumann extension $\mathring{A}_K$ is transversal to A if and only if $\Phi \subset \mathbf{A}^{1/2}\mathfrak{H}_{-1}$. In this case if the operator $C_0 : \Phi \to \mathbf{J}\Phi$ is defined by the equation $(C_0\varphi, \varphi) = \left\|\mathbf{A}^{-1/2}\varphi\right\|^2_{\mathfrak{H}_{-2}}$, $\varphi \in \Phi$, then*

$$\mathring{A}_K = \mathring{A}^* | \left(\mathcal{D}(\mathring{A}) + (A\mathbf{J} + C_0)\Phi\right).$$

Theorem 11.2 (Arlinskiĭ–Tsekanovskiĭ [35]). *Let A be a nonnegative self-adjoint operator. Suppose*

$$\Phi \quad \text{is a subspace in} \quad \mathfrak{H}_{-2},\ \Phi \subset \mathfrak{H}_{-1} \quad \text{and} \quad \Phi \cap \mathfrak{H} = \{0\}.$$

1. *Define the operator $B_0 : \Phi \to \mathbf{J}\Phi$ by the relation*

$$(B_0\varphi, \varphi) = -(A\mathbf{J}\varphi, \varphi) \quad \text{for all} \quad \varphi \in \Phi.$$

Then $(A\mathbf{J} + B_0)\Phi \subset \overset{0}{\mathfrak{H}}_{+1}$ and the Friedrichs extension $\mathring{A}_F$ of $\mathring{A}$ is given by the equality

$$\mathring{A}_F = \mathring{A}^* | \left(\mathcal{D}(\mathring{A}) \dot{+} (A\mathbf{J} + B_0)\Phi\right).$$

2. *Let*

$$\mathfrak{H}_0 = \left\{f \in \mathfrak{H} : \mathbf{A}^{1/2} f \in \Phi\right\}$$

and let P_0 be the orthogonal projection in $\mathfrak{H}$ onto $\mathfrak{H}_0$. Then the Kreĭn-von Neumann extension $\mathring{A}_K$ of $\mathring{A}$ has the following description:

$$\mathcal{D}(\mathring{A}_K) = \left\{u \in \mathfrak{H}_{+1} : (I - P_0)A^{1/2}u \in \mathfrak{H}_{+1}\right\},$$
$$\mathring{A}_K u = A^{1/2}(I - P_0)A^{1/2}u, \quad u \in \mathcal{D}(\mathring{A}_K).$$

3. *The formulas*

$$\mathcal{D}(\widetilde{A}) = \left\{u \in \mathcal{D}(\Omega) : \left(\Omega(u) + \mathbf{A}^{1/2}(I - P_0)A^{1/2}u\right) \cap \mathfrak{H} \neq \{0\}\right\},$$
$$\widetilde{A}u = \Omega(u) + \mathbf{A}^{1/2}(I - P_0)A^{1/2}u, \quad u \in \mathcal{D}(\widetilde{A})$$

establish a one-to-one correspondence between all nonnegative self-adjoint extensions of $\mathring{A}$ and all nonnegative self-adjoint linear relations Ω in $\mathfrak{H}_{+1} \times \mathfrak{H}_{-1}$ such that $\operatorname{Ker}\Omega \supseteq \overset{0}{\mathfrak{H}}_{+1}$. Let $\mathbf{L} = \Omega - \mathbf{A}^{1/2}P_0 A^{1/2}$. Then the resolvent of $\widetilde{A}$ for $\lambda \in \rho(\widetilde{A}) \cap \rho(A)$ takes the form

$$(\widetilde{A} - \lambda I)^{-1} = (A - \lambda I)^{-1} - (\mathbf{A} - \lambda I)^{-1}\left(\mathbf{L}^{-1} + (\mathbf{A} - \lambda I)^{-1}\right)^{-1}(A - \lambda I)^{-1}.$$

12. Sturm-Liouville operators on the semi-axis

Consider Hilbert space $\mathfrak{H} = L^2[a, +\infty)$ and differential operation

$$l(y) = -y'' + q(x)y,$$

where $q(x)$ is real and locally summable function. Let

$$\begin{aligned} Sy = l(y) &= -y'' + q(x)y, \\ y'(a) = y(a) &= 0 \end{aligned} \tag{12.1}$$

be a nonnegative symmetric operator with defect indices $(1,1)$. It is well known that the Friedrichs extension of S is given by

$$S_F y = -y'' + q(x)y, \; y(a) = 0.$$

Let $\varphi_k(x,\lambda)$, $k = 1,2$ be the solutions of the Cauchy problems

$$\begin{aligned} l(\varphi_1) &= \lambda\varphi_1, \quad \varphi_1(a,\lambda) = 0, \qquad \varphi_1'(a,\lambda) = 1, \\ l(\varphi_2) &= \lambda\varphi_2, \quad \varphi_2(a,\lambda) = -1, \quad \varphi_2'(a,\lambda) = 0. \end{aligned}$$

It is known that there exists the Weyl-Titchmarsh (Weyl function) $-m_\infty(\lambda)$ such that

$$\varphi(x,\lambda) = \varphi_2(x,\lambda) + m_\infty(\lambda)\varphi_1(x,\lambda) \in L^2[a, +\infty).$$

The Kreĭn-von Neumann extension in this case has the form

$$\begin{aligned} S_K y &= -y'' + q(x)y, \\ y'(a) &+ m_\infty(-0)y(a) = 0. \end{aligned}$$

The description of all nonnegative self-adjoint extensions of the nonnegative symmetric Sturm-Liouville operator S gives the relation

$$\begin{aligned} \widetilde{S}_u y &= -y'' + q(x)y, \\ y'(a) &= uy(a), \end{aligned}$$

where $u \geq -m_\infty(-0)$. This description as well as the description of all nonnegative self-adjoint extensions for $2n$-order nonnegative minimal differential operator on the semi-axis was obtained by Tsekanovskiĭ [167], [170], [172] by the method of characteristic functions of the Livsic type. The space of boundary values approach for this fact was proposed by Arlinskiĭ [13], [20], Derkach–Malamud [66]. Recently another characterization of positive self-adjoint extensions and its applications to ordinary differential operators was considered by Wei–Jiang in [176].

For the Sturm-Liouville nonnegative symmetric operator of the Bessel type with defect indices $(1,1)$

$$\begin{aligned} Sy &= -y'' + \frac{\nu^2 - \frac{1}{4}}{x^2}y, \quad \nu \geq \frac{1}{2}, \\ y'(1) &= y(1) = 0 \end{aligned}$$

in the Hilbert space $\mathfrak{H} = L^2(1,\infty)$ one has

$$m_{\infty,\nu}(-0) = \nu - \frac{1}{2}.$$

The Friedrichs extension has the form

$$S_F y = -y'' + \frac{\nu^2 - \frac{1}{4}}{x^2} y,$$
$$y(1) = 0.$$

The Kreĭn-von Neumann extension is the following boundary value problem

$$S_K y = -y'' + \frac{\nu^2 - \frac{1}{4}}{x^2} y, \qquad y'(1) + (\nu - \frac{1}{2}) y(1) = 0.$$

When $\nu = \frac{1}{2}$, the Kreĭn-von Neumann extension is simply the Neumann boundary value problem

$$S_K y = -y'',$$
$$y'(1) = 0.$$

The following sharp inequality was established by Gesztesy–Kalton–Makarov–Tsekanovskiĭ [83] on the basis of the properties of the Weyl-Titchmarsh function associated with the Kreĭn-von Neumann extension S_K when $\nu = \frac{1}{2}$

$$\sqrt{2}|y(1)|^2 \le \int_1^\infty (|y(x)|^2 + |y''(x)|^2)dx, \; y(x) \in \mathcal{D}(S_K) = 0.$$

Some new inequalities including mentioned above were established by Arlinskiĭ–Tsekanovskiĭ [38] on the basis of a system theory in triplets of Hilbert spaces and connections with Friedrichs and Kreĭn–von Neumann extensions.

13. Accretive and sectorial operators and the Phillips–Kato extension problems

Let T be closed, densely defined linear operator in Hilbert space $\mathfrak{H}$. Operator T is called an accretive operator if

$$\mathrm{Re}\,(Tx, x) \ge 0 \quad \text{for all } x \in \mathcal{D}(T).$$

Accretive operator T is called maximal accretive (m-accretive) if it does not have accretive extensions in $\mathfrak{H}$.

Let $\theta \in [0, \pi/2)$. An m-accretive operator T is called sectorial with vertex at the origin and a semi-angle θ [104] (θ-sectorial for shorts) if

$$|\mathrm{Im}\,(Tx, x)| \le \tan\theta \, \mathrm{Re}\,(Tx, x), \quad \forall x \in \mathcal{D}(T)$$

(see Section 2). A linear bounded operator B in Hilbert space $\mathfrak{H}$ belongs to the class $C(\theta)$ (θ-co-sectorial contraction) if

$$2\cot\theta |\mathrm{Im}\,(Bx, x)| \le \|x\|^2 - \|Bx\|^2, \quad \forall x \in \mathfrak{H}.$$

The last inequality is equivalent to

$$\|B \pm i \cot\theta I_{\mathfrak{H}}\| \leq \frac{1}{\sin\theta} \iff \|\sin\theta\, B \pm i\cos\theta\, I_{\mathfrak{H}}\| \leq 1.$$

$C(0)$ is the class of all self-adjoint contractions. If T is θ-sectorial then

$$B = (I_{\mathfrak{H}} - T)(I_{\mathfrak{H}} + T)^{-1} \in C(\theta).$$

The problem of a description of all maximal sectorial extension of a given sectorial operator we call the Kato extension problem (see [104, Chapter 6]). One more extension problem is the Phillips problem of a description of all maximal accretive extensions of a given accretive operator [148], [149], [150]. In particular, the following results were established.

Theorem 13.1 (Phillips [149], [150]**).**

1. *Any densely defined accretive operator has an m-accretive extension.*
2. *The following conditions are equivalent for densely defined operator T:*
   (a) *T is m-accretive operator;*
   (b) *T is accretive operator and -1 is a regular point of T;*
   (c) *T and T^* are accretive operators.*

In order to obtain a description of all m-accretive extensions in terms of the abstract "boundary conditions" R. Phillips proposed an approach based on the geometry of Kreĭn spaces with indefinite inner product. His approach has been applied by Evans and Knoweles [77] for some symmetric positive definite ordinary differential operators on the finite interval.

The fractional-linear transformation $A = (I_{\mathfrak{H}} - T)(I_{\mathfrak{H}} + T)^{-1}$ of an accretive operator T is a contraction defined on the subspace $\mathcal{D}(A) = (I_{\mathfrak{H}} + T)\mathcal{D}(T)$. The formula $\widetilde{T} = (I_{\mathfrak{H}} - \widetilde{A})(I_{\mathfrak{H}} + \widetilde{A})^{-1}$ establishes a one-to-one correspondence between the set of all contractive extensions of $\widetilde{A}$ on the space $\mathfrak{H}$ and the set of all m-accretive $\widetilde{T}$ of T. A description of all contractive extensions is given by Theorem 3.2. If T is sectorial with vertex at the origin an semi-angle θ then A satisfies $\|\sin\theta\, A \pm i\cos\theta\, I\| \leq 1$ and is in general nondensely defined (such operator is called $C(\theta)$-suboperator).

Let S be closed, densely defined nonnegative symmetric operator in Hilbert space $\mathfrak{H}$. The Phillips-Kato extension problems [104] in the restricted sense consists of existence and description of m-accretive and θ-sectorial extensions T of S such that

$$S \subset T \subset S^*. \tag{13.1}$$

The operators T satisfying (13.1) are often called *proper extensions* of symmetric operator S.

The existence of the solution of the Phillips-Kato extension problems in restricted sense is presented in the theorem below.

Theorem 13.2 (Tsekanovskiĭ [168], [169]**).** *The nonnegative, closed densely defined symmetric operator S admits proper m-accretive and θ-sectorial non-self-adjoint* *extensions if and only if the Friedrichs extension* S_F *does not coincide with the Kreĭn-von Neumann extension* S_K *of* S*, i.e.,* $S_F \neq S_K$.

The next theorem gives characterization of proper accretive extensions among all accretive extensions of nonnegative S.

Theorem 13.3 (Arlinskiĭ [16]**).** *Let* S *be a nonnegative symmetric operator and let* $\widetilde{S}$ *be its accretive extension. Then the following condition are equivalent:*

(i) $\widetilde{S}$ *is a proper extension of* S*;*

(ii) $\mathcal{D}(\widetilde{S}) \subset \mathcal{D}[S_K]$ *and*

$$\mathrm{Re}\,(\widetilde{S}u, u) \geq S_K[u]$$

for all $u \in \mathcal{D}(\widetilde{S})$*;*

(iii) *the inequalities*

$$|(S\varphi, u)|^2 \leq (S\varphi, \varphi)\,\mathrm{Re}\,(\widetilde{S}u, u)$$

are valid for all $\varphi \in \mathcal{D}(S)$ *and for all* $u \in \mathcal{D}(\widetilde{S})$.

Let A be Hermitian contraction in $\mathfrak{H}$ defined on the subspace $\mathcal{D}(A)$. The operator B in $\mathfrak{H}$ defined on $\mathfrak{H}$ is said to be *quasi-self-adjoint contractive extension* (shortly *qsc*) extension of A if

$$B \supset A, \; B^* \supset A, \quad ||B|| \leq 1.$$

The following theorem gives a solution of the Phillips–Kato extension problems in restricted sense and presents parametrization via fractional-linear transformation of all proper m-accretive and sectorial extensions.

Theorem 13.4 (Arlinskiĭ-Tsekanovskiĭ [30], [32], [34]**).** *Let* A *be a Hermitian contraction defined on the subspace* $\mathcal{D}(A)$ *of* $\mathfrak{H}$*. Then formula*

$$T = \frac{1}{2}(A_M + A_\mu) + \frac{1}{2}(A_M - A_\mu)^{\frac{1}{2}} X (A_M - A_\mu)^{\frac{1}{2}} \tag{13.2}$$

establishes a one-to-one correspondence between contractions X *in* $\overline{\mathcal{R}(A_M - A_\mu)}$ *and all qsc-extensions* T *of* A*. A qsc-extension* T *belongs to the class* $C(\theta)$ *if and only if* $X \in C(\theta)$.

Note that the Kreĭn Theorem 3.1 is being obtained when $\theta = 0$.

For positive definite symmetric operator with finite defect numbers all its m-accretive extensions have been parametrized by means of Phillips approach in the papers of O. Milyo and O. Storozh [141, 142]. It should be pointed out that the Phillips–Kato extension problems have been solved for wide class of initial sectorial operators and even sectorial linear relations (with vertex at the origin) by Yu. Arlinskii [17], [18], [19], [20], [21], [23], [24]. It is established that any closed densely defined sectorial operator S with a semi-angle θ and vertex at the origin admits the m-sectorial extension S_K with the same semi-angle θ (and vertex at the origin) whose properties are very similar to Kreĭn-von Neumann extension of nonnegative operator [17], [18]. Recall that the Friedrichs m-sectorial extension S_F (with the same semi-angle and vertex at the origin) exists by the first representation theorem (see Section 2). It is proved in [17] that for any m-sectorial extension $\widetilde{S}$ of S hold the inclusions

$$\mathcal{D}[S_F] \subseteq \mathcal{D}[\widetilde{S}] \subseteq \mathcal{D}[S_K].$$

In particular, this yields that if $S_F = S_K$, then other m-sectorial extensions (with vertex at the origin) of S do not exist.

We mention that descriptions of all $C(\beta)$-extensions of a given $C(\alpha)$-sub-operator ($\beta \in [\alpha, \pi/2)$) in different forms were obtained in [15], [22], [25], [108], [134], [137], [138]. For the case when A is a Hermitian contraction ($\alpha = 0$) the formula derived in [22] takes the form

$$T = T_0 + D_{T_0}(I + YT_0)^{-1}YD_{T_0} \tag{13.3}$$

with $T_0 = (A_\mu + A_M)/2$ and establishes a one-to-one correspondence between all $Y \in C(\beta)$ in the subspace $\mathfrak{D}_{T_0}$ such that

$$(I + YT_0)^{-1} \in [\mathfrak{D}_{T_0}, \mathfrak{D}_{T_0}], \text{ and } \operatorname{Ker} Y \supset \overline{D_{T_0}\mathcal{D}(A)}$$

and all extension $T \in C(\beta)$ of A. If in additional $\operatorname{Ker} Y^* \supset \overline{D_{T_0}\mathcal{D}(A)}$ then (13.3) can be transformed into (13.2).

Let $\{L_0, M_0\}$ be a dual pair of operators (see Section 4) Suppose L_0 and M_0 are densely defined accretive operators. In [150] and [165] it is established that there exists a maximal accretive extension of the dual pair $\{L_0, M_0\}$. Via the fractional-linear transformation a description of all m-accretive extensions of a dual pair $\{L_0, M_0\}$ is given by (3.5).

Let S be nonnegative, densely defined symmetric operator in Hilbert space $\mathfrak{H}$ and $\{E, \Gamma_1, \Gamma_2\}$ be a boundary value space, $M(z)$ be the corresponding Weyl-Titchmarsh function. Using the boundary value spaces approach, the Phillips–Kato extension problems in restricted sense were considered and solved by Derkach-Malamud-Tselkanovskii [71], [70]. It was shown that operator

$$\mathcal{D}(\widetilde{S}_B) = \operatorname{Ker}(\Gamma_1 - B\Gamma_2), \quad \widetilde{S}_B = S^* | \mathcal{D}(\widetilde{S}_B)$$

is θ-sectorial if and only if $B - M(0)$ is θ-sectorial.

For one-dimensional Schrödinger operator on the semi-axis the Phillips-Kato extension problems in restricted sense has the following form.

Theorem 13.5 (Tsekanovskiĭ [167],[170], [172]**).** *Let S be a nonnegative symmetric Schrödinger operator of the form* (12.1)*with defect indices* $(1,1)$ *with locally summable potential in* $\mathfrak{H} = L^2[a, \infty)$. *Consider operator*

$$T_h y = -y'' + q(x)y,$$
$$y'(a) = hy(a).$$

Then

- *the Friedrichs extension S_F and the Kreĭn-von Neumann extension S_K do not coincide if and only if $m_\infty(-0) < \infty$*
- *operator T_h coincides with the Kreĭn-von Neumann extension if and only if $h = -m_\infty(-0)$*

- *operator* T_h *is* m*-accretive if and only if* $\operatorname{Re} h \geq -m_\infty(-0)$
- *operator* T_h *is* θ*-sectorial if and only if the sharp inequality* $\operatorname{Re} h > -m_\infty(-0)$ *holds.*
- *operator* T_h *is* m*-accretive but not* θ*-sectorial for any* $\theta \in (0, \frac{\pi}{2})$ *if and only if* $\operatorname{Re} h = m_\infty(-0)$
- *If* T_h, $(\operatorname{Im} h > 0)$ *is* θ*-sectorial, then the angle* θ *can be calculated via*

$$\tan\theta = \frac{\operatorname{Im} h}{\operatorname{Re} h + m_\infty(-0)}.$$

For the Schrödinger (Sturm-Liouville) nonnegative symmetric operator of the Bessel type with defect indices $(1,1)$

$$Sy = -y'' + \frac{\nu^2 - \frac{1}{4}}{x^2} y, \quad \nu \geq \frac{1}{2}$$
$$y'(1) = y(1) = 0$$

in the Hilbert space $\mathfrak{H} = L^2[1,\infty)$

$$m_{\infty,\nu}(-0) = \nu - \frac{1}{2}$$

the corresponding operator

$$T_h y = -y'' + \frac{\nu^2 - \frac{1}{4}}{x^2} y, \quad \nu \geq \frac{1}{2}$$
$$y'(1) = hy(1), \quad \operatorname{Im} h \neq 0$$

is θ-sectorial if and only if $\operatorname{Re} h > -(\nu - \frac{1}{2})$. Operator T_h is m-accretive but not θ-sectorial for any $\theta \in (0, \frac{\pi}{2})$ if and only if $\operatorname{Re} h = -(\nu - \frac{1}{2})$. Note that 1 is a regular end point of the interval $[1,+\infty)$ for the coefficients of the operator S. The description of m-accretive and θ-sectorial boundary value problems T_h in $L^2[1,+\infty)$ for the Bessel type of operators for $\nu \geq \frac{1}{2}$ was obtained by Arlinskiĭ-Tsekanovskiĭ.

Consider now the Schrödinger operator in $L^2[0,+\infty)$ of the form

$$\dot{S}y = -y'' + \frac{\nu^2 - \frac{1}{4}}{x^2} y, \quad \nu \in (-1,1)$$
$$\mathcal{D}(\dot{S}) = C_0^\infty.$$

Let S be the closure of $\dot{S}$. Operator S is symmetric, nonnegative operator with defect indices $(1,1)$. Note the the point 0 is a singular point of the potential of S. The Friedrichs extension S_F has the form

$$\mathcal{D}(S_F) = \{y : y \in AC_{loc}(\mathbb{R}_+), \quad y, y' - (\frac{1}{2} + \nu)x^{-1}, -y'' + \frac{\nu^2 - \frac{1}{4}}{x^2} y \in L^2(\mathbb{R}_+)\},$$
$$S_F y = -y'' + \frac{\nu^2 - \frac{1}{4}}{x^2} y.$$

This description was established by Gesztesy-Pittner, Everitt-Kalf [87], [77]. The description of the Kreĭn-von Neumann extension S_K of S was recently obtained by

Makarov-Tsekanovskiĭ and is in preparation for publication. A generalization of the Kreĭn-von Neumann extension for not necessarily semi-bounded symmetric operators with defect indices $(1,1)$ was introduced by Hassi–Kaltenback–de Snoo [101].

In his classical papers Kreĭn discovered the situation when given nonnegative symmetric operator admits only one nonnegative self-adjoint extension or using our terminology when the Friedrichs extension coincides with the Kreĭn-von Neumann extension. Below is the theorem that shows that this case might occur unexpectedly even for the operators of the Bessel type in the situation when 0 is a singular point for the coefficient of differential operator.

Theorem 13.6 (Makarov-Tsekanovskiĭ [133], [168], [169]). *Let S be a closure in $L^2[0,+\infty)$ of the operator*

$$\mathcal{D}(\dot{S}) = C_0^\infty,$$

$$\dot{S}y = -y'' + \frac{\nu^2 - \frac{1}{4}}{x^2}y, \quad \nu \in (-1,1).$$

The operator S admits a unique nonnegative self-adjoint extension, i.e., $S_F = S_K$ if and only if $\nu = 0$. In this case $(\nu = 0)$ operator S does not admit m-accretive and θ-sectorial extensions of S. If $\nu \in (-1,1)$ and $\nu \neq 0$, then $S_F \neq S_K$ and in this case there exist infinitely many non-self-adjoint m-accretive and θ-sectorial extensions.

Note that M-accretivity (θ-sectoriality) of a densely defined in Hilbert space $\mathfrak{H}$ operator T is equivalent to the fact that the solution of the Cauchy problem

$$\frac{dx}{dx} + Tx = 0, \quad x(0) = x_0$$

generates a one-parameter contractive semigroup $U(t)$ having a contractive analytical continuation into a sector $|\arg \varsigma| < \frac{\pi}{2} - \theta$ of a complex plane.

14. The μ-scale invariant operators

Let T be closed, densely defined operator in Hilbert $\mathfrak{H}$. Operator T is called μ-scale invariant $(\mu > 0)$ with respect to the unitary operator U if

$$U^*\mathcal{D}(T) \subseteq \mathcal{D}(T) \qquad \text{and} \qquad UTU^* = \mu T$$

The following theorem describes the role of the Friedrichs and Kreĭn-von Neumann extensions of the given μ-scale invariant nonnegative symmetric operator.

Theorem 14.1 (Makarov-Tsekanovskiĭ [133]). *Let S be a μ-scale invariant, densely defined in Hilbert space $\mathfrak{H}$ nonnegative symmetric operator. Then*

- *The operator S always admits a μ-scale invariant nonnegative self-adjoint extension.*

- *The Friedrichs extension* S_F *and the Kreĭn–von Neumann extension* S_K *are* μ*-scale invariant, i.e.,*
$$US_FU^* = \mu S_F,$$
$$US_KU^* = \mu S_K.$$
- *In the case when* S *has defect indices* $(1,1)$ *the Friedrichs and the Kreĭn–von Neumann extensions are the only ones* μ*-scale invariant.*

Related questions associated with the invariance if Hermitian contractions being fractional-linear transformations of μ-scale invariant nonnegative operators were discussed in [47]

15. Interpolation and system theory

Consider the following linear stationary dynamical system α:
$$(T - zI)x = KJ\varphi_-$$
$$\varphi_+ = \varphi_- - 2iK^*x$$
where $T \in [\mathfrak{H}, \mathfrak{H}]$, $J = J^* = J^{-1} \in E$, $K \in [E, \mathfrak{H}]$, $\operatorname{Im} T = KJK^*$, $\mathfrak{H}$, E are Hilbert spaces, $\varphi_\pm \in E$, φ_- is input vector, φ_+ is output vector and $x \in \mathfrak{H}$ is a state space vector. We call the system α a conservative canonical system of the Livsic type [129] (Brodskii-Livsic colligation [62]). We will use the following notation of the system α as well
$$\alpha = \begin{pmatrix} T & K & J \\ \mathfrak{H} & & E \end{pmatrix}.$$

The operator-valued function
$$W_\alpha(z) = I - 2iK^*(T - zI)^{-1}KJ$$
$$\varphi_+ = W_\alpha(z)\varphi_-$$
is called a transfer function (characteristic function) of the Livsic type of a system α (Brodskii-Livsic colligation). An operator-valued function
$$V_\alpha(z) = i[W_\alpha(z) + I]^{-1}[W_\alpha(z) - I]J$$
is called an impedance function. This function is a Herglotz-Nevanlinna function.

Theorem 15.1 (Derkach-Tsekanovskiĭ [69], [167], [172]**).** *Let*
$$\alpha = \begin{pmatrix} T & K & J \\ \mathfrak{H} & & E \end{pmatrix}$$
be a canonical system of the Livsic type with prime bounded operator T *and finite-dimensional imaginary part. Then* $T \in C(\theta)$ *if and only if*

- V_α *is holomorphic in* $\operatorname{Ext}[-1, 1]$.
- *Operator*

$$K_J = [V_\alpha^{-1}(-1) - V_\alpha^{-1}(1)]^{-\frac{1}{2}}\{2iJ + V_\alpha^{-1}(-1) + V_\alpha^{-1}(1)\}[V_\alpha^{-1}(-1) - V_\alpha^{-1}(1)]^{-\frac{1}{2}}$$
belongs to $C(\theta)$.

The exact value of angle θ can be obtained from the equation

$$\|K_J \pm i\cot\theta\|^2 = 1 + \cot^2\theta.$$

This theorem was discovered as a result of analyzing Kreĭn's paper on semi-bounded operators and its proof is based on parametric representation of all contractive extensions of Hermitian contraction. In scalar case another proof was obtained by Derkach [69]. In another form and in general for infinite dimension of the imaginary part of T the corresponding result has been obtained by Arlinskiĭ in [14].

For the system

$$\alpha = \begin{pmatrix} i\int_x^l f(t)dt & K & 1 \\ L_2[0,l] & & \mathbb{C} \end{pmatrix}$$

where $Kc = cg$, $c \in \mathbb{C}$, $g = g(x) = \frac{1}{\sqrt{2}}$ the main operator

$$T = i\int_x^l f(t)dt$$

is contraction if and only if $0 < l \le \frac{\pi}{2}$.Operator $T \in C(\theta)$if and only if $0 < l < \frac{\pi}{2}$. The exact value of angle θ is $\theta = l$. As a result of this we get inequality

$$\cot l\left|\int_0^l f(t)dt\right|^2 \le \int_0^l |f(t)|^2 dt - \int_0^l\left|\int_x^l f(t)dt\right|^2 dx$$

with the sharp constant $\cot l$.

Consider the classical Nevanlinna-Pick interpolation problem in the class of Herglotz-Nevanlinna matrix-valued functions. The Livsic canonical system α is called an interpolation system for the data $\{z_k\}_1^n$ and $\{V_k\}_1^n$, $\operatorname{Im} z_k > 0$, $\operatorname{Im} V_k \ge 0$, $k = 1,\dots,n$ in Nevanlinna-Pick interpolation problem if the impedance matrix $V_\alpha(z)$ which is the Herglotz-Nevanlinna matrix-valued function satisfies the condition

$$V_\alpha(z_k) = V_k, \quad k = 1,\dots,n.$$

Consider the following classical Pick matrices

$$P = \left\|\frac{V_k - V_j^*}{z_k - z_j^*}\right\|_{k,j=1}^n, \quad Q = \left\|\frac{z_k V_k - z_j^* V_j^*}{z_k - z_j^*}\right\|_{k,j=1}^n.$$

Theorem 15.2 (Alpay-Tsekanovskiĭ [7]). *Let P be strictly positive Pick matrix. Then there exists an interpolation system α of the Livsic type (Brodskii-Livsic colligation) with the state space of dimension n which is an interpolation system for the given data. The main operator T of the system α is θ-sectorial if and only if*

$$Q \ge \cot\theta \left\|V_k V_j^*\right\|_{k,j=1}^n.$$

If P and Q are strictly positive, then the main operator T is θ-sectorial and

$$\tan\theta = (Q^{-1}\varphi, \varphi), \quad \varphi = \begin{pmatrix} V_1^* \\ \vdots \\ V_n^* \end{pmatrix}.$$

From this theorem follows the new sharp inequality involving invertible Pick matrix Q

$$Q \geq \frac{1}{(Q^{-1}\varphi, \varphi)}\varphi\varphi^*.$$

16. Realization problems for Stieltjes functions

Denote by $\mathfrak{S}$ the class of Stieltjes matrix-valued function. It is well known that any matrix-valued function $V(z)$ from the class S admits the following integral representation

$$V(z) = \gamma + \int_0^\infty \frac{d\sigma(t)}{t-z}, \quad \gamma \geq 0, \quad \int_0^\infty \frac{d(\sigma(t)x, x)}{t+1} < \infty, \quad \forall x \in \mathbb{C}^n.$$

In this section we consider realization problems for Stieltjes matrix-valued functions to be represented as the impedance matrix of some conservative, canonical system of the Livsic type and their connections with the Friedrichs and Kreĭn-von Neumann extensions.

Let S be a symmetric, densely defined in Hilbert space $\mathfrak{H}$ (with the inner product $(\cdot,\cdot)$ linear operator with finite and equal defect indices. Consider Hilbert space $\mathfrak{H}_+ = D(S^*)$ with the scalar product

$$(x, y)_+ = (x, y) + (S^*x, S^*y), \quad x, y \in \mathfrak{H}_+$$

and construct triplets of Hilbert spaces (rigged Hilbert space)

$$\mathfrak{H}_+ \subset \mathfrak{H} \subset \mathfrak{H}_-.$$

The geometry of such triplets of Hilbert spaces has been developed by Berezansky [54]. The important role in extension theory (the so-called bi-extensions and $(*)$-extensions) with the exit into triplets of Hilbert spaces plays the Riesz-Berezansky operator [173]. This naturally appearing operator is an isometry that maps Hilbert space $\mathfrak{H}_+$ onto Hilbert space $\mathfrak{H}_-$ [54].

Let T be a quasi-hermitian extension of S with non-empty set of regular points in the lower half-plane ($S \subset T \subset S^*$). An operator $\mathbb{A} \in [\mathfrak{H}_+, \mathfrak{H}_-]$ is called $(*)$-extension of T if

$$\mathbb{A} \supset T, \quad \mathbb{A}^* \supset T^*$$

and

$$\operatorname{Re}\mathbb{A} = \frac{1}{2}(\mathbb{A} + \mathbb{A}^*) \supset \hat{S} \supset S,$$

where

$$\hat{S} = \hat{S}^*, \quad \mathcal{D}(\hat{S}) = \{x \in \mathfrak{H}_+ : \operatorname{Re}\mathbb{A}x \in \mathfrak{H}\}.$$

Let α be a generalized conservative canonical system of the Livsic type (Brodskii-Livsic rigged colligation) of the form

$$(\mathbb{A} - zI)x = KJ\varphi_-$$
$$\varphi_+ = \varphi_- - 2iK^*x,$$

where

$$\operatorname{Im}\mathbb{A} = \frac{\mathbb{A} - \mathbb{A}^*}{2i} = KJK^*, K \in [E, \mathfrak{H}_-], \quad K^* \in [\mathfrak{H}_+, E],$$

E is finite-dimensional Hilbert space and K is invertible. Operator-valued function

$$W_\alpha(z) = I - 2iK^*(\mathbb{A} - zI)^{-1}KJ$$

is a transfer function of the system α that means $\varphi_+ = W_\alpha(z)\varphi_-$. Operator-valued function

$$V_\alpha(z) = K^*(\operatorname{Re}\mathbb{A} - zI)^{-1}K$$

is called an impedance function of the system α. This function is a Heglotz-Nevanlinna function and

$$V_\alpha(z) = i[W_\alpha(z) + I]^{-1}[W_\alpha(z) - I]J.$$

It is known that any Herglotz-Nevanlinna operator-valued function $V(z)$ in finite-dimensional Hilbert space E has the following representation

$$V(z) = Q + Lz + \int_{-\infty}^{+\infty} \left(\frac{1}{t - z} - \frac{t}{1 + t^2}\right) d\Sigma(t),$$

where

$$Q = Q^*, \quad L \geq 0, \quad \int_{-\infty}^{+\infty} \frac{(d\Sigma(t)x, x)_E}{1 + t^2} < \infty, \quad \forall x \in E.$$

The inverse problem when given Herglotz–Nevanlinna operator-valued function in finite-dimensional Hilbert space E can be represented as an impedance function of some conservative, canonical generalized system of the Livsic type was solved by Belyi–Tsekanovskiĭ [50], [51] and the corresponding criterion was established. The general realization theorem for Herglotz–Nevanlinna matrix-valued functions as an impedance of non-canonical generalized conservative systems was considered by Belyi–Hassi–de Snoo–Tsekanovskiĭ [53]. Realizations in terms of transfer matrix-valued functions of canonical conservative systems(characteristic functions) were considered by Tsekanovskiĭ [173] and for operator-valued functions in infinite-dimensional situation of the space E by Arlinskiĭ (see [173]). In terms of different definitions of the characteristic functions of unbounded operators the realization problem was considered by Shtraus, Kuzhel (Sr.), Tsekanovskiĭ, Gubreev, and Derkach–Malamud [125], [173], [100], [67], [66]. Realization problems for rational matrix-valued functions as well as for some classes of transcendental operator-valued functions were considered and studied by Bart–Gohberg–Kaashoek, Arov, Arov–Nudelman, Ball–Staffans, Staffans [46], [43], [42], [45], [162]. Denote over $\mathfrak{S}_0$ the subclass of the Stieltjes class of operator-valued functions in finite-dimensional

Hilbert space E that consists of all Stieltjes operator-valued functions for which the corresponding measure in their integral representation satisfies the condition

$$\int_0^\infty (d\sigma(t)x, x) = \infty, \quad \forall x \neq 0, \quad x \in E.$$

Consider canonical, conservative, generalized system of the Livsic type (Brodskii–Livsic rigged colligation) of the form

$$\alpha = \begin{pmatrix} \mathbb{A} & K & J \\ \mathfrak{H}_+ \subset \mathfrak{H} \subset \mathfrak{H}_- & & E \end{pmatrix}.$$

It was established by Derkach–Tsekanovskiĭ [72] that the impedance function $V_\alpha(z)$ belongs to the class $\mathfrak{S}_0$ if and only if the operator $\mathbb{A}$ is accretive which means that $\mathbb{A}_R = Re\mathbb{A} \geq 0$. If $V(z) \in \mathfrak{S}_0$ and

$$\int_0^\infty \frac{(d\sigma(t)x, x)}{t} = \infty, \quad x \neq 0,\ x \in E,$$

then it can be realized in the form

$$V(z) = V_\alpha(z) = i[W_\alpha(z) + I]^{-1}[W_\alpha(z) - I],$$

where V_α is the impedance function of a dissipative conservative system α of the Livsic type (Brodskii–Livsic rigged colligation) with the accretive and dissipative operator $\mathbb{A}$ ($\operatorname{Im}\mathbb{A} \geq 0$) and the important property

$$\mathbb{A}_R = \operatorname{Re}\mathbb{A} = \frac{1}{2}(\mathbb{A} + \mathbb{A}^*) \supset S_K$$

where S_K is the Kreĭn-von Neumann extension of a nonnegative operator S. At the same time it is impossible to realize any $V(z) \in \mathfrak{S}$ by the conservative system α of the Livsic type with the accretive and dissipative main operator $\mathbb{A}$ such that

$$\mathbb{A}_R = \operatorname{Re}\mathbb{A} = \frac{1}{2}(\mathbb{A} + \mathbb{A}^*) \supset S_F,$$

where S_F is the Friedrichs extension of S. As it is seen, in such realization problem for the Stieltjes function, the Kreĭn–von Neumann extension is involved in the real part of the main operator of a realizing system, but the Friedrichs extension is not. The above-mentioned results were obtained by Dovzhenko–Tsekanovskiĭ [74]. Consider a system of the Livsic type

$$\alpha = \begin{pmatrix} \mathbb{A} & K & J \\ \mathfrak{H}_+ \subset \mathfrak{H} \subset \mathfrak{H}_- & & E \end{pmatrix}$$

with accretive and dissipative main operator $\mathbb{A}$ which is a $(*)$-extension (with the exit into the triplets of Hilbert spaces) of a densely defined quasi-hermitian operator T. Consider the following function

$$Q(z) = i[W_\alpha^{-1}(-1)W_\alpha(z) + I]^{-1}[W_\alpha^{-1}(-1)W_\alpha(z) - I]$$

and the operator

$$\Lambda = [Q^{-1}(-\infty) - Q^{-1}(-0)]^{-\frac{1}{2}} \\ \times [2iI + Q^{-1}(-\infty) + Q^{-1}(-0)][Q^{-1}(-\infty) - Q^{-1}(-0)]^{-\frac{1}{2}}.$$

Operator T is θ-sectorial if and only if Λ is a θ-cosectorial contraction ($\Lambda \in C(\theta)$). This fact is obtained by Tsekanovskiĭ [167], [172].

17. Linear systems with Schrödinger operators

Consider nonnegative symmetric operator in $L_2[0,+\infty)$ with real locally summable potential $q(x)$ and defect indices $(1,1)$

$$y = -y'' + q(x)y \\ y'(a) = y(a) = 0$$

and let

$$T_h y = -y'' + q(x)y \\ y'(a) = hy(a)$$

where nonreal parameter h satisfies the condition $\operatorname{Im} h > 0$. The operator T_h is a quasi-hermitian extension of S ($S \subset T_h \subset S^*$). The corresponding $(*)$-extensions of T_h can be presented in the form

$$\mathbb{A}_{\mu,h} = -y'' + q(x)y + \frac{1}{\mu - h}[hy(a) - y'(a)][\mu\delta(x-a) + \delta'(x-a)],$$

$$\mathbb{A}^*_{\mu,h} = -y'' + q(x)y + \frac{1}{\mu - \bar{h}}[\bar{h}y(a) - y'(a)][\mu\delta(x-a) + \delta'(x-a)],$$

$$\operatorname{Im} \mathbb{A}_{\mu,h} = (\cdot, g)g, \quad g = \frac{(\operatorname{Im} h)^{\frac{1}{2}}}{|\mu - h|}[\mu\delta(x-a) + \delta'(x-a)].$$

Consider the following canonical conservative system of the Livsic type (Brodskii–Livsic rigged colligation) involving Schrödinger operator T_h

$$\alpha = \begin{pmatrix} \mathbb{A}_{\mu,h} & K & 1 \\ \mathfrak{H}_+ \subset L_2[0,+\infty) \subset \mathfrak{H}_- & & \mathbb{C} \end{pmatrix}.$$

In addition to the Stieltjes scalar class $\mathfrak{S}_0$ consider the class of Stieltjes-like scalar functions $\mathfrak{R}_0$ that consists of all functions with the representation

$$V(z) = \gamma + \int_0^{+\infty} \frac{1}{t - z} d\Sigma(t),$$

where γ is an arbitrary real number and the integral term belongs to $\mathfrak{S}_0$. We will consider the following problems:

- When given $V(z)$ belonging to the class $\mathfrak{S}_0$ or to $\mathfrak{N}_0$ can be realized in the form
$$V(z) = V_\alpha(z) = i[W_\alpha(z) + I]^{-1}[W_\alpha(z) - I],$$
where $V_\alpha(z)$ is an impedance function of the system α with Schrödinger operator T_h and its $(*)$-extension $\mathbb{A}_{\mu,h}$ and $W_\alpha(z)$ is a transfer function of this system.
- To find formulas to restor nonreal boundary parameter h and system's real parameter μ from the knowledge of $V(z)$.
- To find conditions on $V(z)$ when the restored operator T_h is a) accretive, b) θ-sectorial, c) extremal (accretive but not θ-sectorial for any $\theta \in (0, \frac{\pi}{2})$.

A non-decreasing function $\sigma(\lambda)$ defined on $[0, +\infty)$ is called a spectral distribution function of an operator pair $\widetilde{S}_\theta$, S, where
$$\widetilde{S}_\theta = -y'' + q(x)y,$$
$$y'(a) = \theta y(a)$$
is a self-adjoint extension of symmetric operator S and if the formulas
$$\varphi(\lambda) = Uf(x),$$
$$f(x) = U^{-1}\varphi(\lambda)$$
establish one-to-one isometric correspondence U between $L_2^\sigma[0,+\infty)$ and $L_2[a,+\infty)$. Moreover, this correspondence is such that the operator $\widetilde{S}_\theta$ is unitarily equivalent to the operator
$$\Lambda_\sigma\varphi(\lambda) = \lambda\varphi(\lambda), \quad (\varphi(\lambda) \in L_2^\sigma[0, +\infty))$$
in $L_2^\sigma[0, +\infty)$ while symmetric operator S is unitarily equivalent to the symmetric operator
$$\Lambda_\sigma\varphi(\lambda) = \lambda\varphi(\lambda),$$
$$\mathcal{D}(\Lambda_\sigma) = \left\{\varphi(\lambda) \in L_2^\sigma[0, +\infty) : \int_0^{+\infty} \varphi(\lambda)d\sigma(\lambda) = 0\right\}.$$

Theorem 17.1 (Belyi–Tsekanovskiĭ [52]). *Let a scalar Stieltjes like function $V(z)$ belongs to the class $\mathfrak{N}_0$ and $\Sigma(t)$ be a spectral function of distribution for nonnegative Schrödinger operator S (with real potential and defect indices $(1,1)$) and its nonnegative self-adjoint extension $\widetilde{S}_\Delta$. Then there exists a unique non-self-adjoint Schrödinger operator $T_h, (\Im h > 0)$ and its unique $(*)$-extension $\mathbb{A}_{\mu,h}$ as a main operator of the conservative system α of the Livsic type such that the function $V(z)$ can be realized in the form*
$$V(z) = V_\alpha(z) = i[W_\alpha(z) + I]^{-1}[W_\alpha(z) - I],$$
where $V_\alpha(z)$ is an impedance function of the system α, $W_\alpha(z)$ is the corresponding transfer function of this system. Operator T_h is accretive if and only if
$$\gamma^2 + \gamma\int_0^\infty \frac{d\Sigma(t)}{t} + 1 \geq 0.$$

The operator T_h is θ-sectorial for some $\theta \in (0, \pi/2)$ if and only if the inequality is strict. In this case the exact value of angle θ can be calculated by the formula

$$\tan\theta = \frac{\int_0^\infty \frac{d\Sigma(t)}{t}}{\gamma^2 + \gamma\int_0^\infty \frac{d\Sigma(t)}{t} + 1}.$$

The operator $\mathbb{A}_{\mu,h}$ is accretive if and only if $V(z)$ is a Stieltjes function from the class $\mathfrak{S}_0$.

We will consider particular case in this paper (for more details in other cases we refer to [52], [74]) when

$$\int_0^\infty \frac{d\Sigma(t)}{t} = \infty.$$

This case profoundly connected with the Kreĭn-von Neumann extension S_K. From the last theorem follows that in this situation operator T_h is extremal if and only if $\gamma = 0$ and operator T_h is θ-sectorial if and only if $\gamma > 0$ with $\tan\theta = \frac{1}{\gamma}$. So, if we have a function

$$V(z) = \int_0^\infty \frac{d\Sigma(t)}{t - z}$$

with the property

$$\int_0^\infty \frac{d\Sigma(t)}{t} = \infty,$$

then the operator T_h in the realizing system is extremal and the following reconstruction formulas take place

$$\begin{aligned} \operatorname{Re} h &= -m_{-\infty}(-0) \\ \operatorname{Im} h &= \frac{1}{C}\int_0^\infty \frac{d\Sigma(t)}{t^2+1} \\ \mu &= \infty \\ C^{\frac{1}{2}} &= \sup \frac{|y(a)|}{(\|y\|_{L_2}^2 + \|S_K y\|_{L_2}^2)^{\frac{1}{2}}}, \quad y \in D(S_K). \end{aligned}$$

Here S_K is the Kreĭn–von Neumann extension of S and has the form

$$\begin{aligned} S_K y &= -y'' + q(x)y \\ y'(a) &+ m_\infty(-0)y(a) = 0. \end{aligned}$$

For the general Stieltjes-like functions from the class $\mathfrak{R}_0$ of the form

$$V(z) = \gamma + \int_0^\infty \frac{d\Sigma(t)}{t - z}, \quad \gamma \in \mathbb{R}$$

in the case of

$$\int_0^\infty \frac{d\Sigma(t)}{t} = \infty$$

the restoration formulas of T_h and $\mathbb{A}_{\mu,h}$ take the form

$$\begin{aligned}\operatorname{Re} h &= -m_\infty(-0) + \frac{\gamma}{1+\gamma^2}\frac{1}{C}\int_0^\infty \frac{d\Sigma(t)}{t^2+1},\\ \operatorname{Im} h &= \frac{1}{1+\gamma^2}\frac{1}{C}\int_0^\infty \frac{d\Sigma(t)}{t^2+1},\\ \mu &= -m_\infty(-0) + \frac{1}{\gamma C}\int_0^\infty \frac{d\Sigma(t)}{t^2+1}.\end{aligned}$$

Restoration formulas as a result of the realization theorem for Stieltjes-like functions were obtained by Belyi–Tsekanovskiĭ [52].

Acknowledgment

We are in debt to Fritz Gesztesy, Konstantin Makarov, and Gerd Grubb for supplying us useful information regarding their work in extension theory. E. Tsekanovskiĭ is grateful to V. Adamyan and Organizing Committee for invitation to make plenary presentation, friendly atmosphere and hospitality.

References

[1] V. Adamyan, *Non-negative perturbations of non-negative self-adjoint operators.* Methods of Functional Anal. and Topology **13** (2007), 103–109.

[2] N. Akhiezer, I. Glazman, *Theory of linear operators.* Pitman advanced publishing Program, 1981.

[3] S. Albeverio, F. Gesztesy, R. Høegh-Krohn, H. Holden, *Solvable models in quantum mechanics.* Springer-Verlag, Berlin, 1988.

[4] S. Albeverio, P. Kurasov, *Singular perturbations of differential operators and solvable Schrödinger type operators.* Cambridge University Press, 2000.

[5] S. Albeverio, J.F. Brasche, M. Malamud, H. Neidhardt, *Inverse spectral theory for symmetric operators with several gaps: scalar-type Weyl functions.* Journ. of Funct. Anal. **228** (2005), no. 1, 144–188.

[6] S. Alonso, B. Simon, *The Birman–Kreĭn–Vishik theory of selfadjoint extensions of semibounded operators.* J. Operator Theory **4** (1980), 251–270.

[7] D. Alpay, E. Tsekanovskiĭ, *Interpolation theory in sectorial Stieltjes classes and explicit system solutions.* Lin. Alg. Appl. **314** (2000), 91–136.

[8] W.N. Anderson, *Shorted operators.* SIAM J. Appl. Math. **20** (1971), 520–525.

[9] W.N. Anderson, G.E. Trapp, *Shorted operators, II.* SIAM J. Appl. Math. **28** (1975), 60–71.

[10] W.N. Anderson, R.J. Duffin, *Series and parallel addition of matrices.* J. Math. Anal. Appl. **26** (1969), 576–594.

[11] T. Ando, *Topics on operator inequalities.* Division of Applied Mathematics, Research Institute of Applied Electricity, Hokkaido University, Sapporo, 1978. ii+44

[12] T. Ando, K. Nishio, *Positive selfadjoint extensions of positive symmetric operators.* Tohóku Math. J. **22** (1970), 65–75.

[13] Yu. Arlinskiĭ, *Positive spaces of boundary values and sectorial extensions of nonnegative operator.* Ukrainian Mat.J. **40** (1988), no. 1, 8–14 (Russian).

[14] Yu. Arlinskiĭ, *Characteristic functions of operators of the class* $C(\alpha)$. Izv. Vyssh. Uchebn. Zaved. Mat. (1991), no. 2, 13–21 (Russian).

[15] Yu. Arlinskiĭ, *On class of extensions of a* $C(\alpha)$*-suboperators.* Dokl. Akad. Nauk Ukraine **no. 8** (1992), 12–16 (Russian).

[16] Yu. Arlinskiĭ, *On proper accretive extensions of positive linear relations.* Ukrainian Mat. J. **47** (1995), no. 6, 723–730.

[17] Yu. Arlinskiĭ, *Maximal sectorial extensions and associated with them closed forms.* Ukrainian Mat. J. **48** (1996), no. 6, 723–739 (Russian).

[18] Yu. Arlinskiĭ, *Extremal extensions of sectorial linear relations.* Matematychnii Studii **7** (1997), no. 1, 81–96.

[19] Yu. Arlinskiĭ, *On functions connected with sectorial operators and their extensions.* Int. Equat. Oper. Theory **33** (1999), no. 2, 125–152.

[20] Yu. Arlinskiĭ, *Abstract boundary conditions for maximal sectorial extensions of sectorial operators*, Math. Nachr. **209** (2000), 5–36.

[21] Yu. Arlinskiĭ, *On m-accretive extensions and restrictions.* Methods of Funct. Anal. and Topol. **4** (1998), no. 3, 1–26.

[22] Yu. Arlinskiĭ, *On a class of nondensely defined contractions on a Hilbert space and their extensions.* Journ. Math. Sci. **97** (1999), no. 5, 4390–4419.

[23] Yu. Arlinskiĭ. *M-accretive extensions of sectorial operators and Kreĭn spaces.* Operator Theory: Advances and Applications **118** (2000), 67–82.

[24] Yu. Arlinskiĭ. *On sectorial block operator matrices.* Matematicheskaya fizika, Analiz, Geometriya **9** (2002), No. 4, 534–573.

[25] Yu. Arlinskiĭ, *Extremal extensions of a* $C(\alpha)$*-suboperator and their representations.* Oper. Theory Adv. Appl. **162** (2006), 47–69.

[26] Yu. Arlinskiĭ, S. Hassi, H.S.V. de Snoo. *Q-functions of quasiselfadjoint contractions.* Operator Theory: Advances and Applications **163** (2005), 23–54.

[27] Yu. Arlinskiĭ, S. Hassi, H.S.V. de Snoo. *Q-functions of Hermitian contractions of Kreĭn–Ovčarenko type.* Int. Eq. and Oper. Theory **53** (2005), 153–189.

[28] Yu. Arlinskiĭ, S. Hassi, Z. Sebestyen, H. de Snoo, *On the class of extremal extensions of a nonnegative operators.* Operator Theory: Advan., and Appl. **127** 2001, v. 127. pp. 41–81.

[29] Yu. Arlinskiĭ, S. Hassi, H. de Snoo, E.Tsekanovskiĭ, *One-dimensional perturbations of self-adjoint operators with finite and discrete spectrum.* Contemporary Mathematics AMS **323** (2003), 419–433.

[30] Yu. Arlinskiĭ, E. Tsekanovskiĭ, *Non-self-adjoint contractive extensions of hermitian contraction and theorem of M.G. Kreĭn.* Uspehi Math. Nauk **1** (1982), 131–132.

[31] Yu. Arlinskiĭ, E. Tsekanovskiĭ, *Generalized resolvents of non-self-adjoint contractive extensions of hermitian contraction*, Ukrainian Math. Journ. **6** (1983), 601–603.

[32] Yu. Arlinskiĭ, E. Tsekanovskiĭ, *On sectorial extensions of positive hermitian operators and their resolvents.* Dokl. Akad. Nauk Armenian SSR **5** (1984), 199–202.

[33] Yu. Arlinskiĭ, E. Tsekanovskiĭ, *On resolvents of m-accretive extensions of symmetric differential operator.* Math. Phys. Nonlin. Mech. **1** (1984), 11–16 (Russian).

[34] Yu. Arlinskiĭ, E. Tsekanovskiĭ, *Quasi-self-adjoint contractive extensions of Hermitian contraction.* Theor. Functions, Funk. Anal. i Prilozhen **50** (1988), 9–16.

[35] Yu. Arlinskiĭ, E. Tsekanovskiĭ, *On the theory of nonnegative extensions of a nonnegative symmetric operator.* Dopov. Nats. Akad. Nauk Ukraini **11** (2002), 30–37.

[36] Yu. Arlinskiĭ, E. Tsekanovskiĭ, *On von Neumann's problem in extension theory of nonnegative operators.* Proc. of AMS **131** 10 (2003), 3143–3154.

[37] Yu. Arlinskiĭ, E. Tsekanovskiĭ. *Some remarks on singular perturbations of selfadjoint operators*, Methods of Functional Analysis and Topology **9** (2003), no. 4, 287–308.

[38] Yu. Arlinskiĭ, E. Tsekanovskiĭ, *Linear systems with Schrödinger operators and their transfer functions.* Oper. Theory, Adv. Appl. **149** (2004), 47–77.

[39] Yu. Arlinskiĭ, E. Tsekanovskiĭ, *The von Neumann problem for nonnegative symmetric operators.* Int. Eq. and Oper. Theory **51** (2005), 319–356.

[40] N. Aronszajn, W.F. Donoghue, *On exponential representations of analytic functions in the upper half-plane with positive imaginary part.* J. Anal. Math. **5** (1957), 321–388.

[41] D. Arov, H. Dym, *Direct and inverse problems for differential systems connected with Dirac systems and related factorization problems.* Indiana Univ. Math. J. **54** 6 (2005), 1769–1815.

[42] D. Arov, M.A. Nudelman, *Passive linear stationary dynamical scattering systems.* Integr. Equ. Oper. Theory **24** (1996), 1–45.

[43] D. Arov, *Passive linear steady-state dynamical systems.* Sibirsk. Math. Zh. **20** (1979), 211–228 (Russian).

[44] G. Arsene, A. Geondea, *Completing matrix contractions.* J. Oper. Theory **7** (1982), no. 1, 179–189.

[45] J.A. Ball, O.J. Staffans, *Conservative state-space realizations of dissipative system behaviors.* Integr. Equ. Oper. Theory **54** (2006), no. 2, 151–213.

[46] H. Bart, I. Gohberg, M.A. Kaashoek, *Minimal Factorizations of Matrix and Operator Functions.* Operator Theory: Advances and Applications, **1**, Birkhäuser, Basel, 1979.

[47] M. Bekker, *On non-densely defined invariant Hermitian contractions.* Methods Funct. Anal. Topol. **13** (2007), no. 3, 223–236.

[48] S. Belyi, G. Menon, E. Tsekanovskiĭ, *On Kreĭn's formula in non-densely defined case.* J. Math. Anal. Appl. **264** (2001), 598–616.

[49] S. Belyi, E. Tsekanovskiĭ, *On Kreĭn's formula in indefinite metric spaces.* Lin. Alg. Appl. **389** (2004), 305–322.

[50] S. Belyi, E. Tsekanovskiĭ, *"Realization theorems for operator-valued R-functions".* Oper. Theory Adv. Appl. **98** (1997), 55–91.

[51] S. Belyi, E. Tsekanovskiĭ, *"On classes of realizable operator-valued R-functions".* Oper. Theory Adv. Appl. **115** (2000), 85–112.

[52] S. Belyi, E. Tsekanovskiĭ, *Stieltjes like functions and inverse problems for systems with Schrödinger operator.* Operators and Matrices. **2** (2008), no. 2, 265–296.

[53] S. Belyi, S. Hassi, H. de Snoo, E. Tsekanovskiĭ, *A general realization theorem for Herglotz-Nevanlinna matrix-valued functions.* Lin. Algbr. Appl. **419** (2006), 331–358.

[54] Yu. Berezansky, *Expansions in eigenfunction of selfadjoint operators.* Amer. Math. Soc. Providence, 1968.

[55] M.S. Birman, *On the selfadjoint extensions of positive definite operators.* Mat. Sbornik **38** (1956), 431–450 (Russian).

[56] J.F. Brasche, V. Koshmanenko, H. Neidhardt, *New aspects of Kreĭn's extension theory.* Ukrain. Mat. Zh. **46** (1994), no. 1–2, 37–54. Translation in Ukrainian Math. J. **46** (1994), no. 1–2, 34–53.

[57] J.F. Brasche, H. Neidhardt, *On the point spectrum of selfadjoint extensions.* Math. Z. **214** (1993), 343–355.

[58] J.F. Brasche, H. Neidhardt, *Some remarks on Kreĭn's extension theory.* Math. Nachr. **165** (1994), 159–181.

[59] J.F. Brasche, H. Neidhardt, *On the absolutely continuous spectrum of self-adjoint extensions.* Journ. of Func. Anal. **131** (1995), no. 2, 364–385.

[60] J.F. Brasche, M. Malamud, H. Neidhardt, *Weyl functions and singular continuous spectra of self-adjoint extensions, Proc. of conference on infinite-dimensional (stochastic) analysis and quantum physics.* Leipzig Germany 18–22, 1999, AMS, CMS Conf. Proc. **29** (2000), 75–84.

[61] M.S. Brodskiĭ, *Triangular and Jordan Representations of Linear Operators.* Nauka, Moscow, 1969 (Russian).

[62] V.M. Bruk, *On one class of boundary value problems with a spectral parameter in the boundary condition.* Mat. Sbornik **100** (1976), No. 2, 210–216 (Russian).

[63] E.A. Coddington, H.S.V. de Snoo, *Positive selfadjoint extensions of positive symmetric subspaces.* Math. Z. **159** (1978), 203–214.

[64] M. Crandall, *Norm preserving extensions of linear transformations on Hilbert spaces.* Proc. Amer. Math. Soc. **21** (1969), no. 2, 335–340.

[65] C. Davis, W.M. Kahan, H.F. Weinberger, *Norm preserving dilations and their applications to optimal error bounds.* SIAM J. Numer. Anal. **19** (1982), no. 3, 445–469.

[66] V. Derkach, M. Malamud, *Weyl function of Hermitian operator and its connection with the characteristic function.* Preprint 85–9, Fiz.-Tekhn. Inst. Akad. Nauk Ukraine (1985), 50 p. (Russian).

[67] V. Derkach, M. Malamud, *Generalized resolvents and the boundary value problems for Hermitian operators with gaps,* J. Funct. Anal. **95** (1991), no. 1, 1–95.

[68] V. Derkach, M. Malamud, *The extension theory of Hermitian operators and the moment problem.* J. of Math. Sci. **73** (1995), no. 2, 141–242.

[69] V. Derkach, E. Tsekanovskiĭ, *On characteristic function of quasi-hermitian contraction.* Izvestia Vysshich Uchebnuch Zavedeni, Math. **6** (1987), 46–51.

[70] V. Derkach, M. Malamud, E. Tsekanovskiĭ, *Sectorial extensions of a positive operator and characteristic function.* Soviet Math. Dokl. **37** 1 (1988), 106–110.

[71] V. Derkach, M. Malamud, E. Tsekanovskiĭ, *Sectorial extensions of positive operator.* Ukrainian Math. J. **41** (1989), no. 2, 151–158 (Russian).

[72] V. Derkach, E. Tsekanovskiĭ, *On characteristic operator-functions of accretive operator colligations.* Ukrainian Math. Dokl., Ser. A, (1981), no. 2, 16–20 (Ukrainian).

[73] W.F. Donoghue, *On the perturbation of spectra.* Commun. Pure Appl. Math. **18** (1965), 559–579.

[74] I. Dovzhenko, E. Tsekanovskiĭ, *Classes of Stieltjes operator-functions and their conservative realizations.* Dokl. Akad. Nauk SSSR, **311** (1990), no. 1, 18–22.

[75] M. Dritschel, J. Rovnyak, *Extension theorems for contraction operators on Kreĭn spaces.* Oper. Theory Adv. Appl. **47** Birkhäuser (1990), 221–305.

[76] W.D. Evans, I. Knoweles, *On the extension problem for accretive differential operators.* Journ. Func. Anal. **63** (1985), no. 3, 276–298.

[77] W.N. Everitt, H. Kalf, *The Bessel differential equation and the Hankel transform.* J. Comp. Appl. Math. **208** (2007), 3–19.

[78] P. Exner, *The absence of the absolutely continuous spectrum for δ's Wannier–Stark ladders.* J. Math. Phys. **36** (1995), 4561–4570.

[79] P. Exner, O. Turek, *Approximations of singular vertex couplings in quantum graphs,* Rev. Math. Phys., **19**, (2007), 571–606.

[80] P.A. Fillmore, J.P. Williams, *On operator ranges.* Advances in Math. **7** (1971), 254–281.

[81] H. Freudental, *Über die Friedrichsche Fortsetzung halbbeschränkter Hermitescher Operatoren.* Proc. Acad. Amsterdam **39** (1936), no. 7, 832–833.

[82] K. Friedrichs, *Spektraltheorie halbbeschränkter Operatoren.* Math. Ann. **109** (1934), 405–487.

[83] F. Gesztesy, N. Kalton, K.A. Makarov, E. Tsekanovskiĭ, *Some applications of operator-valued Herglotz functions.* Oper.Theory, Adv. and Appl. **123** (2001), 271–321.

[84] F. Gesztesy, K.A. Makarov, E. Tsekanovskiĭ, *An addendum to Kreĭn's formula.* Journ. Math. Anal. Appl. **222** (1998), 594–606.

[85] F. Gesztesy, M. Mitrea, *Robin-to-Robin maps and Kreĭn-Type resolvent formulas for Schrödinger operators on bounded Lipschitz domains.* Preprint, arXiv:0803.3072v2 [math.AP] 15 May 2008.

[86] F. Gesztesy, M. Mitrea, *Generalized Robin boundary conditions, Robin-to-Dirichlet maps, and Kreĭn-Type resolvent formulas for Schrödinger operators on bounded Lipschitz domains*, Preprint, arXiv:0803.3179v2 [math.AP] 15 May 2008.

[87] F. Gesztesy and L. Pittner, *On the Friedrichs extension of ordinary differential operators with strongly singular potentials.* Acta Phys. Austriaca **51** (1979), 259–268.

[88] F. Gesztesy, E. Tsekanovskiĭ, *On matrix-valued Herglotz functions.* Math. Nachr. **218** (2000), 61–138.

[89] M.L. Gorbachuk, *Selfadjoint boundary value problems for a second order differential equation with unbounded operator coefficient.* Funct. Anal. and Appl. **5** (1071), no. 1, 10–21 (Russian).

[90] M.L. Gorbachuk, V.I. Gorbachuk, *Boundary value problems for differential-operator equations.* Naukova Dumka, Kiev, 1984 (Russian).

[91] M.L. Gorbachuk, V.I. Gorbachuk, A.N. Kochubeĭ, *Extension theory of symmetric operators and boundary value problems.* Ukrainian Mat. J. **41** (1989), no. 10, 1298–1313 (Russian).

[92] M.L. Gorbachuk, V.A. Mihailets, *Semibounded selfadjoint extensions of symmetric operators.* Dokl. Akad. Nauk SSSR **226** (1976), no. 4, 765–768 (Russian).

[93] G. Grubb, *A characterization of the non-local boundary value problems associated with an elliptic operator.* Ann. Scuola Norm. Sup., Pisa **22** (1968), 425–513.

[94] G. Grubb. *Les problèmes aux limites généraux d'un opérateur elliptique provenant de la théorie variationnelle.* Bull Sci. Math. **91** (1970), 113–157.

[95] G. Grubb. *On coerciveness and semiboundedness of general boundary problems.* Israel Journ. Math. **10** 91971), 32–95.

[96] G. Grubb. *Weakly semibounded boundary problems and sesquilinear forms.* Ann Ins. Fourier **23** (1973), 145–194.

[97] G. Grubb. *Properties of normal boundary problems for elliptic even-order systems.* Ann. Sc. Norm. Sup. Pisa, Ser. IV, **1** (1974), 1–61.

[98] G. Grubb, *Spectral asymptotics for the "soft" self-adjoint extension of a symmetric elliptic differential operator.* J. Operator Theory **10** (1983), 2–20.

[99] G. Grubb, *Known and unknown results on elliptic boundary problems.* Bull. Amer. Math. Soc. **43** 2 (2006), 227–230.

[100] G. Gubreev, *On the characteristic matrix-functions of unbounded non-self-adjoint operators.* Teor. Funkt. Funct. Anal. i Prilozhen. **26** (1976), 12–21 (Russian).

[101] S. Hassi, M. Kaltenback, H. de Snoo, *Generalized Kreĭn-von Neumann extensions and associated operator models.* Acta Sci. Math. (Szeged) **64** (1998), 627–655.

[102] S. Hassi, M. Malamud, H.S.V. de Snoo, *On Kreĭn extension theory of nonnegative operators.* Math. Nach., **274–275** (2004), No. 1, 40–73.

[103] W. Helton, *Systems with infinite-dimensional state space: the Hilbert space approach.* Proceedings of IEEE, **64** 1 (1976), 145–160.

[104] T. Kato, *Perturbation theory for linear operators.* Springer-Verlag, 1966.

[105] Y. Kilpy Y, *Über selbstadjungierte Fortsetzungen symmetrischer Transformationen im Hilbertschen Raum.* Ann. Acad. Fennicae, 1959.

[106] A.N. Kochubeĭ, *On extensions of symmetric operators and symmetric binary relations.* Math. Zametki **17** (1975), no. 1, 41–48 (Russian).

[107] A.N. Kochubeĭ, *On extensions of positive definite symmetric operator.* Dokl. Akad. Nauk Ukr. SSR, Ser. A, No. 3 (1979), 169–171 (Russian).

[108] V. Kolmanovich and M. Malamud, *Extensions of sectorial operators and dual pairs of contractions.* Manuscript No. 4428-85, Deposited at VINITI, (1985), 1–57 (Russian).

[109] V. Koshmanenko, *Singular bilinear forms in perturbations theory of selfadjoint operators.* Kiev, Naukova Dumka, 1993.

[110] V. Koshmanenko, *Singular Operator as a parameter of self-adjoint extensions.* Operator Theory: Advances and Applications **118** (2000), 205–223.

[111] V. Kostrykin, K.A. Makarov, *On Kreĭn's example.* Proc. Amer. Math. Soc. **136** (2008), no. 6, 2067–2071.

[112] M.G. Kreĭn, *The theory of selfadjoint extensions of semibounded Hermitian transformations and its applications, I.* Mat.Sbornik **20** (1947), no. 3, 431–495 (Russian).

[113] M.G. Kreĭn, *The theory of selfadjoint extensions of semibounded Hermitian transformations and its applications, II.* Mat. Sbornik **21** (1947), no. 3) 365–404 (Russian).

[114] M.G. Kreĭn, H. Langer, *On defect subspaces and generalized resolvents of Hermitian operator in the space* Π_κ. Functional Analysis and Appl. **5** (1971), no. 2, 59–71 (Russian).

[115] M.G. Kreĭn, H. Langer, *On defect subspaces and generalized resolvents of Hermitian operator in the space* Π_κ. Functional Analysis and Appl. **5** (1971), no. 3, 54–69 (Russian).

[116] M.G. Kreĭn, H. Langer, *Über die Q-Funktion eines* Π*-Hermiteschen Operators im Raum* Π_κ. Acta Sci. Math. Szeged **34** (1973), 191–230.

[117] M.G. Kreĭn, I.E. Ovčarenko, *On the theory of generalized resolvents for nondensely defined Hermitian contractions.* Dokl.Akad. Nauk Ukr SSR, No. 12, (1976), 881–884 (Russian).

[118] M.G. Kreĭn, I.E. Ovčarenko, *On Q-functions and sc-resolvents of nondensely defined Hermitian contractions.* Siberian Math. Zh. **18** (1977), 728–746 (Russian).

[119] M.G. Kreĭn, I.E. Ovčarenko, *On generalized resolvents and resolvent matrices of positive Hermitian operators.* Sov. Math. Dokl. **231** (1976), no. 5, 1063–1066 (Russian).

[120] M.G. Kreĭn, I.E. Ovčarenko, *Inverse problems for Q-functions and resolvents matrices of positive Hermitian operators.* Sov. Math. Dokl. **242** (1978), no. 3, 521–524 (Russian).

[121] M.G. Kreĭn, Sh.N. Saakyan, *Some new results in the theory of resolvents of Hermitian operators.* Soviet Math. Dokl. **7** (1966), 1086–1089.

[122] M.G. Kreĭn, V.A. Yavryan, *On spectral shift functions appeared under perturbations of positive operator.* J. Oper. Theory **6** (1981), 155–191 (Russian).

[123] P. Kurasov, H_{-n}*-perturbations of self-adjoint operators and Kreĭn's resolvent formula.* Integr. Equ. Oper. Theory **45** (2003), 437–460.

[124] P. Kurasov and B. Pavlov, *Few-body Kreĭn's formula.* Oper. Theory Adv. Appl. **118** (2002), Birkhäuser, 225–254.

[125] A.V. Kuzhel, *On the reduction of unbounded non-self-adjoint operators to triangular form.* Dokl. Akad. Nauk SSSR **119** (1958), 868–871 (Russian).

[126] A.V. Kuzhel, S.A. Kuzhel, *Regular extensions of Hermitian operators.* VSP, the Netherlands, 1998.

[127] A.V. Kuzhel, E. Rotckevich, *Accretive extensions of nonnegative Hermitian operators.* Funct.Anal. Linear Operators, Ul'yanovsk **21** (1983), 94–99 (Russian).

[128] H. Langer, B. Textorius, *Generalized resolvents of contractions.* Acta Sci. Math. **44** (1982), no. 1, 125–131.

[129] M. Livšic, *Operator colligations, waves, open systems.* Transl. Math. Monog, AMS, Providence, R. I., 1973.

[130] V.E. Lyantse, H.B. Majorga, *On selfadjoint extensions of Schrödinger operator with a singular potential.* Lviv university. Deposited in VINITI 15.01.81, N 240-81DEP.

[131] V.E. Lyantse, O.G. Storozh, *Methods of the theory of unbounded operators.* Naukova Dumka, Kiev, 1983 (Russian).

[132] K.A. Makarov, *Survey of new results.* An Addendum to the book of S. Albeverio, F. Gesztesy, R. Høegh-Krohn, H. Holden, Solvable models in quantum mechanics, Springer-Verlag, Berlin, 1988.

[133] K.A. Makarov, E. Tsekanovskiĭ, *On μ-scale invariant operators.* Methods Funct. Anal. Topol. **13** (2007), no. 2, 181–186.

[134] M. Malamud, *On extensions of Hermitian and sectorial operators and dual pairs of contractions.* Dokl. Akad. Nauk SSSR **39** (1989), no. 2, 253–254 (Russian).

[135] M. Malamud, *On some classes of Hermitian operators with gaps.* Ukrainian Mat.J. **44** (1992), no. 2, 215–234 (Russian).

[136] M. Malamud, *On a formula of the generalized resolvents of a nondensely defined hermitian operator.* Ukrain. Math. J. **44** (1992), 1522–1547.

[137] M. Malamud, *On some classes of extensions of sectorial operators and dual pair of contractions.* Operator Theory: Advances and Appl. **124** (2001), 401–448.

[138] M. Malamud, *Operator holes and extensions of sectorial operators and dual pair of contractions.* Math. Nach. **279** (2006), 625–655.

[139] M. Malamud, V.I. Mogilevskii, *Kreĭn type formula for canonical resolvents of dual pair of linear relations.* Methods Funct. Anal. Topol. **8** (2002), no. 4, 72–100.

[140] V.A. Michailets, *Spectral analysis of differential operators.* Sbornik Nauch. Trud., Kiev, Inst. of Math. of Ukrainian Acad. of Sci. (1980), 106–131 (Russian).

[141] O.Ya. Milyo, O.G. Storozh, *On general form of maximal accretive extension of positive definite operator.* Dokl. Akad. Nauk Ukr. SSR, no. 6 (1991), 19–22 (Russian).

[142] O.Ya. Milyo, O.G. Storozh, *Maximal accretive extensions of positive definite operator with finite defect number,* Lviv University, (1993), 31 pages, Deposited in GNTB of Ukraine 28.10.93, no. 2139 Uk93 (Russian).

[143] G. Nenciu, *Applications of the Kreĭn resolvent formula to the theory of self-adjoint extensions of positive symmetric operators.* J. Operator Theory **10** (1983), 209–218.

[144] J. von Neumann, *Allgemeine Eigenwerttheorie Hermitescher Funktionaloperatoren.* Math. Ann. **102** (1929), 49–131.

[145] K. Nishio and T. Ando, *Characterizations of operators derived from network connections.* J. Math. Anal. Appl. **53** (1976), 539–549.

[146] B. Pavlov, *Dilation theory and spectral analysis of non-self-adjoint differential operators.* Proc. of VII Winter School on Mathematical Programming and Related Topics (1976), 3–69.

[147] E. Pekarev, *Shorts of operators and some extremal problems.* Acta Sci. Math. (Szeged) **56** (1992), 147–163.

[148] R. Phillips, *Dissipative parabolic systems.* Trans. Amer. Math. Soc. **86** (1957), 109–173.

[149] R. Phillips, *Dissipative operators and hyperbolic systems of partial differential equations.* Trans. Amer. Math. Soc. **90** (1959), 192–254.

[150] R. Phillips, *On dissipative operators.* Lectures in Differential Equations **3** (1969), 65–113.

[151] A. Posilicano, *A Kreĭn-like formula for singular perturbations of self-adjoint operators and applications.* J. Func. Anal. **183** (2001), 109–147.

[152] V. Prokaj, Z. Sebestyén, *On Friedrichs extensions of operators,* Acta Sci. Math. (Szeged), **62**, (1996), 243–246.

[153] F.S. Rofe-Beketov, *On selfadjoint extensions of differential operators in the space of vector-functions.* Theory of Functions, Functional Anal. and Appl. **8** (1969), 3–24 (Russian).

[154] F.S. Rofe-Beketov, *Selfadjoint extensions of differential operators in the space of vector-functions.* Theory of Functions, Functional Anal. and Appl., Dokl. Akad. Nauk SSSR **184** (1969), no. 5, 1034–1037 (Russian).

[155] F.S. Rofe-Beketov, *Numerical range of a linear relation and maximal relations.* Theory of Functions, Functional Anal. and Appl. **44** (1985), 103–112 (Russian).

[156] Sh.N. Saakyan, *On the theory of resolvents of symmetric operator with infinite defect numbers.* Dokl. acad. Nauk Armenian SSR, **41** (1965), 193–198.

[157] Z. Sebestyén, J. Stochel, *Restrictions of positive self-adjoint operators.* Acta Sci. Math. (Szeged) **55** (1991), 149–154.

[158] Z. Sebestyén, J. Stochel, *Characterizations of positive selfadjoint extensions.* Proceedings of the AMS **135** (2007), no. 5, 1389–1397.

[159] Yu. Shmul'yan, *Hellinger's operator integral.* Mat. Sb. **49** (1959), no. 4, 381–430 (Russian).

[160] Yu. Shmul'yan, P. Yanovskaya, *On blocks of contractive operator matrix.* Izv. Vuzov Math. **7** (1981), 72–75 (Russian).

[161] A.V. Shtraus, *On extensions of semibounded operators.* Dokl. Akad. Nauk SSSR **211** (1973), no. 3, 543–546 (Russian).

[162] O.J. Staffans, *Passive and conservative continuous time impedance and scattering systems, Part I: Well-posed systems.* Math. Control Signals Systems **15** (2002), 291–315.

[163] M. Stone, *Linear transformations in Hilbert spaces and their applications in analysis.* Amer. Math. Soc. Colloquium Publication **15** (1932).

[164] O.G. Storozh, *Extremal extensions of nonnegative operator and accretive boundary problems.* Ukrainian Mat.J. **42** (1990), no. 6, 857–860 (Russian).

[165] B.Sz.-Nagy and C. Foias, *Harmonic analysis of operators on Hilbert space.* North-Holland, New York, 1970.

[166] B. Textorius, *On generalized resolvents of nondensely defined symmetric contractions.* Acta Sci. Math. (Szeged) **49** (1985), 329–338.

[167] E. Tsekanovskiĭ, *Accretive extensions and problems on Stieltjes operator-valued functions relations.* Operator Theory: Adv. and Appl. **59** (1992), 328–347.

[168] E. Tsekanovskiĭ, *Non-self-adjoint accretive extensions of positive operators and theorems of Friedrichs-Kreĭn-Phillips.* Funk. Anal. i Prilozhen. **14** (1980), no. 2, 87–89 (Russian).

[169] E. Tsekanovskiĭ, *The Friedrichs-Kreĭn extensions of positive operators and holomorphic semigroups of contractions.* Funk. Anal. i Prilozhen. **15** (1981), no. 5, 91–93 (Russian).

[170] E. Tsekanovskiĭ, *Characteristic function and description of accretive and sectorial boundary value problems for ordinary differential operators.* Dokl. Akad. Nauk Ukrain. SSR, Ser. A **6** (1985), 21–24.

[171] E. Tsekanovskiĭ, *Triangular models of unbounded accretive operators and regular factorization of their characteristic functions.* Dokl. Akad. Nauk SSSR **297** (1987), 1267–1270.

[172] E. Tsekanovskiĭ, *Characteristic function and sectorial boundary value problems, Proceedings of the Institute of Mathematics.* Novosibirsk, "Nauka" **7** (1987), 180–195 (Russian).

[173] E. Tsekanovskiĭ, Yu. Shmul'yan, *The theory of bi-extensions of operators on rigged Hilbert spaces. Unbounded operator colligations and characteristic functions.* Uspekhi Mat. Nauk **32** (1977), no. 5, 69–124. English translation: Russian Math. Surveys **32:5** (1977), 73–131.

[174] L.I. Vaĭnerman, *On extensions of closed operators in a Hilbert space.* Mat. Zametki **28** (1980), No. 6, 833–841 (Russian).

[175] M.I. Vishik, *On general boundary conditions for elliptic differential equations.* Trudy Moskov. Mat.Obsc. **1** (1952), 187–246 (Russian).

[176] G. Wei, Y. Jiang, *A characterization of positive self-adjoint extensions and its applications to ordinary differential operators.* Proc. Amer. Math. Soc. **133** (2005), 2985–2995.

Yu. Arlinskiĭ
Department of Mathematical Analysis
East Ukrainian National University
91034 Lugansk, Ukraine
e-mail: `yma@snu.edu.ua`

E. Tsekanovskiĭ
Department of Mathematics
P.O. BOX 2044
Niagara University, NY 14109, USA
e-mail: `tsekanov@niagara.edu`

Part 2
Research Papers

Operator Theory:
Advances and Applications, Vol. 190, 115–122

Remarks on the Inverse Spectral Theory for Singularly Perturbed Operators

S. Albeverio, A. Konstantinov and V. Koshmanenko

Abstract. Let A be an unbounded above self-adjoint operator in a separable Hilbert space $\mathcal{H}$ and $E_A(\cdot)$ its spectral measure. We discuss the inverse spectral problem for singular perturbations $\tilde{A}$ of A ($\tilde{A}$ and A coincide on a dense set in $\mathcal{H}$). We show that for any $a \in \mathbb{R}$ there exists a singular perturbation $\tilde{A}$ of A such that $\tilde{A}$ and A coincide in the subspace $E_A((-\infty, a))\mathcal{H}$ and simultaneously $\tilde{A}$ has an additional spectral branch on $(-\infty, a)$ of an arbitrary type. In particular, $\tilde{A}$ may possess prescribed spectral properties in the resolvent set of the operator A on the left of a. Moreover, for an arbitrary self-adjoint operator T in $\mathcal{H}$ there exists $\tilde{A}$ such that T is unitary equivalent to a part of $\tilde{A}$ acting in an appropriate invariant subspace.

Mathematics Subject Classification (2000). Primary 47A10; Secondary 47A55.

Keywords. Singular perturbation, self-adjoint extension, spectrum.

1. Introduction

Let A be a self-adjoint unbounded operator defined on the domain $\mathcal{D}(A) \equiv \mathrm{dom}(A)$ in a separable Hilbert space $\mathcal{H}$ with the inner product $(\cdot, \cdot)$. We shall say that an operator $\tilde{A} \neq A$ in $\mathcal{H}$ is a (pure) singular perturbation of A if the set

$$\mathcal{D} := \{f \in \mathcal{D}(A) \cap \mathcal{D}(\tilde{A}) \mid Af = \tilde{A}f\}$$

is dense in $\mathcal{H}$. In this case, one can define a densely defined symmetric operator $A_0 := A|\mathcal{D} = \tilde{A}|\mathcal{D}$. If, in addition, $\tilde{A}$ is self-adjoint, then A and $\tilde{A}$ are different self-adjoint extensions of A_0.

We shall denote by $\sigma(A)$, $\rho(A)$, and $E_A(\cdot)$ the spectrum, the resolvent set, and the spectral measure of A, respectively. The point, singular continuous, and absolutely continuous spectra of a self-adjoint operator A are denoted by $\sigma_p(A)$, $\sigma_{sc}(A)$, and $\sigma_{ac}(A)$, respectively. For a Borel set $\Delta \subset \mathbb{R}$, we set $A_\Delta := A|E_A(\Delta)\mathcal{H}$. Clearly, A_Δ is as a self-adjoint operator in $\mathcal{H}_{A,\Delta} := \mathrm{Ran}(E_A(\Delta))$.

Assume that an open set $J \subset \mathbb{R}$ is a subset of $\rho(A)$. One can ask the question of whether there exists a singular perturbation $\tilde{A}$ having prescribed spectral properties in J. We show that the answer is positive if A is not semi-bounded above and $J \subset (-\infty, a)$ for some $a \in \mathbb{R}$.

We note that the first detailed investigation of the spectrum of self-adjoint extensions within a gap $J = (a, b)$ $(-\infty \leq a < b < +\infty)$ of a symmetric operator A_0 with finite deficiency indices (n, n) was carried out by M.G. Krein [15]. Namely, he proved that for any auxiliary self-adjoint operator T with the condition $\dim(\mathrm{Ran}(E_T(J))) \leq n$, there exists a self-adjoint extension $\tilde{A}$ such that

$$\tilde{A}_J \simeq T_J. \tag{1.1}$$

Here $A \simeq B$ means that A is unitary equivalent to B. For the operator T with an arbitrary pure point spectrum, this result was generalized in [7] to the case of A_0 with infinite deficiency indices.

Further this problem was intensively studied in a series of papers [2, 8, 9]. The complete solution of it was recently obtained in [10]. It was shown that in the case $J = (a, b)$ and $n \leq \infty$ for any auxiliary self-adjoint operator T there exists a self-adjoint extension $\tilde{A}$ of A_0 satisfying (1.1). In particular, this means that there exists a self-adjoint extension $\tilde{A}$ of A_0 having an arbitrary predetermined structure and type of spectrum in the gap J.

On the other hand, it is known that a similar result is not valid for the much more difficult case of a symmetric operator with several gaps. This problem was studied in [1, 6, 11], where the spectral properties of self-adjoint extensions were described in terms of abstract boundary conditions and the corresponding Weyl functions. In particular, the authors of [1] considered the symmetric operator A_0 of a special structure, namely,

$$A_0 = \bigoplus_{k=1}^{\infty} S_k,$$

where each S_k is unitary equivalent to a fixed densely defined closed symmetric operator S with equal positive deficiency indices. It was assumed that there exists a self-adjoint extension S^0 of S such that the open set $J \subset \rho(S^0) \cap \mathbb{R}$. Then one can associate with the pair $\{S, S^0\}$ a boundary triple $\Pi = \{\mathcal{H}, \Gamma_0, \Gamma_1\}$ (see, [11]) such that $S^0 = S^* | \ker \Gamma_0$. Under the additional assumption that the Weyl function M (see [1, 11]) corresponding to Π is monotone with respect to J, it was shown that for any auxiliary self-adjoint operator T there exists a self-adjoint extension $\tilde{A}$ of A_0 satisfying (1.1).

In the present paper, we consider the above problem from the point of view of singular perturbation theory. Instead of self-adjoint extensions of a fixed symmetric operator A_0, we consider singular perturbations $\tilde{A}$ of a fixed self-adjoint operator A. Therefore, a corresponding symmetric operator A_0 is not unique. This gives us freedom in choosing a dense domain $\mathcal{D} \equiv \mathcal{D}(A_0) = \mathcal{D}(A) \cap \mathcal{D}(\tilde{A})$ and allows us to involve in the consideration a wider class of operators $\tilde{A}$. In particular, instead of an interval (a,b), we can consider an arbitrary open set J which is upper semi-bounded (cf. [10]). We also show that for an an arbitrary self-adjoint operator T in $\mathcal{H}$, there exists a self-adjoint singular perturbation $\tilde{A}$ such that T is unitary equivalent to a certain part of $\tilde{A}$ acting in an appropriate invariant subspace.

The spectral inverse problem in such a formulation (in the case of the point spectrum) was investigated in [3, 13]. In particular, it was shown that for any unbounded self-adjoint operator A and a sequence $\{\lambda_k : k \geq 1\}$ of real numbers, there exists a singular perturbation $\tilde{A}$ of A such that all λ_k are eigenvalues of $\tilde{A}$. Moreover, in [14, 12, 3, 13, 4] the inverse eigenvalue problem of the form

$$\tilde{A}\psi_k = \lambda_k \psi_k, \qquad k = 1, 2, \ldots$$

was studied for a given sequence $\{\lambda_k : k \geq 1\}$ of real numbers and an orthonormal system $\{\psi_k : k \geq 1\}$ satisfying the condition

$$\overline{\operatorname{span}\{\psi_k : k \geq 1\}} \cap \mathcal{D}(A) = \{0\}.$$

Here $\overline{M}$ denotes the closure of the set M.

The aim of this note is to present new observations in the problem of construction of singular perturbations $\tilde{A}$ with prescribed spectral properties, in particular, of the resolvent set of the operator A.

2. Two theorems

In what follows, we assume, without loss of generality, that an unbounded self-adjoint operator A in a separable Hilbert space $\mathcal{H}$ is not semi-bounded above. The main results of this note are formulated in the two following theorems.

Theorem 2.1. *Let A be an unbounded (at least above) self-adjoint operator in a separable Hilbert space $\mathcal{H}$. Then for any fixed $a \in \mathbb{R}$ and an auxiliary self-adjoint operator T in $\mathcal{H}$ there exists a self-adjoint singular perturbation $\tilde{A}$ of A of the form*

$$\tilde{A} = A_{(-\infty,a)} \oplus A', \tag{2.1}$$

where the self-adjoint operator A' in $\mathcal{H}_{[a,\infty)} = E_A([a,\infty))\mathcal{H}$ is such that

$$A'_{(-\infty,a)} \simeq T_{(-\infty,a)}. \tag{2.2}$$

In particular, for an arbitrary open set $J \subset \rho(A) \cap (-\infty, a)$ there exists a self-adjoint singular perturbation $\tilde{A}$ of the form (2.1) *such that*

$$\tilde{A}_J = A'_J \simeq T_J. \tag{2.3}$$

Moreover, we will show that for an arbitrary self-adjoint operator T in $\mathcal{H}$ there exists a self-adjoint singular perturbation $\tilde{A}$ such that T is unitary equivalent to an appropriate part of $\tilde{A}$.

Theorem 2.2. *Let A be an unbounded self-adjoint operator in a separable Hilbert space $\mathcal{H}$ and T be an arbitrary auxiliary self-adjoint operator in $\mathcal{H}$. Then there exists a self-adjoint singular perturbation $\tilde{A}$ of A of the form*

$$\tilde{A} = A' \oplus A'', \tag{2.4}$$

where A' is similar to T,

$$A' \simeq T. \tag{2.5}$$

3. Proofs

Proof of Theorem 2.1. Fix $a \in \mathbb{R}$ and consider the orthogonal decomposition $A = A_{(-\infty,a)} \oplus A_{[a,\infty)}$ where the self-adjoint operators $A_{(-\infty,a)}$ and $A_{[a,\infty)}$ act in the Hilbert spaces $\mathcal{H}_{(-\infty,a)}$ and $\mathcal{H}_{[a,\infty)}$, respectively. Let $\dot{A}$ be an arbitrary densely defined symmetric restriction of $A_{[a,\infty)}$ with infinite deficiency indices. Then $\dot{A} \geq a$ and, according to [10], for any auxiliary self-adjoint operator T there exists a self-adjoint extension A' of $\dot{A}$ (acting in $\mathcal{H}_{[a,\infty)}$) such that $A'_{(-\infty,a)} \simeq T_{(-\infty,a)}$. Define the singular perturbation $\tilde{A}$ of A by

$$\tilde{A} := A_{(-\infty,a)} \oplus A'. \tag{3.1}$$

Clearly, $\tilde{A}$ satisfies (2.2). In particular, for any open subset $J \subset \rho(A) \cap (-\infty, a)$ one can take T_J instead of $T_{(-\infty,a)}$ and get, in the same way, a self-adjoint extension A' of $\dot{A}$ such that

$$A'_{(-\infty,a)} = A'_J \simeq T_J. \tag{3.2}$$

By (3.1) and (3.2),

$$\tilde{A}_{(-\infty,a)} = A_{(-\infty,a)} \oplus A'_{(-\infty,a)} \simeq A_{(-\infty,a)} \oplus T_J. \tag{3.3}$$

Note that $A_{(-\infty,a)} = A_{(-\infty,a)\setminus J}$ since $J \subset \rho(A)$. Therefore (see, (3.3)), $\tilde{A}$ satisfies (2.3). □

Proof of Theorem 2.2. First, suppose additionally that the operator A is not semi-bounded below (recall that we assume throughout the paper that A is not semi-bounded above). Denote $\mathbb{R}_+ := [0, \infty)$, $\mathbb{R}_- := (-\infty, 0)$. In this case, the positive and negative parts $A^\pm := A_{\mathbb{R}_\pm}$ of A are unbounded self-adjoint operators in $\mathcal{H}_\pm := \mathcal{H}_{\mathbb{R}_\pm}$. So, we can apply Theorem 2.1 separately to A^+ and to A^-. Let T be an arbitrary self-adjoint operator in $\mathcal{H}$. Then, by Theorem 2.1, there exist self-adjoint singular perturbations $\widetilde{A^\pm}$ of $A^\pm$ in $\mathcal{H}_\pm$ such that

$$\widetilde{A^+}_{(-\infty,0)} \simeq T^-, \text{ and } \widetilde{A^-}_{[0,\infty)} \simeq T^+.$$

Define the operator

$$\tilde{A} := \tilde{A}^+ \oplus \widetilde{A^-} = \widetilde{A^+}_{(-\infty,0)} \oplus \widetilde{A^+}_{[0,\infty)} \oplus \widetilde{A^-}_{(-\infty,0)} \oplus \widetilde{A^-}_{[0,\infty)}.$$

It has the form (2.4), with

$$A' := \widetilde{A^+}_{(-\infty,0)} \oplus \widetilde{A^-}_{[0,\infty)}, \text{ and } A'' := \widetilde{A^+}_{[0,\infty)} \oplus \widetilde{A^-}_{[0,\infty)}.$$

Clearly, $\tilde{A}$ is a singular perturbation of A such that its part A' satisfies the condition (2.5).

Now, consider the case of a semi-bounded operator A. Suppose that $A \geq a$, $a \in \mathbb{R}$. Then we can decompose $[a, \infty)$ into a union of mutually disjoint Borel sets $\Delta_k \subset [a+k, \infty)$:

$$[a, \infty) = \bigcup_{k=0}^{\infty} \Delta_k,$$

in such a way that each $A^{(k)} := A_{\Delta_k}$ is an unbounded operator in the subspace $\mathcal{H}_k := \mathcal{H}_{\Delta_k}$. Note that

$$A = \bigoplus_{k=0}^{\infty} A^{(k)}.$$

Let T be an arbitrary self-adjoint operator in $\mathcal{H}$. Set $T^{(0)} := T_{(-\infty,a)}$, $T^{(k)} =: T_{[a+k-1,a+k)}$, $k \geq 1$. Then

$$T = \bigoplus_{k=0}^{\infty} T^{(k)}.$$

Applying Theorem 2.1 to $A^{(k)}$, we obtain that there exists a self-adjoint singular perturbation $\widetilde{A^{(k)}}$ of $A^{(k)}$ in $\mathcal{H}_k$ such that

$$\widetilde{A^{(k)}}_{(-\infty,a+k)} \simeq T^{(k)}. \tag{3.4}$$

Define the singular perturbation $\tilde{A}$ of A by

$$\tilde{A} := \bigoplus_{k=0}^{\infty} \widetilde{A^{(k)}}.$$

Clearly, $\tilde{A} = A' \oplus A''$, where

$$A' := \bigoplus_{k=0}^{\infty} \widetilde{A^{(k)}}_{(-\infty,a+k)}, \quad \text{and} \quad A'' := \bigoplus_{k=0}^{\infty} \widetilde{A^{(k)}}_{[a+k,\infty)}.$$

By (3.4) we have that

$$A' \simeq T.$$

□

4. Discussion

We emphasize that the singular perturbations $\tilde{A}$ in Theorems 2.1 and 2.2 are not uniquely defined since in our considerations the symmetric restrictions of $A_{[a,\infty)}$ and $A^{(k)}$ are arbitrary.

Theorem 2.1 shows that the spectral properties of $\tilde{A}$ and T in $J \subset \rho(A) \cap (-\infty, a)$ are the same. In particular,

$$\sigma_\sharp(\tilde{A}) \cap J = \sigma_\sharp(T) \cap J \quad \text{for} \quad \sharp = \text{ac, sc, p}.$$

Besides, on $E_A((-\infty,a)\setminus J)\mathcal{H}$ the operators A and $\tilde{A}$ coincide. If an unbounded self-adjoint operator A is not semi-bounded below, then in Theorem 2.1 one can replace $(-\infty,a)$ by (a,∞).

Note also that Theorem 2.2 shows that for an arbitrary Borel set $\Delta\subset\mathbb{R}$ there exists a self-adjoint singular perturbation $\tilde{A}$ of the form (2.4) such that

$$A'_\Delta \simeq T_\Delta.$$

In particular, one can construct $\tilde{A}$ such that $\sigma_{ac}(\tilde{A})=\sigma_{sc}(\tilde{A})=\overline{\sigma_p(\tilde{A})}=\mathbb{R}$.

We remark that Theorem 2.1 shows that for any fixed $a\in\mathbb{R}$ there exists a singular perturbation $\tilde{A}$ coinciding with A on the subspace $E_A((-\infty,a))\mathcal{H}$ and having any predetermined additional kind of spectra on the left of the point a. Theorem 2.1 can be improved in some sense by using paper [16] and combining it with the results from [10]. The following theorem holds.

Theorem 4.1. *Let A be an unbounded (at least above) self-adjoint operator in a separable Hilbert space $\mathcal{H}$. Then for any fixed $a\in\mathbb{R}$ and an auxiliary self-adjoint operator T in $\mathcal{H}$ there exists a self-adjoint singular perturbation $\tilde{A}$ of A such that*

$$\tilde{A}_{(-\infty,a)} \simeq T_{(-\infty,a)}. \tag{4.1}$$

However, this variant of our main result does not ensure that $\tilde{A}$ coincides with A on the subspace $E_A((-\infty,a))\mathcal{H}$.

Further, taking into account the paper [17], Theorem 2.2 can be given the following stronger form:

Theorem 4.2. *Let A be an unbounded (at least above) self-adjoint operator in a separable Hilbert space $\mathcal{H}$. Then for any auxiliary self-adjoint operator T in $\mathcal{H}$, which is unbounded above, there exists a self-adjoint singular perturbation $\tilde{A}$ of A such that $\tilde{A}\simeq T$.*

One of the aims of our short paper is to show how recent results of the usual spectral theory of self-adjoint extensions imply the corresponding results for singular perturbations. So, Theorem 2.2 can, in particular, be considered as a simple proof of a weak version of the corresponding result from [17].

By the way, Theorem 4.2 shows that the only condition for the existence of a singular perturbation obeying $\tilde{A}\simeq T$ is that both operators A and T together are either semi-bounded above or semi-bounded below. However, if this condition is satisfied, then one can produce by singular perturbation any self-adjoint operator up to unitary equivalence.

Acknowledgment

This work was partly supported by DFG 436 UKR 113/78 grant. The authors would like to thank H. Naidhardt for drawing their attention to [10] and [17].

References

[1] S. Albeverio, J.F. Brasche, M. Malamud, H. Neidhardt, *Inverse spectral theory for symmetric operators with several gaps: scalar-type Weyl functions.* J. Funct. Anal. **228** (2005), 144–188

[2] S. Albeverio, J.F. Brasche, H. Neidhardt, *On inverse spectral theory of self-adjoint extensions: mixed types of spectra.* J. Funct. Anal. **154** (1998), 130–173.

[3] S. Albeverio, M. Dudkin, A. Konstantinov, and V. Koshmanenko, *On the point spectrum of $\mathcal{H}_{-2}$-class singular perturbations.* Math. Nachr. **208** (2007), no. 1-2, 20–27.

[4] S. Albeverio, A. Konstantinov, and V. Koshmanenko, *On inverse spectral theory for singularly perturbed operator: point spectrum.* Inverse Problems **21** (2005), 1871–1878.

[5] S. Albeverio, P. Kurasov, *Singular perturbations of differential operators and solvable Schrödinger type operators.* Cambridge: Univ. Press, 2000.

[6] J.F. Brasche, M.M. Malamud, H. Neidhardt, *Weyl function and spectral properties of self-adjoint extensions.* Integr. Eq. Oper. Theory **43** (2002), 264–289.

[7] J.F. Brasche, H. Neidhardt, J. Weidmann, *On the point spectrum of self-adjoint extensions.* Math. Zeitschr. **214** (1993), 343–355.

[8] J.F. Brasche, H. Neidhardt, *On the absolutely continuous spectrum of self-adjoint extensions.* J. Funct. Anal. **131** (1995), 364–385.

[9] J.F. Brasche, H. Neidhardt, *On the singular continuous spectrum of self-adjoint extensions.* Math. Zeitschr. **222** (1996), 533–542.

[10] J.F. Brasche, *Spectral theory of self-adjoint extensions. Spectral theory of Schrödinger operators.* Contemp. Math. **340** (2004), 51–96.

[11] V.A. Derkach, M.M. Malamud, *General Resolvents and the Boundary Value Problem for Hermitian Operators with Gaps.* J. Funct. Anal. **95** (1991), 1–95.

[12] M.E. Dudkin, V.D. Koshmanenko, *On the point spectrum arising under finite rank perturbations of self-adjoint operators.* Ukrainian Math. J. **55** (2003), no. 9, 1269–1276.

[13] A.Yu. Konstantinov, *Point spectrum of singularly perturbed operators.* Ukrainian Math. J. **57** (2005), no. 5, 776–781.

[14] V. Koshmanenko, *A variant of inverse negative eigenvalues problem in singular perturbation theory.* Methods Funct. Anal. Topology **8** (2002), no. 1, 49–69.

[15] M.G. Krein, *Theory of self-adjoint extensions of semibounded Hermitian operators and its applications. I.* Math. Zbornik **20(62)** (1947), no. 3, 431–495.

[16] H. Neidhardt, V.A. Zagrebnov, *Does each symmetric operator have a stability domain?* Rev. Math. Phys. **10** (1998), no. 6, 829–850.

[17] K. Schmüdgen, *On restrictions of unbounded symmetric operators.* Operator Theory **11** (1984), no. 2, 379–393.

S. Albeverio
Institut für Angewandte Mathematik
Universität Bonn, Wegelerstr. 6
D-53115 Bonn, Germany;
SFB 611, Bonn, Germany;
BiBoS, Bielefeld - Bonn, Germany;
IZKS, Bonn, Germany, CERFIM
Locarno, Switzerland;
Acc Arch., Mendrisio, Switzerland
e-mail: albeverio@uni-bonn.de

A. Konstantinov
Department of Mathematics
Kyiv University
64 Volodymyrs'ka St.
01033 Kyiv, Ukraine
e-mail: konst@faust.kiev.ua

V. Koshmanenko
Institute of Mathematics
National Academy of Science of Ukraine
3 Tereshchenkivs'ka St.
01601 Kyiv, Ukraine
e-mail: kosh@imath.kiev.ua

Operator Theory:
Advances and Applications, Vol. 190, 123–139

An Approach to a Generalization of White Noise Analysis

Yu.M. Berezansky and V.A. Tesko

This paper is dedicated to M.G. Krein

Abstract. In this article, we review some recent developments in white noise analysis and its generalizations. In particular, we describe the main idea of the biorthogonal approach to a generalization of white noise analysis, connected with the theory of hypergroups.

Mathematics Subject Classification (2000). Primary 60H40, 46F25.

Keywords. Fock space, generalized functions, biunitary map, Schefer polynomials.

1. Introduction

The classical white noise analysis (Gaussian white noise analysis) can be understood as a theory of generalized functions of infinitely many variables with a pairing, between test and generalized functions, provided by integration with respect to the Gaussian measure. It is well known that there are several approaches to a construction of such a theory of generalized functions: the Berezansky-Samoilenko approach [19] and the Hida approach [24]. In the Berezansky-Samoilenko approach, the spaces of test and generalized functions are constructed as infinite tensor products of one-dimensional spaces. The Hida approach consists in a construction of some rigging of a Fock space with a subsequent application of the Wiener-Itô-Segal isomorphism to the spaces of this rigging.

After a number of years, it has become clear that the Hida approach is more convenient; in most cases, investigations in white noise analysis and its generalizations are based on it. There exist many works dedicated to white noise analysis developments:

- Works dealing with investigations of spaces of test and generalized functions and operators acting in these spaces, using the Wiener-Itô-Segal isomorphism and various riggings of the Fock space. For more information, see books [24, 12, 25, 39], surveys [40, 41] and the references therein.

- Works dealing with the so-called *Jacobi fields approach* to a generalization of white noise analysis. In these works, the role of the Wiener-Itô-Segal isomorphism is played by a unitary Fourier transform defined by the Jacobi field, i.e., by some family of commuting selfadjoint operators that act in the Fock space and have a Jacobi structure. The theory of Jacobi fields was created by Berezansky under the influence of the works by M.G. Krein (see, e.g., [37, 38]) on Jacobi matrices. A detailed study of general commutative Jacobi fields in the Fock space and of a corresponding spectral measure was carried out in the works by Berezansky and his collaborators (see, e.g., [4], [6]–[10], [15]–[18], [42]–[46]). Note that the Wiener-Itô-Segal isomorphism is possible to understand as the Fourier transform of a certain Jacobi field, the so-called free field. This result was obtained by Koshmanenko and Samoilenko in [36]; see also [12].
- Works devoted to the *biorthogonal approach* to a generalization of white noise analysis. In this approach, one replaces the system of Hermite polynomials, which are orthogonal with respect to the Gaussian measure, with a certain biorthogonal system. The biorthogonal approach was inspired by [22], proposed in [3] and developed in [51, 2, 13, 14, 29, 30, 20, 21] (see survey [20] for the complete bibliography). Note that in [13, 14], it was first observed that the biorthogonal approach is deeply related to the theory of hypergroups.

There exists a deep analogy between the above-mentioned works. In all these works, the spaces of test functions are constructed as images of positive spaces from some rigging of the Fock space. But in the first series of works, the Wiener-Itô-Segal isomorphism is used, in the second series, a Fourier transform is used, and in the third series, a certain biunitary map is used.

This survey is devoted to the biorthogonal approach to a generalization of classical white noise analysis. In the first part of the survey, we recall the main idea of the Hida approach to a construction of classical white noise analysis. In the second part, we give the basic idea of the biorthogonal approach. In order to make the presentation simpler, we first consider the corresponding theory of generalized functions for a model one-dimensional case, and then, briefly, that for the infinite-dimensional case. For the details and proofs, we refer the reader to surveys [20, 21].

2. Gaussian white noise analysis

Let us shortly recall some basic results of Gaussian white noise analysis; for details, see, e.g., [12, 25]. We consider a rigging of the real Hilbert space $L^2(\mathbb{R}) := L^2(\mathbb{R}, dt)$,

$$\mathcal{S}' \supset L^2(\mathbb{R}) \supset \mathcal{S},$$

where $\mathcal{S}$ is the Schwartz space of infinitely differentiable, rapidly decreasing function on $\mathbb{R}$, and $\mathcal{S}'$ is the Schwartz space of distributions dual of $\mathcal{S}$ with respect to the zero space $L^2(\mathbb{R})$. We denote by $\langle \cdot, \cdot \rangle$ the dual pairing between elements of $\mathcal{S}'$ and $\mathcal{S}$ inducted by the scalar product in $L^2(\mathbb{R})$; i.e., for any $f \in L^2(\mathbb{R})$ and any $\varphi \in \mathcal{S}$,

$$\langle f, \varphi \rangle := (f, \varphi)_{L^2(\mathbb{R})}.$$

We will preserve this notation for tensor powers and complexifications of spaces.

Let ρ_G be a probability measure on the Borel σ-algebra $\mathcal{B}(\mathcal{S}')$ such that

$$\int_{\mathcal{S}'} e^{i\langle x, \varphi \rangle} d\rho_G(x) = e^{-\frac{1}{2}\|\varphi\|^2_{L^2(\mathbb{R})}}, \qquad \varphi \in \mathcal{S}. \tag{2.1}$$

By the Minlos theorem, the measure ρ_G is completely characterized by (2.1). This measure ρ_G is called the *Gaussian measure*.

Note that elements $x \in \mathcal{S}'$ can be thought of as paths of the derivative of Brownian motion, i.e., as white noise. More precisely, it follows from (2.1) that

$$\int_{\mathcal{S}'} \langle x, \varphi \rangle^2 d\rho_G(x) = \|\varphi\|^2_{L^2(\mathbb{R})}, \qquad \varphi \in \mathcal{S}.$$

Hence, extending the mapping

$$L^2(\mathbb{R}) \supset \mathcal{S} \ni \varphi \mapsto \langle \cdot, \varphi \rangle \in L^2(\mathcal{S}', \rho_G)$$

by continuity, we obtain a random variable $\langle \cdot, f \rangle \in L^2(\mathcal{S}', \rho_G)$ for each $f \in L^2(\mathbb{R})$. Thus, we can define the stochastic process $\{B_t\}_{t \in \mathbb{R}}$,

$$B_t(\cdot) := \begin{cases} \langle \cdot, \varkappa_{[0,t]} \rangle, & t \geq 0, \\ -\langle \cdot, \varkappa_{[t,0]} \rangle, & t < 0 \end{cases}$$

($\varkappa_\alpha$ is the indicator function of a set α). It is easily seen that $\{B_t\}_{t \in \mathbb{R}}$ is a version of Brownian motion, i.e., finite-dimensional distributions of the process $\{B_t\}_{t \in \mathbb{R}}$ coincide with those of Brownian motion. Now, we informally have, for all $t \in \mathbb{R}$,

$$B_t(x) = \int_0^t x(s) ds, \quad \text{so that} \quad \frac{d}{dt} B_t(x) = x(t).$$

The main technical tool for a construction and study of spaces of test and generalized functions in Gaussian white noise analysis is the *Wiener-Itô-Segal isomorphism*

$$I_G : F(L^2(\mathbb{R})) \to L^2(\mathcal{S}', \rho_G)$$

between the symmetric Fock space $F(L^2(\mathbb{R}))$ and the complex space $L^2(\mathcal{S}', \rho_G)$. Let us recall that the symmetric Fock space $F(L^2(\mathbb{R}))$ over $L^2(\mathbb{R})$ is defined as

$$F(L^2(\mathbb{R})) := \bigoplus_{n=0}^{\infty} \mathcal{F}_n(L^2(\mathbb{R})) n!,$$

where the n-particle Fock space

$$\mathcal{F}_n(L^2(\mathbb{R})) := (L^2_{\mathbb{C}}(\mathbb{R}))^{\widehat{\otimes} n} \qquad ((L^2_{\mathbb{C}}(\mathbb{R}))^{\widehat{\otimes} 0} := \mathbb{C})$$

is equal to the nth symmetric tensor power $\widehat{\otimes}$ of the complexification $L^2_{\mathbb{C}}(\mathbb{R})$ of the real space $L^2(\mathbb{R})$ (henceforth, the subscript $\mathbb{C}$ denotes the complexification of a real space). Thus, for each $f = (f_n)_{n=0}^{\infty} \in F(L^2(\mathbb{R}))$,

$$\|f\|^2_{F(L^2(\mathbb{R}))} = \sum_{n=0}^{\infty} \|f_n\|^2_{\mathcal{F}_n(L^2(\mathbb{R}))} n! < \infty.$$

The isomorphism I_G is completely characterized by its following properties:

1. $I_G : F(L^2(\mathbb{R})) \to L^2(\mathcal{S}', \rho_G)$ is the unitary operator.
2. $I_G(f_0, 0, 0, \ldots) = f_0$ for all $f_0 \in \mathbb{C}$.
3. For each $n \in \mathbb{N}$ and any disjoint Borel sets $\alpha_1, \ldots, \alpha_n \in \mathcal{B}(\mathbb{R})$ of finite Lebesgue measure,

$$\big(I_G(\underbrace{0, \ldots, 0}_{n}, \varkappa_{\alpha_1} \widehat{\otimes} \cdots \widehat{\otimes} \varkappa_{\alpha_n}, 0, 0, \ldots)\big)(\cdot) = \langle \cdot, \varkappa_{\alpha_1} \rangle \ldots \langle \cdot, \varkappa_{\alpha_n} \rangle.$$

There are several equivalent ways of constructing such isomorphism:

- Using multiple stochastic integrals. In this case, I_G is constructed by representing any function from $L^2(\mathcal{S}', \rho_G)$ as an infinite sum of pairwise orthogonal multiple stochastic integrals with respect to the Brownian motion $\{B_t\}_{t \in \mathbb{R}}$; see, e.g., [24, 27, 25].
- Using the Jacobi fields approach. Now, I_G is the Fourier transform of the free field, i.e., a certain family of commuting selfadjoint operators that act in the Fock space $F(L^2(\mathbb{R}))$ and have the Jacobi structure; see, for instance, [36, 12].
- Using the system of infinite-dimensional Hermite polynomials orthogonal (in terms of the Fock space $F(L^2(\mathbb{R}))$, see below) with respect to the Gaussian measure ρ_G; see, e.g., [12, 25].

Our investigation is related to the third way of constructing the Wiener-Itô-Segal isomorphism I_G. Let us take a closer look at it.

We consider the function

$$H(x, \varphi) := e^{\langle x, \varphi \rangle - \frac{1}{2} \|\varphi\|^2_{L^2_{\mathbb{C}}(\mathbb{R})}}, \qquad x \in \mathcal{S}', \quad \varphi \in \mathcal{S}_{\mathbb{C}}.$$

It is well known that H is the *generating function* for the *infinite-dimensional Hermite polynomials* $H_n(x) \in (\mathcal{S}')^{\widehat{\otimes} n}$ which are defined from the decomposition

$$H(x, \varphi) = \sum_{n=0}^{\infty} \frac{1}{n!} \langle \varphi^{\otimes n}, H_n(x) \rangle,$$

where the symbol $\otimes$ denotes the tensor power. The polynomials $H_n(x)$ are *orthogonal in the space* $L^2(\mathcal{S}', \rho_G)$ in terms of the Fock space $F(L^2(\mathbb{R}))$,

$$\int_{\mathcal{S}'} \langle \varphi_n, H_n(x) \rangle \overline{\langle \psi_m, H_m(x) \rangle} \, d\rho_G(x) = \delta_{n,m} n! (\varphi_n, \psi_n)_{\mathcal{F}_n(L^2(\mathbb{R}))}, \tag{2.2}$$

$$\varphi_n \in \mathcal{S}_{\mathbb{C}}^{\widehat{\otimes} n}, \quad \psi_m \in \mathcal{S}_{\mathbb{C}}^{\widehat{\otimes} m}, \qquad n, m \in \mathbb{Z}_+,$$

and the mapping

$$F(L^2(\mathbb{R})) \supset \mathcal{F}_{\text{fin}}(\mathcal{S}) \ni \varphi = (\varphi_n)_{n=0}^{\infty} \mapsto (I_G\varphi)(\cdot) := \sum_{n=0}^{\infty} \langle \varphi_n, H_n(\cdot)\rangle \in L^2(\mathcal{S}', \rho_G)$$

after being extended by continuity to the whole space $F(L^2(\mathbb{R}))$, is the Wiener-Itô-Segal isomorphism. Here, $\mathcal{F}_{\text{fin}}(\mathcal{S})$ denotes the set of all finite sequences $(\varphi_n)_{n=0}^{\infty}$ such that each φ_n belongs to $\mathcal{S}_{\mathbb{C}}^{\widehat{\otimes} n}$.

With the help of the Wiener–Itô–Segal isomorphism I_G, spaces of test and generalized functions are constructed and investigated. These spaces are obtained as the I_G-image of some rigging of the Fock space $F(L^2(\mathbb{R}))$:

$$\begin{array}{ccccc} \mathcal{F}_- & \supset & F(L^2(\mathbb{R})) & \supset & \mathcal{F}_+ \\ & & \big\downarrow I_G & & \big\downarrow I_G \\ \mathcal{H}_- & \supset & L^2(\mathcal{S}', \rho_G) & \supset & \mathcal{H}_+. \end{array}$$

Here, $\mathcal{F}_+$ is a certain Fock space densely and continuously embedded into $F(L^2(\mathbb{R}))$, and $\mathcal{F}_-$ is the negative space with respect to the positive space $\mathcal{F}_+$ and the zero space $F(L^2(\mathbb{R}))$. By definition, the space of test functions $\mathcal{H}_+ := I_G\mathcal{F}_+$ is the I_G-image of the Fock space $\mathcal{F}_+$ with topology inducted by the topology of $\mathcal{F}_+$, and the space of generalized functions $\mathcal{H}_- := (\mathcal{H}_+)'$ is the dual of $\mathcal{H}_+$ with respect to $L^2(\mathcal{S}', \rho_G)$. Note that we can extend the isomorphism $I_G : F(L^2(\mathbb{R})) \to L^2(\mathcal{S}', \rho_G)$ to the isomorphism between the negative Fock space $\mathcal{F}_-$ and the space of generalized functions $\mathcal{H}_-$.

3. Biorthogonal approach

In this section, we give the basic idea of the biorthogonal approach. In order to make the exposition simpler, we first consider the corresponding theory of generalized functions for a model one-dimensional case; then we briefly consider that for the infinite-dimensional case.

3.1. One-dimensional case

At first, we consider a one-dimensional analogue of Gaussian white noise analysis. Then we describe the biorthogonal approach to a generalization of such an analysis.

3.1.1. Gaussian case. Let ρ_G be a *Gaussian measure* on the Borel σ-algebra $\mathcal{B}(\mathbb{R})$. Its Fourier transform has the form

$$\int_{\mathbb{R}} e^{ix\lambda} d\rho_G(x) = e^{-\frac{1}{2}\lambda^2}, \qquad \lambda \in \mathbb{R}.$$

In this case, we have the well-known *generating function*

$$H(x, \lambda) := e^{x\lambda - \frac{1}{2}\lambda^2} = \sum_{n=0}^{\infty} \frac{\lambda^n}{n!} H_n(x), \qquad x \in \mathbb{R}, \quad \lambda \in \mathbb{C},$$

for the *Hermite polynomials* H_n that are orthogonal with respect to ρ_G. More precisely, we have, for all $n, m \in \mathbb{Z}_+$,

$$\int_{\mathbb{R}} H_n(x)\overline{H_m(x)}\, d\rho_G(x) = \delta_{n,m} n!.$$

Now, the role of the Fock space $F(L^2(\mathbb{R}))$ is played by the l^2-space

$$l^2 := \Big\{ f = (f_n)_{n=0}^{\infty},\ f_n \in \mathbb{C} \,\Big|\, \|f\|_{l^2}^2 := \sum_{n=0}^{\infty} |f_n|^2 n! < \infty \Big\}$$

and an analogue of the Wiener-Itô-Segal isomorphism has the form

$$l^2 \ni f = (f_n)_{n=0}^{\infty} \mapsto (I_G f)(\cdot) = \sum_{n=0}^{\infty} f_n H_n(\cdot) \in L^2(\mathbb{R}, \rho_G).$$

By analogy with the infinite-dimensional situation, using the space l^2 instead of the Fock space $F(L^2(\mathbb{R}))$ and the unitary mapping $I_G : l^2 \to L^2(\mathbb{R}, \rho_G)$ instead of the Wiener-Itô-Segal isomorphism, we obtain spaces of test and generalized functions of variables $x \in \mathbb{R}$ as the I_G-image of riggings of the space l^2.

Namely, for fixed $K > 1$ and $q \in \mathbb{N}$ we denote

$$l_+^2(q) := \Big\{ f = (f_n)_{n=0}^{\infty},\ f_n \in \mathbb{C} \,\Big|\, \|f\|_{l_+^2(q)}^2 := \sum_{n=0}^{\infty} |f_n|^2 (n!)^2 K^{qn} < \infty \Big\},$$

$$l_+^2 := \operatorname*{pr\,lim}_{q \in \mathbb{N}} l_+^2(q).$$

Then the dual spaces of $l_+^2(q)$ and l_+^2 with respect to the zero space l^2 are

$$l_-^2(q) := (l_+^2(q))' = \Big\{ f = (f_n)_{n=0}^{\infty},\ f_n \in \mathbb{C} \,\Big|\, \|f\|_{l_-^2(q)}^2 := \sum_{n=0}^{\infty} |f_n|^2 K^{-qn} < \infty \Big\},$$

$$l_-^2 := (l_+^2)' = \operatorname*{ind\,lim}_{q \in \mathbb{N}} l_-^2(q),$$

respectively. Thus, for each $q \in \mathbb{N}$, we get the rigging

$$l_-^2 \supset l_-^2(q) \supset l^2 \supset l_+^2(q) \supset l_+^2.$$

Using the unitary operator I_G, one defines spaces of test functions

$$\mathcal{H}_+(q) := I_G l_+^2(q), \qquad \mathcal{H}_+ := I_G l_+^2 = \operatorname*{pr\,lim}_{q \in \mathbb{N}} \mathcal{H}_+(q),$$

and their dual (with respect to the space $L^2(\mathbb{R}, \rho_G)$) spaces of generalized functions

$$\mathcal{H}_-(q) := (\mathcal{H}_+(q))', \qquad \mathcal{H}_- := (\mathcal{H}_+)' = \operatorname*{ind\,lim}_{q \in \mathbb{N}} \mathcal{H}_-(q).$$

Hence, for each $q \in \mathbb{N}$, we have the rigging

$$\mathcal{H}_- \supset \mathcal{H}_-(q) \supset L^2(\mathbb{R}, \rho_G) \supset \mathcal{H}_+(q) \supset \mathcal{H}_+$$

with pairing, between test and generalized functions, provided by integration with respect to the Gaussian measure ρ_G on $\mathbb{R}$.

3.1.2. Biorthogonal case. Let ρ be a Borel probability measure on $\mathbb{R}$, and $L^2(\mathbb{R},\rho)$ be the corresponding L^2-space. Our purpose is to construct some class of test and generalized functions on $\mathbb{R}$ with pairing for integration with respect to ρ. We try to construct these classes functions on $\mathbb{R}$ in a way parallel to the Gaussian case, but using a certain biunitary mapping instead of the Wiener-Itô-Segal isomorphism.

Let us consider, instead of $H(x,\lambda)$, a fixed function

$$\mathbb{R}\times\mathbb{C}\ni\{x,\lambda\}\mapsto h(x,\lambda)\in\mathbb{C}$$

such that for each λ from some neighborhood B_0 of zero in $\mathbb{C}$, the function $\mathbb{R}\ni x\mapsto h(x,\lambda)\in\mathbb{C}$ is continuous and for every $x\in\mathbb{R}$

$$h(x,\lambda)=\sum_{n=0}^{\infty}\frac{\lambda^n}{n!}h_n(x),\qquad \lambda\in B_0.$$

We additionally assume that $h(\cdot,\lambda)$ is locally bounded, uniformly with respect to λ on any closed ball inside of B_0, and that $h(x,0)=1$ for all x from $\mathbb{R}$. In our consideration, the role of the function $h(x,\lambda)$ is same as the role of the generating function $H(x,\lambda)=e^{x\lambda-\frac{1}{2}\lambda^2}$ for the Hermite polynomials in the Gaussian case.

We denote by $C(\mathbb{R})$ the linear space of all complex-valued locally bounded (i.e., bounded on every ball in $\mathbb{R}$) continuous functions on $\mathbb{R}$. It follows from the properties of h that for every $n\in\mathbb{Z}_+$, the function $\mathbb{R}\ni x\mapsto h_n(x)\in\mathbb{C}$ belongs to the space $C(\mathbb{R})$, and the mapping

$$l^2_+(q)\ni f=(f_n)_{n=0}^{\infty}\mapsto(I^hf)(\cdot):=\sum_{n=0}^{\infty}f_nh_n(\cdot)\in C(\mathbb{R})$$

is well defined for each $q\in\mathbb{N}$ and sufficiently large $K>1$ (we recall that K is used in the definition of the space $l^2_+(q)$). In what follows, we fix such $K>1$.

From the general results, one has (see, e.g., [20])

Theorem 3.1. *Let the above-mentioned function h be such that*

- $\|h_n\|_{L^2(\mathbb{R},\rho)}\le C^nn!$ *for some $C>0$ and all $n\in\mathbb{Z}_+$.*
- *The linear span of the functions $\{h_n\}_{n=0}^{\infty}$ is dense in the space $L^2(\mathbb{R},\rho)$.*
- $\|I^hf\|_{L^2(\mathbb{R},\rho)}=0$ *if and only if $f=0$ in $l^2_+(q)$, $q\in\mathbb{N}$.*

Then the I^h-image

$$\mathcal{H}^h_+(q):=I^h(l^2_+(q))=\Big\{f\in C(\mathbb{R})\,\Big|\,\exists(f_n)_{n=0}^{\infty}\in l^2_+(q),f(x)=\sum_{n=0}^{\infty}f_nh_n(x)\Big\}$$

of the space $l^2_+(q)$, $q\in\mathbb{N}$, is a Hilbert space of continuous functions with topology inducted by the topology of $l^2_+(q)$. Moreover, $\mathcal{H}^h_+(q)$ is densely and continuously embedded in $L^2(\mathbb{R},\rho)$, and we can construct the rigging

$$\mathcal{H}^h_-\supset\mathcal{H}^h_-(q)\supset L^2(\mathbb{R},\rho)\supset\mathcal{H}^h_+(q)\supset\mathcal{H}^h_+,$$

$$\mathcal{H}^h_+:=I^hl^2_+=\operatorname*{pr\,lim}_{q\in\mathbb{N}}\mathcal{H}^h_+(q),\qquad \mathcal{H}^h_-:=(\mathcal{H}^h_+)'=\operatorname*{ind\,lim}_{q\in\mathbb{N}}\mathcal{H}^h_-(q).$$

Let all requirements of Theorem 3.1 be fulfilled. It follows from [5] that for the unitary operator $I^h : l^2_+(q) \to \mathcal{H}^h_+(q)$, there exists a uniquely determined unitary operator $I^h_- : l^2_-(q) \to \mathcal{H}^h_-(q)$ such that

$$(I^h_-\xi, I^h\varphi)_{L^2(\mathbb{R},\rho)} = (\xi,\varphi)_{l^2}, \qquad \xi \in l^2_-(q), \quad \varphi \in l^2_+(q).$$

The pair $\{I^h_-, I^h\}$ is called a *biunitary map.* This biunitary mapping transfers the rigging of the space l^2 to a rigging of the space $L^2(\mathbb{R},\rho)$:

$$\begin{array}{ccccc} l^2_-(q) & \supset & l^2 & \supset & l^2_+(q) \\ \downarrow I^h_- & & & & \downarrow I^h \\ \mathcal{H}^h_-(q) & \supset & L^2(\mathbb{R},\rho) & \supset & \mathcal{H}^h_+(q). \end{array}$$

Thus, in the biorthogonal case the spaces of test and generalized functions are constructed in a way parallel to the Gaussian case (as the image of the rigging of the space l^2), but using the biunitary map $\{I^h_-, I^h\}$ instead of the Wiener-Itô-Segal isomorphism. This give us a possibility to develop the biorthogonal white noise analysis by analogy to the Gaussian analysis. In particular, we can give an inner description of the spaces of test and generalized functions, construct, for general situation, the S-transformation, Wick multiplication etc; see, e.g., [20] for more details.

Note that *the natural question* arises under which conditions on h the biunitary map $\{I^h_-, I^h\}$ is the unitary map, i.e., the system of functions $\{h_n\}_{n=0}^\infty$ constitutes an *orthogonal basis* in the space $L^2(\mathbb{R},\rho)$.

The answer is the following [21].

Theorem 3.2. *The system of functions $\{h_n\}_{n=0}^\infty$ with the generating function h constitutes an orthogonal basis in the space $L^2(\mathbb{R},\rho)$ if and only if the following conditions hold:*

- $\|h_n\|_{L^2(\mathbb{R},\rho)} \le C^n n!$ *for some $C > 0$ and all $n \in \mathbb{Z}_+$.*
- *The linear span of the functions $\{h_n\}_{n=0}^\infty$ is dense in the space $L^2(\mathbb{R},\rho)$.*
- *For each λ, μ from some neighborhood of zero in $\mathbb{C}$,*

$$\int_{\mathbb{R}} h(x,\lambda)\overline{h(x,\mu)}d\rho(x) = e^{\lambda\bar{\mu}}.$$

It is possible to prove that if all conditions of Theorem 3.2 hold, then all conditions of Theorem 3.1 will also hold (see [21]). In other words, the orthogonal situation is a particular case of the biorthogonal situation.

Consider a more special situation.

Example. Let ρ be a Borel probability measure on $\mathbb{R}$ such that

$$\int_{\mathbb{R}} e^{\varepsilon|x|}d\rho(x) < \infty \quad \text{for some} \quad \varepsilon > 0,$$

and $h(x,\lambda)$ be a generating function for the *Schefer polynomials* $h_n(x)$ (in another terminology, the generalized Appel polynomials), that is,

$$h(x,\lambda) := \gamma(\lambda)e^{\alpha(\lambda)x} = \sum_{n=0}^{\infty} \frac{\lambda^n}{n!} h_n(x), \qquad x \in \mathbb{R}, \quad \lambda \in \mathbb{C}, \tag{3.1}$$

where γ and α are fixed analytic functions in some neighborhood of $0 \in \mathbb{C}$ such that $\alpha(0) = 0$, $\alpha^{'}(0) = 1$, and $\gamma(0) = 1$.

In this case, the estimate

$$\|h_n\|_{L^2(\mathbb{R},\rho)} \le C^n n! \quad \text{for some} \quad C > 0 \quad \text{and all} \quad n \in \mathbb{Z}_+$$

is automatically satisfied and the linear span of the functions $\{h_n\}_{n=0}^{\infty}$ is dense in the space $L^2(\mathbb{R},\rho)$, see, e.g., [33, 35]. Hence, if the mapping

$$l^2_+(q) \ni f = (f_n)_{n=0}^{\infty} \mapsto (I^h f)(\cdot) := \sum_{n=0}^{\infty} f_n h_n(\cdot) \in L^2(\mathbb{R},\rho)$$

is injective, then all requirements of Theorem 3.1 are fulfilled, and we can construct the corresponding theory of generalized functions.

The next question is: Which of the Schefer polynomials are orthogonal?

The answer was given by Meixner [47] in 1934 (see also [43, 48] for more details). There exist exactly five types of orthogonal Schefer polynomials: the Hermite, Charlier, Laguerre, Meixner, and Meixner–Pollaczek polynomials, which are orthogonal with respect to the Gaussian, Poissonian, Gamma, Pascal, and Meixner measures respectively.

3.1.3. Some useful tools in biorthogonal analysis. Let ρ be a Borel probability measure on $\mathbb{R}$ and

$$h(x,\lambda) = \sum_{n=0}^{\infty} \frac{\lambda^n}{n!} h_n(x), \qquad x \in \mathbb{R}, \quad \lambda \in \mathbb{C},$$

be a fixed function such that Theorem 3.1 holds.

• *Annihilation and creation operators.* The annihilation operator ∂ acts continuously in the space of test functions $\mathcal{H}^h_+$ by the formula

$$\partial h_n := n h_{n-1}, \qquad \partial h_0 := 0.$$

The creation operator ∂^+ is, by definition, the adjoint to ∂ with respect to the zero space $L^2(\mathbb{R},\rho_G)$ and acts continuously in the space of generalized functions $\mathcal{H}^h_-$.

These operators play an essential role in our considerations. Using them, we investigate the spaces of test and generalized functions, construct generalized translation operators, extended stochastic integral (in the infinite-dimensional case), etc. (see, e.g., [20, 21, 49, 50, 1] and references therein). Note that in the Gaussian case, the annihilation operator ∂ is the derivative and $\partial + \partial^+$ is, as an operator in the space $L^2(\mathbb{R},\rho_G)$, the operator of multiplication by $x \in \mathbb{R}$.

• *S-transform.* For each $\xi \in \mathcal{H}_-^h$, the S-transform is defined by the formula

$$(S\xi)(\lambda) := \langle\!\langle \xi, h(\cdot, \overline{\lambda}) \rangle\!\rangle,$$

where λ belongs to a neighborhood of zero in $\mathbb{C}$, and $\langle\!\langle \cdot, \cdot \rangle\!\rangle$ is the dual pairing between elements of $\mathcal{H}_-^h$ and $\mathcal{H}_+^h$ generated by the scalar product in the space $L^2(\mathbb{R}, \rho)$.

Each generalized function $\xi \in \mathcal{H}_-^h$ is uniquely determined by its S-transform. More exactly, let $\mathrm{Hol}_0(\mathbb{C})$ denote the set of all (germs of) functions which are holomorphic in a neighborhood of zero in $\mathbb{C}$. According to [14], the S-transform is a one-to-one mapping between $\mathcal{H}_-^h$ and $\mathrm{Hol}_0(\mathbb{C})$.

• *Wick multiplication.* Taking into account that $\mathrm{Hol}_0(\mathbb{C})$ is an algebra of analytic functions with ordinary algebraic operations, we can define a Wick product $\xi \lozenge \eta$ of $\xi, \eta \in \mathcal{H}_-^h$ through the formula

$$\xi \lozenge \eta := S^{-1}(S\xi \cdot S\eta)$$

and make $\mathcal{H}_-^h$ an algebra with such multiplication.

Using this multiplication, we can construct the elements of Wick calculus. In Gaussian white noise analysis, such calculus has found numerous applications, in particular, in fluid mechanics and financial mathematics; see, e.g., [23, 26] for more details.

3.2. Infinite-dimensional case

Now, we start with a fixed family $(H_p)_{p\in\mathbb{Z}_+}$ of *real* separable Hilbert spaces H_p such that for all $p \in \mathbb{Z}_+$, the space H_{p+1} is densely embedded in H_p, and this embedding is quasinuclear, i.e., the Hilbert-Schmidt type. So, one can construct the rigging of the space H_0,

$$\Phi' := \operatorname*{ind\,lim}_{p\in\mathbb{Z}_+} H_{-p} \supset H_{-p} \supset H_0 \supset H_p \supset \operatorname*{pr\,lim}_{p\in\mathbb{Z}_+} H_p =: \Phi, \tag{3.2}$$

where H_{-p} is the dual space to H_p with respect to the zero space H_0. We denote by $\langle \cdot, \cdot \rangle$ the dual pairing between elements of H_{-p} and H_p inducted by the scalar product in H_0. As before, we preserve this notation for tensor powers and complexifications of spaces.

For each $p \in \mathbb{Z}$, we introduce a weighted symmetric Fock space $\mathcal{F}(H_p, \tau)$ over H_p with a fixed weight $\tau = (\tau_n)_{n=0}^\infty$, $\tau_n > 0$, by setting

$$\begin{aligned}\mathcal{F}(H_p, \tau) :&= \bigoplus_{n=0}^{\infty} \mathcal{F}_n(H_p)\tau_n \\ :&= \Big\{ f = (f_n)_{n=0}^\infty,\ f_n \in \mathcal{F}_n(H_p) \,\Big|\, \|f\|^2_{\mathcal{F}(H_p,\tau)} = \sum_{n=0}^{\infty} \|f_n\|^2_{\mathcal{F}_n(H_p)} \tau_n < \infty \Big\},\end{aligned}$$

where the n-particle Fock space

$$\mathcal{F}_n(H_p) := H_{p,\mathbb{C}}^{\widehat{\otimes} n} \qquad (H_{p,\mathbb{C}}^{\widehat{\otimes} 0} := \mathbb{C})$$

is equal to the nth symmetric tensor power $\widehat{\otimes}$ of the complexification $H_{p,\mathbb{C}}$ of the real space H_p. Using rigging (3.2) and the weight

$$\tau(q) = ((n!)^2 K^{qn})_{n=0}^{\infty}, \qquad q \in \mathbb{N},$$

with fixed $K > 1$, we construct the rigging of the Fock space $\mathcal{F}(H_0, (n!)_{n=0}^{\infty})$,

$$\mathcal{F}(-p, -q) \supset F(H_0) := \mathcal{F}(H_0, (n!)_{n=0}^{\infty}) \supset \mathcal{F}(p, q),$$

where

$$\mathcal{F}(-p, -q) := \mathcal{F}(H_{-p}, (K^{-qn})_{n=0}^{\infty}), \qquad \mathcal{F}(p, q) := \mathcal{F}(H_p, ((n!)^2 K^{qn})_{n=0}^{\infty})$$

are dual with respect to the zero space $F(H_0)$.

Let ρ be a fixed Borel probability measure on $\Phi^{'}$, and $L^2(\Phi^{'}, \rho)$ be the corresponding space of square integrable functions. Our goal is to construct some class of test and generalized functions on $\Phi^{'}$ with pairing generated by the scalar product in $L^2(\Phi^{'}, \rho)$. We try to construct this class functions on $\Phi^{'}$ in a way parallel to the Gaussian and above one-dimensional cases.

Let B_0 be a neighborhood of zero in the space $\Phi_{\mathbb{C}}$. Instead of the generating function for the infinite-dimensional Hermite polynomials, we take a function

$$\Phi^{'} \times B_0 \ni \{x, \varphi\} \mapsto h(x, \varphi) \in \mathbb{C}$$

such that for each $\varphi \in B_0$, the function $h(\cdot, \varphi)$ is continuous, and for each $x \in \Phi^{'}$, the function $h(x, \cdot)$ is analytic in B_0, i.e., has the representation

$$h(x, \varphi) = \sum_{n=0}^{\infty} \frac{1}{n!} \langle \varphi^{\otimes n}, h_n(x) \rangle, \qquad h_n(x) \in (\Phi_{\mathbb{C}}^{'})^{\widehat{\otimes} n},$$

for all φ from B_0. We additionally assume that $h(\cdot, \varphi)$ is locally bounded, uniformly with respect to φ on any closed ball inside of B_0, and that $h(x, 0) = 1$ for all x from $\Phi^{'}$ (see [20], Sections 2.3, for more details).

Due to such properties of h, one can show that for each $\varphi_n \in \Phi_{\mathbb{C}}^{\widehat{\otimes} n}$, the functions

$$\Phi^{'} \ni x \mapsto \langle \varphi_n, h_n(x) \rangle \in \mathbb{C}$$

belong to the space $C(\Phi^{'})$ of all complex-valued locally bounded continuous functions on $\Phi^{'}$. Moreover, there exist $p, q \in \mathbb{N}$ and $K > 1$ (we recall that K is used in the definition of space $\mathcal{F}(p, q)$) such that the mapping

$$\mathcal{F}(p, q) \ni f = (f_n)_{n=0}^{\infty} \mapsto (I^h f)(\cdot) := \sum_{n=0}^{\infty} \langle f_n, h_n(\cdot) \rangle \in C(\Phi^{'})$$

is well defined. In what follows, we fix such $p, q \in \mathbb{N}$ and $K > 1$.

According to [20], we have

Theorem 3.3. *Let the above-mentioned function h be such that*

- $\| \, \|h_n(\cdot)\|_{\mathcal{F}_n(H_{-p})} \|_{L^2(\Phi', \rho)} \leq C^n n!$ *for some $C > 0$ and all $n \in \mathbb{Z}_+$.*

- *The linear span of the set of functions*

$$\{\langle \varphi_n, h_n(\cdot)\rangle \in L^2(\Phi', \rho) \,|\, \varphi_n \in \Phi_{\mathbb{C}}^{\widehat{\otimes} n},\, n \in \mathbb{Z}_+\}$$

is dense in the space $L^2(\Phi', \rho)$.

- $\|I^h f\|_{L^2(\Phi',\rho)} = 0$ *if and only if* $f = 0$ *in* $\mathcal{F}(p,q)$.

Then the I^h*-image*

$$\mathcal{H}^h(p,q) := I^h(\mathcal{F}(p,q)) = \Big\{ f \in C(\Phi') \,\Big|\, \exists (f_n)_{n=0}^{\infty} \in \mathcal{F}(p,q),\ f(x) = \sum_{n=0}^{\infty} \langle f_n, h_n(x)\rangle \Big\}$$

of the space $\mathcal{F}(p,q)$ *is a Hilbert space of continuous functions with the topology inducted by the topology of* $\mathcal{F}(p,q)$. *Moreover,* $\mathcal{H}^h(p,q)$ *is densely and continuously embedded in* $L^2(\Phi', \rho)$, *and we can construct a rigging*

$$\mathcal{H}^h(-p,-q) \supset L^2(\Phi', \rho) \supset \mathcal{H}^h(p,q)$$

with pairing between the elements of $\mathcal{H}^h(-p,-q) := (\mathcal{H}^h(p,q))'$ *and* $\mathcal{H}^h(p,q)$ *inducted by integration with respect to the measure* ρ.

Under the conditions of Theorem 3.3, for the unitary operator

$$I^h : \mathcal{F}(p,q) \to \mathcal{H}^h(p,q)$$

there exists a uniquely determined unitary operator

$$I^h_- : \mathcal{F}(-p,-q) \to \mathcal{H}^h(-p,-q)$$

such that

$$(I^h_- \xi, I^h \varphi)_{L^2(\Phi',\rho)} = (\xi, \varphi)_{F(H_0)}, \qquad \xi \in \mathcal{F}(-p,-q), \quad \varphi \in \mathcal{F}(p,q).$$

So, we have a *biunitary map* $\{I^h_-, I^h\}$. This mapping transfers the rigging of the space $F(H_0)$ to a rigging of the space $L^2(\Phi', \rho)$:

$$\begin{array}{ccccc} \mathcal{F}(-p.-q) & \supset & F(H_0) & \supset & \mathcal{F}(p,q) \\ \big\downarrow I^h_- & & & & \big\downarrow I^h \\ \mathcal{H}^h(-p,-q) & \supset & L^2(\Phi', \rho) & \supset & \mathcal{H}^h(p,q). \end{array}$$

Thus, in the biorthogonal case, the spaces of test and generalized functions are constructed in a way parallel to the Gaussian case (as the image of the rigging of the Fock space $F(H_0)$), but using the biunitary map $\{I^h_-, I^h\}$ instead of the Wiener-Itô-Segal isomorphism I_G.

The infinite-dimensional analogue of Theorem 3.2 holds; see [20, 21].

Theorem 3.4. *The mapping*

$$F(H_0) \ni f = (f_n)_{n=0}^{\infty} \mapsto (I^h f)(\cdot) := \sum_{n=0}^{\infty} \langle f_n, h_n(\cdot)\rangle \in L^2(\Phi', \rho) \tag{3.3}$$

is well defined and unitary if and only if the following three conditions hold:

- $\| \|h_n(\cdot)\|_{\mathcal{F}_n(H_{-p})}\|_{L^2(\Phi',\rho)} \le C^n n!$ *for some* $C > 0$ *and all* $n \in \mathbb{Z}_+$.
- *The linear span of the set of functions*

$$\{\langle \varphi_n, h_n(\cdot)\rangle \in L^2(\Phi', \rho) \mid \varphi_n \in \Phi_{\mathbb{C}}^{\widehat{\otimes} n},\ n \in \mathbb{Z}_+\}$$

is dense in the space $L^2(\Phi', \rho)$.
- *For each* φ, ψ *from some neighborhood of zero in* $\Phi_{\mathbb{C}}$,

$$\int_{\Phi'} h(x,\varphi)\overline{h(x,\psi)}d\rho(x) = e^{(\varphi,\psi)_{H_{0,\mathbb{C}}}}.$$

Note that under the assumptions of Theorem 3.4, the functions

$$\Phi' \ni x \mapsto \langle \varphi_n, h_n(x)\rangle \in \mathbb{C}, \qquad \Phi' \ni x \mapsto \langle \psi_m, h_m(x)\rangle \in \mathbb{C},$$

$$\varphi_n \in \Phi_{\mathbb{C}}^{\widehat{\otimes} n}, \quad \psi_m \in \Phi_{\mathbb{C}}^{\widehat{\otimes} m}, \qquad n, m \in \mathbb{Z}_+,$$

are orthogonal in the space $L^2(\Phi', \rho)$ in terms of the Fock space $F(H_0)$,

$$\int_{\Phi'} \langle \varphi_n, h_n(x)\rangle \overline{\langle \psi_m, h_m(x)\rangle} d\rho(x) = \delta_{n,m} n! (\varphi_n, \psi_n)_{\mathcal{F}_n(H_0)},$$

and all requirements of Theorem 3.3 are fulfilled.

Let us consider the special case where $\Phi = \mathcal{S}$, $H_0 = L^2(\mathbb{R})$ and, as a consequence, $\Phi' = \mathcal{S}'$. Let the function h satisfy all conditions of Theorem 3.4 and

$$\frac{\partial^n h(x, z_1\varphi_1 + \cdots + z_n\varphi_n)}{\partial z_1 \ldots \partial z_n}\bigg|_{z_1=\ldots=z_n=0\in\mathbb{C}} = \frac{\partial}{\partial z_1} h(x, z_1\varphi_1)\bigg|_{z_1=0\in\mathbb{C}} \cdots \frac{\partial}{\partial z_n} h(x, z_n\varphi_n)\bigg|_{z_n=0\in\mathbb{C}} \tag{3.4}$$

for all $x \in \mathcal{S}'$ and all $\varphi_1, \ldots, \varphi_n \in \mathcal{S}$ such that $\operatorname{supp}\varphi_i \cap \operatorname{supp}\varphi_j = \varnothing$ if $j \neq i$, $i, j \in \{1, \ldots, n\}$, $n \in \mathbb{N}$. Then, according to [1], the mapping (3.3) is completely characterized by the following properties:

1. $I^h : F(L^2(\mathbb{R})) \to L^2(\mathcal{S}', \rho)$ is the unitary operator.
2. $I^h(f_0, 0, 0, \ldots) = f_0$ for all $f_0 \in \mathbb{C}$.
3. For each $n \in \mathbb{N}$ and any disjoint sets $\alpha_1, \ldots, \alpha_n \in \mathcal{B}(\mathbb{R})$ of finite Lebesgue measure,

$$(I^h(\underbrace{0, \ldots, 0}_{n}, \varkappa_{\alpha_1} \widehat{\otimes} \cdots \widehat{\otimes} \varkappa_{\alpha_n}, 0, 0, \ldots))(\cdot) = \langle h_1(\cdot), \varkappa_{\alpha_1}\rangle \ldots \langle h_1(\cdot), \varkappa_{\alpha_n}\rangle.$$

Note that in the case of the Gaussian measure $\rho := \rho_G$ on $\mathcal{B}(\mathcal{S}')$, the function

$$h(x, \varphi) := H(x, \varphi) = e^{\langle x, \varphi\rangle - \frac{1}{2}\|\varphi\|^2_{L^2_{\mathbb{C}}(\mathbb{R})}}, \qquad x \in \mathcal{S}', \quad \varphi \in \mathcal{S}_{\mathbb{C}},$$

satisfies all conditions of Theorem 3.4 and equality (3.4).

Now, we have an analog of the Example from Subsection 3.1.

Example. Let ρ be a Borel probability measure on $\Phi^{'}$ such that

$$\int_{\Phi^{'}} e^{\varepsilon\|x\|_{H_{-p}}} d\rho(x) < \infty \quad \text{for some} \quad \varepsilon > 0 \quad \text{and} \quad p \in \mathbb{N}.$$

It is well known that the Schefer polynomials (3.1) have the corresponding infinite-dimensional counterparts and are defined by the Taylor expansion of the generating function

$$h(x,\varphi) := \gamma(\varphi)e^{\langle\alpha(\varphi),x\rangle} = \sum_{n=0}^{\infty} \frac{1}{n!}\langle\varphi^{\otimes n}, h_n(x)\rangle,$$

where γ and α are fixed analytic functions in some neighborhood of zero in $\Phi_{\mathbb{C}}$ such that $\alpha(0) = 0$, $\gamma(0) = 1$, and α is invertible in a neighborhood of zero.

In this case, the estimate

$$\| \, \|h_n(\cdot)\|_{\mathcal{F}_n(H_{-p})}\|_{L^2(\Phi^{'},\rho)} \le C^n n! \quad \text{for some} \quad C > 0 \quad \text{and all} \quad n \in \mathbb{Z}_+$$

is automatically satisfied, and the linear span of the functions

$$\{\langle\varphi_n, h_n(\cdot)\rangle \in L^2(\Phi^{'},\rho) \,|\, \varphi_n \in \Phi_{\mathbb{C}}^{\widehat{\otimes} n},\, n \in \mathbb{Z}_+\}$$

is dense in the space $L^2(\mathbb{R},\rho)$; see, e.g., [33, 35]. Hence, if the mapping

$$\mathcal{F}(p,q) \ni f = (f_n)_{n=0}^{\infty} \mapsto (I^h f)(\cdot) := \sum_{n=0}^{\infty} \langle f_n, h_n(\cdot)\rangle \in L^2(\Phi^{'},\rho)$$

is injective, then all requirements of Theorem 3.3 are fulfilled, and we can construct the corresponding theory of generalized functions (see [3, 51, 2, 46, 33, 34, 32, 11, 31, 48, 45] for more detailed acount).

For the infinite-dimensional counterparts of the Hermite, Charlier, Laguerre, Meixner, and Meixner–Pollaczek polynomials, the orthogonality persists in the following sense:

- In the Gaussian and Poisson cases, the orthogonality of the Hermite and Charlier polynomials, respectively, is given in terms of the Fock space (relation of type (2.2)); see, e.g., books [24, 12, 25] and articles [28, 51, 34, 10, 31, 21]. Note that the study of Poisson white noise analysis was initiated by Y. Ito and I. Kubo [28] in 1988. They were the first to construct spaces of test and generalized functions of Poisson white noise, to study them and some operators acting in these spaces.
- In the Gamma, Pascal, and Meixner cases, the orthogonality of the corresponding polynomials is more complicated and is given in terms of the so-called "extended Fock space"; see, for instance, [34, 32, 17, 11, 16, 18, 43, 44, 48, 45].

References

[1] S. Albeverio, Yu.M. Berezansky, V. Tesko, *A generalization of an extended stochastic integral.* Ukrainian Math. J. **59** (2007), no. 5, 588–617.

[2] S. Albeverio, Yu.L. Daletsky, Yu.G. Kondratiev, and L. Streit, *Non-Gaussian infinite-dimensional analysis.* J. Funct. Anal. **138** (1996), no. 2, 311–350.

[3] S. Albeverio, Yu.G. Kondratiev, and L. Streit, *How to generalize white noise analysis to non-Gaussian measures.* Proc. Symp. Dynamics of Complex and Irregular Systems (Bielefeld, 1991); Bielefeld Encount. Math. Phys., VIII, World Sci. Publ., River Edge, NJ, 1993, pp. 120–130.

[4] Yu.M. Berezansky, *Spectral approach to white noise analysis.* Proc. Symp. Dynamics of Complex and Irregular Systems (Bielefeld, 1991); Bielefeld Encount. Math. Phys., VIII, World Sci. Publ., River Edge, NJ, 1993, pp. 131–140.

[5] Yu.M. Berezansky, *Infinite-dimensional non-Gaussian analysis connected with generalized translation operators.* Analysis on infinite-dimensional Lie groups and algebras (Marseille, 1997), World Sci. Publ., River Edge, NJ, 1998, pp. 22–46.

[6] Yu.M. Berezansky, *Commutative Jacobi fields in Fock space.* Integr. Equ. Oper. Theory. **30** (1998), no. 2, 163–190.

[7] Yu.M. Berezansky, *Direct and inverse spectral problems for a Jacobi field.* St. Petersburg Math. J. **9** (1998), no. 6, 1053–1071.

[8] Yu.M. Berezansky, *On the theory of commutative Jacobi fields.* Methods Funct. Anal. Topology **4** (1998), no. 1, 1–31.

[9] Yu.M. Berezansky, *Spectral theory of commutative Jacobi fields: direct and inverse problems.* Fields Inst. Conmun. **25** (2000), 211–224.

[10] Yu.M. Berezansky, *Poisson measure as the spectral measure of Jacobi field.* Infin. Dimens. Anal. Quantum Probab. Relat. Top. **3** (2000), no. 1, 121–139.

[11] Yu.M. Berezansky, *Pascal measure on generalized functions and the corresponding generalized Meixner polynomials.* Methods Funct. Anal. Topology **8** (2002), no. 1, 1–13.

[12] Yu.M. Berezansky and Yu.G. Kondratiev, *Spectral Methods in Infinite-dimensional Analysis.* Vols. 1, 2, Kluwer Academic Publishers, Dordrecht–Boston–London, 1995. (Russian edition: Naukova Dumka, Kiev, 1988).

[13] Yu.M. Berezansky and Yu.G. Kondratiev, *Non-Gaussian analysis and hypergroups.* Funct. Anal. Appl. **29** (1995), no. 3, 188–191.

[14] Yu.M. Berezansky and Yu.G. Kondratiev, *Biorthogonal systems in hypergroups: an extension of non-Gaussian analysis.* Methods Funct. Anal. Topology **2** (1996), no. 2, 1–50.

[15] Yu.M. Berezansky, V.O. Livinsky, and E.V. Lytvynov, *A generalization of Gaussian white noise analysis.* Methods Funct. Anal. Topology **1** (1995), no. 1, 28–55.

[16] Yu.M. Berezansky, E. Lytvynov, and D.A. Mierzejewski, *The Jacobi field of a Levy process.* Ukrainian Math. J. **55** (2003), no. 5, 853–858.

[17] Yu.M. Berezansky and D.A. Mierzejewski, *The chaotic decomposition for the Gamma field.* Funct. Anal. Appl. **35** (2001), no. 4, 305–308.

[18] Yu.M. Berezansky and D.A. Mierzejewski, *The Construction of the Chaotic Representation for the Gamma Field.* Infin. Dimens. Anal. Quantum Probab. Relat. Top. **6** (2003), no. 1, 33–56.

[19] Yu.M. Berezansky and Yu.S. Samoilenko, *Nuclear spaces of functions of an infinite number of variables.* Ukrainian Math. J. **25** (1973), no. 6, 723–737 (in Russian).

[20] Yu.M. Berezansky and V.A. Tesko, *Spaces of fundamental and generalized functions associated with generalized translation.* Ukrainian Math. J. **55** (2003), no. 12, 1907–1979.

[21] Yu.M. Berezansky and V.A. Tesko, *An orthogonal approach to the construction of the theory of generalized functions of an infinite number of variables, and Poissonian white noise analysis.* Ukrainian Math. J. **56** (2004), no. 12, 1885–1914.

[22] Yu.L. Daletsky, *A biorthogonal analogy of the Hermite polynomials and the inversion of the Fourier transform with respect to a non-Gaussian measure.* Funct. Anal. Appl. **25** (1991), no. 2, 138–40.

[23] R.J. Elliott and J. van der Hoek, *A general fractional white noise theory and applications to finance.* Math. Finance **13** (2003), no. 2, 301–330.

[24] T. Hida, *Analysis of Brownian functionals.* Carleton Math. Lect. Notes, **13** (1975).

[25] T. Hida, H.-H. Kuo, J. Potthoff, and L. Streit, *White Noise. An Infinite Dimensional Calculus.* Kluwer Academic Publishers, Dordrecht, 1993.

[26] H. Holden, B. Øksendal, J. Ubøe, and T. Zhang, *Stochastic Partial Differential Equations. A Modeling, White Noise Functional Approach.* Birkhäuser, Boston, 1996.

[27] K. Ito, *Multiple Wiener Integral.* J. Math. Soc. Japan **3** (1951), 157–169.

[28] Y. Ito and I. Kubo, *Calculus on Gaussian and Poissonian white noise.* Nagoya Math. J. **111** (1988), 41–84.

[29] N.A. Kachanovsky, *Biorthogonal Appell-like systems in a Hilbert space.* Methods Funct. Anal. Topology **2** (1996), no. 3-4, 36–52.

[30] N.A. Kachanovsky and S.V. Koshkin, *Minimality of Appell-like systems and embeddings of test function spaces in a generalization of white noise analysis.* Methods Funct. Anal. Topology **5** (1999), no. 3, 13–25.

[31] Yu.G. Kondratiev, T. Kuna, M.J. Oliveira, *Analytic aspects of Poissonian white noise analysis.* Methods Funct. Anal. Topology **8** (2002), no. 4, 15–48.

[32] Yu.G. Kondratiev and E.W. Lytvynov, *Operators of Gamma white noise calculus.* Infin. Dimens. Anal. Quantum Probab. Relat. Top. **3** (2000), no. 3, 303–335.

[33] Yu.G. Kondratiev, J.L. Da Silva, and L. Streit, *Generalized Appell systems.* Methods Funct. Anal. Topology **3** (1997), no. 3, 28–61.

[34] Yu.G. Kondratiev, J.L. Da Silva, L. Streit, and G.F. Us, *Analysis on Poisson and Gamma spaces.* Infin. Dimens. Anal. Quantum Probab. Relat. Top. **1** (1998), no. 1, 91–117.

[35] Yu.G. Kondratiev, L. Streit, W. Westerkamp and J. Yan, *Generalized functions in infinite-dimensional analysis.* Hiroshima Math. J. **28** (1998), no. 2, 213–260.

[36] V.D. Koshmanenko and Yu.S. Samoilenko, *The isomorphism between Fock space and a space of functions of infinitely many variables.* Ukrainian Math. J. **27** (1975), no. 5, 669–674 (in Russian).

[37] M.G. Krein, *Fundamental aspects of the representation theory of Hermitian operators with deficiency index* (m, m). Ukrainian Math. J. **1** (1949), no. 2, 3–66 (in Russian).

[38] M.G. Krein, *Infinite J-matrices and a matrix moment problem.* Dokl. Acad. Nauk SSSR **69** (1949), no. 2, 125–128 (in Russian).

[39] H.-H. Kuo, *White noise distribution theory.* CRC Press, 1996.

[40] H.-H. Kuo, *White noise theory.* Handbook of Stochastic Analysis and Applications D. Kannan and V. Lakshmikantham (eds.), 2002, pp. 107–158.

[41] H.-H. Kuo, *A quarter century of white noise theory.* Quantum Information IV, T. Hida and K. Saito (eds.), 2002, pp. 1–37.

[42] E.W. Lytvynov, *Multiple Wiener integrals and non-Gaussian white noises: a Jacobi field approach.* Methods Funct. Anal. Topology **1** (1995), no. 1, 61–85.

[43] E. Lytvynov, *Polynomials of Meixner's type in infinite dimensions – Jacobi fields and orthogonality measures.* J. Funct. Anal. **200** (2003), 118–149.

[44] E. Lytvynov, *Orthogonal decompositions for Levy processes with an applications to the Gamma, Pascal, and Meixner processes.* Infin. Dimens. Anal. Quantum Probab. Relat. Top. **6** (2003), no. 1, 73–102.

[45] E. Lytvynov, *Functional spaces and operators connected with some Levy noises.* arXiv:math.PR/0608380 v1 15 Aug 2006.

[46] E.W. Lytvynov and G.F. Us, *Dual Appell systems in non-Gaussian white noise calculus.* Methods Funct. Anal. Topology **2** (1996), no. 2, 70–85.

[47] J. Meixner, *Orthogonale Polynomsysteme mit einer besonderen Gestalt der erzeugenden Funktion.* J. London Math. Soc. **9** (1934), no. 1, 6–13.

[48] I. Rodionova, *Analysis connected with generating functions of exponential type in one and infinite dimensions.* Methods Funct. Anal. Topology **11** (2005), no. 3, 275–297.

[49] V.A. Tesko, *A construction of generalized translation operators.* Methods Funct. Anal. Topology **10** (2004), no. 4, 86–92.

[50] V.A. Tesko, *On spaces that arise in the construction of infinite-dimensional analysis according to a biorthogonal scheme.* Ukrainian Math. J. **56** (2004), no. 7, 1166–1181.

[51] G.F. Us, *Dual Appel systems in Poissonian analysis.* Methods Funct. Anal. Topology **1** (1995), no. 1, 93–108.

Yu.M. Berezansky and V.A. Tesko
Institute of Mathematics
National Academy of Sciences of Ukraine
3 Tereshchenkivs'ka St.
01601 Kyiv, Ukraine
e-mail: berezan@mathber.carrier.kiev.ua
tesko@imath.kiev.ua

Operator Theory:
Advances and Applications, Vol. 190, 141–155

Some Existence Conditions for the Compact Extrema of Variational Functionals of Several Variables in Sobolev Space W_2^1

E.V. Bozhonok

Abstract. We study sufficient and necessary conditions for the compact extrema of integral functionals in Sobolev space W_2^1 in terms of the Hessian of the integrand. Also, we generalize to the case of compact extrema of variational functionals the classical Legendre necessary condition and the Legendre-Jacobi sufficient condition.

Mathematics Subject Classification (2000). Primary 49K10, 49K20; Secondary 49K24.

Keywords. Compact extremum, integral functional, Sobolev space, Legendre condition, Jacobi condition.

1. Introduction

Analysis [1]–[5] of the extremum problems for variation functionals in Sobolev space W_2^1 leads to the notions of a strong compact extremum, compact continuity, and compact differentiability. In papers [6]–[10], various necessary and sufficient conditions for the compact extrema of the Euler–Lagrange functional in W_2^1 are considered.

In the present work, we consider an application of the compact extremum theory to the case of the Euler–Lagrange functional of several variables. We obtain both sufficient and necessary conditions for a strong compact extremum of the Euler–Lagrange functional in terms of, respectively, a system of inequalities for the positive definiteness and that for the non-negativeness, which are constructed with the help of the strong Hessian of the integrand. Also, we generalize to the case of a strong compact extremum the known Legendre necessary condition and the Legendre–Jacobi sufficient condition. These results are obtained on the basis of a positive definiteness test and a non-negativeness test for the operator matrices and quadratic forms in products of n real Hilbert spaces.

Let us give necessary definitions and results (see [4, 5, 11]). In what follows, E, Y, Z are Banach spaces.

Definition 1.1. We call a functional $\Phi : E \to \mathbb{R}$ *compactly differentiable* at a point $y \in E$ if for every absolutely convex compact set $C \subset E$, the restriction of Φ onto $(y + span\, C)$ is Fréchet differentiable with respect to the norm $\|\cdot\|_C$ generated by C.

We call a K-differentiable functional Φ *twice K-differentiable* at $y \in E$ if for every absolutely convex compact sets C_1 and C_2, there exists a bilinear form $g_{C_1C_2}$ continuous on span $C_1 \times$ span C_2 and such that

$$(\Phi'_K(y+h) - \Phi'_K(y)) \cdot k = g_{C_1C_2}(y) \cdot (h,k) + o(\|h\|_{C_1} \cdot \|k\|_{C_2}).$$

Here Φ'_K and Φ''_K are the first and the second K-derivatives, respectively.

Definition 1.2. We say that a functional Ψ has a *strong compact extremum* (*strong K-extremum*) at a point $y \in E$ if for every absolutely convex compact set $C \subset E$, the restriction of Ψ onto $(y + \text{span } C)$ has a local extremum at y with respect to $\|\cdot\|_C$ in span C.

Remark 1.3. Note that any local extremum of Ψ at a point $y \in E$ is a strong K-extremum.

Definition 1.4. Let Ω be a compact finite Borel measure space. We say that a continuous mapping $\varphi : \Omega \times Y \times Z =: T \to F$ of class C^2 in (y,z) belongs to the class $W^2K_2(z)$ if a representation of φ in the form

$$\varphi(x,y,z) = P(x,y,z) + Q(x,y,z) \cdot \|z\| + R(x,y,z) \cdot \|z\|^2 \tag{1.1}$$

exists such that the mappings P, Q, R, the gradients $\nabla P := \nabla_{yz}P$, $\nabla Q := \nabla_{yz}Q$, $\nabla R := \nabla_{yz}R$, and the Hessians $H(P) := H_{yz}(P)$, $H(Q) := H_{yz}(Q)$, $H(R) := H_{yz}(R)$ are uniformly continuous and bounded on $T_C = \Omega \times C_Y \times Z$ for each compactum $C_Y \subset Y$.

Let us formulate a condition for the twice K-differentiability of a variational functional [4, 11].

Theorem 1.5. *Let $\Omega = [a;b]$, H be a real Hilbert space, and a function $u = f(x,y,z)$, $f : \Omega \times H^2 \to \mathbb{R}$. If $f \in W^2K_2(z)$, then the Euler–Lagrange functional*

$$\Phi(y) = \int_\Omega f(x, y(x), y'(x))dx \quad (y \in W_2^1(\Omega, H)) \tag{1.2}$$

is twice K-differentiable and

$$\Phi''_K(y)(h,k) = \int_\Omega \Big[\frac{\partial^2 f}{\partial y^2}(x,y,y')(h,k) + \frac{\partial^2 f}{\partial y \partial z}(x,y,y')((h',k) + (h,k')) + \frac{\partial^2 f}{\partial z^2}(x,y,y')(h',k')\Big]dx. \tag{1.3}$$

Note that in [4], [5], the classical Euler–Lagrange variational equation is extended to the case as follows:

Theorem 1.6. *If, under the hypothesis of Theorem* 1.5, $y(\cdot) \in \overset{\circ}{W}{}_2^2(\Omega, E)$, *then* $\Phi'_K(y) = 0$ *if and only if the Euler–Lagrange variational equation*

$$\frac{\partial f}{\partial y}(x, y, y') - \frac{d}{dx}\left(\frac{\partial f}{\partial z}(x, y, y')\right) = 0 \tag{1.4}$$

is fulfilled a.e. on Ω.

2. Positive definiteness test and non-negativeness test for the operator matrices and quadratic forms in products of n real Hilbert spaces

Let us first obtain a positive definiteness test for the operator matrix in a product of Hilbert spaces.

Let $H_1, \ldots, H_k$ be separable real Hilbert spaces, $H = H_1 \times \cdots \times H_k$, and operators $B_{ij} : H_j \to H_i$ $(i, j = \overline{1,k})$ be linear and continuous. Define a linear continuous operator $B_k : H \to H$ by the operator matrix $B_k = (B_{ij})_{i,j=1}^k$.

Definition 2.1. Consider the splitting of the matrix B_n into the following four blocks: B_n^{11}, the upper-left block of size $[\frac{n}{2}] \times [\frac{n}{2}]$; B_n^{22}, the lower-right block of size $(n - [\frac{n}{2}]) \times (n - [\frac{n}{2}])$; B_n^{12} and B_n^{21}, the adjacent rectangular blocks of sizes $[\frac{n}{2}] \times (n - [\frac{n}{2}])$ and $(n - [\frac{n}{2}]) \times [\frac{n}{2}]$ (here, $[\,\cdot\,]$ is the integral part of the number).

Assume that a necessary condition of the positive definiteness of a 2×2 operator matrix is fulfilled [12], that is, B_n^{11} and B_n^{22} are continuously invertible operators. On the set of all such matrices B_n $(n = 1, 2, \ldots)$, let us introduce *the first kind Silvester operators*:

$$\triangle_1^1(B_n) = B_n^{11}; \quad \triangle_1^2(B_n) = B_n^{11} - B_n^{12} \cdot (B_n^{22})^{-1} \cdot B_n^{21};$$
$$\triangle_2^2(B_n) = B_n^{22}; \quad \triangle_2^1(B_n) = B_n^{22} - B_n^{21} \cdot (B_n^{11})^{-1} \cdot B_n^{12}.$$

Note that the maximum sizes of the matrices $\triangle_j^i(B_n)$ $(i, j = 1, 2)$ are

$$\left(\left[\frac{n}{2}\right] + 1\right) \times \left(\left[\frac{n}{2}\right] + 1\right) \quad \text{and} \quad \left[\frac{n}{2}\right] \times \left[\frac{n}{2}\right]$$

for odd and even n, respectively.

In [5]–[6], the following positive definiteness test for an operator matrix $B_2 = (B_{ij})_{i,j=1}^2$ in $H_1 \times H_2$ was obtained.

Theorem 2.2. *The operator* B_2 *is positive definite if and only if*

$$B_{11} \gg 0; \quad \triangle_1^2(B_2) = B_{11} - B_{12} \cdot (B_{22})^{-1} \cdot B_{21} \gg 0;$$
$$B_{22} \gg 0; \quad \triangle_2^1(B_2) = B_{22} - B_{21} \cdot (B_{11})^{-1} \cdot B_{12} \gg 0.$$

Applying the first kind Silvester operators repeatedly, we can generalize this result by induction to $n \times n$ operator matrices.

Theorem 2.3. *Let H_i $(i=\overline{1,n})$ be separable real Hilbert spaces, $H = H_1 \times \cdots \times H_n$, and $B_n = (B_{ij})$ where $B_{ij} : H_j \to H_i$ $(i,j=\overline{1,n})$ are linear continuous operators in H. The operator B_n is positive definite if and only if there holds the system of inequalities*

$$\{\triangle^{i_m}_{j_m} \cdots \triangle^{i_2}_{j_2} \triangle^{i_1}_{j_1}(B_n) \gg 0\}^2_{i_l,j_l=1}, \tag{2.1}$$

where $\triangle^{i_l}_{j_l}$ are the first kind Silvester operators and

$$m = \begin{cases} k, & \text{for} \quad n = 2^k \\ k+1, & \text{for} \quad 2^k < n < 2^{k+1}. \end{cases} \tag{2.2}$$

Proof. Consider the splitting of the matrix $B_n = (B_{ij})_{i,j=\overline{1,n}}$ into four blocks according to Definition 2.1. Then $B_n = (B^{ij}_n)_{i,j=1,2}$. Denote $\tilde{H}^1 = H_1 \times \cdots \times H_{[\frac{n}{2}]}$ and $\tilde{H}^2 = H_{[\frac{n}{2}]+1} \times \cdots \times H_n$. Then B_n can be considered as an operator matrix in
$\tilde{H}^1 \times \tilde{H}^2$, $B^{ij}_n : \tilde{H}^j \to \tilde{H}^i$ $(i,j=1,2)$. By Theorem 2.2, B_n is positive definite if and only if:

$$1)\ B^{11}_n = \triangle^1_1(B_n) \gg 0; \quad B^{22}_n = \triangle^2_2(B_n) \gg 0;$$
$$2)\ \triangle^2_1 = \triangle^2_1(B_n) \gg 0; \quad \triangle^1_2 = \triangle^1_2(B_n) \gg 0.$$

Splitting each operator matrix

$$\triangle^1_1(B_n) = B^1_{[\frac{n}{2}]},\ \triangle^2_1(B_n) = B^2_{[\frac{n}{2}]},\ \triangle^2_2(B_n) = B^3_{n-[\frac{n}{2}]},\ \triangle^1_2(B_n) = B^4_{n-[\frac{n}{2}]} \tag{2.3}$$

according to Definition 2.1 and applying Theorem 2.2 to the linear continuous operators (2.3) in $\tilde{H}^1 \times \tilde{H}^1$ and $\tilde{H}^2 \times \tilde{H}^2$ respectively, we obtain 16 inequalities of the form $\{\triangle^{i_2}_{j_2}\triangle^{i_1}_{j_1} \gg 0\}^2_{i_l,j_l=1}$ in spaces

$$\left(H_1 \times \cdots \times H_{\left[\frac{[\frac{n}{2}]}{2}\right]}\right)^2, \quad \left(H_{\left[\frac{[\frac{n}{2}]}{2}\right]+1} \times \cdots \times H_{[\frac{n}{2}]}\right)^2,$$

$$\left(H_{[\frac{n}{2}]+1} \times \cdots \times H_{\left[\frac{n-[\frac{n}{2}]}{2}\right]}\right)^2, \quad \left(H_{\left[\frac{n-[\frac{n}{2}]}{2}\right]+1} \times \cdots \times H_n\right)^2.$$

After the construction is continued by induction p times, the size of the operator matrices is reduced, at least, to $2^{k+1-p} \times 2^{k+1-p}$ as $n < 2^{k+1}$.

Thus, we arrive at the system (2.1) after m steps, where m is defined by (2.2). □

Remark 2.4. Note that the number of inequalities in the system (2.1)–(2.2) can be calculated by the formula

$$V_n = 2^k \cdot (3n - 2^{k+1}) \quad \text{for} \quad 2^k \le n \le 2^{k+1}. \tag{2.4}$$

In addition,

$$n^2 \le V_n \le (n+1)^2. \tag{2.5}$$

Let us consider an application of these results to concrete operator matrices.

Example. Consider a diagonal operator $B_n = \{B_{ij}\}_{i,j=1}^n$ in a product of Hilbert spaces $H_1 \times \cdots \times H_n$, $B_{ij} = 0$ for $i \neq j$. Then

$$\triangle_1^1(B_n) = \triangle_1^2(B_n) = B_n^{11}, \quad \triangle_2^2(B_n) = \triangle_2^1(B_n) = B_n^{22}$$

are diagonal operators, too. Thus, we obtain by induction the diagonal elements

$$\triangle_{j_m}^{i_m} \cdots \triangle_{j_2}^{i_2} \triangle_{j_1}^{i_1}(B_n) = B_{ii}.$$

Hence, the condition for the positive definiteness of the diagonal elements $B_{ii} \gg 0$ is both necessary and sufficient for the positive definiteness of the operator matrix B_n.

Now, we proceed to a non-negativeness test for the $n \times n$ operator matrix in a product of Hilbert spaces.

Definition 2.5. Consider the splitting of B_n into the following four blocks: $\widetilde{B}_n^{11}$, the element B_{11} of the matrix B_n; $\widetilde{B}_n^{22}$, the lower-right block of determinant of size $(n-1) \times (n-1)$; $\widetilde{B}_n^{12}$ and $\widetilde{B}_n^{21}$, the adjacent row and column of sizes $1 \times (n-1)$ and $(n-1) \times 1$.

Assume that $\widetilde{B}_n^{11}$ is continuously invertible. On the set of all such matrices B_n $(n = 1, 2, \ldots)$, let us introduce *the second kind Silvester operators*

$$\widetilde{\triangle_1^1}(B_n) = \widetilde{B}_n^{11}; \quad \widetilde{\triangle_2^1}(B_n) = \widetilde{B}_n^{22} - \widetilde{B}_n^{21} \cdot (\widetilde{B}_n^{11})^{-1} \cdot \widetilde{B}_n^{12}.$$

In [8] (see also [14]), the following non-negativeness test for the operator matrix $B_2 = (B_{ij})_{i,j=1}^2$ in $H_1 \times H_2$ was obtained.

Theorem 2.6. *Let B_{11} be continuously invertible and self-adjoint, $B_{12} = B_{21}^*$. Then the operator B_2 is non-negative if and only if*

(1) $B_{11} \geq 0$,
(2) $\widetilde{\triangle_2^1}(B_2) = B_{22} - B_{21} \cdot (B_{11})^{-1} \cdot B_{12} \geq 0$.

Applying the Silvester operators of the second kind repeatedly, we can generalize the Theorem 2.2 by induction to $n \times n$ operator matrices.

Theorem 2.7. *Let H_i $(i = \overline{1,n})$ be separable real Hilbert spaces, $H = H_1 \times \cdots \times H_n$, $B_n = (B_{ij})$ where $B_{ij} : H_j \to H_i$ $(i, j = \overline{1,n})$ is a linear continuous self-adjoint operator in H, and all operators*

$$\widetilde{\triangle_1^1}(\widetilde{\triangle_2^1})^k(B_n), \qquad k = \overline{0, n-2} \tag{2.6}$$

be continuously invertible. The operator B_n is non-negative if and only if

$$\widetilde{\triangle_1^1}(\widetilde{\triangle_2^1})^k(B_n) \geq 0, \quad k = \overline{0, n-2}; \quad (\widetilde{\triangle_2^1})^{n-1}(B_n) \geq 0, \tag{2.7}$$

where $\widetilde{\triangle_l^1}$, $(l = 1, 2)$ are the second kind Silvester operators.

Proof. Consider the splitting of the matrix $B_n = (B_{ij})_{i,j=\overline{1,n}}$ into four blocks according to Definition 2.5. Then $B_n = (\widetilde{B}_n^{ij})_{i,j=1,2}$. Denote $\tilde{H}^1 = H_1$ and $\tilde{H}^2 = H_2 \times \cdots \times H_k$. Then B_n can be considered as an operator matrix in $\tilde{H}^1 \times \tilde{H}^2$, $\widetilde{B}_n^{ij} : \tilde{H}^j \to \tilde{H}^i$ $(i,j = 1,2)$. By virtue of Theorem 2.6, under the conditions of continuous invertibility of $\widetilde{B}_n^{11} = B_{11} = \widetilde{\triangle_1^1}(B_n)$ and self-adjointness of B_n, B_n is non-negative if and only if

$$\widetilde{\triangle_1^1}(B_n) \geq 0, \qquad \widetilde{\triangle_2^1}(B_n) = \tilde{B}_n^{22} - \tilde{B}_n^{21} \cdot (\tilde{B}_n^{11})^{-1} \cdot \tilde{B}_n^{12} \geq 0.$$

Splitting the operator matrix $\widetilde{\triangle_2^1}(B_n)$ of size $(n-1) \times (n-1)$ according to Definition 2.5 and applying Theorem 2.6 to the linear continuous self-adjoint operator $\widetilde{\triangle_2^1}(B_n)$, under the conditions of continuous invertibility of $\widetilde{\triangle_1^1}(\widetilde{\triangle_2^1}(B_n))$ in $\tilde{H}^2 \times \tilde{H}^2$, we obtain the following necessary and sufficient conditions for the non-negativeness of $\widetilde{\triangle_2^1}(B_n)$:

$$\widetilde{\triangle_1^1}(\widetilde{\triangle_2^1}(B_n)) \geq 0, \quad \widetilde{\triangle_2^1}(\widetilde{\triangle_2^1}(B_n)) \geq 0.$$

After the construction is continued by induction p times, the size of the operator matrices is reduced, at least, to $(n-p) \times (n-p)$. Thus, under the conditions of continuous invertibility of the operators (2.6) and self-adjointness of B_n, we obtain, after $(n-1)$ steps, n inequalities of the form (2.7). □

On the basis of the above-proved tests for the positive definiteness and non-negativeness of operator matrices, we shall obtain appropriate sufficient conditions for the positive definiteness and necessary conditions for the non-negativeness of the quadratic forms in products of n real separable Hilbert spaces.

Theorem 2.8. *Let H_i $(i = \overline{1,n})$ be separable real Hilbert spaces, $H = H_1 \times \cdots \times H_n$, φ be a continuous quadratic form in H generated by a symmetric bilinear form g on H^2, B_n be a linear continuous self-adjoint operator in H associated with g $(g(h,k) = \langle k, B_n h\rangle)$, and $B_n = (B_{ij})$ where $B_{ij} : H_j \to H_i$ $(i,j = \overline{1,n})$. If the system of inequalities* (2.1)–(2.2) *holds, then $\varphi \gg 0$ on H.*

Theorem 2.9. *Let H_i $(i = \overline{1,n})$ be separable real Hilbert spaces, $H = H_1 \times \cdots \times H_n$, φ be a continuous quadratic form on H generated by a symmetric bilinear form g on H^2, B_n be a linear continuous self-adjoint operator in H associated with g $(g(h,k) = \langle k, B_n h\rangle)$, and $B_n = (B_{ij})$ where $B_{ij} : H_j \to H_i$ $(i,j = \overline{1,n})$. If $\varphi \geq 0$ and all operators* (2.6) *are continuously invertible, then the inequalities* (2.7) *hold.*

3. Sufficient and necessary conditions for the strong compact extrema of the Euler–Lagrange functionals of n variables in Sobolev space in terms of the Hessian of the integrand

Now, we use the results proved above to obtain sufficient and necessary conditions for a strong compact extremum of the Euler–Lagrange functionals of n variables in Sobolev space W_2^1.

In [4], the following result was obtained.

Theorem 3.1. *Let $\Omega = [a;b]$, H be a real Hilbert space, and a function $u=f(x,y,z)$, $f : \Omega \times H^2 \to \mathbb{R}$ be of $W^2K_2(z)$. If the equation (1.4) is fulfilled at a point $y(\cdot) \in \overset{\circ}{W}{}_2^2(\Omega, H)$ and the quadratic form (in (y,z)) $f''(x, y(x), y'(x))$ is positive definite on H^2 for any fixed $x \in \Omega$, then the Euler–Lagrange functional (1.2) has a strong K-minimum at $y(\cdot)$.*

A sufficient condition for the strong K-extremum follows from this result and the theorems 1.5, 1.6, 2.7.

Theorem 3.2. *Let $\Omega = [a;b]$, H_i $(i = \overline{1,n})$ be separable real Hilbert spaces, $H = H_1 \times \cdots \times H_n$, and a real function $u = f(x, y_1, \ldots, y_n, z_1, \ldots, z_n)$ be of $W^2K_2(z)$, $z = (z_1 \times \cdots \times z_n)$. Suppose that for some functions $y_m(\cdot) \in \overset{\circ}{W}{}_2^2(\Omega, H_m)$, $m = \overline{1,n}$, the variational Euler–Lagrange equations*

$$\frac{\partial f}{\partial y_m} - \frac{d}{dx}\left(\frac{\partial f}{\partial z_m}\right) \overset{a.e.}{=} 0 \quad (m = \overline{1,n}) \tag{3.1}$$

are fulfilled.

Let $\Gamma_n(f)$ be the Hessian of f in the variables y_i and z_i, $(i = \overline{1,n})$

$$\Gamma_n(f) = \begin{pmatrix} \left(\frac{\partial^2 f}{\partial y_i \partial y_j}\right) & \left(\frac{\partial^2 f}{\partial y_i \partial z_j}\right) \\ \left(\frac{\partial^2 f}{\partial z_i \partial y_j}\right) & \left(\frac{\partial^2 f}{\partial z_i \partial z_j}\right) \end{pmatrix}_{i,j=1}^{n}.$$

If the system of inequalities

$$\{\triangle_{j_m}^{i_m} \cdots \triangle_{j_2}^{i_2} \triangle_{j_1}^{i_1}(\Gamma_n(f)) \gg 0\}_{i_l,j_l=1}^{2}, \tag{3.2}$$

where $\triangle_{j_l}^{i_l}$ are the first kind Silvester operators, $m = \begin{cases} k, & \text{for} \quad n = 2^k \\ k+1, & \text{for} \quad 2^k < n < 2^{k+1}, \end{cases}$ is fulfilled for a K-extremal $y(\cdot) = (y_1(\cdot), \ldots, y_n(\cdot))$ a.e. on Ω, then the Euler–Lagrange functional

$$\Phi(y_1, \ldots, y_n) = \int_a^b f(x, y_1(x), \ldots, y_n(x), y_1'(x), \ldots, y_n'(x))\,dx \tag{3.3}$$

has a strong K-minimum at a point $y(\cdot) = (y_1(\cdot), \ldots, y_n(\cdot))$.

Proof. As is shown in Theorem 2.8, the positive definiteness of $f''(x, y_1(x), \ldots, y_n(x), y_1'(x), \ldots, y_n'(x))$ for all $x \in \Omega$ follows from the condition (3.2). As the system of variational Euler-Lagrange equations (3.1) is equivalent to the equation (1.4), then, by virtue of Theorem 1.6, $y(\cdot) = (y_1(\cdot), \dots, y_n(\cdot))$ is a K–extremal.

In view of Theorem 1.5, under the conditions $f \in C^2(H, \mathbb{R})$ and $f \in W^2K_2(z)$, it follows that $\Phi \in C^2_K(W^1_2(\Omega, H), \mathbb{R})$.

Since $f''\big(x, y_1(x), \dots, y_n(x), y'_1(x), \dots, y'_n(x)\big) \gg 0$ for the K-extremal $y(\cdot) = (y_1(\cdot), \dots, y_n(\cdot))$, it follows from Theorem 3.1 that Φ has a strong K-minimum at the point $y(\cdot) = (y_1(\cdot), \dots, y_n(\cdot))$. □

Corollary 3.3. *If, under the hypothesis of Theorem* 3.2, *the system of inequalities*

$$\{\triangle^{i_m}_{j_m} \cdots \triangle^{i_2}_{j_2} \triangle^{i_1}_{j_1}(\Gamma_n(x)) \ll 0\}^2_{i_l, j_l = 1}$$

is fulfilled for the K-extremal $y(\cdot) = (y_1(\cdot), \dots, y_n(\cdot))$ a.e. on Ω, then the Euler–Lagrange functional (3.3) *has a strong K-maximum at the point $y(\cdot) =$*

$$(y_1(\cdot), \dots, y_n(\cdot)).$$

Now, we obtain, on the basis of Theorem 2.7, a necessary condition for a strong compact extrema of the Euler–Lagrange functional in terms of the Hessian of the integrand.

Theorem 3.4. *If, under the hypothesis of Theorem* 3.2, *the Euler–Lagrange functional* (3.3) *has a strong K-minimum at a point*

$$y(\cdot) = (y_1(\cdot), \dots, y_n(\cdot)) \in \overset{\circ}{W}{}^2_2(\Omega, H),$$

and all operators $\widetilde{\triangle^1_1}(\widetilde{\triangle^1_2})^k(\Gamma_n(f))$, $k = \overline{0, n-2}$ are continuously invertible, then

$$\widetilde{\triangle^1_1}(\widetilde{\triangle^1_2})^k(\Gamma_n(f)) \geq 0, \ (k = \overline{0, n-2}), \quad (\widetilde{\triangle^1_2})^{n-1}(\Gamma_n(f)) \geq 0, \tag{3.4}$$

where $\widetilde{\triangle^1_l}$, $(l = 1, 2)$ are the second kind Silvester operators.

Proof. Since the function $f(x, y_1, \dots, y_n, z_1, \dots, z_n)$ is twice continuously differentiable in (y, z) on $\Omega \times H \times H$ and $f \in W^2K_2(z)$, by Theorem 1.5 the Euler–Lagrange functional (3.3) is twice K-differentiable on $\overset{\circ}{W}{}^1_2(\Omega, H)$. Moreover, the functional (3.3) has a strong K-minimum, whence

$$\Phi''_K(y_1, \dots, y_n)(h, h) \geq 0. \tag{3.5}$$

Suppose that $\exists k_0^2 > 0$ for which on some set $A_0 \subset \Omega$, $\mu A_0 > 0$ and for some fixed $h_0 \in H$,

$$f''(x, y_1(x), \dots, y_n(x), z_1(x), \dots, z_n(x))(h_0, h_0) \leq -k_0^2 < 0.$$

Now, let x_0 be any density point in A_0. Choose a neighborhood $O_{\delta_0}(x_0)$ $(\delta_0 > 0)$ such that for $\delta < \delta_0$,

$$\frac{\mu(A_0 \bigcap O_\delta(x_0))}{2\delta} > 1 - \varepsilon_0 \quad (0 < \varepsilon_0 < 1).$$

Put $h(x) = h_0 \cdot \chi_{(x_0 - \delta; x_0 + \delta)}(x) = \begin{cases} h_0, & x \in (x_0 - \delta; x_0 + \delta); \\ 0, & x \notin (x_0 - \delta; x_0 + \delta). \end{cases}$

Then

$$\begin{aligned}
&\Phi''_K(y_1,\dots,y_n)(h,h)\\
&= \int\limits_{\Omega} f''(x,y_1(x),\dots,y_n(x),z_1(x),\dots,z_n(x))(h(x),h(x))dx\\
&= \int_{x_0-\delta}^{x_0+\delta} f''(x,y_1(x),\dots,y_n(x),z_1(x),\dots,z_n(x))(h_0,h_0)dx\\
&\le -k_0^2\cdot 2\delta\cdot(1-\varepsilon_0)<0
\end{aligned}$$

for $0<\delta<\delta_0$ small enough. Hence $\Phi''_K(y)<0$, which contradicts the inequality (3.5). Thus, if $\Phi''_K(y_1,\dots,y_n)\geq 0$, then $f''(x,y_1(x),\ \dots,\ y_n(x),\ z_1(x),\ \dots,\ z_n(x))\ \geq\ 0$ for the K-extremal $y(\cdot)$. As was shown in Theorem 2.9, the last condition provides the fulfilment of the inequalities (3.4) for the Hessian $\Gamma_n(x)$ associated with $f''(x,y_1(x),\dots,y_n(x),z_1(x),\dots,z_n(x))$. □

Corollary 3.5. *If, under the hypothesis of Theorem* 3.4, *the Euler–Lagrange functional* (3.3) *has a strong K-maximum at a point $y(\cdot)=(y_1(\cdot),\dots,y_n(\cdot))$, then*

$$\widetilde{\triangle_1^1}(\widetilde{\triangle_2^1})^k(\Gamma_n(f))\le 0,\ (k=\overline{0,n-2});\quad (\widetilde{\triangle_2^1})^{n-1}(\Gamma_n(f))\le 0,$$

where $\widetilde{\triangle_l^1}$, $(l=1,2)$ are the second kind Silvester operators.

4. The Legendre condition and the Legendre-Jacobi sufficient condition for a strong K-extremum of the Euler-Lagrange functional of n variables in Sobolev space

Let us extend, to the case of Sobolev space W_2^1 and a strong compact extremum of the Euler–Lagrange functional of n variables, the known Legendre necessary condition and the Legendre–Jacobi sufficient condition [13].

First, we formulate the Legendre necessary condition.

Theorem 4.1. *Let $\Omega=[a;b]$, H_i $(i=\overline{1,n})$ be separable real Hilbert spaces, $H=H_1\times\cdots\times H_n$, a real function $u=f(x,y_1,\dots,y_n,z_1,\dots,z_n)$ be of $W^2K_2(z)$, and all functions $\frac{\partial^2 f}{\partial y_i\partial z_j}(x,y_1,\dots,y_n,y'_1,\dots,y'_n)$ $(i,j=\overline{1,n})$ be absolutely continuous on Ω, where the functions $y_m(\cdot)\in \overset{\circ}{W}{}_2^2(\Omega,H)$ satisfy the Euler-Lagrange variational equations*

$$\frac{\partial f}{\partial y_m}-\frac{d}{dx}\left(\frac{\partial f}{\partial z_m}\right)\overset{\text{a.e.}}{=}0\quad (m=\overline{1,n}).$$

Set $P_n(x)=\left(\frac{\partial^2 f}{\partial z_i\partial z_j}\right)_{i,j=1}^n$. If the Euler-Lagrange functional (3.3) *has a strong K-minimum at $y(\cdot)=(y_1(\cdot),\dots,y_n(\cdot))\in\overset{\circ}{W}{}_2^2(\Omega,H)$, and all operators $\widetilde{\triangle_1^1}(\widetilde{\triangle_2^1})^k(P_n)$, $k=\overline{0,n-2}$ are continuously invertible, then the system of differential inequalities*

$$\widetilde{\triangle_1^1}(\widetilde{\triangle_2^1})^k(P_n)\ge 0,\quad k=\overline{0,n-2};\quad (\widetilde{\triangle_2^1})^{n-1}(B_n)\ge 0,\tag{4.1}$$

where $\widetilde{\triangle_l^1}$ $(l = 1,2)$ *are the second kind Silvester operators, holds true for the extremal* $y(\cdot)$ *a.e. on* Ω.

Proof. Since the functional (3.3) has a strong K-minimum at the point $y(\cdot) = (y_1(\cdot), \dots, y_n(\cdot))$, the Legendre necessary condition ([10], Theorem 1.1)

$$\frac{\partial^2 f}{\partial z^2}(x, y_1(x), \dots, y_n(x), y_1'(x), \dots, y_n'(x)) \geq 0,$$

where $\frac{\partial^2 f}{\partial z^2}(x, y_1(x), \dots, y_n(x), y_1'(x), \dots, y_n'(x))$ is the symmetric bilinear form associated with $P_n(x)$, holds true for a.e. $x \in \Omega$.

Then, by Theorem 2.9, under the conditions of continuous invertibility of $\widetilde{\triangle_1^1}(\widetilde{\triangle_2^1})^k(P_n)$, $k = \overline{0, n-2}$, the system of inequalities (4.1) is satisfied. □

Note that the representation (1.3) can be rewritten in the following form:

$$\Phi_K''(y)(h,k) = \int_\Omega \left(\left[-\frac{d}{dx} \left(\frac{\partial^2 f}{\partial y \partial z}(x, y, y') \right) + \frac{\partial^2 f}{\partial y^2}(x, y, y') \right] (h, k) + \frac{\partial^2 f}{\partial z^2}(x, y, y')(h', k') \right) dx. \tag{4.2}$$

For the Euler–Lagrange functional (1.2), the second K-derivative is the quadratic functional (1.3), (4.2). To extend the Legendre-Jacobi sufficient condition to the case of a compact extremum of a variational functional in Sobolev space, consider some properties of the quadratic integral functional

$$\widehat{\Phi}(h) = \int_\Omega (P(h', h') + Q(h, h))\, dx, \tag{4.3}$$

where $h \in \overset{\circ}{W}{}_2^1(\Omega, H)$, $H = H_1 \times \dots \times H_n$, H_i $(i = \overline{1,n})$ are real separable Hilbert spaces, and for each $x \in \Omega = [a; b]$, $P(x)$ and $Q(x)$ are symmetric bilinear continuous forms on $H \times H$, strongly measurable and essentially bounded with respect to x as mappings from Ω into $(H, H)^* \cong (H, H^*) = (H, H)$. Note that the forms $P(x)$ and $Q(x)$ can be considered as self-adjoint linear continuous operators $P = (P_{ij})$, $Q = (Q_{ij})$ where $P_{ij} : H_j \to H_i$, $Q_{ij} : H_j \to H_i$, $(i, j = \overline{1,n})$.

Define the Jacobi condition for the functional $\widehat{\Phi}(h)$.

Definition 4.2. Consider the Jacobi equation

$$-\frac{d}{dx}(PU') + QU \overset{\text{a.e.}}{=} 0, \quad U(a) = 0, \quad U'(a) = I_H, \tag{4.4}$$

in the class of mappings $U(\cdot) \in W_2^1(\Omega, \mathcal{L}(H))$. We say that the functional (4.3) satisfies the Jacobi condition if any solution of the Jacobi equation (4.4) is such that the operators $U(x)$ are continuously invertible for $a < x \leq b$.

The Jacobi condition can be rewritten in the coordinate form.

Definition 4.3. Consider the linear continuous operator $U = (U_{ij})$ where

$$U_{ij} : H_j \to H_i, \quad (i, j = \overline{1,n}).$$

Then it is possible to consider the system of Jacobi equations

$$\sum_{i=1}^{n} \left(-\frac{d}{dx}(P_{ki}U'_{im}) + Q_{ki}U_{im} \right) \overset{\text{a.e.}}{=} 0,$$

$$U_{km}(a) = 0, \;\; U'_{km}(a) = \begin{cases} I_{H_k}, & k = m \\ 0, & k \neq m \end{cases}, \;\; (k, m = \overline{1,n}) \tag{4.5}$$

in the class of mappings $U(\cdot) \in W_2^1(\Omega, \mathcal{L}(H))$. We say that the functional (4.3) satisfies the Jacobi condition if any solution of the system of Jacobi equations (4.5) is such that the operators $U(x)$ are continuously invertible for $a < x \leq b$.

Theorem 4.4. *If the system of inequalities*

$$\{\triangle_{j_m}^{i_m} \cdots \triangle_{j_2}^{i_2} \triangle_{j_1}^{i_1}(P) \gg 0\}_{i_l, j_l = 1}^{2}, \tag{4.6}$$

where $\triangle_{j_l}^{i_l}$ *are the first kind Silvester operators and*

$$m = \begin{cases} k, & \text{for} \quad n = 2^k \\ k+1, & \text{for} \quad 2^k < n < 2^{k+1}, \end{cases} \tag{4.7}$$

holds true for a.e. $x \in \Omega$ *and the Jacobi condition (Definitions 4.2–4.3) is fulfilled, then the quadratic functional (4.3) is positive definite on* $\overset{\circ}{W}{}_2^1(\Omega, H)$.

Proof. If the system of inequalities (4.6)–(4.7) is satisfied, then, by Theorem 2.8, the bilinear continuous form $P(x) \gg 0$ for a.e. $x \in \Omega$. Hence, by Theorem 2.1 [10], $\widehat{\Phi}(h) \gg 0$ on $\overset{\circ}{W}{}_2^1(\Omega, H)$. □

Applying the sufficient condition for the positive definiteness of the quadratic functional (4.3) (Theorem 4.4), we obtain the Legendre–Jacobi sufficient condition for a compact extremum of the Euler–Lagrange functional of several variables in Sobolev space.

Theorem 4.5. *Suppose that the assumptions of Theorem 4.1 are fulfilled. Set*

$$P_n(x) = \left(\frac{\partial^2 f}{\partial z_i \partial z_j} \right)_{i,j=1}^{n}, \quad Q_n(x) = \left(\frac{d}{dx}\left(\frac{\partial^2 f}{\partial y_i \partial z_j} \right) + \frac{\partial^2 f}{\partial y_i \partial y_j} \right)_{i,j=1}^{n}.$$

If for a K*-extremal* $y(\cdot) = (y_1(\cdot), \ldots, y_n(\cdot)) \in \overset{\circ}{W}{}_2^2(\Omega, H)$*:*

1) *The system of inequalities*

$$\{\triangle_{j_m}^{i_m} \cdots \triangle_{j_2}^{i_2} \triangle_{j_1}^{i_1}(P_n) \gg 0\}_{i_l, j_l = 1}^{2}, \tag{4.8}$$

where $\triangle_{j_l}^{i_l}$ *are the first kind Silvester operators and*

$$m = \begin{cases} k, & for \quad n = 2^k \\ k+1, & for \quad 2^k < n < 2^{k+1}, \end{cases} \tag{4.9}$$

is fulfilled for a.e. $x \in \Omega$*;*

2) *For the functional* (4.2), *the Jacobi condition is fulfilled, i.e., any solution of the Jacobi equation*

$$-\frac{d}{dx}\left(\frac{\partial^2 f}{\partial z^2}U'\right) + \left[-\frac{d}{dx}\left(\frac{\partial^2 f}{\partial y \partial z}\right) + \frac{\partial^2 f}{\partial y^2}\right] U \overset{a.e.}{=} 0,$$

$$U(a) = 0, \quad U'(a) = I_H,$$

(*where* $\frac{\partial^2 f}{\partial z^2}$ *and* $-\frac{d}{dx}\left(\frac{\partial^2 f}{\partial y \partial z}\right) + \frac{\partial^2 f}{\partial y^2}$ *are the bilinear continuous forms associated with* $P_n(x)$ *and* $Q_n(x)$*, respectively) in the class* $W_2^1(\Omega, \mathcal{L}(H))$ *is such that the operators* $U(x)$ *are continuously invertible for* $a < x \leq b$*,*

then the Euler-Lagrange functional (3.3) *has a strong* K*-minimum at* $y(\cdot)$*.*

Proof. Since the system of inequalities (4.8)–(4.9) for the operator $P_n(x)$, associated with the form $\frac{\partial^2 f}{\partial z^2}$, is fulfilled, $\frac{\partial^2 f}{\partial z^2} \gg 0$ almost everywhere on Ω (Theorem 2.8). Under the Jacobi condition, the Euler-Lagrange functional (3.3) has a strong K–minimum at $y(\cdot)$ ([10], Theorem 3.1). □

Corollary 4.6. *If, under the hypothesis of Theorem* 4.5*,*

1) *The system of inequalities*

$$\{\triangle_{j_m}^{i_m} \cdots \triangle_{j_2}^{i_2} \triangle_{j_1}^{i_1}(P_n) \ll 0\}_{i_l, j_l = 1}^{2}, \tag{4.10}$$

where $\triangle_{j_l}^{i_l}$ *are the first kind Silvester operators and*

$$m = \begin{cases} k, & for \quad n = 2^k \\ k+1, & for \quad 2^k < n < 2^{k+1}, \end{cases} \tag{4.11}$$

is fulfilled for a.e. $x \in \Omega$*;*

2) *For the functional* (4.2), *the Jacobi condition is fulfilled,*

then the Euler-Lagrange functional (3.3) *has a strong* K*-maximum at* $y(\cdot)$*.*

Remark 4.7. According to Definition 4.3, the Jacobi condition for the functional (4.2) can be rewritten in the coordinate form, i.e., any solution $U(x)$, where $U = (U_{ij})$, $U_{ij} : H_j \to H_i$, $(i, j = \overline{1, n})$, of the system of Jacobi equations

$$\sum_{j=1}^{n}\left(-\frac{d}{dx}\left(\frac{\partial^2 f}{\partial z_i \partial z_j}U'_{jm}\right) + \left[-\frac{d}{dx}\left(\frac{\partial^2 f}{\partial y_i \partial z_j}\right) + \frac{\partial^2 f}{\partial y_i \partial y_j}\right] U_{jm}\right) \overset{\text{a.e.}}{=} 0, \tag{4.12}$$

$$U_{im}(a) = 0, \quad U'_{im}(a) = \begin{cases} I_{H_i}, & i = m \\ 0, & i \neq m \end{cases}, \quad (i, m = \overline{1, n}), \tag{4.13}$$

in the class $W_2^1(\Omega, \mathcal{L}(H))$ is such that the operators $U(x)$ are continuously invertible for $a < x \leq b$.

5. Example of a K-extremum of the Euler–Lagrange functional of n variables that is not a local extremum

1) Let us consider a real function $\varphi \in C^2[0,+\infty)$ such that $\varphi(t) = t$ for $0 \le t \le 1-\delta$, $\varphi(t) = 2-t$ for $1+\delta \le t \le +\infty$, $\varphi \nearrow$ as $0 \le t \le 1$, and $\varphi \searrow$ as $1 \le t < \infty$ ($\delta > 0$ is small enough). Assume

$$f(x, y_1, \ldots, y_n, z_1, \ldots, z_n) = \varphi\left(\sum_{i=1}^{n} \|y_i\|_i^2 + \|z_i\|_i^2\right),$$

where $y_i, z_i \in H_i$, $H = H_1 \times \ldots \times H_n$, and H_i are the separable real Hilbert spaces with norms $\|\cdot\|_i$ generated by inner products $\langle\cdot,\cdot\rangle_i$ $(i = \overline{1,n})$. Consider the functional

$$\Phi(y_1, \ldots, y_n) = \int_0^1 \varphi\left(\sum_{i=1}^{n} \|y_i(x)\|_i^2 + \|y_i'(x)\|_i^2\right) dx,$$

where $y(\cdot) \in \overset{\circ}{W}{}_2^1([0;1], H)$, $(i = \overline{1,n})$.

Then the variational Euler–Lagrange equation (1.4) takes the form

$$\frac{d\varphi}{dt}\langle y_i - y_i'', \cdot\rangle_i - 2\frac{d^2\varphi}{dt^2}\langle y_i + y_i'', y_i'\rangle_i \cdot \langle y_i', \cdot\rangle_i = 0 \quad (i = \overline{1,n}). \tag{5.1}$$

Thus, the function $y_0(x) = (y_{01}(x), \ldots, y_{0n}(x)) \equiv 0$ is a K-extremal and

$$\Phi(y_{01}, \ldots, y_{0n}) = 0.$$

2) Direct verification shows that Φ has a compact minimum at the point $y_0(x) \equiv 0$, but no local minimum at this point [7].

3) Let us show that the sufficient condition for a K-minimum (Theorem 3.2) is fulfilled on the K-extremal $y_0 \equiv 0$.

Consider

$$\Gamma_n(f) = \begin{pmatrix} \frac{\partial^2 f}{\partial y_i \partial y_j} & \frac{\partial^2 f}{\partial y_i \partial z_j} \\ \frac{\partial^2 f}{\partial z_i \partial y_j} & \frac{\partial^2 f}{\partial z_i \partial z_j} \end{pmatrix}_{i,j=1}^{n}.$$

Direct verification shows that

$$\frac{\partial^2 f}{\partial y_i^2} = 4\frac{d^2\varphi}{dt^2}\langle y_i, \cdot\rangle_i \cdot \langle y_i, \cdot\rangle_i + 2\frac{d\varphi}{dt}\langle\cdot,\cdot\rangle_i; \qquad \frac{\partial^2 f}{\partial y_i \partial y_j} = 4\frac{d^2\varphi}{dt^2}\langle y_i, \cdot\rangle_i \cdot \langle y_j, \cdot\rangle_j;$$

$$\frac{\partial^2 f}{\partial z_i^2} = 4\frac{d^2\varphi}{dt^2}\langle z_i, \cdot\rangle_i \cdot \langle z_i, \cdot\rangle_i + 2\frac{d\varphi}{dt}\langle\cdot,\cdot\rangle_i; \qquad \frac{\partial^2 f}{\partial z_i \partial z_j} = 4\frac{d^2\varphi}{dt^2}\langle z_i, \cdot\rangle_i \cdot \langle z_j, \cdot\rangle_j;$$

$$\frac{\partial^2 f}{\partial z_i \partial y_j} = 4\frac{d^2\varphi}{dt^2}\langle z_i, \cdot\rangle_i \cdot \langle y_j, \cdot\rangle_j; \qquad \frac{\partial^2 f}{\partial y_i \partial z_j} = 4\frac{d^2\varphi}{dt^2}\langle y_i, \cdot\rangle_i \cdot \langle z_j, \cdot\rangle_j;$$

$$(i, j = \overline{1,n})$$

Then

$$\Gamma_n(x) = 4\frac{d^2\varphi}{dt^2}\begin{pmatrix} \Gamma_n^{11} & \Gamma_n^{12} \\ \Gamma_n^{21} & \Gamma_n^{22} \end{pmatrix} + 2\frac{d\varphi}{dt}\begin{pmatrix} D_1 & 0 \\ 0 & D_1 \end{pmatrix},$$

where

$$\Gamma_n^{11} = \left(\langle y_i, \cdot\rangle_i \cdot \langle y_j, \cdot\rangle_j\right)_{i,j=1}^n, \quad \Gamma_n^{12} = \left(\langle y_i, \cdot\rangle_i \cdot \langle z_j, \cdot\rangle_j\right)_{i,j=1}^n,$$
$$\Gamma_n^{21} = \left(\langle z_i, \cdot\rangle_i \cdot \langle y_j, \cdot\rangle_j\right)_{i,j=1}^n, \quad \Gamma_n^{22} = \left(\langle z_i, \cdot\rangle_i \cdot \langle z_j, \cdot\rangle_j\right)_{i,j=1}^n,$$
$$D_1 = \left(\delta_{ij}\langle \cdot, \cdot\rangle_i\right)_{i,j=1}^n.$$

Since $\langle \cdot, \cdot\rangle \gg 0$, $i = \overline{1,n}$, by Example in Section 2 we obtain

$$\Gamma_n(x) = 2\begin{pmatrix} D_1 & 0 \\ 0 & D_1 \end{pmatrix} \gg 0$$

for the extremal $(y_{01}, \dots, y_{0n}) \equiv 0$.

5) Let us show that the Legendre–Jacobi sufficient condition for a K–minimum (Theorem 4.5) is fulfilled on the K–extremal $y_0 \equiv 0$.

a) For the K-extremal $y_0 \equiv 0$, $P_n = 2\left(\delta_{ij}\langle \cdot, \cdot\rangle_i\right)_{i,j=1}^n$. Since $\langle \cdot, \cdot\rangle \gg 0$, $i = \overline{1,n}$, by Example in Section 2 we obtain $P_n(x) \gg 0$.

b) The system of equations (4.12)–(4.13) can be rewritten in the form

$$-U_{ij}'' + U_{ij} \overset{\text{a.e.}}{=} 0,$$

$$U_{ij}(0) = 0, \quad U_{ij}'(0) = \begin{cases} I_{H_i}, & i = j \\ 0, & i \neq j \end{cases}, \quad (i, j = \overline{1,n}).$$

The solution of this system is $U(x) = \left(\delta_{ij} sh(xI_{H_i})\right)_{i,j=1}^n$, where

$$sh(xI_{H_i}) := \frac{e^{xI_{H_i}} - e^{xI_{H_i}}}{2}.$$

As the operators $sh(xI_{H_i})$ $(i = \overline{1,n})$ are continuously invertible for $0 < x \leq 1$, the operator $U(x)$ is continuously invertible for $0 < x \leq 1$.

References

[1] M.M. Vainberg, *Variational method and method of monotone operators.* Nauka, Moskow, 1972. (In Russian)

[2] I.V. Skrypnik, *Nonlinear elliptic high order equations.* Naukova Dumka, Kiev, 1973. (In Russian)

[3] B. Ricceri, *Integral functionals on Sobolev spaces having multiple local minima.* arXiv:math.OC/0402445, **1** (2004).

[4] I.V. Orlov, *K-differentiability and K-extrema.* Ukrainian Mathematical Bulletin **3** (2006), no. 1, 97–115. (In Russian)

[5] I.V. Orlov, *Extremum Problems and Scales of the Operator Spaces.* North-Holland Math. Studies. Funct. Anal. & Appl., Elsevier, Amsterdam–Boston–..., **197** (2004), 209–228.

[6] I.V. Orlov, *Sufficient extremum and -extremum conditions in product of two nuclear locally convex spaces.* Scientific Notes of Taurida National University, **17(56)** (2004), 68–77. (In Russian)

[7] E.V. Bozhonok, *The sufficient conditions for an extremum of integral functionals in product of nuclear spaces.* Dynamic Systems **19** (2005), 100–117. (In Russian)

[8] E.V. Bozhonok, *Sufficient and necessary functional extremum conditions in nuclear locally convex space (case of several variables).* Scientific Notes of Taurida National University **18(57)** (2005), no. 1, 3–26. (In Russian)

[9] K.V. Bozhonok, I.V. Orlov, *Legendre–Jacobi conditions for compact extrema of integral functionals.* Reports NAS of Ukraine, (2006), no. 11, 11–15. (In Ukrainian)

[10] E.V. Bozhonok, I.V. Orlov, *Legendre and Jacobi conditions for compact extrema of variation functionals in Sobolev spaces.* Proceedings of Institute Math. NAS of Ukraine **3** (2006), no.1, 97–115. (In Russian)

[11] I.V. Orlov, E.V. Bozhonok, *Well-definiteness, K-continuity and K-differentiability conditions for the Euler–Lagrange functional in Sobolev space W_2^1.* Scientific Notes of Taurida National University **19(58)** (2006), no. 2, 121–136. (In Russian)

[12] N.D. Kopachevskii, S.G. Krein, Ngo Zuy Kan, *Operator methods in linear hydrodynamics.* Nauka, Moscow, 1989. (In Russian)

[13] I.M. Gelfand, S.V. Fomin, *Calculus of variations.* Fizmatgiz, Moscow, 1961. (In Russian)

[14] S. Hassi, M. Malamud, H. De Snoo, *On Krein's extension theory of nonnegative operators.* Math. Nachr **274–275** (2004), 40–73.

E.V. Bozhonok
Taurida National University
4 V. Vernadsky Ave.
95007 Simferopol, Ukraine
e-mail: `katboz@tnu.crimea.ua`

Operator Theory:
Advances and Applications, Vol. 190, 157–169

The Moment Problem for Rational Measures: Convexity in the Spirit of Krein

Christopher I. Byrnes and Anders Lindquist

In memory of Mark Grigoryevich Krein on the occasion of the 100th anniversary of his birth

Abstract. The moment problem as formulated by Krein and Nudel'man is a beautiful generalization of several important classical moment problems, including the power moment problem, the trigonometric moment problem and the moment problem arising in Nevanlinna-Pick interpolation. Motivated by classical applications and examples, in both finite and infinite dimensions, we recently formulated a new version of this problem that we call the moment problem for positive rational measures. The formulation reflects the importance of rational functions in signals, systems and control. While this version of the problem is decidedly nonlinear, the basic tools still rely on convexity. In particular, we present a solution to this problem in terms of a nonlinear convex optimization problem that generalizes the maximum entropy approach used in several classical special cases.

Mathematics Subject Classification (2000). Primary 30E05; Secondary 44A60.

Keywords. Moment problems, interpolation, rational positive measures, convex optimization.

1. Introduction

The moment problem for positive measures is the synthesis, over the course of more than 70 years by Krein and his collaborators (see [1, 15] and references therein), of many important classical problems in pure and applied mathematics. This paper is devoted to the study of a class of moment problems, which we refer to as the moment problem for positive *rational* measures, whose formulation reflects the importance of rational functions in signals, systems and control. This class of problems abstracts the recent work of a number of authors [3, 4, 5, 6, 8, 9, 11, 12,

This research was supported in part by grants from AFOSR, Swedish Research Council, Swedish Foundation for Strategic Research, and the Göran Gustafsson Foundation.

13] who incorporated various complexity constraints into the refinements of the moment problem for arbitrary positive measures [15].

We refer to this problem as the moment problem for positive *rational* measures. In this paper we develop some basic results for this problem, closely following the approach outlined in [15]. Indeed, in Section 2 we recall the fundamental result on the generalized moment problem as derived in [15] using convex cones in finite-dimensional function spaces and properties of positive measures. In Section 3 we derive similar basic results for the moment problem for positive rational measures using a topological proof that mirrors the steps in the convexity proof in [15].

While this version of the problem is decidedly nonlinear, one can still develop an approach based entirely on convexity. In particular, in Section 4 we present a synthesis of our topological approach with a nonlinear convex optimization problem that we discovered in the context of interpolation problems [4, 5] and generalized to the case of moment problems with complexity constraints [6, 8, 9, 13]. In fact, the topological approach developed in Section 3 allows us to significantly streamline our previous proofs concerning the convex functional and its extrema. Naturally, the optimization problem itself generalizes the maximum entropy approach. Indeed, for cases where the space of test functions lie in the Hardy space on the unit circle, we provide in Section 5 a succinct closed form for the maximum entropy solution. In Section 6 we describe some amplifications of our basic results using differentiable maps and manifolds, a methodology upon which we based an alternative approach to this problem in [8, 9] and which is also streamlined by our topological arguments.

2. The moment problem following Krein and Nudel'man

The fundamental result on the generalized moment problem derived in [15] is based on two results, one about properties of convex cones in finite-dimensional spaces of continuous functions and the other about properties of positive measures.

The first result concerns a subspace $\mathfrak{P}$ of the Banach space $C[a,b]$ of complex-valued continuous functions defined on the real interval $[a,b]$ and a choice of basis $(u_0, u_1, \dots, u_n)$ of $\mathfrak{P}$. If $p \in \mathfrak{P}$ we denote by P its real part $P := \mathrm{Re}(p)$. Following [15], we define the subset $\mathfrak{P}_+$ of those elements $p \in \mathfrak{P}$ such that $P \geq 0$. The space $\mathfrak{P}_+ \subset \mathfrak{P}$ is a closed, convex cone. In terms of the basis (u_i), every $\phi \in \mathfrak{P}^*$ corresponds to a complex sequence $c = (c_0, c_1, \dots, c_n) \in \mathbb{C}^{n+1}$. Since every ϕ is determined by its real part as a linear functional on $\mathfrak{P}$ as a real vector space, we can characterize elements of the dual cone $\mathfrak{P}_+^\top$ as those sequences c satisfying

$$\langle c, p \rangle := \mathrm{Re}\left\{ \sum_{k=0}^{n} p_k c_k \right\} \geq 0 \tag{2.1}$$

for all $p \in \mathfrak{P}_+$. Such a sequence is classically called *positive*, and the space of positive sequences is denoted by $\mathfrak{C}_+$. In particular, $\mathfrak{C}_+$ is a closed, convex cone with $\mathfrak{C}_+^\top = \mathfrak{P}_+$.

Following [15], consider the curve

$$U(t) = \begin{pmatrix} u_0(t) \\ u_1(t) \\ \vdots \\ u_n(t) \end{pmatrix}, \quad a \le t \le b.$$

We then define the subset $U = \{U(t) : t \in [a, b]\} \subset \mathbb{C}^{n+1}$, and let $K(U)$ denote its convex conic hull. Clearly, $K(U)^{\mathsf{T}} = \mathfrak{P}_+$, from which follows:

Theorem 2.1 ([15]). $K(U) = \mathfrak{C}_+$.

We now turn to some results concerning positive measures. Given $c \in \mathbb{C}^{n+1}$, the generalized moment problem (see [15]) is to find a positive measure $d\mu$ such that

$$\int_a^b u_k(t) d\mu(t) = c_k, \quad k = 0, 1, \dots, n. \tag{2.2}$$

For the sake of brevity, from now on we shall refer to this problem as simply the moment problem, omitting the adjective "generalized". More generally, let

$$\mathfrak{M} : C[a, b]^* \to \mathbb{C}^{n+1} \tag{2.3}$$

be the continuous mapping defined via (2.2) for an arbitrary bounded measure $d\mu \in C[a, b]^*$ and consider the subset $\mathcal{M}_+ \subset C[a, b]^*$ of positive measures.

Lemma 2.2 ([15]). $\mathfrak{M}(\mathcal{M}_+) \subset \mathfrak{C}_+$.

Proof. If $p \in \mathfrak{P}_+ = \mathfrak{C}_+^{\mathsf{T}}$, then

$$\langle c, p \rangle := \mathrm{Re}\left\{ \sum_{k=0}^{n} p_k c_k \right\} = \int_a^b P \, d\mu \ge 0, \tag{2.4}$$

so that $c \in \mathfrak{C}_+$. □

By Theorem 2.1 and Lemma 2.2, we have that $\mathfrak{M}(\mathcal{M}_+)$ is a convex subset of $K(U)$. On the other hand, by choosing $d\mu = \delta_{t_0}$ for each $t_0 \in [a, b]$, it follows that $U \subset \mathfrak{M}(\mathcal{M}_+)$. In particular, to say that $\mathfrak{M}(\mathcal{M}_+)$ is closed is to say that $K(U) \subset \mathfrak{M}(\mathcal{M}_+)$.

In [15], the Helly Selection Theorem is used to show that $\mathfrak{M}(\mathcal{M}_+)$ is closed in $\mathfrak{C}_+$ under the following hypothesis.

Hypothesis 2.3. There exists $p \in \mathfrak{P}_+$ such that $P > 0$ on $[a, b]$.

Theorem 2.4 ([15]). *Whenever Hypothesis* 2.3 *holds,* $K(U) = \mathfrak{M}(\mathcal{M}_+)$. *In particular,*

$$\mathfrak{C}_+ = \mathfrak{M}(\mathcal{M}_+). \tag{2.5}$$

Of course, in order for the moment equations to hold it is necessary that c_k be real whenever u_k is real. Moreover, a purely imaginary moment condition can always be reduced to a real one.

Convention 2.5. Henceforth we shall assume that $u_0, \dots, u_{r-1}$ are real functions and $u_r, \dots, u_n$ are complex-valued functions whose real and imaginary parts, taking together with $u_0, \dots, u_{r-1}$, are linearly independent over $\mathbb{R}$.

In particular, we may regard $\mathfrak{P}$ as the real vector space $\mathbb{R}^{2n-r+2}$ and $\mathfrak{C}_+ \subset \mathbb{R}^{2n-r+2}$. Therefore, it follows [15] that $\mathfrak{C}_+$ is a closed convex cone of dimension $2n - r + 2$, with interior $\mathring{\mathfrak{C}}_+$ consisting of *strictly positive* sequences c; i.e., those sequences c satisfying

$$\langle c, p \rangle := \operatorname{Re} \left\{ \sum_{k=0}^{n} p_k c_k \right\} > 0 \tag{2.6}$$

for all $p \in \mathfrak{P}_+ \setminus \{0\}$. Assuming Hypothesis 2.3, it then follows that $\mathfrak{P}_+$ is also a closed convex cone of dimension $2n-r+2$, with a nonempty interior $\mathring{\mathfrak{P}}_+$ consisting of those $p \in \mathfrak{P}_+$ for which $\operatorname{Re}(p) > 0$.

3. The main results

In the power and the trigonometric moment problems, the elements of the subspace $\mathfrak{P}$ are polynomials and trigonometric polynomials, respectively. In part for this reason, the elements of the subspace $\mathfrak{P}$ in an arbitrary moment problem are referred to as "polynomials in $\mathfrak{P}$". Following this precedent, we shall refer to the ratio p/q with $p, q \in \mathfrak{P}$ as a "rational function". For the classical Nevanlinna-Pick interpolation problem, it turns out that $\mathfrak{P}$ is a coinvariant subspace of H^2 so that the "polynomials" are rational functions σ/τ, where τ is fixed. This of course implies that the rational functions in $\mathfrak{P}$ are rational in the usual sense.

Definition 3.1. The functions $P := \operatorname{Re}(p)$, for $p \in \mathfrak{P}$ in the moment problem are referred to as *real polynomials for* $\mathfrak{P}$. We shall refer to the ratio P/Q with $p, q \in \mathfrak{P}$ as a *real rational functions for* $\mathfrak{P}$.

Remark 3.2. Under Convention 2.5,

$$p := \sum_{k=0}^{n} p_k u_k \in \mathfrak{P} \tag{3.1}$$

corresponds to an $(n+1)$-tuple of points $(p_0, p_1, \dots, p_n)$, where $p_0, p_1, \dots, p_{r-1}$ are real and $p_r, p_{r+1}, \dots, p_n$ are complex. Moreover, p is determined by P [8, p. 165].

The moment problem is about measures and combining these two concepts leads us to following definition.

Definition 3.3. Any measure of the form

$$d\mu = \frac{P(t)}{Q(t)} dt, \tag{3.2}$$

where P, Q are positive real polynomials for $\mathfrak{P}$, is a *rational positive measure.* Let $\mathcal{R}_+ \subset \mathcal{M}_+$ denote the subset of rational positive measures.

Problem 3.4. Given a sequence of complex numbers $c_0, c_1, \dots, c_n$ and a subspace $\mathfrak{P} = \operatorname{span}(u_0, \dots, u_n) \subset C[a,b]$, the *moment problem for rational measures* is to parameterize all positive rational measures $\frac{P(t)}{Q(t)}dt$ such that

$$\int_a^b u_k(t)\frac{P(t)}{Q(t)}dt = c_k, \quad k = 0, 1, \dots, n. \tag{3.3}$$

We shall need an additional hypothesis to accommodate the restriction to rational positive measures.

Hypothesis 3.5. The space $\mathfrak{P}$ consists of Lipschitz continuous functions.

Remark 3.6. To the best of our knowledge, all instances of the generalized moment problem that arise in systems and control involve subspaces of $C[a,b]$ consisting of Lipschitz continuous functions. Moreover, we recall the classical result that, if $\mathfrak{P}$ is spanned by a Chebyshev system (or T-system) and contains a constant function, then after a reparameterization $\mathfrak{P}$ consists of Lipschitz continuous functions [15, p. 37].

In the setting of Section 2, our first result is the following.

Theorem 3.7. *If Hypotheses* 2.3 *and* 3.5 *hold, then*

$$\mathfrak{M}(\mathcal{R}_+) = \mathring{\mathfrak{C}}_+.$$

In other words, the moment problem for rational measures is solvable if, and only if, the sequence c is strictly positive.

For any $d\mu \in \mathcal{R}_+$, consider the sequence c defined by (2.2) and any $p = \sum_{k=0}^n p_k u_k \in \mathfrak{P}_+ \setminus \{0\}$. Then

$$\langle c, p \rangle := \operatorname{Re}\left\{\sum_{k=0}^n p_k c_k\right\} = \int_a^b P(t)d\mu > 0, \tag{3.4}$$

so that $c \in \mathring{\mathfrak{C}}_+$. This observation yields the rational analogue of Lemma 2.2.

Lemma 3.8. *If Hypothesis* 2.3 *holds, then* $\mathfrak{M}(\mathcal{R}_+) \subset \mathring{\mathfrak{C}}_+$.

The following result implies the reverse inclusion.

Theorem 3.9. *If Hypotheses* 2.3 *and* 3.5 *hold, then* $\mathfrak{M}(\mathcal{R}_+)$ *contains a set which is both open and closed in the convex set* $\mathring{\mathfrak{C}}_+$.

Remark 3.10. The assertions in Theorem 3.9 are the topological analogue, for the case of rational measures, of the convexity assertions used in the proof of Theorem 2.4 for the generalized moment problem, where it was shown that the convex subset $\mathfrak{M}(\mathcal{M}_+) \subset \mathfrak{C}_+ = K(U)$ both contains U and is closed. In light of Lemma 3.8, a point mass δ_{t_0} cannot be realized on $\mathfrak{P}$ by a positive rational measure so that $U \not\subset \mathfrak{M}(\mathcal{R}_+)$. Nonetheless, there exists $\mathcal{P}_+ \subset \mathcal{R}_+$ such that $\mathfrak{M}(\mathcal{P}_+)$ is both open and closed in $\mathring{\mathfrak{C}}_+$.

Indeed, for a fixed $P \in \mathring{\mathfrak{P}}_+$ consider the set

$$\mathcal{P}_+ = \{d\mu \in \mathcal{R}_+ : d\mu = \frac{P}{Q}dt, \quad Q \in \mathring{\mathfrak{P}}_+\} \tag{3.5}$$

and the restriction of the moment mapping $\mathfrak{M}|_{\mathcal{P}_+} : \mathcal{P}_+ \to \mathring{\mathfrak{C}}_+$.

Proposition 3.11. *If Hypothesis* 2.3 *holds, then* $\mathfrak{M}(\mathcal{P}_+) \subset \mathring{\mathfrak{C}}_+$ *is open.*

Proof. For simplicity, we view $\mathfrak{P}$ and $\mathfrak{C}$ as real vector spaces, so that $\mathfrak{P}$ is spanned by the real basis (u_i), where we have replaced a complex-valued (u_k) by its real and imaginary parts. We shall also parameterize $d\mu \in \mathcal{P}_+$ by $q \in \mathring{\mathfrak{P}}_+$. The Jacobian, $\text{Jac}(\mathfrak{M}|_{\mathcal{P}_+})_{q_0}$, of $\mathfrak{M}|_{\mathcal{P}_+}$ at a point $q_0 \in \mathring{\mathfrak{P}}_+$ is a square matrix M_q whose (i,j)th entry is

$$(M_q)_{(i,j)} = -\int_a^b u_i(t)u_j(t)\frac{P(t)}{Q^2(t)}dt \tag{3.6}$$

evaluated at the point q_0. Thus, $-M_q$ is the gramian matrix of the real basis (u_i) with respect to the positive definite inner product defined by $P(t)/Q^2(t)dt$ on $C[a,b]$. Therefore, $\text{Jac}(\mathfrak{M}|_{\mathcal{P}_+})_q$ has rank $2n-r+2$ at each point $q \in \mathring{\mathfrak{P}}_+$ so that, by the Implicit Function Theorem, $\mathfrak{M}(\mathcal{P}_+)$ is open. □

Proposition 3.12. *If Hypotheses* 2.3 *and* 3.5 *hold, then* $\mathfrak{M}(\mathcal{P}_+) \subset \mathring{\mathfrak{C}}_+$ *is closed.*

Proof. Suppose

$$\mathfrak{M}(\frac{P}{Q_j}dt) = c^j \in \mathring{\mathfrak{C}}_+ \tag{3.7}$$

and $\lim_{j\to\infty} c^j = c_0 \in \mathring{\mathfrak{C}}_+$. We claim that there exists an $M > 0$ such that $\|Q_j\|_\infty \le M$. To see this note that

$$\lim_{j\to\infty} \int_a^b \frac{P^2(t)}{Q_j(t)}dt = \lim_{j\to\infty}\langle c^j, p\rangle = \langle c, p\rangle > 0. \tag{3.8}$$

Setting $\widetilde{Q}_j = Q_j/\|Q_j\|_\infty$, we have

$$\lim_{j\to\infty}\|Q_j\|_\infty \int_a^b \frac{P^2(t)}{\widetilde{Q}_j(t)}dt = \langle c, p\rangle > 0. \tag{3.9}$$

Since $\int_a^b P^2(t)/\widetilde{Q}_j(t)\,dt \ge \epsilon$ for some $\epsilon > 0$, we must have $\|Q_j\|_\infty \le M < \infty$ for some $M > 0$. Therefore, there is a convergent subsequence in $\mathfrak{P}$,

$$\lim_{k\to\infty} q_k = q_0, \quad \text{with } q_0 \in \mathfrak{P}_+ \tag{3.10}$$

for which

$$0 < \int_a^b \frac{P^2(t)}{Q_0(t)}dt = \langle c, p\rangle < \infty.$$

We claim that $q_0 \in \mathring{\mathfrak{P}}_+$.

Suppose, on the contrary, that $Q_0(t_0) = 0$ for some $t_0 \in [a,b]$, Then, since Q_0 is Lipschitz continuous at t_0, there exists an $\varepsilon > 0$ and an $L > 0$ such that $Q_0(t) \le L|t - t_0|$ whenever $|t - t_0| < \varepsilon$ and $t \in [a,b]$. In particular, if $t_0 \in (a,b)$,

$$\int_a^b \frac{P^2}{Q_0}\,dt \ge \frac{1}{L}\int_{t_0-\varepsilon}^{t_0+\varepsilon} \frac{P^2}{|t-t_0|}\,dt = +\infty,$$

contrary to assumption. If $t_0 = a$ or $t_0 = b$, a similar estimate holds. Hence, $q_0 \in \mathring{\mathfrak{P}}_+$, as claimed. □

Corollary 3.13. *If Hypotheses* 2.3 *and* 3.5 *hold, the moment mapping* $\mathfrak{M}|_{\mathcal{P}_+} : \mathcal{P}_+ \to \mathring{\mathfrak{C}}_+$ *is surjective.*

4. A Dirichlet principle for the moment problem with rational positive measures

In the course of proving Theorem 3.7 we showed that, for any $c \in \mathring{\mathfrak{C}}_+$ and any choice of $P \in \mathring{\mathfrak{P}}_+$, the moment problem for rational positive measures always has a solution in the set

$$\mathcal{P}_+ = \{d\mu \in \mathcal{R}_+ : d\mu = \frac{P}{Q}dt,\ \ Q \in \mathring{\mathfrak{P}}_+\}. \tag{4.1}$$

In this section, using a convex optimization argument, we show that the surjection $\mathfrak{M}|_{\mathcal{P}_+}$ is injective, and we characterize the unique rational measure as the solution of a variational problem. In fact, we derive both a primal optimization problem and its dual. Remarkably, the moment problem for rational positive measures *is* the set of critical point equations for the dual variational problem. In this classical sense, a nonlinear convex optimization provides an illustration of the Dirichlet Principle for this class of moment problems.

Let $\mathbb{I}_p : C_+[a,b] \to \mathbb{R} \cup \{-\infty\}$ be the relative entropy functional

$$\mathbb{I}_p(\Phi) = \int_a^b P(t)\log\Phi(t)dt, \tag{4.2}$$

which is a generalization of the entropy functional obtained by setting $P = 1$. From Jensen's inequality we see that $\mathbb{I}_p(\Phi) \le \log\left(\int_a^b P\Phi dt\right) \le \int_a^b P\Phi dt < \infty$.

Theorem 4.1. *Assume that Hypotheses* 2.3 *and* 3.5 *hold, and let* $c \in \mathring{\mathfrak{C}}_+$. *Then, for any* $P \in \mathring{\mathfrak{P}}_+$, *the constrained optimization problem to maximize* (4.2) *over* $C_+[a,b]$ *subject to the moment constraints*

$$\int_a^b u_k(t)\Phi(t)dt = c_k, \quad k = 0,1,\dots,n, \tag{4.3}$$

has a unique solution, and it has the form

$$\Phi = \frac{P}{Q}, \quad Q := \mathrm{Re}\{q\}, \tag{4.4}$$

where $q \in \mathring{\mathfrak{P}}_+$.

The optimization problem of Theorem 4.1, to which we shall refer as the *primal problem*, can be solved by Lagrange relaxation. In fact, we have the Lagrangian

$$L(\Phi, q) = \mathbb{I}(\Phi) + \mathrm{Re} \sum_{k=0}^{n} q_k \left[c_k - \int_a^b u_k \Phi dt \right],$$

where $(q_0, q_1, \dots, q_n) \in \mathbb{R}^r \times \mathbb{C}^{n-r+1}$ are Lagrange multipliers. Then,

$$L(\Phi, q) = \int_a^b P \log \Phi \, dt + \langle c, q \rangle - \int_a^b Q \Phi dt,$$

where $Q = \mathrm{Re}\{q\}$ with $q := \sum_{k=0}^{n} q_k u_k \in \mathfrak{P}$. Clearly, comparing linear and logarithmic growth, we see that the dual functional

$$\psi(q) = \sup_{\Phi \in C_+[a,b]} L(\Phi, q)$$

takes finite values only if $q \in \mathfrak{P}_+$, so we may restrict our attention to such Lagrange multipliers. For any $q \in \mathfrak{P}_+$ and any $\Phi \in C_+[a, b]$ such that P/Φ is integrable, the directional derivative

$$d_{(\Phi,q)} L(h) = \int_a^b \left[\frac{P}{\Phi} - Q \right] h \, dt = 0$$

for all $h \in C[a, b]$ if and only if $\Phi = \frac{P}{Q} \in C_+[a, b]$, which inserted into the dual functional yields

$$\psi(q) = \mathbb{J}_P(q) + \int_a^b P(\log P - 1) dt, \tag{4.5}$$

where $\mathbb{J}_P : \mathfrak{P}_+ \to \mathbb{R} \cup \{\infty\}$ is the strictly convex functional

$$\mathbb{J}_P(q) = \langle c, q \rangle - \int_a^b P \log Q dt. \tag{4.6}$$

As the last term in (4.5) is constant, the dual problem to minimize $\psi(q)$ over $\mathfrak{P}_+$ is equivalent to the convex optimization problem

$$\min_{q \in \mathfrak{P}_+} \mathbb{J}(q). \tag{4.7}$$

Since

$$\frac{\partial \mathbb{J}_P}{\partial q_k} = c_k - \int_a^b u_k \frac{P}{Q} dt, \quad k = 0, 1, \dots, n,$$

it follows from Corollary 3.13 that the optimization problem (4.7) has an optimal solution $\hat{q} \in \mathring{\mathfrak{P}}_+$ satisfying the moment equations (3.3). Moreover, since the functional (4.6) is strictly convex, this optimum is unique.

Consequently,

$$\hat{\Phi} := \frac{P}{\hat{Q}} \in C_+[a,b] \tag{4.8}$$

is the unique optimal solution of the primal problem. To see this, observe that $\Phi \mapsto L(\Phi, \hat{q})$ is strictly concave and that $dL_{(\hat{\Phi},\hat{q})}(h) = 0$ for all $h \in C_+[a,b]$. Therefore,

$$L(\Phi, \hat{q}) \leq L(\hat{\Phi}, \hat{q}), \quad \text{for all } \Phi \in C_+[a,b] \tag{4.9}$$

with equality if and only if $\Phi = \hat{\Phi}$. However, $L(\Phi, \hat{q}) = \mathbb{I}_p(\Phi)$ for all Φ satisfying the moment conditions (4.3). In particular, since (4.3) holds with $\Phi = \hat{\Phi}$, $L(\hat{\Phi}, \hat{q}) = \mathbb{I}_p(\hat{\Phi})$. Consequently, (4.9) implies that $\mathbb{I}_p(\Phi) \leq \mathbb{I}_p(\hat{\Phi})$ for all $\Phi \in C_+[a,b]$ satisfying the moment conditions, with equality if and only if $\Phi = \hat{\Phi}$. Hence, $\mathbb{I}_p$ has a unique maximum in the space of all $\Phi \in C_+[a,b]$ satisfying the constraints (4.3), and it is given by (4.8).

This concludes the proof of Theorem 4.1, but we have also proven the following theorem.

Theorem 4.2. *Assume that Hypotheses* 2.3 *and* 3.5 *hold. Let* $(c,p) \in \mathring{\mathfrak{C}}_+ \times \mathring{\mathfrak{P}}_+$, *and set* $P := \mathrm{Re}\{p\}$. *Then the functional* (4.6) *has a unique minimizer* $\hat{q} \in \mathring{\mathfrak{P}}_+$ *and* $\hat{Q} := \mathrm{Re}\{\hat{q}\}$ *is the unique solution to the moment equations*

$$\int_a^b u_k \frac{P}{Q} dt = c_k, \quad k = 0, 1, \ldots, n. \tag{4.10}$$

Corollary 4.3. *If Hypotheses* 2.3 *and* 3.5 *hold, the moment mapping* $\mathfrak{M}|_{\mathcal{P}_+} : \mathcal{P}_+ \to \mathring{\mathfrak{C}}_+$ *is a bijection.*

5. Moment problems in a Hardy space setting

Some important special cases of the moment problem is when

$$u_k(t) = g_k(e^{it}) \quad \text{where } g_k \in H^2(\mathbb{D}), \quad k = 0, 1, \ldots, n, \tag{5.1}$$

and $[a,b] = [-\pi, \pi]$. A case in point is the trigonometric moment problem when $g_k(z) = \frac{1}{2\pi} z^k$; another is Nevanlinna-Pick interpolation when $g_k(z) = \frac{1}{2\pi}\frac{z+z_k}{z-z_k}$, where $z_0, z_1, \ldots, z_n$ are the (distinct) interpolation points. In both of these cases, $g := (g_0, g_1, \ldots, g_n)^{\mathsf{T}}$ can be represented as

$$g(z) = (I - zA)^{-1}B, \tag{5.2}$$

where A is a $n \times n$ stability matrix and B an n-vector such that (A, B) is a reachable pair; i.e.,

$$G = \int_{-\pi}^{\pi} g(e^{it}) g(e^{it})^* dt > 0. \tag{5.3}$$

Indeed, positive definiteness of G follows readily from the fact that the basis functions are linearly independent. This condition also insures that there is a unique function of the form

$$w(z) = \sum_{k=0}^{n} w_k g_k(e^{it})^*$$

that satisfies

$$\int_{-\pi}^{\pi} g_k(e^{it}) w(e^{it}) dt = c_k, \quad k = 0, 1, \dots, n,$$

namely the one provided by the unique solution of the system of linear equations

$$\sum_{k=0}^{n} G_{kj} w_j = c_k, \quad k = 0, 1, \dots, n.$$

Consequently, for any $q \in \mathfrak{P}$,

$$\langle c, q \rangle = \int_{-\pi}^{\pi} Q(t) w(e^{it}) dt. \tag{5.4}$$

It can be shown that $g_0, g_1, \dots, g_n$ span the coinvariant subspace $\mathcal{K} := H^2 \ominus \phi H^2$, where ϕ is the inner function

$$\phi(z) = \frac{\det(zI - A^*)}{\det(I - zA)}.$$

In view of (5.1), $\mathcal{K}$ is a Hardy space model of $\mathfrak{P}$. Moreover, for any $\psi \in \mathcal{K}$, there is a $v \in \mathcal{K}$ such that $\Psi := \text{Re}\{\psi\} = vv^*$ [7, Proposition 9]. Therefore, for any $q \in \mathfrak{P}_+$, there is an $\mathbf{a} \in \mathbb{C}^n$ such that $Q(t) = a(e^{it})^* a(e^{it})$ where $a(z) := g(z)^* \mathbf{a}$. Then, by (5.4),

$$\langle c, q \rangle = \mathbf{a}^* \int_{-\pi}^{\pi} w(e^{it}) g(e^{it}) g(e^{it})^* dt \, \mathbf{a} = \mathbf{a}^* \mathbf{P} \mathbf{a}, \tag{5.5}$$

where

$$\mathbf{P} := \frac{1}{2} \int_{-\pi}^{\pi} g(e^{it}) [w(e^{it}) + w(e^{it})^*] g(e^{it})^* dt. \tag{5.6}$$

Consequently, $c \in \mathfrak{C}_+$ if and only if $\mathbf{P} \geq 0$, and $c \in \mathring{\mathfrak{C}}_+$ if and only if $\mathbf{P} > 0$. In the trigonometric moment problem $\mathbf{P}$ is the Toeplitz matrix, and in the Nevanlinna-Pick case $\mathbf{P}$ is the the Pick matrix.

If $\mathfrak{P}$ contains constants, then we may determine the maximum-entropy solution, corresponding to setting $P = 1$ in (4.2), in closed form.

Proposition 5.1. *Suppose that the basis functions in* $\mathfrak{P}$ *satisfy* (5.1) *and* $\mathfrak{P}$ *contains constants. Then the maximum-entropy solution is*

$$\hat{\Phi}(t) = \frac{g(0)^* \mathbf{P}^{-1} g(0)}{|g(e^{it})^* \mathbf{P}^{-1} g(0)|^2}, \tag{5.7}$$

where $\mathbf{P}$ *is given by* (5.6).

Proof. We proceed as in [14, 16]. Since, by Jensen's formula [2, p. 184], the last term in the dual functional (4.6) (with $P = 1$) can be written $2\log|a(0)|$, (4.6) becomes

$$J(\mathbf{a}) := \mathbb{J}_p(a^*a) = \mathbf{a}^*\mathbf{P}\mathbf{a} - 2\log|\mathbf{a}^*g(0)|.$$

Setting the gradient of $J(\mathbf{a})$ equal to zero, we obtain $\mathbf{a} = \mathbf{P}^{-1}g(0)/|a(0)|$ and hence $a(z) = g(z)^*\mathbf{P}^{-1}g(0)/|a(0)|$. Then $|a(0)|^2 = g(0)^*\mathbf{P}^{-1}g(0)$, and therefore the optimal $\mathbf{a}$ becomes

$$a(z) = \frac{g(z)^*\mathbf{P}^{-1}g(0)}{\sqrt{g(0)^*\mathbf{P}^{-1}g(0)}}. \tag{5.8}$$

Moreover, in view of Theorems 4.1 and 4.2,

$$\hat{\Phi}(t) = \frac{1}{Q(t)} = \frac{1}{|a(e^{it})|^2},$$

and therefore (5.7) follows from (5.8). □

In the trigonometric moment problem, modulo normalization,

$$\varphi_n(z) := g(z)^*\mathbf{P}^{-1}g(0)$$

reduces to the Szegö polynomial orthogonal on the unit circle of degree n (cf. [10]).

6. Amplifications and conclusions

In this paper we showed that the moment problem for rational positive measures is solvable for all strictly positive sequences, provide Hypotheses 2.3 and 3.5 hold for $\mathfrak{P}$. In the language of functions and spaces, we showed that the moment map $\mathfrak{M}$ defined by (2.3) restricts to a surjection

$$\mathfrak{M}|_{\mathcal{R}_+} : \mathcal{R}_+ \to \mathring{\mathfrak{C}}_+ \tag{6.1}$$

by proving that the restriction

$$\mathfrak{M}|_{\mathcal{P}_+} : \mathcal{P}_+ \to \mathring{\mathfrak{C}}_+ \tag{6.2}$$

is surjective. Indeed, using the strict convexity of the dual functional, we were able to conclude in Corollary 4.3 that (6.2) is a bijection.

In this section we briefly discuss these maps in more detail. Following Hadamard, the problem of solving, for $c \in \mathring{\mathfrak{C}}_+$, the equations

$$\mathfrak{M}|_{\mathcal{P}_+}(d\mu(q)) = c, \ \text{for } q \in \mathring{\mathfrak{P}}_+, \tag{6.3}$$

is *well posed* provided a solution q exists, is unique and varies continuously with c (in some reasonable topology). As elements of open convex subsets of Euclidean space, the choice of topology is clear. Existence and uniqueness is the essence of Corollary 4.3. Moreover, our proof of Proposition 3.11 reposed on the observation that $\mathrm{Jac}(\mathfrak{M}|_{\mathcal{P}_+})_q$ is nonsingular at each $q \in \mathring{\mathfrak{P}}_+$ so that, by the Inverse Function

Theorem, $\mathfrak{M}|_{\mathcal{P}_+}$ is a smooth bijection with a smooth inverse. Since $\mathfrak{M}|_{\mathcal{P}_+}^{-1}$ is differentiable, it is continuous, so that q is a continuous function of c and this restricted moment problem is well posed.

Our second amplification concerns the map (6.1). Here, $\mathfrak{M}|_{\mathcal{R}_+}$ is not injective and one would instead like a continuous or smooth parameterization of the solutions, for $c \in \mathring{\mathfrak{C}}_+$, to the equations

$$\mathfrak{M}|_{\mathcal{R}_+}(d\mu) = c, \text{ for } d\mu \in \mathcal{R}_+. \tag{6.4}$$

As before, one can compute the Jacobian $\mathrm{Jac}(\mathfrak{M}|_{\mathcal{R}_+})_{d\mu}$ and show [9] that

$$\text{rank } \mathrm{Jac}(\mathfrak{M}|_{\mathcal{R}_+})_{d\mu} = 2n - r + 2,$$

for all $d\mu \in \mathcal{R}_+$. In fact, in [9] we prove that the solution space $\mathfrak{M}|_{\mathcal{R}_+}^{-1}(c)$ is a smooth manifold, smoothly parameterized by $p \in \mathring{\mathfrak{P}}_+$ as described in Theorem 4.1.

References

[1] N.I. Ahiezer and M. Krein, *Some Questions in the Theory of Moments.* American Mathematical Society, Providence, Rhode Island, 1962.

[2] L.V. Ahlfors, *Complex Analysis.* McGraw-Hill, 1953.

[3] C.I. Byrnes, A. Lindquist, S.V. Gusev, and A.S. Matveev, *A complete parameterization of all positive rational extensions of a covariance sequence.* IEEE Trans. Automat. Control **40** (1995), 1841–1857.

[4] C.I. Byrnes, S.V. Gusev, and A. Lindquist, *From finite covariance windows to modeling filters: A convex optimization approach.* SIAM Review **43** (2001) 645–675.

[5] C.I. Byrnes, T.T. Georgiou and A. Lindquist, *A generalized entropy criterion for Nevanlinna-Pick interpolation with degree constraint.* IEEE Trans. Automatic Control **AC-46** (2001) 822–839.

[6] C.I. Byrnes and A. Lindquist, *A convex optimization approach to generalized moment problems.* In "Control and Modeling of Complex Systems: Cybernetics in the 21st Century: Festschrift in Honor of Hidenori Kimura on the Occasion of his 60th Birthday", K. Hashimoto, Y. Oishi and Y. Yamamoto, Editors, Birkhäuser, 2003, 3–21.

[7] C.I. Byrnes, T.T. Georgiou, A. Lindquist, and A. Megretski, *Generalized interpolation in H^∞ with a complexity constraint.* Trans. American Mathematical Society **358** (2006), 965–987.

[8] C.I. Byrnes and A. Lindquist, *The generalized moment problem with complexity constraint.* Integral Equations and Operator Theory **56** (2006) 163–180.

[9] C.I. Byrnes and A. Lindquist, *Important moments in systems and control.* SIAM J. Control and Optimization **47(5)** (2008), 2458–2469.

[10] P. Enqvist, *A homotopy approach to rational covariance extension with degree constraint.* Intern. J. Applied Mathematics and Computer Science **11(5)** (2001), pp. 1173–1201.

[11] T.T. Georgiou, *Partial Realization of Covariance Sequences.* Ph.D. thesis, CMST, University of Florida, Gainesville 1983.

[12] T.T. Georgiou, *Realization of power spectra from partial covariance sequences.* IEEE Trans. Acoustics, Speech and Signal Processing **35** (1987), 438–449.

[13] T.T. Georgiou, *Solution of the general moment problem via a one-parameter imbedding.* IEEE Trans. Automatic Control **AC-50** (2005) 811–826.

[14] T.T. Georgiou and A. Lindquist, *Kullback-Leibler approximation of spectral density functions.* IEEE Trans. on Information Theory **49(11)**, November 2003.

[15] M.G. Krein and A.A. Nudelman, *The Markov Moment Problem and Extremal Problems.* American Mathematical Society, Providence, Rhode Island, 1977.

[16] A. Lindquist, *Prediction-error approximation by convex optimization, in Modeling, Estimation and Control: Festschrift in honor of Giorgio Picci on the occasion of his sixty-fifth Birthday.* A. Chiuso, A. Ferrante and S. Pinzoni (eds.), Springer-Verlag, 2007, 265–275.

Christopher I. Byrnes
Department of Electrical and Systems Engineering
Washington University
St. Louis, MO 63130, USA
e-mail: `chrisbyrnes@wustl.edu`

Anders Lindquist
Department of Mathematics
Optimization and Systems Theory
Royal Institute of Technology
SE-100 44 Stockholm, Sweden
e-mail: `alq@math.kth.se`

Operator Theory:
Advances and Applications, Vol. 190, 171–190

Almost Periodic Factorization of Some Triangular Matrix Functions

M.C. Câmara, Yu.I. Karlovich and I.M. Spitkovsky

To the esteemed memory of Mark Grigorevich Krein, with our deep respect and gratitude for everything he has taught us, directly or indirectly.

Abstract. The paper is devoted to matrices of the form $G(x) = \begin{bmatrix} e^{i\lambda x} & 0 \\ f(x) & e^{-i\lambda x} \end{bmatrix}$, with almost periodic off-diagonal entry f. Some new cases are found, in terms of the Bohr-Fourier spectrum of f, in which G is factorable. Formulas for the partial indices are derived and, under additional constraints, the factorization itself is constructed explicitly. Some a priori conditions on the Bohr-Fourier spectra of the factorization factors (provided that a canonical factorization exists) are also given.

Mathematics Subject Classification (2000). Primary 47A68; Secondary 42A75, 47A10, 47B35.

Keywords. Almost periodic factorization, Toeplitz operator.

1. Introduction

First, let us fix some notation. Throughout the paper, we denote by e_λ the functions

$$e_\lambda(x) = e^{i\lambda x}, \quad x \in \mathbb{R},$$

with the parameter λ also being real: $\lambda \in \mathbb{R}$. Finite linear combinations of $\{e_\lambda \colon \lambda \in \mathbb{R}\}$ form the algebra APP of *almost periodic polynomials*. It may be considered as a (non-closed) subalgebra of the algebra $C(\mathbb{R})$ of all functions continuous on $\mathbb{R}$. The closure of APP with respect to the uniform norm $\|\cdot\|$ is the Bohr algebra AP of all almost periodic functions. On the other hand, the closure of APP with

The work on this paper was started during the visits of Karlovich and Spitkovsky to Instituto Superior Técnico, supported by the FCT Projects POCTI/MAT/59972/2004 and PTDC/MAT/81385/2006 (Portugal). The authors were also partially supported by FCT through the Program POCI 2010/FEDER (Portugal), the SEP-CONACYT Project No. 25564 (México), and the NSF grant DMS-0456625 (USA), respectively.

respect to the stronger norm

$$\|f\|_W := \sum |c_j|, \quad f = \sum c_j e_{\lambda_j}, \; c_j \in \mathbb{C},$$

is the Banach algebra APW. Of course, APW is a dense subalgebra of AP with respect to the uniform norm.

For any $f \in AP$ there exists the *Bohr mean value*

$$\mathbf{M}(f) := \lim_{T\to\infty} \frac{1}{2T} \int_{-T}^{T} f(x)\, dx.$$

Consequently, the *Bohr-Fourier coefficients*

$$\widehat{f}(\lambda) := \mathbf{M}(e_{-\lambda} f)$$

are also defined. The set

$$\Omega(f) := \{\lambda \in \mathbb{R}\colon \widehat{f}(\lambda) \neq 0\}$$

is at most countable; it is called the *Bohr-Fourier spectrum* of f. Functions $f \in AP$ are defined uniquely by their formal *Bohr-Fourier series*

$$f \sim \sum_{\lambda\in\Omega(f)} \widehat{f}(\lambda) e_\lambda.$$

Naturally, for $f \in APW$ these series converge to f uniformly and absolutely on $\mathbb{R}$.

For $X = AP$, APW, or APP, we let

$$X^{\pm} = \{f \in X\colon \Omega(f) \subset \mathbb{R}_{\pm}\},$$

where of course $\mathbb{R}_{\pm} = \{x \in \mathbb{R}\colon \pm x \geq 0\}$. All these classes also are algebras, with $APP^{\pm}$ being dense in $APW^{\pm}$ and $AP^{\pm}$, in their respective norms.

We refer interested readers to the books [9, 11], Chapter 1 in [12] and Section 1.4 in [5] for a more detailed treatment of AP functions, in particular, proofs of the above-mentioned results.

Finally (as far as the notation goes), we will denote by X_n ($X_{n\times n}$) the set of all n-columns (respectively, $n \times n$ matrices) with elements in X. To save space, it is convenient to write a column $x \in X_n$ in the form of an n-tuple $(x_1, \dots, x_n)$.

An *AP factorization* of the $n \times n$ matrix function G is by definition its representation in the form

$$G = G_+ \Lambda G_-, \tag{1.1}$$

where

$$G_+^{\pm 1} \in AP^+_{n\times n}, \quad G_-^{\pm 1} \in AP^-_{n\times n}, \tag{1.2}$$

and the diagonal matrix Λ has the simplest AP functions $e_{\lambda_1}, \dots, e_{\lambda_n}$ as its diagonal entries. The numbers $\lambda_1, \dots, \lambda_n \in \mathbb{R}$ are defined uniquely (up to the order), provided that the factorization (1.1) exists, and are called the *partial AP indices* of G. The AP factorization (1.1) is *canonical* provided that $\lambda_1 = \cdots = \lambda_n = 0$, in which case the middle multiple Λ can be dropped:

$$G = G_+ G_-. \tag{1.3}$$

Canonical factorization of G exists exactly when the Toeplitz (or Wiener-Hopf) operator with the symbol G is invertible, and is therefore stable. An important characteristic of the canonical factorization is the *geometric mean* of G,

$$\mathbf{d}(G) := \mathbf{M}(G_+)\mathbf{M}(G_-), \tag{1.4}$$

where $\mathbf{M}(G_\pm)$ are computed entry-wise. The geometric mean is defined by G uniquely, in spite of the fact that the canonical factorization is not unique, and depends on G continuously ([14], see also [5, Section 18.5]).

The representation (1.1) is called an *APW* (respectively, *APP*) factorization of G if in condition (1.2) the classes $AP^\pm$ are replaced by $APW^\pm$ (respectively, $APP^\pm$). Of course, G must be an invertible element of *AP* (*APW*, *APP*) in order to admit an *AP* (respectively, *APW*, *APP*) factorization.

The notion of the *AP* factorization may be thought of as a natural generalization of the classical Wiener-Hopf factorization of purely periodic matrix functions. Various applications of it, along with the state of the existence and construction problem as of 2002, can be found in [5]. For instance, it is known that a canonical *AP* factorization of an *APW* matrix function, if it exists, automatically must be an *APW* factorization.

In this paper, we will work with 2×2 matrices of the form

$$G = \begin{bmatrix} e_\lambda & 0 \\ f & e_{-\lambda} \end{bmatrix}, \tag{1.5}$$

$f \in AP$. Such matrix functions arise naturally in the consideration of convolution type equation (in particular, difference equations) on finite intervals, with λ being the length of these intervals and f describing the behavior of the Fourier transform of the kernel at infinity. In spite of their seemingly simple structure, even for such matrices the *AP* factorability criteria are known only under some, rather restrictive, additional conditions on f. See [10, 1, 6] for pertinent recent results.

In Section 2 we describe the factorization criterion and provide explicit factorization formulas in the so-called big gap case. The existence part is then used in Section 3 to single out some new factorability cases. These cases happen to be new even when the off-diagonal entry f contains only three terms, and in Section 5 we provide explicit factorization formulas for some of them. The derivation of these formulas is based on solving appropriate Riemann-Hilbert problems, as described in Section 4. Finally, in Section 6 we make some general observations regarding the spectral structure of the factorization multiples for matrix functions of the type (1.5).

2. The big gap case

Matrices (1.5) have constant determinant: $\det G = 1$. This property easily implies that the partial *AP* indices sum up to zero; in what follows we will denote them by $\delta(\geq 0)$ and $-\delta$. The factorization (1.1) therefore takes the form

$$G = G_+ \operatorname{diag}[e_{-\delta}, e_\delta] G_-. \tag{2.1}$$

Since, according to (2.1), $\det G_+ = \det G_-^{-1}$, both determinants must be constant. A simple adjustment of the factors by a constant scalar multiple allows us to assume without loss of generality that

$$\det G_+ = \det G_- \equiv 1.$$

It is known (and easy to prove, see [5, Section 13.2]) that $\delta \leq \lambda$. Also, in case $f \in APW$, a simple transformation of G allows without loss of generality to suppose that

$$\Omega(f) \subset (-\lambda, \lambda). \tag{2.2}$$

If the set $\Omega(f)$ contains two points α, β such that $d := \alpha - \beta \geq \lambda$ and the rest of the Bohr-Fourier spectrum of $f \in APW$ lies either to the left of β or to the right of α, then the matrix (1.5) is APW factorable with the partial AP indices equal $\pm(d - \lambda)$. This result, under the name "big gap case", can be found in [5, Section 14.2]. In this section we show that basically the same factorization result holds when portions of $\Omega(f)$ are present both to the left of β and to the right of α.

Theorem 2.1. *Let $f \in APW$ be such that its Bohr-Fourier spectrum has a gap of length (at least) λ within $(-\lambda, \lambda)$:*

$$\Omega(f) \cap (\alpha - \lambda, \alpha) = \emptyset \tag{2.3}$$

for some $\alpha \in (0, \lambda)$. Then the matrix (1.5) *admits a canonical AP factorization if and only if $\widehat{f}(\alpha), \widehat{f}(\alpha - \lambda) \neq 0$.*

Proof. Under condition (2.3), f can be represented as

$$f = f_+ e_\alpha - f_- e_{\alpha-\lambda},$$

where $f_\pm \in APW^\pm$. In this representation, $\widehat{f}(\alpha) = \mathbf{M}(f_+)$ and $\widehat{f}(\alpha - \lambda) = -\mathbf{M}(f_-)$.

Sufficiency. Suppose first that $c_\pm := \mathbf{M}(f_\pm) \neq 0$. The functions f_+ and $e_{\lambda-\alpha}$ then satisfy the corona condition

$$\inf(|f_+(z)| + |e_{\lambda-\alpha}(z)|) > 0$$

in the upper half-plane $\mathbb{C}_+$. By the corona theorem for APW^+, there exist $\widetilde{f_+}, \widetilde{g_+} \in APW^+$ such that

$$f_+\widetilde{f_+} + e_{\lambda-\alpha}\widetilde{g_+} = 1. \tag{2.4}$$

Similarly, there exist $\widetilde{f_-}, \widetilde{g_-} \in APW^-$ such that

$$f_-\widetilde{f_-} + e_{-\alpha}\widetilde{g_-} = 1. \tag{2.5}$$

Now let

$$F = \widetilde{f_+} f \widetilde{f_-} + \widetilde{g_+}\widetilde{f_-} e_\lambda - \widetilde{f_+}\widetilde{g_-} e_{-\lambda}$$

and denote by $F_\pm$ the functions in $APW^\pm$ such that $F_+ + F_- = F$ (the latter are defined up to constant summands).

Finally, introduce

$$G_+ = \begin{bmatrix} \widetilde{f_+} + e_{\lambda-\alpha}F_+ & e_{\lambda-\alpha} \\ -\widetilde{g_+} + f_+F_+ & f_+ \end{bmatrix}, \quad G_- = \begin{bmatrix} f_- & -e_{-\alpha} \\ \widetilde{g_-} + f_-F_- & \widetilde{f_-} - e_{-\alpha}F_- \end{bmatrix}. \tag{2.6}$$

Obviously, $G_\pm \in APW^\pm$. Direct computations show that $\det G_\pm = 1$ (so that $G_\pm^{-1} \in APW^\pm$ as well) and

$$G = G_+ G_-. \tag{2.7}$$

This gives the canonical *AP* factorization of G.

This concludes the proof of sufficiency. However, we will make an additional observation which will become useful later. From (2.4) and (2.5) it follows, respectively, that

$$\mathbf{M}(\widetilde{f_+}) = 1/c_+, \quad \mathbf{M}(\widetilde{f_-}) = 1/c_-.$$

This yields the following information regarding the geometric mean $\mathbf{d}(G)$ of the matrix function (1.5):

$$\mathbf{d}(G) = \begin{bmatrix} 1/c_+ & 0 \\ * & c_+ \end{bmatrix} \begin{bmatrix} c_- & 0 \\ * & 1/c_- \end{bmatrix} = \begin{bmatrix} c_-/c_+ & 0 \\ * & c_+/c_- \end{bmatrix}. \tag{2.8}$$

Necessity. Suppose that at least one of the means $\mathbf{M}(f_\pm)$ is zero. By an arbitrarily small perturbation of f preserving property (2.3) these coefficients can be made non-zero. The resulting perturbation of the matrix function (1.5) admits a canonical *AP* factorization. According to (2.8), by choosing sufficiently small $\mathbf{M}(f_+)$ or $\mathbf{M}(f_-)$ for the perturbed f, the norm of $\mathbf{d}(G)$ for the perturbed G can be made arbitrarily large. This would contradict the continuity of the geometric mean, if the matrix (1.5) with f satisfying (2.3) and vanishing $\widehat{f}(\alpha)$ or $\widehat{f}(\alpha - \lambda)$ had a canonical *AP* factorization. □

Our proof of sufficiency follows the lines of [4, 3] and [2] (also see [5, Corollary 23.2]). The statement of Theorem 2.1 can be extracted as well from Theorem 4.1 and Corollary 4.5 of [7]. Moreover, it is contained in [5, Theorem 22.16]) as its very particular case corresponding to $n = 0$ in formula (22.53).

It follows from Theorem 2.1 that if the gap in the Bohr-Fourier spectrum of f is strictly wider than λ, then the matrix (1.5) does *not* admit a canonical *AP* factorization. Our next results implies that a (naturally, non-canonical) *AP* factorization of such matrices does exist, provided that both endpoints of the gap belong to the spectrum.

Theorem 2.2. *Let $f \in APW$ be such that for some $\alpha, \beta \in (-\lambda, \lambda)$ with $\alpha - \beta := d \geq \lambda$,*

$$\alpha, \beta \in \Omega(f) \text{ and } (\beta, \alpha) \cap \Omega(f) = \emptyset. \tag{2.9}$$

Then the matrix (1.5) *admits an APW factorization with the partial AP indices equal $\pm(d - \lambda)$.*

Proof. Observe first of all that the case $d = \lambda$ follows from Theorem 2.1. It remains therefore to consider the situation when

$$\delta := d - \lambda > 0.$$

Along with (1.5), let us introduce the auxiliary matrix

$$H = \begin{bmatrix} e_d & 0 \\ f & e_{-d} \end{bmatrix} \tag{2.10}$$

having the same structure as (1.5) but different exponents in the diagonal positions. Due to (2.9), the matrix (2.10) satisfies the conditions of Theorem 2.1 (with λ changed to d) and therefore admits a canonical AP factorization

$$H = H_+ H_-. \tag{2.11}$$

But then

$$G = \begin{bmatrix} e_{-\delta} & 0 \\ 0 & 1 \end{bmatrix} H \begin{bmatrix} 1 & 0 \\ 0 & e_\delta \end{bmatrix} = G_+ \begin{bmatrix} e_{-\delta} & 0 \\ 0 & e_\delta \end{bmatrix} G_-, \tag{2.12}$$

where

$$G_+ = \begin{bmatrix} e_{-\delta} & 0 \\ 0 & 1 \end{bmatrix} H_+ \begin{bmatrix} e_\delta & 0 \\ 0 & 1 \end{bmatrix} \text{ and } G_- = \begin{bmatrix} 1 & 0 \\ 0 & e_{-\delta} \end{bmatrix} H_- \begin{bmatrix} 1 & 0 \\ 0 & e_\delta \end{bmatrix}.$$

Matrix functions $G_\pm$ are unimodular along with $H_\pm$. Moreover, the property $H_\pm \in APW^\pm_{2\times 2}$ implies that three out of four entries of $G_\pm$ also belong to $APW^\pm$, with the dubious ones located in $(1,2)$-positions. From the formulas analogous to (2.6) it follows however that the $(1,2)$ entries of H_+ and H_- equal $e_{d-\alpha}$ and $-e_{-\alpha}$, respectively. The corresponding entries of $G_\pm$ are therefore $e_{-\delta}e_{d-\alpha} = e_{\lambda-\alpha} \in APW^+$ and $-e_\delta e_{-\alpha} = -e_{-\beta-\lambda} \in APW^-$. Thus, (2.12) delivers the desired APW factorization of G. □

Remark 2.3. If, in the setting of Theorem 2.2, $f \in APP$, then the matrix (1.5) actually admits an APP factorization.

For $d = \lambda$, when the factorization is canonical, this follows from [5, formulas (13.42), (13.43)] for the solutions of corona problem with corona data as in (the proof of) Theorem 2.1. In this case, by the way, the factors of all AP factorizations are automatically in APP. For $d > \lambda$ the construction of the proof of Theorem 2.2 yields an APP factorization.

Of course, the above-mentioned APP factorization can be obtained explicitly, by first finding the canonical AP factorization of the matrix (2.10) with the use of [5, Theorem 22.15] (and its analogue for the corona problem in APW^-) or Theorems 4.2, 5.3 from [7], and then proceeding to the factorization of (1.5) as described in the proof of Theorem 2.2.

Consider for example the case of a trinomial

$$f = c_{-\nu}e_{-\nu} + c_\mu e_\mu + c_\alpha e_\alpha, \quad -\lambda < -\nu < 0 < \mu < \alpha < \lambda, \; c_{-\nu}c_\mu c_\alpha \neq 0, \tag{2.13}$$

with $\mu + \nu := d \geq \lambda$.

Direct computations as outlined above show that a canonical AP factorization (2.11) is delivered by

$$H_+ = \begin{bmatrix} h_{1+} & e_\nu \\ -h_{2+} & c_\mu + c_\alpha e_{\alpha-\mu} \end{bmatrix}, \quad H_- = \begin{bmatrix} -c_{-\nu} & -e_{-\mu} \\ h_{2-} & h_{1-} \end{bmatrix}$$

with

$$h_{1+} = \frac{1}{c_\mu}\sum_{j=0}^{N}\left(-\frac{c_\alpha}{c_\mu}\right)^j e_{j(\alpha-\mu)} - \frac{1}{c_{-\nu}}\, e_d,$$

$$h_{2+} = \frac{c_\mu}{c_{-\nu}} e_\mu + \left(-\frac{c_\alpha}{c_\mu}\right)^{N+1} e_{(N+1)(\alpha-\mu)-\nu} + \frac{c_\alpha}{c_{-\nu}}\, e_\alpha,$$

$$h_{1-} = e_{-d}h_{1+}, \; h_{2-} = \frac{c_{-\nu}}{c_\mu}\sum_{j=0}^{N}\left(-\frac{c_\alpha}{c_\mu}\right)^j e_{j(\alpha-\mu)-\nu}.$$

Here N is the greatest integer such that $N(\alpha-\mu) < d-\mu = \nu$, i.e.,[1] $N = \left\lceil \frac{d-\alpha}{\alpha-\mu} \right\rceil$. An *APP* factorization of G is then given by (2.12) with $\delta = d-\lambda = \mu+\nu-\lambda$ and

$$G_+ = \begin{bmatrix} h_{1+} & e_{\lambda-\mu} \\ -h_{2+}e_{\mu+\nu-\lambda} & c_\mu + c_\alpha e_{\alpha-\mu} \end{bmatrix}, \quad G_- = \begin{bmatrix} -c_{-\nu} & -e_{\nu-\lambda} \\ h_{2-}e_{\lambda-\mu-\nu} & h_{1-} \end{bmatrix}.$$

3. Not so big gap. Existence

We now move to the consideration of matrices (1.5) with a more delicate structure of the gap in the Bohr-Fourier spectrum of f. Namely, let us suppose that f is an *AP* polynomial, the negative portion of $\Omega(f)$ consists of one point $-\nu$, and its positive portion has the endpoints α and $\mu(<\alpha)$. Denote the respective Bohr-Fourier coefficients of f by $c_{-\nu}$, c_α and c_μ. Also, let

$$n = \left\lceil \frac{\lambda}{\mu+\nu} \right\rceil$$

and suppose in addition that

$$\frac{\lambda}{\alpha+\nu} \geq n-1, \tag{3.1}$$

$$\mu \geq \alpha\left(1-\frac{1}{n}\right). \tag{3.2}$$

Theorem 3.1. *Let f be as described above. Then the matrix* (1.5) *admits an APP factorization. This factorization is canonical if and only if $\frac{\lambda}{\mu+\nu}$ is an integer or the equality is attained in* (3.1) *or* (3.2).

Observe that $n=1$ if and only if $\mu+\nu \geq \lambda$ which is a particular case of the situation of Theorem 2.2. Conditions (3.1) and (3.2) then hold automatically, and the equality in either of them is not possible. The statement of Theorem 3.1 therefore implies that G is *APP* factorable, and the factorization is canonical if and only if $\lambda = \mu+\nu$ (which is in line with Theorem 2.2).

---

[1]Here and in what follows we use the standard notation $\lceil x \rceil$ for the smallest integer upper bound of $x \in \mathbb{R}$.

Proof. We invoke the Portuguese transformation (see [5, Chapter 13]) according to which the matrix (1.5) is APP factorable only simultaneously (and has the same set of the partial AP indices) with

$$\begin{bmatrix} e_\nu & 0 \\ f_1 & e_{-\nu} \end{bmatrix}. \tag{3.3}$$

Here f_1 is an AP polynomial constructed from f explicitly as described in [5, Sections 13.3–13.4]. By this construction, $\Omega(f_1)$ lies in the set

$$-\lambda + \sum_j k_j \gamma_j, \tag{3.4}$$

where k_j are non-negative integers and γ_j are the distances from $-\nu$ to the remaining points of $\Omega(f)$. Moreover, property (2.2) for the matrix (3.3) means that the terms of $\Omega(f_1)$ lying outside $(-\nu, \nu)$ can be dropped.

Apparently, for our structure of $\Omega(f)$

$$\mu + \nu \le \gamma_j \le \alpha + \nu.$$

Denoting $\sum k_j = k$, we conclude from here that

$$-\lambda + \sum_j k_j \gamma_j \le -\lambda + k(\alpha + \nu) \le (k - n + 1)(\alpha + \nu)$$

due to (3.1), and

$$-\lambda + \sum_j k_j \gamma_j \ge -\lambda + k(\mu + \nu) \ge (k - n)(\mu + \nu)$$

by the definition of n. So, there are only two values of k for which $-\lambda + \sum_j k_j \gamma_j$ may lie between $-\nu$ and ν: $k = n - 1$ and $k = n$.

The biggest exponent corresponding to $k = n-1$ is $\beta_1 = -\lambda + (n-1)(\alpha+\nu)$, and the smallest exponent corresponding to $k = n$ is $\alpha_1 = -\lambda + n(\mu + \nu)$. The respective coefficients of f_1 can be computed by formulas (13.40), (13.41) from [5] and are equal to

$$c_{-\nu}^{-1}(-c_\alpha/c_{-\nu})^{n-1} \text{ and } c_{-\nu}^{-1}(-c_\mu/c_{-\nu})^n.$$

It is important in what follows that they are different from zero.

Observe also that $\beta_1 \le 0$ due to (3.1) while $\alpha_1 \ge 0$ due to the definition of n. Moreover,

$$\alpha_1 - \beta_1 = -\lambda + n(\mu + \nu) - (-\lambda + (n-1)(\alpha + \nu)) = n\mu - (n-1)\alpha + \nu \ge \nu$$

because of (3.2).

If both α_1 and β_1 fall between $-\nu$ and ν, the matrix (3.3) satisfies the conditions of Theorem 2.2, with an obvious change of notation. By Remark 2.3, it is APP factorable. Its partial AP indices equal $\pm(n\mu - (n-1)\alpha)$ according to Theorem 2.2. The same is then true for the matrix (1.5), so that in particular its AP factorization is canonical (in this case) if and only if the equality holds in (3.2).

If $\beta_1 \le -\nu$, then $\Omega(f_1)$ lies to the right of α_1. This situation falls into the so-called *one-sided* case (see [5, Theorem 14.1]) in which the matrix (3.3) is APP

factorable with the partial *AP* indices equal $\pm\min(\alpha_1, \nu)$. Consequently, the matrix (1.5) also is *APP* factorable, and the factorization is canonical if and only if $\alpha_1 = 0$, that is, $\lambda/(\mu+\nu)$ is an integer.

In a similar fashion, the matrices (3.3) and (1.5) remain *APP* factorable if $\alpha_1 \geq \nu$, and the partial *AP* indices in this case equal $\pm\min(-\beta_1, \nu)$. The canonical *AP* factorization takes place if and only if $\beta_1 = 0$, that is, when equality holds in (3.1). □

4. Auxiliary result

Theorem 3.1 delivers the *AP* factorability criterion and formulas for the partial *AP* indices of the matrix function (1.5) under conditions (3.1), (3.2). In principle, the *AP* factorization itself can also be produced by carrying out the Portuguese transformation and then using the explicit construction of the *AP* factorization in the big gap case (implied by Theorems 2.1, 2.2) or in the one-sided case.

However, there is a more direct approach to the factorization construction for matrix functions of the type (1.5) with a priori known partial *AP* indices. We now present this approach.

Let us continue using notation (2.1) for the *AP* factorization of matrices (1.5), with

$$G_+ = \left[g_{ij}^+\right]_{i,j=1,2}, \quad \det G_+ = 1, \tag{4.1}$$

$$G_-^{-1} = \left[g_{ij}^-\right]_{i,j=1,2}, \quad \det G_- = 1. \tag{4.2}$$

It is easy to see that, on the one hand,

$$\Phi_+ = e_\delta(g_{12}^+, g_{22}^+), \quad \Phi_- = (g_{12}^-, g_{22}^-)$$

satisfy the Riemann-Hilbert problem

$$G\Phi_- = \Phi_+, \quad \Phi_\pm \in APW_2^\pm; \tag{4.3}$$

on the other hand

$$\Psi_+ = (g_{11}^+, g_{21}^+), \quad \Psi_- = (g_{11}^-, g_{21}^-)$$

satisfy the Riemann-Hilbert problem

$$G\Psi_- = e_{-\delta}\Psi_+, \quad \Psi_\pm \in APW_2^\pm. \tag{4.4}$$

Since $\det G_\pm = 1$, we also have

$$\lim_{y\to+\infty} \det\left[\Psi_+(iy),\ (e_{-\delta}\Phi_+)(iy)\right] = \lim_{y\to-\infty} \det\left[\Psi_-(iy),\ \Phi_-(iy)\right] = 1.$$

In the following theorem we show that the converse is true.

Theorem 4.1. *Let* $G \in APW_{2\times 2}$ *of the form* (1.5) *admit an AP factorization with partial indices* $\pm\delta$. *Let also* Φ_+, Φ_- *be a non-trivial solution to* (4.3) *such that*

$$\widetilde{\Phi}_+ = e_{-\delta}\Phi_+ \in APW_2^+, \tag{4.5}$$

and Ψ_+, Ψ_- *be a solution to* (4.3) *such that*

$$\lim_{y\to+\infty} \det\left[\Psi_+(iy),\ \widetilde{\Phi}_+(iy)\right] = \lim_{y\to-\infty} \det\left[\Psi_-(iy),\ \Phi_-(iy)\right] = 1. \tag{4.6}$$

Then an AP factorization of G is given by (2.1), *where*

$$G_+ = \left[\Psi_+, \widetilde{\Phi}_+\right], \quad G_-^{-1} = \left[\Psi_-, \Phi_-\right]. \tag{4.7}$$

Proof. From (4.3), (4.4) and (4.5) it follows that $G_+ \in APW^+_{2\times 2}$, $G_-^{-1} \in APW^-_{2\times 2}$ and

$$GG_-^{-1}\operatorname{diag}[e_\delta, e_{-\delta}] = G_+. \tag{4.8}$$

In particular,

$$\det G_+ = \det G_-^{-1},$$

so that the latter determinants are actually constant. From (4.6) it follows that this constant is 1. Consequently, (4.8) can be rewritten as (2.1), with $G_\pm$ satisfying (4.1), (4.2). □

5. Factorization construction for some trinomials

Let us return to matrix functions (1.5) with trinomial off-diagonal entry f given by (2.13). In this section we will apply Theorem 4.1 to obtain an AP factorization of G provided that conditions (3.1), (3.2) hold with $n = 2$, that is,

$$\lambda/2 \le \mu + \nu < \lambda, \quad \alpha + \nu \le \lambda, \text{ and } \alpha > \mu \ge \alpha/2. \tag{5.1}$$

Recall that the case $n = 1$ was disposed of in Section 2. The case of general n will be considered elsewhere.

We denote by $\mathcal{T}$ the class of all matrix functions (1.5) with the trinomial f given by (2.13) and satisfying (5.1).

First we determine a solution to the Riemann-Hilbert problem (4.3). Let

$$\widetilde{\nu} = \min\left\{\nu, \frac{\lambda-\alpha}{2}\right\}, \quad \widetilde{\mu} = \frac{\lambda}{2} - \widetilde{\nu}. \tag{5.2}$$

Then

$$f = e_{-\widetilde{\nu}}\widetilde{f}_- + e_{\widetilde{\mu}}\widetilde{f}_+,$$

where

$$\widetilde{f}_- = c_{-\nu}e_{-\nu+\widetilde{\nu}}, \qquad \widetilde{f}_+ = c_\mu e_{\mu-\widetilde{\mu}} + c_\alpha e_{\alpha-\widetilde{\mu}}.$$

Denoting

$$g = f\left(e_{-\widetilde{\nu}}\widetilde{f}_- - e_{\widetilde{\mu}}\widetilde{f}_+\right) = e_{-2\widetilde{\nu}}\widetilde{f}_-^2 - e_{2\widetilde{\mu}}\widetilde{f}_+^2,$$

where $\widetilde{f}_\pm \in APW^\pm$, observe that the Riemann-Hilbert problem

$$G\left(\Phi_{1-}, \Phi_{2-}\right) = \left(\Phi_{1+}, \Phi_{2+}\right)$$

is equivalent to

$$G_1\left(\frac{\Phi_{1-}}{e_{-\widetilde{\nu}}\widetilde{f}_- - e_{\widetilde{\mu}}\widetilde{f}_+}, \Phi_{2-}\right) = \left(\frac{\Phi_{1+}}{e_{-\widetilde{\nu}}\widetilde{f}_- - e_{\widetilde{\mu}}\widetilde{f}_+}, \Phi_{2+}\right),$$

where

$$G_1 = \begin{bmatrix} e_\lambda & 0 \\ g & e_{-\lambda} \end{bmatrix}.$$

It can be easily checked (see also [7, Theorem 4.2] for a similar reasoning) that a solution of the Riemann-Hilbert problem

$$G_1\eta_- = \eta_+; \quad \eta_\pm = (\eta_{1\pm}, \eta_{2\pm}) \in APW_2^\pm$$

is given by

$$\eta_{1+} = e_{2\widetilde{\nu}}, \ \eta_{2+} = -\widetilde{f}_+^2, \ \eta_{1-} = e_{-2\widetilde{\mu}}, \ \eta_{2-} = -\widetilde{f}_-^2.$$

Therefore, it is clear that

$$\Phi_{1\pm} = \eta_{1\pm}\left(e_{-\widetilde{\nu}}\widetilde{f}_- - e_{\widetilde{\mu}}\widetilde{f}_+\right), \ \Phi_{2\pm} = \eta_{2\pm} \tag{5.3}$$

give a solution to (4.3) if $\Phi_{1\pm} \in APW^\pm$.

We have

$$\eta_{1+}e_{-\widetilde{\nu}}\widetilde{f}_- = c_{-\nu}e_{2\widetilde{\nu}-\nu}.$$

If $\widetilde{\nu} = \nu$, then $2\widetilde{\nu} - \nu = \nu > 0$; if $\widetilde{\nu} = \frac{\lambda-\alpha}{2}$, then $2\widetilde{\nu} - \nu = \lambda - (\alpha + \nu) \geq 0$ due to the second inequality in (5.1). Thus, $\eta_{1+}e_{-\widetilde{\nu}}\widetilde{f}_-$ lies in APW^+. By (5.3), so does Φ_{1+}.

As for Φ_{1-}, we have

$$\eta_{1-}e_{\widetilde{\mu}}\widetilde{f}_+ = c_\mu e_{\mu-2\widetilde{\mu}} + c_\alpha e_{\alpha-2\widetilde{\mu}} = c_\mu e_{\mu+2\widetilde{\nu}-\lambda} + c_\alpha e_{\alpha+2\widetilde{\nu}-\lambda}.$$

If $\widetilde{\nu} = \nu$, then by (5.2) $\nu \leq \frac{\lambda-\alpha}{2}$ and it follows that $\alpha + 2\widetilde{\nu} - \lambda = \alpha + 2\nu - \lambda \leq 0$; if $\widetilde{\nu} = \frac{\lambda-\alpha}{2}$, then $\alpha + 2\widetilde{\nu} - \lambda = 0$. Either way, $\eta_{1-}e_{\widetilde{\mu}}\widetilde{f}_+ \in APW^-$ and therefore $\Phi_{1-} \in APW^-$ as well.

We have thus proved the following.

Theorem 5.1. *Let $G \in \mathcal{T}$ and let $\widetilde{\nu}$ be defined by* (5.2). *Then the Riemann-Hilbert problem* (4.3) *has a solution $\Phi_\pm = (\Phi_{1\pm}, \Phi_{2\pm})$ with*

$$\begin{aligned}
\Phi_{1+} &= c_{-\nu}e_{2\widetilde{\nu}-\nu} - c_\mu e_{2\widetilde{\nu}+\mu} - c_\alpha e_{2\widetilde{\nu}+\alpha}, \\
\Phi_{2+} &= -\left(c_\mu e_{\mu+\widetilde{\nu}-\lambda/2} + c_\alpha e_{\alpha+\widetilde{\nu}-\lambda/2}\right)^2, \\
\Phi_{1-} &= e_{-\lambda}\Phi_{1+}, \\
\Phi_{2-} &= -c_{-\nu}^2 e_{2(\widetilde{\nu}-\nu)}.
\end{aligned}$$

Since matrix functions $G \in \mathcal{T}$ are *AP* factorable (by Theorem 3.1) and the values of their partial *AP* indices $\pm\delta$ are known, we can now follow Theorem 4.1 to compute the factors $G_\pm$. To this end, it is convenient to introduce the notation

$$\gamma_1 = \mu + \nu, \quad \gamma_2 = \alpha + \nu. \tag{5.4}$$

First we derive formulas for the second columns of G_+, G_-^{-1}, sticking to the notation of (4.7).

<u>Case 1</u>: $\alpha + \nu - \lambda \leq -\nu < 0 < \nu \leq 2(\mu + \nu) - \lambda$. As is shown in the proof of Theorem 3.1, then $\delta = \nu$.

Definition (5.2) implies that in this case $\widetilde{\nu} = \nu$. From (4.5) therefore:

$$\begin{aligned} \widetilde{\Phi}_{1+} &= c_{-\nu} - c_\mu e_{\gamma_1} - c_\alpha e_{\gamma_2}, \\ \widetilde{\Phi}_{2+} &= -\left(c_\mu^2 e_{2\gamma_1-\gamma_2+\alpha-\lambda} + 2c_\mu c_\alpha e_{\gamma_1+\alpha-\lambda} + c_\alpha^2 e_{\gamma_2+\alpha-\lambda}\right), \end{aligned} \tag{5.5}$$

while Φ_- is given by

$$\begin{aligned} \Phi_{1-} &= c_{-\nu} e_{\gamma_2-\alpha-\lambda} - c_\mu e_{\gamma_1+\gamma_2-\lambda-\alpha} - c_\alpha e_{2\gamma_2-\lambda-\alpha}, \\ \Phi_{2-} &= -c_{-\nu}^2. \end{aligned} \tag{5.6}$$

<u>Case 2</u>: $\alpha + \nu - \lambda \leq -\nu < 0 \leq 2(\mu+\nu) - \lambda < \nu$. As shown in the proof of Theorem 3.1, then $\delta = 2(\mu+\nu) - \lambda$.

As in Case 1, $\widetilde{\nu} = \nu$. Formulas for the elements of $\widetilde{\Phi}_+$ change to

$$\begin{aligned} \widetilde{\Phi}_{1+} &= c_{-\nu} e_{-2\gamma_1+\gamma_2-\alpha+\lambda} - c_\mu e_{-\gamma_1+\gamma_2-\alpha+\lambda} - c_\alpha e_{-2\gamma_1+2\gamma_2-\alpha+\lambda}, \\ \widetilde{\Phi}_{2+} &= -\left(c_\mu^2 + 2c_\mu c_\alpha e_{\gamma_2-\gamma_1} + c_\alpha^2 e_{2(\gamma_2-\gamma_1)}\right), \end{aligned} \tag{5.7}$$

while formulas for Φ_- remain the same as in Case 1.

<u>Case 3</u>: $-\nu < \alpha + \nu - \lambda \leq 0 < \nu \leq 2(\mu+\nu) - \lambda$. As shown in the proof of Theorem 3.1, then $\delta = \lambda - (\alpha+\nu)$.

In this case $\widetilde{\nu} = \frac{\lambda-\alpha}{2}$ and from Theorem 5.1 we find that elements of $\widetilde{\Phi}_+$ are given by the same formulas (5.5) as in Case 1 while

$$\begin{aligned} \Phi_{1-} &= c_{-\nu} e_{-\gamma_2} - c_\mu e_{\gamma_1-\gamma_2} - c_\alpha, \\ \Phi_{2-} &= -c_{-\nu}^2 e_{-2\gamma_2+\lambda+\alpha}. \end{aligned} \tag{5.8}$$

<u>Case 4</u>: $-\nu < \alpha + \nu - \lambda \leq 0 \leq 2(\mu+\nu) - \lambda < \nu$. As shown in the proof of Theorem 3.1, then $\delta = 2\mu - \alpha$.

As in Case 3, $\widetilde{\nu} = \frac{\lambda-\alpha}{2}$. Formulas for Φ_- remain the same as in this case, while $\widetilde{\Phi}_+$ are given by the formulas from Case 2.

Having determined the second column of G_+ and G_-^{-1}, we now move on to computing the first column, that is, to constructing a solution of (4.4) under the additional condition (4.6). To this end, it is convenient to rewrite (4.4) in the scalar form: determine $\Psi_{1+}, \Psi_{2+} \in APW^+$, $\Psi_{2-} \in APW^-$ such that

$$(c_{-\nu} + c_\mu e_{\gamma_1} + c_\alpha e_{\gamma_2})\Psi_{1+} = e_{\lambda+\nu}\Psi_{2+} - e_{\delta+\nu}\Psi_{2-} \tag{5.9}$$

while

$$\Omega(\Psi_{1+}) \subset [0, \lambda+\delta]. \tag{5.10}$$

The latter condition is equivalent to the requirement $\Psi_{1-}\ (= e_{-\lambda-\delta}\Psi_{1+}) \in APW^-$.

Once again, we consider separately the four cases corresponding to different formulas for the partial AP indices.

<u>Case 1</u>. Since $\delta = \nu$, conditions (5.9) and (5.10) take the form

$$(c_{-\nu} + c_\mu e_{\gamma_1} + c_\alpha e_{\gamma_2})\Psi_{1+} = e_{\lambda+\gamma_2-\alpha}\Psi_{2+} - e_{2(\gamma_2-\alpha)}\Psi_{2-}$$

and

$$\Omega(\Psi_{1+}) \subset [0, \lambda + \gamma_2 - \alpha],$$

respectively. From (5.6) we see that $0 \in \Omega(\Phi_{2-})$ while $0 \notin \Omega(\Phi_{1-})$. For (4.6) to be satisfied it is therefore necessary that $0 \in \Omega(\Psi_{1-})$, that is,

$$\lambda + \nu \ (= \lambda + \gamma_2 - \alpha) \in \Omega(\Psi_{1+}).$$

In fact, setting

$$\Psi_{1+} = e_{\lambda+\gamma_2-\alpha},\ \Psi_{2+} = c_{-\nu} + c_\mu e_{\gamma_1} + c_\alpha e_{\gamma_2},\ \Psi_{1-} = 1,\ \Psi_{2-} = 0 \tag{5.11}$$

in this case yields the desired solution.

Case 2. Since $\delta = 2\gamma_1 - \lambda$, conditions (5.9) and (5.10) now take the form

$$(c_{-\nu} + c_\mu e_{\gamma_1} + c_\alpha e_{\gamma_2})\Psi_{1+} = e_{\lambda+\gamma_2-\alpha}\Psi_{2+} - e_{2\gamma_1+\gamma_2-\lambda-\alpha}\Psi_{2-} \tag{5.12}$$

and

$$\Omega(\Psi_{1+}) \subset [0, 2\gamma_1]. \tag{5.13}$$

As in Case 1, from (5.6) we conclude that $0 \in \Omega(\Psi_{1-})$. In the current case it is equivalent to $2\gamma_1 \in \Omega(\Psi_{1+})$.

On the other hand, from (5.7) we see that $0 \in \Omega(\widetilde{\Phi}_{2+})$, $0 \notin \Omega(\widetilde{\Phi}_{1+})$ (unless $2\mu + \nu = \lambda$), so that we need $0 \in \Omega(\Psi_{1+})$ in order to satisfy (4.6).

Taking this as a starting point we can use a method of recursive construction, following the ideas of [7, Section 4.2], to determine a solution to (5.12) under condition (5.13). Denoting by M the smallest integer such that

$$\gamma_1 + M(\gamma_2 - \gamma_1) \geq \lambda - \alpha, \tag{5.14}$$

i.e., letting

$$M = \left\lceil \frac{\lambda - \alpha - \gamma_1}{\gamma_2 - \gamma_1} \right\rceil = \left\lceil \frac{\lambda - (\alpha + \mu + \nu)}{\alpha - \mu} \right\rceil, \tag{5.15}$$

we thus obtain

$$\Psi_{1+} = \frac{c_\mu^2}{c_{-\nu}^2} e_{2\gamma_1} + \sum_{j=0}^{M} (-1)^j (j+1) \left(\frac{c_\alpha}{c_\mu}\right)^j e_{j(\gamma_2-\gamma_1)} \tag{5.16}$$

$$+ \sum_{j=0}^{M} \frac{(-1)^{j+1} c_\alpha^j}{c_{-\nu} c_\mu^{j-1}} e_{\gamma_1+j(\gamma_2-\gamma_1)} + (-1)^{M+1}(M+1) \frac{c_\alpha^{M+1}}{c_\mu^M c_{-\nu}} e_{\gamma_1+(M+1)(\gamma_2-\gamma_1)},$$

$$\Psi_{2+} = \frac{c_\mu^3}{c_{-\nu}^2} e_{3\gamma_1-\gamma_2+\alpha-\lambda} + \frac{c_\mu^2 c_\alpha}{c_{-\nu}^2} e_{2\gamma_1+\alpha-\lambda} \tag{5.17}$$

$$+ (-1)^{M+1} \frac{(M+2) c_\alpha^{M+1}}{c_{-\nu} c_\mu^{M-1}} e_{\gamma_1+M(\gamma_2-\gamma_1)+\alpha-\lambda}$$

$$+ (-1)^{M+1} \frac{(M+1) c_\alpha^{M+2}}{c_{-\nu} c_\mu^{M}} e_{\gamma_1+(M+1)(\gamma_2-\gamma_1)+\alpha-\lambda},$$

$$\Psi_{1-} = e_{-2\gamma_1}\Psi_{1+}, \tag{5.18}$$

$$\Psi_{2-} = -\sum_{j=0}^{M} (-1)^j (j+1) \frac{c_\alpha^j c_{-\nu}}{c_\mu^j} e_{-3\gamma_1+(j-1)(\gamma_2-\gamma_1)+\alpha+\lambda}. \tag{5.19}$$

Simple computations show that, along with (5.12), also (4.6) holds. It remains to check that $\Psi_{2\pm} \in APW^{\pm}$ and that (5.13) holds.

In (5.17), the exponent $3\gamma_1 - \gamma_2 + \alpha - \lambda$ is non-negative because $2\gamma_1 \geq \lambda$ and $\gamma_1 - \gamma_2 + \alpha = \mu \geq 0$; $\gamma_1 + M(\gamma_2 - \gamma_1) + \alpha - \lambda$ is non-negative due to (5.15). The other two exponents are obtained from these two by adding a positive quantity $\gamma_2 - \gamma_1$, so that indeed $\Psi_{2+} \in APW^{+}$.

Now observe that (5.15) implies

$$\gamma_1 + (M-1)(\gamma_2 - \gamma_1) < \lambda - \alpha. \tag{5.20}$$

Consequently, the biggest exponent in the right-hand side of (5.19), $-3\gamma_1 + \alpha + \lambda + (M-1)(\gamma_2 - \gamma_1)$, does not exceed $-4\gamma_1 + 2\lambda$, and is therefore non-positive. Finally, (5.20) also implies that

$$\gamma_1 + (M+1)(\gamma_2 - \gamma_1) \leq \lambda - \alpha + 2(\gamma_2 - \gamma_1) = \lambda + \alpha - 2\mu \leq \lambda.$$

Thus, all the exponents in the right-hand side of (5.16) lie between 0 and λ.

Case 3. In this case $\delta = \lambda - \gamma_2$, so that (5.9), (5.10) take the form

$$(c_{-\nu} + c_\mu e_{\gamma_1} + c_\alpha e_{\gamma_2})\Psi_{1+} = e_{\lambda+\gamma_2-\alpha}\Psi_{2+} - e_{\lambda-\alpha}\Psi_{2-},$$

$$\Omega(\Psi_{1+}) \subset [0, 2\lambda - \gamma_2].$$

From (5.5) we have $0 \in \Omega(\widetilde{\Phi}_{1+})$, $0 \notin \Omega(\widetilde{\Phi}_{2+})$ (unless $2\mu + \nu = \lambda$), so that $0 \in \Omega(\Psi_{2+})$. Similarly, from (5.8) we have that $0 \in \Omega(\Phi_{1-})$, $0 \notin \Omega(\Phi_{2-})$ (unless $\alpha + 2\nu = \lambda$), and therefore $0 \in \Omega(\Psi_{2-})$. In any case, we conclude that $\lambda - \alpha \in \Omega(\Psi_{1+})$.

Proceeding as in Case 2, we obtain

$$\Psi_{1+} = \sum_{j=0}^{N} \left(-\frac{c_\mu}{c_\alpha}\right)^j e_{\lambda-\alpha-j(\gamma_2-\gamma_1)} + \frac{(-1)^{N+1} c_\mu^{N+1}}{c_\alpha^N c_{-\nu}} e_{\lambda-\alpha+\gamma_1-N(\gamma_2-\gamma_1)} \tag{5.21}$$

$$\Psi_{2+} = c_\alpha - \frac{(-1)^N c_\mu^{N+2}}{c_\alpha^N c_{-\nu}} e_{\gamma_1-(N+1)(\gamma_2-\gamma_1)} - \frac{(-1)^N c_\mu^{N+1}}{c_\alpha^{N-1} c_{-\nu}} e_{\gamma_1-N(\gamma_2-\gamma_1)} \tag{5.22}$$

$$\Psi_{1-} = e_{-2\lambda+\gamma_2}\Psi_{1+}, \quad \Psi_{2-} = -\sum_{j=0}^{N} \left(-\frac{c_\mu}{c_\alpha}\right)^j c_{-\nu} e_{-j(\gamma_2-\gamma_1)}, \tag{5.23}$$

where N is the smallest integer such that

$$2\gamma_1 - (N-1)(\gamma_2 - \gamma_1) \leq \lambda + \alpha,$$

i.e.,

$$N = \left\lceil \frac{-\lambda - \alpha + \gamma_1 + \gamma_2}{\gamma_2 - \gamma_1} \right\rceil = \left\lceil \frac{\mu + 2\nu - \lambda}{\alpha - \mu} \right\rceil. \tag{5.24}$$

To check that $\Psi_{j\pm} \in APW^{\pm}$ $(j = 1, 2)$ we reason as follows.

The smallest exponent in the right-hand side of (5.21) is $\lambda - \alpha - N(\gamma_2 - \gamma_1)$ which is non-negative because

$$N(\gamma_2 - \gamma_1) = \gamma_2 - \gamma_1 + (N-1)(\gamma_2 - \gamma_1) < -\lambda - \alpha + 2\gamma_2$$

due to (5.24), and $\gamma_2 < \lambda$. Thus, $\Psi_{1+} \in APW^+$. On the other hand,

$$\lambda - \alpha - j(\gamma_2 - \gamma_1) \le \lambda - \alpha < 2\lambda - \gamma_2 \text{ for } j = 0, \ldots, N,$$

while

$$\lambda - \alpha + \gamma_1 - N(\gamma_2 - \gamma_1) \le 2\lambda - \gamma_2,$$

again due to (5.24). Consequently, $\Omega(\Psi_{1+}) \subset [0, 2\lambda - \gamma_2]$, and $\Psi_{1-} \in APW^-$.

The smallest exponent in the right-hand side of (5.22) is $\gamma_1 - (N+1)(\gamma_2 - \gamma_1)$, and due to (5.24) it is bigger than $\lambda + \alpha + (2\gamma_1 - \gamma_2) - 2\gamma_2$. But in the case under consideration $2\gamma_1 - \gamma_2 \ge \lambda - \alpha$ (because $2\mu + \nu \ge \lambda$) and $\lambda - \gamma_2 \ge 0$. Hence, $\Psi_{2+} \in APW^+$.

Of course, from (5.23) it is clear that $\Psi_{2-} \in APW^-$.

Case 4. We have now $\delta = \alpha - 2(\gamma_2 - \gamma_1)$, and (5.9), (5.10) take the form

$$(c_{-\nu} + c_\mu e_{\gamma_1} + c_\alpha e_{\gamma_2})\Psi_{1+} = e_{\lambda+\gamma_2-\alpha}\Psi_{2+} - e_{2\gamma_1-\gamma_2}\Psi_{2-}, \tag{5.25}$$

$$\Omega(\Psi_{1+}) \subset [0, \lambda + \alpha - 2(\gamma_2 - \gamma_1)]. \tag{5.26}$$

From (5.7) we see that $0 \in \Omega(\widetilde{\Phi}_{2+})$, $0 \notin \Omega(\widetilde{\Phi}_{1+})$ (unless $2\mu + \nu = \lambda$), and from (5.8) we see that $0 \in \Omega(\Phi_{1-})$, $0 \notin \Omega(\Phi_{2-})$ (unless $\alpha + 2\nu = \lambda$). Thus we should look for a solution to (4.4) satisfying

$$0, 2\gamma_1 - \gamma_2 \in \Omega(\Psi_{1+}).$$

The recursive construction yields the following:

$$\begin{aligned}\Psi_{1+} =& \sum_{j=1}^{K} \frac{(-1)^{j+1} c_\mu^{j+1}}{c_{-\nu} c_\alpha^j} e_{\gamma_1 - j(\gamma_2-\gamma_1)} + \frac{(-1)^K c_\mu^{K+2}}{c_{-\nu}^2 c_\alpha^K} e_{2\gamma_1 - K(\gamma_2-\gamma_1)} \\ &+ (-1)^{M+1} \frac{(M+1) c_\alpha^{M+1}}{c_\mu^M c_{-\nu}} e_{\gamma_1 + (M+1)(\gamma_2-\gamma_1)} \\ &+ \sum_{j=0}^{M} \frac{(-1)^{j+1} c_\alpha^j}{c_{-\nu} c_\mu^{j-1}} e_{\gamma_1 + j(\gamma_2-\gamma_1)} + \sum_{j=0}^{M} (-1)^j (j+1) \frac{c_\alpha^j}{c_\mu^j} e_{j(\gamma_2-\gamma_1)},\end{aligned} \tag{5.27}$$

$$\begin{aligned}\Psi_{2+} =& e_{2\gamma_1-\gamma_2-\lambda+\alpha} \left[\frac{(-1)^K c_\mu^{K+3}}{c_{-\nu}^2 c_\alpha^K} e_{\gamma_1 - K(\gamma_2-\gamma_1)} \right. \\ &+ \frac{(-1)^K c_\mu^{K+2}}{c_{-\nu}^2 c_\alpha^{K-1}} e_{\gamma_1 - (K-1)(\gamma_2-\gamma_1)} \\ &+ (-1)^{M+1} \frac{(M+2) c_\alpha^{M+1}}{c_\mu^{M-1} c_{-\nu}} e_{(M+1)(\gamma_2-\gamma_1)} \\ &\left. + (-1)^{M+1} \frac{(M+1) c_\alpha^{M+2}}{c_\mu^M c_{-\nu}} e_{(M+2)(\gamma_2-\gamma_1)} \right],\end{aligned} \tag{5.28}$$

$$\Psi_{1-} = e_{-\lambda-\alpha+2(\gamma_2-\gamma_1)} \Psi_{1+}, \tag{5.29}$$

$$\begin{aligned}\Psi_{2-} = &- e_{\gamma_2-\gamma_1}\left[\sum_{j=0}^{M}(-1)^j(j+1)\frac{c_\alpha^j}{c_\mu^j}e_{-\gamma_1+j(\gamma_2-\gamma_1)}\right.\\ &\left.+\sum_{j=1}^{K}(-1)^{j+1}\frac{c_\mu^{j+1}}{c_\alpha^j}e_{-j(\gamma_2-\gamma_1)}\right].\end{aligned} \tag{5.30}$$

Here M is given by (5.15) while K is the smallest integer such that

$$2\gamma_2 - K(\gamma_2-\gamma_1) \le \lambda+\alpha, \tag{5.31}$$

i.e.,

$$K = \left\lceil \frac{2\gamma_2-\lambda-\alpha}{\gamma_2-\gamma_1}\right\rceil.$$

As in the preceding case, we now need to check that $\Psi_{j\pm} \in APW^{\pm}$, $j = 1, 2$.

From the definition of K it follows in particular that

$$(K-1)(\gamma_2-\gamma_1) < 2\gamma_2-\lambda-\alpha. \tag{5.32}$$

Consequently,

$$\begin{aligned}\gamma_1 - K(\gamma_2-\gamma_1) &= 2\gamma_1-\gamma_2-(K-1)(\gamma_2-\gamma_1) > \lambda-\gamma_2+(\alpha-2(\gamma_2-\gamma_1))\\ &= (\lambda-\gamma_2)+(2\mu-\alpha)\end{aligned}$$

is non-negative. It follows then that all the exponents in the right-hand side of (5.27) are non-negative, that is, $\Psi_{1+} \in APW^+$.

Inequality (5.31) implies directly that

$$2\gamma_1 - K(\gamma_2-\gamma_1) \le \lambda+\alpha-2(\gamma_2-\gamma_1).$$

On the other hand, from the definition of M we have:

$$\gamma_1+(M+1)(\gamma_2-\gamma_1) < \lambda-\alpha+2(\gamma_2-\gamma_1),$$

and therefore

$$\gamma_1+(M+1)(\gamma_2-\gamma_1) < \lambda+\alpha-2(\gamma_2-\gamma_1),$$

because $\alpha-2(\gamma_2-\gamma_1) = 2\mu-\alpha \ge 0$ by the last inequality in (5.1). Thus, all the exponents in the right-hand side of (5.27) do not exceed $\lambda+\alpha-2(\gamma_2-\gamma_1)$. Consequently, (5.29) defines a function in APW^-.

Property $\Psi_{2+} \in APW^+$ boils down to

$$3\gamma_1-\gamma_2-\lambda+\alpha-K(\gamma_2-\gamma_1) \ge 0, \quad 2\gamma_1-\gamma_2-\lambda+\alpha+(M+1)(\gamma_2-\gamma_1) \ge 0.$$

The first of the exponents in question equals $4\gamma_1-2\gamma_2-\lambda+\alpha-(K-1)(\gamma_2-\gamma_1)$, and due to (5.32) is bounded below by $2(\alpha-2(\gamma_2-\gamma_1)) = 2(2\mu-\alpha) \ge 0$; the second is non-negative simply because of (5.14).

Finally, $\Psi_{2-} \in APW^-$ because the biggest exponent in the right-hand side of (5.30) is

$$\gamma_2-2\gamma_1+M(\gamma_2-\gamma_1) < \lambda-2\gamma_1+2(\gamma_2-\gamma_1)-\alpha = (\lambda-2\gamma_1)+(\alpha-2\mu) \le 0.$$

We have thus proved the following.

Theorem 5.2. *Let $G \in \mathcal{T}$. Then G admits an APP factorization*

$$G = G_+ \operatorname{diag}[e_{-\delta}, e_\delta] G_-,$$

where the factors

$$G_+ = \begin{bmatrix} \Psi_{1+} & \widetilde{\Phi}_{1+} \\ \Psi_{2+} & \widetilde{\Phi}_{2+} \end{bmatrix}, \quad G_- = \begin{bmatrix} \Phi_{2-} & -\Phi_{1-} \\ -\Psi_{2-} & \Psi_{1-} \end{bmatrix}$$

and the partial AP indices $\pm\delta$ are computed according to the rule:

(i) *If $\alpha + \nu - \lambda \le -\nu < 0 < \nu \le 2(\mu + \nu) - \lambda$, then $\delta = \nu$ and elements of $G_\pm$ are given by* (5.11), (5.5), (5.6).

(ii) *If $\alpha + \nu - \lambda \le -\nu < 0 \le 2(\mu + \nu) - \lambda < \nu$, then $\delta = 2(\mu + \nu) - \lambda$ and elements of $G_\pm$ are given by* (5.16)–(5.19), (5.6), (5.7).

(iii) *If $-\nu < \alpha + \nu - \lambda \le 0 < \nu \le 2(\mu + \nu) - \lambda$, then $\delta = \lambda - (\alpha + \nu)$ and elements of $G_\pm$ are given by* (5.5), (5.8) and (5.21)–(5.23).

(iv) *If $-\nu < \alpha + \nu - \lambda \le 0 \le 2(\mu + \nu) - \lambda < \nu$, then $\delta = 2\mu - \alpha$ and elements of $G_\pm$ are given by* (5.7), (5.8) *and* (5.27)–(5.30).

In particular, Theorem 5.2 gives explicit formulas for the canonical *AP* factorization of G in the cases when it exists, that is $\mu + \nu = \lambda/2$, $\alpha + \nu = \lambda$, or $2\mu = \alpha$.

6. Bohr-Fourier spectra of the factorization factors

It is known [3, 13] that if G is an arbitrary $n \times n$ matrix function admitting a canonical *AP* factorization (1.3) and the Bohr-Fourier spectrum $\Omega(G)$ of G lies in some additive subgroup Σ of $\mathbb{R}$, then also

$$\Omega(G_+^{\pm 1}), \Omega(G_-^{\pm 1}) \subset \Sigma.$$

This point is illustrated in particular by formulas from Theorem 5.2, corresponding to the cases $\mu + \nu = \lambda/2$, $\alpha + \nu = \lambda$, or $2\mu = \alpha$. A closer look, however, suggests that there is an additional structure to the Bohr-Fourier spectra of the individual entries. Namely, the Bohr-Fourier spectra of the diagonal elements of $G_+^{\pm 1}$ and the off diagonal entries of $G_-^{\pm 1}$ actually belong to a smaller subgroup Σ_0 generated just by $\Omega(f)$, while the remaining entries of $G_\pm^{\pm 1}$ have their Bohr-Fourier spectra in $\pm\lambda + \Sigma_0$. The next theorem shows that this is no coincidence but rather a general property of the matrices (1.5).

Theorem 6.1. *Let f be an AP function with the Bohr-Fourier spectrum $\Omega(f)$ contained in the subgroup Σ_0 of $\mathbb{R}$. Suppose that the matrix function* (1.5) *admits a canonical factorization* (1.3). *Then this factorization can be chosen in such a way that: the diagonal entries of $G_+^{\pm 1}$ and the off diagonal entries of $G_-^{\pm 1}$ also have the Bohr-Fourier spectra contained in Σ_0; the* (1,2) *entry of $G_+^{\pm 1}$,* (1,1) *entry of G_- and* (2,2) *entry of G_-^{-1} have the Bohr-Fourier spectra located in $\lambda + \Sigma_0$; the* (2,1) *entry of* $G_+^{\pm 1}$, (2,2) *entry of* G_- *and* (1,1) *entry of* G_-^{-1} *have the Bohr-Fourier spectra located in* $-\lambda + \Sigma_0$.

Proof. Factorization (1.3), when it exists, is defined up to multiplying G_+ by a constant invertible matrix on the right and G_- by the inverse of this constant matrix on the left. Therefore, we may choose $\mathbf{M}(G_+)$ to be any invertible matrix. We will show that the desired spectral properties of the multiples are associated with the choice $\mathbf{M}(G_+) = I$. Observe that for this choice $\det G_\pm = 1$.

Consider first the case when $f \in APW$. Then [5, Corollary 10.7] also $G_\pm \in APW_{2\times 2}$. Moreover, the Bohr-Fourier spectra of $G_\pm$ lie in the group $\Sigma = \lambda\mathbb{Z} + \Sigma_0$. Represent the latter as the union of pairwise disjoint subsets

$$\Sigma_k = k\lambda + \Sigma_0, \quad k \in \mathcal{J},$$

where $\mathcal{J} = \mathbb{Z}$ if λ is rationally independent from Σ_0 and $\mathcal{J} = \{0, \dots, p-1\}$ otherwise. Here p is the smallest natural p for which $p\lambda \in \Sigma_0$.

Now write G_-^{-1} as

$$G_-^{-1} = \sum_{k \in \mathcal{J}} F_k, \tag{6.1}$$

where (in obvious notation)

$$\Omega(F_k) \subset \begin{bmatrix} \Sigma_{k-1} & \Sigma_k \\ \Sigma_k & \Sigma_{k+1} \end{bmatrix}, \tag{6.2}$$

the right-hand side of (6.1) converges in $\| \ \|_W$ norm if $\mathcal{J}$ is infinite, and $k \pm 1$ in (6.2) are understood (mod p) if $\mathcal{J}$ is finite.

Then, from (1.5) and (6.1),

$$GG_-^{-1} = \sum_{k \in \mathcal{J}} GF_k = \sum_{k \in \mathcal{J}} G_k,$$

where

$$\Omega(G_k) \subset \begin{bmatrix} \Sigma_k & \Sigma_{k+1} \\ \Sigma_{k-1} & \Sigma_k \end{bmatrix}. \tag{6.3}$$

Since the sets Σ_k corresponding to different values of $k \in \mathcal{J}$ are disjoint, the non-negativity of $\Omega(GG_-^{-1})$ implies that each of $\Omega(GF_k)$ must be non-negative. In particular,

$$G_0 = GF_0 \in AP^+_{2\times 2},$$

so that $\det G_0 = \det F_0$ must be constant. Since the only set Σ_k containing 0 is Σ_0, conditions (6.3) imply that

$$\mathbf{M}(G_0) = \mathbf{M}(G_+) = I;$$

therefore, the constant in question is equal to 1. Consequently,

$$G = G_0 F_0^{-1}$$

is a canonical factorization of G. The desired spectral properties of its multiples follow from (6.2), (6.3) with $k = 0$.

To cover the general case when $f \in AP$ does not (necessarily) lie in APW, introduce a sequence of functions $f_n \in APW$ converging uniformly to f and such that $\Omega(f_n) \subset \Sigma_0$. For n large enough, the matrix functions

$$\begin{bmatrix} e_\lambda & 0 \\ f_n & e_{-\lambda} \end{bmatrix}$$

all admit canonical AP factorizations $G_+^{(n)} G_-^{(n)}$. Normalizing these factorizations by setting $\mathbf{M}(G_+^{(n)}) = I$, from the already proved part of the theorem we conclude for example that the Bohr-Fourier spectra of the diagonal entries of $G_+^{(n)}$ lie in Σ_0. But, according to [5, Section 21.3], $G_+^{(n)}$ converge to the factorization multiple G_+ of (1.5) in the topology of the Besicovitch space B_2. Therefore, the Bohr-Fourier spectra of the diagonal entries of the latter matrix also lie in Σ_0. The statements about the remaining entries can be treated similarly. □

In relation with Theorem 6.1, let us mention the following question posed in [8]: given rationally independent $\alpha, \beta \in (0,1)$, characterize all such $g \in APP$ with $\Omega(g) \subset \Gamma := \alpha\mathbb{Z} + \beta\mathbb{Z}$ for which there exists $h \in AP^+$ with $\Omega(h) \subset \Gamma$ and $\Omega(gh) \cap (0,1) = \emptyset$. According to Theorem 6.1, g has the desired property whenever the matrix function $\begin{bmatrix} e_1 & 0 \\ g & e_{-1} \end{bmatrix}$ admits a canonical factorization. Moreover, this observation carries over to Γ being an arbitrary additive subgroup of $\mathbb{R}$, not just those with two generators as in [8].

References

[1] S. Avdonin, A. Bulanova, and W. Moran, *Construction of sampling and interpolating sequences for multi-band signals. The two-band case.* Int. J. Appl. Math. Comput. Sci. **17** (2007), no. 2, 143–156.

[2] M.A. Bastos, Yu.I. Karlovich, and A.F. dos Santos, *Oscillatory Riemann-Hilbert problems and the corona theorem.* J. Functional Analysis **197** (2003), 347–397.

[3] M.A. Bastos, Yu.I. Karlovich, A.F. dos Santos, and P.M. Tishin, *The corona theorem and the canonical factorization of triangular AP-matrix functions – Effective criteria and explicit formulas.* J. Math. Anal. Appl. **223** (1998), 523–550.

[4] ———, *The corona theorem and the existence of canonical factorization of triangular AP-matrix functions.* J. Math. Anal. Appl. **223** (1998), 494–522.

[5] A. Böttcher, Yu.I. Karlovich, and I.M. Spitkovsky, *Convolution operators and factorization of almost periodic matrix functions.* OT 131, Birkhäuser Verlag, Basel and Boston, 2002.

[6] M.C. Câmara and C. Diogo, *Invertibility of Toeplitz operators and corona conditions in a strip.* J. Math. Anal. Appl. **342** (2008), no. 2, 1297–1317.

[7] M.C. Câmara and A.F. dos Santos, *Wiener-Hopf factorization for a class of oscillatory symbols.* Integral Equations and Operator Theory **36** (2000), no. 4, 409–432.

[8] M.C. Câmara, A.F. dos Santos, and M.C. Martins, *A new approach to factorization of a class of almost-periodic triangular symbols and related Riemann-Hilbert problems.* J. Funct. Anal. **235** (2006), no. 2, 559–592.

[9] C. Corduneanu, *Almost periodic functions.* J. Wiley & Sons, 1968.

[10] Yu.I. Karlovich, *Approximation approach to canonical APW factorability.* Izv. Vuzov., Sev.-Kavk. Region, 2005, pp. 143–151.

[11] B.M. Levitan and V.V. Zhikov, *Almost periodic functions and differential equations.* Cambridge University Press, 1982.

[12] A.A. Pankov, *Bounded and almost periodic solutions of nonlinear differential operator equations.* Kluwer, Dordrecht/Boston/London, 1990.

[13] L. Rodman, I.M. Spitkovsky, and H.J. Woerdeman, *Carathéodory-Toeplitz and Nehari problems for matrix valued almost periodic functions.* Trans. Amer. Math. Soc. **350** (1998), 2185–2227.

[14] I.M. Spitkovsky, *On the factorization of almost periodic matrix functions.* Math. Notes **45** (1989), no. 5–6, 482–488.

M.C. Câmara
Departamento de Matemática
Instituto Superior Técnico
Technical University of Lisbon (TULisbon), Portugal
e-mail: `cristina.camara@math.ist.utl.pt`

Yu.I. Karlovich
Facultad de Ciencias
Universidad Autónoma del Estado de Morelos
Av. Universidad 1001, Col. Chamilpa
C.P. 62209 Cuernavaca, Morelos, México
e-mail: `karlovich@buzon.uaem.mx`

I.M. Spitkovsky
Department of Mathematics
College of William and Mary
Williamsburg, VA 23187, USA
e-mail: `ilya@math.wm.edu`

Operator Theory:
Advances and Applications, Vol. 190, 191–195

On Generalized Operator-valued Toeplitz Kernels

Olga Chernobai

This paper is dedicated to M.G. Krein

Abstract. In this report, we sketch the proof of an integral representation for operator-valued Toeplitz kernels. The proof is based on the spectral theory for the corresponding differential operator constructed from this kernel and acting in Hilbert space. The report also contains references to other new results concerning such Toeplitz kernels.

Keywords. Positive definite functions, Toeplitz kernels, generalized Toeplitz kernels, quasinuclear riggings of spaces, projection spectral theorem.

1. Introduction

In 1979, M. Cotlar and C. Sadosky [1] introduced an essential generalization of positive definite functions on the axis, called generalized Toeplitz kernels, and gave their integral representation. Later, R. Bruzual [2] developed this construction for the case of a finite interval. In 1988–1999, M. Bekker investigated matrix positive definite Toeplitz kernels (see [3]).

In this talk, we present a survey of our results [4]–[7] concerning an integral representation of positive definite Toeplitz kernels, whose values are bounded operators in fixed separable Hilbert space.

The proof is based on the methods of construction of a representation for positive definite kernels, developed by Yu.M. Berezansky [8] in 1956, and generalizes some M.G. Krein's ideas [9]. In particular, it exploits the theory of eigenvector expansions for a selfadjoint operator acting in Hilbert space constructed by such a kernel.

2. Results

Let H be a full separable complex Hilbert space with involution $H \ni f \mapsto \overline{f} \in H$, scalar product $(\cdot\,,\cdot)_H$, and norm $\|\cdot\|$. Let $\mathcal{L}(H)$ be the set of all bounded operators in H. For $I = (-l,l),\ \ 0 < l \leq \infty$, we denote: $I_1 = I \cap [0,\infty)$, $I_2 = I \cap (-\infty, 0]$, and $\forall \alpha, \beta = 1,2$

$$I_{\alpha\beta} = \{t = x - y | x \in I_\alpha, y \in I_\beta\}.$$

Thus, $I_{11} = I_{22} = (-l,l),\ \ I_{12} = (0,2l)$.

Consider the operator-valued kernel K

$$I \times I \ni (x,y) \mapsto K(x,y) \in \mathcal{L}(H).$$

By definition, this kernel is a generalized Toeplitz kernel if it is positive definite and there exist four continuous operator-valued functions $I_{\alpha\beta} \ni t \longmapsto k_{\alpha\beta}(t) \in \mathcal{L}(H)$ such that

$$K(x,y) = k_{\alpha\beta}(x-y), \qquad (x,y) \in I_\alpha \times I_\beta,\ \alpha,\beta = 1,2.$$

Remind that an operator-valued kernel K is called positive definite if for an arbitrary continuous bounded vector-valued function $I \ni x \mapsto f(x) \in H$, there holds the inequality

$$\int_I \int_I (K(x,y)f(y), f(x))_H \, dxdy \geq 0.$$

Theorem 2.1. *For every generalized operator-valued Toeplitz kernel, the following integral representation holds:*

$$K(x,y) = \int_{\mathbb{R}^1} e^{i\lambda(x-y)} \sum_{\alpha,\beta=1}^{2} \kappa_\alpha(x)\kappa_\beta(y) d\sigma_{\alpha\beta}(\lambda), \qquad (x,y) \in I \times I. \tag{2.1}$$

Here, κ_α are the characteristic functions of I_α, $\alpha = 1,2$, and $(d\sigma_{\alpha\beta}(\Delta))_{\alpha,\beta=1}^2$ is a positive definite matrix-valued Borel measure on $\mathbb{R}^1$.

Conversely, every kernel of form (2.1) *is a generalized Toeplitz kernel.*

Sketch of the proof. Let $C(H,I)$ be the set of all continuous vector-valued functions $I \ni x \mapsto f(x) \in H$. Introduce the (quasi)scalar product

$$(f,g)_{H_K} = \int_I \int_I (K(x,y)f(y), g(x))_H dxdy.$$

Denote by H_K the completion of the space $C(H,I)$ wrt the scalar product introduced above.

Let us construct some quasinuclear rigging of Hilbert space H_K. For this purpose, we take any quasinuclear rigging of the space H :

$$H_- \supset H \supset H_+,$$

and construct the tensor product

$$W_2^{-1}(I) \otimes H_- \supset L^2(H,I) = L^2(I) \otimes H \supset W_2^1(I) \otimes H_+,$$

where $W_2^1(I)$ is the ordinary Sobolev space and $W_2^{-1}(I)$ is the corresponding negative space. The imbedding $L^2(H,I) \hookrightarrow H_K$ is continuous and imbedding $W_2^1(I) \hookrightarrow L^2(I)$ is quasinuclear; therefore, the imbedding $W_2^1(H,I) \otimes H_+ \hookrightarrow H_K$ is quasinuclear. So, we have constructed the required rigging of Hilbert space H_K :

$$H_{K,-} \supset H_K \supset H_{K,+}.$$

In the space H_K, we consider the operator

$$\operatorname{Dom}(A') = C_0^\infty(H,I) \ni f(x) \mapsto -i\frac{d}{dx}f(x) = (\mathcal{L}f)(x);$$
$$\mathcal{L}^+ = \mathcal{L}.$$

Here, $C_0^\infty(H,I)$ denotes the set of all continuously differentiable vector-valued functions finite near the points $-l$, 0, l. The operator A' is Hermitian wrt the scalar product in the space H_K :

$$(A'f,g)_{H_K} = (f,A'g)_{H_K}, \qquad f,g \in C_0^\infty(H,I).$$

It has equal defect numbers, because it is real wrt the corresponding involution in the space H_K. Therefore, it has a selfadjoint extension A in the space H_K.

This operator A is standardly connected with the quasinuclear chain introduced above:

$$H_{K,-} \supset H_K \supset H_{K,+} \supset D = C_0^\infty(H_+,I).$$

Therefore, we can apply the results connected with the spectral theorem for a selfadjoint operator. In particular, the Parseval equality can be written now:

$$(u,v)_{H_K} = \int_{\mathbb{R}^1} (P(\lambda)u,v)_{H_K}\,d\rho(\lambda), \qquad u,v \in H_{K,+} = W_2^1(I) \otimes H_+.$$

Here, $P(\lambda) : H_{K,+} \to H_{K,-}$ is the corresponding operator of generalized projection.

From the Parseval equality, it is possible to show that the following lemma holds:

Lemma 2.2. *For the kernel K, there holds the representation*

$$K = \int_{\mathbb{R}^1} \Omega(\lambda)\,d\rho(\lambda). \tag{2.2}$$

Here,

$$\Omega(\lambda) \in \left(W_2^{-1}(I) \otimes H_-\right) \otimes \left(W_2^{-1}(I) \otimes H_-\right)$$

is an elementary positive definite kernel with bounded norm, $d\rho(\lambda)$ is a positive Borel bounded measure on $\mathbb{R}^1$. This kernel $\Omega(\lambda)$ is definite for ρ-almost every $\lambda \in \mathbb{R}^1$. The positive definiteness of $\Omega(\lambda), \lambda \in \mathbb{R}^1$, means that for every $u \in H_{K,+}$,

$$(\Omega(\lambda), u \otimes \overline{u})_{L^2(H,I)\otimes L^2(H,I)} \geq 0.$$

An elementary character of this kernel means that the following equalities holds:

$$\begin{aligned}(\Omega(\lambda), v \otimes \overline{A_0 u})_{L^2(H,I)\otimes L^2(H,I)} &= (\Omega(\lambda), (A_0 v) \otimes \overline{u})_{L^2(H,I)\otimes L^2(H,I)} \\ &= \lambda(\Omega(\lambda), v \otimes \overline{u}), \qquad u,v \in C_0^\infty(H_+,I).\end{aligned}$$

Denote by $\Omega_{\alpha\beta}(\lambda)$ $(\alpha,\beta=1,2)$ the restriction of $\Omega(\lambda)$ on the space $H_{+,\alpha}\otimes H_{+,\beta}$ where $H_{+,\alpha}$ is the subspace of H_+ consisting from all functions equal to zero on $I\setminus I_\alpha$. For arbitrary $l,m\in H_+$, we introduce the kernel $\Omega^{lm}_{\alpha\beta}(\lambda)=\Big(\Omega_{\alpha\beta}(\lambda),(\cdot\otimes l)\otimes\overline{(\cdot,\otimes m)}\Big)$. Using Lemma 2.2 and the theorem concerning the smoothness of generalized solutions of the equation $-i\frac{du}{dx}=\lambda u$, we can prove the following representation:

$$\Omega^{lm}_{\alpha\beta}(\lambda)=\Omega^{lm}_{\alpha\beta}(\lambda;x,y)=\tau^{lm}_{\alpha\beta}(\lambda)e^{i\lambda(x-y)},\qquad x\in I_\alpha,\ y\in I_\beta,\ \alpha,\beta=1,2,\tag{2.3}$$

where $\tau^{lm}_{\alpha\beta}$ are some coefficients. From (2.3), it is possible to obtain the following representation:

$$\Omega_{\alpha\beta}(\lambda)=\tau_{\alpha\beta}(\lambda)e^{i\lambda(x-y)};$$

$$\Omega(\lambda;x,y)=\sum_{\alpha,\beta=1}^{2}e^{i\lambda(x-y)}\kappa_\alpha(x)\kappa_\beta(y)\tau_{\alpha\beta}(\lambda),\qquad x,y\in I.\tag{2.4}$$

Putting the expression (2.4) into (2.2), we get the representation

$$\begin{aligned}K(x,y)&=\int_{\mathbb{R}^1}\Omega(\lambda;x,y)d\rho(\lambda)\\&=\int_{\mathbb{R}^1}\left(\sum_{\alpha,\beta=1}^{2}e^{i\lambda(x-y)}\kappa_\alpha(x)\kappa_\beta(y)\tau_{\alpha\beta}(\lambda)\right)d\rho(\lambda).\end{aligned}\tag{2.5}$$

Denoting $\tau_{\alpha\beta}(\lambda)d\rho(\lambda)$ by $d\sigma_{\alpha\beta}(\lambda)$, we get from (2.5):

$$K(x,y)=\int_{\mathbb{R}^1}e^{i\lambda(x-y)}\sum_{\alpha,\beta=1}^{2}\kappa_\alpha(x)\kappa_\beta(y)d\sigma_{\alpha\beta}(\lambda),\qquad (x,y)\in I\times I.$$

As a result, we get the representation (2.1).

To prove the second part of theorem, it remains only to check that the kernel (2.1) is positive definite.

Theorem 2.1 for a scalar-valued kernel was published, together with the proofs, in [4], and that for the general case in [6]. The articles [4]–[6] also contain some additional facts concerning generalized Toeplitz kernels, namely: the conditions for the uniqueness of the measure $d\sigma_{\alpha\beta}(\lambda)$ in the representation (2.1), investigation of the Fourier transform connected with the representation (2.1) and selfadjointness of the above operator A. In the article [7], we prove a theorem concerning the smoothness of generalized solutions for simple differential equations with operator-valued coefficients, which is necessary to prove Theorem 2.1.

References

[1] M. Cotlar, C. Sadoscy, *On the Helson-Szegö theorem and related class of modified Toeplitz kernels.* Proc. Symp. Pure Math. **35** (1979) -PtI, 383–407.

[2] R. Bruzual, *Local semigroups of contractions and some applications to Fourier representations theorems.* Integr. Equat. and Operator Theory **10** (1987), 780–801.

[3] M. Bekker, *On the extension problem for continuous positive definite generalized Toeplitz kernels definite on a finite interval.* Integr. Equat. and Operator Theory **35** (1999), no. 4, 379–397.

[4] Yu.M. Berezansky, O.B. Chernobai, *On the theory of generalized Toeplitz kernels.* Ukr. Math. J. **52** (2000), no. 11, 1458–1472.

[5] O. Chernobai, *On spectral theory of generalized Toeplitz kernels.* Ukr. Math. J. **55** (2003), no. 6, 850–857.

[6] O. Chernobai, *Spectral representation for generalized operator-valued Toeplitz kernels.* Ukr. Math. J. **57** (2005). no. 12, 1698-1710.

[7] O. Chernobai, *On generalized solutions of differential equations with operatorial coefficients.* Ukr. Math. J. **58** (2006), no. 5, 715–720.

[8] Yu.M. Berezanskii, *Expansions in Eigenfunctions of Selfadjoint Operators.* – Amer. Math. Soc., Providence, R.I:AMS, 1968. (Russian edition: Kiev: Naukova Dumka, 1965.)

[9] M.G. Krein, *On Hermitian operators with directed functionals.* Zbirnyk prac' Inst. Math. Akad. Nauk Ukrain. RSR (1948), no. 10, 83–106.

Olga Chernobai
National University State Tax Service of Ukraine
31 Karl Marx St.
08200 Irpin, Ukraine
e-mail: kneu_6103@mail.ru

Operator Theory:
Advances and Applications, Vol. 190, 197–236

Abstract Interpolation Problem in Nevanlinna Classes

Vladimir Derkach

Abstract. The abstract interpolation problem (AIP) in the Schur class was posed V. Katznelson, A. Kheifets and P. Yuditskii in 1987. In the present paper an analog of the AIP for Nevanlinna classes is considered. The description of solutions of the AIP is reduced to the description of $\mathcal{L}$-resolvents of some model symmetric operator associated with the AIP. The latter description is obtained by using the M.G. Kreĭn's theory of $\mathcal{L}$-resolvent matrices. Both regular and singular cases of the AIP are treated. The results are illustrated by the following examples: bitangential interpolation problem, full and truncated moment problems. It is shown that each of these problems can be included into the general scheme of the AIP.

Mathematics Subject Classification (2000). Primary 47A57; Secondary 30E05, 47A06, 47B25, 47B32.

Keywords. Symmetric relation, selfadjoint extension, reproducing kernel, boundary triplet, resolvent matrix, interpolation problem, moment problem.

1. Introduction

The abstract interpolation problem (AIP) was posed by V. Katznelson, A. Kheifets and P. Yuditskii in [30] as an extension of the V.P. Potapov's approach to interpolation problems [34]. A description of the set of all solutions of the AIP was reduced in [30] to the description of all scattering matrices of unitary extensions of a given partial isometry V ([8]). In a number of papers it was shown that many problems of analysis, such that the bitangential interpolation problem [31], moment problem [33], lifting problem [39], and others can be included into the general scheme of the AIP.

In the present paper we consider a parallel version of the AIP for the Nevanlinna class $N[\mathcal{L}]$. The class $N[\mathcal{L}]$ consists of all operator-valued functions (ovf's)

This research has been done partially while the author was visiting the Department of Mathematics of Weizmann Institute of Science as a Weston Visiting Scholar.

holomorphic in the upper half-plane $\mathbb{C}_+$ with values in the set $[\mathcal{L}]$ of bounded linear operators in a Hilbert space $\mathcal{L}$, such that the kernel

$$\mathsf{N}_\omega^m(\lambda) = \frac{m(\lambda) - m(\omega)^*}{\lambda - \bar{\omega}} \tag{1.1}$$

is nonnegative on $\mathbb{C}_+$. If m is extended to $\mathbb{C}_-$ by the symmetry $m(\bar{\lambda}) = m(\lambda)^*$ then the kernel $\mathsf{N}_\omega^m(\lambda)$ is also nonnegative on $\mathbb{C} \setminus \mathbb{R}$.

In introduction we restrict ourselves to the case when $\dim \mathcal{L} < \infty$ and identify $\mathcal{L}$ with the space $\mathbb{C}^d$, where the standard basis is chosen. Then every ovf $m \in N^{d\times d} := N[\mathbb{C}^d]$ can be considered as a $d \times d$ matrix-valued function (mvf). Let $\mathcal{H}(m)$ be the reproducing kernel Hilbert space of vector-valued functions holomorphic in $\mathbb{C} \setminus \mathbb{R}$ (see [17], [4]), which is characterized by the properties:

(1) $\mathsf{N}_\omega^m(\cdot)u \in \mathcal{H}(m)$ for all $\omega \in \mathbb{C} \setminus \mathbb{R}$ and $u \in \mathbb{C}^d$;
(2) for every $f \in \mathcal{H}(m)$ the following identity holds

$$\langle f(\cdot), \mathsf{N}_\omega^m(\cdot)u\rangle_{\mathcal{H}(m)} = u^* f(\omega), \quad \omega \in \mathbb{C} \setminus \mathbb{R}, u \in \mathbb{C}^d. \tag{1.2}$$

The AIP in the class $N^{d\times d}$ can be formulated as follows.

Let $\mathcal{X}$ be a complex linear space, let B_1, B_2 be linear operators in $\mathcal{X}$, let C_1, C_2 be linear operators from $\mathcal{X}$ to $\mathcal{L} := \mathbb{C}^d$, and let K be a nonnegative sesquilinear form on $\mathcal{X}$ which satisfies the following identity

(A1) $K(B_2h, B_1g) - K(B_1h, B_2g) = (C_1h, C_2g)_{\mathbb{C}^d} - (C_2h, C_1g)_{\mathbb{C}^d}$

for every $h, g \in \mathcal{X}$. Consider the following

Problem $AIP(B_1, B_2, C_1, C_2, K)$. Let the data set (B_1, B_2, C_1, C_2, K) satisfies the assumption $(A1)$. Find a mvf $m \in N^{d\times d}$, such that for some linear mapping F from $\mathcal{X}$ to $\mathcal{H}(m)$ the following conditions hold for all $h \in \mathcal{X}$:

(C1) $(FB_2h)(\lambda) - \lambda(FB_1h)(\lambda) = \begin{bmatrix} I & -m(\lambda) \end{bmatrix} \begin{bmatrix} C_1h \\ C_2h \end{bmatrix}$;

(C2) $\|Fh\|^2_{\mathcal{H}(m)} \le K(h,h)$.

Clearly $\ker K = \{h \in \mathcal{X} : K(h,h) = 0\}$ is a linear subspace of $\mathcal{X}$. Let $\mathcal{H}$ be the completion of the factor-space $\widehat{\mathcal{X}} = \mathcal{X}/\ker K$ endowed with the scalar product

$$(\widehat{h}, \widehat{g})_{\mathcal{H}} = K(h,g), \quad \widehat{h} = h + \ker K, \widehat{g} = g + \ker K, \ h, g \in \mathcal{X}. \tag{1.3}$$

In the present paper we will use the notion of a *linear relation* in a Hilbert space $\mathfrak{H}$. Recall, that a subspace T of $\mathfrak{H}^2$ is called the linear relation in $\mathfrak{H}$, [11]. For a linear relation T in $\mathfrak{H}$ the symbols $\operatorname{dom} T$, $\ker T$, $\operatorname{ran} T$, and $\operatorname{mul} T$ stand for the domain, kernel, range, and the multivalued part, respectively. The adjoint T^* is the closed linear relation in $\mathfrak{H}$ defined by (see [11], [19])

$$T^* = \{\{h,k\} \in \mathfrak{H}^2 : (k,f)_{\mathfrak{H}} = (h,g)_{\mathfrak{H}}, \{f,g\} \in T\}. \tag{1.4}$$

Recall that a linear relation T in $\mathfrak{H}$ is called *symmetric* (*selfadjoint*) if $T \subset T^*$ ($T = T^*$, respectively).

It follows from (A1) that the linear relation

$$\widehat{A} = \left\{\left\{\begin{bmatrix} \widehat{B_1 h} \\ C_1 h \end{bmatrix}, \begin{bmatrix} \widehat{B_2 h} \\ C_2 h \end{bmatrix}\right\} : h \in \mathcal{X}\right\}$$

is symmetric in $\mathcal{H} \oplus \mathbb{C}^d$. The main result of the paper is the following description of all the AIP solutions m.

Theorem 1. *Let the data set (B_1, B_2, C_1, C_2, K) satisfies the assumption (A1) and let $\operatorname{ran} C_2 = \mathcal{L}$. Then the Problem $AIP(B_1, B_2, C_1, C_2, K)$ is solvable and the set of its solutions is parametrized by the formula*

$$m(\lambda) = P_{\mathcal{L}}(\widetilde{A} - \lambda)^{-1}|_{\mathcal{L}}(I_{\mathcal{L}} + \lambda P_{\mathcal{L}}(\widetilde{A} - \lambda)^{-1}|_{\mathcal{L}})^{-1}, \tag{1.5}$$

where $\widetilde{A}$ ranges over the set of all selfadjoint extensions of $\widehat{A}$ with the exit in a Hilbert space $\widetilde{\mathcal{H}} \oplus \mathcal{L} \supset \mathcal{H} \oplus \mathcal{L}$. The corresponding linear mapping $F : \mathcal{X} \to \mathcal{H}(\varphi, \psi)$ is given by

$$(Fh)(\lambda) = P_{\mathcal{L}}(\widetilde{A} - \lambda)^{-1}\widehat{h}, \quad h \in \mathcal{X}. \tag{1.6}$$

Due to Theorem 1 the description of all solutions m of the $AIP(B_1, B_2, C_1, C_2, K)$ is reduced to the problem of description of all $\mathcal{L}$-resolvents of the linear relation $\widehat{A}$. The latter description has been obtained by M.G. Kreĭn in [36] (see also [38]). In order to apply this theory to the linear relation $\widehat{A}$ we will impose additional assumptions on the data set (B_1, B_2, C_1, C_2, K):

(A2) $\dim \ker K < \infty$ and $\mathcal{X}$ admits the representation

$$\mathcal{X} = \mathcal{X}_0 \dotplus \ker K, \tag{1.7}$$

such that $B_j \mathcal{X}_0 \subseteq \mathcal{X}_0$ $(j = 1, 2)$.

(A3) $B_2 = I_{\mathcal{X}}$ and the operators $B_1|_{\mathcal{X}_0} : \mathcal{X}_0 \subset \mathcal{H} \to \mathcal{H}$, $C_1|_{\mathcal{X}_0}$, $C_2|_{\mathcal{X}_0} : \mathcal{X}_0 \subset \mathcal{H} \to \mathcal{L}$ are bounded.

Notice that the decomposition (1.7) was used by V. Dubovoj in [25] to study the degenerate matrix Schur problem. The continuations of the operators $B_1|_{\mathcal{X}_0}$, $C_1|_{\mathcal{X}_0}$, $C_2|_{\mathcal{X}_0}$ will be denoted by $\widetilde{B}_1 \in [\mathcal{H}]$, $\widetilde{C}_1$, $\widetilde{C}_2 \in [\mathcal{H}, \mathcal{L}]$. Here $[\mathcal{H}, \mathcal{L}]$ stands for the set of bounded linear operators from $\mathcal{H}$ to $\mathcal{L}$.

Denote by $\widetilde{N}^{d\times d}$ the set of Nevanlinna pairs $\{p, q\}$ of $d \times d$ mvf's $p(\cdot)$, $q(\cdot)$ holomorphic on $\mathbb{C} \setminus \mathbb{R}$ such that:

(i) the kernel $\mathsf{N}_\omega^{p,q}(\lambda) = \dfrac{q(\bar{\lambda})^* p(\bar{\omega}) - p(\bar{\lambda})^* q(\bar{\omega})}{\lambda - \bar{\omega}}$ is nonnegative on $\mathbb{C}_+$;

(ii) $q(\bar{\lambda})^* p(\lambda) - p(\bar{\lambda})^* q(\lambda) = 0$, $\lambda \in \mathbb{C} \setminus \mathbb{R}$;

(iii) $0 \in \rho(p(\lambda) - \lambda q(\lambda))$, $\lambda \in \mathbb{C}_\pm$.

In the regular case ($\ker K = \{0\}$) the set of $\mathcal{L}$-resolvents $P_{\mathcal{L}}(\widetilde{A} - \lambda)^{-1}|_{\mathcal{L}}$ of $\widehat{A}$ can be described by the formula (see [35], [38])

$$P_{\mathcal{L}}(\widetilde{A} - \lambda)^{-1} I_{\mathcal{L}} = (w_{11}(\lambda)q(\lambda) + w_{12}(\lambda)p(\lambda))(w_{21}(\lambda)q(\lambda) + w_{22}(\lambda)p(\lambda))^{-1}, \tag{1.8}$$

where $\{p, q\} \in \widetilde{N}^{d\times d}$ and $W = [w_{ij}(\lambda)]_{i,j=1}^2$ is an $\mathcal{L}$-resolvent matrix of $\widehat{A}$ which can be calculated explicitly in terms of the data set (see (4.38)).

Combining (1.5) and (1.8) one obtains the following

Theorem 2. *Let the AIP data set* (B_1, B_2, C_1, C_2, K) *satisfy* $(A1)$, $(A3)$, $\ker K = \{0\}$, $\operatorname{ran} C_2 = \mathcal{L} = \mathbb{C}^d$, *and let*

$$\Theta(\lambda) = \begin{bmatrix} \theta_{11}(\lambda) & \theta_{12}(\lambda) \\ \theta_{21}(\lambda) & \theta_{22}(\lambda) \end{bmatrix} = I_{\mathcal{L}\oplus\mathcal{L}} - \lambda \begin{bmatrix} \widetilde{C}_1 \\ \widetilde{C}_2 \end{bmatrix} (I_{\mathcal{H}} - \lambda \widetilde{B}_1)^{-1} \begin{bmatrix} -\widetilde{C}_2^+ & \widetilde{C}_1^+ \end{bmatrix}. \quad (1.9)$$

Then the formula

$$m(\lambda) = (\theta_{11}(\lambda)q(\lambda) + \theta_{12}(\lambda)p(\lambda))(\theta_{21}(\lambda)q(\lambda) + \theta_{22}(\lambda)p(\lambda))^{-1}, \quad (1.10)$$

establishes the one-to-one correspondence between the set of all solutions m *of the AIP and the set of all equivalence classes of Nevanlinna pairs* $\{p, q\} \in \widetilde{N}^{d\times d}$.

The operators $\widetilde{C}_1^+$, $\widetilde{C}_2^+$ in (1.9) are adjoint operators to $\widetilde{C}_1$, $\widetilde{C}_2 : \mathcal{H} \to \mathcal{L}$, i.e.,

$$(\widetilde{C}_j^+ u, h)_{\mathcal{H}} = (u, \widetilde{C}_j h)_{\mathcal{L}} \quad \text{for all } h \in \mathcal{H}, u \in \mathcal{L} \quad (j = 1, 2).$$

Let the matrix $J \in \mathbb{C}^{2d\times 2d}$ be given by

$$J = \begin{bmatrix} 0 & -iI_d \\ iI_d & 0 \end{bmatrix}$$

The mvf $\Theta(\lambda)$ belongs to the Potapov class $\mathcal{P}(J)$ (see [43]), i.e., $\Theta(\lambda)$ is meromorphic in $\mathbb{C}_+$ and has the following J-property

$$\frac{J - W(\lambda)JW(\lambda)^*}{\lambda - \bar{\lambda}} \geq 0 \quad \text{for all } \lambda \in \mathfrak{H}_\Theta \cap \mathbb{C}_+,$$

where $\mathfrak{H}_\Theta$ is the set of holomorphy of Θ.

In the singular case ($\ker K \neq \{0\}$) the formula (1.8) gives a description of all $\mathcal{L}$-resolvents of the linear relation

$$A_0 = \left\{ \left\{ \begin{bmatrix} B_1 h \\ C_1 h \end{bmatrix}, \begin{bmatrix} B_2 h \\ C_2 h \end{bmatrix} \right\} : h \in \mathcal{X}_0 \right\}.$$

To obtain a description of all $\mathcal{L}$-resolvents of $\widehat{A}(\supset A_0)$ one should consider in (1.8) only those Nevanlinna pairs $\{p, q\} \in \widetilde{N}(\mathcal{L})$ for which $\widetilde{A} \supset \widehat{A}(\supset A_0)$. We show that after the replacement of $\Theta(\lambda)$ in (1.10) by the $\Theta(\lambda)V$, where V is an appropriate J-unitary matrix, the formula (1.10) gives a description of all the solutions of the AIP when $\{p, q\}$ ranges over the set of all Nevanlinna pairs of the form

$$p(\lambda) = \begin{bmatrix} \widetilde{0}_\nu & 0 \\ 0 & p_1(\lambda) \end{bmatrix}, \quad q(\lambda) = \begin{bmatrix} \widetilde{I}_\nu & 0 \\ 0 & q_1(\lambda) \end{bmatrix}, \quad \{p_1, q_1\} \in \widetilde{N}^{d-\nu}, \quad (1.11)$$

where $\nu = \dim C\ker K$.

All these results are formulated in the paper in a more general situation, when the mvf m is replaced by a Nevanlinna pair $\{\varphi, \psi\}$. Moreover, we do not suppose, in general, that $\dim \mathcal{L} < \infty$.

The paper is organized as follows. In Section 2 we recall the definition of the class $\widetilde{N}(\mathcal{L})$ of Nevanlinna pairs from [7], [23]. To each selfadjoint linear relation $\widetilde{A}$

and a scale spaces $\mathcal{L}$ we associate a Nevanlinna pair $\{\varphi, \psi\}$ by the formula

$$\psi(\lambda) := P_{\mathcal{L}}(\widetilde{A} - \lambda)^{-1}|_{\mathcal{L}}, \quad \varphi(\lambda) := I_{\mathcal{L}} + \lambda P_{\mathcal{L}}(\widetilde{A} - \lambda)^{-1}|_{\mathcal{L}}, \quad \lambda \in \rho(\widetilde{A}). \tag{1.12}$$

Conversely, given a Nevanlinna pair $\{\varphi, \psi\}$ normalized by the condition $\varphi(\lambda) - \lambda\psi(\lambda) = I_{\mathcal{L}}$ we construct a functional model for a selfadjoint linear relation $\widetilde{A} = A(\varphi, \psi)$, such that the pair $\{\varphi, \psi\}$ is related to $A(\varphi, \psi)$ via (1.12). In the case when the pair $\{\varphi, \psi\}$ is equivalent to a pair $\{I_d, m\}$ with $m \in N^{d\times d}$, functional model $A(m)$ for a selfadjoint linear relation was given in [7] (see also [22]). Conditions when the model $A(\varphi, \psi)$ is reduced to $A(m)$ are discussed. In Sections 3 and 4 we formulate the AIP in the classes $N^{d\times d}$ and $\widetilde{N}^{d\times d}$ and give a complete description of its solutions under some additional restrictions on the data set both in the regular and singular cases. The results of the paper are illustrated in Section 5 with an example of bitangential interpolation problems in the classes $N^{d\times d}$ and $\widetilde{N}^{d\times d}$, reduced there to the AIP with appropriately chosen data set. These problems have been studied earlier in [42], [7], [32], [26], [27], [16].

Mention, that the Arov and Grossman's description of scattering matrices of unitary extensions of an isometry V in [8] used in the Schur type AIP is an analog of M.G. Kreĭn's description (1.8) of $\mathcal{L}$-resolvents of a symmetric operator [35]. One of the goals of this paper is to formulate the AIP, where the M.G. Kreĭn's formula (1.8) works directly. In particular, we use the example of the full moment problem to show that the reduction of this problem to the Nevanlinna type AIP is more natural and simpler than that in [33], where the reduction of the moment problem to the Schur type AIP was performed.

Another goal of the paper is to elaborate the operator approach to singular AIP's. This approach is illustrated with an example of singular truncated moment problem, where we discuss the results of [15] and explain them from our point of view.

The paper is dedicated to the centennial of M.G. Kreĭn.

2. Functional model of a selfadjoint linear relation

2.1. Nevanlinna pairs

Let $\mathcal{L}$ be a Hilbert space.

Definition 2.1. A pair $\{\Phi, \Psi\}$ of $[\mathcal{L}]$-valued functions $\Phi(\cdot)$, $\Psi(\cdot)$ holomorphic on $\mathbb{C} \setminus \mathbb{R}$ is said to be a *Nevanlinna pair* if:

(i) the kernel

$$\mathsf{N}_{\omega}^{\Phi\Psi}(\lambda) = \frac{\Psi(\bar{\lambda})^*\Phi(\bar{\omega}) - \Phi(\bar{\lambda})^*\Psi(\bar{\omega})}{\lambda - \bar{\omega}}, \quad \lambda, \omega \in \mathbb{C}_+$$

is nonnegative on $\mathbb{C}_+$;

(ii) $\Psi(\bar{\lambda})^*\Phi(\lambda) - \Phi(\bar{\lambda})^*\Psi(\lambda) = 0$, $\lambda \in \mathbb{C} \setminus \mathbb{R}$;

(iii) $0 \in \rho(\Phi(\lambda) - \lambda\Psi(\lambda))$, $\lambda \in \mathbb{C}_{\pm}$.

Two Nevanlinna pairs $\{\Phi, \Psi\}$ and $\{\Phi_1, \Psi_1\}$ are said to be *equivalent*, if $\Phi_1(\lambda) = \Phi(\lambda)\chi(\lambda)$ and $\Psi_1(\lambda) = \Psi(\lambda)\chi(\lambda)$ for some operator function $\chi(\cdot) \in [\mathcal{H}]$, which is holomorphic and invertible on $\mathbb{C}_+ \cup \mathbb{C}_-$. The set of all equivalence classes of Nevanlinna pairs in $\mathcal{L}$ will be denoted by $\widetilde{N}(\mathcal{L})$. We will write, for short, $\{\Phi, \Psi\} \in \widetilde{N}(\mathcal{L})$ for the Nevanlinna pair $\{\Phi, \Psi\}$.

A Nevanlinna pair $\{\Phi, \Psi\}$ will be said to be *normalized* if $\Phi(\lambda) - \lambda\Psi(\lambda) = I_{\mathcal{H}}$. Clearly, every Nevanlinna pair $\{\Phi, \Psi\}$ is equivalent to the unique normalized Nevanlinna pair $\{\varphi, \psi\}$ given by

$$\varphi(\lambda) = \Phi(\lambda)(\Phi(\lambda) - \lambda\Psi(\lambda))^{-1}, \quad \psi(\lambda) = \Psi(\lambda)(\Phi(\lambda) - \lambda\Psi(\lambda))^{-1}. \tag{2.1}$$

The set $\widetilde{N}(\mathcal{L})$ can be identified with the set of Nevanlinna families (see [23])

$$\tau(\lambda) = \{\{\Phi(\lambda)u, \Psi(\lambda)u\} : u \in \mathcal{L}\}, \quad \{\Phi, \Psi\} \in \widetilde{N}(\mathcal{L}). \tag{2.2}$$

Define the class $N(\mathcal{L})$ as the set of all Nevanlinna pairs $\{\Phi, \Psi\} \in \widetilde{N}(\mathcal{L})$ such that $\ker \Phi(\lambda) = \{0\}$ for some (and hence for all) $\lambda \in \mathbb{C} \setminus \mathbb{R}$. Then $\Psi(\lambda)\Phi(\lambda)^{-1}$ is an ovf with values in the set of maximal dissipative operators in $\mathcal{L}$ for $\lambda \in \mathbb{C}_+$. The set $N[\mathcal{L}]$ can be embedded in $\widetilde{N}(\mathcal{L})$ via the mapping

$$m \in N[\mathcal{L}] \mapsto \{I_{\mathcal{L}}, m\} \in N(\mathcal{L}).$$

2.2. Nevanlinna pair corresponding to a selfadjoint linear relation and a scale

Let $\mathfrak{H}$, $\mathcal{L}$ be Hilbert spaces, let $\widetilde{A}$ be a selfadjoint linear relation in $\mathfrak{H} \oplus \mathcal{L}$ and let $P_{\mathcal{L}}$ be the orthogonal projection onto the scale space $\mathcal{L}$. Define the operator-valued functions

$$\varphi(\lambda) := I_{\mathcal{L}} + \lambda P_{\mathcal{L}}(\widetilde{A} - \lambda)^{-1}|_{\mathcal{L}}, \quad \psi(\lambda) := P_{\mathcal{L}}(\widetilde{A} - \lambda)^{-1}|_{\mathcal{L}}, \quad \lambda \in \rho(\widetilde{A}). \tag{2.3}$$

Proposition 2.2. *The pair of ovf's $\{\varphi, \psi\}$, associated with a selfadjoint linear relation $\widetilde{A}$ and the scale $\mathcal{L}$ via* (2.3) *is a normalized Nevanlinna pair.*

Proof. Consider the kernel

$$\mathsf{N}_\omega^{\varphi\psi}(\lambda) = \frac{\psi(\bar{\lambda})^*\phi(\bar{\omega}) - \phi(\bar{\lambda})^*\psi(\bar{\omega})}{\lambda - \bar{\omega}}, \quad \lambda, \omega \in \mathbb{C}_+ \cup \mathbb{C}_-. \tag{2.4}$$

It follows from (2.3)–(2.4) that

$$\begin{aligned} \mathsf{N}_\omega^{\varphi\psi}(\lambda) &= \frac{\psi(\lambda) - \psi(\omega)^*}{\lambda - \bar{\omega}} - \psi(\lambda)\psi(\omega)^* \\ &= P_{\mathcal{L}} \frac{R_\lambda - R_{\bar{\omega}}}{\lambda - \bar{\omega}}|_{\mathcal{L}} - P_{\mathcal{L}} R_\lambda P_{\mathcal{L}} R_{\bar{\omega}}|_{\mathcal{L}} \\ &= P_{\mathcal{L}} R_\lambda P_{\mathcal{H}} R_{\bar{\omega}}|_{\mathcal{L}} \end{aligned} \tag{2.5}$$

and hence the kernel $\mathsf{N}_\omega^{\varphi\psi}(\lambda)$ is nonnegative.

The property (ii) is easily checked, $\varphi(\lambda) - \lambda\psi(\lambda) \equiv I_{\mathcal{L}}$ and, hence, the pair $\{\phi, \psi\}$ is a normalized Nevanlinna pair. □

Definition 2.3. The pair of ovf's $\{\varphi, \psi\}$ determined by (2.3) will be called the Nevanlinna pair corresponding to the selfadjoint linear relation $\widetilde{A}$ and the scale $\mathcal{L}$.

Remark 2.4. Definition 2.3 is inspired by the notion of the Weyl family of a symmetric operator corresponding to a boundary relation, see [24]. Namely, the Nevanlinna pair $\{\varphi, \psi\}$ determined by (2.3) generates via (2.2) the Weyl family of the symmetric linear relation $S = \widetilde{A} \cap (\mathcal{H} \oplus \mathcal{H})$, corresponding to the boundary relation

$$\Gamma = \left\{ \left\{ \begin{bmatrix} f \\ f' \end{bmatrix}, \begin{bmatrix} h' \\ h \end{bmatrix} \right\} : \left\{ \begin{bmatrix} f \\ h \end{bmatrix}, \begin{bmatrix} f' \\ h' \end{bmatrix} \right\} \in \widetilde{A} \right\}.$$

The proof of Proposition 2.2 is contained in [24, Theorem 3.9]. Moreover, it is shown in [24] that the converse is also true, every Nevanlinna family generates via (2.2) the Weyl family of a symmetric linear operator S. In the case when the pair $\{\varphi, \psi\}$ is equivalent to a pair $\{I_{\mathcal{L}}, m\}$ with $m \in N[\mathcal{L}]$, functional model $A(m)$ for this symmetric operator S was given in [6] (see also [22] and [41]).

In the following theorem we give another functional model of a selfadjoint linear relation $\widetilde{A}$ recovered from a Nevanlinna pair. The author was aware that the same result was obtained independently in [14] by another method. Consider the reproducing kernel Hilbert space $\mathcal{H}(\Phi, \Psi)$, which is characterized by the properties:

(1) $\mathsf{N}_\omega^{\Phi\Psi}(\lambda)u \in \mathcal{H}(\Phi, \Psi)$ for all $\omega \in \mathbb{C} \setminus \mathbb{R}$ and $u \in \mathcal{L}$;
(2) for every $f \in \mathcal{H}(\Phi, \Psi)$ the following identity holds

$$\langle f(\cdot), \mathsf{N}_\omega^{\Phi\Psi}(\lambda)u \rangle_{\mathcal{H}(\Phi,\Psi)} = (f(\omega), u)_{\mathcal{L}}, \quad \omega \in \mathbb{C} \setminus \mathbb{R}, u \in \mathcal{L}. \tag{2.6}$$

It follows from (2.6) that the evaluation operator $E(\lambda) : f \mapsto f(\lambda)$ ($f \in \mathcal{H}(\Phi, \Psi)$) is a bounded operator from $\mathcal{H}(\Phi, \Psi)$ to $\mathcal{L}$.

Theorem 2.5. *Let $\{\Phi, \Psi\} \in \widetilde{N}(\mathcal{L})$. Then the linear relation*

$$A(\Phi, \Psi) = \left\{ \left\{ \begin{bmatrix} f \\ u \end{bmatrix}, \begin{bmatrix} f' \\ u' \end{bmatrix} \right\} : \begin{array}{c} f, f' \in \mathcal{H}(\Phi, \Psi), \ u, u' \in \mathcal{L}, \\ f'(\lambda) - \lambda f(\lambda) = \Phi(\bar{\lambda})^* u - \Psi(\bar{\lambda})^* u' \end{array} \right\} \tag{2.7}$$

is a selfadjoint linear relation in $\mathcal{H}(\Phi, \Psi) \oplus \mathcal{L}$ and the normalized pair $\{\varphi, \psi\}$ given by (2.1) *is the Nevanlinna pair corresponding to $\widetilde{A}(\Phi, \Psi)$ and $\mathcal{L}$.*

Proof. Step 1. Let us show that $A(\Phi, \Psi)$ contains vectors of the form

$$\{F_\omega u, F'_\omega u\} := \left\{ \begin{bmatrix} \mathsf{N}_\omega(\cdot)u \\ \Psi(\bar{\omega})u \end{bmatrix}, \begin{bmatrix} \bar{\omega}\mathsf{N}_\omega(\cdot)u \\ \Phi(\bar{\omega})u \end{bmatrix} \right\}, \quad u \in \mathcal{L}, \ \omega \in \mathbb{C}_+ \cup \mathbb{C}_-, \tag{2.8}$$

where $\mathsf{N}_\omega(\cdot) := \mathsf{N}_\omega^{\Phi\Psi}(\cdot)$ and the restriction $\widetilde{A}'$ of $A(\Phi, \Psi)$ to the span of vectors $\{F_\omega u, F'_\omega u\}$ is a symmetric linear relation.

Indeed, it follows from (2.7) and the equality

$$(\bar{\omega} - \lambda)\mathsf{N}_\omega(\lambda)u = \Phi(\bar{\lambda})^* \Psi(\bar{\omega}) - \Psi(\bar{\lambda})^* \Phi(\bar{\omega})$$

that $\{F_\omega u, F'_\omega u\} \in A(\Phi, \Psi)$.

For arbitrary $\omega_j \in \mathbb{C}\setminus\mathbb{R}$, $u_j \in \mathcal{L}$ $(j=1,2)$ one obtains

$$\begin{aligned}
&\langle\bar\omega_1\ \mathsf{N}_{\omega_1}(\cdot)u_1, \mathsf{N}_{\omega_2}(\cdot)u_2\rangle_{\mathcal{H}(\Phi,\Psi)} - \langle \mathsf{N}_{\omega_1}(\cdot)u_1, \bar\omega_2\mathsf{N}_{\omega_2}(\cdot)u_2\rangle_{\mathcal{H}(\Phi,\Psi)}\\
&\quad + (\Phi(\bar\omega_1)u_1, \Psi(\bar\omega_2)u_2)_{\mathcal{L}} - (\Psi(\bar\omega_1)u_1, \Phi(\bar\omega_2)u_2)_{\mathcal{L}}\\
&\quad = (\bar\omega_1-\omega_2)(\mathsf{N}_{\omega_1}(\omega_2)u_1,u_2)_{\mathcal{L}} - ((\Phi(\bar\omega_2)^*\Psi(\bar\omega_1) - \Psi(\bar\omega_2)^*\Phi(\bar\omega_1))u_1,u_2)_{\mathcal{L}}\\
&\quad = 0,
\end{aligned}$$

therefore, $\widetilde{A}'$ is symmetric in $\mathcal{H}(\Phi,\Psi)\oplus\mathcal{L}$.

Step 2. Let us show that ran $(\widetilde{A}'-\lambda)$ is dense in $\mathcal{H}(\Phi,\Psi)\oplus\mathcal{L}$ for $\lambda\in\mathbb{C}\setminus\mathbb{R}$. Choose the vector $\{F_\omega u, F'_\omega u\}$ with $\omega=\bar\lambda$. Then

$$\{F_{\bar\lambda}u, F'_{\bar\lambda}u-\lambda F_{\bar\lambda}u\} = \left\{\begin{bmatrix}\mathsf{N}_{\bar\lambda}(\cdot)u\\ \Psi(\lambda)u\end{bmatrix}, \begin{bmatrix}0\\ \Phi(\lambda)u-\lambda\Psi(\lambda)u\end{bmatrix}\right\}\in \widetilde{A}'-\lambda. \tag{2.9}$$

Since ran $(\Phi(\lambda)u-\lambda\Psi(\lambda))=\mathcal{L}$ one obtains $0\oplus\mathcal{L}\subset$ ran $(\widetilde{A}'-\lambda)$. Taking $\widehat{F}_\omega u$ with $\omega\ne\bar\lambda$ one obtains from (2.8)

$$\left\{\begin{bmatrix}\mathsf{N}_\omega(\cdot)u\\ \Psi(\bar\omega)u\end{bmatrix}, \begin{bmatrix}(\bar\omega-\lambda)\mathsf{N}_\omega(\cdot)u\\ \Phi(\bar\omega)u-\lambda\Psi(\bar\omega)u\end{bmatrix}\right\}\in \widetilde{A}'-\lambda \tag{2.10}$$

and, hence, $\begin{bmatrix}\mathsf{N}_\omega(\cdot)u\\ 0\end{bmatrix}\in$ ran $(\widetilde{A}'-\lambda)$ for all $\omega\ne\bar\lambda$. Due to the properties 1) and 2) of $\mathcal{H}(\Phi,\Psi)$ one obtains the statement.

Step 3. Let us show that $A(\Phi,\Psi)=(\widetilde{A}')^*$. Indeed, for every vector

$$\widehat{F}=\{F,F'\}=\left\{\begin{bmatrix}f(\cdot)\\ u\end{bmatrix}, \begin{bmatrix}f'(\cdot)\\ u'\end{bmatrix}\right\}\in A(\Phi,\Psi),\quad f,f'\in\mathcal{H}(\Phi,\Psi),\ u,u'\in\mathcal{L},$$

and $\omega\in\mathbb{C}\setminus\mathbb{R}$, $v\in\mathcal{L}$ it follows from (2.7) that

$$\begin{aligned}
\langle F', F_\omega v\rangle_{\mathcal{H}(\Phi,\Psi)} - \langle F, F'_\omega v\rangle_{\mathcal{H}(\Phi,\Psi)} &= \langle f', \mathsf{N}_\omega(\cdot)v\rangle_{\mathcal{H}(\Phi,\Psi)} - \langle f, \bar\omega\mathsf{N}_\omega(\cdot)v\rangle_{\mathcal{H}(\Phi,\Psi)}\\
&\quad + (u',\Psi(\bar\omega)v)_{\mathcal{L}} - (u,\Phi(\bar\omega)v)_{\mathcal{L}}\\
&= (f'(\omega)-\omega f(\omega)+\Psi(\bar\omega)^*u'-\Phi(\bar\omega)^*u, v)_{\mathcal{L}}=0.
\end{aligned}$$

Hence $\widehat{F}\in(\widetilde{A}')^*$ and $A(\Phi,\Psi)\subset(\widetilde{A}')^*$. Conversely, if

$$\langle f', \mathsf{N}_\omega(\cdot)h\rangle_{\mathcal{H}(\Phi,\Psi)} - \langle f,\bar\omega\mathsf{N}_\omega(\cdot)v\rangle_{\mathcal{H}(\Phi,\Psi)} + (u',\Psi(\bar\omega)v)_{\mathcal{L}} - (u,\Phi(\bar\omega)v)_{\mathcal{L}}=0$$

for some $f,f'\in\mathcal{H}(\varphi,\psi)$, $u,u'\in\mathcal{L}$ and all $\omega\in\mathbb{C}\setminus\mathbb{R}$, $v\in\mathcal{L}$, then

$$f'(\omega)-\omega f(\omega)-(\Phi(\bar\omega)^*u-\Psi(\bar\omega)^*u')=0$$

and, hence, $\widehat{F}\in A(\Phi,\Psi)$. This proves that $(\widetilde{A}')^*\subset A(\Phi,\Psi)$, and, hence, $(\widetilde{A}')^*=A(\Phi,\Psi)$.

Step 4. Finally, in view of (2.9) and Definition 2.1 (iii) one obtains

$$P_{\mathcal{L}}(\widetilde{A}(\Phi,\Psi)-\lambda)^{-1}|_{\mathcal{L}} = \Psi(\lambda)(\Phi(\lambda)-\lambda\Psi(\lambda))^{-1} = \psi(\lambda),$$
$$I_{\mathcal{L}} + \lambda\psi(\lambda) = \Phi(\lambda)(\Phi(\lambda)-\lambda\Psi(\lambda))^{-1} = \varphi(\lambda).$$

Therefore, the pair $\{\varphi,\psi\}$ is a normalized Nevanlinna pair corresponding to the linear relation $A(\Phi,\Psi)$ and the scale $\mathcal{L}$. □

Remark 2.6. For every normalized Nevanlinna pair $\{\varphi,\psi\}$ and $h \in \mathcal{H}(\varphi,\psi)$ the following identity holds

$$P_{\mathcal{L}}(A(\varphi,\psi)-\lambda)^{-1}\begin{bmatrix} h \\ 0 \end{bmatrix} = h(\lambda), \quad (\lambda \in \mathbb{C}\setminus\mathbb{R}). \tag{2.11}$$

Indeed, it follows from (2.9) that for every $u \in \mathcal{L}$ one obtains

$$\left(P_{\mathcal{L}}(A(\varphi,\psi)-\lambda)^{-1}\begin{bmatrix} h \\ 0 \end{bmatrix}, u\right)_{\mathcal{L}} = \left(\begin{bmatrix} h \\ 0 \end{bmatrix}, (A(\varphi,\psi)-\bar{\lambda})^{-1}\begin{bmatrix} 0 \\ u \end{bmatrix}\right)_{\mathcal{H}(\varphi,\psi)\oplus\mathcal{L}}$$
$$= (h, \mathsf{N}_\lambda(\cdot)u)_{\mathcal{H}(\varphi,\psi)} = (h(\lambda), u)_{\mathcal{L}}.$$

Remark 2.7. In notations of [24] the pair $\{\Phi,\Psi\}$ generates via (2.2) the Weyl family of the symmetric operator

$$S(\Phi,\Psi) = \{\{f,f'\} : f, f' \in \mathcal{H}(\Phi,\Psi), f'(\lambda) - \lambda f(\lambda) = 0\}$$

corresponding to the boundary relation

$$\Gamma = \left\{\left\{\begin{bmatrix} f \\ f' \end{bmatrix}, \begin{bmatrix} u' \\ u \end{bmatrix}\right\} : \begin{array}{c} f, f' \in \mathcal{H}(\Phi,\Psi),\ u, u' \in \mathcal{L}, \\ f'(\lambda) - \lambda f(\lambda) = \Phi(\bar{\lambda})^* u - \Psi(\bar{\lambda})^* u' \end{array}\right\}.$$

Remark 2.8. Mention that the linear space

$$\mathfrak{N}_{\bar{\omega}}(T) := \{\mathsf{N}_\omega(\cdot)u : u \in \mathcal{L}\}$$

in general is not closed, since

$$(\mathsf{N}_\omega(\cdot)u, \mathsf{N}_\omega(\cdot)u)_{\mathcal{H}(\varphi,\psi)} = (\mathsf{N}_\omega(\omega)u, u)_{\mathcal{L}}$$

and the operator $\mathsf{N}_\omega(\omega)$ not necessarily is boundedly invertible. If, however, $0 \in \rho(\mathsf{N}_\omega(\omega))$ then $\mathfrak{N}_{\bar{\omega}}(T)$ is closed. Recall that in this case $0 \in \rho(\mathsf{N}_\lambda(\lambda))$ for all $\lambda \in \mathbb{C}\setminus\mathbb{R}$ and, hence, all the subspaces $\mathfrak{N}_\lambda(\bar{\lambda})$ are closed for all $\lambda \in \mathbb{C}\setminus\mathbb{R}$.

Let, the ovf $\gamma(\lambda) : \mathcal{L} \to \mathcal{H}$ be defined by

$$\gamma(\lambda) := P_{\mathcal{H}}(\widetilde{A}-\lambda)^{-1}|_{\mathcal{L}} \quad (\lambda \in \rho(\widetilde{A})). \tag{2.12}$$

Proposition 2.9. *Let $\widetilde{A}$ be a selfadjoint linear relation in $\mathcal{H}\oplus\mathcal{L}$ and let $\{\varphi,\psi\}$ be the normalized Nevanlinna pair given by (2.3). Then the following identity holds*

$$\mathsf{N}_\omega^{\varphi\psi}(\lambda) = \gamma(\bar{\lambda})^*\gamma(\bar{\omega}). \tag{2.13}$$

Proof. Indeed, it follows from (2.5) that the kernel $\mathsf{N}_\omega^{\varphi\psi}(\lambda)$ takes the form

$$\mathsf{N}_\omega^{\varphi\psi}(\lambda) = (P_{\mathcal{L}}R_\lambda P_{\mathcal{H}})(P_{\mathcal{H}}R_{\bar{\omega}}|_{\mathcal{L}}) = \gamma(\bar{\lambda})^*\gamma(\bar{\omega}).$$ □

In general case one obtains

$$\begin{aligned} \mathsf{N}_\omega^{\Phi\Psi}(\lambda) &= (\Phi(\bar\lambda) - \bar\lambda\Psi(\bar\lambda))^* \mathsf{N}_\omega^{\varphi\psi}(\lambda)(\Phi(\bar\omega) - \bar\omega\Psi(\bar\omega)) \\ &= (\Phi(\bar\lambda) - \bar\lambda\Psi(\bar\lambda))^* \gamma(\bar\lambda)^*\gamma(\bar\omega)(\Phi(\bar\omega) - \bar\omega\Psi(\bar\omega)). \end{aligned}$$

The following statement formulated in terms of boundary relations can be found in [23, Lemma 4.1].

Lemma 2.10. *Let $\widetilde{A}$ be a selfadjoint linear relation in $\mathcal{H} \oplus \mathcal{L}$, let $\{\varphi, \psi\}$ be the normalized Nevanlinna pair given by* (2.3), *and let* $\dim \mathcal{L} < \infty$. *Then:*

(i) $\ker \psi(\lambda) = \{0\}$ *for $\lambda \in \mathbb{C} \setminus \mathbb{R}$ iff $P_\mathcal{L} \operatorname{dom} \widetilde{A}$ is dense in $\mathcal{L}$;*

(ii) $\ker \varphi(\lambda) = \{0\}$ *for $\lambda \in \mathbb{C} \setminus \mathbb{R}$ iff $P_\mathcal{L} \operatorname{ran} \widetilde{A}$ is dense in $\mathcal{L}$.*

Proof. Let us prove the first statement. The set $P_\mathcal{L} \operatorname{dom} \widetilde{A}$ consists of the vectors $u \in \mathcal{L}$ such that

$$\left\{ \begin{bmatrix} f \\ u \end{bmatrix}, \begin{bmatrix} f' \\ u' \end{bmatrix} \right\} \in \widetilde{A} \quad \text{for some} \quad f, f' \in \mathcal{H}, u' \in \mathcal{L}.$$

If there is a vector $v \in \mathcal{L}$ such that $v \perp u$ for all $u \in P_\mathcal{L} \operatorname{dom} \widetilde{A}$ then

$$\left\{ \begin{bmatrix} 0 \\ 0 \end{bmatrix}, \begin{bmatrix} 0 \\ v \end{bmatrix} \right\} \in \widetilde{A},$$

and then $\psi(\lambda)v = 0$, $\varphi(\lambda)v = v$, due to (2.3).

Conversely, if $\psi(\lambda)v = 0$ for some $v \neq 0$, then in view of (2.3)

$$\left\{ \begin{bmatrix} 0 \\ 0 \end{bmatrix}, \begin{bmatrix} 0 \\ v \end{bmatrix} \right\} \in \widetilde{A},$$

and hence $v \perp P_\mathcal{L} \operatorname{dom} \widetilde{A}$.

The proof of the second statement is similar. □

Remark 2.11. If $\{\Phi, \Psi\}$ is an arbitrary Nevanlinna pair and $\{\varphi, \psi\}$ is the corresponding normalized Nevanlinna pair, then the mapping

$$\mathcal{U} : f \in \mathcal{H}(\varphi, \psi) \mapsto (\Phi(\bar\lambda)^* - \lambda\Psi(\bar\lambda)^*) f(\lambda) \tag{2.14}$$

maps isometrically $\mathcal{H}(\varphi, \psi)$ onto $\mathcal{H}(\Phi, \Psi)$. The linear relations $A(\varphi, \psi)$ and $A(\Phi, \Psi)$ are unitarily equivalent under this mapping, that is

$$\{f, f'\} \in A(\varphi, \psi) \Leftrightarrow \{\mathcal{U}f, \mathcal{U}f'\} \in A(\Phi, \Psi).$$

If the Nevanlinna pair $\{\varphi, \psi\}$ satisfies the first condition (i) in Lemma 2.10 then it is equivalent to an ovf $m \in N(\mathcal{L})$. If, additionally, $m(\lambda)$ takes values in $[\mathcal{L}]$ for $\lambda \in \mathbb{C} \setminus \mathbb{R}$ then $m \in N[\mathcal{L}]$ and the reproducing kernel Hilbert space $\mathcal{H}(\varphi, \psi)$ is unitary equivalent to the reproducing kernel Hilbert space $\mathcal{H}(m)$ with the kernel $\mathsf{N}_\omega^m(\lambda)$ (see (1.1)) via the mapping

$$\mathcal{U} : f \in \mathcal{H}(\varphi, \psi) \to (I - \lambda m(\lambda)) f(\lambda) \in \mathcal{H}(m).$$

These spaces have been introduced by L. de Branges [17]. The following statement can be derived from [5], however we will give a proof for the convenience of the reader.

Lemma 2.12. *Let* $m \in N[\mathcal{L}]$*, and let* $\dim \mathcal{L} < \infty$*. Then:*

(i) $f(\lambda) = O(1)$ $(\lambda \widehat{\to} \infty)$ *for all* $f \in \mathcal{H}(m)$*;*

(ii) *If, additionally,* $m \in N_0[\mathcal{L}]$ *then* $f(\lambda) = O(\frac{1}{\lambda})$ $(\lambda \widehat{\to} \infty)$ *for all* $f \in \mathcal{H}(m)$*.*

Proof. The reproducing kernel property and Schwartz inequality yield

$$\begin{aligned} |(f(\lambda), u)_{\mathcal{L}}| &= |\langle f(\cdot), \mathsf{N}_\lambda(\cdot)u\rangle_{\mathcal{H}(m)}| \\ &\le ||f(\cdot)||_{\mathcal{H}(m)} ||\mathsf{N}_\lambda(\cdot)u||_{\mathcal{H}(m)} \\ &= ||f||_{\mathcal{H}(m)} \left(\frac{\Im m(\lambda)}{\Im \lambda} u, u\right)^{1/2} = O(1) \end{aligned}$$

for all $f \in \mathcal{H}(m)$ and $u \in \mathcal{L}$. If, additionally, $m \in N_0[\mathcal{L}]$, then m admits the integral representation

$$m(\lambda) = \int_{\mathbb{R}} \frac{d\sigma(t)}{t - \lambda} \tag{2.15}$$

with a nondecreasing function $\sigma(t)$ such that

$$\lim_{t \to -\infty} \sigma(t) = 0, \qquad \lim_{t \to +\infty} \sigma(t) = \Sigma \in [\mathcal{L}].$$

It follows from (2.15) that

$$\left(\frac{\Im m(\lambda)}{\Im \lambda} u, u\right) = \int \frac{d(\sigma(t)u, u)}{|t - \lambda|^2} = O(\frac{1}{\lambda^2})$$

Therefore, $(f(\lambda), u)_{\mathcal{L}} = O(1/\lambda)$ for all $u \in \mathcal{L}$. □

2.3. Generalized Fourier transform

Definition 2.13. A linear relation $\widetilde{A} = \widetilde{A}^*$ in $\mathcal{H} \oplus \mathcal{L}$ is said to be $\mathcal{L}$-minimal if

$$\mathcal{H}_0 = \overline{\operatorname{span}}\{P_{\mathcal{H}}(\widetilde{A} - \lambda)^{-1}\mathcal{L} : \lambda \in \rho(\widetilde{A})\}. \tag{2.16}$$

In this section we show that every $\mathcal{L}$-minimal selfadjoint linear relation A is unitarily equivalent to its functional model $A(\varphi, \psi)$, constructed in Theorem 2.5. The operator $\mathcal{F} : \mathcal{H} \to \mathcal{H}(\varphi, \psi)$ given by the formula

$$h \mapsto (\mathcal{F}h)(\lambda) = \gamma(\bar{\lambda})^* h = P_{\mathcal{L}}(\widetilde{A} - \lambda)^{-1} h \quad (h \in \mathcal{H}) \tag{2.17}$$

is called the *generalized Fourier transform* associated with $\widetilde{A}$ and the scale $\mathcal{L}$. In the discrete case this definition was given in [30].

Theorem 2.14. *Let* $\widetilde{A}$ *be a selfadjoint linear relation in* $\mathcal{H} \oplus \mathcal{L}$ *and let* $\{\varphi, \psi\}$ *be the corresponding Nevanlinna pair given by* (2.3)*. Then:*

1) *The generalized Fourier transform* $\mathcal{F}$ *maps isometrically the subspace* $\mathcal{H}_0$ *onto* $\mathcal{H}(\varphi, \psi)$ *and* $\mathcal{F}$ *is identically equal to* 0 *on* $\mathcal{H} \ominus \mathcal{H}_0$*;*

2) *For every* $\left\{\begin{bmatrix} f \\ u \end{bmatrix}, \begin{bmatrix} f' \\ u' \end{bmatrix}\right\} \in \widetilde{A}$ *the following identity holds*

$$E(\lambda)\mathcal{F}(f' - \lambda f) = \begin{bmatrix}\varphi(\lambda) & -\psi(\lambda)\end{bmatrix}\begin{bmatrix} u \\ u' \end{bmatrix}. \tag{2.18}$$

Proof. 1) For every vector $h = \gamma(\bar{\omega})v$ ($\omega \in \rho(\widetilde{A})$, $v \in \mathcal{L}$) it follows from Proposition 2.9 that

$$(\mathcal{F}h)(\lambda) = \gamma(\bar{\lambda})^*\gamma(\bar{\omega})v = \mathsf{N}_\omega^{\varphi\psi}(\lambda)v.$$

Therefore, $\mathcal{F}$ maps the linear space $\operatorname{span}\{\gamma(\omega)\mathcal{L} : \omega \in \rho(\widetilde{A})\}$ dense in $\mathcal{H}_0$ onto the linear space $\operatorname{span}\{\mathsf{N}_\omega^{\varphi\psi}(\cdot)\mathcal{L} : \omega \in \rho(\widetilde{A})\}$ which is dense in $\mathcal{H}(\varphi,\psi)$. Moreover, this mapping is isometric, since

$$\begin{aligned}(\mathcal{F}h, \mathcal{F}h)_{\mathcal{H}(\varphi,\psi)} &= (\mathsf{N}_\omega^{\varphi\psi}(\cdot)v, \mathsf{N}_\omega^{\varphi\psi}(\cdot)v)_{\mathcal{H}(\varphi,\psi)} \\ &= (\mathsf{N}_\omega^{\varphi\psi}(\omega)v, v)_{\mathcal{L}} = (h,h)_{\mathcal{H}}.\end{aligned} \tag{2.19}$$

This proves the first statement. It is clear from (2.17) that $\mathcal{F}h \equiv 0$ for $h \in \mathcal{H} \ominus \mathcal{H}_0$.

2) Let $h = \gamma(\bar{\omega})v = P_{\mathcal{H}}(\widetilde{A} - \bar{\omega})^{-1}v$, $\quad v \in \mathcal{L}$. It follows from (2.3), (2.12) that

$$\begin{bmatrix} h \\ \psi(\bar{\omega})v \end{bmatrix} = (\widetilde{A} - \bar{\omega})^{-1}\begin{bmatrix} 0 \\ v \end{bmatrix}, \quad \begin{bmatrix} \bar{\omega}h \\ \varphi(\bar{\omega})v \end{bmatrix} = (I + \bar{\omega}(\widetilde{A} - \bar{\omega})^{-1})\begin{bmatrix} 0 \\ v \end{bmatrix}$$

and hence

$$\left\{\begin{bmatrix} h \\ \psi(\bar{\omega})v \end{bmatrix}, \begin{bmatrix} \bar{\omega}h \\ \varphi(\bar{\omega})v \end{bmatrix}\right\} \in \widetilde{A}.$$

Since $\widetilde{A} = \widetilde{A}^*$ one obtains for all $\left\{\begin{bmatrix} f \\ u \end{bmatrix}, \begin{bmatrix} f' \\ u' \end{bmatrix}\right\} \in \widetilde{A}$

$$(f', h)_{\mathcal{H}} - (f, \bar{\omega}h)_{\mathcal{H}} + (u', \psi(\bar{\omega}))_{\mathcal{L}} - (u, \varphi(\bar{\omega})v)_{\mathcal{L}} = 0, \quad v \in \mathcal{L}.$$

This implies

$$\gamma(\bar{\omega})^*(f' - \bar{\omega}f) = \varphi(\omega)u - \psi(\omega)u', \quad \omega \in \mathbb{C} \setminus \mathbb{R}. \tag{2.20}$$

This proves the identity (2.18). □

Remark 2.15. In the case, when the linear relation $\widetilde{A}$ is $\mathcal{L}$-minimal it is unitary equivalent to the linear relation $A(\varphi,\psi)$ via the formula

$$\widetilde{A}(\varphi,\psi) = \left\{\left\{\begin{bmatrix} \mathcal{F}f \\ u \end{bmatrix}, \begin{bmatrix} \mathcal{F}f' \\ u' \end{bmatrix}\right\} : \left\{\begin{bmatrix} f \\ u \end{bmatrix}, \begin{bmatrix} f' \\ u' \end{bmatrix}\right\} \in \widetilde{A}\right\}. \tag{2.21}$$

The operator $\mathcal{F} \oplus I_{\mathcal{L}}$ establishes this unitary equivalence.

Remark 2.16. It follows from (2.11) that the Fourier transform $\mathcal{F}$ associated with the operator $A(\varphi,\psi)$ is identical, since

$$(\mathcal{F}h)(\lambda) = P_{\mathcal{L}}(A(\varphi,\psi) - \lambda)^{-1}\begin{bmatrix} h \\ 0 \end{bmatrix} = h(\lambda) \quad \text{for every } h \in \mathcal{H}(\varphi,\psi).$$

Remark 2.17. The Fourier transform $\mathcal{F}(\Phi,\Psi)$ associated with an arbitrary Nevanlinna pair $\{\Phi,\Psi\}$ can be defined by the formula

$$\mathcal{F}(\Phi,\Psi)h := \mathcal{U}\mathcal{F}(\varphi,\psi)h, \quad h \in \mathcal{H},$$

where $\{\varphi,\psi\}$ is the corresponding normalized Nevanlinna pair and $\mathcal{U}$ is the mapping from $\mathcal{H}(\varphi,\psi)$ to $\mathcal{H}(\Phi,\Psi)$ given by (2.14). Then $\mathcal{F}(\Phi,\Psi)$ maps isometrically the subspace $\mathcal{H}_0$ onto $\mathcal{H}(\Phi,\Psi)$ and

$$E(\lambda)\mathcal{F}(\Phi,\Psi)(f'-\lambda f) = \left[\Phi(\bar\lambda)^* - \Psi(\bar\lambda)^*\right]\begin{bmatrix} u \\ u' \end{bmatrix} \text{ for every } \left\{\begin{bmatrix} f \\ u \end{bmatrix}, \begin{bmatrix} f' \\ u' \end{bmatrix}\right\} \in \widetilde{A}.$$

3. Abstract interpolation problem

Let $\mathcal{X}$ be a complex linear space, let $\mathcal{L}$ be a Hilbert space, let B_1, B_2 be linear operators in $\mathcal{X}$, let C_1, C_2 be linear operators from $\mathcal{X}$ to $\mathcal{L}$, and let K be a nonnegative sesquilinear form on $\mathcal{X}$ which satisfies the identity

(A1) $K(B_2h, B_1g) - K(B_1h, B_2g) = (C_1h, C_2g)_{\mathcal{L}} - (C_2h, C_1g)_{\mathcal{L}}$

for every $h, g \in \mathcal{X}$. Consider the following continuous analog of the AIP considered in [30].

Problem $AIP(B_1,B_2,C_1,C_2,K)$**.** Let the data set (B_1,B_2,C_1,C_2,K) satisfies the assumption $(A1)$. Find a normalized Nevanlinna pair $\{\varphi,\psi\} \in \widetilde{N}(\mathcal{L})$ such that for some linear mapping $F : \mathcal{X} \to \mathcal{H}(\varphi,\psi)$ the following conditions hold:

(C1) $(FB_2h)(\lambda) - \lambda(FB_1h)(\lambda) = \left[\ \varphi(\lambda) \quad -\psi(\lambda)\ \right]\begin{bmatrix} C_1h \\ C_2h \end{bmatrix};$

(C2) $\|Fh\|^2_{\mathcal{H}(\varphi,\psi)} \le K(h,h)$

for all $h \in \mathcal{X}$.

Consider the factor space $\widehat{\mathcal{X}} = \mathcal{X}/\ker K$ and denote by $\hat h$ the equivalence class $h + \ker K$ in $\widehat{\mathcal{X}}$, $h \in \mathcal{X}$. Let $\widehat{\mathcal{X}}$ be endowed with the scalar product (1.3) and let $\mathcal{H}$ be the completion of $\widehat{\mathcal{X}}$ with respect to the norm $\|h\|_{\mathcal{H}}$.

In some examples (see Section 5) the linear space $\mathcal{X}$ has its own inner product. In order to avoid an ambiguity we denote by B^+ the adjoint to the operator $B : \mathcal{H} \to \mathcal{H}$ and by B^* the adjoint to $B : \mathcal{X} \to \mathcal{X}$.

Proposition 3.1. *Let the data set* (B_1,B_2,C_1,C_2,K) *satisfies the assumption* $(A1)$*. Then the linear relation*

$$\widehat{A} = \left\{\left\{\begin{bmatrix} \widehat{B_1h} \\ C_1h \end{bmatrix}, \begin{bmatrix} \widehat{B_2h} \\ C_2h \end{bmatrix}\right\} : h \in \mathcal{X}\right\} \tag{3.1}$$

is symmetric in $\mathcal{H} \oplus \mathcal{L}$*.*

Proof. The statement is implied by (A1) since

$$\left\langle \begin{bmatrix} \widehat{B_2h} \\ C_2h \end{bmatrix}, \begin{bmatrix} \widehat{B_1h} \\ C_1h \end{bmatrix} \right\rangle_{\mathcal{H}\oplus\mathcal{L}} - \left\langle \begin{bmatrix} \widehat{B_1h} \\ C_1h \end{bmatrix}, \begin{bmatrix} \widehat{B_2h} \\ C_2h \end{bmatrix} \right\rangle_{\mathcal{H}\oplus\mathcal{L}}$$
$$= K(B_2h, B_1h) - K(B_1h, B_2h) - (C_1h, C_2h)_{\mathcal{L}} + (C_2h, C_1h)_{\mathcal{L}} = 0. \qquad \square$$

Remark 3.2. The linear relation $\widehat{A}$ is called the AIP symmetry. In general, the AIP symmetry $\widehat{A}$ need not be simple and its deficiency indices not necessarily coincide.

Theorem 3.3. *Let the data set* (B_1, B_2, C_1, C_2, K) *satisfies the assumption* $(A1)$. *Then the Problem* $AIP(B_1, B_2, C_1, C_2, K)$ *is solvable and the set of its normalized solutions is parametrized by the formula*

$$\begin{bmatrix} \psi(\lambda) \\ \varphi(\lambda) \end{bmatrix} = \begin{bmatrix} I_{\mathcal{L}} & 0 \\ \lambda & I_{\mathcal{L}} \end{bmatrix} \begin{bmatrix} P_{\mathcal{L}}(\widetilde{A} - \lambda)^{-1}|_{\mathcal{L}} \\ I_{\mathcal{L}} \end{bmatrix}, \tag{3.2}$$

where $\widetilde{A}$ *ranges over the set of all selfadjoint extensions of* $\widehat{A}$ *with the exit in a Hilbert space* $\widetilde{\mathcal{H}} \oplus \mathcal{L} \supset \mathcal{H} \oplus \mathcal{L}$. *The corresponding linear mapping* $F : \mathcal{X} \to \mathcal{H}(\varphi, \psi)$ *is given by*

$$(Fh)(\lambda) = P_{\mathcal{L}}(\widetilde{A} - \lambda)^{-1}\widehat{h}, \quad h \in \mathcal{X}. \tag{3.3}$$

Proof. Sufficiency. Let $\widetilde{A}$ be a selfadjoint extension of $\widehat{A}$ and let $\{\varphi, \psi\}$ be the normalized Nevanlinna pair corresponding to $\widetilde{A}$ and the scale $\mathcal{L}$, and let $\mathcal{F} : \widetilde{\mathcal{H}} \to \mathcal{H}(\varphi, \psi)$ be the corresponding generalized Fourier transform given by (2.17). Then the formula (3.2) is implied by (2.3) and in view of (2.17) the linear mapping $F : \mathcal{X} \to \mathcal{H}(\varphi, \psi)$ given by (3.3) is connected to $\mathcal{F}$ via the formula

$$Fh = \mathcal{F}\widehat{h}, \quad (h \in \mathcal{X}). \tag{3.4}$$

Since $\mathcal{F}$ satisfies the identity (2.18) and

$$\left\{ \begin{bmatrix} \widehat{B_1 h} \\ C_1 h \end{bmatrix}, \begin{bmatrix} \widehat{B_2 h} \\ C_2 h \end{bmatrix} \right\} \in \widehat{A} \subset \widetilde{A}$$

one obtains from (2.18)

$$\begin{aligned} (FB_2h)(\lambda) - \lambda(FB_1h)(\lambda) &= (\mathcal{F}\widehat{B_2h})(\lambda) - \lambda(\mathcal{F}\widehat{B_1h})(\lambda) \\ &= \begin{bmatrix} \varphi(\lambda) & -\psi(\lambda) \end{bmatrix} \begin{bmatrix} C_1 h \\ C_2 h \end{bmatrix} \quad \forall h \in \mathcal{H}. \end{aligned} \tag{3.5}$$

Next, it follows from (1.3) and Theorem 2.14 that

$$\|Fh\|^2_{\mathcal{H}(\varphi,\psi)} = \|\mathcal{F}\widehat{h}\|^2_{\mathcal{H}(\varphi,\psi)} \le \|\widehat{h}\|^2_{\mathcal{H}} = K(h, h).$$

This proves $(C1)$, $(C2)$ and, hence, $\{\varphi, \psi\}$ is a solution of the AIP.

Necessity. Let a normalized Nevanlinna pair $\{\varphi, \psi\}$ be a solution of the AIP and let the mapping $F : \mathcal{X} \to \mathcal{H}(\varphi, \psi)$ satisfies (C1), (C2). We will construct a selfadjoint exit space extension $\widetilde{A}$ of $\widehat{A}$ such that (3.2) and (3.3) hold.

Step 1. *Isometric embedding of* $\mathcal{H}$ *into a Hilbert space.* It follows from $(C2)$ that $(Fh)(\lambda) \equiv 0$ for all $h \in \ker K$. Thus F induces the mapping $\widehat{F} : \widehat{\mathcal{X}} \to \mathcal{H}(\varphi, \psi)$, which is well defined by

$$\hat{h} \to (\widehat{F}\widehat{h})(\lambda) = (Fh)(\lambda), \quad h \in \mathcal{H} \tag{3.6}$$

and is contractive due to $(C2)$

$$\|(\widehat{F}\widehat{h})(\lambda)\|^2_{\mathcal{H}(\varphi,\psi)} = \|(Fh)(\lambda)\|^2_{\mathcal{H}(\varphi,\psi)} \le K(h, h) = \|\widehat{h}\|^2_{\mathcal{H}}.$$

We will keep the same notation for the continuous extension of $\widehat{F}$ to $\mathcal{H}$.

Let $D = D^*(\geq 0)$ be the defect operator of the contraction $\widehat{F}$ defined by

$$D^2 = I - \widehat{F}^+\widehat{F} : \mathcal{H} \to \mathcal{H} \tag{3.7}$$

and let $\mathcal{D} = \overline{\operatorname{ran}}\, D$ be the defect subspace of $\widehat{F}$ in $\mathcal{H}$. Consider the column extension $\widetilde{F}$ of the operator $\widehat{F}$ to the isometric mapping from $\mathcal{H}$ to $\mathcal{D} \oplus \mathcal{H}(\varphi, \psi)$ by the formula

$$\widetilde{F}h = \begin{bmatrix} Dh \\ \widehat{F}h \end{bmatrix}, \quad h \in \mathcal{H}. \tag{3.8}$$

Step 2. *Construction of a selfadjoint linear relation* $\widetilde{A}$. Let $A_{\mathcal{D}}$ be a linear relation in $\mathcal{D}$ defined by

$$A_{\mathcal{D}} = \left\{ \left\{ D\widehat{B}_1 h, D\widehat{B}_2 h \right\} : h \in \mathcal{X} \right\}$$

and let us show that $A_{\mathcal{D}}$ is symmetric in $\mathcal{D}$. Indeed, it follows from (3.6), (3.7) that

$$\begin{aligned}(D\widehat{B_2h}, D\widehat{B_1h})_{\mathcal{H}} &- (D\widehat{B_1h}, D\widehat{B_2h})_{\mathcal{H}} \\ &= ((I - \widehat{F}^+\widehat{F})\widehat{B_2h}, \widehat{B_1h})_{\mathcal{H}} - ((I - \widehat{F}^+\widehat{F})\widehat{B_1h}, \widehat{B_2h})_{\mathcal{H}} \\ &= K(B_2h, B_1h) - K(B_1h, B_2h) \\ &- (\widehat{F}\widehat{B_2h}, \widehat{F}\widehat{B_1h})_{\mathcal{H}(\varphi,\psi)} + (\widehat{F}\widehat{B_1h}, \widehat{F}\widehat{B_2h})_{\mathcal{H}(\varphi,\psi)}.\end{aligned} \tag{3.9}$$

As follows from (C1)

$$\begin{aligned}(\widehat{F}\widehat{B_2h})(\lambda) - \lambda(\widehat{F}\widehat{B_1h})(\lambda) &= (FB_2h)(\lambda) - \lambda(FB_1h)(\lambda) \\ &= \begin{bmatrix} \varphi(\lambda) & \psi(\lambda) \end{bmatrix} \begin{bmatrix} C_1h \\ C_2h \end{bmatrix} \quad \forall h \in \mathcal{X}.\end{aligned} \tag{3.10}$$

In view of (2.7) this implies

$$\left\{ \begin{bmatrix} \widehat{F}\widehat{B_1h} \\ C_1h \end{bmatrix}, \begin{bmatrix} \widehat{F}\widehat{B_2h} \\ C_2h \end{bmatrix} \right\} \in A(\varphi, \psi).$$

Therefore, the right-hand part of (3.9) can be rewritten as

$$K(B_2h, B_1h) - K(B_1h, B_2h) - (C_1h, C_2h)_{\mathcal{L}} + (C_2h, C_1h)_{\mathcal{L}}$$

and hence it is vanishing due to (A1).

Let $\widetilde{A}_{\mathcal{D}}$ be a selfadjoint extension of $A_{\mathcal{D}}$ in a Hilbert space $\widetilde{\mathcal{D}} \supset \mathcal{D}$ and let

$$\widetilde{A} = \widetilde{A}_{\mathcal{D}} \oplus A(\varphi, \psi). \tag{3.11}$$

Step 3. *Linear relation* $\widetilde{A}$ *satisfies* (3.2) *and* (3.3). Under the identification of the vector $h \in \mathcal{H}$ with $\widetilde{F}h$ the symmetric linear relation $\widehat{A}$ in $\mathcal{H} \oplus \mathcal{L}$ can be identified with the symmetric linear relation

$$A_1 = (\widetilde{F} \oplus I_{\mathcal{L}})\widehat{A}(\widetilde{F} \oplus I_{\mathcal{L}})^{-1}$$
$$= \left\{ \left\{ \begin{bmatrix} D\widehat{B_1 h} \\ \widehat{F}\widehat{B_1 h} \\ C_1 h \end{bmatrix}, \begin{bmatrix} D\widehat{B_2 h} \\ \widehat{F}\widehat{B_2 h} \\ C_2 h \end{bmatrix} \right\} : h \in \mathcal{X} \right\}$$

in $\widetilde{\mathcal{H}} := \widetilde{\mathcal{D}} \oplus \mathcal{H}(\varphi,\psi) \oplus \mathcal{L}$. Moreover A_1 is contained in the selfadjoint linear relation $\widetilde{A} = \widetilde{A}_{\mathcal{D}} \oplus A(\varphi,\psi)$, since $\{D\widehat{B_1 h}, D\widehat{B_2 h}\} \in A_{\mathcal{D}} \subset \widetilde{A}_{\mathcal{D}}$ and

$$\left\{ \begin{bmatrix} \widehat{F}\widehat{B_1 h} \\ C_1 h \end{bmatrix}, \begin{bmatrix} \widehat{F}\widehat{B_2 h} \\ C_2 h \end{bmatrix} \right\} \in A(\varphi,\psi).$$

The formula (3.2) is implied by the analogous formula for $A(\varphi,\psi)$

$$\begin{bmatrix} \psi(\lambda) \\ \varphi(\lambda) \end{bmatrix} = \begin{bmatrix} I_{\mathcal{L}} & 0 \\ \lambda I_{\mathcal{L}} & I_{\mathcal{L}} \end{bmatrix} \begin{bmatrix} P_{\mathcal{L}}(A(\varphi,\psi)-\lambda)^{-1}|_{\mathcal{L}} \\ I_{\mathcal{L}} \end{bmatrix}$$

since

$$P_{\mathcal{L}}(A(\varphi,\psi)-\lambda)^{-1}|_{\mathcal{L}} = P_{\mathcal{L}}(\widetilde{A}-\lambda)^{-1}|_{\mathcal{L}}.$$

In view of Remark 2.6 one obtains for every $h \in \mathcal{X}$

$$P_{\mathcal{L}}(\widetilde{A}-\lambda)^{-1} \begin{bmatrix} \widetilde{F}\widehat{h} \\ 0 \end{bmatrix} = P_{\mathcal{L}}(\widetilde{A}-\lambda)^{-1} \begin{bmatrix} \widetilde{F}\widehat{h} \\ 0 \end{bmatrix}$$
$$= P_{\mathcal{L}}(A(\varphi,\psi)-\lambda)^{-1} \begin{bmatrix} Fh \\ 0 \end{bmatrix} = (Fh)(\lambda).$$

This proves the formula (3.3), since $\widetilde{F}\widehat{h}$ is identified with $\widehat{h}$. □

Remark 3.4. If (B_1, B_2, C_1, C_2, K) is a data set for the AIP in the Nevanlinna classes then the data set (T_1, T_2, M_1, M_2, K) given by

$$T_1 = B_1 - iB_2, \quad T_2 = B_1 + iB_2;$$
$$M_1 = C_1 - iC_2, \quad M_2 = C_1 + iC_2,$$

is a data set for the AIP in the Schur class. Remind (see [30]), that a contractive $[\mathcal{L}]$-valued function $\omega(\zeta)$ is said to be a solution of the problem $AIP(T_1, T_2, M_1, M_2, K)$, if there exists a map Φ from $\mathcal{X}$ to the de Branges–Rovnyak space $\mathcal{H}^\omega$, such that

$$(\Phi T_1 h)(t) - t(\Phi T_2 h)(t) = \begin{bmatrix} I & -\omega(t) \\ -\omega^*(t) & I \end{bmatrix} \begin{bmatrix} M_2 h \\ M_1 h \end{bmatrix},$$

and $\|\Phi h\|^2_{\mathcal{H}^\omega} \le K(h,h)$ for all $h \in \mathcal{X}$. One can check, that solutions of the problems $AIP(B_1, B_2, C_1, C_2, K)$ and $AIP(T_1, T_2, M_1, M_2, K)$ are related via some linear fractional transformation and the result of Theorem 3.3 can be derived from the corresponding result in [30]. However, we prefer to give a direct proof based on the space $\mathcal{H}(\varphi,\psi)$, especially since these spaces will be useful in applications to some interpolation problems.

In general, the mapping $F : \mathcal{X} \to \mathcal{H}(\varphi, \psi)$ in (C1), (C2) is not uniquely defined by the solution $\{\varphi, \psi\}$ of the $AIP(B_1, B_2, C_1, C_2, K)$. We impose an additional assumption on the data set:

(U) $B_2 - \lambda B_1$ is an isomorphism in $\mathcal{X}$ for λ in nonempty domains $\mathcal{O}_\pm \subset \mathbb{C}_\pm$,

which ensures the uniqueness of F. Let us set

$$G(\lambda) = \begin{bmatrix} C_1 \\ C_2 \end{bmatrix} (B_2 - \lambda B_1)^{-1} \quad (\lambda \in \mathcal{O}_\pm).$$

Proposition 3.5. *Let the data set (B_1, B_2, C_1, C_2, K) satisfies the assumptions (A1), (U). Then the mapping $F : \mathcal{X} \to \mathcal{H}(\varphi, \psi)$ in (C1), (C2) is uniquely defined by the solution $\{\varphi, \psi\}$ of the $AIP(B_1, B_2, C_1, C_2, K)$ by the formula*

$$(Fh)(\lambda) = \begin{bmatrix} \varphi(\lambda) & -\psi(\lambda) \end{bmatrix} G(\lambda)h \quad (\lambda \in \mathcal{O}_\pm). \tag{3.12}$$

Proof. Applying (C1) to the vector

$$h = h_\mu := (B_2 - \mu B_1)^{-1} g \quad (\mu \in \mathcal{O}_\pm,\, g \in \mathcal{X}),$$

one obtains

$$\begin{aligned} (Fg)(\lambda) &= (FB_2 h_\mu)(\lambda) - \mu(FB_1 h_\mu)(\lambda) \\ &= (FB_2 h_\mu)(\lambda) - \lambda(FB_1 h_\mu)(\lambda) + (\lambda - \mu)(FB_1 h_\mu)(\lambda) \\ &= \begin{bmatrix} \varphi(\lambda) & -\psi(\lambda) \end{bmatrix} G(\mu) g + (\lambda - \mu)(FB_1 h_\mu)(\lambda). \end{aligned} \tag{3.13}$$

Setting in (3.13) $\lambda = \mu$, one obtains $(Fg)(\mu)) = \begin{bmatrix} \varphi(\mu) & -\psi(\mu) \end{bmatrix} G(\mu) g$. □

4. Description of solutions of abstract interpolation problem

In view of Theorem 3.3 a description of the set of solutions of the AIP is reduced to the description of $\mathcal{L}$-resolvents of the linear relation $\widehat{A}$. The latter problem can be solved within the theory of $\mathcal{L}$-resolvent matrix [35, 38] (see also [21], [22]). In this section we will treat both the nondegenerate ($\ker K \neq \{0\}$) and the degenerate case ($\ker K \neq \{0\}$). In the case when the form $K(\cdot, \cdot)$ is degenerate it is more convenient to calculate the resolvent matrix of some auxiliary linear relation A_0 which is a restriction of $\widehat{A}$.

4.1. Symmetric linear relation A

Let us impose some additional assumptions on the data set (B_1, B_2, C_1, C_2, K):

(A2) $\dim \ker K < \infty$ and $\mathcal{X}$ admits the representation

$$\mathcal{X} = \mathcal{X}_0 \dotplus \ker K, \tag{4.1}$$

such that $B_j \mathcal{X}_0 \subseteq \mathcal{X}_0$ ($j = 1, 2$);

(A3) $B_2 = I_{\mathcal{X}}$ and the operators $B_1|_{\mathcal{X}_0} : \mathcal{X}_0 \subseteq \mathcal{H} \to \mathcal{H}$, $C_1|_{\mathcal{X}_0}$, $C_2|_{\mathcal{X}_0} : \mathcal{X}_0 \subseteq \mathcal{H} \to \mathcal{L}$ are bounded.

Due to the assumption (A2) one can identify $\mathcal{X}_0$ with $\widehat{\mathcal{X}} = \mathcal{X}/\ker K$ and consider the space $\mathcal{H}$ as a completion of $\mathcal{X}_0$. The continuations of the operators $B_1|_{\mathcal{X}_0}$, $C_1|_{\mathcal{X}_0}$, $C_2|_{\mathcal{X}_0}$ will be denoted by $\widetilde{B}_1 \in [\mathcal{H}]$, $\widetilde{C}_1$, $\widetilde{C}_2 \in [\mathcal{H}, \mathcal{L}]$.

Define a linear relation

$$A_0 = \left\{ \left\{ \begin{bmatrix} B_1 h \\ C_1 h \end{bmatrix}, \begin{bmatrix} B_2 h \\ C_2 h \end{bmatrix} \right\} : h \in \mathcal{X}_0 \right\} \tag{4.2}$$

in a Hilbert space $\mathcal{H} \oplus \mathcal{L}$. Clearly, A_0 is a restriction of the symmetric linear relation $\widehat{A}$ which can be rewritten as

$$\widehat{A} = \left\{ \left\{ \begin{bmatrix} B_1 h + \widehat{B_1 u} \\ C_1 h + C_1 u \end{bmatrix}, \begin{bmatrix} B_2 h + \widehat{B_2 u} \\ C_2 h + C_2 u \end{bmatrix} \right\} : h \in \mathcal{X}_0, u \in \ker K \right\}. \tag{4.3}$$

In view of (A3) the closures of A_0 and $\widehat{A}$ take the form

$$A := \operatorname{clos} A_0 = \left\{ \left\{ \begin{bmatrix} \widetilde{B}_1 h \\ \widetilde{C}_1 h \end{bmatrix}, \begin{bmatrix} h \\ \widetilde{C}_2 h \end{bmatrix} \right\} : h \in \mathcal{H} \right\}, \tag{4.4}$$

$$\operatorname{clos} \widehat{A} = \left\{ \left\{ \begin{bmatrix} \widetilde{B}_1 h + \widehat{B_1 u} \\ \widetilde{C}_1 h + C_1 u \end{bmatrix}, \begin{bmatrix} h \\ \widetilde{C}_2 h + C_2 u \end{bmatrix} \right\} : h \in \mathcal{H}, u \in \ker K \right\}. \tag{4.5}$$

A point $\lambda \in \mathbb{C}$ is said to be a *regular type point* for a closed symmetric linear relation A if $\operatorname{ran}(A - \lambda)$ is closed in $\mathcal{H} \oplus \mathcal{L}$. It is well known that the set $\widehat{\rho}(A)$ of regular type points for symmetric linear relation A contains $\mathbb{C}_+ \cup \mathbb{C}_-$ and the defect subspaces

$$\mathfrak{N}_\lambda(A) := (\mathcal{H} \oplus \mathcal{L}) \ominus \operatorname{ran}(A - \bar{\lambda})$$

have the same dimensions $n_+(A)$ and $n_-(A)$ for $\lambda \in \mathbb{C}_+$ and $\lambda \in \mathbb{C}_-$, respectively, which are called the *defect numbers* of the symmetric linear relation A. In the following proposition we show, that the symmetric linear relation A in (4.4) has equal defect numbers $n_+(A) = n_-(A) = \dim \mathcal{L}$ and, moreover, $0 \in \widehat{\rho}(A)$.

Proposition 4.1. *Let the data set (B_1, B_2, C_1, C_2, K) satisfy the assumptions (A1)–(A3). Then:*

(1) *the adjoint linear relation A^+ takes the form*

$$A^+ = \left\{ \widehat{g} = \left\{ \begin{bmatrix} g \\ v \end{bmatrix}, \begin{bmatrix} g' \\ v' \end{bmatrix} \right\} : \begin{array}{c} v, v' \in \mathcal{L}, g' \in \mathcal{H}; \\ g = \widetilde{B}_1^+ g' + \widetilde{C}_1^+ v' - \widetilde{C}_2^+ v \end{array} \right\}; \tag{4.6}$$

(2) *the set $\widehat{\rho}(A)$ of regular type points for symmetric linear relation A contains the resolvent set of the linear relation $\widetilde{B}_1^{-1}$*

$$\rho(\widetilde{B}_1^{-1}) = \{\lambda \in (\mathbb{C} \setminus \{0\}) : 1/\lambda \in \rho(\widetilde{B}_1)\} \cup \{0\} \text{ if } \widetilde{B}_1 \in [\mathcal{H}]$$

and the defect subspace $\mathfrak{N}_\lambda(A)$ for $\lambda \in \rho(\widetilde{B}_1^{-1})$ consists of the vectors

$$\begin{bmatrix} -F(\bar{\lambda})^+ u \\ u \end{bmatrix}, \quad u \in \mathcal{L}, \tag{4.7}$$

where $F(\lambda) = (\widetilde{C}_2 - \lambda \widetilde{C}_1)(I_{\mathcal{H}} - \lambda \widetilde{B}_1)^{-1}$.

Proof. 1) Let

$$\widehat{g} = \left\{ \begin{bmatrix} g \\ v \end{bmatrix}, \begin{bmatrix} g' \\ v' \end{bmatrix} \right\} \in A^+ \quad (g, g' \in \mathcal{H}; v, v' \in \mathcal{L}).$$

Then it follows from (4.4) that

$$(g', \widetilde{B}_1 h)_{\mathcal{H}} - (g, h)_{\mathcal{H}} + (v', \widetilde{C}_1 h)_{\mathcal{L}} - (v, \widetilde{C}_2 h)_{\mathcal{L}} = 0$$

for all $h \in \mathcal{H}$ and, therefore, $\widetilde{B}_1^+ g' - g + \widetilde{C}_1^+ v' - \widetilde{C}_2^+ v = 0$. This gives the equality

$$g = \widetilde{B}_1^+ g' + \widetilde{C}_1^+ v' - \widetilde{C}_2^+ v. \tag{4.8}$$

2) It follows from (4.4) that

$$A - \lambda = \left\{ \left\{ \begin{bmatrix} \widetilde{B}_1 h \\ \widetilde{C}_1 h \end{bmatrix}, \begin{bmatrix} (I_{\mathcal{H}} - \lambda \widetilde{B}_1) h \\ (\widetilde{C}_2 - \lambda \widetilde{C}_1) h \end{bmatrix} \right\} : h \in \mathcal{H} \right\}, \tag{4.9}$$

and, hence

$$\operatorname{ran}(A - \lambda) = \left\{ \begin{bmatrix} h \\ F(\lambda) h \end{bmatrix} : h \in \mathcal{H} \right\}.$$

Therefore, $\operatorname{ran}(A - \lambda)$ is closed for all $\lambda \in \rho(\widetilde{B}_1^{-1})$.

If $\lambda \in \rho(\widetilde{B}_1^{-1})$ and $\widehat{g} \in \widehat{\mathfrak{N}}_\lambda(\widehat{A})$ then $g' = \lambda g$, $v' = \lambda v$. Substituting these equalities in (4.8) one obtains

$$(I_{\mathcal{H}} - \lambda \widetilde{B}_1^+) g = -(\widetilde{C}_2^+ - \lambda \widetilde{C}_1^+) v.$$

This proves the second statement since $g = -F(\bar{\lambda})^+ v$. □

4.2. $\mathcal{L}$-resolvent matrix

Let $\widetilde{A}$ be a selfadjoint extension of A in a Hilbert space $\widetilde{\mathcal{H}} \oplus \mathcal{L}$. The compression $\mathbf{R}_\lambda = P_{\mathcal{L}}(\widetilde{A} - \lambda)^{-1} \upharpoonright \mathcal{L}$ of the resolvent of $\widetilde{A}$ to $\mathcal{L}$ is said to be an $\mathcal{L}$-*resolvent* of A.

Recall some facts from M.G. Kreĭn's representation theory ([36], [22]) for symmetric linear relations in Hilbert spaces.Let us say that $\lambda \in \rho(A, \mathcal{L})$ for a symmetric linear relation A in $\mathcal{H} \oplus \mathcal{L}$, if λ is a regular type point and

$$\mathcal{H} \oplus \mathcal{L} = \operatorname{ran}(A - \lambda) \dot{+} \mathcal{L}. \tag{4.10}$$

For every $\lambda \in \rho(A, \mathcal{L})$ the operator-valued function $\mathcal{P}(\lambda) : \mathcal{H} \to \mathcal{L}$ is defined as a skew projection onto $\mathcal{L}$ in the decomposition (4.10) and $\mathcal{Q}(\lambda) : \mathcal{H} \to \mathcal{L}$ is given by

$$\mathcal{Q}(\lambda) = P_{\mathcal{L}}(A - \lambda)^{-1}(I - \mathcal{P}(\lambda)), \quad \lambda \in \rho(A, \mathcal{L}). \tag{4.11}$$

Let the block mvf $W(\lambda)$ be defined by the formula

$$W(\lambda) = [w_{ij}(\lambda)]_{i,j=1}^2 = I + i\lambda \mathcal{V}(\lambda)\mathcal{V}(0)^+ J \quad (\lambda \in \rho(A, \mathcal{L})), \tag{4.12}$$

where $\mathcal{V}(\lambda) = \begin{bmatrix} -\mathcal{Q}(\lambda) \\ \mathcal{P}(\lambda) \end{bmatrix}$ and the block structure of $W(\lambda)$ is conformal with that of $\mathcal{V}(\lambda)$. The ovf $W(\lambda)$ is called the $\mathcal{L}$-*resolvent matrix* for the linear relation A.

Theorem 4.2. [35, 38]. *The set of $\mathcal{L}$-resolvents of A is parametrized by the formula*

$$P_{\mathcal{L}}(\widetilde{A}-\lambda)^{-1}|_{\mathcal{L}} = (w_{11}(\lambda)q(\lambda)+w_{12}(\lambda)p(\lambda))(w_{21}(\lambda)q(\lambda)+w_{22}(\lambda)p(\lambda))^{-1} \quad (4.13)$$

where $\{p,q\}$ ranges over the set $\widetilde{N}(\mathcal{L})$ of all Nevanlinna pairs.

In the following theorem we calculate $\mathcal{P}(\lambda)$, $\mathcal{Q}(\lambda)$ and the $\mathcal{L}$-resolvent matrix $W(\lambda)$ for the linear relation A in terms of the AIP data set.

Theorem 4.3. *Let B_1, B_2, C_1, C_2, K satisfy the assumptions* (A1)–(A3). *Then:*

1. $\rho(A,\mathcal{L}) = \rho(B_1^{-1})$ *and for $\lambda \in \rho(A,\mathcal{L})$ the ovf's $\mathcal{P}(\lambda)$, $\mathcal{Q}(\lambda)$ are given by*

$$\mathcal{P}(\lambda)\begin{bmatrix} f \\ u \end{bmatrix} = u - F(\lambda)f, \quad f\in\mathcal{H}, \quad u\in\mathcal{L}, \quad (4.14)$$

$$\mathcal{Q}(\lambda)\begin{bmatrix} f \\ u \end{bmatrix} = \widetilde{C}_1(I_{\mathcal{H}}-\lambda\widetilde{B}_1)^{-1}f, \quad f\in\mathcal{H}; \quad (4.15)$$

2. *The adjoint operators to $\mathcal{P}(\lambda), \mathcal{Q}(\lambda) : \begin{bmatrix} \mathcal{H} \\ \mathcal{L} \end{bmatrix} \to \mathcal{L}$ take the form*

$$\mathcal{P}(\lambda)^+ u = \begin{bmatrix} -F(\lambda)^+ u \\ u \end{bmatrix} \quad u\in\mathcal{L},\ \lambda\in\rho(A,\mathcal{L}), \quad (4.16)$$

$$\mathcal{Q}(\lambda)^+ u = \begin{bmatrix} (I_{\mathcal{H}}-\bar{\lambda}\widetilde{B}_1^+)^{-1}\widetilde{C}_1^+ u \\ 0 \end{bmatrix} \quad u\in\mathcal{L},\ \lambda\in\rho(A,\mathcal{L}); \quad (4.17)$$

3. *The $\mathcal{L}$-resolvent matrix $W(\lambda)$ in* (4.12) *takes the form*

$$W_{\Pi\mathcal{L}}(\lambda) = \begin{bmatrix} I_{\mathcal{L}} & 0 \\ -\lambda & I_{\mathcal{L}} \end{bmatrix}\left(I + i\lambda\begin{bmatrix} \widetilde{C}_1 \\ \widetilde{C}_2 \end{bmatrix}(I_{\mathcal{H}}-\lambda\widetilde{B}_1)^{-1}\begin{bmatrix} \widetilde{C}_1 \\ \widetilde{C}_2 \end{bmatrix}^+ J\right). \quad (4.18)$$

Proof. 1) Assume that $\lambda\in\rho(A,\mathcal{L})$ and the decomposition (4.10) holds. Then for $f\in\mathcal{H}$, $u\in\mathcal{L}$ there are unique $h\in\mathcal{H}$ and $v\in\mathcal{L}$ such that

$$(I_{\mathcal{H}}-\lambda\widetilde{B}_1)h = f, \quad (\widetilde{C}_2-\lambda\widetilde{C}_1)h + v = u. \quad (4.19)$$

This implies, in particular, that $\lambda\in\rho(B_1^{-1})$. Conversely, if $\lambda\in\rho(B_1^{-1})$, then the equations (4.19) have unique solutions $h\in\mathcal{H}$ and $v\in\mathcal{L}$ and, hence, $\lambda\in\rho(A,\mathcal{L})$. In view of (4.19) these solutions take the form

$$h = (I_{\mathcal{H}}-\lambda\widetilde{B}_1)^{-1}f, \quad v = \mathcal{P}(\lambda)\begin{bmatrix} f \\ u \end{bmatrix} = u - F(\lambda)f. \quad (4.20)$$

It follows from (4.11), (4.20) and (4.9) that

$$\mathcal{Q}(\lambda)\begin{bmatrix} f \\ u \end{bmatrix} = P_{\mathcal{L}}(A-\lambda)^{-1}\begin{bmatrix} f \\ F(\lambda)f \end{bmatrix} = P_{\mathcal{L}}\begin{bmatrix} \widetilde{B}_1 h \\ \widetilde{C}_1 h \end{bmatrix}$$
$$= \widetilde{C}_1(I_{\mathcal{H}}-\lambda\widetilde{B}_1)^{-1}f.$$

2) The formulas (4.16), (4.17) are implied by

$$\left\langle \mathcal{P}(\lambda)^+ v, \begin{bmatrix} f \\ u \end{bmatrix} \right\rangle_{\mathcal{H}\oplus\mathcal{L}} = (v, u - F(\lambda)f)_{\mathcal{L}} = \left\langle \begin{bmatrix} -F(\lambda)^+ v \\ v \end{bmatrix}, \begin{bmatrix} f \\ u \end{bmatrix} \right\rangle_{\mathcal{H}\oplus\mathcal{L}},$$

$$\begin{aligned} \left\langle \mathcal{Q}(\lambda)^+ v, \begin{bmatrix} f \\ u \end{bmatrix} \right\rangle_{\mathcal{H}\oplus\mathcal{L}} &= (v, \widetilde{C}_1(I_{\mathcal{H}} - \lambda\widetilde{B}_1)^{-1} f)_{\mathcal{L}} \\ &= \left\langle \begin{bmatrix} (I_{\mathcal{H}} - \lambda\widetilde{B}_1^+)^{-1}\widetilde{C}_1^+ v \\ 0 \end{bmatrix}, \begin{bmatrix} f \\ u \end{bmatrix} \right\rangle_{\mathcal{H}\oplus\mathcal{L}}. \end{aligned}$$

3) For every $u = \begin{bmatrix} u_1 \\ u_2 \end{bmatrix} \in \mathcal{L}^2$ one obtains from (4.12) and (4.14)–(4.17)

$$\begin{aligned} W(\lambda)u &= \left(I + \lambda \begin{bmatrix} -\mathcal{Q}(\lambda) \\ \mathcal{P}(\lambda) \end{bmatrix} \begin{bmatrix} -\mathcal{Q}(0)^+ & \mathcal{P}(0)^+ \end{bmatrix} \begin{bmatrix} 0 & I \\ I & 0 \end{bmatrix} \right) u \\ &= \begin{bmatrix} u_1 \\ u_2 \end{bmatrix} + \lambda \begin{bmatrix} -\mathcal{Q}(\lambda) \\ \mathcal{P}(\lambda) \end{bmatrix} \begin{bmatrix} -\widetilde{C}_1^+ u_2 + \widetilde{C}_2^+ u_1 \\ -u_1 \end{bmatrix} \\ &= \begin{bmatrix} u_1 \\ u_2 - \lambda u_1 \end{bmatrix} + \lambda \begin{bmatrix} \widetilde{C}_1(I_{\mathcal{H}} - \lambda\widetilde{B}_1)^{-1}(\widetilde{C}_1^+ u_2 - \widetilde{C}_2^+ u_1) \\ (\widetilde{C}_2 - \lambda\widetilde{C}_1)(I_{\mathcal{H}} - \lambda\widetilde{B}_1)^{-1}(\widetilde{C}_1^+ u_2 - \widetilde{C}_2^+ u_1) \end{bmatrix} \\ &= \begin{bmatrix} I_{\mathcal{L}} & 0 \\ -\lambda & I_{\mathcal{L}} \end{bmatrix} \left(I_{\mathcal{L}\oplus\mathcal{L}} + i\lambda \begin{bmatrix} \widetilde{C}_1 \\ \widetilde{C}_2 \end{bmatrix} (I_{\mathcal{H}} - \lambda\widetilde{B}_1)^{-1} \begin{bmatrix} \widetilde{C}_1 \\ \widetilde{C}_2 \end{bmatrix}^+ J \right) u. \quad \square \end{aligned}$$

The $\mathcal{L}$-resolvent matrix $W(\lambda)$ satisfies the identity

$$W(\lambda)JW(\mu)^* = J + i(\lambda - \bar{\mu})\mathcal{V}(\lambda)\mathcal{V}(\mu)^+ \quad (\lambda, \mu \in \rho(A, \mathcal{L})). \tag{4.21}$$

Originally, the identity (4.21) has been used by M.G. Kreĭn as a definition of the $\mathcal{L}$-resolvent matrix. It follows from (4.21) that $W(\lambda)$ belongs to the Potapov class $\mathcal{P}(J)$, that is

$$\frac{W(\lambda)JW(\lambda)^* - J}{i(\lambda - \bar{\lambda})} \geq 0 \quad (\lambda \in \rho(A, \mathcal{L})). \tag{4.22}$$

Corollary 4.4. *Let the data set* (B_1, B_2, C_1, C_2, K) *satisfy* (A1),(A2) *and*

(A3′) *the operator* $D = B_2 - \mu B_1$ *is an isomorphism in* $\mathcal{X}$ *for some* $\mu \in \mathbb{R}$, *and the operators* $B_1 D^{-1}|\mathcal{X}_0 : \mathcal{X}_0 \to \mathcal{X}_0$, $G(\mu)|\mathcal{X}_0 : \mathcal{X}_0 \to \mathcal{L}^2$ *are bounded.*

Then one of the $\mathcal{L}$-*resolvent matrices can be found from*

$$\begin{aligned} &\begin{bmatrix} I & 0 \\ -\lambda & I \end{bmatrix}^{-1} W^{\mu}(\lambda) \\ &= I + i(\lambda - \mu) \begin{bmatrix} \widetilde{C}_1 \\ \widetilde{C}_2 \end{bmatrix} (\widetilde{B}_2 - \lambda\widetilde{B}_1)^{-1}(\widetilde{B}_2^+ - \mu\widetilde{B}_1^+)^{-1} \begin{bmatrix} \widetilde{C}_1 \\ \widetilde{C}_2 \end{bmatrix}^+ J. \end{aligned} \tag{4.23}$$

Proof. The data set

$$(B_1(B_2 - \mu B_1)^{-1}, I_{\mathcal{X}}, C_1(B_2 - \mu B_1)^{-1}, (C_2 - \mu C_1)(B_2 - \mu B_1)^{-1}, K)$$

satisfies the assumptions (A1)–(A3). Consider the linear relation $A-\mu$

$$A-\mu=\left\{\left\{\begin{bmatrix}\widetilde{B}_1(\widetilde{B}_2-\mu\widetilde{B}_1)^{-1}h\\ \widetilde{C}_1(\widetilde{B}_2-\mu\widetilde{B}_1)^{-1}h\end{bmatrix},\begin{bmatrix}h\\ (\widetilde{C}_2-\mu\widetilde{C}_1)(\widetilde{B}_2-\mu\widetilde{B}_1)^{-1}h\end{bmatrix}\right\}:h\in\mathcal{H}\right\}.$$

Due to (4.18) its $\mathcal{L}$-resolvent matrix $W(\lambda)$ satisfies the equality

$$\begin{bmatrix}-I & 0\\ \lambda & -I\end{bmatrix}^{-1}W(\lambda)=$$

$$=I_{\mathcal{L}\oplus\mathcal{L}}+i\lambda\begin{bmatrix}\widetilde{C}_1\\ \widetilde{C}_2-\mu\widetilde{C}_1\end{bmatrix}(\widetilde{B}_2-(\lambda+\mu)\widetilde{B}_1)^{-1}(\widetilde{B}_2^+-\mu\widetilde{B}_1^+)^{-1}\begin{bmatrix}\widetilde{C}_1\\ \widetilde{C}_2-\mu\widetilde{C}_1\end{bmatrix}^+J.$$

Then the matrix $W^\mu(\lambda)=W(\lambda-\mu)$ is the $\mathcal{L}$-resolvent matrix of A and, hence,

$$\begin{bmatrix}I & 0\\ -\lambda & I\end{bmatrix}^{-1}W^\mu(\lambda)=\begin{bmatrix}I & 0\\ -\lambda & I\end{bmatrix}^{-1}W(\lambda-\mu)$$

$$=\left(I_{\mathcal{L}\oplus\mathcal{L}}+i(\lambda-\mu)\begin{bmatrix}\widetilde{C}_1\\ \widetilde{C}_2\end{bmatrix}(\widetilde{B}_2-\lambda\widetilde{B}_1)^{-1}(\widetilde{B}_2^+-\mu\widetilde{B}_1^+)^{-1}\begin{bmatrix}\widetilde{C}_1\\ \widetilde{C}_2\end{bmatrix}^+J\right)\begin{bmatrix}I & 0\\ \mu & I\end{bmatrix}.$$

This prove (4.23) since the class of $\mathcal{L}$-resolvent matrices is invariant under the multiplication by a right J-unitary factor. □

4.3. Boundary triplets for A^+ and the $\mathcal{L}$-resolvent matrix

Definition 4.5. ([28], [40]) A triplet $\Pi=\{\mathcal{L},\Gamma_1,\Gamma_2\}$, where $\Gamma_i:A^+\to\mathcal{L}$, $i=1,2$, is said to be a *boundary triplet* for A^+, if for all $\widehat{f}=\{f,f'\}$, $\widehat{g}=\{g,g'\}\in A^+$;

$$(f',g)_{\mathcal{H}\oplus\mathcal{L}}-(f,g')_{\mathcal{H}\oplus\mathcal{L}}=(\Gamma_1\widehat{f},\Gamma_2\widehat{g})_{\mathcal{L}}-(\Gamma_2\widehat{f},\Gamma_1\widehat{g})_{\mathcal{L}}$$

and the mapping $\Gamma:=\begin{bmatrix}\Gamma_1\\ \Gamma_2\end{bmatrix}:A^+\to\begin{bmatrix}\mathcal{L}\\ \mathcal{L}\end{bmatrix}$ is surjective.

The set of all selfadjoint extensions $\widetilde{A}$ of A can be parametrized by the set of selfadjoint linear relations τ in $\mathcal{L}$ via the formula

$$\widehat{f}\in\widetilde{A}\Leftrightarrow\Gamma\widehat{f}\in\tau.$$

Let the operator-valued functions $\widehat{\mathcal{P}}(\lambda)^+$ and $\widehat{\mathcal{Q}}(\lambda)^+$ be given by

$$\widehat{\mathcal{P}}(\lambda)^+u=\{\mathcal{P}(\lambda)^+u,\bar{\lambda}\mathcal{P}(\lambda)^+u\},\quad u\in\mathcal{L},\tag{4.24}$$

$$\widehat{\mathcal{Q}}(\lambda)^+u=\{\mathcal{Q}(\lambda)^+u,u+\bar{\lambda}\mathcal{Q}(\lambda)^+u\},\quad u\in\mathcal{L},\tag{4.25}$$

where $\widehat{\mathcal{P}}(\lambda)^+$, $\widehat{\mathcal{Q}}(\lambda)^+:\mathcal{L}\to\mathcal{H}$ are adjoint operators to $\mathcal{P}(\lambda)$, $\mathcal{Q}(\lambda):\mathcal{H}\to\mathcal{L}$. The following statements were proved in [21] and [22].

Theorem 4.6. ([21], [22]) *Under the above assumptions the linear relation A^+ can be decomposed in the following direct sum*

$$A^+=A\dotplus\widehat{\mathcal{P}}(\lambda)^+\mathcal{L}\dotplus\widehat{\mathcal{Q}}(\lambda)^+\mathcal{L},\quad\lambda\in\rho(A,\mathcal{L}),\tag{4.26}$$

and one of the $\mathcal{L}$-resolvent matrices of A corresponding to the boundary triplet Π can be found by

$$W_{\Pi\mathcal{L}}(\lambda) = \begin{bmatrix} -\Gamma_2\widehat{\mathcal{Q}}(\lambda)^+ & \Gamma_2\widehat{\mathcal{P}}(\lambda)^+ \\ -\Gamma_1\widehat{\mathcal{Q}}(\lambda)^+ & \Gamma_1\widehat{\mathcal{P}}(\lambda)^+ \end{bmatrix}^*, \quad \lambda \in \rho(A, \mathcal{L}). \tag{4.27}$$

Mention that every $\mathcal{L}$-resolvent matrix $W(\lambda)$ of A can be found in this way by a suitable choice of the boundary triplet Π and this correspondence between $\mathcal{L}$-resolvent matrices and boundary triplets is one-to-one, [22].

Proposition 4.7. *Let the data set* $(B_1,\ B_2,\ C_1,\ C_2,\ K)$ *satisfy the assumptions* (A1)–(A3). *Then:*

1) *A boundary triplet* $\Pi = \{\mathcal{L}, \Gamma_1, \Gamma_2\}$ *for* A^+ *can be defined by*

$$\Gamma_1\widehat{g} = v - \widetilde{C}_1 g', \quad \Gamma_2\widehat{g} = -v' + \widetilde{C}_2 g'; \tag{4.28}$$

2) *The $\mathcal{L}$-resolvent matrix of A corresponding to the boundary triplet Π coincides with that in* (4.18).

Proof. 1) For two vectors

$$\widehat{f} = \left\{ \begin{bmatrix} f \\ u \end{bmatrix}, \begin{bmatrix} f' \\ u' \end{bmatrix} \right\}, \quad \widehat{g} = \left\{ \begin{bmatrix} g \\ v \end{bmatrix}, \begin{bmatrix} g' \\ v' \end{bmatrix} \right\} \in A^+$$

one obtains

$$\begin{aligned} (f', g)_{\mathcal{H}} - (f, g')_{\mathcal{H}} + (u', v)_{\mathcal{L}} - (u, v')_{\mathcal{L}} &= (u', v)_{\mathcal{L}} - (u, v')_{\mathcal{L}} \\ + (f', \widetilde{B}_1^+ g' + \widetilde{C}_1^+ v' - \widetilde{C}_2^+ v)_{\mathcal{H}} &- (\widetilde{B}_1^+ f' + \widetilde{C}_1^+ u' - \widetilde{C}_2^+ u, g')_{\mathcal{H}}. \end{aligned} \tag{4.29}$$

Then the right-hand part of (4.29) takes the form

$$\begin{aligned} &(\widetilde{B}_1 f', g')_{\mathcal{H}} - (f', \widetilde{B}_1 g')_{\mathcal{H}} + (\widetilde{C}_1 f', v')_{\mathcal{L}} - (\widetilde{C}_2 f', v)_{\mathcal{L}} \\ &\quad - (u', \widetilde{C}_1 g')_{\mathcal{L}} + (u, \widetilde{C}_2 g')_{\mathcal{L}} + (u', v)_{\mathcal{L}} - (u, v')_{\mathcal{L}} \\ &= (\widetilde{C}_2 f', \widetilde{C}_1 g')_{\mathcal{L}} - (\widetilde{C}_1 f', \widetilde{C}_2 g')_{\mathcal{L}} + (\widetilde{C}_1 f', v')_{\mathcal{L}} - (\widetilde{C}_2 f', v)_{\mathcal{L}} \\ &\quad - (u', \widetilde{C}_1 g')_{\mathcal{L}} + (u, \widetilde{C}_2 g')_{\mathcal{L}} + (u', v)_{\mathcal{L}} - (u, v')_{\mathcal{L}} \\ &= (\widetilde{C}_2 f' - u', \widetilde{C}_1 g' - v)_{\mathcal{L}} - (\widetilde{C}_1 f' - u, \widetilde{C}_2 g' - v')_{\mathcal{L}}. \end{aligned}$$

Since $0 \in \rho(A, \mathcal{L})$ one can rewrite the formula (4.26) in the form

$$A^+ = A \dotplus \widehat{\mathcal{P}}(0)^+\mathcal{L} \dotplus \widehat{\mathcal{Q}}(0)^+\mathcal{L}.$$

Due to (4.16), (4.17), (4.24), (4.25) one obtains for every $u \in \mathcal{L}$

$$\widehat{\mathcal{P}}(0)^+ u = \left\{ \begin{bmatrix} -\widetilde{C}_2^+ u \\ u \end{bmatrix}, \begin{bmatrix} 0 \\ 0 \end{bmatrix} \right\}, \quad \widehat{\mathcal{Q}}(0)^+ u = \left\{ \begin{bmatrix} \widetilde{C}_1^+ u \\ 0 \end{bmatrix}, \begin{bmatrix} 0 \\ u \end{bmatrix} \right\}$$

It follows from (4.28) that

$$\Gamma\widehat{\mathcal{P}}(0)^+ u = \begin{bmatrix} u \\ 0 \end{bmatrix}, \quad \Gamma\widehat{\mathcal{Q}}(0)^+ u = \begin{bmatrix} 0 \\ -u \end{bmatrix}$$

and hence the mapping $\Gamma : A^+ \to \mathcal{L}\oplus\mathcal{L}$ is surjective and $\{\mathcal{L},\Gamma_1,\Gamma_2\}$ is a boundary triplet for A^+.

2) Now one obtains from (4.24), (4.16) and (4.28) that

$$\Gamma_2\widehat{\mathcal{P}}(\lambda)^+v = -\bar{\lambda}v - \bar{\lambda}\widetilde{C}_2F(\lambda)^+v, \tag{4.30}$$

$$\Gamma_1\widehat{\mathcal{P}}(\lambda)^+v = v + \bar{\lambda}\widetilde{C}_1F(\lambda)^+v. \tag{4.31}$$

Similarly (4.25), (4.17) and (4.28) imply

$$-\Gamma_2\widehat{\mathcal{Q}}(\lambda)^+v = v - \bar{\lambda}\widetilde{C}_2(I_{\mathcal{H}} - \lambda\widetilde{B}_1^+)^{-1}\widetilde{C}_1^+v, \tag{4.32}$$

$$-\Gamma_1\widehat{\mathcal{Q}}(\lambda)^+v = \bar{\lambda}\widetilde{C}_1(I_{\mathcal{H}} - \lambda\widetilde{B}_1^+)^{-1}\widetilde{C}_1^+v. \tag{4.33}$$

It follows from (4.30)–(4.33) and (4.27) that

$$W_{\Pi\mathcal{L}}(\lambda)^* = \begin{bmatrix} I_{\mathcal{L}} & -\bar{\lambda} \\ 0 & I_{\mathcal{L}} \end{bmatrix} - \bar{\lambda}\begin{bmatrix} \widetilde{C}_2 \\ -\widetilde{C}_1 \end{bmatrix}(I_{\mathcal{H}} - \bar{\lambda}\widetilde{B}_1^+)^{-1}\begin{bmatrix} \widetilde{C}_1^+ & \widetilde{C}_2^+ - \bar{\lambda}\widetilde{C}_1^+ \end{bmatrix}$$

and hence

$$\begin{aligned} W_{\Pi\mathcal{L}}(\lambda) &= \begin{bmatrix} I & 0 \\ -\lambda & I \end{bmatrix} - \lambda\begin{bmatrix} \widetilde{C}_1 \\ \widetilde{C}_2 - \lambda\widetilde{C}_1 \end{bmatrix}(I_{\mathcal{H}} - \lambda\widetilde{B}_1)^{-1}\begin{bmatrix} \widetilde{C}_2^+ & -\widetilde{C}_1^+ \end{bmatrix} \\ &= \begin{bmatrix} I & 0 \\ -\lambda & I \end{bmatrix}\left(I_{\mathcal{L}\oplus\mathcal{L}} - \lambda\begin{bmatrix} \widetilde{C}_1 \\ \widetilde{C}_2 \end{bmatrix}(I_{\mathcal{H}} - \lambda\widetilde{B}_1)^{-1}\begin{bmatrix} \widetilde{C}_2^+ & -\widetilde{C}_1^+ \end{bmatrix}\right). \qquad \square \end{aligned}$$

4.4. $\mathcal{L}$-resolvents of $\widehat{A}$

In the case when $\ker P$ is nontrivial we calculated the $\mathcal{L}$-resolvent matrix of the linear relation $A_0(\subset \widehat{A})$. A description of $\mathcal{L}$-resolvents of A is given in Theorem 4.2. In order to obtain a description of $\mathcal{L}$-resolvents of $\widehat{A}$ we will use the same formula and specify the set of parameters $\{p,q\} \in \widetilde{N}(\mathcal{L})$ which correspond to $\mathcal{L}$-resolvents of $\widehat{A}$ via (4.13).

Recall (see [44]) that every generalized resolvent $P_{\mathcal{H}\oplus\mathcal{L}}(\widetilde{A} - \lambda)^{-1}|_{\mathcal{H}\oplus\mathcal{L}}$ of A corresponding to an exit space selfadjoint extension $\widetilde{A}$ in a Hilbert space $\widetilde{\mathcal{H}} \oplus \mathcal{L}$, where $\mathcal{H} \subset \widetilde{\mathcal{H}}$, can be represented as

$$P_{\mathcal{H}\oplus\mathcal{L}}(\widetilde{A} - \lambda)^{-1}|_{\mathcal{H}\oplus\mathcal{L}} = (T(\lambda) - \lambda)^{-1}, \quad \lambda \in \mathbb{C}_+, \tag{4.34}$$

where $T(\lambda)$ ($\lambda \in \mathbb{C}_+$) is the *Strauss family* of maximal dissipative linear relations in $\mathcal{H}$ defined by

$$T(\lambda) = \{\{Pf, Pf'\} : \{f, f'\} \in \widetilde{A},\ f' - \lambda f \in \mathcal{H}\oplus\mathcal{L}\} \tag{4.35}$$

and P is the orthogonal projection onto $\mathcal{H}\oplus\mathcal{L}$.

Proposition 4.8. ([22]) *Let $\widetilde{A}$ be an exit space selfadjoint extension of A, let $T(\lambda)$ be the Strauss family of maximal dissipative linear relations defined by (4.35), let $\{\mathcal{L},\Gamma_1,\Gamma_2\}$ be a boundary triplet for A^+. Then the pair $\{p,q\} \in \widetilde{N}(\mathcal{L})$ is the*

Nevanlinna pair corresponding to $\widetilde{A}$ *via* (4.13) *if and only if the pair* $\{p,q\}$ *is related to* $T(\lambda)$ *via the formula*

$$\Gamma T(\lambda) = \operatorname{ran} \begin{bmatrix} q(\lambda) \\ p(\lambda) \end{bmatrix}.$$

We will need the following simple statements.

Lemma 4.9. *Let under the assumptions of Proposition* 4.8 $\widehat{A}$ *be a symmetric extension of* A *in* $\mathcal{H} \oplus \mathcal{L}$. *Then:*

(i) $\widehat{A} \subset \widetilde{A}$ *if and only if* $\widehat{A} \subset T(\lambda)$ *for some* $\lambda \in \mathbb{C}_+$;

(ii) $\widehat{A} \subset \widetilde{A}$ *if and only if* $\Gamma\widehat{A} \subset \operatorname{ran} \begin{bmatrix} q(\lambda) \\ p(\lambda) \end{bmatrix}$ *for some* $\lambda \in \mathbb{C}_+$.

Proof. (i) The implication $\Rightarrow$ is immediate from (4.35). Conversely, assume that $\widehat{A} \subset T(\lambda)$. In view of (4.35) for every $\{g, g'\} \in \widehat{A}$ there are $\{f, f'\} \in \widetilde{A}$ and $g_1 \in \mathcal{H}_1 := \widetilde{\mathcal{H}} \ominus \mathcal{H}$ such that

$$f = g + g_1, \quad f' = g' + \lambda g_1. \tag{4.36}$$

Hence

$$(f', f)_{\widetilde{\mathcal{H}}} = (g', g)_{\mathcal{H}} + \lambda (g_1, g_1)_{\mathcal{H}_1}.$$

Since $\widehat{A}$ and $\widetilde{A}$ are symmetric this implies $g_1 = 0$. Therefore, $\{g, g'\} \in \widetilde{A}$ and hence $\widehat{A} \subset \widetilde{A}$.

2) The statement (ii) is implied by (i) since the inclusion $\widehat{A} \subset T(\lambda)$ is equivalent to $\Gamma\widehat{A} \subset \Gamma T(\lambda) = \operatorname{ran} \begin{bmatrix} q(\lambda) \\ p(\lambda) \end{bmatrix}$. □

It follows from Lemma 4.9 that all Nevanlinna pairs corresponding to generalized resolvents of $\widehat{A}$ have a common constant part $\Gamma\widehat{A}$.

Lemma 4.10. *Let* $\widehat{A}$ *be the symmetric linear relation* (4.3), *and let* $\{\mathcal{L}, \Gamma_1, \Gamma_2\}$ *be a boundary triplet for* A^+. *Then*

(i) $\Gamma\widehat{A}$ *is a neutral subspace in* $(\mathcal{L}^2, J_{\mathcal{L}})$ *of dimension* $\nu := \dim C\ker K$;

(ii) *There is a subspace* $\mathcal{L}_0 \subset \mathcal{L}$ *and a* $J_{\mathcal{L}}$*-unitary operator* $V \in [\mathcal{L}^2]$ *such that*

$$V(\{0\} \times \mathcal{L}_0) = \Gamma\widehat{A}.$$

Proof. 1) It follows from (4.5) and (4.28) that

$$\Gamma\widehat{A} = \left\{ \begin{bmatrix} \Gamma_1 \widehat{g} \\ \Gamma_2 \widehat{g} \end{bmatrix} : \widehat{g} \in \widehat{A} \right\} = \left\{ \begin{bmatrix} C_1 u \\ -C_2 u \end{bmatrix} : u \in \ker K \right\}.$$

Clearly, the subspace $\Gamma\widehat{A}$ is finite-dimensional and $\dim \Gamma\widehat{A} = \dim C\ker K$. The subspace $\Gamma\widehat{A}$ is neutral since for every $u \in \ker K$ one has

$$\begin{aligned}(\Gamma_1\widehat{g}, \Gamma_2\widehat{g}) - (\Gamma_1\widehat{g}, \Gamma_2\widehat{g}) &= (C_1 u, C_2 u)_{\mathcal{L}} - (C_2 u, C_1 u)_{\mathcal{L}} \\ &= K(u, B_1 u) - K(B_1 u, u) = 0.\end{aligned}$$

2) Let us decompose $\mathcal{L}$ into the orthogonal sum of two subspaces $\mathcal{L}_0$ and $\mathcal{L}_1$

$$\mathcal{L} = \mathcal{L}_0 \oplus \mathcal{L}_1,$$

with $\dim \mathcal{L}_0 = \nu$. The subspace $\{0\} \times \mathcal{L}_0$ is $J_{\mathcal{L}}$-neutral and hence there exists a $J_{\mathcal{L}}$-unitary operator $V \in [\mathcal{L}^2]$ such that $V(\{0\} \times \mathcal{L}_0) = \Gamma \widehat{A}$. □

Let V be the $J_{\mathcal{L}}$-unitary operator, constructed in Lemma 4.10. Then

$$\widehat{W}(\lambda) = (\widehat{w}_{ij}(\lambda))_{i,j=1}^2 := W_{\Pi\mathcal{L}}(\lambda)V. \tag{4.37}$$

is also the $\mathcal{L}$-resolvent matrix of A_0 with the advantage that the $\mathcal{L}$-resolvents of $\widehat{A}$ can be easily described in its terms.

Proposition 4.11. *The set of all $\mathcal{L}$-resolvents of $\widehat{A}$ is parametrized by the formula*

$$P_{\mathcal{L}}(\widetilde{A} - \lambda)^{-1}|_{\mathcal{L}} = (\widehat{w}_{11}(\lambda)q(\lambda) + \widehat{w}_{12}(\lambda)p(\lambda))(\widehat{w}_{21}(\lambda)q(\lambda) + \widehat{w}_{22}(\lambda)p(\lambda))^{-1} \tag{4.38}$$

where $\{p, q\}$ ranges over the set $\widetilde{N}(\mathcal{L})$ of Nevanlinna pairs of the form

$$p(\lambda) = \begin{bmatrix} I_{\mathcal{L}_0} & 0 \\ 0 & p_1(\lambda) \end{bmatrix}, \quad q(\lambda) = \begin{bmatrix} 0_{\mathcal{L}_0} & 0 \\ 0 & q_1(\lambda) \end{bmatrix}, \quad \{p_1, q_1\} \in \widetilde{N}(\mathcal{L}_1). \tag{4.39}$$

Proof. Let $\{\widetilde{p}, \widetilde{q}\}$ be a Nevanlinna pair defined by

$$\begin{bmatrix} \widetilde{q}(\lambda) \\ \widetilde{p}(\lambda) \end{bmatrix} = V \begin{bmatrix} q(\lambda) \\ p(\lambda) \end{bmatrix}, \quad \{p, q\} \in \widetilde{N}(\mathcal{L}).$$

It follows from Lemma 4.9 that the formula

$$P_{\mathcal{L}}(\widetilde{A} - \lambda)^{-1}|_{\mathcal{L}} = (w_{11}(\lambda)\widetilde{q}(\lambda) + w_{12}(\lambda)\widetilde{p}(\lambda))(w_{21}(\lambda)\widetilde{q}(\lambda) + w_{22}(\lambda)\widetilde{p}(\lambda))^{-1}$$

establishes a one-to-one correspondence between the set of all $\mathcal{L}$-resolvents of $\widehat{A}$ and the set of Nevanlinna families $\{\widetilde{p}, \widetilde{q}\}$ such that

$$\Gamma \widehat{A} \subset \operatorname{ran} \begin{bmatrix} \widetilde{q}(\lambda) \\ \widetilde{p}(\lambda) \end{bmatrix}, \quad \lambda \in \mathbb{C}_+. \tag{4.40}$$

Since

$$\Gamma \widehat{A} = V \begin{bmatrix} 0 \\ \mathcal{L}_0 \end{bmatrix}, \quad \text{and } \operatorname{ran} \begin{bmatrix} \widetilde{q}(\lambda) \\ \widetilde{p}(\lambda) \end{bmatrix} = V \operatorname{ran} \begin{bmatrix} q(\lambda) \\ p(\lambda) \end{bmatrix},$$

the inclusion (4.40) is equivalent to the inclusion

$$\begin{bmatrix} 0 \\ \mathcal{L}_0 \end{bmatrix} \subset \operatorname{ran} \begin{bmatrix} q(\lambda) \\ p(\lambda) \end{bmatrix}, \quad \lambda \in \mathbb{C}_+, \tag{4.41}$$

which, in turn, means that the pair $\{p, q\}$ admits the representation (4.39). □

4.5. Description of AIP solutions

To describe solutions of the AIP it remains to combine Theorem 3.3 and Proposition 4.11. Let the ovf $\Theta(\lambda)$ be defined by

$$\Theta(\lambda) = \begin{bmatrix} I & 0 \\ \lambda & I \end{bmatrix} \widehat{W}(\lambda) = \left(I - \lambda \begin{bmatrix} \widetilde{C}_1 \\ \widetilde{C}_2 \end{bmatrix} (I_{\mathcal{H}} - \lambda \widetilde{B}_1)^{-1} \begin{bmatrix} \widetilde{C}_2^+ & -\widetilde{C}_1^+ \end{bmatrix} \right) V \quad (4.42)$$

Theorem 4.12. *Let the AIP data set satisfy* (A1)–(A3). *Then the formula*

$$\begin{bmatrix} \psi(\lambda) \\ \phi(\lambda) \end{bmatrix} = \Theta(\lambda) \begin{bmatrix} q(\lambda) \\ p(\lambda) \end{bmatrix} (\widehat{w}_{21}(\lambda) q(\lambda) + \widehat{w}_{22}(\lambda) p(\lambda))^{-1} \quad (4.43)$$

establishes the one-to-one correspondence between the set of all normalized solutions $\{\varphi, \psi\}$ *of the* $AIP(B_1, B_2, C_1, C_2, K)$ *and the set of all equivalence classes of Nevanlinna pairs* $\{p, q\} \in \widetilde{N}(\mathcal{L})$ *of the form* (4.39). *The corresponding mapping* $F : \mathcal{X} \to \mathcal{H}(\varphi, \psi)$ *in* (C1), (C2) *is uniquely defined by the solution* $\{\varphi, \psi\}$:

$$(Fg)(\mu) = \begin{bmatrix} \varphi(\mu) & -\psi(\mu) \end{bmatrix} \widetilde{G}(\mu) P_{\mathcal{X}_0} g \quad (\mu \in \mathcal{O},\, g \in \mathcal{X}), \quad (4.44)$$

where $\mathcal{O}$ *is a neighborhood of 0,*

$$\widetilde{G}(\mu) = \begin{bmatrix} \widetilde{C}_1 \\ \widetilde{C}_2 \end{bmatrix} (I_{\mathcal{H}} - \mu \widetilde{B}_1)^{-1} \quad (\mu \in \mathcal{O}),$$

and $P_{\mathcal{X}_0}$ *is the skew projection onto* $\mathcal{X}_0$ *in the decomposition* (4.1).

Proof. Indeed, the description (4.43) is implied by (3.2), (4.38) and (4.42).

Let $\mathcal{O}$ is a neighborhood of 0, such that $(I_{\mathcal{H}} - \mu \widetilde{B}_1)$ is invertible in $\mathcal{H}$ for $\mu \in \mathcal{O}$ and let $g \in (I - \mu B_1)\mathcal{X}_0$ $(\mu \in \mathcal{O})$. Applying (C1) to the vector $h = h_\mu := (I - \mu B_1)^{-1} g$, one obtains

$$\begin{aligned} (Fg)(\lambda) &= (Fh_\mu)(\lambda) - \mu (F B_1 h_\mu)(\lambda) \\ &= \begin{bmatrix} \varphi(\lambda) & -\psi(\lambda) \end{bmatrix} G(\mu) g + (\lambda - \mu)(F B_1 h_\mu)(\lambda). \end{aligned} \quad (4.45)$$

Setting in (4.45) $\lambda = \mu$, one obtains

$$(Fg)(\mu) = \begin{bmatrix} \varphi(\mu) & -\psi(\mu) \end{bmatrix} G(\mu) g \quad (\mu \in \mathcal{O}, g \in (I - \mu B_1)\mathcal{X}_0). \quad (4.46)$$

Let $g \in \mathcal{X}_0$, let $g_n \in (I - \mu B_1)\mathcal{X}_0$ and $g_n \to g$. Then taking the limit in

$$(F g_n)(\mu) = \begin{bmatrix} \varphi(\mu) & -\psi(\mu) \end{bmatrix} \widetilde{G}(\mu) g_n$$

one obtains (4.44) for $g \in \mathcal{X}_0$. To complete the proof of (4.44) it remains to notice that $Fg \equiv 0$ for all $g \in \ker K$. □

Theorem 4.13. *Let the AIP data set satisfy* (*A*1), (*A*2), (*A*3′) *and let*

$$\begin{aligned} \Theta^\mu(\lambda) &= \begin{bmatrix} I & 0 \\ \lambda & I \end{bmatrix} \widehat{W}^\mu(\lambda) \\ &= \left(I + i(\lambda - \mu) \begin{bmatrix} \widetilde{C}_1 \\ \widetilde{C}_2 \end{bmatrix} (\widetilde{B}_2 - \lambda \widetilde{B}_1)^{-1} (\widetilde{B}_2^+ - \mu \widetilde{B}_1^+)^{-1} \begin{bmatrix} \widetilde{C}_1 \\ \widetilde{C}_2 \end{bmatrix}^+ J \right) V. \end{aligned}$$

Then the formula

$$\begin{bmatrix} \psi(\lambda) \\ \varphi(\lambda) \end{bmatrix} = \Theta^\mu(\lambda) \begin{bmatrix} q(\lambda) \\ p(\lambda) \end{bmatrix} (\widehat{w}_{21}^\mu(\lambda)q(\lambda) + \widehat{w}_{22}^\mu(\lambda)p(\lambda))^{-1} \tag{4.47}$$

establishes the one-to-one correspondence between the set of all normalized solutions $\{\varphi, \psi\}$ *of the* $AIP(B_1, B_2, C_1, C_2, K)$ *and the set of all equivalence classes of Nevanlinna pairs* $\{p, q\} \in \widetilde{N}(\mathcal{L})$ *of the form* (4.39).

Corollary 4.14. *Let the AIP data set satisfies* (A1)–(A3), *let* $\Theta(\lambda)$ *be given by* (4.42), *and let* $\operatorname{ran} C_2 = \mathcal{L}$. *Then the formula*

$$m(\lambda) = (\theta_{11}(\lambda)q(\lambda) + \theta_{12}(\lambda)p(\lambda))(\theta_{21}(\lambda)q(\lambda) + \theta_{22}(\lambda)p(\lambda))^{-1} \tag{4.48}$$

establishes the one-to-one correspondence between the set of all solutions $m(\lambda)$ *of the* $AIP(B_1, B_2, C_1, C_2, K)$ *and the set of all equivalence classes of Nevanlinna pairs* $\{p, q\} \in \widetilde{N}(\mathcal{L})$ *of the form* (4.39).

In the case when the AIP data set satisfy $(A1)$, $(A2)$, and $(A3')$ similar formula can be written in terms of the mvf $\Theta^\mu(\lambda)$.

5. Examples

5.1. Tangential interpolation problem

Let $\lambda_j \in \mathbb{C}_+$, $\xi_j \in \mathbb{C}^d$, $\eta_j \in \mathbb{C}^d$ $(1 \le j \le n)$. Consider the following problem. Find $m \in N^{d\times d}$ such that

$$m(\lambda_j)\eta_j = \xi_j \quad (1 \le j \le n). \tag{5.1}$$

The problem (5.1) is called *tangential* (or one-sided) interpolation problem and was considered first by I. Fedchina [29] in the Schur class. In [30], [32] the inclusion of this (and more general bitangential) problem into the scheme of the AIP was demonstrated.

For the case of Nevanlinna class let us set $B_1 = I_n$, $B_2 = \operatorname{diag}(\lambda_1 \dots \lambda_n)$, $C_1 = \begin{bmatrix} \xi_1 & \cdots & \xi_n \end{bmatrix}$, $C_2 = \begin{bmatrix} \eta_1 & \cdots & \eta_n \end{bmatrix}$, and let

$$P = \left[\frac{\eta_k^* \xi_j - \xi_k^* \eta_j}{\lambda_j - \bar{\lambda}_k} \right]_{j,k=1}^n \tag{5.2}$$

be the unique solution of the Lyapunov equation

$$PB_2 - B_2^* P = C_2^* C_1 - C_1^* C_2. \tag{5.3}$$

Assume that P is nonnegative and nondegenerate and that $\operatorname{ran} C_2 = \mathbb{C}^d$. Then the data set (B_1, B_2, C_1, C_2, P) satisfies the assumptions (A1)–(A3). Consider the AIP corresponding to this data set. Due to Lemma 2.10 and Proposition 3.5 every AIP solution is equivalent to a mvf $m(\cdot) \in N^{d\times d}$ and the mapping $F : \mathcal{H} \to \mathcal{H}(m)$

in (C1), (C2) is uniquely defined by (4.44). Therefore, the conditions (C1), (C2) can be rewritten as

$$(Fu)(\lambda) := \begin{bmatrix} I_d & -m(\lambda) \end{bmatrix} \begin{bmatrix} C_1 \\ C_2 \end{bmatrix} (B_2 - \lambda)^{-1} u \in \mathcal{H}(m) \quad (u \in \mathbb{C}^n); \tag{5.4}$$

$$\|Fu\|^2_{\mathcal{H}(m)} \le (Pu, u) \quad (u \in \mathbb{C}^n). \tag{5.5}$$

We claim that the problem (5.1) is equivalent to the problem (5.4), (5.5). Indeed, the condition (C1) takes the form

$$\left[\frac{\xi_j - m(\lambda)\eta_j}{\lambda - \lambda_j} \right]_{j=1}^n u \in \mathcal{H}(m), \quad u \in \mathbb{C}^n,$$

which implies (5.1). Moreover, if $m(\cdot)$ has the integral representation

$$m(\lambda) = \alpha + \beta\lambda + \int_{\mathbb{R}} \left(\frac{1}{t-\lambda} - \frac{t}{1+t^2} \right) d\sigma(t), \tag{5.6}$$

where $\alpha, \beta \in \mathbb{C}^{d\times d}$, $\alpha = \alpha^*$, $\beta \ge 0$ and $\sigma(t)$ is a nondecreasing $d \times d$-valued mvf, then the vvf $(Fu)(\lambda)$ takes the form

$$(Fu)(\lambda) = \left[\beta\eta_j + \int_{\mathbb{R}} \frac{d\sigma(t)\eta_j}{(t-\lambda)(t-\lambda_j)} \right]_{j=1}^n u.$$

Due to [4, Theorem 2.5] $Fu \in \mathcal{H}(m)$ and

$$\begin{aligned} \|Fu\|^2_{\mathcal{H}(m)} &= u^* \left[\eta_k^* \left(\beta + \int_{\mathbb{R}} \frac{d\sigma(t)}{(t-\lambda_j)(t-\bar{\lambda}_k)} \right) \eta_j \right]_{j,k=1}^n u \\ &= u^* \left[\eta_k^* \frac{m(\lambda_j) - m(\lambda_k)^*}{\lambda_j - \bar{\lambda}_k} \eta_j \right]_{j,k=1}^n u. \end{aligned} \tag{5.7}$$

Thus, (5.5) is implied by (5.7) and (5.1).

More general bitangential interpolation problems in the classes of Nevanlinna pairs with multiple points can be included in the AIP by using the data set (B_1, B_2, C_1, C_2, P):

1) $B_1 = I_N$, $B_2 = \operatorname{diag}\,(J(\lambda_1) \dots J(\lambda_\ell))$, where $J(\lambda_j)$ is a Jordan cell, corresponding to an eigenvalue $\lambda_j \in \mathbb{C}\setminus\mathbb{R}$ of order n_j $(1 \le j \le \ell)$, $n = n_1 + n_2 + \dots + n_\ell$.
2) $C_1 = \begin{bmatrix} \xi_1 & \cdots & \xi_n \end{bmatrix}$, $C_2 = \begin{bmatrix} \eta_1 & \cdots & \eta_n \end{bmatrix}$;
3) P is a nonnegative solution of (5.3).

If the set $\{\lambda_j\}_{j=1}^\ell$ contains symmetric points then the solution P of the Lyapunov equation (5.3) not necessarily exists and is not unique. Assume that there is a nonnegative nondegenerate solution P of (5.3). If $\operatorname{ran} C_2 \ne \mathbb{C}^d$ then the AIP corresponding to the data set (B_1, B_2, C_1, C_2, P) can be formulated as follows.

Find a normalized Nevanlinna pair $\{\varphi, \psi\}$ such that:

$$(Fu)(\lambda) := \begin{bmatrix}\varphi(\lambda) & -\psi(\lambda)\end{bmatrix} \begin{bmatrix} C_1 \\ C_2 \end{bmatrix} (B_2 - \lambda)^{-1} u \in \mathcal{H}(\varphi, \psi) \quad (u \in \mathbb{C}^n); \tag{5.8}$$

$$\|Fu\|^2_{\mathcal{H}(\varphi,\psi)} \leq (Pu, u) \quad (u \in \mathbb{C}^n). \tag{5.9}$$

One can show that every solution $\{\varphi, \psi\}$ of the problem (5.8), (5.9) satisfies the Parseval equality

$$\|Fu\|^2_{\mathcal{H}(\varphi,\psi)} = (Pu, u) \quad (u \in \mathbb{C}^n).$$

Regular bitangential interpolation problems in the Schur and Nevanlinna classes were studied in [42], [10], [26], [32], [9], [7]. Singular tangential and bitangential interpolation problems considered in [29], [42], [26], [18], [16], [27] can be also included in the above consideration by imposing the assumption (A2).

5.2. Hamburger moment problem

Let $s_j \in \mathbb{C}^{d\times d}$, $j \in \mathbb{N}$ and let S_n be the Hankel block matrix

$$S_n = (s_{i+j})^n_{i,j=0}.$$

A $\mathbb{C}^{d\times d}$-valued nondecreasing right continuous function $\sigma(t)$ is called a solution of the Hamburger moment problem if

$$\int t^j d\sigma(t) = s_j \quad (j \in \mathbb{N}). \tag{5.10}$$

It is known (see [2], [12], [36]) that the Hamburger moment problem (5.10) is solvable iff $S_n \geq 0$ for all $n \in \mathbb{N}$. Due to Hamburger-Nevanlinna theorem a function $\sigma(t)$ is a solution of the Hamburger moment problem (5.10) if and only if the associated mvf

$$m(\lambda) = \int_{-\infty}^{\infty} \frac{d\sigma(t)}{t - \lambda}$$

has the following nontangential asymptotic at ∞

$$m(\lambda) \sim -\frac{s_0}{\lambda} - \frac{s_1}{\lambda^2} - \cdots - \frac{s_{2n}}{\lambda^{2n+1}} + O\left(\frac{1}{\lambda^{2n+2}}\right) \quad (\lambda \widehat{\to} \infty) \tag{5.11}$$

for every $n \in \mathbb{N}$.

Let $\mathcal{L} = \mathbb{C}^d$, let $\mathcal{X}$ be the space of all vector polynomials

$$h(X) = \sum_{j=0}^{n} h_j X^j, \quad h_j \in \mathcal{L}, \tag{5.12}$$

and let the nonnegative form $K(h, h)$ be defined by

$$K(h, h) = \sum_{j,k=0}^{n} (s_{j+k} h_j, h_k)_{\mathcal{L}}. \tag{5.13}$$

Assume that all the matrices S_n, $n \in \mathbb{N}$ are nondegenerate and consider the closure $\mathcal{H}$ of the space $\mathcal{X}$ endowed with the inner product $K(\cdot, \cdot)$. Then the closure M of the multiplication operator M_0 in $\mathcal{X}$ is a symmetric operator in $\mathcal{H}$. The moment problem (5.10) is called *indeterminate* if the defect numbers of M are equal to d.

As was shown in [13] the scalar moment problem ($d = 1$) is indeterminate if and only if there exists $\delta > 0$ such that

$$S_n \geq \delta > 0 \quad \text{for all } n \in \mathbb{N}. \tag{5.14}$$

Slight modification of the proof of this statement in [13] shows that the condition (5.14) is also necessary and sufficient for the moment problem (5.10) to be indeterminate for arbitrary d.

Let us consider the abstract interpolation problem in the class $N^{d\times d}$ corresponding to the indeterminate moment problem (5.10). Define the operators $B_1, B_2 : \mathcal{X} \to \mathcal{X}$ and $C_1, C_2 : \mathcal{X} \to \mathcal{L}$ by the equalities

$$B_1 h = \frac{h(X) - h_0}{X}, \quad B_2 h = h, \quad C_1 h = \sum_{j=1}^{n} s_{j-1} h_j, \quad C_2 h = -h(0). \tag{5.15}$$

Then the data set (B_1, B_2, C_1, C_2, K) satisfies the assumption (A1). Clearly, (A2) is in force, since $\ker K = \{0\}$. Let us show that (A3) is fulfilled.

Proposition 5.1. *The operators B_1, C_1, C_2 admit continuous extensions to the operators $\widetilde{B}_1 \in [\mathcal{H}]$, and $\widetilde{C}_1, \widetilde{C}_2 \in [\mathcal{H}, \mathcal{L}]$.*

Proof. Let $\widetilde{B}_1$ be the closure of the graph of the operator B_1. Then

$$\widetilde{B}_1^{-1} = \{\{h, Mh + u\} : h \in \operatorname{dom} M, u \in \mathcal{L}\}.$$

As was shown in [36] $\rho(M, \mathcal{L}) = \mathbb{C}$ in the case of indeterminate moment problem (5.10). In particular, $0 \in \rho(M, \mathcal{L})$, that is

$$\operatorname{ran} \widetilde{B}_1^{-1} = \operatorname{ran} M \dotplus \mathcal{L} = \mathcal{H}, \quad \ker \widetilde{B}_1^{-1} = \operatorname{ran} M \cap \mathcal{L} = \{0\}.$$

Therefore $\widetilde{B}_1$ is the graph of a bounded operator in $\mathcal{H}$ for which we will keep the same notation.

It follows from (5.14) that for every polynomial $h \in \mathcal{X}$

$$\|h\|_{\mathcal{H}}^2 = K(h, h) = \sum_{j,k=0}^{n} h_k^* s_{j+k} h_j \geq \delta \sum_{j,k=0}^{n} \|h_j\|^2 \geq \delta \|h_0\|^2.$$

Therefore,

$$\|C_2 h\|^2 \leq \frac{1}{\delta} \|h\|_{\mathcal{H}}^2$$

and, hence, the operator $C_2 : \mathcal{X} \subset \mathcal{H} \to \mathcal{L}$ is bounded. Let us note that the boundedness of C_2 is implied also by the fact that $0 \in \rho(M, \mathcal{L})$, since $C_2 h = \mathcal{P}_{M,\mathcal{L}}(0)h$, where $\mathcal{P}_{M,\mathcal{L}}(\lambda)$ is the skew projection onto $\mathcal{L}$ in the decomposition $\mathcal{H} = \operatorname{ran}(M - \lambda) \dotplus \mathcal{L}$.

The boundedness of $C_1 : \mathcal{X} \subset \mathcal{H} \to \mathcal{L}$ is implied by the equality

$$(C_1 h, u)_{\mathcal{L}} = K\left(\frac{h(X) - h(0)}{X}, u\right) = K(B_1 h, u). \tag{5.16}$$

Indeed, it follows from (5.16) that

$$\begin{aligned}|(C_1h,u)_{\mathcal{L}}| &\le K(B_1h,B_1h)^{1/2}K(u,u)^{1/2}\\ &= \|B_1h\|_{\mathcal{H}}(s_0u,u)^{1/2}\\ &\le \|B_1\|\|s_0^{1/2}\|\|h\|_{\mathcal{H}}\|u\|_{\mathcal{L}}\end{aligned}$$

and hence C_1 is bounded and $\|C_1\| \le \|B_1\|\|s_0^{1/2}\|$. □

Remark 5.2. The definition (5.15) of C_1 can be rewritten as

$$C_1h = \widetilde{h}(0), \tag{5.17}$$

where the adjacent polynomial $\widetilde{h}$ is defined by

$$(\widetilde{h}(\lambda),u)_{\mathcal{L}} = K\left(\frac{h(X)-h(\lambda)}{X-\lambda},u\right), \quad u \in \mathcal{L}. \tag{5.18}$$

Let us consider the $[\mathcal{H},\mathcal{L}^2]$-valued operator function

$$G(\lambda) = \begin{bmatrix} C_1 \\ C_2 \end{bmatrix}(I-\lambda B_1)^{-1}, \quad \lambda \in \mathbb{C}. \tag{5.19}$$

Recall some useful formulas (see [33])

$$(I-\lambda B_1)^{-1}h = \frac{Xh(X)-\lambda h(\lambda)}{X-\lambda}, \quad (h \in \mathcal{X}). \tag{5.20}$$

$$C_1(I-\lambda B_1)^{-1}h = \widetilde{h}(\lambda), \quad C_2(I-\lambda B_1)^{-1}h = -h(\lambda). \tag{5.21}$$

The corresponding abstract interpolation problem can be formulated as follows. Find a Nevanlinna pair $\{\varphi,\psi\} \in \widetilde{N}(\mathbb{C}^d)$, such that

$$Fh := \begin{bmatrix}\varphi(\lambda) & -\psi(\lambda)\end{bmatrix} G(\lambda)h \in \mathcal{H}(\varphi,\psi); \tag{5.22}$$

$$\|Fh\|^2_{\mathcal{H}(\varphi,\psi)} \le K(h,h) \tag{5.23}$$

for all $h \in \mathcal{X}$.

Since the operators B_1, $B_2 = I$ satisfy the assumption (U) the mapping $F : \mathcal{X} \to \mathcal{H}(\varphi,\psi)$ corresponding to the solution $\{\varphi,\psi\}$ of the AIP is uniquely defined (see Proposition 4.44). Moreover, since $\operatorname{ran} C_2 = \mathcal{L}$ it follows from Lemma 2.10 that any solution $\{\varphi,\psi\}$ of the AIP is equivalent to a pair $\{I_d, m(\lambda)\}$, where $m \in N^{d\times d}$.

Theorem 5.3. *Let m be a solution of the $AIP(B_1,B_2,C_1,C_2,K)$, which assumes the integral representation* (5.6)*, and let $F : \mathcal{X} \to \mathcal{H}(m)$ be the mapping corresponding to m via the formula $(Fh)(\lambda) := \begin{bmatrix}I_d & -m(\lambda)\end{bmatrix} G(\lambda)h$, $h \in \mathcal{X}$. Then:*

1) *σ is a solution of the Hamburger moment problem* (5.10)*;*
2) *for every polynomial $h \in \mathcal{X}$ one has*

$$(Fh)(\lambda) = \int_{-\infty}^{\infty} \frac{d\sigma(t)h(t)}{t-\lambda}, \tag{5.24}$$

$$\|Fh\|^2_{\mathcal{H}(m)} = K(h,h). \tag{5.25}$$

Proof. For a monic polynomial $h = uX^j$ $(u \in \mathcal{L})$ one obtains from (5.15)–(5.19)

$$(Gh)(\lambda) = \begin{bmatrix} C_1 \\ C_2 \end{bmatrix} (X^j + \lambda X^{j-1} + \cdots + \lambda^j)u$$
$$= \begin{bmatrix} (\lambda^{j-1}s_0 + \cdots + \lambda s_{j-2} + s_{j-1})u \\ -\lambda^j u \end{bmatrix}$$

and hence

$$\begin{aligned}(Fh)(\lambda) &= \begin{bmatrix} I_d & -m(\lambda) \end{bmatrix} G(\lambda)h \\ &= (\lambda^j m(\lambda) + \lambda^{j-1}s_0 + \cdots + s_{j-1})u \\ &= \int_{-\infty}^{\infty} \frac{t^j}{t-\lambda} d\sigma(t)u.\end{aligned} \tag{5.26}$$

When $j = 0$ it follows from (5.26) and Lemma 2.12 that $m(\lambda) = O(1)$ if $\lambda \widehat{\rightarrow} \infty$. Setting $j = 1$ one derives from (5.26) that

$$\lambda m(\lambda) + s_0 = O(1) \quad (\lambda \widehat{\rightarrow} \infty).$$

Therefore, $m(\lambda) = O(\frac{1}{\lambda})$ and applying Lemma 2.12, (ii) gives

$$m(\lambda) + \frac{s_0}{\lambda} = O(\frac{1}{\lambda^2}).$$

Similarly, for $h = uX^n$ one obtains from (C1) $(\lambda^n m(\lambda) + s_0\lambda^{n-1} + \cdots + s_{n-1})u \in \mathcal{H}(m)$, or by Lemma 2.12, (ii)

$$m(\lambda) + \frac{s_0}{\lambda} + \cdots + \frac{s_{n-1}}{\lambda^n} = O(\frac{1}{\lambda^{n+1}})$$

for arbitrary $n \in \mathbb{N}$. In view of the Hamburger-Nevanlinna theorem ([2]) this implies that σ is a solution of the Hamburger moment problem (5.10).

For arbitrary polynomial $h = \sum_{j=0}^{n} u_j X^j \in \mathcal{X}$ the formula (5.26) can be rewritten as (5.24). Applying the formula for the inner product in $\mathcal{H}(m)$ (see [4, Theorem 2.5]) one obtains

$$\|Fh\|^2_{\mathcal{H}(m)} = \int_{-\infty}^{\infty} (d\sigma(t)h(t), h(t)) = \sum_{j,k=0}^{n} (s_{j+k}u_j, u_k)_{\mathcal{L}} = K(h,h).$$

This proves (5.25).

Conversely, let σ be a solution of the Hamburger moment problem (5.10). Then it follows from (5.24) and Theorem 2.5 from [4] that $Fh \in \mathcal{H}(m)$ for arbitrary polynomial $h \in \mathcal{X}$. This proves (C1). (C2) is implied by the equality (5.25). □

To calculate the $\mathcal{L}$-resolvent matrix let us introduce a system of matrix polynomials $\{P_n(\lambda)\}_{n=0}^{\infty}$ orthogonal with respect to the form K:

$$K(P_j u, P_k v) = v^* u \delta_{jk}, \quad u, v \in \mathbb{C}^d;\ j, k \in \mathbb{N}$$

and a system of adjacent polynomials $\{\widetilde{P}_k(\lambda)\}_{n=0}^{\infty}$

$$v^* \widetilde{P}_k(\lambda) u = K\left(\frac{P_k(t) - P_k(\lambda)}{t-\lambda} u, v\right), \quad u, v \in \mathbb{C}^d;\ k \in \mathbb{N}.$$

Proposition 5.4. *For every $u \in \mathcal{L}$ the following formulas hold*

$$C_1^+ u = \sum_{k=1}^{\infty} P_k(t)\widetilde{P}_k(0)^* u, \quad C_2^+ u = -\sum_{k=0}^{\infty} P_k(t)P_k(0)^* u. \tag{5.27}$$

Proof. Indeed, for every $j \in \mathbb{N} \cup \{0\}$, $u, v \in \mathcal{L}$ one obtains from (5.15), (5.17)

$$\begin{aligned}
(C_1^+ u, P_j v)_{\mathcal{H}} = (u, C_1 P_j v)_{\mathcal{L}} \quad &= (u, \widetilde{P}_j(0)v)_{\mathcal{L}} \\
= (\widetilde{P}_j(0)^* u, v)_{\mathcal{H}} \quad &= \left(\sum_{k=1}^{\infty} P_k(t)\widetilde{P}_k(0)^* u, P_j v \right)_{\mathcal{H}}, \\
(C_2^+ u, P_j v)_{\mathcal{H}} = (u, C_2 P_j v)_{\mathcal{L}} \quad &= -(u, P_j(0)v)_{\mathcal{L}} \\
= -(P_j(0)^* u, v)_{\mathcal{H}} &= -\left(\sum_{k=1}^{\infty} P_k(t)P_k(0)^* u, P_j v \right)_{\mathcal{H}}.
\end{aligned}$$

□

Applying the formulas (4.18), (5.21) and (5.27) one obtains the resolvent matrix $\Theta(\lambda)$:

$$\begin{aligned}
\theta_{11}(\lambda)u &= u + \lambda\widetilde{C}_1(I - \lambda\widetilde{B}_1)^{-1}\left(\sum_{k=0}^{\infty} P_k(t)P_k(0)^* u\right) = \left(I + \lambda\sum_{k=0}^{\infty}\widetilde{P}_k(\lambda)P_k(0)^*\right)u, \\
\theta_{12}(\lambda)u &= \lambda\widetilde{C}_1(I - \lambda\widetilde{B}_1)^{-1}\left(\sum_{k=0}^{\infty} P_k(t)\widetilde{P}_k(0)^* u\right) = \lambda\sum_{k=0}^{\infty}\widetilde{P}_k(\lambda)\widetilde{P}_k(0)^* u, \\
\theta_{21}(\lambda)u &= \lambda\widetilde{C}_2(I - \lambda\widetilde{B}_1)^{-1}\left(\sum_{k=0}^{\infty} P_k(t)P_k(0)^* u\right) = -\lambda\sum_{k=0}^{\infty} P_k(\lambda)P_k(0)^* u, \\
\theta_{22}(\lambda)u &= u + \lambda\widetilde{C}_2(I - \lambda\widetilde{B}_1)^{-1}\left(\sum_{k=1}^{\infty} P_k(t)\widetilde{P}_k(0)^* u\right) = \left(I - \lambda\sum_{k=1}^{\infty} P_k(\lambda)\widetilde{P}_k(0)^*\right)u.
\end{aligned}$$

Application of general result in Corollary 4.14 gives the well-known description of solutions of the moment problem (5.10)

$$\int_{\mathbb{R}} \frac{d\sigma(t)}{t - \lambda} = (\theta_{11}(\lambda)q(\lambda) + \theta_{12}(\lambda)p(\lambda))(\theta_{21}(\lambda)q(\lambda) + \theta_{22}(\lambda)p(\lambda))^{-1}, \tag{5.28}$$

when the pair $\{p, q\}$ ranges over the class $\widetilde{N}^{d\times d}$.

5.3. Truncated Hamburger moment problem

Let $s_0, s_1, \ldots, s_{2n} \in \mathbb{C}^{d\times d}$. A $\mathbb{C}^{d\times d}$-valued nondecreasing right continuous function $\sigma(t)$ is called a solution of the truncated Hamburger moment problem if

$$\int t^j d\sigma(t) = s_j \quad (j = 0, 1, \ldots, 2n - 1) \tag{5.29}$$

$$\int t^{2n} d\sigma(t) \leq s_{2n}. \tag{5.30}$$

It is known that the problem (5.29)–(5.30) is solvable if and only if $S_n \geq 0$. A solution σ of the problem (5.29)–(5.30) is called "exact", if $\int t^{2n} d\sigma(t) = s_{2n}$. Singular truncated Hamburger moment problem has been studied in [20], [15], [1].

Theorem 5.5. ([15]) *Let* $S_n = (s_{i+j})_{i,j=0}^n \in \mathbb{C}^{d(n+1)\times d(n+1)})$ – *be a nonnegative block Hankel matrix. The following assertions are equivalent:*

1) *The problem* (5.29)–(5.30) *has an "exact" solution;*
2) S_n *admits a nonnegative block Hankel extension* S_{n+1}.

The equivalence (1) ⇔ (2) was proved in [15]. Moreover, it was shown in [15] that if the conditions 1)–2) in Theorem 5.5 fail to hold then one can replace the right lower block in the matrix S_n in such a way that the new matrix $S_n{}' = (s'_{i+j})_{i,j=0}^n$ satisfies 1)–2) in Theorem 5.5 and the sets $\mathcal{Z}(S_n)$ and $\mathcal{Z}(S'_n)$ coincide.

In what follows it is supposed that S_n satisfies the assumptions 1)–2) of Theorem 5.5. We will need also the following statement from [15].

Lemma 5.6. *Let a block Hankel matrix* $S_n = (s_{i+j})_{i,j=0}^n$ *satisfy the assumptions of Theorem* 5.5 *and let the matrix* $T \in \mathbb{C}^{N\times N}$ $(N = (n+1)d)$ *be given by*

$$T = \begin{bmatrix} 0_d & I_d & & \\ & \ddots & \ddots & \\ & & 0_d & I_d \\ & & & 0_d \end{bmatrix}.$$

Then there exists a matrix $X = X^* \in \mathbb{C}^{N\times N}$ *of* $\operatorname{rank} X = \operatorname{rank} S_n$ *such that*

$$XS_nX = X, \quad S_nXS_n = S_n, \quad T\operatorname{ran} X \subseteq \operatorname{ran} X. \tag{5.31}$$

Let $\mathcal{X}$ be the space of vector polynomials $h(X)$ of the form (5.12) of formal degree n and let the form $K(\cdot,\cdot)$ be given by (5.13). Define the operators $B_1, B_2 : \mathcal{X} \to \mathcal{X}$ and $C_1, C_2 : \mathcal{X} \to \mathcal{L}$ by (5.15). Then the data set (B_1, B_2, C_1, C_2, K) satisfies the assumption (A1). Choosing the basis $1, X, \dots, X^n$ in $\mathcal{X}$ one can identify $\mathcal{X}$ with $\mathbb{C}^N$ and then the form $K(\cdot,\cdot)$ is given by

$$K(h,g) = (S_nh, g), \quad h, g \in \mathbb{C}^N, \quad N = (n+1)d.$$

The operators B_1, B_2 and C_1, C_2 can be identified with their matrix representations in this basis

$$B_1 = T, \quad B_2 = I_N, \tag{5.32}$$

$$C_1 = \begin{bmatrix} 0 & s_0 & \dots & s_{n-1} \end{bmatrix}, \quad C_2 = \begin{bmatrix} -I_d & 0 & \dots & 0 \end{bmatrix}. \tag{5.33}$$

Then $B_2 - \lambda B_1 = I_N - \lambda T$ is invertible for all $\lambda \in \mathbb{C} \setminus \{0\}$ and

$$G(\lambda) = \begin{bmatrix} C_1 \\ C_2 \end{bmatrix} (I_N - \lambda T)^{-1} = \begin{bmatrix} C_1 \\ C_2 \end{bmatrix} \begin{bmatrix} I_d & \dots & \lambda^n I_d \\ & \ddots & \vdots \\ & & I_d \end{bmatrix}. \tag{5.34}$$

Since the operators $B_1 = T$, $B_2 = I_N$ satisfy the assumption (U) and $\operatorname{ran} C_2 = \mathbb{C}^d$ the mapping $F : \mathcal{X} \to \mathcal{H}(\varphi, \psi)$ corresponding to the solution $\{\varphi, \psi\}$ of the

$AIP(B_1, B_2, C_1, C_2, K)$ is uniquely defined and any solution $\{\varphi, \psi\}$ of the AIP is equivalent to a pair $\{I_d, m(\lambda)\}$, where $m \in N^{d\times d}$. The corresponding AIP can be formulated as follows:

Find a Nevanlinna mvf $m \in N^{d\times d}$ such that:

(C1) $Fh = \begin{bmatrix} I_d & -m(\lambda) \end{bmatrix} G(\lambda)h \in \mathcal{H}(m)$ for all $h \in \mathcal{X}$;
(C2) $\|Fh\|^2_{\mathcal{H}(m)} \le (S_n h, h)$ for all $h \in \mathcal{X}$.

Proposition 5.7. *Let m be a solution of the $AIP(B_1, B_2, C_1, C_2, S_n)$. Then m admits the integral representation*

$$m(\lambda) = \int_{-\infty}^{\infty} \frac{d\sigma(t)}{t-\lambda} \tag{5.35}$$

where $\sigma \in \mathcal{Z}(S_n)$. Conversely, if $\sigma \in \mathcal{Z}(S_n)$, then m is a solution of the AIP.

Proof. Necessity. The same arguments as in the proof of Theorem 5.3 show that (C1) implies

$$m(\lambda) + \frac{s_0}{\lambda} + \cdots + \frac{s_{n-1}}{\lambda^n} = O\left(\frac{1}{\lambda^{n+1}}\right) \quad (\lambda \widehat{\to} \infty). \tag{5.36}$$

Let $m(\lambda)$ have the integral representation (5.35) and let us set

$$s_j^{(m)} = \int_{\mathbb{R}} t^j d\sigma(t) \quad (j = 0, 1, \dots, 2n). \tag{5.37}$$

It follows from (5.36) that

$$s_j^{(m)} = s_j \text{ for } j = 0, 1, \dots, n-1. \tag{5.38}$$

The rest of the equalities (5.29) and the inequality (5.30) are implied by (C2).

Let us show that for every polynomial $h(X) = \sum_{j=0}^n u_j X^j, \quad u_j \in \mathbb{C}^d$, the following equality holds

$$\|Fh\|^2_{\mathcal{H}(m)} = \sum_{j,k=0}^{n} u_k^* s_{j+k}^{(m)} u_j. \tag{5.39}$$

Indeed, it follows from (5.26) that

$$\begin{aligned}(Fh)(\lambda) &= \begin{bmatrix} m(\lambda) & \lambda m(\lambda) + s_0 & \dots & \lambda^n m(\lambda) + \lambda^{n-1} s_0 + \cdots + s_{n-1} \end{bmatrix} u \\ &= \begin{bmatrix} \int_{\mathbb{R}} \frac{d\sigma(t)}{t-\lambda} & \int_{\mathbb{R}} \frac{t d\sigma(t)}{t-\lambda} & \dots & \int_{\mathbb{R}} \frac{t^n d\sigma(t)}{t-\lambda} \end{bmatrix} u \in \mathcal{H}(m),\end{aligned} \tag{5.40}$$

where $u = \operatorname{col}(u_0, u_1, \dots, u_n) \in \mathbb{C}^{(n+1)d}$. Due to [4, Theorem 2.5]

$$\|Fh\|^2_{\mathcal{H}(m)} = u^* S_n^{(m)} u, \tag{5.41}$$

where

$$S_n^{(m)} = \left[s_{i+j}^{(m)} \right]_{i,j=0}^{n}. \tag{5.42}$$

Then the inequality $S_n^{(m)} \le S_n$ in (C2) and (5.38) imply that (5.29) and (5.30) hold.

Sufficiency. Let $\sigma \in \mathcal{Z}(S_n)$ and let m be defined by (5.35). Then it follows from (5.40) that (C1) holds.

The condition (C2) is implied by (5.41), (5.29) and (5.30), since

$$\|Fh\|^2_{\mathcal{H}(m)} = u^* S_n^{(m)} u \le u^* S_n u, \quad u \in \mathbb{C}^N. \qquad \square$$

In the regular case (when $\det S_n \neq 0$) the solution matrix $\Theta(\lambda)$ can be calculated by (4.42). Since $C_1^+ = S_n^{-1} C_1^*$ and $C_2^+ = S_n^{-1} C_2^*$ one obtains from (4.42)

$$\Theta(\lambda) = I_{2d} - \lambda \begin{bmatrix} C_1 \\ C_2 \end{bmatrix} (I_{\mathcal{H}} - \lambda T)^{-1} S_n^{-1} \begin{bmatrix} C_2^* & -C_1^* \end{bmatrix}.$$

Then by Corollary 4.14 the formula (5.28) establishes the one-to-one correspondence between the set of all solutions σ of the truncated moment problem (5.29)–(5.30) and the set of all equivalence classes of Nevanlinna pairs $\{p, q\} \in \widetilde{N}^{d\times d}$.

In the singular case ($\det S_n = 0$) let us consider the matrix $X \in \mathbb{C}^{N\times N}$ which satisfies (5.31). Then the decomposition

$$\mathcal{X} = \operatorname{ran} X \dotplus \ker S_n$$

satisfies the assumptions (A2), (A3), since $T \operatorname{ran} X \subseteq \operatorname{ran} X$, and the solution matrix $\Theta(\lambda)$ takes the form (4.42). Now, let us calculate the operators $C_1^+, C_2^+ : \mathcal{L} \to \mathcal{H}$. For arbitrary $h = Xg \in \operatorname{ran} X$, $u \in \mathcal{L}$ and $j = 1, 2$ one obtains

$$\begin{aligned} (C_j X g, u)_{\mathbb{C}^d} &= (Xg, C_j^* u)_{\mathbb{C}^N} \\ &= (Xg, S_n X C_j^* u)_{\mathbb{C}^N} \\ &= (Xg, X C_j^* u)_{\mathcal{H}}. \end{aligned}$$

Therefore, $C_1^+ = XC_1^*$, $C_2^+ = XC_2^*$, and the mvf $\Theta(\lambda)$ takes the form

$$\Theta(\lambda) = \left(I_{\mathcal{L}\oplus\mathcal{L}} - \lambda \begin{bmatrix} C_1 \\ C_2 \end{bmatrix} (I_{\mathcal{H}} - \lambda T)^{-1} X \begin{bmatrix} C_2^* & -C_1^* \end{bmatrix} \right) V, \tag{5.43}$$

where $V \in \mathbb{C}^{2d\times 2d}$ is a unitary matrix such that

$$V(\{0\} \times \mathbb{C}^\nu) = \begin{bmatrix} C_1 \\ -C_2 \end{bmatrix} \ker S_n = \begin{bmatrix} 0 & s_1 & \dots & s_{n-1} \\ I_d & 0 & \dots & 0 \end{bmatrix} \ker S_n$$

and

$$\nu = \dim \begin{bmatrix} 0 & s_1 & \dots & s_{n-1} \\ -I_d & 0 & \dots & 0 \end{bmatrix} \ker S_n.$$

Then by Corollary 4.14 the formula (5.28) establishes the one-to-one correspondence between the set of all solutions σ of the truncated moment problem (5.29)–(5.30) and the set of all equivalence classes of pairs $\{p, q\} \in \widetilde{N}^{d\times d}$ of the form (4.39).

References

[1] V.M. Adamjan, I.M. Tkachenko, *Solutions of the truncated matrix Hamburger moment problem according to M.G. Kreĭn.* Oper. Theory: Adv. Appl. **118** (2000), Birkhäuser Verlag, Basel, 35–51.

[2] N.I. Achieser, *The classical moment problem.* Moscow, 1961.

[3] D. Alpay, A. Dijksma, J. Rovnyak, and H.S.V. de Snoo, *Schur functions, operator colligations, and reproducing kernel Pontryagin spaces.* Oper. Theory: Adv. Appl. **96**, Birkhäuser Verlag, Basel, 1997.

[4] D. Alpay, I. Gohberg, *Pairs of selfadjoint operators and their invariants.* St. Petersburg Math. J. **16** (2005), no. 1, 59–104.

[5] D. Alpay, H. Dym, *Hilbert spaces of analytic functions, inverse scattering and operator models. I.* Integral Equations and Operator Theory **7** (1984), 589–641.

[6] D. Alpay, P. Bruinsma, A. Dijksma, and H.S.V. de Snoo, *A Hilbert space associated with a Nevanlinna function.* Proceedings MTNS Meeting, Amsterdam (1989), 115–122.

[7] D. Alpay, P. Bruinsma, A. Dijksma, and H.S.V. de Snoo, *Interpolation problems, extensions of symmetric operators and reproducing kernel spaces. I.* Oper. Theory: Adv. Appl. **50** (1991), Basel: Birkhäuser Verlag, 35–82.

[8] D.Z. Arov and L.Z. Grossman, *Scattering matrices in the theory of unitary extensions of isometric operators.* Math. Nachr. **157** (1992), 105–123.

[9] J. Ball, I. Gohberg, and L. Rodman, *Interpolation of rational matrix functions.* OT45, Birkhäuser Verlag, 1990.

[10] J.A. Ball, H.W. Helton, *A Beurling–Lax theorem for the Lie group $U(m, n)$ which contains most classical interpolation theory.* J. Operator Theory **9** (1983), 107–142.

[11] C. Bennewitz, *Symmetric relations on a Hilbert space.* Lect. Notes Math. **280** (1972), 212–218.

[12] Yu.M. Berezanskii, *Expansions in eigenfunctions of selfadjoint operators.* Naukova Dumka, Kiev, 1965 [English translation: Amer. Math. Soc., Providence, RI, 1968].

[13] Ch. Berg, Y. Chen, M.E.H. Ismail, *Small eigenvalues of large Hankel matrices: The indeterminate case.* Math. Scand. **91** (2002), no. 1, 67–81.

[14] J. Berndt, S. Hassi, H.S.V. de Snoo, *Boundary relations, unitary colligations and functional models.* Complex Analysis Operator Theory, 3 (2009), 57–98.

[15] V. Bolotnikov, On degenerate Hamburger moment problem and extensions of nonnegative Hankel block matrices, Integral Equations and Operator Theory 25 (1996), no. 3, 253–276.

[16] V. Bolotnikov, H. Dym, *On degenerate interpolation, entropy and extremal problems for matrix Schur functions.* Integral Equations Operator Theory **32** (1998), no. 4, 367–435.

[17] L. de Branges, *Perturbation theory.* J. Math. Anal.Appl. **57** (1977), 393–415.

[18] P. Bruinsma, *Degenerate interpolation problems for Nevanlinna pairs.* Indag Math. N.S. **2** (1991), 179–200.

[19] E.A. Coddington, *Extension theory of formally normal and symmetric subspaces.* Mem. Amer. Math. Soc. **134** (1973), 1–80.

[20] R.E. Curto, L.A. Fialkow, *Recursiveness, positivity, and truncated moment problems.* Houston J. Math. **17** (1991), 603–635.

[21] V.A. Derkach and M.M. Malamud, *Generalized resolvents and the boundary value problems for hermitian operators with gaps.* J. Funct. Anal. **95** (1991), 1–95.

[22] V.A. Derkach, M.M. Malamud, *The extension theory of hermitian operators and the moment problem.* J. of Math.Sci. **73** (1995), no. 2, 141–242.

[23] V.A. Derkach, S. Hassi, M.M. Malamud, H.S.V. de Snoo, *Generalized resolvents of symmetric operators and admissibility.* Methods of Functional Analysis and Topology **6** (2000), 24–55.

[24] V.A. Derkach, S. Hassi, M.M. Malamud, H.S.V. de Snoo, *Boundary relations and their Weyl families*, Trans. Amer. Math. Soc. **358** (2006), 5351–5400.

[25] V.K. Dubovoj, Indefinite metric in the interpolation problem of Schur for analytic matrix functions, IV, Theor. Funkts. Func. Anal. i Prilozen., 42 (1984), 46–57 (Russian) [English transl. in: Topics in Interpolation Theory, Oper. Theory: Adv. Appl., OT 95, Birkhäuser Verlag, Basel, 1997, 93–104].

[26] H. Dym, *J-contractive matrix functions, reproducing kernel Hilbert spaces and interpolation*, CBMS Regional Series in Math. **71**, Providence, RI, 1989.

[27] H. Dym, Riccati equations and bitangential interpolation problems with singular Pick matrices. Fast algorithms for structured matrices: theory and applications (South Hadley, MA, 2001), 361–391, Contemp. Math., 323, Amer. Math. Soc., Providence, RI, 2003.

[28] V.I. Gorbachuk and M.L. Gorbachuk, *Boundary value problems for operator differential equations.* **48**. Kluwer, Dordrecht, 1991. xii+347 pp.

[29] I.P. Fedchina, *Criteria for the solvability of Nevanlinna-Pick tangent problem.* Matem. Issl. **7** (1972), no. 4 (26), 213–227.

[30] V.E. Katsnelson, A.Ya. Kheifets and P.M. Yuditskii, *The abstract interpolation problem and extension theory of isometric operators.* Operators in Spaces of Functions and Problems in Function Theory, Kiev, Naukova Dumka (1987), 83–96 (Russian).

[31] A.Ya. Kheifets and P.M. Yuditskii, *An analysis and extension of V.P. Potapov's approach to interpolation problems with applications to the generalized bi-tangential Schur–Nevanlinna–Pick problem and J-inner-outer factorization.* Operator Theory: Adv. Appl. **72** (1994), Birkhäuser, Basel, 133–161.

[32] A. Kheifets, *Generalized bitangential Schur-Nevanlinna-Pick problem and related with it Parseval equality.* Teor. funkcij i pril., Kharkov **54** (1990), 89–96.

[33] A. Kheifets, *Hamburger Moment problem: Parseval equality and A-singularity.* J. Funct. Analysis **141** (1996), 374–420.

[34] I.V. Kovalishina and V.P. Potapov, *Indefinite metric in Nevanlinna-Pick problem.* Dokl. Akad. Nauk Armjan. SSR, ser. mat. **59** (1974), 17–22.

[35] M.G. Kreĭn, *On Hermitian operators with defect indices equal to one.* Dokl. Akad. Nauk SSSR **43** (1944), no. 8, 339–342.

[36] M.G. Kreĭn, *Fundamental aspects of the representation theory of Hermitian operators with deficiency index (m, m).* Ukrain. Math. Zh. **1** (1949), 3–66 (Russian); (English translation: Amer. Math. Soc. Transl. **(2) 97** (1970), 75–143).

[37] M.G. Kreĭn and H.Langer, *Über die verallgemeinerten Resolventen und die characteristische Function eines isometrischen Operators im Raume* Π_κ. Hilbert space Operators and Operator Algebras (Proc. Intern. Conf., Tihany, 1970); Colloq. Math. Soc. Janos Bolyai, **5** (1972), North–Holland, Amsterdam, 353–399.

[38] M.G. Kreĭn and Sh.N. Saakyan, *Some new results in the theory of resolvent matrices.*Dokl. Akad. Nauk SSSR **169** (1966), no. 1, 657–660.

[39] S. Kupin, *Lifting theorem as a special case of abstract interpolation problem.* J. Anal. Appl. **15** (1996), no. 4, 789–798.

[40] M.M. Malamud, *On the formula of generalized resolvents of a nondensely defined Hermitian operator*, Ukr. Mat. Zh. **44** (1992), no. 2, 1658–1688.

[41] M.M. Malamud and S.M. Malamud, *Spectral theory of operator measures in Hilbert space.* St.-Petersburg Math. Journal **15** (2003), no. 3, 1–77.

[42] A.A. Nudel'man, *On a new problem of moment type.* Dokl. Akad. Nauk SSSR **233** (1977), 792–795; Soviet Math. Dokl. **18** (1977), 507–510.

[43] V.P. Potapov, *Multiplicative structure of J-nonexpanding matrix functions.* Trudy Mosk. Matem. Obsch. **4** (1955), 125–236.

[44] A.V. Štraus, *Extensions and generalized resolvents of a symmetric operator which is not densely defined.* Izv. Akad. Nauk. SSSR, Ser. Mat. **34** (1970), 175–202 (Russian) [English translation: Math. USSR-Izvestija **4** (1970), 179–208].

Vladimir Derkach
Department of Mathematics
Donetsk National University
24 Universitetskaya str.
83055 Donetsk, Ukraine
e-mail: derkach.v@gmail.com

Operator Theory:
Advances and Applications, Vol. 190, 237–242

On the Strong Kreiss Resolvent Condition in the Hilbert Space

Alexander Gomilko and Jaroslav Zemánek

Abstract. It is proved that if an operator T on a Hilbert space satisfies the strong Kreiss resolvent condition then so does the operator T^m for any $m \in \mathbb{N}$.

Mathematics Subject Classification (2000). Primary 47A10; Secondary 47D03.

Keywords. Hilbert space, resolvent, Carleson embedding theorem.

1. Introduction

Let H be a Hilbert space with inner product $(\cdot,\cdot)$ and norm $\|\cdot\|$, and let $L(H)$ be the algebra of bounded linear operators on H. We denote the spectrum of $T \in L(H)$ by $\sigma(T)$, the identity operator on H by I, and the resolvent of T by $R(T,\lambda) = (T-\lambda I)^{-1}$, $\lambda \notin \sigma(T)$.

Let us recall (see, e.g., [1], [2]) that an operator T with spectrum in the unit disc is said to satisfy the strong Kreiss resolvent condition with constant $M \geq 1$ if

$$\|R^k(T,\lambda)\| \leq \frac{M}{(|\lambda|-1)^k} \quad \text{for all } |\lambda| > 1, \ \text{ and } k = 1,2,\dots\,. \qquad \text{[SR]}$$

We recall that the condition [SR] is equivalent to the condition

$$\|e^{zT}\| \leq Me^{|z|}, \ \text{ for all } z \in \mathbb{C}. \tag{1.1}$$

In this article we prove that if the condition [SR] holds for an operator T then it also holds for the operator T^m, for any integer $m \geq 2$ (with constant depending on m). It means that if the condition (1.1) holds then there exists $M_m \geq 1$ such that

$$\|e^{zT^m}\| \leq M_m e^{|z|} \ \text{ for all } z \in \mathbb{C}. \tag{1.2}$$

Research partially supported by "TODEQ" MTKD-CT-2005-030042. The first author partially supported by grant 0107U000937, Ukraine.

2. Auxiliary results

We intend to use the Carleson embedding theorem [3, Ch. 2, § 3]. To this end, we need the following assertion.

Lemma 2.1. *Let $\alpha \in (0,1)$, $\sigma > 1$ and let γ be the Jordan curve in the complex plane defined by*

$$\gamma = \{\lambda^\alpha : \lambda = \sigma + is,\ s \in (-\infty, \infty)\},$$

where $\lambda^\alpha = |\lambda|^\alpha e^{i\alpha \arg \lambda}$, $\arg \lambda \in (-\pi/2, \pi/2)$. Then for any $s_0 \in \mathbb{R}$ and $h > 0$ the length $l(\gamma_{s_0}(h))$ of the curve $\gamma_{s_0}(h) = \gamma \cap \{\operatorname{Im} z \in (s_0, s_0 + h)\}$ satisfies the estimate

$$l(\gamma_{s_0}(h)) \le \frac{2\sqrt{2}}{\alpha} h. \tag{2.1}$$

Proof. We note that $\gamma = \{f(s) + ig(s) : s \in (-\infty, \infty)\}$, where the real-valued functions

$$f(s) = (\sigma^2 + s^2)^{\alpha/2} \cos(\alpha \arctan(s/\sigma)),$$
$$g(s) = (\sigma^2 + s^2)^{\alpha/2} \sin(\alpha \arctan(s/\sigma)), \tag{2.2}$$

satisfy $f(-s) = f(s)$, $g(-s) = -g(s)$, and for $s > 0$ they are positive and monotonically increasing.

It follows directly from the definition (2.2) that

$$g'(s) = \alpha(\sigma^2 + s^2)^{\alpha/2-1}[s \sin(\alpha \arctan(s/\sigma)) + \sigma \cos(\alpha \arctan(s/\sigma))],$$

and if $z(s) = f(s) + ig(s)$, then

$$|z'(s)| = (|f'(s)|^2 + |g'(s)|^2)^{1/2} = \alpha(\sigma^2 + s^2)^{\alpha/2-1/2}.$$

Thus we have the following equality

$$\frac{|z'(s)|}{g'(s)} = \frac{(\sigma^2 + s^2)^{1/2}}{[s \sin(\alpha \arctan(s/\sigma)) + \sigma \cos(\alpha \arctan(s/\sigma))]}.$$

From this and the inequalities $\cos(\alpha\pi/4) \ge 1/\sqrt{2}$, $\sin(\alpha\pi/4) \ge \alpha/2$, $\alpha \in (0,1)$, one derives the estimates

$$\frac{|z'(s)|}{g'(s)} \le \frac{\sqrt{2}}{\cos(\alpha\pi/4)} \le 2, \quad |s| \le \sigma, \quad \frac{|z'(s)|}{g'(s)} \le \frac{\sqrt{2}}{\sin(\alpha\pi/4)} \le \frac{2\sqrt{2}}{\alpha}, \quad |s| \ge \sigma,$$

so that the inequality

$$\frac{|z'(s)|}{g'(s)} \le \frac{2\sqrt{2}}{\alpha}, \qquad s \in \mathbb{R}, \tag{2.3}$$

is true.

Let $s_0 \in \mathbb{R}$ and let $s_2 > s_1$ be such that $g(s_1) = s_0$, $g(s_2) = s_0 + h$, and that the curve γ intersects the lines $\operatorname{Im} z = s_0$ and $\operatorname{Im} z = s_0 + h$ at the points $z(s_1)$ and $z(s_2)$, respectively. Then, in view of (2.3), we have

$$l(\gamma_{s_0}(h)) = \int_{s_1}^{s_2} |z'(s)| ds = \int_{s_1}^{s_2} \frac{|z'(s)|}{g'(s)} g'(s) ds \le \frac{2\sqrt{2}}{\alpha} \int_{s_1}^{s_2} g'(s) ds = \frac{2\sqrt{2}}{\alpha} h.$$

The proof of the lemma is finished. □

Suppose the plane Jordan curve γ belongs to the half-plane $\operatorname{Re} z > 0$. Recall [3, Ch. 1, § 5] that γ is called a Carleson curve, if there exists a constant $N(\gamma) > 0$ such that

$$l(\gamma \cap Q_{s_0}(h)) \leq N(\gamma)h$$

for all squares

$$Q_{s_0}(h) = \{z \in \mathbb{C} : \operatorname{Re} \in (0, h), \operatorname{Im} z \in (s_0, s_0 + h)\}.$$

From Lemma 2.1 we obtain the following assertion.

Corollary 2.2. *For any $m \in \mathbb{N}$, $m \geq 2$, $\sigma > 1$ the curve*

$$\Gamma_{m,\sigma} = \{\lambda^{\alpha_m} - \sigma^{\alpha_m} : \lambda = \sigma + is,\ s \in (-\infty, \infty)\}, \quad \alpha_m = 1/m,$$

is a Carleson curve such that

$$l(\Gamma_{m,\sigma} \cap Q_{s_0}(h)) \leq 2\sqrt{2}m\, h.$$

In the next considerations we need the following assertion (see [4]).

Lemma 2.3. *For an operator $S \in L(H)$ with the spectrum in the half-plane $\operatorname{Re} \lambda \leq 1$, the following assertions are equivalent:*

1) *the estimate*

$$\|e^{tS}\| \leq Me^t, \qquad t \geq 0, \tag{2.4}$$

is true;

2) *there exists $C > 0$ such that for all $x, y \in H$ the inequality*

$$\sup_{\sigma>1} (\sigma - 1) \int_{\sigma-i\infty}^{\sigma+i\infty} |(R^2(S, \lambda)x, y)|\, |d\lambda| \leq C\|x\|\|y\| \tag{2.5}$$

holds.

Furthermore, if the assertion 1) holds, then the estimate (2.5) is true with the constant $C = 2\pi M^2$, and if the assertion 2) holds, then the estimate (2.4) is true with the constant $M = eC/\pi$.

From Corollary 2.2 and Lemma 2.3, we deduce the following statement.

Lemma 2.4. *Let $S \in L(H)$ satisfy the estimate (2.4). Then its spectrum lies in the half-plane $\operatorname{Re} \lambda \leq 1$, and there exists an absolute constant $C_1 > 1$ such that for any $m \in \mathbb{N}$, $m \geq 2$, and for all $\sigma > 1$ we have*

$$J_{m,\sigma}(x; y) := (\sigma - 1) \int_{\sigma-i\infty}^{\sigma+i\infty} \frac{|(R^2(S, \lambda^{1/m})x, y)|}{|\lambda|^{2(1-1/m)}} |d\lambda|$$
$$\leq 8\pi M^2 C_1 m^3 \|x\|\|y\|, \qquad \forall\, x, y \in H, \tag{2.6}$$

$$G_{m,\sigma} := (\sigma - 1) \int_{\sigma-i\infty}^{\sigma+i\infty} \frac{\|(R(S, \lambda^{1/m})\|}{|\lambda|^{2-1/m}} |d\lambda| \leq \sqrt{2}\pi M(m + 1). \tag{2.7}$$

Proof. The Hille-Yosida theorem [5, Chap. 9] and (2.4) imply the first (spectral) assertion. If $m = 1$, then by Lemma 2.3

$$J_{1,\sigma}(x;y) = (\sigma - 1)\int_{\sigma-i\infty}^{\sigma+i\infty} |(R^2(S,\lambda)x,y)|\,|d\lambda| \le 2\pi M^2\|x\|\|y\|, \quad \sigma > 1. \quad (2.8)$$

Let now $m \ge 2$. By the change of the integration variable in (2.6):

$$z = \lambda^{1/m} = |\lambda|^{1/m} e^{im^{-1}\arg\lambda}, \ \ \arg\lambda \in (-\pi/2, \pi/2),$$

we get

$$\begin{aligned} J_{m,\sigma}(x) &= m(\sigma-1)\int_{\gamma_{m,\sigma}} \frac{|(R^2(S,z)x,y)|}{|z|^{m-1}}\,|dz| \\ &\le m\frac{(\sigma-1)\sigma^{1/m}}{\sigma}\int_{\gamma_{m,\sigma}} |(R^2(S,z)x,y))\,|dz|, \end{aligned} \quad (2.9)$$

where $\gamma_{m,\sigma} = \{z = \lambda^{\alpha_m} : \lambda = \sigma + is, \ s \in (-\infty,\infty)\}$, $\alpha_m = 1/m$. From (2.8) it follows that the function $(R^2(S, z + \sigma^{\alpha_m})x, y)$ is analytic for $\operatorname{Re} z > 0$ and belongs to the Hardy space H^1 in the half-plane $\operatorname{Re} z > 0$. Moreover

$$\|(R^2(S,z{+}\sigma^{\alpha_m})x,y)\|_{H^1} := \sup_{\beta>0}\int_{\beta-i\infty}^{\beta+i\infty} |(R^2(S,z{+}\sigma^{\alpha_m})x,y)|\,|dz| \le \frac{2\pi M^2\|x\|\|y\|}{\sigma^{\alpha_m}-1}.$$

Then, using Corollary 2.2, the Carleson embedding theorem and the estimate (2.9), we obtain

$$\begin{aligned} J_{m,\sigma}(x) &\le C_1 2\sqrt{2}m^2\frac{(\sigma-1)\sigma^{1/m}}{\sigma}\|(R^2(S,z+\sigma^{\alpha_m})x,y)\|_{H^1} \\ &\le 4\sqrt{2}\pi M^2 C_1\frac{m^2(\sigma-1)\sigma^{1/m}}{\sigma(\sigma^{1/m}-1)}\|x\|\|y\| \\ &\le 8\pi M^2 C_1 m^3\|x\|\|y\|, \qquad \sigma > 1, \ m \ge 2, \end{aligned}$$

where $C_1 > 1$ is the constant from the Carleson embedding theorem for the space H^1 on a half-plane. Hence the inequality (2.6) holds.

To derive (2.7), we note that the resolvent of S admits the estimate

$$\|R(S,\lambda)\| \le \frac{M}{\operatorname{Re}\lambda - 1}, \qquad \operatorname{Re}\lambda > 1.$$

Then, using the inequalities

$$\cos(\pi/(2m)) \ge 1/\sqrt{2}, \quad \sigma - 1 \le m\sigma(\sigma^{1/m}-1),$$

$$(s^2+\sigma^2)^{1/(2m)}\cos(m^{-1}\arctan(s/\sigma)) \ge \sigma^{1/m},$$

where $\sigma > 1$ and $m \geq 2$, we have for $\sigma > 1$:

$$\begin{aligned}G_{m,\sigma} &\leq M(\sigma-1)\int_{\sigma-i\infty}^{\sigma+i\infty}\frac{|d\lambda|}{(\operatorname{Re}\lambda^{1/m}-1)|\lambda|^{2-1/m}}\\ &\leq \frac{2M(\sigma-1)}{\cos(\pi/(2m))}\int_0^\infty \frac{(s^2+\sigma^2)^{1/(2m)}\cos(m^{-1}\arctan(s/\sigma))\,ds}{((s^2+\sigma^2)^{1/(2m)}\cos(m^{-1}\arctan(s/\sigma))-1)(s^2+\sigma^2)}\\ &\leq \frac{2M(\sigma-1)}{\cos(\pi/(2m))}\left(1+\frac{1}{\sigma^{1/m}-1}\right)\int_0^\infty\frac{ds}{s^2+\sigma^2}\\ &= M\pi\frac{(\sigma-1)}{\sigma\cos(\pi/(2m))}\left(1+\frac{1}{\sigma^{1/m}-1}\right) \leq M\sqrt{2}\pi(m+1).\end{aligned}$$

The proof of the lemma is complete. □

3. Main result

Theorem 3.1. *Let $m \in \mathbb{N}$, $m \geq 2$ be fixed. Let the operator $S \in L(H)$ satisfy the conditions:*

$$\|e^{\nu_k tS}\| \leq Me^t, \qquad t \geq 0,\ k=1,2,\dots,m, \tag{3.1}$$

where $\nu_k = e^{2\pi ik/m}$ are the m-th roots of the unity. Then there exists such a constant $M_m \geq 1$ that

$$\|e^{tS^m}\| \leq M_m e^t,\ t \geq 0. \tag{3.2}$$

Proof. From (3.1) it follows that the spectrum $\sigma(S)$ is contained in $\Omega_m := \{\lambda \in \mathbb{C} : \operatorname{Re}\nu_k\lambda \leq 1,\ k=1,2,\dots,m\}$, that $\Omega_2 = \{\lambda \in \mathbb{C} : |\operatorname{Re}\lambda| \leq 1\}$, and for $m > 2$ the set Ω_m is the regular m-angle polygon circumscribed to the circle $|\lambda| = 1$ with points of contact $\lambda = \nu_k$, $k = 1,2,\dots,m$. Hence $\mu^m \in \sigma(S^m)$ if and only if $\mu \in \sigma(S) \subset \Omega_m$.

Observe further that

$$R(S^m,\mu^m) = \mu^{-m}((S/\mu)^m - I)^{-1} = \frac{1}{m\mu^{m-1}}\sum_{k=1}^m R(\nu_k S,\mu), \qquad \mu \notin \Omega_m. \tag{3.3}$$

In particular, (3.3) holds for $\operatorname{Re}\mu^m > 1$. From (3.3) we obtain

$$\begin{aligned}R^2(S^m,\mu^m) &= \frac{1}{m\mu^{m-1}}\frac{d}{d\mu}R(S^m,\mu^m) \qquad (3.4)\\ &= \frac{1}{m^2\mu^{2(m-1)}}\left\{\sum_{k=1}^m R^2(\nu_k S,\mu) - \frac{m-1}{\mu}\sum_{k=1}^m R(\nu_k S,\mu)\right\},\quad \operatorname{Re}\mu^m > 1.\end{aligned}$$

Thus, using (3.4), we have

$$\begin{aligned}\int_{\sigma-i\infty}^{\sigma+i\infty}|(R^2(S^m,\lambda)x,y)||d\lambda| &\leq \frac{1}{m^2}\sum_{k=1}^m\int_{\sigma-i\infty}^{\sigma+i\infty}\frac{|(R^2(\nu_k S,\lambda^{1/m})x,y)|}{|\lambda|^{2(1-1/m)}}|d\lambda|\\ &+\frac{(m-1)}{m^2}\|x\|\|y\|\sum_{k=1}^m\int_{\sigma-i\infty}^{\sigma+i\infty}\frac{\|(R(\nu_k S,\lambda^{1/m})\|}{|\lambda|^{2-1/m}}|d\lambda|, \quad \sigma>1.\end{aligned}$$

Then, using Lemma 2.4 for the operators $\nu_k S$, $k = 1, 2, \ldots, m$, we obtain

$$(\sigma - 1) \int_{\sigma - i\infty}^{\sigma + i\infty} |(R^2(S_m, \lambda)x, y)||d\lambda| \leq \frac{2\pi M}{m}(4MC_1 m^3 + (m^2 - 1))\|x\|\|y\|$$
$$\leq 2\pi m^2 M(4MC_1 + 1)\|x\|\|y\|.$$

Then, according to Lemma 2.3, we conclude that the estimate (3.2) is true with

$$M_m = 2em^2 M(4MC_1 + 1). \tag{3.5}$$

□

The following theorem is the main result of the paper. It is an immediate corollary of Theorem 3.1.

Theorem 3.2. *Let T be a bounded linear operator on a Hilbert space H satisfying the strong Kreiss resolvent condition $[SR]$. Then for any $m \in \mathbb{N}$, $m \geq 2$, the operator T^m satisfies the condition $[SR]$ with the constant M_m defined by* (3.5).

Added in proof. A different approach, in the Banach space setting, also concerning the converse implication, is in preparation.

References

[1] J.C. Strikwerda, B.A. Wade, *A survey of the Kreiss matrix theorem for power bounded families of matrices and its extensions.* Linear Operators (eds. J. Janas, F.H. Szafraniec and J. Zemánek), Banach Center Publications **38** (Institute of Mathematics, Polish Academy of Sciences, Warsaw, 1997), 339–360.

[2] O. Nevanlinna, *Resolvent conditions and powers of operators.* Studia Mathematica. **145** (2001), no. 2, 113–134.

[3] J.B. Garnett, *Bounded Analytic Functions.* Academic Press, New York–London, 1981.

[4] A.M. Gomilko, *Conditions on the generator of uniformly bounded C_0-semigroup.* Funkts. Anal. Prilozhen. **33** (1999), no. 4, 66–69 (in Russian); English transl. in Funct. Anal. Appl. **33** (1999), no. 4, 294–296.

[5] T. Kato, *Perturbation Theory for Linear Operators.* Springer-Verlag, Berlin, 1966.

Alexander Gomilko
Institute of Telecommunications and Global Information Space
National Academy of Sciences of Ukraine
3, Chokolivsky Blvd., 03186 Kiev, Ukraine
e-mail: `alex@gomilko.com`

Jaroslav Zemánek
Institute of Mathematics
Polish Academy of Sciences
P.O. Box 21, 00-956 Warsaw, Poland
e-mail: `zemanek@impan.gov.pl`

Operator Theory:
Advances and Applications, Vol. 190, 243–265

Anatomy of the C^*-algebra Generated by Toeplitz Operators with Piece-wise Continuous Symbols

Sergei Grudsky and Nikolai Vasilevski

Abstract. We study the structure of the C^*-algebra generated by Toeplitz operators with piece-wise continuous symbols, putting a special emphasis on Toeplitz operators with unbounded symbols. We show that none of a finite sum of finite products of the initial generators is a compact perturbation of a Toeplitz operator. At the same time the uniform closure of the set of such sum of products contains a huge number of Toeplitz operators with bounded and unbounded symbols drastically different from symbols of the initial generators.

Mathematics Subject Classification (2000). Primary 47B35; Secondary 47C15.

Keywords. Toeplitz operator, Bergman space, piece-wise continuous symbol, unbounded symbol, C^*-algebra.

1. Preliminaries

In the paper we continue a detailed study of the C^*-algebra generated by Toeplitz operators T_a with piece-wise continuous symbols a acting on the Bergman space $\mathcal{A}^2(\mathbb{D})$ on the unit disk $\mathbb{D}$ in $\mathbb{C}$, which was initiated in [4, 6].

We start by recalling the necessary definitions and results of [4].

Let $\mathbb{D}$ be the unit disk on the complex plane and $\gamma = \partial\mathbb{D}$ be its boundary. Consider the space $L_2(\mathbb{D})$ with the standard Lebesgue plane measure $dv(z) = dxdy$, $z = x + iy \in \mathbb{D}$, and its Bergman subspace $\mathcal{A}^2(\mathbb{D})$ which consists of all functions analytic in $\mathbb{D}$. It is well known that the orthogonal Bergman projection B of $L_2(\mathbb{D})$ onto $\mathcal{A}^2(\mathbb{D})$ has the form

$$(B\varphi)(z) = \frac{1}{\pi}\int_{\mathbb{D}} \frac{\varphi(\zeta)\,dv(\zeta)}{(1 - z\overline{\zeta})^2}.$$

This work was partially supported by CONACYT Project 80503, México.

Given a function $a \in L_\infty$, the Toeplitz operator T_a with symbol a is defined as follows:

$$T_a : \varphi \in \mathcal{A}^2(\mathbb{D}) \longmapsto B(a\varphi) \in \mathcal{A}^2(\mathbb{D}).$$

As was already mentioned in [6], considering Toeplitz operators with piece-wise continuous symbols, it turns out that neither the curves supporting symbol discontinuities nor the number of such curves meeting at a boundary point of discontinuity play actually any essential role for the Toeplitz operator algebra studied. We can start from very different sets of symbols and obtain exactly the same operator algebra as a result. Thus, without loss of generality, we will use the same setup as in [4].

We fix a finite number of distinct points $T = \{t_1, \ldots, t_m\}$ on the boundary γ of the unit disk $\mathbb{D}$, and let

$$\delta = \min_{k \neq j}\{|t_k - t_j|, 1\}.$$

Denote by ℓ_k, $k = 1, \ldots, m$, the part of the radius of $\mathbb{D}$ starting at t_k and having length $\delta/3$; and let $\mathcal{L} = \bigcup_{k=1}^m \ell_k$. We denote by $PC(\overline{\mathbb{D}}, T)$ the set (algebra) of all functions piece-wise continuous on $\mathbb{D}$ which are continuous in $\overline{\mathbb{D}} \setminus \mathcal{L}$ and have one-sided limit values at every point of $\mathcal{L}$. In particular, every function $a \in PC(\overline{\mathbb{D}}, T)$ has at each point $t_k \in T$ two (different, in general) limit values:

$$a^-(t_k) = a(t_k - 0) = \lim_{\gamma \ni t \to t_k,\, t \prec t_k} a(t) \quad \text{and} \quad a^+(t_k) = a(t_k + 0) = \lim_{\gamma \ni t \to t_k,\, t \succ t_k} a(t),$$

where the signs $\pm$ correspond to the standard orientation of the boundary γ of $\mathbb{D}$.

For each $k = 1, \ldots, m$, we denote by $\chi_k = \chi_k(z)$ the characteristic function of the half-disk obtained by cutting $\mathbb{D}$ by the diameter passing through $t_k \in T$, and such that $\chi_k^+(t_k) = 1$, and thus $\chi_k^-(t_k) = 0$.

In [4] we define the functions $v_k = v_k(z)$, $k = 1, \ldots, m$, as follows. For each $k = 1, \ldots, m$, we introduce two neighborhoods of the point t_k:

$$V_k' = \{z \in \overline{\mathbb{D}} : |z - t_k| < \frac{\delta}{6}\} \quad \text{and} \quad V_k'' = \{z \in \overline{\mathbb{D}} : |z - t_k| < \frac{\delta}{3}\},$$

and fix a continuous function $v_k = v_k(z) : \overline{\mathbb{D}} \to [0, 1]$ such that

$$v_k|_{\overline{V_k'}} \equiv 1, \qquad v_k|_{\overline{\mathbb{D}} \setminus V_k''} \equiv 0.$$

But in this paper we need to make the functions $v_k = v_k(z)$ more specific. For each $k = 1, \ldots, m$, we introduce the Möbius transformation

$$\alpha_k(z) = i\frac{t_k - z}{z + t_k}, \tag{1.1}$$

which maps the unit disk $\mathbb{D}$ onto the upper half-plane Π, sending the point t_k to 0 and the opposite point $-t_k$ to ∞. We assume now that each $v_k = v_k(z)$ is a C^∞-function and that the function $v_k(\alpha_k^{-1}(z)) = \widehat{v}_k(r)$ depends only on r, the radial part of a point $z = re^{i\theta} \in \Pi$.

We denote by $\mathcal{T}(PC(\overline{\mathbb{D}},T))$ the C^*-algebra generated by all Toeplitz operators T_a whose symbols a belong to $PC(\overline{\mathbb{D}},T)$. It is well known that this algebra is irreducible and contains the entire ideal $\mathcal{K}$ of all compact on $\mathcal{A}^2(\mathbb{D})$ operators.

Recall that the main reason causing a quite complicated structure of the algebra $\mathcal{T}(PC(\overline{\mathbb{D}},T))$ was that the semi-commutator $[T_a,T_b) = T_aT_b - T_{ab}$, for $a,\, b \in PC(\overline{\mathbb{D}},T)$, is not compact in general (while the commutator $[T_a,T_b] = T_aT_b - T_bT_a$ is always compact).

This implies that the algebra $\mathcal{T}(PC(\overline{\mathbb{D}},T))$, apart from its initial generators T_a with $a \in PC(\overline{\mathbb{D}},T)$, contains all elements of the form

$$\sum_{k=1}^{p}\prod_{j=1}^{q_k} T_{a_{j,k}} \tag{1.2}$$

and the uniform limits of sequences of such elements.

In what follows we will need the description of the (Fredholm) symbol algebra $\operatorname{Sym}\mathcal{T}(PC(\overline{\mathbb{D}},T)) = \mathcal{T}(PC(\overline{\mathbb{D}},T))/\mathcal{K}$ of the algebra $\mathcal{T}(PC(\overline{\mathbb{D}},T))$, which we now proceed to characterize.

Let $\widehat{\gamma}$ be the boundary γ, cut at the points $t_k \in T$. The pair of points of $\widehat{\gamma}$ which correspond to the point $t_k \in T$, $k = 1,\dots,m$, will be denoted by $t_k - 0$ and $t_k + 0$, following the positive orientation of γ. Let $\overline{X} = \bigsqcup_{k=1}^{m} \Delta_k$ be the disjoint union of segments $\Delta_k = [0,1]_k$. Denote by Γ the union $\widehat{\gamma} \cup \overline{X}$ with the following point identification:

$$t_k - 0 \equiv 0_k, \qquad t_k + 0 \equiv 1_k,$$

where $t_k \pm 0 \in \widehat{\gamma}$, 0_k and 1_k are the boundary points of Δ_k, $k = 1,\dots,m$.

Theorem 1.1 ([7, 8, 9]). *The symbol algebra* $\operatorname{Sym}\mathcal{T}(PC(\overline{\mathbb{D}},T)) = \mathcal{T}(PC(\overline{\mathbb{D}},T))/\mathcal{K}$ *of the algebra* $\mathcal{T}(PC(\overline{\mathbb{D}},T))$ *is isomorphic and isometric to the algebra* $C(\Gamma)$. *The homomorphism*

$$\mathrm{sym}\,:\; \mathcal{T}(PC(\overline{\mathbb{D}},T)) \longrightarrow \operatorname{Sym}\mathcal{T}(PC(\overline{\mathbb{D}},T)) \cong C(\Gamma)$$

is generated by the mapping of generators of $\mathcal{T}(PC(\overline{\mathbb{D}},T))$

$$\mathrm{sym}\,: T_a \longmapsto \begin{cases} a(t), & t \in \widehat{\gamma} \\ a(t_k - 0)(1-x) + a(t_k+0)x, & x \in [0,1]_k \end{cases},$$

where $t_k \in T$, $k = 1,2,\dots,m$.

The following results were, in particular, obtained in [4].

Theorem 1.2. *Each operator $A \in \mathcal{T}(PC(\overline{\mathbb{D}},T))$ admits the canonical representations*

$$\begin{aligned} A &= T_{s_A} + \sum_{k=1}^{m} T_{v_k} f_{A,k}(T_{\chi_k}) T_{v_k} + K \\ &= T_{s_A} + \sum_{k=1}^{m} T_{u_k} f_{A,k}(T_{\chi_k}) + K' \\ &= T_{s_A} + \sum_{k=1}^{m} f_{A,k}(T_{\chi_k}) T_{u_k} + K'', \end{aligned}$$

where $u_k(z) = v_k(z)^2$; K, K', K'' are compact operators,

$$f_{A,k}(x) = (\operatorname{sym} A)|_{\Delta_k}, \qquad x \in [0,1]_k, \qquad k = 1,\dots,m,$$

$$s_A(t) = (\operatorname{sym} A)(t) - \sum_{k=1}^{m} v_k^2(t)[f_{A,k}(0)(1-\chi_k(t)) + f_{A,k}(1)\chi_k(t)]. \tag{1.3}$$

We mention that $s_A(t)$ is a function continuous on γ and that $s_A(t_k) = 0$ for all $t_k \in T$.

The next two theorems characterize Toeplitz operators with bounded measurable symbols in the algebra $\mathcal{T}(PC(\overline{\mathbb{D}},T))$.

Theorem 1.3. *An operator $A \in \mathcal{T}(PC(\overline{\mathbb{D}},T))$ is a compact perturbation of a Toeplitz operator if and only if each operator $f_{A,k}(T_{\chi_k})$, $k = 1,\dots,m$, is a Toeplitz operator.*

Theorem 1.4. *Let $A = T_a + K \in \mathcal{T}(PC(\overline{\mathbb{D}},T))$, thus, for each $k = 1,\dots,m$, the operator $f_{A,k}(T_{\chi_k})$ is Toeplitz, i.e., $f_{A,k}(T_{\chi_k}) = T_{a_k}$, for some $a_k \in L_\infty(\mathbb{D})$. Then the symbol a of the operator T_a is*

$$a(z) = s_A(z) + \sum_{k=1}^{m} a_k(z) v_k^2(z),$$

where $s_A(z)$ is given by (1.3).

The functions $f_{A,k}(x)$, $x \in [0,1]$, and the symbols $a_k(z)$ of the Toeplitz operators $T_{a_k} = f_{A,k}(T_{\chi_k})$, $k = 1,\dots,m$, are connected by the formula

$$f_{A,k}(x) = \frac{2x^2}{\pi} \frac{\ln(1-x) - \ln x}{(1-x) - x} \int_0^\pi \widehat{a}_k(\theta) \left(\frac{1-x}{x}\right)^{\frac{2\theta}{\pi}} d\theta,$$

where $\widehat{a}_k(\theta) = a_k(\alpha_k^{-1}(e^{i\theta}))$, where θ is the angular part of $z = re^{i\theta} \in \Pi$, and α_k is given by (1.1).

Anticipating and motivating a further study we give an example showing how monstrous the symbols of Toeplitz operators from $\mathcal{T}(PC(\overline{\mathbb{D}},T))$ can be.

Example. Consider the algebra $\mathcal{T}(PC(\overline{\mathbb{D}}, T_0))$ for the special case of the discontinuity set $T_0 = \{t_1, t_2\}$, where $t_2 = -t_1$. Then the Toeplitz operator

$$T_{\chi_1} = T_s + T_{\chi_1 v_1^2} + T_{(1-\chi_2)v_2^2} + K,$$

where $s(z)$ is a function continuous on $\mathbb{D}$ whose restriction on γ coincides with

$$\chi_1(t) - \chi_1(t)v_1^2(t) - (1-\chi_2(t))v_2^2(t) = \chi_1(t)(1 - v_1^2(t) - v_2^2(t))$$

and K is a compact operator, obviously belongs to the algebra $\mathcal{T}(PC(\overline{\mathbb{D}}, T_0))$. And thus for each function $f(x) \in C[0,1]$ the operator $f(T_{\chi_1})$ belongs to the algebra $\mathcal{T}(PC(\overline{\mathbb{D}}, T_0))$ as well.

Introduce the space $L_2(\Pi)$, with the usual Lebesgue plane measure, and its Bergman subspace $\mathcal{A}^2(\Pi)$ which consists of all functions analytic in Π. For each $t_k \in T$, the operator

$$(V_k\varphi)(z) = -\frac{2it_k}{(z+t_k)^2}\,\varphi\left(\alpha_k(z)\right) \tag{1.4}$$

is obviously the unitary operator both from $L_2(\Pi)$ onto $L_2(\mathbb{D})$, and from $\mathcal{A}^2(\Pi)$ onto $\mathcal{A}^2(\mathbb{D})$, and its inverse (and adjoint) has the form

$$(V_k^{-1}\varphi)(w) = -\frac{2it_k}{(w+i)^2}\,\varphi\left(\alpha_k^{-1}(w)\right).$$

It is obvious that

$$V_k T_{\chi_k} V_k^{-1} = T_{\chi_+},$$

where χ_+ is the characteristic function of the right quarter-plane in Π, and that this unitary equivalence implies that

$$f(T_{\chi_k}) = V_k^{-1} f(T_{\chi_+}) V_k. \tag{1.5}$$

Now for $t_1 \in T_0$, let $a_0(z)$ be a function on the unit disk such that

$$\widehat{a}_0(\theta) = a_0(\alpha_1^{-1}(e^{i\theta})) = (\sin\theta)^{-\beta}\sin(\sin\theta)^{-\alpha},$$

where $0 \le \beta < 1$ and $\alpha > 0$.

By Example 6.4 of [4] the Toeplitz operator $T_{\widehat{a}_0}$ is bounded on $\mathcal{A}^2(\Pi)$ and belongs to the algebra generated by T_{χ_+} Moreover for the function

$$f_0(x) = \frac{2x^2}{\pi}\,\frac{\ln(1-x) - \ln x}{(1-x) - x}\int_0^\pi (\sin\theta)^{-\beta}\sin(\sin\theta)^{-\alpha}\left(\frac{1-x}{x}\right)^{\frac{2\theta}{\pi}} d\theta,$$

which belongs to $C[0,1]$ and obeys the property $f_0(0) = f_0(1) = 0$, we have that $T_{\widehat{a}_0} = f_0(T_{\chi_+})$. Thus the Toeplitz operator

$$T_{a_0} = f_0(T_{\chi_1}) = V_1^{-1} f_0(T_{\chi_+}) V_1 = V_1^{-1} T_{\widehat{a}_0} V_1$$

belongs to the algebra $\mathcal{T}(PC(\overline{\mathbb{D}}, T_0))$.

We note that the symbol $a_0(z)$ is quite horrible, being *unbounded and oscillating* near every point of $\gamma \setminus T$ and having quite a complicated angular behavior approaching the points of T. At the same time the (Fredholm) symbol of the operator T_{a_0} has quite a respectable form:

$$\operatorname{sym} T_{a_0} = \begin{cases} 0, & t \in \widehat{\gamma} \\ f_0(x), & x \in \Delta_1 = [0,1] \\ f_0(1-x), & x \in \Delta_2 = [0,1] \end{cases}.$$

We describe now some results of [10] which we will use in the paper. Passing to polar coordinates on the upper half-plane Π we have

$$L_2(\Pi) = L_2(\mathbb{R}_+, rdr) \otimes L_2([0,\pi], d\theta) := L_2(\mathbb{R}_+, rdr) \otimes L_2(0,\pi).$$

We introduce two operators: the unitary operator

$$U = M \otimes I : L_2(\mathbb{R}_+, rdr) \otimes L_2(0,\pi) \longrightarrow L_2(\mathbb{R}) \otimes L_2(0,\pi),$$

where the Mellin transform $M : L_2(\mathbb{R}_+, rdr) \longrightarrow L_2(\mathbb{R})$ is given by

$$(M\psi)(\lambda) = \frac{1}{\sqrt{2\pi}} \int_{\mathbb{R}_+} r^{-i\lambda}\, \psi(r)\, dr,$$

and the isometric imbedding $R_0 : L_2(\mathbb{R}) \longrightarrow \mathcal{A}_1^2 \subset L_2(\mathbb{R} \times [0,\pi])$, which is given by

$$(R_0 f)(\lambda, \theta) = f(\lambda) \cdot \sqrt{\frac{2\lambda}{1 - e^{-2\pi\lambda}}}\; e^{-(\lambda+i)\theta}.$$

The adjoint operator $R_0^* : L_2(\mathbb{R} \times [0,\pi]) \longrightarrow L_2(\mathbb{R})$ has the form

$$(R_0^*\psi)(\lambda) = \sqrt{\frac{2\lambda}{1 - e^{-2\pi\lambda}}} \int_0^\pi \psi(\lambda,\theta)\, e^{-(\lambda-i)\theta}\, d\theta.$$

Now the operator $R = R_0^* U$ maps the space $L_2(\Pi)$ onto $L_2(\mathbb{R})$, and its restriction

$$R|_{\mathcal{A}^2(\Pi)} : \mathcal{A}^2(\Pi) \longrightarrow L_2(\mathbb{R})$$

is an isometric isomorphism. The adjoint operator

$$R^* = U^* R_0 : L_2(\mathbb{R}) \longrightarrow \mathcal{A}^2(\Pi) \subset L_2(\Pi)$$

is an isometric isomorphism of $L_2(\mathbb{R})$ onto the Bergman subspace $\mathcal{A}^2(\Pi)$ of the space $L_2(\Pi)$.

We have

$$RR^* = I : \; L_2(\mathbb{R}) \longrightarrow L_2(\mathbb{R}) \qquad \text{and} \qquad R^*R = B_\Pi : \; L_2(\Pi) \longrightarrow \mathcal{A}^2(\Pi),$$

where B_Π is the orthogonal Bergman projection of $L_2(\Pi)$ onto $\mathcal{A}^2(\Pi)$.

Denote by $H(L_1(0,\pi))$ the space of all functions homogeneous of zero order on the upper half-plane whose restrictions onto the upper half of the unit circle (angle parameterized by $\theta \in (0,\pi)$) belong to $L_1(0,\pi)$. Writing $a = a(\theta)$ we will often mean both a function from $L_1(0,\pi)$ and its homogeneous extension on the upper half-plane.

Theorem 1.5 ([10]). *Let $a = a(\theta) \in H(L_1(0,\pi))$ such that the Toeplitz operator T_a is bounded. Then T_a, acting on $\mathcal{A}^2(\Pi)$, is unitary equivalent to the multiplication operator $\gamma_a I = RT_aR^*$, acting on $L_2(\mathbb{R})$. The function $\gamma_a(\lambda)$ is given by*

$$\gamma_a(\lambda) = \frac{2\lambda}{1-e^{-2\pi\lambda}} \int_0^\pi a(\theta)\, e^{-2\lambda\theta}\, d\theta, \qquad \lambda \in \mathbb{R}. \tag{1.6}$$

In particular, for $a = \chi_+(\theta)$, we have (see [6])

$$\gamma_{\chi_+}(\lambda) = \frac{1}{e^{-\pi\lambda}+1}, \qquad \lambda \in \mathbb{R}, \tag{1.7}$$

and

$$T_{\chi_+} = R^* \gamma_{\chi_+}(\lambda) R.$$

We mention as well, see for details [6], that the C^*-algebra with identity $\mathcal{T}_+$ generated by the Toeplitz operator T_{χ_+} is isomorphic and isomorphic to $C(\overline{\mathbb{R}})$, and that this isomorphism is generated by the assignment

$$T_{\chi_+} \longmapsto \gamma_{\chi_+}(\lambda).$$

In particular this implies that for every Toeplitz operator T_a, with $a = a(\theta) \in H(L_1(0,\pi))$ in the algebra $\mathcal{T}_+$ the corresponding function $\gamma_a(\lambda)$, given by (1.6), must belong to $C(\overline{\mathbb{R}})$, where $\overline{\mathbb{R}} = \mathbb{R} \cup \{\pm\infty\}$ is the two-point compactification of $\mathbb{R}$.

We note that for general symbols $c = c(r,\theta)$ the Toeplitz operator T_c is no longer unitary equivalent to a multiplication operator. The operator RT_cR^* now has a much more complicated structure: it turns out to be a pseudodifferential operator with a certain compound (or double) symbol. The next theorem clarifies this statement for bounded symbols of a special and important case: $c = c(r,\theta) = a(\theta)v(r)$. The case of unbounded $a(\theta)$ will be treated in Theorem 2.2.

Theorem 1.6. *Given a bounded symbol $a(\theta)v(r)$, the Toeplitz operator T_{av} acting on $\mathcal{A}^2(\Pi)$ is unitary equivalent to the pseudodifferential operator $A_1 = RT_{av}R^*$, acting on $L_2(\mathbb{R})$. The operator A_1 is given by*

$$(A_1 f)(\lambda) = \frac{1}{2\pi} \int_{\mathbb{R}} d\xi \int_{\mathbb{R}} a_1(x,y,\xi) e^{i(x-y)\xi} f(y)dy, \qquad x \in \mathbb{R}, \tag{1.8}$$

where its compound symbol $a_1(x,y,\xi)$ has the form

$$a_1(x,y,\xi) = c(x,y)\, \gamma_a\left(\frac{x+y}{2}\right) \widetilde{v}(\xi)$$

with

$$c(x,y) = \frac{1-e^{-\pi(x+y)}}{x+y} \sqrt{\frac{2x}{1-e^{-2\pi x}}} \sqrt{\frac{2y}{1-e^{-2\pi y}}}, \tag{1.9}$$

and $\widetilde{v}(\xi) = v(e^{-\xi})$.

Proof. We have

$$\begin{aligned}
(A_1 f)(\lambda) &= (RT_{a(\theta)v(r)}R^* f)(\lambda) = (R(R^*R)a(\theta)v(r)(R^*R)R^* f)(\lambda)\\
&= ((RR^*)Ra(\theta)v(r)R^*(RR^*)f)(\lambda) = (Ra(\theta)v(r)R^* f)(\lambda)\\
&= (R_0^* a(\theta)(M\otimes I)v(r)(M^{-1}\otimes I)R_0 f)(\lambda)\\
&= \sqrt{\frac{2\lambda}{1-e^{-2\pi\lambda}}}\int_0^\pi e^{-(\lambda-i)\theta}a(\theta)d\theta\,\frac{1}{\sqrt{2\pi}}\int_{\mathbb{R}_+} r^{-i\lambda}v(r)dr\\
&\quad\cdot\frac{1}{\sqrt{2\pi}}\int_{\mathbb{R}} r^{i\alpha-1}f(\alpha)\sqrt{\frac{2\alpha}{1-e^{-2\pi\alpha}}}\,e^{-(\alpha+i)\theta}d\alpha\\
&= \frac{1}{\sqrt{2\pi}}\int_{\mathbb{R}_+} r^{-i\lambda}v(r)dr\,\frac{1}{\sqrt{2\pi}}\int_{\mathbb{R}} r^{i\alpha-1}f(\alpha)d\alpha\\
&\quad\cdot\sqrt{\frac{2\lambda}{1-e^{-2\pi\lambda}}}\sqrt{\frac{2\alpha}{1-e^{-2\pi\alpha}}}\int_0^\pi e^{-(\lambda+\alpha)\theta}\,a(\theta)d\theta.
\end{aligned}$$

The last integral gives

$$\int_0^\pi e^{-(\lambda+\alpha)\theta}\,a(\theta)d\theta = \frac{1-e^{-\pi(\lambda+\alpha)}}{\lambda+\alpha}\,\gamma_a\left(\frac{\lambda+\alpha}{2}\right),$$

and thus we have

$$\begin{aligned}
(A_1 f)(\lambda) &= (RT_{a(\theta)v(r)}R^* f)(\lambda)\\
&= \frac{1}{2\pi}\int_{\mathbb{R}_+} dr\int_{\mathbb{R}} c(\lambda,\alpha)\,\gamma_a\left(\frac{\lambda+\alpha}{2}\right)\,v(r)\,r^{-i(\lambda-\alpha)-1}f(\alpha)d\alpha,
\end{aligned}$$

where

$$c(\lambda,\alpha) = \frac{1-e^{-\pi(\lambda+\alpha)}}{\lambda+\alpha}\sqrt{\frac{2\lambda}{1-e^{-2\pi\lambda}}}\sqrt{\frac{2\alpha}{1-e^{-2\pi\alpha}}}. \tag{1.10}$$

Changing variables, $\lambda = x$, $\alpha = y$, and $r = e^{-\xi}$, we finally have

$$(A_1 f)(x) = \frac{1}{2\pi}\int_{\mathbb{R}} d\xi\int_{\mathbb{R}} a_1(x,y,\xi)e^{i(x-y)\xi}f(y)dy, \qquad x\in\mathbb{R},$$

with

$$a_1(x,y,\xi) = c(x,y)\,\gamma_a\left(\frac{x+y}{2}\right)\,\widetilde{v}(\xi),$$

where $c(x,y)$ is given by (1.10), and $\widetilde{v}(\xi) = v(e^{-\xi})$. □

2. Semi-commutators involving unbounded symbols

The following classical semi-commutator property

$$[T_a, T_b) = T_aT_b - T_{ab} \in \mathcal{K}, \qquad \text{for all} \quad a\in L_\infty(\mathbb{D}),\ b\in C(\overline{\mathbb{D}}),$$

played an essential role in [4]. The question on compactness of the semi-commutator for *unbounded* a is quite delicate and does not have any universal answer. In the next two examples we show that the compactness result is not valid for general, and even special, unbounded symbols a and arbitrary $b \in C(\overline{\mathbb{D}})$. At the same time we will prove it for a certain special (and important for us) case of unbounded symbols a and a special choice of $b \in C(\overline{\mathbb{D}})$.

The first example is a minor modification of Example 7 from [3], to which we refer for further details.

Example. Let

$$a(z) = a(r) = (1 - r^2)^{-\beta} \sin(1 - r^2)^{-\alpha} \in L_1(\mathbb{D})$$

and

$$b(z) = b(r) = (1 - r^2)^{\varepsilon} \sin(1 - r^2)^{-\alpha} \in C(\overline{\mathbb{D}})$$

where $z = re^{i\theta}$, $0 < \varepsilon < \beta < 1$. Then both T_a and T_b are bounded and compact.

The product ab has the form

$$a(r)b(r) = \frac{(1 - r^2)^{-(\beta-\varepsilon)}}{2} - \frac{(1 - r^2)^{-(\beta-\varepsilon)} \cos 2(1 - r^2)^{-\alpha}}{2} = c_1(r) - c_2(r).$$

Then the operator T_{c_1} is unbounded, while the operator T_{c_2} is compact. That is, the operator T_{ab} is not bounded, and the (unbounded) semi-commutator is not compact.

In what follows we will deal with the class of unbounded symbols which, considered in the upper half-plane setting, are the functions $a(\theta) \in H(L_1(0, \pi))$, where $z = re^{i\theta} \in \Pi$, for which the corresponding Toeplitz operators T_a are bounded.

The second example shows that even for such specific symbols $a(\theta)$ the semi-commutator is not compact for each $b(z) \in C(\overline{\Pi})$.

Example. Let

$$a(z) = a(\theta) = \theta^{-\beta} \sin \theta^{-\alpha}$$

and

$$b(z) = w(r)\, \theta^{\varepsilon} \sin \theta^{-\alpha},$$

where $z = re^{i\theta}$, $0 < \varepsilon < \beta < 1$, $\alpha > 0$, and $w(r)$ is a $[0, 1]$-valued C^∞-function such that

$$w(r) \equiv \begin{cases} 0, & r \in [0, \delta_1] \\ 1, & r \in [\delta_2, \delta_3] \\ 0, & r \in [\delta_4, +\infty] \end{cases},$$

and $0 < \delta_1 < \delta_2 < \delta_3 < \delta_4 < +\infty$.

The operator T_a is bounded by results of Example 6.3 of [4]; the operator T_b is bounded as well because of $b(z) \in C(\overline{\Pi})$. The product ab has the form

$$a(\theta)b(z) = \frac{w(r)\theta^{-\delta}}{2} - \frac{w(r)\theta^{-\delta} \cos 2\theta^{-\alpha}}{2} = c_1(z) - c_2(z),$$

where $\delta = \beta - \varepsilon \in (0, 1)$.

The Toeplitz operator T_{c_2} is bounded by Theorem 2.2. To prove that the semi-commutator $[T_a, T_b)$ is not compact, it is sufficient to show, for example, that the operator T_{c_1} is unbounded. Let $a_\delta(\theta) = \theta^{-\delta}$, then

$$\gamma_{a_\delta}(\lambda) = \frac{2\lambda}{1-e^{-2\pi\lambda}} \int_0^\pi \theta^{-\delta} e^{-2\lambda\theta} d\theta = \frac{(2\lambda)^\delta}{1-e^{-2\pi\lambda}} \int_0^{2\pi\lambda} u^{-\delta} e^{-u} du.$$

It is clear that if $\lambda \to +\infty$ then we have the asymptotics

$$\gamma_{a_\delta}(\lambda) = c_0\lambda^\delta + o(1), \tag{2.1}$$

$$\frac{\partial\gamma_{a_\delta}(\lambda)}{\partial\lambda} = \delta c_0 \lambda^{\delta-1} + o(1), \tag{2.2}$$

where $c_0 = 2^\delta\Gamma(1-\delta)$.

We will use now the representation (1.8) for the operator $A_1 = RT_{c_1}R^*$. Denoting

$$\widehat{w}(x-y) = \frac{1}{2\pi}\int_{\mathbb{R}} \widetilde{w}(\xi) e^{i(x-y)\xi}\, d\xi,$$

where $\widetilde{w}(\xi) = w(e^{-\xi})$, we have

$$(A_1 f)(x) = \int_{\mathbb{R}} c(x,y)\gamma_{a_\delta}\left(\frac{x+y}{2}\right)\widehat{w}(x-y) f(y)dy,$$

where the function $c(x,y)$ is given by (1.9).

We show now that the operator A_1 is unbounded on $L_2(\mathbb{R})$. Introduce the family of functions

$$f_{x_0}(y) = \begin{cases} \varepsilon^{-1/2}, & y \in I_\varepsilon = [x_0 - \varepsilon/2, x_0 + \varepsilon/2] \\ 0, & y \in \mathbb{R}\setminus I_\varepsilon \end{cases},$$

where $\varepsilon = \varepsilon(x_0) = x_0^{-\delta/2}$. It is clear that $\|f_{x_0}\|_{L_2(\mathbb{R})} = 1$.

Let $x \in I_\varepsilon$; denoting

$$K(x,y) = c(x,y)\gamma_{a_\delta}\left(\frac{x+y}{2}\right)\widehat{w}(x-y)$$

we have

$$\begin{aligned}(A_1 f_{x_0})(x) &= \varepsilon^{-1/2}\int_{x_0-\varepsilon/2}^{x_0+\varepsilon/2} K(x,y)dy \\ &= \varepsilon^{1/2}K(x,x) + \varepsilon^{-1/2}\int_{x_0-\varepsilon/2}^{x_0+\varepsilon/2} (K(x,y)-K(x,x))dy \\ &= I_1(x) + I_2(x).\end{aligned}$$

When $x_0 \to +\infty$, for the first summand we have

$$\begin{aligned}I_1(x) &= 1\cdot\gamma_{a_\delta}(x)\cdot\widehat{w}(0)\cdot\varepsilon^{1/2}(x_0) \\ &= \widehat{w}(0)c_0\left(x^\delta\cdot x_0^{-\delta/4} + o(1)\right) = \widehat{w}(0)c_0\left(x_0^{3\delta/4} + o(1)\right).\end{aligned} \tag{2.3}$$

As $\widetilde{w}(\xi) \geq 0$, we have that $\widetilde{w}(0) > 0$.

Now for the second summand we have

$$|I_2(x)| \le \varepsilon^{3/2} \sup_{y\in I_\varepsilon} \left|\frac{\partial K}{\partial y}(x,y)\right|.$$

Both functions $\frac{\partial c}{\partial y}(x,y)$ and $\frac{\partial \widehat{w}}{\partial y}(x-y)$ are uniformly bounded on x. The former is bounded by Theorem 4.2, while the latter is bounded as the Fourier transform of a function with a compact support. Thus we have that

$$|I_2(x)| \le \text{const}\, \varepsilon^{3/2} \sup_{y\in I_\varepsilon} \left(\left|\frac{\partial \gamma_{a_\delta}}{\partial y}\left(\frac{x+y}{2}\right)\right| + \left|\gamma_{a_\delta}\left(\frac{x+y}{2}\right)\right| \right).$$

Asymptotics (2.1) and (2.2) imply that for $x_0 \to +\infty$ we have

$$|I_2(x)| \le \text{const}\, \varepsilon^{3/2} x^\delta \le \text{const}\, \left(x_0^{-\delta/2}\right)^{3/2} x_0^\delta = \text{const}\, x_0^{\delta/4}. \tag{2.4}$$

Comparing (2.3) and (2.4), for sufficiently large x_0 and $x \in I_\varepsilon$, we have that

$$|(A_1 f_{x_0})(x)| \ge \frac{|\widehat{w}(0)|\, c_0}{2}\, x_0^{3\delta/4}.$$

Thus

$$\|A_1 f_{x_0}\|_{L_2(\mathbb{R})} \ge \left(\left(\frac{|\widehat{w}(0)|\, c_0}{2}\, x_0^{3\delta/4} \right)^2 \int_{x_0-\varepsilon/2}^{x_0+\varepsilon/2} dx \right)^{1/2}$$

$$\ge \text{const}\, \left(x_0^{3\delta/2} \cdot \varepsilon(x_0) \right)^{1/2} = \text{const}\, x_0^{\delta/2}.$$

This obviously yields unboundedness of the operator A_1, which in turn implies unboundedness of $T_a b$.

Now as a special choice of functions continuous on $\overline{\mathbb{D}}$ we select any $v_k(z)$, $k = 1, 2, \ldots, m$, considered in the upper half-plane setting as a function $v = v(r)$, where $z = re^{i\theta} \in \Pi$, as introduced in Section 1. That is, v is a $[0,1]$-valued C^∞-function such that for some $0 < \delta_1 < \delta_2 < +\infty$, we have

$$v(r) \equiv \begin{cases} 1, & r \in [0, \delta_1] \\ 0, & r \in [\delta_2, +\infty] \end{cases}. \tag{2.5}$$

Our aim is to prove that for each $a(\theta) \in H(L_1(0,\pi))$, for which the corresponding Toeplitz operator T_a is bounded, the semi-commutator $T_a T_v - T_{av}$ is compact. To do this we first represent the operators $T_a T_v$ and T_{av} in the form of pseudodifferential operators with certain compound (or double) symbols and then use the next result, which can be found, for example, in [5, Theorem 4.2 and Theorem 4.4].

Denote by $V(\mathbb{R})$ the set of all absolutely continuous functions on $\mathbb{R}$ of bounded total variation, and by $C_b(\mathbb{R}^2, V(\mathbb{R}))$ the set of all functions $a: \mathbb{R}^2 \times \mathbb{R} \to \mathbb{C}$ such that $u \mapsto a(u,\cdot)$ is a bounded continuous $V(\mathbb{R})$-valued function on $\mathbb{R}^2$. Then, for $a \in C_b(\mathbb{R}^2, V(\mathbb{R}))$, we define

$$cm_u^C(a) = \max\left\{ \|a(u+\Delta u,\cdot) - a(u,\cdot)\|_C :\ \Delta u \in \mathbb{R}^2,\ \|\Delta u\| \le 1 \right\},$$

and denote by $\mathcal{E}_2^C$ the subset of all functions in $C_b(\mathbb{R}^2, V(\mathbb{R}))$ such that the $V(\mathbb{R})$-valued function $u \mapsto a(u, \cdot)$ is uniformly continuous on $\mathbb{R}^2$ and the following conditions hold,

$$\lim_{\|u\|\to\infty} cm_u^C(a) = 0 \quad \text{and} \quad \lim_{|h|\to 0} \sup_{u\in\mathbb{R}^2} \|a(u,\cdot) - a^h(u,\cdot)\|_V = 0, \tag{2.6}$$

where $a^h(u,\cdot) = a(u, \xi + h)$, for all $(u, \xi) \in \mathbb{R}^2 \times \mathbb{R}$.

Theorem 2.1 ([5]). *If $\partial_\xi^j \partial_y^k a(x, y, \xi) \in C_b(\mathbb{R}\times\mathbb{R}, V(\mathbb{R}))$ for all $k, j = 0, 1, 2$, then the pseudodifferential operator A with compound symbol $a(x, y, \xi)$ defined on functions $f \in C_0^\infty(\mathbb{R})$ by the iterated integral*

$$(Af)(x) = \frac{1}{2\pi} \int_{\mathbb{R}} d\xi \int_{\mathbb{R}} a(x, y, \xi) e^{i(x-y)\xi} f(y) dy, \qquad x \in \mathbb{R}, \tag{2.7}$$

extends to a bounded linear operator on every Lebesgue space $L_p(\mathbb{R})$, $p \in (1, \infty)$.

If $\partial_\xi^j \partial_y^k a(x, y, \xi) \in \mathcal{E}_2^C$ for all $k, j = 0, 1, 2$, then the pseudodifferential operator (2.7) *with compound symbol*

$$r(x, y, \xi) = a(x, y, \xi) - a(x, x, \xi)$$

is compact on every Lebesgue space $L_p(\mathbb{R})$, $p \in (1, \infty)$.

Considering semi-commutators, we prove first that for our selection of symbols $a(\theta)$ and $v(r)$ the Toeplitz operator T_{av} is bounded.

Theorem 2.2. *For each $a(\theta) \in H(L_1(0, \pi))$ such that the Toeplitz operator T_a is bounded and the $[0, 1]$-valued C^∞-function $v = v(r)$ of the form* (2.5)*, the Toeplitz operator T_{av} is bounded on $\mathcal{A}^2(\Pi)$.*

Proof. We mention first that the boundedness of T_a is equivalent (by Theorem 1.5) to the boundedness of the corresponding function

$$\gamma_a(\lambda) = \frac{2\lambda}{1 - e^{-2\pi\lambda}} \int_0^\pi a(\theta)\, e^{-2\lambda\theta}\, d\theta, \qquad \lambda \in \mathbb{R}.$$

The C^∞-functions with compact support in $\mathbb{R}_+$ obviously form a dense set in $L_2(\mathbb{R}_+)$. Taking any such function f we consider

$$(A_1 f)(\lambda) = \frac{1}{2\pi} \int_{\mathbb{R}} d\xi \int_{\mathbb{R}} a_1(x, y, \xi) e^{i(x-y)\xi} f(y) dy, \qquad x \in \mathbb{R},$$

where the compound symbol $a_1(x, y, \xi)$ has the form

$$a_1(x, y, \xi) = c(x, y)\, \gamma_a\left(\frac{x+y}{2}\right) \widetilde{v}(\xi)$$

with

$$c(x, y) = \frac{1 - e^{-\pi(x+y)}}{x + y} \sqrt{\frac{2x}{1 - e^{-2\pi x}}} \sqrt{\frac{2y}{1 - e^{-2\pi y}}},$$

and $\widetilde{v}(\xi) = v(e^{-\xi})$. We note that $c(x, x) \equiv 1$.

The boundedness of the operator A_1 follows from Theorem 2.1, Theorems 4.1–4.4, and the fact that $\widetilde{v}(\xi)$ is a C^∞-function with a compact support.

By the calculations of Theorem 1.6 we have that $T_{av} = R^* A_1 R$. Thus the Toeplitz operator T_{av} is bounded on $\mathcal{A}^2(\Pi)$. □

Now we are ready to prove that the semi-commutator $T_a T_v - T_{av}$ is compact.

Theorem 2.3. *For each $a(\theta) \in H(L_1(0,\pi))$ such that the Toeplitz operator T_a is bounded and the $[0,1]$-valued C^∞-function $v = v(r)$ of the form (2.5), the semi-commutator $T_a T_v - T_{av}$ is compact.*

Proof. Calculation analogous to that of Theorem 1.6 yields

$$\begin{aligned}(A_2 f)(x) &= R T_a T_v R^* f = (RaR^*)(RvR^*)f \\ &= \gamma_a(x)\,(RvR^*)f = \frac{1}{2\pi}\int_{\mathbb{R}} d\xi \int_{\mathbb{R}} a_2(x,y,\xi) e^{i(x-y)\xi} f(y)dy, \qquad x \in \mathbb{R},\end{aligned}$$

with

$$a_2(x,y,\xi) = c(x,y)\,\gamma_a(x)\,\widetilde{v}(\xi),$$

where $c(x,y)$ is given by (1.9), and $\widetilde{v}(\xi) = v(e^{-\xi})$.

Thus the operator $R^*(T_{av} - T_a T_v)R = A_1 - A_2$ can be represented as a difference of two pseudodifferential operators having the compound symbols

$$\begin{aligned}r_1(x,y,\xi) &= a_1(x,y,\xi) - a_1(x,x,\xi) \\ &= c(x,y)\,\gamma_a\!\left(\frac{x+y}{2}\right)\widetilde{v}(\xi) - \gamma_a(x)\,\widetilde{v}(\xi)\end{aligned}$$

and

$$\begin{aligned}r_2(x,y,\xi) &= a_2(x,y,\xi) - a_2(x,x,\xi) \\ &= c(x,y)\,\gamma_a(x)\,\widetilde{v}(\xi) - \gamma_a(x)\,\widetilde{v}(\xi).\end{aligned}$$

The compactness of each of the last pseudodifferential operators easily follows from Theorem 2.1, Theorems 4.1–4.4, and the fact that $\widetilde{v}(\xi)$ is a C^∞-function with a compact support. Indeed, the above property of $\widetilde{v}(\xi)$ guarantees that both $a_1(x,y,\xi)$ and $a_2(x,y,\xi)$, as well as their two consecutive derivatives on ξ satisfy the second property in (2.6); while the properties

$$\lim_{(x,y)\to\infty} \frac{\partial^k d_{1,2}}{\partial y^k}(x,y) = 0, \qquad \text{for} \quad k = 1,2,$$

where $d_1(x,y) = c(x,y)\,\gamma_a\left(\frac{x+y}{2}\right)$ and $d_2(x,y) = c(x,y)\,\gamma_a(x)$ imply the first equality in (2.6). □

The above result leads directly to the following extension (of the sufficient part) of Theorem 1.3.

Corollary 2.4. *Let the operator $A \in \mathcal{T}(PC(\overline{\mathbb{D}},T))$ be such that in its canonical representation*

$$A = T_{s_A} + \sum_{k=1}^{m} T_{v_k} f_{A,k}(T_{\chi_k}) T_{v_k} + K$$

all operators $f_{A,k}(T_{\chi_k})$ *are Toeplitz with possibly unbounded symbols* a_k, $k = 1,\dots,m$, *correspondingly. Then* $A = T_a + K_A$ *is a compact perturbation of the Toeplitz operator* T_a, *where*

$$a(z) = s_A(z) + \sum_{k=1}^{m} a_k(z)v_k^2(z),$$

where $s_A(z)$ *is given by (1.3).*

We note that Corollary 2.4 immediately reveals, via property (1.5), many Toeplitz operators in $\mathcal{T}(PC(\overline{\mathbb{D}},T))$ having unbounded symbols. Indeed, recall in this connection the following result ([4, Theorem 6.2]).

For any L_1-symbol $a(\theta) \in H(L_1(0,\pi))$ we define the following averaging functions, corresponding to the endpoints of $[0,\pi]$,

$$C_a^{(1)}(\theta) = \int_0^{\theta} a(u)du, \qquad D_a^{(1)}(\theta) = \int_{\pi-\theta}^{\pi} a(u)du$$

and

$$C_a^{(p)}(\theta) = \int_0^{\theta} C_a^{(p-1)}(u)du, \qquad D_a^{(p)}(\theta) = \int_{\pi-\theta}^{\pi} D_a^{(p-1)}(u)du,$$

for each $p = 2,3,\dots$.

The next statement gives the conditions on some regular behavior of L_1-symbols near endpoints 0 and π guaranteeing that the corresponding Toeplitz operator is a certain continuous function of $T_{\chi+}$, and thus belongs to the algebra $\mathcal{T}_+$.

Theorem 2.5. *Let* $a(\theta) \in H(L_1(0,\pi))$ *and for some* $p,q \in \mathbb{N}$,

$$\lim_{\theta\to 0} \theta^{-p} C_a^{(p)}(\theta) = c_p \ (\in \mathbb{C}) \qquad \text{and} \qquad \lim_{\theta\to\pi} \theta^{-q} D_a^{(q)}(\theta) = d_q \ (\in \mathbb{C}). \quad (2.8)$$

Then $\gamma_a(\lambda) \in C(\overline{\mathbb{R}})$, *and thus* $T_a \in \mathcal{T}_+$.

The conditions (2.8) are obviously satisfied, with $p = q = 1$, for example, for any function $a(\theta) \in H(L_1(0,\pi))$ which has limits at the endpoints of $[0,\pi]$. Of course, the existence of symbol limits at the endpoints by no means is necessary for the Toeplitz operator T_a to be an element of $\mathcal{T}_+$. As the example on page 247 shows, the corresponding symbol can even be unbounded near each of the endpoints 0 and π. Many further particular symbols can be given, for example, by combining polynomial growth with logarithmic and iterated logarithmic growth, then by considering linear combinations of different symbols, etc. The following symbol may serve as an illustrative example,

$$a(\theta) = \sum_{k=1}^{n} c_k \theta^{-\beta_k} \ln^{\lambda_k} \theta^{-1} \sin\left(\theta^{-\alpha_k} \ln^{\mu_k} \theta^{-1}\right),$$

where $c_k \in \mathbb{C}$, $0 < \beta_k < 1$, $\alpha_k > 0$, $\lambda_k \in \mathbb{R}$, $\mu_k \in \mathbb{R}$, $k = 1,\dots,n$.

We mention especially that when speaking about a compact perturbation of a Toeplitz operator, say T_a, one should always remember that the coset $T_a + \mathcal{K}$

contains many Toeplitz operators of the form T_{a+k} for which the Toeplitz operator T_k is compact; and that all such operators have the same image $\operatorname{sym} T_{a+k} = \operatorname{sym} T_a$ in the (Fredholm) symbol algebra $\operatorname{Sym} \mathcal{T}(PC(\overline{\mathbb{D}}, T))$. At the same time the properties of the functions a and $a+k$ can be extremely different. Indeed, even having as nice as possible a, say $a \in C(\overline{\mathbb{D}})$, one can always add, for example, the function

$$k(z) = (1-r^2)^{-\beta} \sin(1-r^2)^{-\alpha} + (1-r)\chi_Q(z), \qquad z = re^{i\theta},$$

where the first summand is taken from the first example on page 251 and Q is the set of all points $z = r_1 + ir_2 \in \overline{\mathbb{D}}$ with rational r_1 and r_2. This converts the initial symbol a to the symbol $a+k$, which does not have a limit at every point of $\overline{\mathbb{D}}$, and moreover is unbounded near every point of the boundary.

That is, when speaking about the representation $A = T_a + K$ it is preferable to have a symbol a with fewer unnecessary singularities. It seems that the option given by Theorem 1.4 and Corollary 2.4 may be optimal in this respect.

3. Toeplitz or not Toeplitz

The key question in the description of Toeplitz operators in $\mathcal{T}(PC(\overline{\mathbb{D}}, T))$ is whether the operators of the form $f(T_{\chi_k})$, where $f(x) \in C[0,1]$ and $k = 1, 2, \dots, m$, are Toeplitz or not. By (1.5) this question reduces to the following question in the upper half-plane setting: given $f(x) \in C[0,1]$, whether the operator $f(T_{\chi_+})$ is Toeplitz or not. The last question is in turn equivalent to: whether the function $\gamma(\lambda) \in C(\overline{\mathbb{R}})$, which is connected with $f(x) \in C[0,1]$ by (see (1.7))

$$\gamma(\lambda) = f\left(\frac{1}{e^{-\pi\lambda}+1}\right),$$

admits the representation (1.6) for some $a(\theta) \in L_1(0,\pi)$, i.e.,

$$\gamma(\lambda) = \gamma_a(\lambda) = \frac{2\lambda}{1-e^{-2\pi\lambda}} \int_0^\pi a(\theta)\, e^{-2\lambda\theta}\, d\theta, \qquad \lambda \in \overline{\mathbb{R}}. \tag{3.1}$$

The statements of the next theorem are necessary for the existence of the above representation for a given function $\gamma(\lambda) \in C(\overline{\mathbb{R}})$.

Theorem 3.1. *Let $a(\theta) \in L_1(0,\pi)$. Then the function $\gamma_a(\lambda)$ is analytic in the whole complex plane with the exception of the points $\lambda_n = in$, where $n = \pm 1, \pm 2, \dots$, where $\gamma_a(\lambda)$ has simple poles. Moreover, for any fixed and sufficiently small δ the function $\gamma_a(\lambda)$ admits, on the set*

$$\mathbb{C} \setminus \bigcup_{\mathbb{Z}\setminus\{0\}} K_n(\delta), \qquad \text{where} \qquad K_n(\delta) = \{\lambda \in \mathbb{C} : \ |\lambda - in| < \delta\},$$

the estimate

$$|\gamma_a(\lambda)| \le \mathrm{const}\, |\lambda|,$$

where const *depends on δ.*

Proof. The function

$$\beta_a(\lambda) = \int_0^\pi a(\theta)\, e^{-2\lambda\theta}\, d\theta, \qquad \lambda = x + iy,$$

is obviously analytic in $\mathbb{C}$, and for large $|\lambda|$ admits the estimate

$$|\beta_a(\lambda)| \leq \int_0^\pi |a(\theta)|\, e^{-2x\theta}\, d\theta.$$

Thus for $x > 0$ we have

$$|\beta_a(\lambda)| \leq \|a(\theta)\|_{L_1},$$

while for $x < 0$ we have

$$\begin{aligned}|\beta_a(\lambda)| &\leq e^{-2\pi x} \int_0^\pi |a(\theta)|\, e^{2x(\pi-\theta)}\, d\theta \\ &= e^{-2\pi x} \int_0^\pi |a(\theta)|\, d\theta \leq e^{-2\pi x}\, \|a(\theta)\|_{L_1}.\end{aligned}$$

The theorem statements now follow from

$$\gamma_a(\lambda) = \frac{2\lambda}{1 - e^{-2\pi\lambda}}\, \beta_a(\lambda).$$

□

To give a sufficient condition for the representation (3.1) we start with some definitions (see [1] for details).

An entire function $\varphi(\lambda)$ is called a function of exponential type if it obeys an estimate

$$|\varphi(\lambda)| \leq A e^{B|\lambda|},$$

where the positive constants A and B do not depend on $\lambda \in \mathbb{C}$. The infimum of all constants B for which this estimate holds is called the type of the function $\varphi(\lambda)$.

We denote by $\mathcal{L}_2^\sigma$ the set of all functions of exponential type less than or equal to σ whose restrictions to $\mathbb{R}$ belong to $L_2(\mathbb{R})$.

An analytic function on the upper half-plane $\varphi(\lambda)$ is said to belong to the Hardy space $H^2(\mathbb{R})$ if

$$\sup_{y>0} \int_{\mathbb{R}} |\varphi(x+iy)|^2 dx < \infty.$$

The proof of the next theorem can be found, for example, in [1, Theorem 1.4].

Theorem 3.2. *Let $\varphi(z) \in \mathcal{L}_2^{2\pi} \cap H^2(\mathbb{R})$. Then there exists a function $a(\theta) \in L_2(0, 2\pi)$ such that*

$$\varphi(z) = \int_0^{2\pi} a(\theta)\, e^{iz\theta}\, d\theta, \qquad \lambda \in \mathbb{C}.$$

As $L_2(0, 2\pi) \subset L_1(0, 2\pi)$, the theorem can be used as a sufficient condition for the existence of representation (3.1). Indeed, given a function $\gamma(\lambda)$, introduce

$$\varphi(z) = i \frac{1 - e^{i\pi z}}{z}\, \gamma\left(-\frac{iz}{2}\right).$$

If this function $\varphi(z)$ belongs to $\mathcal{L}_2^{2\pi} \cap H^2(\mathbb{R})$ then $\gamma(\lambda)$ does admit representation (3.1). That is, there is a function $a(\theta) \in L_1(0, 2\pi)$ such that $\gamma(\lambda) = \gamma_a(\lambda)$ and

$$T_a = R^*\gamma(\lambda)R = f(T_{\chi+}),$$

where

$$f(x) = \gamma\left(\gamma_{\chi+}^{-1}(x)\right) = \gamma\left(-\frac{1}{\pi}\ln\frac{1-x}{x}\right).$$

Theorem 3.3. *Let*

$$p(x) = \sum_{k=1}^{n} a_k x^k, \qquad a_n \neq 0,$$

be a polynomial of degree $n \geq 2$ with complex coefficients. Then the bounded operator $p(T_{\chi+})$ is not a Toeplitz operator.

Proof. The operator $p(T_{\chi+})$ belongs to the algebra generated by all Toeplitz operators on the upper half-plane with homogeneous L_∞-symbols $a(\theta)$ of zero order. Thus by [2] the operator $p(T_{\chi+})$ being Toeplitz must have a symbol which belongs to $H(L_1(0, \pi))$. The corresponding function $\gamma(\lambda)$, that is, such that $p(T_{\chi+}) = R^*\gamma(\lambda)R$, obviously has the form

$$\gamma(\lambda) = p\left(\gamma_{\chi+}(\lambda)\right) = p\left(\frac{1}{e^{-\pi\lambda}+1}\right).$$

But this function has poles of order n at the points $\lambda_n = i(2n-1)$, where $n \in \mathbb{Z}$. Thus by Theorem 3.1 there is no function $a(\theta) \in H(L_1(0, \pi))$ for which the representation (3.1) holds. □

Corollary 3.4. *Let A be an operator of the algebra $\mathcal{T}(PC(\overline{\mathbb{D}}, T))$ having the form*

$$A = \sum_{i=1}^{p} \prod_{j=1}^{q_i} T_{a_{i,j}},$$

where all $a_{i,j} \in PC(\overline{\mathbb{D}}, T)$. Then A is a compact perturbation of a Toeplitz operator if and only if A is a compact perturbation of one of the initial generators of $\mathcal{T}(PC(\overline{\mathbb{D}}, T))$, which is a Toeplitz operator T_a with $a \in PC(\overline{\mathbb{D}}, T)$.

Proof. By Corollary 4.3 of [4], or Theorem 1.2 of this paper, the operator A admits the canonical representation

$$A = \sum_{i=1}^{p} \prod_{j=1}^{q_i} T_{a_{i,j}} = T_{s_A} + \sum_{k=1}^{m} T_{v_k} p_{A,k}(T_{\chi_k}) T_{v_k} + K_A,$$

where $s_A = s_A(z) \in C(\overline{\mathbb{D}})$, $p_{A,k} = p_{A,k}(x)$, $k = 1, \dots, m$, are some polynomials, and K_A is a compact operator. Thus by Theorem 1.3, A is a compact perturbation of a Toeplitz operator if and only if each $p_{A,k}(T_{\chi_k})$, $k = 1, \dots, m$, is a Toeplitz operator, or by (1.5) if and only if each $p_{A,k}(T_{\chi+})$, $k = 1, \dots, m$, is a Toeplitz operator. By Theorem 3.3 the last statement is equivalent to the fact that the degree of each polynomial $p_{A,k}(x)$, $k = 1, \dots, m$, must be less than or equal to one, which in turn is equivalent to the fact that A is a compact perturbation of a Toeplitz operator T_a with $a \in PC(\overline{\mathbb{D}}, T)$. □

We summarize now the results obtained on Toeplitz operators of the algebra $\mathcal{T}(PC(\overline{\mathbb{D}}, T))$. By its construction, the C^*-algebra $\mathcal{T}(PC(\overline{\mathbb{D}}, T))$ consists of its initial generators, Toeplitz operators T_a with symbols $a \in PC(\overline{\mathbb{D}}, T)$, then of all elements of the form

$$\sum_{i=1}^{p} \prod_{j=1}^{q_i} T_{a_{i,j}},$$

forming thus a nonclosed algebra, and finally of all elements of the uniform closure of the nonclosed algebra. The information on Toeplitz operators is as follows.

- All initial generators are Toeplitz operators.
- None of the elements of the nonclosed algebra which does not reduce to a compact perturbation of an initial generator can be (a compact perturbation of) a Toeplitz operator. Thus at this stage we have not increased the quantity of Toeplitz operators.
- The uniform closure of the nonclosed algebra contains a huge amount of Toeplitz operators, with bounded and even unbounded symbols, which are drastically different from the initial generators. All these Toeplitz operators are uniform limits of sequences of non-Toeplitz operators.
- The uniform closure, apart from Toeplitz operators, contains many more non-Toeplitz operators (this is a consequence of Theorem 3.1).

At the same time each operator in the C^*-algebra $\mathcal{T}(PC(\overline{\mathbb{D}}, T))$ admits a very transparent canonical representation (given in Theorem 1.2).

4. Appendix: Technical statements

We prove here several statements whose results were used in Theorems 2.2 and 2.3.

We start with some properties of the function (see (1.9))

$$c(x,y) = \frac{1-e^{-\pi(x+y)}}{x+y}\sqrt{\frac{2x}{1-e^{-2\pi x}}}\sqrt{\frac{2y}{1-e^{-2\pi y}}}, \qquad x, y \in \mathbb{R}.$$

Theorem 4.1. *The function $c(x,y)$ is bounded in $\mathbb{R}^2$; i.e.,*

$$\sup_{(x,y)\in\mathbb{R}^2} |c(x,y)| < \infty.$$

Proof. Introduce the function

$$f(u) = \sqrt{\frac{u}{1-e^{-u}}}.$$

Then

$$c(x,y) = \frac{f(2\pi x)\, f(2\pi y)}{f^2(\pi(x+y))}.$$

Let $D_1 = [1, +\infty)$ and $D_{-1} = (-\infty, -1]$. We obviously have the following asymptotics in the above domains:

$$f(u) = u^{1/2}\left(1 + O(e^{-u})\right), \qquad u \in D_1, \tag{4.1}$$

$$f(u) = |u|^{1/2} e^{u/2}\left(1 + O(e^{u})\right), \qquad u \in D_{-1}, \tag{4.2}$$

$$f^{-2}(u) = u^{-1}\left(1 + O(e^{-u})\right), \qquad u \in D_1, \tag{4.3}$$

$$f^{-2}(u) = |u|^{-1} e^{-u}\left(1 + O(e^{u})\right), \qquad u \in D_{-1}. \tag{4.4}$$

In what follows the relation $\varphi(u) \sim \psi(u)$ means that

$$0 < c \leq \frac{\varphi(u)}{\psi(u)} \leq C < \infty,$$

for all u in the domain under consideration. We note as well that if u belongs to any bounded domain in $\mathbb{R}$, then

$$f(u) \sim 1 \qquad \text{and} \qquad f^{-2}(u) \sim 1. \tag{4.5}$$

We will prove the statement of the theorem considering successively all possible locations of x and y on $\mathbb{R}$. The symmetry of $c(x, y)$ with respect to its arguments implies that it is sufficient to consider only the following cases:

1. $x, y \in [-1, 1]$. Then $x + y \in [-2, 2]$, and by (4.5) we have that $c(x, y) \sim 1$.
2. $x \in [-1, 1]$, $y \in D_1$. Then either $x + y \in D_1$ and thus by (4.2) and (4.4) we have
$$c(x, y) \sim \frac{1 \cdot (2\pi y)^{1/2}}{\pi(x + y)} \sim y^{-1/2},$$
or $y \in [1, 2]$ and thus, as in the first case, $c(x, y) \sim 1$.
3. $x \in [-1, 1]$, $y \in D_{-1}$. Then either $x + y \in D_{-1}$ and thus by (4.1) and (4.3) we have
$$c(x, y) \sim \frac{1 \cdot |2\pi y|^{1/2} e^{\pi y}}{|\pi y| e^{\pi(x+y)}} \sim y^{-1/2},$$
or $y \in [-2, -1]$ and again $c(x, y) \sim 1$.
4. $x \in D_1$, $y \in D_{-1}$. Then we have the following three possibilities for $x + y$:
   (a) $x + y \in [-1, 1]$. Then by (4.1), (4.2), and (4.5), assuming that $x + y = \delta \in [-1, 1]$, we have
$$c(x, y) \sim (2\pi x)^{1/2} \cdot (2\pi |y|)^{1/2} e^{\pi y} \cdot 1 \sim (\delta + y)^{1/2} |y|^{1/2} e^{\pi y} \sim |y| e^{\pi y}.$$
   (b) $x + y \in D_1$. Then by (4.1), (4.2), and (4.3) we have
$$c(x, y) \sim \frac{(2\pi x)^{1/2} \cdot (2\pi |y|)^{1/2} e^{\pi y}}{\pi(x + y)} \sim \begin{cases} \frac{x^{1/2} |y|^{1/2} e^{\pi y}}{x}, & x \geq 2|y| \\ \frac{|y|^{1/2} |y|^{1/2} e^{\pi y}}{1}, & x < 2|y| \end{cases}$$
$$\sim \begin{cases} \frac{|y|^{1/2} e^{\pi y}}{x^{1/2}}, & x \geq 2|y| \\ |y| e^{\pi y}, & x < 2|y| \end{cases}.$$

(c) $x+y \in D_{-1}$. Then by (4.1), (4.2), and (4.4) we have

$$c(x,y) \sim \frac{(2\pi x)^{1/2} \cdot (2\pi|y|)^{1/2} e^{\pi y}}{\pi|x+y|e^{\pi(x+y)}} \sim \frac{x^{1/2}|y|^{1/2}e^{-\pi x}}{|x+y|}$$
$$\sim \begin{cases} \frac{|x|^{1/2}e^{-\pi x}}{y^{1/2}}, & |y| \geq 2x \\ xe^{-\pi x}, & |y| < 2x \end{cases}.$$

5. $x \in D_1$, $y \in D_1$. Then $x+y \in D_1$ and thus by (4.1) and (4.3) we have

$$c(x,y) = \frac{(2\pi x)^{1/2}(2\pi y)^{1/2}}{\pi(x+y)} \left(1 + O(e^{-x}) + O(e^{-y})\right)$$
$$= \frac{2x^{1/2}y^{1/2}}{x+y} \left(1 + O(e^{-x}) + O(e^{-y})\right).$$

As $2x^{1/2}y^{1/2} \leq x+y$, the boundedness of $c(x,y)$ is obvious.

6. $x \in D_{-1}$, $y \in D_{-1}$. Then $x+y \in D_{-1}$ and thus by (4.2) and (4.4) we have

$$c(x,y) = \frac{(2\pi|x|)^{1/2}e^{\pi x}(2\pi|y|)^{1/2}e^{\pi y}}{\pi|x+y|e^{\pi(x+y)}} \left(1 + O(e^{-|x|}) + O(e^{-|y|})\right)$$
$$= \frac{2|x|^{1/2}|y|^{1/2}}{|x|+|y|} \left(1 + O(e^{-|x|}) + O(e^{-|y|})\right).$$

The theorem is proved. □

Theorem 4.2. *Both functions $\frac{\partial c}{\partial x}(x,y)$ and $\frac{\partial c}{\partial y}(x,y)$ are bounded in $\mathbb{R}^2$, and moreover*

$$\lim_{(x,y)\to\infty} \frac{\partial c}{\partial x}(x,y) = 0 \qquad \text{and} \qquad \lim_{(x,y)\to\infty} \frac{\partial c}{\partial y}(x,y). \tag{4.6}$$

Proof. We start with the asymptotics of the derivatives of $f(u)$. We have

$$f'(u) = \frac{1}{2}\sqrt{\frac{1-e^{-u}}{u}} \cdot \frac{1-e^{-u}+ue^{-u}}{(1-e^{-u})^2},$$

thus, as is easy to see,

$$f'(u) = u^{-1/2}\left(1 + O(ue^{-u})\right), \qquad u \in D_1, \tag{4.7}$$
$$f'(u) = |u|^{1/2}e^{u/2}\left(1 + O(e^{u})\right), \qquad u \in D_{-1}. \tag{4.8}$$

Then

$$\frac{\partial c}{\partial y}(x,y) = f(2\pi x)\frac{2\pi f'(2\pi y)f^2(\pi(x+y)) - 2\pi f(2\pi y)f(\pi(x+y))f'(\pi(x+y))}{f^4(\pi(x+y))}$$
$$= 2\pi\, c(x,y)\left(\frac{f'(2\pi y)}{f(2\pi y)} - \frac{f'(\pi(x+y))}{f(\pi(x+y))}\right). \tag{4.9}$$

We check now the boundedness of the logarithmic derivative of f. By (4.1) and (4.7) we have

$$\begin{aligned}\frac{f'(2\pi y)}{f(2\pi y)} &= \frac{(2\pi y)^{-1/2}}{(2\pi y)^{1/2}}\left(1+O(ye^{-2\pi y})\right) \\ &= (2\pi y)^{-1}\left(1+O(ye^{-2\pi y})\right), \qquad y\in D_1,\end{aligned}$$

and by (4.2) and (4.8),

$$\frac{f'(2\pi y)}{f(2\pi y)} = \frac{(2\pi|y|)^{1/2}e^{\pi y}}{(2\pi|y|)^{1/2}e^{\pi y}}\left(1+O(e^{2\pi y})\right) = 1+O(e^{2\pi y}), \qquad y\in D_{-1}.$$

Thus the function $\frac{\partial c}{\partial y}(x,y)$ is bounded in $\mathbb{R}^2$.

To prove the second equality in (4.6) we note that if both $|y|\to\infty$, and $|x+y|\to\infty$, and moreover $\operatorname{sign} y = \operatorname{sign}(x+y)$ then the result follows from (4.9) and boundedness of $c(x,y)$. If both $|y|\to\infty$, and $|x+y|\to\infty$, but $\operatorname{sign} y = -\operatorname{sign}(x+y)$ then by case 4.b of Theorem 4.1 we have that $c(x,y)\to 0$. If $(x,y)\to\infty$ while y belongs to a bounded domain, then $x+y$ is unbounded and we are in the situation of the cases 2 or 3 of Theorem 4.1, when $c(x,y)\to 0$. Finally, if $(x,y)\to\infty$ but $x+y$ is bounded, then as in the case 4.a of Theorem 4.1 we have that $c(x,y)\to 0$. In the last three cases the result follows from (4.9), boundedness of the logarithmic derivatives, and $c(x,y)\to 0$.

Boundedness of $\frac{\partial c}{\partial x}(x,y)$ and the first equality in (4.6) follow from the above and the symmetry of $c(x,y)$ with respect to x and y. □

Theorem 4.3. *The function $\frac{\partial^2 c}{\partial y^2}(x,y)$ is bounded in $\mathbb{R}^2$, and moreover*

$$\lim_{(x,y)\to\infty}\frac{\partial^2 c}{\partial y^2}(x,y) = 0.$$

Proof. For the second derivative of f, after elementary calculations, we have

$$f''(u) = u^{-3/2}\left(1+O(ue^{-u})\right), \qquad u\in D_1, \tag{4.10}$$

$$f''(u) = |u|^{1/2}e^{u/2}\left(1+O(e^{u})\right), \qquad u\in D_{-1}. \tag{4.11}$$

Differentiating (4.9) we have

$$\begin{aligned}\frac{\partial^2 c}{\partial y^2}(x,y) &= 2\pi\frac{\partial c}{\partial y}(x,y)\left(\frac{f'(2\pi y)}{f(2\pi y)} - \frac{f'(\pi(x+y))}{f(\pi(x+y))}\right) \\ &\quad - 2\pi^2\, c(x,y)\left[2\left(\frac{f''(2\pi y)}{f(2\pi y)} - \left(\frac{f'(2\pi y)}{f(2\pi y)}\right)^2\right)\right. \\ &\quad \left. - \left(\frac{f''(\pi(x+y))}{f(\pi(x+y))} - \left(\frac{f'(\pi(x+y))}{f(\pi(x+y))}\right)^2\right)\right].\end{aligned}$$

We note that by Theorem 4.2 the first summand is bounded and tends to 0 as $(x,y)\to\infty$. Considering the second summand we have that if $|x+y|\to\infty$ and $y\to\infty$ then formulas (4.1), (4.2), (4.7), (4.8), (4.10), and (4.11), for $u=2\pi y$ or

$u = \pi(x+y)$, yield

$$\frac{f''(u)}{f(u)} - \left(\frac{f'(u)}{f(u)}\right)^2 = \begin{cases} u^{-2}\left(1+O(ue^{-u})\right) - u^{-2}\left(1+O(ue^{-u})\right), & u \in D_1 \\ \left(1+O(e^{u})\right) - \left(1+O(e^{u})\right), & u \in D_{-1} \end{cases}$$
$$= \begin{cases} O(u^{-1}e^{-u}), & u \in D_1 \\ O(e^{u}), & u \in D_{-1} \end{cases}.$$

If $(x,y) \to \infty$, but $|x+y|$ is bounded, then we are in the situation of the case 4.a of Theorem 4.1, and thus $c(x,y) \to 0$ as $(x,y) \to \infty$. Finally, if $(x,y) \to \infty$, but y is bounded, then we are in the situation of the cases 2 or 3 of Theorem 4.1, and thus $c(x,y) \to 0$ as $(x,y) \to \infty$. □

Theorem 4.4. *Let $a(\theta) \in L_1(0,\pi)$ be such that*

$$\gamma_a(\lambda) = \frac{2\lambda}{1-e^{-2\pi\lambda}} \int_0^\pi a(\theta) e^{-2\lambda\theta}\, d\theta \in L_\infty(\mathbb{R}).$$

Then, for each $j = 1, 2, \ldots,$

$$\lim_{\lambda \to \pm\infty} \frac{d^j \gamma_a}{d\lambda^j}(\lambda) = 0.$$

Proof. Let $j = 1$. Then

$$\frac{d\gamma_a}{d\lambda}(\lambda) = \frac{2}{1-e^{-2\pi\lambda}} \int_0^\pi a(\theta) e^{-2\lambda\theta}\, d\theta + \frac{4\pi\lambda e^{-2\pi\lambda}}{(1-e^{-2\pi\lambda})^2} \int_0^\pi a(\theta) e^{-2\lambda\theta}\, d\theta$$
$$- \frac{4\lambda}{1-e^{-2\pi\lambda}} \int_0^\pi \theta a(\theta) e^{-2\lambda\theta}\, d\theta = I_1(\lambda) + I_2(\lambda) + I_3(\lambda).$$

We consider first the behaviour of the derivative when $\lambda \to +\infty$. We have

$$I_1(\lambda) = \frac{2}{\lambda}\, \gamma_a(\lambda) \qquad \text{and} \qquad I_2(\lambda) = \frac{2\pi e^{-2\pi\lambda}}{1-e^{-2\pi\lambda}}\, \gamma_a(\lambda),$$

and thus the first two summands tend to 0 as $\lambda \to +\infty$.

Consider now the last summand,

$$|I_3(\lambda)| \le 2 \int_0^\delta |a(\theta)| (2\lambda\theta) e^{-2\lambda\theta} d\lambda + 4\pi\lambda e^{-2\lambda\delta} \int_\delta^\pi |a(\theta)| d\theta$$
$$\le \int_0^\delta |a(\theta)| d\theta + 4\pi\lambda e^{-2\lambda\delta} \int_0^\pi |a(\theta)| d\theta.$$

Then for any $\varepsilon > 0$ we can select both δ small enough and $\lambda_0 = \lambda_0(\delta)$ large enough, such that

$$\int_0^\delta |a(\theta)| d\theta < \varepsilon/2 \qquad \text{and} \qquad 4\pi\lambda e^{-2\lambda\delta} \int_0^\pi |a(\theta)| d\theta < \varepsilon/2,$$

for all $\lambda \ge \lambda_0(\delta)$. That is, $\lim_{\lambda \to +\infty} I_3(\lambda) = 0$.

The case when $\lambda \to -\infty$ follows from the above and the equality

$$\gamma_{a(\theta)}(\lambda) = \gamma_{a(\pi-\theta)}(-\lambda).$$

The cases when $j > 1$ are considered analogously. □

References

[1] V. Dybin and S. Grudsky, *Inrtoduction to the theory of Toeplitz operators with infinite index.* Birkhäuser Verlag, Basel, 2002.

[2] S. Grudsky, R. Quiroga-Barranco, and N. Vasilevski, *Commutative C^*-algebras of Toeplitz operators and quantization on the unit disk.* J. Funct. Anal. **234** (2006), no. 1, 1–44.

[3] S. Grudsky and N. Vasilevski, *Bergman-Toeplitz operators: Radial component influence.* Integr. Equat. Oper. Th. **40** (2001), no. 1, 16–33.

[4] S. Grudsky and N. Vasilevski, *On the structure of the C^*-algebra generated by Toeplitz operators with piece-wise continuous symbols.* Complex Analysis Operator Theory, v. 2, no. 4, 525–548, 2008.

[5] Yu.I. Karlovich, *Pseudodifferential operators with compound slowly oscillating symbols.* In: "The extended field of operator theory". Operator Theory, Advances and Applications **171**, Birkhäuser, Basel-Boston-Berlin (2007), 189–224.

[6] N. Vasilevski, *On Toeplitz operators with piecewise continuous symbols on the Bergman space.* In: "Modern Operator Theory and Applications". Operator Theory, Advances and Applications **170**, Birkhäuser, Basel-Boston-Berlin (2007), 229–248.

[7] N.L. Vasilevski, *Banach algebras that are generated by certain two-dimensional integral operators. II.* (Russian). Math. Nachr. **99** (1980), 135–144.

[8] N.L. Vasilevski, *Banach algebras generated by two-dimensional integral operators with a Bergman kernel and piecewise continuous coefficients. I.* Soviet Math. (Izv. VUZ) **30** (1986), no. 3, 14–24.

[9] N.L. Vasilevski, *Banach algebras generated by two-dimensional integral operators with a Bergman kernel and piecewise continuous coefficients. II.* Soviet Math. (Izv. VUZ) **30** (1986), no. 3, 44–50.

[10] N.L. Vasilevski, *Bergman space structure, commutative algebras of Toeplitz operators and hyperbolic geometry.* Integr. Equat. Oper. Th. **46** (2003), 235–251.

Sergei Grudsky and Nikolai Vasilevski
Departamento de Matemáticas
CINVESTAV del I.P.N.
México, D.F., México
e-mail: grudsky@math.cinvestav.mx
nvasilev@math.cinvestav.mx

Operator Theory:
Advances and Applications, Vol. 190, 267–277

Block Tridiagonal Matrices and a Beefed-up Version of the Ehrenfest Urn Model

F. Alberto Grünbaum

To the memory of Prof. Mark G. Krein, with gratitude.

Abstract. The very classical Ehrenfest urn model can be solved exactly in terms of Krawtchouk polynomials. I consider a natural extension of this model which goes beyond "nearest neighbours" random walks and whose analysis benefits from the study of a family of matrix-valued orthogonal polynomials. This subject was started by M.G. Krein around 1949. This paper shows one more example of his vast and lasting legacy in the never-ending task of finding new mathematical tools to analyze the physical world.

Mathematics Subject Classification (2000). 33C45, 22E45.

Keywords. Matrix-valued orthogonal polynomials, Karlin-McGregor representation, Krawtchouk polynomials.

1. Introduction

We establish some general results for finite block tridiagonal matrices which are motivated by the study of a concrete example of interest in mathematical physics. Most of the discussion is devoted to this important special example, but as we observe many of the results are valid in general.

2. The Ehrenfest urn model

Consider two urns where a total of $2N$ balls, each one carrying a label $1, 2, \ldots, 2N$ is distributed in these two urns. The state of the system is described by an integer i denoting the number of balls in one of the urns. At integer values of time one picks a label $1, 2, \ldots, 2N$ with the uniform distribution and the ball whose label

The author was supported in part by NSF Grant # DMS 0603901.

is chosen in this fashion is removed from the urn where it is located and placed in the other urn.

This gives rise to a Markov chain in the finite state space $0, 1, 2, \ldots, 2N$ where the matrix $\mathbb{P}$ given by

$$\begin{pmatrix} 0 & 1 & & & & & \\ \frac{1}{2N} & 0 & \frac{2N-1}{2N} & & & & \\ & \frac{2}{2N} & 0 & \frac{2N-2}{2N} & & & \\ & & \ddots & 0 & \ddots & & \\ & & & \ddots & \ddots & \ddots & \\ & & & & \ddots & 0 & \frac{1}{2N} \\ & & & & & \frac{2N}{2N} & 0 \end{pmatrix}.$$

This situation arises in a model introduced by P. and T. Ehrenfest, see [9], in an effort to illustrate the issue that irreversibility and recurrence can coexist. The background here is, of course, the famous H-theorem of L. Boltzmann, that claims to start from the time-reversible equations of Newton and obtain the result that a quantity like entropy increases monotonically in time. On the face of it, this is a flagrant contradiction.

For a more detailed discussion of the model see [11, 14]. This model has also been considered in dealing with a quantum mechanical version of a discrete harmonic oscillator by Schrödinger himself, see [23].

In this case the corresponding orthogonal polynomials (on a finite set) can be given explicitly. Consider the so-called Krawtchouk polynomials, given by means of the (truncated) Gauss series

$${}_2\tilde{F}_1\left(\begin{matrix} a, b \\ c \end{matrix}; z\right) = \sum_0^{2N} \frac{(a)_n (b)_n}{n!(c)_n} z^n$$

with

$$(a)_n \equiv a(a+1)\ldots(a+n-1), \quad (a)_0 = 1.$$

The polynomials are given by

$$K_i(x) = {}_2\tilde{F}_1\left(\begin{matrix} -i, -x \\ -2N \end{matrix}; 2\right)$$
$$x = 0, 1, \ldots, 2N; \quad i = 0, 1, \ldots, 2N.$$

Observe that

$$K_0(x) \equiv 1, K_i(2N) = (-1)^i.$$

The orthogonality measure is read off from

$$\sum_{x=0}^{2N} K_i(x) K_j(x) \frac{\binom{2N}{x}}{2^{2N}} = \frac{(-1)^i i!}{(-2N)_i} \delta_{ij} \equiv \pi_i^{-1} \delta_{ij} \quad 0 \le i, j \le 2N.$$

These polynomials satisfy the three term recursion relation

$$\frac{1}{2}(2N-i)K_{i+1}(x) - \frac{1}{2}2NK_i(x) + \frac{i}{2}K_{i-1}(x) = -xK_i(x)$$

and this has the consequence that

$$\begin{pmatrix} 0 & 1 & & & & & \\ \frac{1}{2N} & 0 & \frac{2N-1}{2N} & & & & \\ & \frac{2}{2N} & 0 & \frac{2N-2}{2N} & & & \\ & & \ddots & 0 & \ddots & & \\ & & & \ddots & \ddots & \ddots & \\ & & & & \ddots & 0 & \frac{1}{2N} \\ & & & & & \frac{2N}{2N} & 0 \end{pmatrix} \begin{pmatrix} K_0(x) \\ K_1(x) \\ \\ \\ \\ \\ K_{2N}(x) \end{pmatrix} = \left(1-\frac{x}{N}\right)\begin{pmatrix} K_0(x) \\ K_1(x) \\ \vdots \\ K_{2N}(x) \end{pmatrix}$$

any time that x is one of the values $x = 0, 1, \ldots, 2N$. This means that the eigenvalues of the matrix $\mathbb{P}$ above are given by the values of $1 - \frac{x}{N}$ at these values of x, i.e.,

$$1, 1 - \frac{1}{N}, \ldots, -1$$

and that the corresponding eigenvectors are the values of

$$[K_0(x), K_1(x), \ldots, K_{2N}(x)]^T$$

at these values of x.

Since the matrix $\mathbb{P}$ above is the one step transition probability matrix for our urn model we conclude that the Karlin-McGregor representation is given by

$$(\mathbb{P}^n)_{ij} = \pi_j \sum_{x=0}^{2N} \left(1 - \frac{x}{N}\right)^n K_i(x)K_j(x)\frac{\binom{2N}{x}}{2^{2N}}.$$

We can use these expressions to rederive some results given in [14].

We have

$$(\mathbb{P}^n)_{00} = \sum_{x=0}^{2N} \left(1 - \frac{x}{N}\right)^n \frac{\binom{2N}{x}}{2^{2N}}$$

and the "generating function" for these probabilities, defined by

$$U(z) \equiv \sum_{n=0}^{\infty} z^n (\mathbb{P}^n)_{00}$$

becomes

$$U(z) = \sum_{x=0}^{2N} \frac{N}{N(1-z)+xz}\frac{\binom{2N}{x}}{2^{2N}}.$$

In particular $U(1) = \infty$ and then the familiar "renewal equation", see [11] given by

$$U(z) = F(z)U(z) + 1$$

where $F(z)$ is the generating function for the probabilities f_n of returning from state 0 to state 0 for the first time in n steps

$$F(z) = \sum_{n=0}^{\infty} z^n f_n$$

gives

$$F(z) = 1 - \frac{1}{U(z)}.$$

Therefore we have $F(1) = 1$, indicating that one returns to state 0 with probability one in finite time. The same is true for any other state, and the physically important issue has to do with the time that it takes our random walk to return to state i for the first time if we take this as the initial state.

The results above allow us to compute the expected time to return to state 0. This expected value is given by $F'(1)$, and we have

$$F'(z) = \frac{U'(z)}{U^2(z)}.$$

Since

$$U'(z) = \sum_{x=0}^{2N} \frac{N(N-x)}{(N(1-z)+xz)^2} \frac{\binom{2N}{x}}{2^{2N}}$$

we get $F'(1) = 2^{2N}$. The same method shows that any state $i = 0, \ldots, 2N$ is also recurrent and that the expected time to return to it is given by

$$\frac{2^{2N}}{\binom{2N}{i}}.$$

The moral of this is clear: if $i = 0$ or $2N$, or close to these values, i.e., we start from a state where most balls are in one urn it will take on average a huge amount of time to get back to this state.

For $N = 10000$ and a repetition rate of 1 second M. Kac gives about 10^{6000} years as the expected time.

If on the other hand $i = N$, i.e., we are starting from a very balanced state, then we will (on average) return to this state fairly soon.

Again, for $N = 10000$ and a repetition rate of 1 second M. Kac gives about 175 seconds as the expected time.

Thus we see how the issues of irreversibility and recurrence are rather subtle and that certain models show how they can coexist.

It is worth pointing out that the complete explicit solution in terms of the Krawtchouk polynomials came up much later that the original work of P. and

T. Ehrenfest and even after the important contributions of M. Kac. Two papers dealing with this point are [2] and [17].

3. Tridiagonal matrices and nearest neighbours random walks

The example above is a special case of a class of random walks of the birth-and-death kind, where the one step transition probability matrix is a tridiagonal one. Starting with the work of S. Karlin and J. McGregor, [18], one knows that the corresponding family of orthogonal polynomials and their spectral measure can be of use in analyzing properties of the random walk. This work is a discrete version of similar analysis in the continuous case done by W. Feller and H.P. McKean, see [10, 22].

The Eherenfest model is very special in that all the ingredients that make up the spectral analysis have been eventually found in great detail.

4. Block tridiagonal matrices: going beyond nearest neighbours

In [12, 13] and [5] one sees examples of so-called Quasi-birth-and-death-processes, see [21], where the theory of matrix-valued orthogonal polynomials can be used to study these processes whose one step transition probability matrix is block diagonal. This theory was, as it is well known, started by M.G. Krein in [19, 20].

One natural way to get processes of this type is to consider Markov chains either on a finite set or on the non-negative integers where the the transition probability matrix is pentadiagonal.

5. A beefed-up version of the Ehrenfest urn model

Here we consider the transition probability matrix for a beefed-up Ehrenfert model based on N balls, (N odd), with $N+1$ states $0, 1, \ldots, N$. Notice that we have made an unimportant change from the standard model where the number of balls was even. In the previous case this was only a matter of notational convenience, but now it will be important that we have an odd number of balls and thus an even number of states.

The state of the system, denoted by i, is the number of balls in the left urn. From state i you can make a transition to state $i-2, i-1, i, i+1, i+2$ with probabilities

$$\frac{\binom{i}{2}}{2\binom{N}{2}}, \frac{1}{2}\frac{i}{N}, \frac{i(N-i)}{2\binom{N}{2}}, \frac{1}{2}\frac{N-i}{N}, \frac{1}{2}\frac{\binom{N-i}{2}}{\binom{N}{2}}.$$

Since the number of states is even and $\mathbb{P}$ is pentadiagonal, it is natural to think of it as a block tridiagonal matrix with 2×2 blocks

$$\mathbb{P} = \begin{bmatrix} B_0 & A_0 & & & \\ C_1 & B_1 & A_1 & & \\ & \ddots & \ddots & \ddots & \\ & & C_{\frac{N-3}{2}} & B_{\frac{N-3}{2}} & A_{\frac{N-3}{2}} \\ & & & C_{\frac{N-1}{2}} & B_{\frac{N-1}{2}} \end{bmatrix}.$$

The matrices A_i are all invertible. We construct, following Krein, a finite set of polynomials

$$P_1(\lambda), P_1(\lambda), \ldots, P_{\frac{N-1}{2}}(\lambda)$$

given by the recursion relation

$$C_i P_{i-1}(\lambda) + B_i P_i(\lambda) + A_i P_{i+1}(\lambda) = \lambda P_i(\lambda)$$

for $i = 0, 1, \ldots, \frac{N-3}{2}$, where we put $P_{-1}(\lambda) = 0$, $P_0(\lambda) = I$.

One can also consider the so-called polynomials of the second kind, Q_i defined by the same recursion relation with initial conditions given by $Q_0(\lambda) = 0$ and $Q_1(\lambda) = A_0^{-1}$.

From the expression given above we have

$$B_i = \frac{1}{2} \begin{pmatrix} \dfrac{2i(N-2i)}{\binom{N}{2}} & \dfrac{N-2i}{N} \\ \dfrac{2i+1}{N} & \dfrac{(2i+1)(N-2i-1)}{\binom{N}{2}} \end{pmatrix}$$

$$A_i = \frac{1}{2} \begin{pmatrix} \dfrac{\binom{N-2i}{2}}{\binom{N}{2}} & 0 \\ \dfrac{N-2i-1}{N} & \dfrac{\binom{N-2i-1}{2}}{\binom{N}{2}} \end{pmatrix}$$

$$C_i = \frac{1}{2} \begin{pmatrix} \dfrac{\binom{2i}{2}}{\binom{N}{2}} & \dfrac{2i}{N} \\ 0 & \dfrac{\binom{2i+1}{2}}{\binom{N}{2}} \end{pmatrix}.$$

Since we are dealing with a finite matrix this recursion relation cannot be used for $i = \frac{N-1}{2}$ for the simple reason that we do not have $A_{\frac{N-1}{2}}$ to be able to define $P_{\frac{N+1}{2}}(\lambda)$.

We can however consider the polynomial

$$\left(\lambda I - B_{\frac{N-1}{2}}\right) P_{\frac{N-1}{2}}(\lambda) - C_{\frac{N-1}{2}} P_{\frac{N-3}{2}}$$

which would be the value of $P_{\frac{N+1}{2}}(\lambda)$ given by the recursion relation above if we had $A_{\frac{N-1}{2}} = I$. By a slight abuse of notation we denote this polynomial by $P_{\frac{N+1}{2}}(\lambda)$. We define its zeros as the zeros of the determinant of $P_{\frac{N+1}{2}}(\lambda)$. These happen to be the eigenvalues of $\mathbb{P}$ considered as a pentadiagonal $(N+1) \times (N+1)$ matrix.

In the same fashion, the polynomial $Q_{\frac{N+1}{2}}(\lambda)$ requires a special definition and this is handled as before: one pretends that the missing matrix $A_{\frac{N-1}{2}} = I$.

As we will see later on these two "last polynomials" play an important role.

For the matrix $\mathbb{P}$ given above we find out the following situation

a) These eigenvalues are given by the values of

$$\frac{(i - N - 1)\left(i - \frac{N+1}{2}\right)}{\binom{N}{2}} = \lambda_i$$

for $i = 1, 2, \ldots, \tau(N)$ where the value of $\tau(N)$ depends on the nature of the odd number N modulo 4. If $N = 4i + 3$ then $\tau(N) = 3(i+1)$, while of $N = 4i + 1$ then $\tau(N) = 3i + 1$.

b) Of these $\tau(N)$ eigenvalues, the positive ones are simple eigenvalues and there are $s(N)$ of them. The value of $s(N)$ depends again on the value of $\mod(N, 4)$. If $N = 4i+3$, then $s(N) = 2i+1$, if $N = 4i+1$ then $s(N) = 2i$. In the first case there is an extra negative simple eigenvalue corresponding to $i = \tau(N) = 3(i+1)$. When this extra simple eigenvalue is present it sits at the bottom of the spectrum.

c) Of these $\tau(N)$ eigenvalues there is finally a number of nonpositive double eigenvalues starting with the value 0. The number of them is $d(N)$ and its value depends, once again, on the value of $\mod(N, 4)$. If $N = 4i + 3$ then $d(N) = i + 1$, if $N = 4i + 1$ then $d(N) = 2i$.

In summary we may (or may not) have one simple negative eigenvalue at the bottom of the spectrum, then $d(N)$ negative eigenvalues of which the largest one is zero and then $s(N)$ simple positive eigenvalues of which the largest is one.

For the considerations below it is convenient to replace the non-symmetric matrix $\mathbb{P}$ by one obtained by conjugating it by a diagonal scalar matrix so as to get a symmetric one. Since $\mathbb{P}$ is no longer tridiagonal this is not automatic, but it is still possible to do such an adjustment.

From now on the polynomials P_i refer to the ones obtained going with this symmetrized version of the original $\mathbb{P}$. We retain the same symbol for this new version of $\mathbb{P}$.

It may be worth pointing out that if we put together the $\left(\frac{N+1}{2}\right) \times \tau(N)$ matrix made up of 2×2 blocks

$$P_i(\lambda_j),\ i = 0, \dots, \frac{N-1}{2},\ j = 1, 2, \dots, \tau(N)$$

and denote this matrix by X, we *do not* quite have

$$\mathbb{P}X = X\Lambda$$

with Λ a block diagonal matrix with 2×2 diagonal blocks $\lambda_i I$. Note that when one views all these matrices as made up to scalars, we have that $\mathbb{P}$ is $(N+1) \times (N+1)$, X is $(N+1) \times 2\tau(N)$ and finally Λ is $2\tau(N) \times 2\tau(N)$.

If one considers $\mathbb{P}X - X\Lambda$, then all of its 2×2 blocks except those in the last row vanish. In this last row the only blocks that are zero are those that correspond to double eigenvalues. The other 2×2 blocks are singular nonzero matrices. Of course this phenomenon (the distinction between vanishing and being singular) is absent in the case of a scalar tridiagonal matrix.

We find that there exists a unique collection of symmetric nonnegative matrices 2×2 W_k, $k = 1, \dots, \tau(N)$ such that

$$\sum_{k=1}^{\tau(N)} P_i(\lambda_k) W_k P_j^+(\lambda_k) = \delta_{ij} I$$

for $0 \le i, j \le \frac{N-1}{2}$.

Moreover we have for $r = 0, 1, 2, \dots$

$$\sum_{k=1}^{\tau(N)} \lambda_k^r P_i(\lambda_k) W_k P_j^+(\lambda_k) = (\mathbb{P}^r)_{ij}.$$

A way of seeing this relation is to give a formula for the resolvent of $\mathbb{P}$, which is of course a matrix version of the Stieltjes transform of the discrete measure in question. The formula is

$$(\mathbb{P} - \lambda I)_{ij}^{-1} = \sum_{k=1}^{\tau(N)} \frac{P_i(\lambda_k) W_k P_j^+(\lambda_k)}{\lambda_k - \lambda}.$$

This is not the only useful expression for the inverse of this block tridiagonal matrix. There is a well-known construction in the scalar case that can be adapted to the block situation at hand. This is recalled below. A reader familiar with the construction of the Green's function for a second-order differential operator from two solutions of the homogeneous equation, each one satisfying one of the two boundary conditions will recognize the analogy of the construction below with its this classical counterpart, see [26].

One can give a general expression for the corresponding weight matrix at any one of the λ_k, namely

$$W_k = \lim_{\lambda \to \lambda_k} (\lambda_k - \lambda)(\mathbb{P} - \lambda I)^{-1}_{00}.$$

This formula follows directly from the one given above.

We have found that a formula of A. Duran, developed for a different purpose, see [6], works nicely in this case to give a general expression for the weight matrices W_k. For $k = 1, \dots, \tau(N)$, we have

$$W_k = \frac{l_k}{\left(\det(P_{\frac{N+1}{2}}(t)\right)^{(l_k)}(\lambda_k)} \left(\left(Adj P_{\frac{N+1}{2}}(t)\right)\right)^{(l_k-1)}(\lambda_k) Q_{\frac{N+1}{2}}(\lambda_k).$$

Here l_k is the multiplicity of λ_k, which can be one or two.

In the particular case when λ_k is double, but only in this case, we get the familiar looking expression

$$W_k = \left(\sum_{i=0}^{\frac{N-1}{2}} P_i^*(\lambda_k) P_i(\lambda_k)\right)^{-1}.$$

We have not found an explicit treatment of this finite tridiagonal case in the literature, but there are several references that touch upon several of these issues in the context of Gaussian quadrature formulas. For a collection of some of these papers see [8, 7, 6, 4, 24]. In particular in [8] there are references to other papers on this topic.

The careful reader will have noticed that we have not given here an explicit expression for the polynomials $P_i(\lambda)$ themselves. For the classical Ehrenfest model this was achieved, as noticed earlier, much later than the time the model was proposed. One can only conjecture than in the matrix-valued case at hand such an expression is possible in terms of the matrix-valued hypergeometric function introduced in [25].

6. Beyond the Ehrenfest model

It may be worth noticing that all the considerations above, with the exception of the detailed form of the spectrum of the finite block tridiagonal matrix $\mathbb{P}$ applies anytime we are dealing with such a matrix and the matrix blocks A_i are invertible.

In particular one can consider situations where the original model is modified in fancier ways allowing, for instance, picking not only one or two pairs of indices but also triplets, etc. This leads to scalar banded matrices which can be seen as block tridiagonal ones with larger blocks. The study of these simple models already exhibits situations where the multiplicity of the eigenvalues can range all the way from one to the maximum possible value, given by the size of the blocks.

References

[1] G. Andrews, R. Askey, and R. Roy, *Special functions.* Encyclopedia of Mathematics and its applications, Cambridge University Press, 1999.

[2] R. Askey, *Evaluation of Sylvester type determinants using orthogonal polynomials.* Advances in Analysis, Proceed. 4th international ISAAC Congress, ed. H.G.W. Begehr et al., World Scientific, Singapore (2005), 1–16.

[3] S. Basu, and N.K. Bose, *Matrix Stieltjes series and network models.* SIAM J. Matrix Anal. Applic. **14** (1983), no. 2, 209–222.

[4] H. Dette, W. Studden, W., *Quadrature formulas for matrix measures, a geometric approach.* Linear Algebra Appl. **364** (2003), 33–64.

[5] H. Dette, B. Reuther, W. Studden, and M. Zygmunt, *Matrix measures and random walks with a block tridiagonal transition matrix.* SIAM J. Matrix Anal. Applic. **29** (2006), no. 1, 117–142.

[6] A. Duran *Markov's theorem for orthogonal matrix polynomials.* Can. J. Math. **48** (1996), 1180–1195.

[7] A.J. Duran, and P. Lopez Rodriguez, *Orthogonal matrix polynomials: zeros and Blumenthal theorem.* J. Approx. Theory **84** (1996), 96–118.

[8] A.J. Duran, and B. Polo, *Gauss quadrature formulae for orthogonal matrix polynomials.* Linear Alg. Appl. **355** (2002), 119–146.

[9] P. Ehrenfest, and T. Eherenfest, *Über zwei bekannte Einwände gegen das Boltzmannsche H-Theorem.* Physikalische Zeitschrift **8** (1907), 311–314.

[10] W. Feller, *On second order differential operators.* Ann. of Math. **61** (1955), no. 1, 90–105.

[11] W. Feller, *An introduction to Probability Theory and its Applications.* vol. 1, 3rd edition, Wiley 1967.

[12] F.A. Grünbaum, *Random walks and orthogonal polynomials: some challenges.* to appear in Probability, Geometry and Integrable Systems, MSRI Publication **55** (2007), see arXiv math.PR/0703375.

[13] F.A. Grünbaum, and M.D. de la Iglesia, *Matrix valued orthogonal polynomials arising from group representation theory and a family of quasi-birth-and-death processes.* submitted to SIAM J. Matrix Anal. Applic.

[14] M. Kac, *Random walk and the theory of Brownian motion.* American Math. Monthly **54** (1947), 369–391.

[15] S. Karlin, *A first course in stochastic process.* Academic Press, 1966.

[16] S. Karlin, and H. Taylor, *A second course in stochastic processes.* Academic Press, 1981.

[17] S. Karlin, and J. McGregor, *Ehrenfest urn models.* J. Appl. Prob **2** (1965) 352–376.

[18] S. Karlin, and J. McGregor, *Random walks.* Illinois J. Math. **3** (1959), 66–81.

[19] M.G. Krein, *Fundamental aspects of the representation theory of hermitian operators with deficiency index* (m, m). AMS Translations, Series 2, vol. 97, Providence, Rhode Island (1971), 75–143.

[20] M.G. Krein, *Infinite J-matrices and a matrix moment problem.* Dokl. Akad. Nauk SSSR **69** (1949), no. 2, 125–128.

[21] G. Latouche, and V. Ramaswami, *Introduction to Matrix Analytic Methods in Stochastic Modeling.* ASA-SIAM Series on Statistics and Applied Probability, 1999.

[22] H.P. McKean jr., *Elementary solutions for certain parabolic partial differential equations.* Trans. Amer. Math. Soc. **82** (1956), 519–548.

[23] E. Schroedinger, and F. Kohlrausch, *Das Ehrenfestsche Model der H-Kurve.* Phys. Zeit. **27** (1926), 306–313.

[24] A. Sinap, and W. van Assche, *Polynomial Interpolation and Gaussian Quadrature for Matrix Valued Functions.* Linear Algebra Appl. **207** (1994), 71–114.

[25] J. Tirao, *The matrix valued hypergeometric equation.* Proc. Nat. Acad. Sci. U.S.A. **100** (2003), no. 14, 8138–8141.

[26] E.C. Titchmarsh, *Eigenfunction expansions associated with second order differential equations.* Oxford at the Clarendon Press, 1946.

F. Alberto Grünbaum
Department of Mathematics
University of California
Berkeley, CA 94720, USA
e-mail: grunbaum@math.berkeley.edu

Operator Theory:
Advances and Applications, Vol. 190, 279–294

On the Unitarization of Linear Representations of Primitive Partially Ordered Sets

Roman Grushevoy and Kostyantyn Yusenko

Abstract. We describe all weights which are appropriated for the unitarization of indecomposable linear representations of primitive partially ordered sets of finite type.

Mathematics Subject Classification (2000). Primary 65N30; Secondary 65N15.

Keywords. Unitarizations, quivers, partially ordered sets, Dynkin graphs, subspaces, *-algebras, representations.

0. Introduction

The representation theory of partially ordered sets (posets) in linear vector spaces has become one of the classical fields in linear algebra (see [1, 2, 3, 4] and others). See Section 1 for details.

On the other hand when one tries to develop a similar theory for Hilbert spaces, one will be faced with difficulties even with three subspaces two of which are orthogonal: it is impossible to obtain their description in a reasonable way (it is the so-called $*$-wild problem) see [5, 6]. But adding linear relation

$$\alpha_1 P_1 + \cdots + \alpha_n P_n = \gamma I,$$

between the projections P_i on corresponding subspaces, in some cases gives nice answers which have deep interconnections with the linear case (see Section 2 for details).

The aim of this article is to show that each indecomposable linear representation of a primitive poset of finite type can be unitarized with some weight $\chi = (\alpha_1, \ldots, \alpha_n, \gamma)$ and to describe all possible weights χ appropriated to the unitarization for a given indecomposable linear representation of the primitive poset (Section 3).

This work has been partially supported by the Scientific Program of National Academy of Sciences of Ukraine, Project No. 0107U002333.

1. Posets and their linear representations

In this section we will briefly recall some results concerning partially ordered sets and their linear representation.

1.1. Posets and Hasse diagrams

Let $(\mathcal{N}, \prec)$ be a finite partially ordered set (or poset for short) which for us will be $\{1, \dots, n\}$. By the width of the poset $\mathcal{N}$ we will understand the maximal number of two-by-two incomparable elements of this set.

The poset $\mathcal{N}$ of width m is called primitive and denoted by $(k_1, \dots, k_m)$ if this set is the cardinal sum of m linearly ordered sets $\mathcal{N}_1, \dots, \mathcal{N}_m$ with orders $k_1, \dots, k_m$.

We will use the standard graphic representations for the poset $(\mathcal{N}, \prec)$ called a Hasse diagram. This representation associates to each point $x \in \mathcal{N}$ the vertex v_x and the arrow $v_x \to v_y$ if $x \prec y$ and there is no $z \in \mathcal{N}$, such that $x \prec z \prec y$. For example let $\mathcal{N} = \{1, 2, 3, 4, 5, 6\}$ with the following order:

$$2 \prec 1, \quad 2 \prec 3, \quad 4 \prec 3, \quad 6 \prec 5,$$

then the corresponding Hasse diagram is the following:

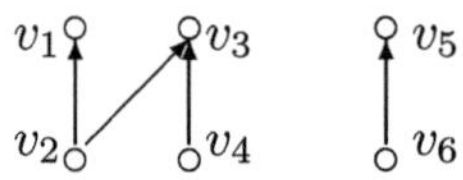

For each primitive poset $\mathcal{N} = (k_1, \dots, k_m)$ its Hasse diagram has the following form:

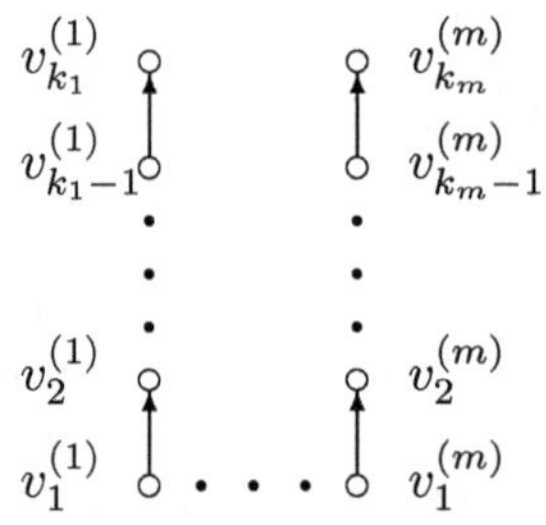

1.2. Linear representations of posets. Indecomposability and bricks

By a linear representation π of the poset $\mathcal{N}$ in some complex vector space V, we will understand a relationship such that, for each element $i \in \mathcal{N}$ corresponding to the subspace $V_i \subset V$, if $i \prec j$, then $V_i \subseteq V_j$. We will denote the representation π in the following way: $\pi = (V; V_1, \dots, V_n)$, where n is the cardinality of $\mathcal{N}$, and for the primitive posets $(k_1, \dots, k_m)$ we will use the notation $\pi = (V; V^{(1)}_1, \dots, V^{(1)}_{k_1}; \dots; V^{(m)}_1, \dots, V^{(m)}_{k_m})$.

By the dimension vector d_π of given representation π we will understand the vector $d_\pi = (d_0; d_1, \dots, d_n)$, where $d_0 = \dim V$, $d_i = \dim V_i$ (correspondingly, for representations of primitive posets the dimensional vector will be denoted by $d_\pi = (d_0; d^{(1)}_1, \dots, d^{(1)}_{k_1}; \dots; d^{(m)}_1, \dots, d^{(m)}_{k_m})$).

In fact the linear representations of a poset $\mathcal{N}$ form the additive category $\mathrm{Rep}(\mathcal{N})$, where the set of morphisms $\mathrm{Mor}(\pi_1, \pi_2)$ between two representations $\pi_1 = (V; V_1, \dots, V_n)$ and $\pi_2 = (W; W_1, \dots, W_n)$ consists of such linear maps $C : V \to W$, that $C(V_i) \subset W_i$. Two representations π_1 and π_2 of a poset $\mathcal{N}$ are isomorphic (or equivalent) if there exists an invertible morphism $C \in \mathrm{Mor}(\pi_1, \pi_2)$, i.e., there exists an invertible linear map $C : V \to W$ such that $C(V_i) = W_i$.

One can define a direct sum $\pi = \pi_1 \oplus \pi_2$ of two objects $\pi_1, \pi_2 \in \mathrm{Rep}(\mathcal{N})$ in the following way:

$$\pi = (V \oplus W; V_1 \oplus W_1, \dots, V_n \oplus W_n).$$

Using the definition of a direct sum it is natural to define *indecomposable* representations as the representations that are not isomorphic to the direct sum of two non-zero representations. Otherwise representations are called *decomposable.* It is easy to show that a representation π is indecomposable iff there exist no non-trivial idempotents in $\mathrm{End}(\pi)$. A representation π is called *brick* if there is no non-trivial endomorphism of this representation. One can show that *brick* implies *indecomposability.* But there exist indecomposable representations of posets which are not brick: let us take

$$A_\alpha = \begin{pmatrix} 1 & \alpha \\ 0 & 1 \end{pmatrix}, \qquad \alpha \in \mathbb{R} \backslash \{0\},$$

and consider the representation $\pi_\alpha = (V; V_1, V_2, V_3, V_4)$ of the poset $\mathcal{N} = (1, 1, 1, 1)$,

$$V = \mathbb{C}^2 \oplus \mathbb{C}^2, \qquad V_1 = \mathbb{C}^2 \oplus 0, \quad V_2 = 0 \oplus \mathbb{C}^2,$$
$$V_3 = \{(x, x) \in \mathbb{C}^4 | x \in \mathbb{C}^2\}, \quad V_4 = \{(x, A_\alpha x) \in \mathbb{C}^4 | x \in \mathbb{C}^2\};$$

this representation is indecomposable but is not brick.

We call the representation π *non-degenerated* if the components of dimension vector d_π satisfy the following conditions:

- $d_i \neq 0$;
- if $i \prec j$ then $d_i < d_j$;
- $d_i < d_0$,

otherwise the representation is called *degenerated.* Note that in works [3, 4], by the term *non-degenerated,* the authors meant only the first and second of the above conditions.

1.3. Posets and quivers

In studying the representations of the primitive posets it is helpful to use the well-developed representation theory of quivers (for the representations of quivers see for example [7, 8]).

A quiver $Q = (Q_0, Q_1, s, t : Q_1 \to Q_0)$ is given by a set of vertices Q_0, which for us will be $\{v_0, \dots, v_n\}$, and a set Q_1 of arrows. An arrow $\rho \in Q_1$ starts at the vertex $s(\rho)$ and terminates at $t(\rho)$. A representation X of Q is given by a vector space X_{v_i} for each vertex $v_i \in Q_0$ and linear operator $X_\rho : X_{s(\rho)} \to X_{t(\rho)}$. In a similar way as for the posets one can define the set of morphisms between two representations X and Y, *indecomposable* and *brick* representations (see [7] for the details).

To each poset $\mathcal{N}$ one can associate the quiver $Q^{\mathcal{N}}$ in a similar way as the Hasse diagram was associated, the only difference being that we add the extra vertex v_0 to the quiver and connect by arrows all vertices that correspond to maximal elements of $\mathcal{N}$ with the vertex v_0. More formally, $Q_0^{\mathcal{N}} = \{v_x \mid x \in \mathcal{N}\} \cup \{v_0\}$ and $Q_1^{\mathcal{N}}$ consists of the following arrows:

- $v_x \to v_y$ if $x \prec y$ and there are no such z that $x \prec z \prec y$, and
- $v_x \to v_0$ if there no such z that $x \prec z$.

For example, for each primitive poset $\mathcal{N} = (k_1, \dots, k_m)$ the corresponding quiver $Q^{\mathcal{N}}$ has the following form:

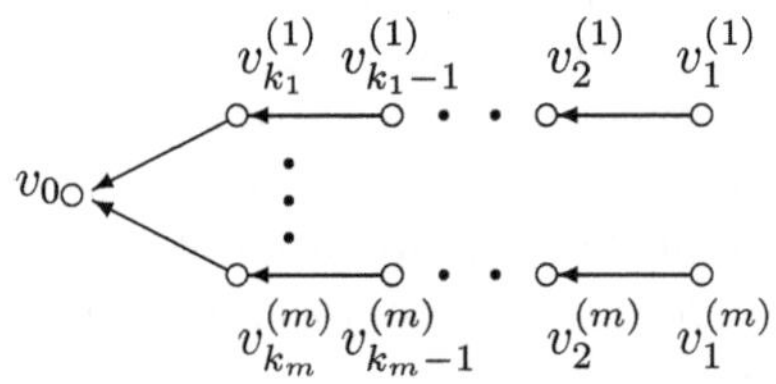

Remark 1.1. Notice that the underlying graph of the quiver $Q_{\mathcal{N}}$ that corresponds to primitive poset $\mathcal{N} = (k_1, \dots, k_m)$ is the tree which is usually denoted by $T_{k_1+1,\dots,k_m+1}$.

It is obvious that each linear representation π of the poset $\mathcal{N} = (k_1, \dots, k_m)$ defines some representation X_π of the corresponding quiver $Q^{\mathcal{N}}$. Indeed, let be given the representation

$$\pi = (V; V_1^{(1)}, \dots, V_{k_1}^{(1)}; \dots; V_1^{(m)}, \dots, V_{k_m}^{(m)})$$

of the poset $\mathcal{N}$, then the corresponding representation X_π of quiver $Q_{\mathcal{N}}$ could be built in the following way: to each vertex $v_i^{(j)}$ we associate the space $V_i^{(j)}$, $X_{v_i^{(j)}} = V_i^{(j)}$ (to the vertex v_0 we associate the space V, $X_{v_0} = V$) and to each arrow $v_i^{(j)} \to v_{i+1}^{(j)}$ we associate the embedding $V_i^{(j)} \hookrightarrow V_{i+1}^{(j)}$ (to the arrows $v_{k_i}^{(j)} \to v_0$ we associate the embedding $V_{k_i}^{(j)} \hookrightarrow V$). And vice versa every representation X of quiver Q with underlaying graph $T_{k_1+1,\dots,k_m+1}$, such that for all $\rho \in Q_1$, X_ρ is a monomorphism, defines a representation of the poset $\mathcal{N} = (k_1, \dots, k_n)$:

$$\pi_X = (V; V_1^{(1)}, \dots, V_{k_1}^{(1)}; \dots; V_1^{(n)}, \dots, V_{k_n}^{(n)}),$$

where $V = X_{v_0}$ is a representation of the vertex v_0, and $V_j^{(i)} = X_{v_j^{(i)} \to v_{j+1}^{(i)}} \cdot \ldots \cdot X_{v_{k_i}^{(i)} \to v_0}(X_{v_j^{(i)}})$ is an image of space corresponding to the vertex $v_j^{(i)}$ under the maps that correspond to the path from vertex $v_j^{(i)}$ to vertex v_0.

In fact, having chosen a representation of the poset, then using the above construction, we build a representation of the corresponding quiver and finally, after building a representation of the corresponding poset, we will get the representation that is isomorphic to the one with which we started, i.e., $\pi \simeq \pi_{X_\pi}$ in the category $\mathrm{Rep}(\mathcal{N})$.

Proposition 1.2.

1) *The representations π_1 and π_2 of the poset $\mathcal{N}$ are isomorphic if and only if the corresponding representation X_{π_1} and X_{π_2} of the quiver $Q^{\mathcal{N}}$ are isomorphic.*
2) *The representation π is indecomposable if and only if the corresponding representation X_π of the quiver $Q^{\mathcal{N}}$ is indecomposable.*

Proof. 1) The morphism $C : X_{\pi_1} \to X_{\pi_2}$, establishes an isomorphism between the representations of the quiver $Q_{\mathcal{N}}$ if and only if the linear map C_{v_0} is an isomorphism between the representations π_1 and π_2 of $\mathcal{N}$.

2) If the representation π of primitive poset $\mathcal{N}$ is decomposable, i.e., $\pi \simeq \pi_1 \oplus \pi_2$, then the corresponding representation X_π of the quiver $Q^{\mathcal{N}}$ has the following form: for all $v_i \in Q_0^{\mathcal{N}}$, $X_{v_i} = U_i \oplus W_i$, for all $v_i \to v_j \in Q_1^{\mathcal{N}}$, $X_{v_i \to v_j} = \hat{\Gamma}_{i,j} \oplus \tilde{\Gamma}_{i,j}$, where $\hat{\Gamma}_{i,j} : U_j \to U_i$ and $\tilde{\Gamma}_{i,j} : W_j \to W_i$, i.e., X is decomposable. □

1.4. Finite type posets

Recall that a poset $\mathcal{N}$ has *finite (linear) representation type* if there exist only finitely many indecomposable representations of $\mathcal{N}$ in $\mathrm{Rep}(\mathcal{N})$. A result obtained by M.M. Kleiner [2] gives a complete description of the posets of finite type. In case of primitive posets using connections between posets and quivers and analogous results for quivers (see [7], for instance) one can obtain description of the primitive posets of finite type.

Proposition 1.3. *Each primitive poset of finite type has one of the following forms:*

- (k) *for all* $k \in \mathbb{N}$;
- (k_1, k_2) *for all* $k_1, k_2 \in \mathbb{N}$;
- $(k, 1, 1)$, *for all* $k \in \mathbb{N}$;
- $(2, 2, 1)$;
- $(3, 2, 1)$;
- $(4, 2, 1)$.

Proof. if. As it was shown in the previous subsection that for every representation of primitive posets one can build a representation of the corresponding quiver, non-isomorphic representations of the poset correspond to non-isomorphic representations of the quiver. The quivers A_{k+1}, $A_{k_1+k_2+1}$, D_{k+3}, E_6, E_7, E_8 are the corresponding quivers for the posets listed in the statement. Each of these quivers has finitely many non-isomorphic indecomposable representations (see [7]), hence the listed posets also have finitely many non-isomorphic indecomposable representations.

only if. One can show that each primitive poset that is not included in the list contains a subposet with corresponding extended Dynkin quiver. But this quiver has infinitely many non-isomorphic indecomposable representations (see [9, 10] and others for description). We list infinite series of non-isomorphic indecomposable representations of the corresponding poset:

- for the poset $(1,1,1,1)$ (the corresponding quiver is $\tilde{D}_4$):

$$V = \mathbb{C}^2 = \langle e_1, e_2\rangle;\ V_1 = \langle e_1\rangle,\ V_2 = \langle e_2\rangle,$$
$$V_3 = \langle e_1 + e_2\rangle,\ V_4 = \langle e_1 + \lambda e_2\rangle, \quad \lambda \neq 0, 1;$$

- for the poset $(2,2,2)$ (the corresponding quiver is $\tilde{E}_6$):

$$V = \mathbb{C}^3 = \langle e_1, e_2, e_3\rangle,$$
$$V_1 = \langle e_1\rangle, V_2 = \langle e_1, e_2\rangle, V_3 = \langle e_3\rangle, V_4 = \langle e_2, e_3\rangle,$$
$$V_5 = \langle e_1 + e_2 + e_3\rangle, V_6 = \langle e_1 + e_2 + e_3, \lambda e_1 + e_2\rangle, \quad \lambda \neq 0, 1;$$

- for the poset $(3,3,2)$ (the corresponding quiver is $\tilde{E}_7$):

$$V = \mathbb{C}^4 = \langle e_1, e_2, e_3, e_4\rangle,$$
$$V_1 = \langle e_1\rangle, V_2 = \langle e_1, e_2\rangle, V_3 = \langle e_1, e_2, e_3\rangle,$$
$$V_4 = \langle e_4\rangle, V_5 = \langle e_3, e_4\rangle, V_6 = \langle e_2, e_3, e_4\rangle,$$
$$V_7 = \langle e_1 + e_2 + e_3, \lambda e_1 + e_3 + e_4\rangle, \quad \lambda \neq 0, 1;$$

- for the poset $(5,2,1)$ (the corresponding quiver is $\tilde{E}_8$):

$$V = \mathbb{C}^6 = \langle e_1, e_2, e_3, e_4, e_5, e_6\rangle,$$
$$V_1 = \langle e_1\rangle, V_2 = \langle e_1, e_2\rangle, V_3 = \langle e_1, e_2, e_3\rangle, V_4 = \langle e_1, e_2, e_3, e_4\rangle,$$
$$V_5 = \langle e_1, e_2, e_3, e_4, e_5\rangle, V_6 = \langle e_5, e_6\rangle, V_7 = \langle e_3, e_4, e_5, e_6\rangle,$$
$$V_8 = \langle e_1 + e_3 + e_4 + e_5, \lambda e_1 + e_3 + e_5 + e_6, e_2 + e_4\rangle, \quad \lambda \neq 0, 1.$$

Hence the theorem is proved. □

1.5. Coxeter functors for representations of posets in linear spaces

Coxeter functors for representations of quivers were studied in [12] for proving Gabriel's theorem. Similar Coxeter functors F^+, F^- act in the category $\operatorname{Rep}(\mathcal{N})$ of linear representations of posets that was constructed in [4]. We will not give their full construction here, but we recall some basic necessary facts.

There are functors σ, ρ, ρ^* ($\sigma^2 = \mathrm{id}$, $\rho \circ \rho^* = \mathrm{id}$, $\rho^* \circ \rho = \mathrm{id}$) which act between categories $\operatorname{Rep}(\mathcal{N})$ and $\operatorname{Rep}(\mathcal{N}^*)$ (where $\mathcal{N}^*$ is dual to poset $\mathcal{N}$). These functors change the dimensions of the representations by the following formulas:

$$(d_0; d_1^{(1)}, \ldots, d_{k_1}^{(1)}; \ldots; d_1^{(m)}, \ldots, d_{k_m}^{(m)}) \overset{\rho}{\longmapsto} (\sum_{j=1}^{m} d_{k_j}^{(j)} - d_0; d_{k_1}^{(1)}, d_{k_1}^{(1)} - d_1^{(1)}, \ldots,$$
$$d_{k_1}^{(1)} - d_{k_1-1}^{(1)}; \ldots; d_{k_1}^{(m)}, \ldots, d_{k_m}^{(m)} - d_{k_m-1}^{(m)}), \quad \text{for } \rho : \operatorname{Rep}(\mathcal{N}) \to \operatorname{Rep}(\mathcal{N}^*);$$
$$(d_0; d_1^{(1)}, \ldots, d_{k_1}^{(1)}; \ldots; d_1^{(m)}, \ldots, d_{k_m}^{(m)}) \overset{\rho^*}{\longmapsto} (\sum_{j=1}^{m} d_1^{(j)} - d_0; d_1^{(1)} - d_2^{(1)},$$
$$d_1^{(1)} - d_3^{(1)}, \ldots, d_1^{(1)}; \ldots; d_1^{(m)} - d_2^{(m)}, \ldots, d_1^{(m)}), \quad \text{for } \rho^* : \operatorname{Rep}(\mathcal{N}^*) \to \operatorname{Rep}(\mathcal{N}).$$

$$(d_0; d_1^{(1)}, \dots, d_{k_1}^{(1)}; \dots; d_1^{(m)}, \dots, d_{k_m}^{(m)}) \xrightarrow{\sigma} (d_0; d_0 - d_1^{(1)}, \dots, d_0 - d_{k_1}^{(1)}; \dots; d_0 - d_1^{(m)}, \dots, d_0 - d_{k_m}^{(m)}).$$

By definition $F^- = \rho \circ \sigma : \mathrm{Rep}(\mathcal{N}) \to \mathrm{Rep}(\mathcal{N})$, $F^+ = \sigma \circ \rho = \mathrm{Rep} : (\mathcal{N}) \to \mathrm{Rep}(\mathcal{N})$, hence their action on generalized dimensions could be written as follows:

$$(d_0; d_1^{(1)}, \dots, d_{k_1}^{(1)}; \dots; d_1^{(m)}, \dots, d_{k_m}^{(m)}) \xrightarrow{F^+} (\sum_{j=1}^{m} d_{k_j}^{(j)} - d_0; \sum_{j=2}^{m} d_{k_j}^{(j)} - d_0,$$
$$\sum_{j=2}^{m} d_{k_j}^{(j)} - d_0 + d_1^{(1)}, \dots, \sum_{j=2}^{m} d_{k_j} - d_0 + d_{k_1-1}^{(1)}; \dots; \sum_{j=1}^{m-1} d_{k_j}^{(j)} - d_0 + d_{k_m-1}^{(m)}); \quad (1.1)$$

$$(d_0; d_1^{(1)}, \dots, d_{k_1}^{(1)}; \dots; d_1^{(m)}, \dots, d_{k_m}^{(m)}) \xrightarrow{F^-} ((m-1)d_0 - \sum_{j=1}^{m} d_{k_j}^{(j)}; d_2^{(1)} - d_1^{(1)},$$
$$d_3^{(1)} - d_1^{(1)}, \dots, d_{k_1-1}^{(1)} - d_1^{(1)}, d_0 - d_1^{(1)}; \dots; d_{k_l-1}^{(m)} - d_1^{(m)}, d_0 - d_1^{(m)}). \quad (1.2)$$

Importance of these functors is due to the following fact (see [4]):

Any non-degenerate representation of finite type poset $\mathcal{N}$ is the following: $(F^+)^k\pi$, where π is degenerate.

2. Representations of posets in Hilbert spaces

In this section we will consider the unitary representation of the posets. We will show how these representations are connected with $*$-representation of certain $*$-algebras that are generated by projections and linear relations. The representations of these algebras associated with primitive posets could be studied using the Coxeter functors, which will allow us to study unitary representations of the posets, using a similar technique.

2.1. Unitary representations of posets

Denote by $\mathrm{Rep}(\mathcal{N}, \mathrm{H})$ a sub-category in $\mathrm{Rep}(\mathcal{N})$. Its set of objects consists of representations in finite-dimensional Hilbert spaces and two objects π and $\tilde{\pi}$ are equivalent in $\mathrm{Rep}(\mathcal{N})$ if there exists a morphism between them that is a unitary operator $U : H \to \tilde{H}$ such that $U(H_i) = \tilde{H}_i$ (unitary equivalent). Representation $\pi \in \mathrm{Rep}(\mathcal{N}, \mathrm{H})$ is called *irreducible* iff the C^*-algebra generated by a set of orthogonal projections $\{P_i\}$ on the subspaces $\{H_i\}$ is irreducible. Let us remark that indecomposability of representation π in $\mathrm{Rep}(\mathcal{N})$ implies irreducibility of $C^*(\{P_i, i \in \mathcal{N}\})$ but the converse is not true (see [11] for details).

The problem of classifying all irreducible objects in category $\mathrm{Rep}(\mathcal{N}, \mathrm{H})$ becomes much harder. Even for primitive poset $\mathcal{N} = (1, 2)$ it is impossible to classify in a reasonable way all irreducible representations: indeed this leads us to classify up to unitary equivalence three subspaces in Hilbert space, two of which are orthogonal, but it is well known [5] that such a task is $*$-wild. Hence it is natural to consider some additional relation.

Let us consider those objects $\pi \in \mathrm{Rep}(\mathcal{N}, \mathrm{H})$, $\pi = (H; H_1, \ldots, H_2)$, which satisfy the following linear relation:

$$\alpha_1 P_1 + \cdots + \alpha_n P_n = \gamma I, \tag{2.1}$$

where α_i, γ are some positive real numbers and P_i is an orthogonal projection on space H_i. These objects form a category (in the sequel this category will be also denoted by $\mathrm{Rep}(\mathcal{N}, \mathrm{H})$, of course we will consider only those representations in Hilbert spaces which satisfy the required relation). We can also fix the weight $\chi = (\alpha_1, \ldots, \alpha_n; \gamma) \in \mathbb{R}_+^{n+1}$ and consider those representations that satisfy (2.1)). The category of such representations (which is a subcategory of $\mathrm{Rep}(\mathcal{N}, \mathrm{H})$) we will denote by $\mathrm{Rep}_\chi(\mathcal{N}, \mathrm{H})$.

The surprising result of [13] is that if there is an indecomposable $\pi \in \mathrm{Rep}(\mathcal{N})$ that is equivalent to some $\pi' \in \mathrm{Rep}(\mathcal{N}, \mathrm{H})$, then π is brick.

2.2. Certain $*$-algebras and their representation in the category of Hilbert spaces

Let be given the weight $\chi = (\alpha_1, \ldots, \alpha_n; \gamma) \in \mathbb{R}_+^{n+1}$ and consider the following $*$-algebra $\mathcal{A}_{\mathcal{N},\chi}$, which is generated by n projections and corresponds to poset $\mathcal{N}$ of order n:

$$A_{\mathcal{N},\chi} := \mathbb{C}\langle p_1, p_2, \ldots, p_n \mid p_i = p_i^2 = p_i^*,\ p_i p_j = p_j p_i = p_i,\ i \prec j,$$
$$\alpha_1 p_1 + \cdots + \alpha_n p_n = \gamma e\rangle. \tag{2.2}$$

It is easy to see that every $*$-representation π of $*$-algebra $\mathcal{A}_{\mathcal{N},\chi}$ gives us a representation of poset $\mathcal{N}$ in category $\mathrm{Rep}_\chi(\mathcal{N}, \mathrm{H})$ ($H_i = \pi(p_i)$), and vice versa every representation of poset $\mathcal{N}$ in category $\mathrm{Rep}_\chi(\mathcal{N}, \mathrm{H})$ generates the $*$-representation of $\mathcal{A}_{\mathcal{N},\chi}$. This means that categories $\mathrm{Rep}_\chi(\mathcal{N}, \mathrm{H})$ and $\mathrm{Rep}(\mathcal{A}_{\mathcal{N},\chi})$ are equivalent.

For primitive poset $\mathcal{N} = (k_1, \ldots, k_m)$, weights, for convenience, will be denoted by $\chi = (\alpha_1^{(1)}, \ldots, \alpha_{k_1}^{(1)}; \ldots; \alpha_1^{(m)}, \ldots, \alpha_{k_m}^{(m)}; \gamma)$ and correspondingly

$$\mathcal{A}_{\mathcal{N},\chi} = \mathcal{C}\langle p_1^{(1)}, \ldots, p_{k_1}^{(1)}, \ldots, p_1^{(m)}, \ldots, p_{k_m}^{(m)} \mid p_i^{(j)2} = p_i^{(j)*} = p_i^{(j)},$$
$$p_i^{(j)} p_k^{(j)} = p_k^{(j)} p_i^{(j)} = p_{\min\{i,k\}}^{(j)},\ \sum_{j=1}^{m} \sum_{i=1}^{k_l} \alpha_i^{(j)} p_i^{(j)} = \gamma e\rangle. \tag{2.3}$$

Under some conditions there exist isomorphisms between these algebras and $*$-algebras $\mathcal{A}_{\Gamma,\hat{\chi}}$ which are associated with star-shaped graphs $\Gamma = T_{k_1+1,\ldots,k_m+1}$ and character $\hat{\chi} = (\beta_1^{(1)}, \ldots, \beta_{k_1}^{(1)}; \ldots; \beta_1^{(m)}, \ldots, \beta_{k_m}^{(m)}; \gamma)$. $*$-Algebras $\mathcal{A}_{\Gamma,\hat{\chi}}$ defined in the following way:

$$\mathcal{A}_{\Gamma,\hat{\chi}} = \mathcal{C}\langle q_1^{(1)}, \dots, q_{k_1}^{(1)}, \dots, q_1^{(m)}, \dots, q_{k_m}^{(m)} \mid q_i^{(j)*} = q_i^{(j)},$$
$$q_i^{(j)} q_k^{(j)} = q_k^{(j)} q_i^{(j)} = \delta_{i,k} q_k^{(j)}, \sum_{j=1}^{m} \sum_{i=1}^{k_l} \beta_i^{(j)} q_i^{(j)} = \gamma e\rangle. \quad (2.4)$$

Proposition 2.1. *Let $\mathcal{N}$ be a primitive poset and*

$$\chi = (\alpha_1^{(1)}, \dots, \alpha_{k_1}^{(1)}; \dots; \alpha_1^{(m)}, \dots, \alpha_{k_m}^{(m)}; \gamma)$$

and

$$\hat{\chi} = (\beta_1^{(1)}, \dots, \beta_{k_1}^{(1)}; \dots; \beta_1^{(m)}, \dots, \beta_{k_m}^{(m)}; \gamma)$$

be two given weights such that for all $j = 1, \dots, m$, $i = 1, \dots, k_j$ the following conditions hold:

$$\beta_i^{(j)} = \sum_{s=i}^{k_j} \alpha_s^{(j)},$$

then there exists an isomorphism ϕ between $*$-algebras $\mathcal{A}_{\mathcal{N},\chi}$ and $\mathcal{A}_{\Gamma,\hat{\chi}}$.

Proof. One can easily prove this map, which is given on generators of algebras, in the following way:

$$p_i^{(j)} \stackrel{\phi}{\longmapsto} \sum_{s=1}^{i} q_s^{(j)}, \quad j = 1, \dots, l, \ \ i = 1, \dots, k_j,$$

establishes the isomorphism between $\mathcal{A}_{\mathcal{N},\chi}$ and $\mathcal{A}_{\Gamma,\hat{\chi}}$. □

The isomorphism ϕ can be extended to an equivalence functor between categories of representations of the algebras. Let π be a representation of $*$-algebra $\mathcal{A}_{\Gamma,\hat{\chi}}$ then $\phi \circ \pi$ is a representation of $\mathcal{A}_{\mathcal{N},\chi}$. Furthermore if π_1 and π_2 are equivalent objects in $\mathrm{Rep}(\mathcal{A}_{\Gamma,\hat{\chi}})$ then $\phi \circ \pi_1$ and $\phi \circ \pi_2$ are equivalent objects in $\mathrm{Rep}(\mathcal{A}_{\mathcal{N},\chi})$.

Let us recall some definitions from [15]:

Definition 2.2. An irreducible finite-dimensional $*$-representation π of the algebra $\mathcal{A}_{\Gamma,\hat{\chi}}$ such that

$$\pi(q_i^{(j)}) \neq 0, \qquad j = 1, \dots, m, \ \ i = 1, \dots, k_j \quad \text{and}$$
$$\sum_{i=1}^{k_j} \pi(q_i^{(j)}) \neq I, \qquad j = 1, \dots, m$$

will be called *non-degenerate*. By $\overline{\mathrm{Rep}}(\mathcal{A}_{\Gamma,\hat{\chi}})$ we will denote a full subcategory of non-degenerate representations in the category $\mathrm{Rep}(\mathcal{A}_{\Gamma,\hat{\chi}})$.

Objects that corresponds to the subcategory $\overline{\mathrm{Rep}}(\mathcal{A}_{\Gamma,\hat{\chi}})$ in $\mathrm{Rep}\mathcal{A}_{\mathcal{N},\chi})$ will be called non-degenerated as well. Non-degenerate representations of $*$-algebra $\mathcal{A}_{\mathcal{N},\chi}$ correspond to non-degenerate representations of poset $\mathcal{N}$.

For the Hilbert representation $(H; H_1, \dots, H_n)$ of a poset $\mathcal{N}$ to be degenerated means that one of the following conditions holds:

- there exist i such that $H_i = 0$, i.e., $P_i = 0$;
- there exist i such that $H_i = H_{i-1}$, i.e., $P_i = P_{i-1}$;
- there exist i such that $H_i = H$, i.e., $P_i = I$.

In this case one can obtain a non-degenerate representation of some subposet of $\mathcal{N}$ by the following procedure. If for some i, $P_i = 0$, then one automatically has a representation of poset $\mathcal{N}' = \mathcal{N} \setminus \{i\}$. If $P_i = P_{i-1}$, then one can consider π as a representation of poset $\mathcal{N}'$ where $\pi(i-1) = H_i = H_{i-1}$ and $\sum_{j \in \mathcal{N} \setminus \{i-1,i\}} \alpha_j P_j + (\alpha_{i-1} + \alpha_i) P_{i-1} = \gamma I$. And finally, if for some i $\pi(i) = H$ one can consider representation π as representation of $\mathcal{N}'$ for which holds $\sum_{j \in \mathcal{N}'} \alpha_j P_j = (\gamma - \alpha_i) I$.

2.3. Coxeter functors for representations of $*$-algebras $\mathcal{A}_{\mathcal{N},\chi}$ in case of primitive poset $\mathcal{N}$

In [14] there were developed Coxeter functors for representations of quivers in Hilbert spaces which preserve the so-called 'orthoscalarity' condition. It was proved that any irreducible representation of a quiver whose underlying graph is a Dynkin diagram can be obtained from the simplest ones by using these Coxeter functors.

These functors allow us to construct some other Coxeter functors Φ^+ and Φ^-, which act on representations of algebras $\mathcal{A}_{\Gamma,\hat{\chi}}$. Recall that $*$-algebras $\mathcal{A}_{\Gamma,\hat{\chi}}$ and their involutive representations have been studied in many recent works (see [15, 17] and others). Using the isomorphism ϕ between $*$-algebras $\mathcal{A}_{\Gamma,\hat{\chi}}$ and $\mathcal{A}_{\mathcal{N},\chi}$ we obtain two new functors which we will also denote by Φ^+ and Φ^- and call Coxeter functors. These functors act from category $\operatorname{Rep}(\mathcal{A}_{\mathcal{N},\chi})$ to $\operatorname{Rep}(\mathcal{A}_{\mathcal{N},\chi_{\Phi^+}})$ and $\operatorname{Rep}(\mathcal{A}_{\mathcal{N},\chi_{\Phi^-}})$ respectively, where

$$\chi_{\Phi^+} = (\gamma - \sum_{i=1}^{k_1} \alpha_i^{(1)}, \alpha_1^{(1)}, \dots, \alpha_{k_1-1}^{(1)}; \dots; \\ \gamma - \sum_{i=1}^{k_m} \alpha_i^{(m)}, \alpha_1^{(m)}, \dots, \alpha_{k_m-1}^{(m)}; (m-1)\gamma - \sum_{j=1}^{m} \alpha_{k_j}^{(j)});$$

$$\chi_{\Phi^-} = (\alpha_2^{(1)}, \alpha_3^{(1)}, \dots, \alpha_{k_1}^{(1)}, \sum_{j=2}^{m} \sum_{i=1}^{k_j} \alpha_i^{(j)} - \gamma; \dots; \\ \alpha_2^{(k_m)}, \dots, \sum_{j=1}^{m-1} \sum_{i=1}^{k_j} \alpha_i^{(j)} - \gamma; \sum_{j=1}^{m} \sum_{i=1}^{k_j} \alpha_i^{(j)} - \gamma).$$

In other words, these functors can be considered as functors from $\operatorname{Rep}_\chi(\mathcal{N}, \mathrm{H})$ to $\operatorname{Rep}_{\chi_{\Phi^\pm}}(\mathcal{N}, \mathrm{H})$, and these functors change the dimensions of representations by formulas similar to (1.1) and (1.2).

3. Unitarization of linear representations of primitive posets

This section is devoted to the interconnection between linear and unitary representation of the primitive posets. Actually for primitive posets of finite type we will obtain a complete list of possible weights that are appropriate to given indecomposable linear representations.

3.1. Unitarization

Let be given the representation $\pi \in \mathrm{Rep}(\mathcal{N})$, $\pi = (V; V_1, \dots, V_n)$ of the poset $\mathcal{N}$. We say that π can be unitarized if there exists an appropriate choice of hermitian structure $\langle \cdot, \cdot \rangle_{\mathbb{C}}$ in V, such that the corresponding projection $P_i : V \to V_i$ satisfies the following relation:

$$\alpha_1 P_1 + \cdots + \alpha_n P_n = \gamma I,$$

for some weight $\chi = (\alpha_1, \dots, \alpha_n; \gamma) \in \mathbb{R}_+^{n+1}$, and correspondingly we say that π can be unitarized with weight χ if this weight is fixed.

Recall that a similar notion of *unitarization* was provided in article [14] for the unitarization of representations of a given quiver Q. In fact that work studied the question when it is possible to define hermitian structure on each space X_i, $i \in Q_0$, in such a way that in each vertex i the following condition is satisfied:

$$\sum_{j \longrightarrow i} X_{j \longrightarrow i} X^*_{i \longrightarrow j} + \sum_{i \longrightarrow j} X^*_{i \longrightarrow i} X_{j \longrightarrow j} = \alpha_i I_{X_i},$$

where $X^*_{i \longrightarrow j}$ denotes the adjoint map to $X_{j \longrightarrow i}$ with respect to hermitian structure in X_i and X_j. One of the results of paper [14] is that if Q is a Dynkin quiver then every representation could be unitarizable, and if Q is an extended Dynkin quiver then there are representations that cannot be unitarized.

One can obtain an analogous fact for representations of a primitive poset of finite type using Coxeter functors. Noticed that it can be obtained using results from [14].

Proposition 3.1. *Every indecomposable linear representation of a primitive poset of finite type can be unitarized with some weight.*

Proof. To start with, remark that it is obvious that any one-dimensional representation with dimension vector $d_\pi = (1; d_1, \dots, d_n)$ of any poset can be unitarized with weight $\chi = (\alpha_1, \dots, \alpha_n; \gamma)$ satisfying *trace* condition $\sum_{s \in \mathcal{N}} d_s \alpha_s = \gamma$. All indecomposable representations of primitive posets $\mathcal{N} = (k), k \in \mathbb{N}$ and $\mathcal{N} = (k_1, k_2)$, $k_1, k_2 \in \mathbb{N}$ are one-dimensional hence they could be unitarized.

Let π be some representation of a primitive poset $\mathcal{N}$ of finite type. There are two possibilities: π is degenerate or π is non-degenerate. In the first case the representation π can be considered as one of some subposets $\mathcal{N}'$ of $\mathcal{N}$. Then by induction it can be unitarized and a corresponding unitary representation of $\mathcal{N}'$ can be restricted to a unitary representation of $\mathcal{N}$ which is a unitarization of π.

In the latter case representation π has the form $(F^-)k\pi = \pi'$ where π' is degenerate ([4] and Section 1.5), hence $(F^-)k\pi = \pi'$. Each degenerate representation

π' can be unitarized with some weight χ'. Applying functor $(\Phi^+)^k$ to its corresponding unitary representation, we obtain a unitary representation with some weight χ equivalent to π. Therefore π unitarizes with weight χ. As a result we obtain a statement of the theorem and even more: this gives the algorithm which allows us to describe all possible weights that are appropriated for the unitarization of the given representation of a finite type primitive poset. □

The next theorem gives a complete list of all possible weights for non-degenerated representations of primitive posets of finite type. In other words it describes the set of weights χ for every primitive poset $\mathcal{N}$ such that $*$-algebra $\mathcal{A}_{\mathcal{N},\chi}$ has irreducible representation in fixed dimension D. Analogous results for $*$-algebras $\mathcal{A}_{\Gamma,\hat{\chi}}$ were obtained in [15] and the following theorem can be obtained using those results but we did it independently. The list of weights is organized in the following way: for each representation of primitive poset $\mathcal{N} = (k_1, \dots, k_n)$ (which for us is given by generalized dimensional $d = (d_1^{(1)}, \dots, d_{k_1}^{(1)}; \dots; d_1^{(n)}, \dots, d_{k_n}^{(n)}; d_0)$, since this gives a representation up to isomorphism) we state the condition on weights $\chi = (\alpha_1, \dots, \alpha_{k_1}; \beta_1, \dots; \gamma)$, under which it is possible to unitarize the linear representation. A complete list (for all representations including degenerated) is available online (http://arxiv.org/pdf/0807.0155).

Theorem 3.2. *For primitive poset $\mathcal{N}$ its non-degenerate linear representation with dimension D unitarizes with every weight χ for which conditions C are satisfied:*

1. *Poset* $\mathcal{N} = (1; 1; 1)$

Dimensions D	*Conditions C*
$(1;1;1;2)$	$\alpha < \gamma, \beta < \gamma, \delta < \gamma, \alpha + \beta + \delta = 2\gamma$

2. *Poset* $\mathcal{N} = (2; 1; 1)$

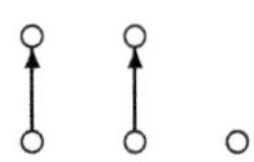

Dimensions D	*Conditions C*
$(1,2;1,2;1;3)$	$\alpha_1 + \alpha_2 < \gamma, \alpha_1 + \alpha_2 + \beta_2 + \delta < 2\gamma, \beta_1 + \beta_2 < \gamma, \alpha_2 + \beta_1 + \beta_2 + \delta < 2\gamma, \alpha_2 + \beta_2 < \gamma, \alpha_1 + 2\alpha_2 + \beta_1 + 2\beta_2 + \delta = 3\gamma$
$(1,2;1,2;2;3)$	$\beta_2 + \delta < \gamma, \alpha_2 + \beta_1 + 2\beta_2 + \delta < 2\gamma, \alpha_2 + \delta < \gamma, \alpha_1 + 2\alpha_2 + \beta_2 + \delta < 2\gamma, \alpha_1 + \alpha_2 + \beta_1 + \beta_2 + \delta < 2\gamma, \alpha_1 + 2\alpha_2 + \beta_1 + 2\beta_2 + 2\delta = 3\gamma$

3. *Poset* $\mathcal{N} = (3; 2; 1)$

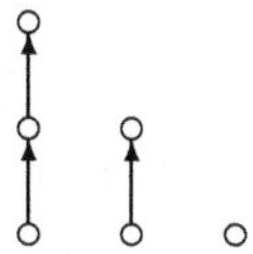

Dimensions D	*Conditions C*
$(1,2,3;1,2;2;4)$	$\alpha_1+\alpha_2+\alpha_3<\gamma, \alpha_1+\alpha_2+\alpha_3+\beta_2+\delta<2\gamma, \alpha_1+\alpha_2+2\alpha_3+\beta_1+2\beta_2+\delta<3\gamma, \alpha_3+\delta<\gamma, \alpha_2+2\alpha_3+\beta_2+\delta<2\gamma, \alpha_2+\alpha_3+\beta_1+\beta_2+\delta<2\gamma, \alpha_1+2\alpha_2+3\alpha_3+\beta_1+2\beta_2+2\delta=4\gamma$
$(1,2,3;1,3;2;4)$	$\beta_2+\delta<\gamma, \alpha_3+\beta_1+2\beta_2+\delta<2\gamma, \alpha_2+2\alpha_3+\beta_1+2\beta_2+2\delta<3\gamma, \alpha_2+\alpha_3+\beta_2<\gamma, \alpha_1+2\alpha_2+2\alpha_3+\beta_1+2\beta_2+\delta<3\gamma, \alpha_1+\alpha_2+2\alpha_3+\beta_2+\delta<2\gamma, \alpha_1+2\alpha_2+3\alpha_3+\beta_1+3\beta_2+2\delta=4\gamma$
$(1,2,3;2,3;2;4)$	$\alpha_3+\beta_1+\beta_2<\gamma, \alpha_2+2\alpha_3+\beta_1+\beta_2+\delta<2\gamma, \alpha_1+2\alpha_2+3\alpha_3+\beta_1+2\beta_2+\delta<3\gamma, \alpha_1+\alpha_2+\alpha_3+\beta_1+\beta_2+\delta<2\gamma, \alpha_1+\alpha_2+2\alpha_3+\beta_1+2\beta_2+2\delta<3\gamma, \alpha_2+\alpha_3+\beta_1+2\beta_2+\delta<2\gamma, \alpha_1+2\alpha_2+3\alpha_3+2\beta_1+3\beta_2+2\delta=4\gamma$

4. *Poset* $\mathcal{N} = (4; 2; 1)$

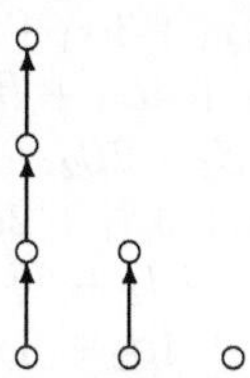

Dimensions D	*Conditions C*
$(1,2,3,4;1,3;2;5)$	$\alpha_1+\alpha_2+\alpha_3+\alpha_4<\gamma, \alpha_1+\alpha_2+\alpha_3+\alpha_4+\beta_2+\delta<2\gamma, \alpha_1+\alpha_2+\alpha_3+2\alpha_4+\beta_1+2\beta_2+\delta<3\gamma, \alpha_1+\alpha_2+2\alpha_3+3\alpha_4+\beta_1+2\beta_2+2\delta<4\gamma, \alpha_3+\alpha_4+\beta_2<\gamma, \alpha_2+2\alpha_3+2\alpha_4+\beta_1+2\beta_2+\delta<3\gamma, \alpha_2+\alpha_3+2\alpha_4+\beta_2+\delta<2\gamma, \alpha_1+2\alpha_2+3\alpha_3+4\alpha_4+\beta_1+3\beta_2+2\delta=5\gamma$
$(1,2,3,4;2,3;2;5)$	$\alpha_1+\alpha_2+\alpha_3+\alpha_4<\gamma, \alpha_4+\beta_1+\beta_2<\gamma, \alpha_3+2\alpha_4+\beta_1+\beta_2+\delta<2\gamma, \alpha_2+2\alpha_3+3\alpha_4+\beta_1+2\beta_2+\delta<3\gamma, \alpha_2+\alpha_3+\alpha_4+\beta_1+\beta_2+\delta<2\gamma, \alpha_1+2\alpha_2+2\alpha_3+3\alpha_4+\beta_1+2\beta_2+2\delta<4\gamma, \alpha_1+\alpha_2+2\alpha_3+2\alpha_4+\beta_1+2\beta_2+\delta<3\gamma, \alpha_1+2\alpha_2+3\alpha_3+4\alpha_4+2\beta_1+3\beta_2+2\delta=5\gamma$

$(1,2,3,4;2,4;2;5)$	$\alpha_4+\beta_1+\beta_2<\gamma, \alpha_2+\alpha_3+\alpha_4+\beta_2<\gamma, \alpha_1+2\alpha_2+2\alpha_3+2\alpha_4+\beta_1+2\beta_2+\delta<3\gamma, \alpha_1+2\alpha_2+2\alpha_3+3\alpha_4+\beta_1+3\beta_2+2\delta<4\gamma, \alpha_3+\alpha_4+\beta_1+2\beta_2+\delta<2\gamma, \alpha_2+2\alpha_3+3\alpha_4+2\beta_1+3\beta_2+2\delta<4\gamma, \alpha_1+\alpha_2+2\alpha_3+3\alpha_4+\beta_1+2\beta_2+\delta<3\gamma, \alpha_1+2\alpha_2+3\alpha_3+4\alpha_4+2\beta_1+4\beta_2+2\delta=5\gamma$
$(1,2,3,4;1,3;3;5)$	$\beta_2+\delta<\gamma, \alpha_3+\alpha_4+\delta<\gamma, \alpha_2+2\alpha_3+2\alpha_4+\beta_2+\delta<2\gamma, \alpha_1+2\alpha_2+3\alpha_3+3\alpha_4+\beta_1+2\beta_2+2\delta<4\gamma, \alpha_1+\alpha_2+\alpha_3+2\alpha_4+\beta_2+\delta<2\gamma, \alpha_1+\alpha_2+2\alpha_3+3\alpha_4+\beta_1+3\beta_2+2\delta<4\gamma, \alpha_2+\alpha_3+2\alpha_4+\beta_1+2\beta_2+2\delta<3\gamma, \alpha_1+2\alpha_2+3\alpha_3+4\alpha_4+\beta_1+3\beta_2+3\delta=5\gamma$
$(1,2,3,4;2,3;3;5)$	$\alpha_3+\alpha_4+\delta<\gamma, \alpha_1+\alpha_2+\alpha_3+\alpha_4+\beta_1+\beta_2+\delta<2\gamma, \alpha_1+\alpha_2+\alpha_3+2\alpha_4+\beta_1+2\beta_2+2\delta<3\gamma, \alpha_1+\alpha_2+2\alpha_3+3\alpha_4+2\beta_1+3\beta_2+2\delta<4\gamma, \alpha_2+\alpha_3+2\alpha_4+\beta_1+\beta_2+\delta<2\gamma, \alpha_1+2\alpha_2+3\alpha_3+4\alpha_4+\beta_1+2\beta_2+2\delta<4\gamma, \alpha_2+2\alpha_3+2\alpha_4+\beta_1+2\beta_2+2\delta<3\gamma, \alpha_1+2\alpha_2+3\alpha_3+4\alpha_4+2\beta_1+3\beta_2+3\delta=5\gamma$
$(1,2,3,4;2,4;3;5)$	$\alpha_4+\beta_2+\delta<\gamma, \alpha_3+2\alpha_4+\beta_1+2\beta_2+\delta<2\gamma, \alpha_2+2\alpha_3+3\alpha_4+\beta_1+2\beta_2+2\delta<3\gamma, \alpha_1+2\alpha_2+3\alpha_3+4\alpha_4+\beta_1+3\beta_2+2\delta<4\gamma, \alpha_2+\alpha_3+\alpha_4+\beta_1+2\beta_2+\delta<2\gamma, \alpha_1+2\alpha_2+2\alpha_3+3\alpha_4+2\beta_1+3\beta_2+2\delta<4\gamma, \alpha_1+\alpha_2+2\alpha_3+2\alpha_4+\beta_1+2\beta_2+2\delta<3\gamma, \alpha_1+2\alpha_2+3\alpha_3+4\alpha_4+2\beta_1+4\beta_2+3\delta=5\gamma$
$(1,2,3,4;2,4;3;6)$	$\beta_2+\delta<\gamma, \alpha_4+\beta_1+2\beta_2+\delta<2\gamma, \alpha_3+2\alpha_4+\beta_1+2\beta_2+2\delta<3\gamma, \alpha_2+2\alpha_3+3\alpha_4+\beta_1+3\beta_2+2\delta<4\gamma, \alpha_2+\alpha_3+\alpha_4+\beta_1+\beta_2+\delta<2\gamma, \alpha_1+2\alpha_2+2\alpha_3+3\alpha_4+\beta_1+2\beta_2+2\delta<4\gamma, \alpha_1+\alpha_2+2\alpha_3+2\alpha_4+\beta_1+2\beta_2+\delta<3\gamma, \alpha_1+2\alpha_2+3\alpha_3+4\alpha_4+2\beta_1+4\beta_2+3\delta=6\gamma$
$(1,2,3,5;2,4;3;6)$	$\alpha_4+\beta_1+\beta_2<\gamma, \alpha_3+2\alpha_4+\beta_1+\beta_2+\delta<2\gamma, \alpha_2+2\alpha_3+3\alpha_4+\beta_1+2\beta_2+\delta<3\gamma, \alpha_1+2\alpha_2+3\alpha_3+4\alpha_4+2\beta_1+3\beta_2+2\delta<5\gamma, \alpha_1+\alpha_2+\alpha_3+2\alpha_4+\beta_2+\delta<2\gamma, \alpha_1+\alpha_2+2\alpha_3+3\alpha_4+\beta_1+3\beta_2+2\delta<4\gamma, \alpha_2+\alpha_3+2\alpha_4+\beta_1+2\beta_2+2\delta<3\gamma, \alpha_1+2\alpha_2+3\alpha_3+5\alpha_4+2\beta_1+4\beta_2+3\delta=6\gamma$

$(1,2,4,5;2,4;3;6)$	$\alpha_3+\alpha_4+\delta<\gamma, \alpha_2+2\alpha_3+2\alpha_4+\beta_2+\delta<2\gamma, \alpha_1+2\alpha_2+3\alpha_3+3\alpha_4+\beta_1+2\beta_2+2\delta<4\gamma, \alpha_1+2\alpha_2+3\alpha_3+4\alpha_4+\beta_1+3\beta_2+3\delta<5\gamma, \alpha_3+\alpha_4+\beta_1+2\beta_2+\delta<2\gamma, \alpha_2+2\alpha_3+3\alpha_4+2\beta_1+3\beta_2+2\delta<4\gamma, \alpha_1+\alpha_2+2\alpha_3+3\alpha_4+\beta_1+2\beta_2+\delta<3\gamma, \alpha_1+2\alpha_2+4\alpha_3+5\alpha_4+2\beta_1+4\beta_2+3\delta=6\gamma$
$(1,3,4,5;2,4;3;6)$	$\alpha_2+\alpha_3+\alpha_4+\beta_2<\gamma, \alpha_1+2\alpha_2+2\alpha_3+2\alpha_4+\beta_1+2\beta_2+\delta<3\gamma, \alpha_1+2\alpha_2+2\alpha_3+3\alpha_4+\beta_1+3\beta_2+2\delta<4\gamma, \alpha_1+2\alpha_2+3\alpha_3+4\alpha_4+2\beta_1+4\beta_2+2\delta<5\gamma, \alpha_2+\alpha_3+2\alpha_4+\beta_1+\beta_2+\delta<2\gamma, \alpha_1+2\alpha_2+3\alpha_3+4\alpha_4+\beta_1+2\beta_2+2\delta<4\gamma, \alpha_2+2\alpha_3+2\alpha_4+\beta_1+2\beta_2+2\delta<3\gamma, \alpha_1+3\alpha_2+4\alpha_3+5\alpha_4+2\beta_1+4\beta_2+3\delta=6\gamma$
$(2,3,4,5;2,4;3;6)$	$\alpha_1+\alpha_2+\alpha_3+\alpha_4+\beta_1+\beta_2+\delta<2\gamma, \alpha_1+\alpha_2+\alpha_3+2\alpha_4+\beta_1+2\beta_2+2\delta<3\gamma, \alpha_1+\alpha_2+2\alpha_3+3\alpha_4+2\beta_1+3\beta_2+2\delta<4\gamma, \alpha_1+2\alpha_2+3\alpha_3+4\alpha_4+2\beta_1+3\beta_2+3\delta<5\gamma, \alpha_1+\alpha_2+2\alpha_3+2\alpha_4+\beta_2+\delta<2\gamma, \alpha_1+2\alpha_2+3\alpha_3+3\alpha_4+\beta_1+3\beta_2+2\delta<4\gamma, \alpha_1+2\alpha_2+2\alpha_3+3\alpha_4+\beta_1+2\beta_2+\delta<3\gamma, 2\alpha_1+3\alpha_2+4\alpha_3+5\alpha_4+2\beta_1+4\beta_2+3\delta=6\gamma$

Acknowledgments

The authors are deeply grateful to Prof. Yu.S. Samoilenko for the statement of the problem and for his constant attention to this work.

References

[1] L.A. Nazarova, A.V. Roiter, *Representations of the partially ordered sets.* Investigations on the theory of representations. Zap. Naučn. Sem. Leningrad. Otdel. Mat. Inst. Steklov. (LOMI) (Russian) **28** (1972), 5–31.

[2] M.M. Kleiner, *Partially ordered sets of finite type.* Investigations on the theory of representations. Zap. Naučn. Sem. Leningrad. Otdel. Mat. Inst. Steklov. (LOMI) (Russian) **28** (1972), 32–41.

[3] M.M. Kleiner, *On the faithful representations of partially ordered sets of finite type.* Investigations on the theory of representations. Zap. Naučn. Sem. Leningrad. Otdel. Mat. Inst. Steklov. (LOMI) (Russian) **28** (1972), 42–59.

[4] Ju.A. Drozd, *Coxeter transformations and representations of partially ordered sets.* Funkcional. Anal. i Prilozen. (Russian) **8** (1974), no. 3, 34–42.

[5] S.A. Krugljak, Yu.S. Samoilenko, *Unitary equivalence of sets of selfadjoint operators.* Funkcional. Anal. i Prilozen. (Russian) **14**, (1980), no. 1, 60–62.

[6] S.A. Krugljak, Yu.S. Samoilenko, *On the complexity of description of representations of $*$-algebras generated by idempotents.* Proceedings of the American Mathematical Society **128**, (2000), no. 6, 1655–1664.

[7] P. Gabriel, A.V. Roiter, *Representations of Finite-Dimensional Algebras.* Enc. of Math. Sci., **73**, Algebra VIII, (Springer, 1992).

[8] I. Assem, D. Simpson, A. Skowronskiy, *Elements of the Representation Theory of Associative Algebras. Volume* 1 *Techniques of Representation Theory.* London Mathematical Society Student Texts **65**, 2006.

[9] V. Dlab and C.M. Ringel, *Indecomposable representations of graphs and algebras.* Memories Amer. Math. Soc **6** (1976), no. 176, 1–72.

[10] P. Donovan and M. Freislich, *The Representation Theory of Finite Graphs and Associated Algebras.* Carleton Mathematical Lecture Notes **5** 1973, 1–187.

[11] M. Enomoto and Ya. Watatani, *Relative position of four subspaces in a Hilbert space.* Adv. Math. **201** (2006), no. 2, 263–317.

[12] I.N. Berstein, I.M. Gelfand, V.A. Ponomarev, *Coxeter functors, and Gabriel's theorem.* Uspehi Mat. Nauk (Russian) **2** (1973), no. 170, pp. 19–33.

[13] S.A. Kruglyak, L.A. Nazarova, A.V. Roiter, *Orthoscalar representations of quivers in the category of Hilbert spaces.* J. Math. Sci. (N.Y.) **145** (2007), no. 1, 4793–4804.

[14] S.A. Kruglyak, A.V. Roiter, *Locally scalar representations of graphs in the category of Hilbert spaces.* Funct. Anal. Appl. **39** (2005), no. 2. 91–105.

[15] S.A. Krugljak, S.V. Popovych, Yu.S. Samoilenko, *The spectral problem and $*$-representations of algebras associated with Dynkin graphs.*, J. Algebra Appl. **4** (2005), no. 6, 761–776.

[16] W. Crawley-Boevey and Ch. Geiss, *Horn's problem and semi-stability for quiver representations.* Representations of Algebras, V.I.Poc. IX Intern. Conf. Benjing (2000), 40–48.

[17] V.L. Ostrovskyi, Yu.S. Samoilenko, *On spectral theorems for families of linearly connected selfadjoint operators with prescribed spectra associated with extended Dynkin graphs.* Ukrain. Mat. Zh. (Ukrainian) **58** (2006), no. 11, 1556–1570.

Roman Grushevoy and Kostyantyn Yusenko
Institute of Mathematics
National Academy of Science of Ukraine
3 Tereshchenkivs'ka St.
01601 Kyiv, Ukraine
e-mail: grushevoy@imath.kiev.ua
kay@imath.kiev.ua

Operator Theory:
Advances and Applications, Vol. 190, 295–314

Direct and Inverse Theorems in the Theory of Approximation of Banach Space Vectors by Exponential Type Entire Vectors

Yaroslav Grushka and Sergiy Torba

Abstract. An arbitrary operator A on a Banach space $\mathfrak{X}$ such that either A or iA generates the C_0-group with certain growth condition at infinity is considered. The direct and inverse theorems on connection between the degree of smoothness of a vector $x \in \mathfrak{X}$ with respect to the operator A, the rate of convergence to zero of the best approximation of x by exponential type entire vectors for the operator A, and the k-module of continuity are established. These results allow to obtain Jackson-type and Bernstein-type inequalities in weighted L_p spaces.

Mathematics Subject Classification (2000). Primary 41A25, 41A27, 41A17, 41A65.

Keywords. Direct and inverse theorems, Module of continuity, Banach space, Non-quasianalytic operators, Entire vectors of exponential type.

1. Introduction

Direct and inverse theorems which establish the relationship between the degree of smoothness of a function with respect to the differentiation operator and the rate of convergence to zero of its best approximation by trigonometric polynomials are well known in the theory of approximation of periodic functions. Jackson's and Bernstein's inequalities are ones among such results.

N.P. Kuptsov proposed a generalized notion of module of continuity, expanded onto C_0-groups in a Banach space [1]. Using this notion, N.P. Kuptsov [1] and A.P. Terekhin [2] proved the generalized Jackson's inequalities for the cases of

This work was partially supported by the Ukrainian State Foundation for Fundamental Research (project N14.1/003).

bounded group and s-regular group. Recall that group $\{U(t)\}_{t\in\mathbb{R}}$ is called s-regular if resolvent of its generator A satisfies the condition

$$\exists\theta\in\mathbb{R}: \ \|R_\lambda(e^{i\theta}A^s)\| \le \frac{C}{\mathrm{Im}\lambda}.$$

G.V. Radzievsky studied direct and inverse theorems [3, 4], using the notion of K-functional instead of module of continuity, but it should be noted that the K-functional has two-sided estimates with regard to the module of continuity at least for bounded C_0-groups.

In the papers [5, 6] and [7] authors investigated the case of a group of unitary operators in a Hilbert space and established Jackson-type inequalities in Hilbert spaces and their rigs. These inequalities are used to estimate the rate of convergence to zero of the best approximation of both finite and infinite smoothness vectors for operator A by exponential type entire vectors.

We consider C_0-groups, generated by so-called *non-quasianalytic operators* [8], i.e., groups satisfying

$$\int_{-\infty}^{\infty}\frac{\ln\|U(t)\|}{1+t^2}dt<\infty. \tag{1.1}$$

We recall that the belonging of group to the C_0 class means that for every $x\in\mathfrak{X}$ vector-function $U(t)x$ is continuous on $\mathbb{R}$ with respect to the norm of space $\mathfrak{X}$.

As it was shown in [5], the set of exponential type entire vectors for the non-quasianalytic operator A is dense in $\mathfrak{X}$, so the problem of approximation by exponential type entire vectors is correct. On the other hand, it was shown in [9] that condition (1.1) is close to the necessary one, so in the case when (1.1) doesn't hold, the class of entire vectors isn't necessary dense in $\mathfrak{X}$, and the corresponding approximation problem loses its meaning.

The purpose of this work is to obtain Jackson- and Bernstein-type inequalities and inverse theorems in the case where a vector of a Banach space is approximated by exponential type entire vectors for a non-quasianalytic operator, and to give some applications of these results to weighted L_p spaces. First, in Section 2, all necessary definitions and the statement of approximation problem are given, next, in Section 3 we describe required apparatus, in Section 4 direct and inverse theorems are established, and, in the last Section 5, we give some examples of using the results in weighted L_p spaces.

2. Preliminaries

Let A be a closed linear operator with dense domain of definition $\mathcal{D}(A)$ in the Banach space $(\mathfrak{X},\|\cdot\|)$ over the field of complex numbers.

Let $C^\infty(A)$ denotes the set of all infinitely differentiable vectors of operator A, i.e.,

$$C^\infty(A)=\bigcap_{n\in\mathbb{N}_0}\mathcal{D}(A^n), \quad \mathbb{N}_0=\mathbb{N}\cup\{0\}.$$

For number $\alpha > 0$ we set

$$\mathfrak{E}^\alpha(A) = \left\{x \in C^\infty(A) \mid \exists c = c(x) > 0 \; \forall k \in \mathbb{N}_0 \; \left\|A^k x\right\| \le c\alpha^k\right\}.$$

The set $\mathfrak{E}^\alpha(A)$ is a Banach space with respect to the norm

$$\|x\|_{\mathfrak{E}^\alpha(A)} = \sup_{n\in\mathbb{N}_0} \frac{\|A^n x\|}{\alpha^n}.$$

Then $\mathfrak{E}(A) = \bigcup_{\alpha>0} \mathfrak{E}^\alpha(A)$ is a linear locally convex space with respect to the topology of inductive limit of Banach spaces $\mathfrak{E}^\alpha(A)$:

$$\mathfrak{E}(A) = \liminf_{\alpha\to\infty} \mathfrak{E}^\alpha(A).$$

Elements of space $\mathfrak{E}(A)$ are called exponential type entire vectors of operator A. The type $\sigma(x, A)$ of vector $x \in \mathfrak{E}(A)$ is defined as the number

$$\sigma(x, A) = \inf\{\alpha > 0 \,:\, x \in \mathfrak{E}^\alpha(A)\} = \limsup_{n\to\infty} \|A^n x\|^{\frac{1}{n}}. \tag{2.1}$$

Denote by $\Xi^\alpha(A)$ the following set

$$\Xi^\alpha(A) = \left\{x \in \mathfrak{E}(A) \mid \sigma(x) \le \alpha\right\}. \tag{2.2}$$

It is easy to see that

$$\mathfrak{E}^\alpha(A) \subset \Xi^\alpha(A) = \bigcap_{\epsilon>0} \mathfrak{E}^{\alpha+\epsilon}(A). \tag{2.3}$$

The question if the inclusion in the equation (2.3) is strict naturally arises. If the operator A is a generator of bounded C_0-group $U(t)$, then, as follows from [2, Theorem 1], $\mathfrak{E}^\alpha(A) = \Xi^\alpha(A)$. But if the group $U(t)$ is unbounded (as is for non-quasianalytic operators A), it is possible that $\mathfrak{E}^\alpha(A) \neq \Xi^\alpha(A)$, that is there exists such vector x that $\sigma(x) = r$, but $x \notin \mathfrak{E}^r(A)$. We will return to this question later, discussing the generalization of the Bernstein inequality in Section 4 and in Section 5, where the example of such operator A and vector x is constructed.

For arbitrary $x \in \mathfrak{X}$ we set, according to [7, 6],

$$\mathcal{E}_r(x, A) = \inf_{y\in\Xi^r(A)} \|x - y\|, \quad r > 0,$$

i.e., $\mathcal{E}_r(x, A)$ is the best approximation of element x by exponential type entire vectors y of operator A for which $\sigma(y, A) \le r$. For fixed x, $\mathcal{E}_r(x, A)$ does not increase and $\mathcal{E}_r(x, A) \to 0$, $r \to \infty$ for every $x \in \mathfrak{X}$ if and only if the set $\mathfrak{E}(A)$ of exponential type entire vectors is dense in $\mathfrak{X}$. Particularly, as indicated above, the set $\mathfrak{E}(A)$ is dense in $\mathfrak{X}$ if A generates C_0-group $\{U(t) \,:\, t \in \mathbb{R}\}$ and this group belongs to non-quasianalytic class (that is, it satisfies (1.1)).

Example. Let $\mathfrak{X}$ is one of the $L_p(2\pi)$ $(1 \le p < \infty)$ spaces of integrable in pth degree over $[0, 2\pi]$, 2π-periodical functions or the space $C(2\pi)$ of continuous 2π-periodical functions (the norm in $\mathfrak{X}$ is defined in a standard way), and let A is the differentiation operator in the space $\mathfrak{X}$ ($\mathcal{D}(A) = \{x \in \mathfrak{X} \cap AC(\mathbb{R}) \,:\, x' \in \mathfrak{X}\}$; $(Ax)(t) = \frac{dx}{dt}$, where $AC(\mathbb{R})$ denotes the space of absolutely continuous functions over $\mathbb{R}$). It can be proved that in such case $\mathfrak{E}(A)$ coincides with the set of all trigonometric polynomials, and for $y \in \mathfrak{E}(A)$ $\sigma(y, A) = \deg(y)$, where $\deg(y)$ is the degree of trigonometric polynomial y.

Note that all previous definitions do not change if we replace the operator A by any operator of form $e^{i\theta}A$, $\theta \in \mathbb{R}$. Moreover, main results of this article do not depend on which operator generates group $U(t)$ – either A or iA. So, in what follows, we always assume that the operator iA is the generator of C_0-group of linear continuous operators $\{U(t) : t \in \mathbb{R}\}$ on $\mathfrak{X}$. Moreover, we suppose that the operator A is non-quasianalytic.

For $t \in \mathbb{R}_+$, we set

$$M_U(t) := \sup_{\tau \in \mathbb{R},\, |\tau| \le t} \|U(\tau)\| . \tag{2.4}$$

The estimation $\|U(t)\| \le Me^{\omega t}$ for some $M, \omega \in \mathbb{R}$ implies $M_U(t) < \infty$ (for all $t \in \mathbb{R}_+$). It is easy to see that the function $M_U(\cdot)$ has the following properties:

1) $M_U(t) \ge 1$, $t \in \mathbb{R}_+$;
2) $M_U(\cdot)$ is monotonically non-decreasing on $\mathbb{R}_+$;
3) $M_U(t_1 + t_2) \le M_U(t_1)M_U(t_2)$, $t_1, t_2 \in \mathbb{R}_+$.

According to [1], for $x \in \mathfrak{X}$, $t \in \mathbb{R}_+$ and $k \in \mathbb{N}$ we set as a generalization of module of smoothness,

$$\omega_k(t, x, A) = \sup_{0 \le \tau \le t} \left\|\Delta_\tau^k x\right\|, \qquad \text{where} \tag{2.5}$$

$$\Delta_h^k = (U(h) - \mathbb{I})^k = \sum_{j=0}^{k} (-1)^{k-j} \binom{j}{k} U(jh), \quad k \in \mathbb{N}_0,\, h \in \mathbb{R} \quad (\Delta_h^0 \equiv 1). \tag{2.6}$$

Moreover, let

$$\widetilde{\omega}_k(t, x, A) = \sup_{|\tau| \le t} \left\|\Delta_\tau^k x\right\| . \tag{2.7}$$

Remark 2.1. It is easy to see that in the case of bounded group $\{U(t)\}$ ($\|U(t)\| \le M$, $t \in \mathbb{R}$) the quantities $\omega_k(t, x, A)$ and $\widetilde{\omega}_k(t, x, A)$ are equivalent within constant factor ($\omega_k(t, x, A) \le \widetilde{\omega}_k(t, x, A) \le M\, \omega_k(t, x, A)$), and in the case of isometric group ($\|U(t)\| \equiv 1$, $t \in \mathbb{R}$) these quantities coincide.

It is immediate from the definition of $\widetilde{\omega}_k(t, x, A)$ that for $k \in \mathbb{N}$:

1) $\widetilde{\omega}_k(0, x, A) = 0$;
2) for fixed x the function $\widetilde{\omega}_k(t, x, A)$ is non-decreasing and is continuous by the variable t on $\mathbb{R}_+$;
3) $\widetilde{\omega}_k(nt, x, A) \le \big(1 + (n-1)M_U((n-1)t)\big)^k \widetilde{\omega}_k(t, x, A)$ ($n \in \mathbb{N}$, $t > 0$);
4) $\widetilde{\omega}_k(\mu t, x, A) \le \big(1 + \mu M_U(\mu t)\big)^k \widetilde{\omega}_k(t, x, A)$ ($\mu, t > 0$);
5) for fixed $t \in \mathbb{R}_+$ the function $\widetilde{\omega}_k(t, x, A)$ is continuous in x.

3. Integration kernels and spectral subspaces

In contrast to the Hilbert space case and selfadjoint operators, we do not have spectral decomposition of operator as the convenient tool to deal with entire vectors of exponential type, we need another instruments. This section devoted to two such things – the investigation of properties of generalizations of classic integral kernels, such as

$$\frac{1}{2\pi}\left(\frac{\sin\frac{x}{2}}{\frac{x}{2}}\right)^2 \quad \text{and} \quad \frac{1}{K_n}\left(\frac{\sin\frac{x}{2n}}{\frac{x}{2n}}\right)^{2n}, \tag{3.1}$$

and to determining the relationship between entire vectors of exponential type and spectral subspaces of non-quasianalytic operators, constructed in [8].

3.1. One class of integration kernels

Integral kernels, generalizing (3.1), were first constructed in [10]. They are entire functions, bounded and fast decreasing on real axis. They will be used later for constructing of approximation. Now we describe how these kernels are constructed and state their properties.

In what follows we denote as $\mathfrak{Q}$ the class of functions $\alpha : \mathbb{R} \mapsto \mathbb{R}$, satisfying the following conditions:

I) $\alpha(\cdot)$ is measurable and bounded on any segment $[-T, T] \subset \mathbb{R}$.
II) $\alpha(t) > 0,\ t \in \mathbb{R}$.
III) $\alpha(t_1 + t_2) \leq \alpha(t_1)\alpha(t_2), \quad t_1, t_2 \in \mathbb{R}$.
IV) $\int_{-\infty}^{\infty} \frac{|\ln(\alpha(t))|}{1+t^2}\,dt < \infty$.

Note that these conditions are properties of norm of non-quasianalytic group. Without lost of generality we may assume that the function $\alpha(t)$ satisfies additional conditions:

V) $\alpha(t) \geq 1,\ t \in \mathbb{R}$; [1]
VI) $\alpha(t)$ is even on $\mathbb{R}$ and is monotonically increasing on $\mathbb{R}_+$;
VII) $\|\alpha^{-1}\|_{L_1(\mathbb{R})} = \int_{-\infty}^{\infty} |\alpha^{-1}(t)|dt < \infty$.

Really, it is easy to verify that assumptions V), VII) and condition that the function $\alpha(t)$ is even in VI) don't confine the general case if one examine the function $\alpha_1(t) = \widetilde{\alpha}(t)\widetilde{\alpha}(-t)$, where $\widetilde{\alpha}(t) = (1 + \alpha(t))(1 + t^2)$. In [11, Theorems 1 and 2] it has been proved that the monotony condition on $\alpha(t)$ in VI) doesn't confine the general case too.

It follows from VII) that

$$\alpha(t) \to \infty,\ t \to \infty. \tag{3.2}$$

[1]As shown in [8], for non-quasianalytic groups the condition $\|U(t)\| \geq 1$ always holds, therefore in this paper the condition V) automatically takes place.

Let $\beta(t) = \ln \alpha(t),\ t \in \mathbb{R}$. Conditions III)–VII) and (3.2) lead to conclusion that

$$\beta(t) > 0, \qquad \beta(-t) = \beta(t), \qquad \beta(t) \to \infty,\ t \to \infty;$$

$$\beta(t_1 + t_2) \le \beta(t_1) + \beta(t_2),\ t_1, t_2 \in \mathbb{R} \tag{3.3}$$

$$\int_1^\infty \frac{\beta(t)}{t^2} dt < \infty \tag{3.4}$$

Because of (3.3) there exists limit $\lim_{t\to\infty} \frac{\beta(t)}{t}$. And, by virtue of (3.4):

$$\lim_{t\to\infty} \frac{\beta(t)}{t} = 0. \tag{3.5}$$

Also, using monotony of $\beta(t)$, for $t \in [k, k+1],\ k \in \mathbb{N}$ one gets

$$\frac{\beta(t)}{t^2} \ge \frac{\beta(k)}{(k+1)^2} \ge \frac{1}{4}\frac{\beta(k)}{k^2},$$

thus using (3.4) it is easy to obtain that

$$\sum_{k=1}^{\infty} \frac{\beta(k)}{k^2} < \infty, \tag{3.6}$$

moreover, all terms of series (3.6) are positive. From the convergence of series (3.6) it follows the existence of such sequence $\{Q_n\}_{n=1}^{\infty} \subset \mathbb{R}$ that $Q_n > 1,\quad Q_n \to \infty,\ n \to \infty$ and

$$\sum_{k=1}^{\infty} \frac{\beta(k)}{k^2} Q_k = S < \infty. \tag{3.7}$$

We set

$$a_k := \frac{\beta(k) Q_k}{S\, k^2}, \qquad k \in \mathbb{N}.$$

The definition of a_k and (3.7) result in equality

$$\sum_{k=1}^{\infty} a_k = 1. \tag{3.8}$$

We construct the sequence of functions, which, obviously, are entire for every $n \in \mathbb{N}$:

$$f_n(z) := \prod_{k=1}^{n} P_k(z), \quad \text{where } P_k(z) = \left(\frac{\sin \frac{a_k z}{2}}{\frac{a_k z}{2}} \right)^2,\ z \in \mathbb{C},\ n \in \mathbb{N}.$$

Similarly to the proof of Denjoy-Carleman theorem [12, p. 378] it can be concluded that the sequence of (entire) functions $f_n(z)$ converges uniformly to the function

$$f(z) = \prod_{k=1}^{\infty} \left(\frac{\sin \frac{a_k z}{2}}{\frac{a_k z}{2}} \right)^2, \qquad z \in \mathbb{C}$$

in every disk $\{z \in \mathbb{C} \,|\, |z| \le R\}$. Thus, by Weierstrass theorem, the function $f(z)$ is entire.

It can be verified that $0 < \|f\|_{L_1(\mathbb{R})} < \infty$. Thus let

$$\mathcal{K}_\alpha(z) := \frac{1}{\|f\|_{L_1(\mathbb{R})}} f(z), \quad z \in \mathbb{C}. \tag{3.9}$$

Functions $\mathcal{K}_\alpha(z)$ are generalizations of integral kernels (3.1). The following two lemmas state properties of these kernels.

Lemma 3.1 ([13]). *Let $\alpha \in \mathfrak{Q}$. Then the function $\mathcal{K}_\alpha(z)$, constructed above, satisfies*

1) $\mathcal{K}_\alpha(t) \geq 0,\ t \in \mathbb{R}$;
2) $\int_{-\infty}^{\infty} \mathcal{K}_\alpha(t)\,dt = 1$;
3) $\forall r > 0\ \exists c_r = c_r(\alpha) > 0\ \forall z \in \mathbb{C} \quad |\mathcal{K}_\alpha(rz)| \leq c_r \frac{e^{r|\operatorname{Im} z|}}{\alpha(|z|)}$

Let $\alpha \in \mathfrak{Q}$, and $\mathcal{K}_\alpha : \mathbb{C} \mapsto \mathbb{C}$ is the function constructed by the function α as above. We set

$$\mathcal{K}_{\alpha,r}(z) := r\mathcal{K}_\alpha(rz), \quad z \in \mathbb{C},\ r \in (0,\infty).$$

Lemma 3.1 ensures us that the function $\mathcal{K}_{\alpha,r}$ has the following properties:

1) $\mathcal{K}_{\alpha,r}(t) \geq 0, \quad t \in \mathbb{R}$;
2) $\int_{-\infty}^{\infty} \mathcal{K}_{\alpha,r}(t)\,dt = 1$;
3) $\forall z \in \mathbb{C}\ |\mathcal{K}_{\alpha,r}(z)| \leq rc_r \frac{e^{r|\operatorname{Im} z|}}{\alpha(|z|)}; \quad r > 0.$

Next lemma describes the behavior of derivatives of $\mathcal{K}_{\alpha,r}(t)$.

Lemma 3.2 ([13]). *For every $r \in (0,\infty)$ there exists a constant $\widetilde{c}_r = \widetilde{c}_r(\alpha) > 0$ such that for each $n \in \mathbb{N}$ the following inequality holds*

$$|\mathcal{K}_{\alpha,r}^{(n)}(t)| \leq \widetilde{c}_r \frac{\sqrt{2\pi n}\,\alpha\left(\frac{n}{r}\right)}{\alpha(|t|)} r^n, \qquad t \in \mathbb{R}.$$

Remark 3.3. If the function $\alpha(t)$ satisfies the conditions of lemma 3.1, but, moreover, has polynomial order of growth at infinity, i.e., $\exists m \in \mathbb{N}_0$, $\exists M > 0$:

$$\alpha(t) \leq M(1+|t|)^{2m}, \quad t \in \mathbb{R}, \tag{3.10}$$

another integral kernel may be used:

$$\tilde{K}_\alpha(z) = \frac{1}{\mathrm{K}_m}\left(\frac{\sin\frac{z}{2m}}{\frac{z}{2m}}\right)^{2m}, \qquad \mathrm{K}_m = \int_{-\infty}^{\infty}\left(\frac{\sin\frac{x}{2m}}{\frac{x}{2m}}\right)^{2m} dx.$$

In much the same way as in Lemmas 3.1 and 3.2 one can show that

$$\big|\tilde{K}_\alpha(rz)\big| \leq \tilde{C}_r \frac{e^{r|\operatorname{Im} z|}}{\alpha(|z|)}, \qquad \text{where } \tilde{C}_r = \frac{M}{\mathrm{K}_m}\left(1+\frac{2m}{r}\right)^{2m},$$

and

$$\big|\tilde{K}_{\alpha,r}^{(n)}(t)\big| \leq \tilde{c}_r \frac{\sqrt{2\pi n}\,\alpha(\frac{n}{r})}{\alpha(|t|)} r^n, \qquad \text{where } \tilde{c}_r < 2r\tilde{C}_r,$$

that is to say, defined in such a way integral kernel satisfies Lemmas 3.1 and 3.2.

3.2. Spectral subspaces of non-quasianalytic operators

Main instrument for proving generalized Bernstein inequality is the theory of spectral subspaces of non-quasianalytic operator A, constructed in [8]. Recall that spectral subspaces (denoted by $\mathcal{L}(\Delta)$) are defined for all segments $\Delta \subset \mathbb{R}$ and are characterized by the following properties:

1) The operator A is defined on whole $\mathcal{L}(\Delta)$ and is bounded on it;
2) $\mathcal{L}(\Delta)$ is invariant with respect to A;
3) the spectrum of part A_Δ of operator A, induced in $\mathcal{L}(\Delta)$, consists of intersection of spectrum of A with the interior of segment Δ and, perhaps, the endpoints of segment Δ. And at that, if the endpoint of segment Δ does not belong to the spectrum of A, it does not belong to the spectrum of A_Δ either;
4) if there is some subspace $\mathcal{L}$ on which the operator A is defined everywhere and is bounded, and this subspace is invariant with respect to A, and at the same time the spectrum of $\mathcal{L}$-induced part of A is included in Δ, then $\mathcal{L} \subset \mathcal{L}(\Delta)$.

Now we describe the construction of spectral subspaces and their main properties, and later state the relationship with the entire vectors of exponential type. Let $\theta(t)$ $(-\infty < t < \infty)$ is the entire function of order 1 with zeroes on the positive imaginary ray:

$$\theta(t) = C \prod_{k=1}^{\infty} \left(1 - \frac{t}{it_k}\right), \quad \text{where } 0 < t_1 \le t_2 \le \dots, \quad \sum_{k=1}^{\infty} \frac{1}{t_k} < \infty, \tag{3.11}$$

C is a constant. Define by $E_\theta^{(\infty)}$ the class of entire functions $\phi(t)$ of finite type and order 1 which satisfies for all $m = 0, 1, \dots$ and for all $a > 0$ the condition

$$M_\theta^{(m,a)}(\phi) := \int_{-\infty}^{\infty} |t^m \theta(at)\phi(t)|\, dt < \infty. \tag{3.12}$$

As shown in [8, Lemma 1.1.1], the Fourier transform of functions from $E_\theta^{(\infty)}$ is non-quasianalytic, that is the following property takes place:

Proposition 3.4. *For any segment Δ of real axis and for any open finite interval $I \supset \Delta$ there exists $\phi(t) \in E_\theta^{(\infty)}$ such that its Fourier transform equals one in Δ and equals zero outside I.*

Moreover, the class $E_\theta^{(\infty)}$ is linear and is closed under convolutions and differentiation.

Next step is the construction of finite functions of operator A. For C_0-group with non-quasianalytic generator there exists [11] such entire function $\theta(t)$ of order 1 with zeroes on the positive imaginary ray that

$$\|U(t)\| \le |\theta(t)| \qquad \forall t \in \mathbb{R}.$$

Let's consider arbitrary $\phi(t) \in E_\theta^{(\infty)}$ and construct linear operator

$$P_\phi = \int_{-\infty}^{\infty} \phi(t) U(t)\, dt. \tag{3.13}$$

The operator, defined by (3.13), is bounded due to (3.12). Next, consider arbitrary segment Δ of real axis and denote by $E_\theta^{(\infty)}(\Delta)$ the set of such functions $\phi(t) \in E_\theta^{(\infty)}$ that the Fourier transform $\tilde{\phi}(\lambda) = 1$ in some interval containing Δ. Denote by $\mathcal{L}(\Delta)$ the subspace of vectors x such that

$$P_\phi x = x \tag{3.14}$$

for all $\phi(t) \in E_\theta^{(\infty)}(\Delta)$.

Operators P_ϕ are useful to studying of vectors $A^n x$ and to proving of the Bernstein inequality because of properties (3.13), (3.14) and property [8, p. 445]

$$AP_\phi = \overline{P_\phi A} = P_{-i\phi'}, \tag{3.15}$$

which allow to deal with derivatives of some entire functions instead of Banach-space operators and vectors.

The following theorem shows the close relationship between spectral subspaces and entire vectors of exponential type.

Theorem 3.5 ([14]). *For all* $\alpha > 0$

$$\mathfrak{E}^\alpha(A) \subset \Xi^\alpha(A) = \mathcal{L}([-\alpha, \alpha]),$$

moreover, $\Xi^\alpha(A)$ *is the closed subspace of* $\mathfrak{X}$.

4. Direct and inverse theorems

In this section we establish generalizations of well-known Jackson's inequality (direct theorem), Bernstein inequality and the inverse theorem.

4.1. Abstract Jackson-type inequality

Direct theorem can be proven in several ways, one of which is indirect – that is, we can show that there exists a vector or an approximation method giving good estimate for the best approximation. But from the practical point of view more interesting is the method giving direct expression for the approximation, even if this approximation is not the best.

We show how to construct an approximation of a Banach-space vector by entire vectors of exponential type and state the main results (an analogue of the Jackson's inequality) for it.

Let the group $\{U(t) : t \in \mathbb{R}\}$ satisfies (1.1). Then it follows from [11, Theorems 1 and 2] that

$$\int_{-\infty}^{\infty} \frac{\ln(M_U(|t|))}{1+t^2} dt < \infty. \tag{4.1}$$

We fix arbitrary $k \in \mathbb{N}$ and set

$$\alpha(t) := \big(M_U(|t|)\big)^k (1+|t|)^{k+2}, \qquad t \in \mathbb{R}.$$

The function α is, obviously, even on $\mathbb{R}$. Condition (4.1) and properties of function $M_U(\cdot)$ imply $\alpha \in \mathfrak{Q}$, and, moreover,

$$\int_{-\infty}^{\infty} \frac{\big((1+|t|)M_U(|t|)\big)^k}{\alpha(t)} dt = \int_{-\infty}^{\infty} \frac{dt}{(1+|t|)^2} = 2. \tag{4.2}$$

Using lemma 3.1 (or remark 3.3 if $\alpha(t) \le M(1+|t|)^m$) for the function $\alpha(t)$, we construct the family of kernels $K_{\alpha,r}$.

In what follows, we assume $x \in \mathfrak{X}$, $r \in (0,\infty)$ and $n \in \{1,\dots,k\}$. We define

$$x_{r,n} := \int_{-\infty}^{\infty} \mathcal{K}_{\alpha,r}(t)U(nt)x\,dt.$$

Let $\nu \in \mathbb{N}_0$. It can be shown that $x_{r,n} \in C^\infty(A) = \bigcap_{\nu\in\mathbb{N}_0} \mathcal{D}(A^\nu)$ and

$$A^\nu x_{r,n} = \frac{(-1)^\nu}{n^\nu} \int_{-\infty}^{\infty} \mathcal{K}_{\alpha,r}^{(\nu)}(t)U(nt)x\,dt. \tag{4.3}$$

Moreover, for the relation (4.3) we can get [13]

$$\limsup_{\nu\to\infty} \big(\,\|A^\nu x_{r,n}\|\,\big)^{1/\nu} \le \frac{r}{n}.$$

The last inequality brings us to the conclusion that

$$x_{r,n} \in \mathfrak{E}(A) \quad \text{and} \quad \sigma(x_{r,n}, A) \le \frac{r}{n}, \tag{4.4}$$

that is $x_{r,n}$, as the entire vector of exponential type, constructed from vector x, is a candidate for constructing of the approximation.

For arbitrary $x \in \mathfrak{X}$ we define

$$\begin{aligned}\widetilde{x}_{r,k} &:= \int_{-\infty}^{\infty} \mathcal{K}_{\alpha,r}(t)(x + (-1)^{k-1}(U(t)-\mathbb{I})^k x)dt \\ &= \int_{-\infty}^{\infty} \mathcal{K}_{\alpha,r}(t) \sum_{n=1}^{k} (-1)^{n+1}\binom{k}{n} U(nt)x\,dt \end{aligned}\tag{4.5}$$

Using definition (4.5) one can get

$$\widetilde{x}_{r,k} = \sum_{n=1}^{k}(-1)^{n+1}\binom{k}{n}\int_{-\infty}^{\infty}\mathcal{K}_{\alpha,r}(t)U(nt)x dt = \sum_{n=1}^{k}(-1)^{n+1}\binom{k}{n} x_{r,n}.$$

Therefore, accordingly to (4.4),

$$\widetilde{x}_{r,k} \in \mathfrak{E}(A) \quad \text{and} \quad \sigma(\widetilde{x}_{r,k}, A) \le r.$$

Hence for an arbitrary $x \in \mathfrak{X}$ we have

$$\mathcal{E}_r(x, A) = \inf_{y\in\mathfrak{E}(A)\,:\,\sigma(y,A)\le r} \|x-y\| \le \|x - \widetilde{x}_{r,k}\|.$$

$\widetilde{x}_{r,k}$ is the approximation of vector x. Remark that the rule by which we set the vector $\widetilde{x}_{r,k}$ in correspondence to vector x is the linear operator

$$L_{r,k}x = \int_{-\infty}^{\infty} \mathcal{K}_{\alpha,r}(t) \sum_{n=1}^{k} (-1)^{n+1} \binom{n}{k} U(nt)x\, dt,$$

defined on all $\mathfrak{X}$, that is we constructed linear approximation method and gave the direct expression for it.

Next we state the generalizations of Jackson's inequality.

Theorem 4.1 ([13]). *Suppose that* $\{U(t) : t \in \mathbb{R}\}$ *satisfies condition* (1.1). *Then for every* $k \in \mathbb{N}$ *there exists a constant* $\mathbf{m}_k = \mathbf{m}_k(A) > 0$, *such that for all* $x \in \mathfrak{X}$ *the following inequality holds*

$$\mathcal{E}_r(x, A) \leq \mathbf{m}_k \cdot \tilde{\omega}_k \left(\frac{1}{r}, x, A\right), \quad r \geq 1. \tag{4.6}$$

Remark 4.2. If, additionally, the group $\{U(t)\}$ is bounded ($M_U(t) \leq \widetilde{M} < \infty$, $t \in \mathbb{R}$), then the assumption $r \geq 1$ can be changed to $r > 0$.

Theorem 4.1 allows us to prove the analogue of classic Jackson's inequality for m times differentiable functions.

Corollary 4.3 ([13]). *Let* $x \in \mathcal{D}(A^m)$, $m \in \mathbb{N}_0$. *Then for every* $k \in \mathbb{N}_0$

$$\mathcal{E}_r(x, A) \leq \mathbf{m}_{k+m} \frac{M_U\left(\frac{m}{r}\right)}{r^m} \tilde{\omega}_k \left(\frac{1}{r}, A^m x, A\right), \quad r \geq 1, \tag{4.7}$$

where constants $\mathbf{m}_n$ $(n \in \mathbb{N})$ *are the same as in Theorem* 4.1.

By setting in corollary 4.3 $k = 0$ and taking into account that $\tilde{\omega}_0\left(\cdot, A^m x, A\right) \equiv \|A^m x\|$, one can conclude the following inequality:

Corollary 4.4 ([13]). *Let* $x \in \mathcal{D}(A^m)$, $m \in \mathbb{N}_0$. *Then*

$$\mathcal{E}_r(x, A) \leq \frac{\mathbf{m}_m}{r^m} \left(M_U(1/r)\right)^m \|A^m x\| \qquad r \geq 1, \tag{4.8}$$

where the constants $\mathbf{m}_n$ $(n \in \mathbb{N})$ *are the same as in Theorem* 4.1.

4.2. Generalized Bernstein inequality

One of the well-known inequalities in approximation theory is the Bernstein inequality. If $f(x)$ is an entire function of exponential type $\sigma > 0$, and

$$|f(x)| \leq M, \qquad -\infty < x < \infty,$$

then

$$|f'(x)| \leq \sigma M, \qquad -\infty < x < \infty. \tag{4.9}$$

This inequality was generalized for exponential type entire vectors in [14], but in contrast to the classical one, the generalized inequality $\|A^n x\| \leq c_n \alpha^n \|x\|$ has the constants c_n in it. In this section we show that the constants c_n grows slowly than exponent and that in the general case of non-quasianalytic group $U(t)$ it is impossible to prove an analogue of Bernstein inequality with $c_n \le c$ uniformly for all $n \in \mathbb{N}$.

Theorem 4.5 (Bernstein-type inequality). *For every $x \in \mathfrak{E}(A)$ with type, not exceeding some $\alpha \ge 1$, the following inequality holds*

$$\|A^n x\| \le c_n \alpha^n \|x\|, \tag{4.10}$$

where constants c_n do not depend on x and α and are growing slowly than exponent, i.e., for any $\epsilon > 0$ there exists such c_ϵ that $c_n \le c_\epsilon (1+\epsilon)^n$ for all positive integer n.

Proof. Let's consider majorant $\theta(t)$ for the function $\|U(t)\|$, constructed in [11]. Remark that $\theta(t)$ is of the form (3.11). As it was done in (3.9) in Section 2, by the function $\theta(t)$ one can construct the entire function $K(t)$ of exponential type.

Let's consider arbitrary $\epsilon > 0$ and such function $\phi_\alpha(t)$ that its Fourier transform equals 1 in $[-\alpha, \alpha]$ and equals 0 outside $\big(-\alpha(1+2\epsilon), \alpha(1+2\epsilon)\big)$. According to [8, Lemma 1.1.1], one can use as $\phi_\alpha(t)$ the function

$$\phi_\alpha(t) = \frac{K^2(\alpha\epsilon t)\sin\big(\alpha(1+\epsilon)t\big)}{\pi t}. \tag{4.11}$$

Denote by

$$\phi(t) := \frac{K^2(\epsilon t)\sin\big((1+\epsilon)t\big)}{\pi t}.$$

Then $\phi_\alpha(t) = \alpha\phi(\alpha t)$. As it follows from (3.13) and (3.15), it is enough to estimate the quantity

$$\int_{-\infty}^{\infty} |\phi_\alpha^{(n)}(t)\theta(t)|\, dt.$$

to prove the theorem. For $\alpha \ge 1$ we have $|\theta(t)| \le |\theta(\alpha t)|$ and

$$\int_{-\infty}^{\infty} |\phi_\alpha^{(n)}(t)\theta(t)|\, dt \le \int_{-\infty}^{\infty} |\phi^{(n)}(\alpha t)\theta(\alpha t)|\, \alpha dt. \tag{4.12}$$

The change of variables $\tau = \alpha \cdot t$ gives

$$\frac{d^n\phi(\alpha t)}{dt^n} = \frac{d^n\phi(\tau)}{d\tau^n} \cdot \alpha^n,$$

thus

$$\int_{-\infty}^{\infty} |\phi^{(n)}(\alpha t)\theta(\alpha t)|\, \alpha dt = \alpha^n \cdot \int_{-\infty}^{\infty} |\phi^{(n)}(\tau)\theta(\tau)|\, d\tau.$$

It is easy to see that the last integral exists and does not depend on α. Let it equals $c_n > 0$. Then

$$\|A^n x\| = \|P_{(-i)^n\phi^{(n)}} x\| \le c_n \alpha^n \|x\|.$$

Let's show that c_n are growing slowly than exponent. Using the estimate [14, Proof of Theorem 1]

$$|\phi_{r,\epsilon}^{(n)}(t)| \le c^{(!)} c_\epsilon^2 \sqrt{2\pi n}(r+2\epsilon)^n \left|\frac{\theta\big(\frac{n}{r+2\epsilon}\big)}{\theta(t)}\right|^2$$

on derivatives of $\phi_{r,\epsilon}(t) = K^2(\epsilon t)\frac{\sin rt}{rt}$ we obtain

$$|\phi^{(n)}(t)| \leq \frac{c^{(!)} c_\epsilon^2 \sqrt{2\pi n}}{\pi}(1+\epsilon)(1+3\epsilon)^n \left|\frac{\theta(n)}{\theta(t)}\right|^2.$$

Using [14, Proof of Theorem 1]

$$\lim_{n\to\infty} \left(\left|\theta^2\left(\frac{n}{\alpha}\right)\right|\right)^{1/n} = 1, \qquad \alpha \in \mathbb{R}_+,$$

and the arbitrary of $\epsilon > 0$, from the last inequality we can get that c_n grows slowly then the exponent. □

We see that the inequality (4.10) differs from direct analogue of classic one (4.9) by the constants c_n, which equals 1 in the classic case. The question if these constants c_n are the characteristic property of our generalization or only the imperfection of proof, naturally arises. It is shown in [2, Theorem 1] that for bounded C_0-group $U(t)$ the direct analogue of Bernstein inequality (with $c_n \equiv 1$) can be proved. But for non-quasianalytic operators the situation is different. In Section 5.2 we construct the example of operator A and vector x such that the type of x equals some $\alpha > 1$, but $x \notin \mathfrak{E}^\alpha(A)$, which means that there is no such constant c that

$$\forall k \in \mathbb{N}_0 \quad \|A^k x\| \leq c\alpha^k,$$

that is there exists a counterexample for any inequality of the form (4.10) with any bounded sequence c_n.

As the consequence of Theorem 4.5 we get the following estimate for operator Δ_h^k:

Corollary 4.6 ([14]). *Let $x \in \mathfrak{E}(A)$ and $\sigma(x) \leq \alpha$, $\alpha \geq 1$. Then for all $k \in \mathbb{N}$*

$$\|\Delta_h^k x\| \leq c_k (h\alpha)^k M_U(kh)\|x\|, \tag{4.13}$$

where constants c_k are the same as in Theorem 4.5 and the function $M_U(t)$ is defined by (2.4).

4.3. Inverse theorem

Following results generalize classical Bernstein theorem (also known as inverse theorem).

Theorem 4.7 ([14]). *Let $\omega(t)$ is the function of type of a module of continuity for which the following conditions are satisfied:*

1. *$\omega(t)$ is continuous and nondecreasing for $t \in \mathbb{R}_+$.*
2. *$\omega(0) = 0$.*
3. *$\exists c > 0\ \forall t \in [0,1] \quad \omega(2t) \leq c\omega(t)$.*
4. *$\int_0^1 \frac{\omega(t)}{t} dt < \infty$.*

If, for $x \in \mathfrak{X}$, there exist $n \in \mathbb{N}$ and $m > 0$ such that

$$\mathcal{E}_r(x, A) \leq \frac{m}{r^n}\omega\left(\frac{1}{r}\right), \qquad r \geq 1, \tag{4.14}$$

then $x \in \mathcal{D}(A^n)$ *and for every* $k \in \mathbb{N}$ *there exists a constant* $m_k > 0$ *such that*

$$\omega_k(t, A^n x, A) \leq m_k \left(t^k \int_t^1 \frac{\omega(u)}{u^{k+1}} du + \int_0^t \frac{\omega(u)}{u} du \right), \quad 0 < t \leq 1/2. \tag{4.15}$$

The following lemma is used for the proof of theorem.

Lemma 4.8 ([14]). *Suppose that the function* $\omega(t)$ *satisfies conditions* 1–3 *of theorem* 4.7. *If, for* $x \in \mathfrak{X}$, *there exists* $m > 0$ *such that*

$$\mathcal{E}_r(x, A) \leq m\omega\left(\frac{1}{r}\right), \qquad r \geq 1, \tag{4.16}$$

then, for every $k \in \mathbb{N}$, *there exists a constant* $\tilde{c}_k > 0$ *such that*

$$\omega_k(t, x, A) \leq \tilde{c}_k t^k \int_k^1 \frac{\omega(\tau)}{\tau^{k+1}} d\tau, \quad 0 < t \leq 1/2. \tag{4.17}$$

5. Examples of application of abstract direct and inverse theorems in particular spaces

In this section we discuss two applications of presented theory. First is the approximation of continuous functions by entire functions in the weighted $L_p(\mathbb{R}, \mu)$ spaces with growing at the infinity weight (for example, $L_1(\mathbb{R}, x^n)$ spaces). Similar problems studied in several papers (see the review [15]). Second is the approximation of continuous functions by entire functions in the weighted $L_p(\mathbb{R}, \mu)$ spaces with decreasing at the infinity weight. Many papers, started from the works of S. Bernstein [16, 17], concerned to this problem.

5.1. Direct and inverse theorems of approximation by exponential type entire functions in the space $L_p(\mathbb{R}, \mu^p)$

We consider a real-valued function $\mu(t)$ satisfying the following conditions:

1) $\mu(t) \geq 1, \quad t \in \mathbb{R}$;
2) $\mu(t)$ is even, monotonically non-decreasing when $t > 0$;
3) $\mu(t)$ satisfies naturally occurring in many applications condition $\mu(t+s) \leq \mu(t) \cdot \mu(s), \ s, t \in \mathbb{R}$.
4) $\int_{-\infty}^{\infty} \frac{\ln \mu(t)}{1+t^2} dt < \infty$,

or alternatively, instead of 4), the equivalent condition holds:

4′) $\sum_{k=1}^{\infty} \frac{\ln \mu(k)}{k^2} < \infty$.

Let's consider several important classes of functions satisfying conditions 1)–4).

1. Constant function $\mu(t) \equiv 1, \quad t \in \mathbb{R}$.

2. Functions with polynomial order of growth at the infinity. It is easy to check that for such functions following estimate holds:

$$\exists k \in \mathbb{N}, \ \exists M \geq 1 \qquad \mu(t) \leq M(1 + |t|)^k, \quad t \in \mathbb{R}.$$

3. Functions of the form

$$\mu(t) = e^{|t|^\beta}, \quad 0 < \beta < 1, \; t \in \mathbb{R}.$$

4. $\mu(t)$ represented as a power series for $t > 0$. I.e.,

$$\mu(t) = \sum_{n=0}^{\infty} \frac{|t|^n}{m_n},$$

where $\{m_n\}_{n\in\mathbb{N}}$ is the sequence of positive real numbers satisfying two conditions:

- $m_0 = 1, \; m_n^2 \le m_{n-1} \cdot m_{n+1}, \; n \in \mathbb{N}$;
- $\forall k, l \in \mathbb{N} \;\; \frac{(k+l)!}{m_{k+l}} \le \frac{k!}{m_k}\frac{l!}{m_l}$.

The function $\mu(t)$, defined above, obviously satisfies conditions 1) and 2). The condition $\forall k, l \in \mathbb{N} \;\; \frac{(k+l)!}{m_{k+l}} \le \frac{k!}{m_k}\frac{l!}{m_l}$ implies

$$\sum_{k=0}^{n} \frac{t^k s^{n-k} n!}{k!(n-k)!\, m_n} \le \sum_{k=0}^{n} \frac{t^k s^{n-k}}{m_k m_{n-k}}, \tag{5.1}$$

and it is easy to see that condition 3) follows from inequality (5.1). Denjoy-Carleman theorem [12, p. 376] asserts that the following conditions are equivalent:

a) $\mu(t)$ satisfies condition 4);

b) $\sum_{n=1}^{\infty} \left(\frac{1}{m_n}\right)^{1/n} < \infty$;

c) $\sum_{n=1}^{\infty} \frac{m_{n-1}}{m_n} < \infty$.

5. $\mu(t)$ as a module of an entire function with zeroes on the imaginary axis. We consider

$$\omega(t) = C \prod_{k=1}^{\infty} \left(1 - \frac{t}{it_k}\right), \quad t \in \mathbb{R},$$

where $C \ge 1, \; 0 < t_1 \le t_2 \le \cdots, \; \sum_{k=1}^{\infty} \frac{1}{t_k} < \infty$. We set $\mu(t) := |\omega(t)|$. Then $\mu(t)$ satisfies conditions 1)–3), and, as shown in [8], $\mu(t)$ satisfies condition 4) also.

Let's proceed to the description of spaces $L_p(\mathbb{R}, \mu^p)$. Let the function $\mu(t)$ satisfies conditions 1)–4). We consider the space $L_p(\mathbb{R}, \mu^p)$, $1 \le p \le \infty$ of functions $x(s)$, $s \in \mathbb{R}$, integrable in pth degree with the weight μ^p:

$$\|x\|^p_{L_p(\mathbb{R},\mu^p)} = \int_{-\infty}^{\infty} |x(s)|^p \mu^p(s)\, ds.$$

$L_p(\mathbb{R}, \mu^p)$ is the Banach space. We consider the differential operator

$$(Ax)(t) = \frac{dx}{dt}, \qquad \mathcal{D}(A) = \{x \in L_p(\mathbb{R}, \mu^p) \cap AC(\mathbb{R}) : \; x' \in L_p(\mathbb{R}, \mu^p)\}.$$

The operator A generates the group of shifts $\{U(t)\}_{t\in\mathbb{R}}$ in the space $L_p(\mathbb{R}, \mu^p)$. This group isn't bounded. As shown in [13],

$$\|U(t)\|_{L_p(\mathbb{R},\mu^p)} \le \mu(|t|), \quad t \in \mathbb{R}.$$

To obtain meaningful results of approximation by entire functions, we need to establish how the space $\mathfrak{E}(A)$ and the space of entire functions of exponential type are connected. Denote by B_σ the set of exponential functions of entire type σ. We showed [14] that the following embedding holds

$$\Xi^\sigma(A) \subset B_\sigma \cap L_p(\mathbb{R}, \mu^p). \tag{5.2}$$

It was shown in [14] that for all functions $\mu(t)$, satisfying

$$\mu(t) \geq 1 + R|t|^q$$

for some $q > 1 - \frac{1}{p}$, $R > 0$ and for all $t > t_0 \geq 0$, the reverse embedding holds. In this paper we improve this result and show that $B_\sigma \cap L_p(\mathbb{R}, \mu^p) \subset \Xi^\sigma(A)$ for all weights $\mu(t)$, satisfying conditions 1)–4). Let $f \in B_\sigma \cap L_p(\mathbb{R}, \mu^p)$. Obviously $f \in L_p(\mathbb{R})$. f is entire function, so (see [18, p. 191]) it is bounded on the real axis. One can conclude that $f \in \mathcal{S}'$ (remind that $\mathcal{S}'$ is the Schwartz space of tempered distributions, see [19]).

To prove that $f \in \Xi^\sigma(A)$ by the use of Theorem 3.5, let's consider majorant $\theta(t)$ for the function $\mu(t)$, constructed as in the proof of Theorem 4.5. We need to show that for all $\phi \in E_\theta^{(\infty)}([-\sigma, \sigma])$

$$f = P_\phi f = \int_{-\infty}^{\infty} \phi(t) U(t) f dt.$$

Note that it follows from the definition of class $E_\theta^{(\infty)}$ that every such $\phi \in \mathcal{S}$. Let's consider $\phi \in E_\theta^{(\infty)}([-\sigma, \sigma])$. As $\phi \in \mathcal{S}$ and $f \in \mathcal{S}'$, the convolution $\phi * f$ is well defined [19, p. 324]. But for $\phi_1(t) = \phi(-t)$

$$(P_{\phi_1} f)(x) = \int_{-\infty}^{\infty} \phi_1(t) U(t) f(x) dt = \int_{-\infty}^{\infty} \phi(t) f(x - t) dt = (\phi * f)(x).$$

The Fourier transform of $\phi * f$ equals to

$$\widetilde{\phi * f} = \tilde{\phi} \cdot \tilde{f} = \tilde{f},$$

because by the definition of $E_\theta^{(\infty)}([-\sigma, \sigma])$ we have $\tilde{\phi} = 1$ on $[-\sigma, \sigma]$, and by the Paley-Wiener theorem $\operatorname{supp} f \subset [-\sigma, \sigma]$. Thus,

$$P_{\phi_1} f = f \qquad \forall \phi_1 \in E_\theta^{(\infty)}([-\sigma, \sigma]),$$

(since $\phi(-t)$ also runs over all $E_\theta^{(\infty)}([-\sigma, \sigma])$ so $f \in \mathcal{L}([-\sigma, \sigma])$ and by the means of Theorem 3.5, $f \in \Xi^\sigma(A)$.

We have shown that $\Xi^\sigma(A) = B_\sigma \cap L_p(\mathbb{R}, \mu^p)$ for all weights $\mu(t)$, satisfying conditions 1)–4). Note that $\|f - g_\sigma\|_{L_p(\mathbb{R}, \mu^p)}$ is defined only for those functions that belongs to $L_p(\mathbb{R}, \mu^p)$ (because of $\|g_\sigma\|_{L_p(\mathbb{R}, \mu^p)} \leq \|f - g_\sigma\|_{L_p(\mathbb{R}, \mu^p)} + \|f\|_{L_p(\mathbb{R}, \mu^p)}$), thus the best approximation by exponential type entire vectors is the same as the best approximation by entire functions of exponential type.

By applying Theorems 4.1, 4.5, 4.7 and Corollary 4.3 we get several results for the approximation theory in $L_p(\mathbb{R}, \mu^p)$ spaces.

Corollary 5.1. *For every $k \in \mathbb{N}$ there exists a constant $\mathbf{m}_k(p,\mu) > 0$ such that for all $f \in L_p(\mathbb{R},\mu^p)$*

$$\mathcal{E}_r(f) \le \mathbf{m}_k \cdot \tilde{\omega}_k\left(\frac{1}{r}, f\right), \quad r \ge 1.$$

Corollary 5.2. *Let $f \in W_p^m(\mathbb{R},\mu^p)$, $m \in \mathbb{N}_0$. Then for all $k \in \mathbb{N}_0$*

$$\mathcal{E}_r(f) \le \mathbf{m}_{k+m} \frac{\mu\left(\frac{m}{r}\right)}{r^m} \widetilde{\omega}_k\left(\frac{1}{r}, f^{(m)}\right), \quad r \ge 1,$$

where constants $\mathbf{m}_n$ $(n \in \mathbb{N})$ are the same as in Corollary 5.1.

Corollary 5.3. *Let $f \in L_p(\mathbb{R},\mu^p) \cap B_\sigma$, $\sigma \ge 1$. Then for all $n \in \mathbb{N}$ there exist such constants $c_n > 0$, not depending on σ and growing slower than exponent, that*

$$\|f^{(n)}\|_{L_p(\mathbb{R},\mu^p)} \le c_n \sigma^n \|f\|_{L_p(\mathbb{R},\mu^p)}.$$

Corollary 5.4. *Let $\omega(t)$ be a function of type of a module of continuity for which the following conditions are satisfied:*

1. *$\omega(t)$ is continuous and nondecreasing for $t \in \mathbb{R}_+$.*
2. *$\omega(0) = 0$.*
3. *$\exists c > 0\ \forall t \in [0,1]\quad \omega(2t) \le c\omega(t)$.*
4. *$\int_0^1 \frac{\omega(t)}{t} dt < \infty$.*

If, for $f \in L_p(\mathbb{R},\mu^p)$ there exist such $n \in \mathbb{N}$ and $m > 0$ that

$$\mathcal{E}_r(f) \le \frac{m}{r^n}\omega\left(\frac{1}{r}\right), \qquad r \ge 1,$$

then $f \in W_p^n(\mathbb{R},\mu^p)$ and for every $k \in \mathbb{N}$ there exists a constant $m_k > 0$ such that

$$\omega_k(t, f^{(n)}) \le m_k\left(t^k \int_t^1 \frac{\omega(u)}{u^{k+1}} du + \int_0^t \frac{\omega(u)}{u} du\right), \quad 0 < t \le 1/2.$$

5.2. Exponential type entire functions in the space $L_p(\mathbb{R},\mu^{-p})$ and constants in the Bernstein inequality

We consider the same real-valued function $\mu(t)$ as in the previous subsection, but another functional space. Consider the space $L_p(\mathbb{R},\mu^{-p})$ of functions $x(s)$, $s \in \mathbb{R}$, integrable in pth degree with the weight μ^{-p}:

$$\|x\|^p_{L_p(\mathbb{R},\mu^{-p})} = \int_{-\infty}^{\infty} |x(s)|^p \mu^{-p}(s)\, ds.$$

$L_p(\mathbb{R},\mu^{-p})$ is the Banach space. We consider the same differential operator as in the previous subsection

$$(Ax)(t) = \frac{dx}{dt}, \qquad \mathcal{D}(A) = \{x \in L_p(\mathbb{R},\mu^{-p}) \cap AC(\mathbb{R}) : \ x' \in L_p(\mathbb{R},\mu^{-p})\}.$$

The operator A generates the group of shifts $\{U(t)\}_{t\in\mathbb{R}}$ in the space $L_p(\mathbb{R},\mu^{-p})$. This group isn't bounded for unbounded $\mu(t)$. Let's show this and that the group is non-quasianalytic. Indeed, let's consider for arbitrary $t \geq 1$

$$x(s) = \begin{cases} 1, & s \in [t, t+1], \\ 0, & s \notin [t, t+1]. \end{cases}$$

Obviously, $x(s) \in L_p(\mathbb{R},\mu^{-p})$. We have

$$\|U(t)x\|^p = \int_{-\infty}^{\infty} |x(t+s)|^p \mu^{-p}(s)\, ds = \int_0^1 \mu^{-p}(s)\, ds = C > 0,$$

the last quantity does not depend on t. But

$$\|x\|^p = \int_{-\infty}^{\infty} |x(s)|^p \mu^{-p}(s)\, ds = \int_t^{t+1} \mu^{-p}(s)\, ds,$$

and from the monotony of $\mu(s)$ for $s \in [0,1]$ we get

$$\frac{1}{\mu(t+1)} \leq \frac{1}{\mu(t+s)} \leq \frac{1}{\mu(t)},$$

so $\|x\|^p \leq \frac{1}{\mu^p(t)}$. For unbounded $\mu(t)$ this shows that

$$\|U(t)\| \geq \frac{\|U(t)x\|}{\|x\|} \geq C\mu(t) \to \infty,$$

that is the group $U(t)$ is unbounded.

From the other hand, for every function x, because of the property 3),

$$\begin{aligned} \|U(t)x\|^p &= \int_{-\infty}^{\infty} \left| \frac{x(t+s)}{\mu(s)} \right|^p ds \\ &= \int_{-\infty}^{\infty} \left| \frac{x(t+s)\cdot\mu(t)}{\mu(s)\cdot\mu(t)} \right|^p ds \\ &\leq \int_{-\infty}^{\infty} \left| \frac{x(t+s)\cdot\mu(t)}{\mu(t+s)} \right|^p ds \\ &= \big(\mu(t)\big)^p \|x\|^p, \end{aligned}$$

so $\|U(t)\|_{L_p(\mathbb{R},\mu^{-p})} \leq \mu(t)$, that is the group $U(t)$ is non-quasianalytic.

Now we construct an example of vector such that the type of x equals some $\alpha > 1$, but $x \notin \mathfrak{E}^\alpha(A)$. Consider $\mu(t) = (1+|t|)^3$ and (for the simplicity)

$$x(t) = t \cdot e^{ikt}, \quad k \in \mathbb{N}.$$

Then we can calculate

$$A^n x(t) = \left(t \cdot e^{ikt}\right)^{(n)} = n(ik)^{n-1} e^{ikt} + (ik)^n t \cdot e^{ikt}.$$

Let

$$a = \int_{\infty}^{\infty} |e^{ikt}|^p \mu^{-p}(t)\, dt, \qquad b = \int_{\infty}^{\infty} |t \cdot e^{ikt}|^p \mu^{-p}(t)\, dt.$$

It is easy to see that $0 < a < \infty$ and $0 < b < \infty$, and from the property $|y| - |z| \leq |y + z| \leq |y| + |z|$ of the absolute value we obtain

$$nk^{n-1}a - k^n b \leq \|A^n x\|_{L_p(\mathbb{R},\mu^{-p})} \leq nk^{n-1}a + k^n b. \tag{5.3}$$

It can be seen from the inequality (5.3) that there is no such constant $c > 0$ that $\|A^n x\|_{L_p(\mathbb{R},\mu^{-p})} \leq c \cdot k^n$ for all $n \in \mathbb{N}$, that is $x \notin \mathfrak{E}^k(A)$, but, on the other hand, for any $\epsilon > 0$ there exists such constant c_ϵ that $\|A^n x\|_{L_p(\mathbb{R},\mu^{-p})} \leq c_\epsilon \cdot (k+\epsilon)^n$, that is $x \in \Xi^k(A)$ and $\sigma(x) = k$.

References

[1] N.P. Kupcov, *Direct and inverse theorems of approximation theory and semigroups of operators.* Uspekhi Mat.Nauk. **23** (1968), No. 4, 118–178. (Russian)

[2] A.P. Terehin, *A bounded group of operators and best approximation.* Differencial'nye Uravneniya i Vyçisl.Mat., Vyp. 2, 1975, 3–28. (Russian)

[3] G.V. Radzievskii, *On the best approximations and the rate of convergence of decompositions in the root vectors of an operator.* Ukrain. Math. Zh. **49** (1997), no. 6, 754–773. (Russian); English trans. in Ukrainian Math. J. **49** (1997), no. 6, 844–864.

[4] G.V. Radzievskii, *Direct and converse theorems in problems of approximation by vectors of finite degree.* Math. Sb. **189** (1998), no. 4, 83–124.

[5] M.L. Gorbachuk and V.I. Gorbachuk, *On the approximation of smooth vectors of a closed operator by entire vectors of exponential type.* Ukrain. Mat. Zh. **47** (1995), no. 5, 616–628. (Ukrainian); English transl. in Ukrainian Math. J. **47** (1995), no. 5, 713–726.

[6] M.L. Gorbachuk and V.I. Gorbachuk, *Operator approach to approximation problems.* St. Petersburg Math. J. **9** (1998), no. 6, 1097–1110.

[7] M.L. Gorbachuk, Ya.I. Grushka and S.M. Torba, *Direct and inverse theorems in the theory of approximations by the Ritz method.* Ukrain. Mat. Zh. **57** (2005), no. 5, 633–643. (Ukrainian); English transl. in Ukrainian Math. J. **57** (2005), no. 5, 751–764.

[8] Ju.I. Ljubic and V.I. Macaev, *Operators with separable spectrum* (Russian). Mat. Sb. **56 (98)** (1962), no. 4, 433–468. (Russian)

[9] M.L. Gorbachuk, *On analytic solutions of operator-differential equations.* Ukrain. Mat. Zh. **52** (2000), no. 5, 596–607. (Ukrainian); English transl. in Ukrainian Math. J. **52** (2000), no. 5, 680–693.

[10] V.A. Marchenko, *On some questions of the approximation of continuous functions on the whole real axis.* Zap. Mat. Otd. Fiz-Mat. Fak. KhGU i KhMO **22** (1951), no. 4, 115–125. (Russian)

[11] O.I. Inozemcev and V.A. Marchenko, *On majorants of genus zero.* Uspekhi Mat. Nauk **11** (1956), 173–178. (Russian)

[12] Walter Rudin, *Real and Complex Analysis.* McGrow-Hill, New York, 1970.

[13] Ya. Grushka and S. Torba, *Direct theorems in the theory of approximation of vectors in a Banach space with exponential type entire vectors.* Methods Func. Anal. Topology **13** (2007), no. 3, pp. 267–278.

[14] S. Torba, *Inverse theorems in the theory of approximation of vectors in a Banach space with exponential type entire vectors.* Methods Func. Anal. Topology **15** (2009), no. 4 (to appear).

[15] Ganzburg M.I., *Limit theorems and best constants in approximation theory // Anastassiou, George (ed.), Handbook of analytic-computational methods in applied mathematics.* Boca Raton, FL: Chapman & Hall/CRC, 2000, pp. 507–569.

[16] S.N. Bernstein, *Collected Works, vol. I.* Akad. Nauk SSSR, Moscow, 1952.

[17] S.N. Bernstein, *Collected Works, vol. II.* Akad. Nauk SSSR, Moscow, 1954.

[18] N.I. Akhiezer, *Lectures on Approximation Theory, 2nd edition.* Nauka, Moscow, 1965.

[19] M. Reed, B. Simon, *Methods of Modern Mathematical Physics: Functional Analysis I, Revised and enlarged edition.* Academic Press, 1980.

Yaroslav Grushka and Sergiy Torba
Institute of Mathematics
National Academy of Science of Ukraine
3 Tereshchenkivs'ka St.
01601 Kyiv, Ukraine
e-mail: grushka@imath.kiev.ua
sergiy.torba@gmail.com

Operator Theory:
Advances and Applications, Vol. 190, 315–327

Results on Convergence in Norm of Exponential Product Formulas and Pointwise of the Corresponding Integral Kernels

Takashi Ichinose and Hideo Tamura

Abstract. For the last one and a half decades it has been known that the exponential product formula holds also *in norm* in nontrivial cases. In this note, we review the results on its convergence in norm as well as pointwise of the integral kernels in the case for Schrödinger operators, with error bounds. Optimality of the error bounds is elaborated.

Mathematics Subject Classification (2000). Primary 47D08; Secondary 47D06, 41A80, 81Q10, 35J10.

Keywords. Exponential product formula, Trotter product formula, Lie–Trotter product formula, Trotter–Kato product formula, pointwise integral kernel convergence, evolution group and semigroup, Schrödinger operator.

1. Introduction

The Trotter product formula, Trotter–Kato product formula or exponential product formula is usually a product formula which in strong operator topology approximates the group/semigroup with generator being a sum of two operators. It is often a useful tool to study Schrödinger evolution groups/semigroups in quantum mechanics and to study Gibbs semigroups in statistical mechanics.

To think of a typical case, let A and B be selfadjoint operators in a Hilbert space $\mathcal{H}$ with domains $D[A]$ and $D[B]$ and $H := A + B$ their operator sum with domain $D[H] = D[A] \cap D[B]$. Assume that H is selfadjoint or essentially selfadjoint on $D[H]$ and denote its closure by the same H. Then Trotter [44] proved the unitary product formula

$$[e^{-itB/2n}e^{-itA/n}e^{-itB/2n}]^n - e^{-itH} \to 0, \quad \text{strongly},$$
$$[e^{-itA/n}e^{-itB/n}]^n - e^{-itH} \to 0, \quad \text{strongly}, \quad n \to \infty,$$

and also, when A and B are nonnegative, the selfadjoint product formula

$$[e^{-tB/2n}e^{-tA/n}e^{-tB/2n}]^n - e^{-tH} \to 0, \quad \text{strongly},$$

$$[e^{-tA/n}e^{-tB/n}]^n - e^{-tH} \to 0, \quad \text{strongly.} \quad n \to \infty,$$

The convergence is *locally uniform*, i.e., uniform on compact t-intervals, respectively in the real line $\mathbf{R}$ and in the closed half-line $[0, \infty)$. Kato [29] discovered the latter selfadjoint product formula to hold also for the form sum $H := A \dot{+} B$ with form domain $D[H^{1/2}] = D[A^{1/2}] \cap D[B^{1/2}]$, which we assume for simplicity is dense in $\mathcal{H}$. However, it remains to be an open problem whether the unitary product formula for the form sum holds.

However, since around 1993 we have begun to know that selfadjoint product formulas converge even in (*operator*) *norm*, though in some special cases, by the following two first results. Rogava [37] proved, when B is A-bounded and $H = A + B$ is selfadjoint, among others, the abstract product formula that

$$\|[e^{-tA/n}e^{-tB/n}]^n - e^{-tH}\| = O(n^{-1/2}\log n), \quad n \to \infty,$$

locally uniformly in $[0, \infty)$. Helffer [13] proved, when $H := -\Delta + V(x)$ is a Schrödinger operator in $L^2(\mathbf{R}^d)$ with nonnegative potential $V(x)$ satisfying $|\partial_x^\alpha V(x)| \leq C_\alpha$ ($|\alpha| \geq 2$) so that H is selfadjoint on the domain $D[-\Delta] \cap D[V]$, the symmetric product formula that

$$\|[e^{-tV/2n}e^{-t(-\Delta)/n}e^{-tV/2n}]^n - e^{-tH}\| = O(n^{-1}), \quad n \to \infty,$$

locally uniformly in $[0, \infty)$. Many works were done to extend these results before 2000, e.g., in [5, 20, 22, 32, 33, 35] for the abstract product formula, [9, 10, 17, 18, 19, 41] for the Schrödinger operators, and after that, e.g., in [23, 27, 16], [3, 4, 6, 7] for the abstract product formula. In most of them, use was made of operator-theoretic methods, though of a probabilistic method in [17, 18, 19, 41].

In this note, we want to describe more recent results on convergence in norm for exponential product formulas and also pointwise of the corresponding integral kernels, mainly based on our works since around 2000, [23, 27, 24, 25, 26]. As for the error bounds, although it is easy to see by the Baker–Campbell–Hausdorff formula (e.g., [45], [40]) that with both operators A and B being bounded, the *nonsymmetric* product formula has an optimal error bound $O(n^{-1})$ while the *symmetric* one does $O(n^{-2})$, it was shown in [27] that even the *symmetric* product formula has an optimal error bound $O(n^{-1})$ in general, if both A and B are unbounded. However, in [25] (cf. [26]), a better upper sharp error bound $O(n^{-2})$ has been obtained for the *symmetric* product formula with the Schrödinger operator $-\Delta + V(x)$ having nonnegative potentials $V(x)$ growing polynomially at infinity, in spite that both $-\Delta$ and V are unbounded operators. In this note we mention, with a sketch of proof, a latest complementary result [2] which settles the sharp optimal error bound is in fact $O(n^{-2})$ with the symmetric product formula for the harmonic oscillator, by estimating the error not only from above but also from below, in norm as well as pointwise.

Theorems are described in Section 2. Optimality of error bounds is discussed separately in Section 3. The idea of proof is briefly mentioned in Section 4. In Section 5 we give concluding remarks, and also refer to a connection of the exponential product formula with the Feynman path integral.

It should be also noted that in almost the same context with the notion of norm ideals (e.g., [12], [38]) we are able to deal with the trace norm convergence as in [46, 30, 31, 34, 14, 21, 42]. For an extensive literature on this we refer to [47].

The content of this note is an expanded version of the lecture entitled "On convergence pointwise of integral kernels and in norm for exponential product formulas" given by T.I. at the International Conference "*Modern Analysis and Applications* (MAA 2007)", Odessa, Ukraine, April 9–14, 2007, which is a slightly extended version of of the lecture (unpublished) given at the Conference on "*Heat Kernel in Mathematical Physics*", Blaubeuren, Germany, November 28–December 2, 2006.

2. Theorems

We begin with our result which extends ultimately Rogava and Helffer's.

Theorem 2.1 (Ichinose-Tamura-Tamura-Zagrebnov 2001 [23, 27]**).** *Let A and B be nonnegative selfadjoint operators, and assume $H = A + B$ is selfadjoint on $D[H] = D[A] \cap D[B]$. Then as $n \to \infty$,*

$$\|[e^{-tB/2n}e^{-tA/n}e^{-tB/2n}]^n - e^{-tH}\| = O(n^{-1}), \tag{2.1}$$

$$\|[e^{-tA/n}e^{-tB/n}]^n - e^{-tH}\| = O(n^{-1}). \tag{2.2}$$

The convergence is locally uniform in the closed half-line $[0, \infty)$, while on the whole half-line $[0, \infty)$, if H is strictly positive, i.e., $H \geq \eta I$ for some $\eta > 0$. The error bound $O(n^{-1})$ in (2.1) and (2.1) is optimal.

We can go beyond this result. First, focussing on the Schrödinger operator $-\Delta + V(x)$, we ask whether norm convergence implies pointwise convergence of integral kernels. The answer is yes, though strong convergence does not. This problem is discussed for Schrödinger operators with potentials of polynomial growth (Theorem 2.2), with positive Coulomb potential (Theorem 2.3), and also for the Dirichlet Laplacian (Theorem 2.4). Pointwise convergence of integral kernels for Schrödinger semigroups is important, because it gives a time-sliced approximation to the imaginary-time Feynman path integral.

Next, we ask, for the unitary exponential/Trotter product formula, whether there are nontrivial cases where it converges in norm, though it does not in general hold (see [15]). The answer is yes. In fact, it holds for the Dirac operator and relativistic Schrödinger operator (Theorem 2.5).

Let $H = H_0 + V := -\Delta + V(x)$ with $V(x)$ a real-valued function. By $K^{(n)}(t, x, y)$ we denote the integral kernel of $[e^{-tH_0/2n}e^{-tV/n}e^{-tH_0/2n}]^n$, and by $e^{-tH}(x, y)$ that of e^{-tH}.

Theorem 2.2 (Ichinose-Tamura 2004 [25] (positive potential of polynomial growth)). *Assume that $V(x)$ is in $C^\infty(\mathbf{R}^d)$, bounded below and satisfies*

$$|\partial_x^\alpha V(x)| \le C_\alpha \langle x\rangle^{m-\delta|\alpha|}, \qquad \langle x\rangle = (1+x^2)^{1/2}$$

with some $0 < \delta \le 1$.

(i) (*In norm*)

$$\|[e^{-tH_0/2n}e^{-tV/n}e^{-tH_0/2n}]^n - e^{-tH}\|_{L^2} = O(n^{-2}), \tag{2.3}$$

locally uniformly in the open half-line $(0,\infty)$.

(ii) (*Integral kernel*)

$$\begin{aligned}&[K^{(n)}(t,x,y) - e^{-tH}(x,y)] = O(n^{-2}),\\ &\text{in } C^\infty(\mathbf{R}^d\times\mathbf{R}^d)\text{-topology, locally uniformly in } (0,\infty),\end{aligned} \tag{2.4}$$

i.e., together with all x, y-derivatives.

This theorem improves the result of Takanobu [41], who used a probabilistic method with the Feynman–Kac formula (see Sect. 5) to show uniform pointwise convergence of the integral kernels, roughly speaking, with error bound $O(n^{-\rho/2})$, if $V(x)$ satisfies $V(x) \ge C(1+|x|^2)^{\rho/2}$ and $|\partial_x^\alpha V(x)| \le C_\alpha(1+|x|^2)^{(\rho-\delta|\alpha|)_+/2}$ for some constants C, $C_\alpha \ge 0$ and $\rho \ge 0$, $0<\delta\le 1$. The claim of Theorem 2.2 is a little bit sharpened in Theorems 3.1 and 3.2, in the next section, in the case of the harmonic oscillator.

Theorem 2.3 (Ichinose-Tamura 2006 [26] (positive Coulomb potential)). *Let $H := -\Delta + V(x)$ with $V(x) \ge 0$. Assume that $V(-\Delta+1)^{-\alpha}$: $L^2(\mathbf{R}^d) \to L^2(\mathbf{R}^d)$ is bounded for some $0<\alpha<1$, and that $V \in C^\infty$ near a neighbourhood U of both p and q (after $p, q \in \mathbf{R}^d$ taken). Then*

$$\begin{aligned}&[K(t/n)^n(x,y) - e^{-tH}(x,y)] = O(n^{-1}),\\ &\text{in } C^\infty(U)\text{-topology, locally uniformly in } (0,\infty).\end{aligned} \tag{2.5}$$

The condition is satisfied if V is in $L^2(\mathbf{R}^3)+L^\infty(\mathbf{R}^3)$, in particular, if V is the positive Coulomb potential $1/|x|$. We don't know what happens at the singularities of $V(x)$.

Theorem 2.4 (Ichinose-Tamura 2006 [26] (Dirichlet Laplacian)). *Let $\Omega \subset \mathbf{R}^d$ be a bounded domain with smooth boundary and χ_Ω the indicator function of Ω. Let $H_0 := -\Delta$ in $L^2(\mathbf{R}^d)$, and $H := -\Delta_D$ the Dirichlet Laplacian in Ω with domain $D[H] = H^2(\Omega)\cap H_0^1(\Omega)$. Then for $0<\sigma<\frac{1}{6}$,*

$$\begin{aligned}&\left(\chi_\Omega\, e^{-tH_0/n}\chi_\Omega\right)^n(x,y) - e^{-tH}(x,y) = O(n^{-\sigma}),\\ &\text{locally uniformly in } (t,x,y)\in(0,\infty)\times\Omega\times\Omega.\end{aligned} \tag{2.6}$$

We don't know what happens when x or y approaches the boundary of Ω.

Corollary.

$$\|[\chi_\Omega\, e^{-tH_0/n}\chi_\Omega]^n f - e^{-tH}f\|_{L^2} \to 0, \quad f\in L^2(\Omega).$$

Consequently, Theorem 2.4 is a *stronger* statement than this corollary, though the latter is also obtained by Kato [29] as an abstract result: If A is a nonnegative selfadjoint operator and P an orthogonal projection in a Hilbert space $\mathcal{H}$, then $(Pe^{-tA/n}P)^n \to e^{-tA_P}$, *strongly*, as $n \to \infty$, where $A_P := (A^{1/2}P)^*(A^{1/2}P)$. In passing, however, it is an open question whether it holds that $(Pe^{-itA/n}P)^n \to e^{-itA_P}P$, strongly (*Zeno product formula*). A partial answer was given in [11].

All Theorems 2.2–2.4 hold with order of products exchanged, e.g., in Theorem 2.2, $[e^{-tV/2n}e^{-tH_0/n}e^{-tV/2n}]^n$ instead of $[e^{-tH_0/2n}e^{-tV/n}e^{-tH_0/2n}]^n$.

Theorem 2.5 (Ichinose-Tamura 2004 [24] (Unitary Trotter in norm)). *Let A and B be selfadjoint, and assume $H := A + B$ to be essentially selfadjoint in a Hilbert space $\mathcal{H}$. Assume that there exists a dense subspace $\mathcal{D}$ of $\mathcal{H}$ with $\mathcal{D} \subset D[A] \cap D[B]$ such that e^{-itA}, e^{-itB} : $\mathcal{D} \to \mathcal{D}$. Further assume that the commutators $[A, B]$, $[A, [A, B]]$ and $[B, [A, B]]$ are bounded on $\mathcal{H}$. Then*

$$\|\big(e^{-itB/2n}e^{-itA/n}e^{-itB/2n}\big)^n - e^{-itH}\| = O(n^{-2}), \quad n \to \infty, \tag{2.7}$$

locally uniformly in the real line $\mathbf{R}$.

As important applications we have ones to the Dirac operator $H = H_0 + V = (i\alpha \cdot \nabla + m\beta) + V(x)$ in $L^2(\mathbf{R}^3)^4$, where $\alpha = (\alpha_1, \alpha_2, \alpha_3)$ and β are the 4 Dirac matrices, with $\partial^\gamma V(x)(|\gamma| = 2)$ being bounded, as well as to the relativistic Schrödinger operator $H = H_0 + V = \sqrt{-\Delta + m^2} + V(x)$ on $L^2(\mathbf{R}^d)$ with $\partial_x^\gamma V(x)$ being bounded for $1 \leq |\gamma| \leq 4$ ($0 \leq |\gamma| \leq 4$, if $m = 0$). In these cases, H are essentially selfadjoint, and satisfy the conditions in the theorem. So it holds that

$$\|[e^{-itV/2n}e^{-itH_0/n}e^{-itV/2n}]^n - e^{-itH}\|_{L^2} = O(n^{-2}), \quad n \to \infty, \tag{2.8}$$

locally uniformly in $\mathbf{R}$.

However, this theorem does not apply to Schrödinger operators except for the *Stark Hamiltonian* $(-\Delta + V(x)) + a \cdot x$ in $L^2(\mathbf{R}^d)$, where a is a constant real vector in $\mathbf{R}^d$.

Finally it should be noted that we have shown in Theorems 2.2–2.4 that the convergence is uniform only on compact t-intervals which are away from 0, though in Theorems 2.1 and 2.5, on ones which are allowed to be not away from 0.

3. Optimality of error bounds

In this section we discuss optimality of error bounds. The error bound $O(1/n)$ in Theorem 2.1 is optimal, because if both A and B are bounded operators, by the Baker–Campbell–Hausdorff formula we know

$$\begin{aligned} [e^{-tA/n}e^{-tB/n}]^n - e^{-tH} &= R'_n \cdot n^{-1}, \\ [e^{-tB/2n}e^{-tA/n}e^{-tB/2n}]^n - e^{-tH} &= R_n \cdot n^{-2}, \end{aligned}$$

for some R'_n and R_n being uniformly bounded operators which in general are not the zero operator. From this, optimality in the former non-symmetric case is evident. But even in the symmetric case it is optimal. Indeed, there exist unbounded nonnegative selfadjoint operators A, B such that $H = A + B$ is selfadjoint and

$$\|[e^{-tB/2n}e^{-tA/n}e^{-tB/2n}]^n - e^{-tH}\| \geq c(t)n^{-1}$$

for some continuous function $c(t)$ with $c(t) > 0,\ t > 0$ and $c(0) = 0$ ([27]).

However, further in some special symmetric case in Theorem 2.2 where $-\Delta$, V are taken as A, B, we have seen the symmetric product formula hold with a *sharp* error bound $O(n^{-2})$. We can make more precise this result with the 1-dimensional harmonic oscillator $H := H_0 + V := \frac{1}{2}(-\partial_x^2 + x^2)$ in $L^2(\mathbf{R})$.

Theorem 3.1 (Azuma-Ichinose 2007 [2]**).** *There exists bounded continuous functions $C(t) \geq 0$ and $c(t) \geq 0$ in $t \geq 0$, which are positive except $t = 0$ with $C(0) = c(0) = 0$, independent of n, such that for $n = 1, 2, \ldots,$*

$$c(t)n^{-2} \leq \|[e^{-\frac{t}{2n}V}e^{-\frac{t}{n}H_0}e^{-\frac{t}{2n}V}]^n - e^{-tH}\| \leq C(t)n^{-2}, \qquad t \geq 0. \tag{3.1}$$

This theorem mentions an error bound from below, extending the harmonic oscillator case of Theorem 2.2 which treats only the right-half inequality with $C(t) = C$ being a positive constant depending on each compact t-interval in the *open* half-line $(0, \infty)$.

It is anticipated that the same is true for the Schrödinger operator $H = -\Delta + V(x)$ with growing potentials like $V(x) = |x|^{2m}$ treated in Theorem 2.2.

Theorem 3.1 is obtained as a corollary from the following theorem of its integral kernel version. Here one calculates explicitly the integral kernel $K^{(n)}(t, x, y)$ of $[e^{-tV/2n}e^{-tH_0/n}e^{-tV/2n}]^n$ to estimate its difference from the integral kernel $e^{-tH}(x, y)$ of e^{-tH}.

Theorem 3.2 (Azuma-Ichinose 2007 [2]**).** *There exists a bounded operator $R(t)$ and uniformly bounded operators $\{Q^{(n)}(t)\}_{n=1}^{\infty}$ with integral kernels $R(t, x, y)$ and $Q^{(n)}(t, x, y)$ being uniformly bounded continuous functions in $(0, \infty) \times \mathbf{R} \times \mathbf{R}$ such that*

$$K^{(n)}(t, x, y) - e^{-tH}(x, y) = \big[R(t, x, y) + Q^{(n)}(t, x, y)n^{-1}\big]n^{-2}; \tag{3.2}$$

they satisfy

$$\sup_{x,y}|R(t, x, y)|,\ \sup_n \sup_{x,y}|Q^{(n)}(t, x, y)| \to 0,\ t \to 0\,; \quad \sup_{x,y}|R(t, x, y)| \to 0,\ t \to \infty.$$

$R(t, x, y)$ is explicitly given by

$$\begin{aligned} R(t, x, y) = e^{-tH}(x, y)\frac{t^2}{12}\Big[t\Big(\frac{1}{4}\frac{e^t + e^{-t}}{e^t - e^{-t}} + \frac{(e^t + e^{-t})xy - (x^2 + y^2)}{(e^t - e^{-t})^2}\Big) \\ + \frac{1}{16}\Big(1 + \frac{4xy - (e^t + e^{-t})(x^2 + y^2)}{e^t - e^{-t}}\Big)\Big]. \end{aligned} \tag{3.3}$$

If $t > 0$, $R(t, x, y)$ can become positive and negative.

Lemma 3.3.

$$K^{(n)}(t,x,y)$$
$$= \frac{1}{\sqrt{\pi}}\left(\frac{\sqrt{1+\frac{t^2}{4n^2}}}{\left(1+\frac{t}{n}\sqrt{1+\frac{t^2}{4n^2}}+\frac{t^2}{2n^2}\right)^n-\left(1-\frac{t}{n}\sqrt{1+\frac{t^2}{4n^2}}+\frac{t^2}{2n^2}\right)^n}\right)^{1/2}$$
$$\times \exp\left[\frac{2\sqrt{1+\frac{t^2}{4n^2}}}{\left(1+\frac{t}{n}\sqrt{1+\frac{t^2}{4n^2}}+\frac{t^2}{2n^2}\right)^n-\left(1-\frac{t}{n}\sqrt{1+\frac{t^2}{4n^2}}+\frac{t^2}{2n^2}\right)^n}xy\right]$$
$$\times \exp\left\{\left[-\frac{t}{4n}-\frac{n}{2t}\left(1-\right.\right.\right. \tag{3.4}$$
$$\left.\left.\left.\frac{\left(1+\frac{t}{n}\sqrt{1+\frac{t^2}{4n^2}}+\frac{t^2}{2n^2}\right)^{n-1}-\left(1-\frac{t}{n}\sqrt{1+\frac{t^2}{4n^2}}+\frac{t^2}{2n^2}\right)^{n-1}}{\left(1+\frac{t}{n}\sqrt{1+\frac{t^2}{4n^2}}+\frac{t^2}{2n^2}\right)^n-\left(1-\frac{t}{n}\sqrt{1+\frac{t^2}{4n^2}}+\frac{t^2}{2n^2}\right)^n}\right)\right](x^2+y^2)\right\}.$$

Proof. Calculate the Gaussian integral

$$K^{(n)}(t,x,y) \equiv \left(\frac{n}{2\pi t}\right)^{\frac{n}{2}} \overbrace{\int_{\mathbf{R}}\cdots\int_{\mathbf{R}}}^{n-1} \prod_{j=1}^{n}\left[e^{-\frac{t}{4n}x_j^2}e^{-\frac{(x_j-x_{j-1})^2}{2t/n}}e^{-\frac{t}{4n}x_{j-1}^2}\right]dx_1\cdots dx_{n-1},$$

where $x = x_n$, $y = x_0$. We shall encounter with continued fraction to lead to the final expression (3.4) of the lemma.

To show Theorem 3.2, we simply calculate the difference $K^{(n)}(t,x,y) - e^{-tH}(x,y)$, though it is not so simple.

Here we mention what the operator with $R(t,x,y)$ as its integral kernel is. By the Baker–Campbell–Hausdorff formula (e.g., [45], [40]), if A and B are bounded operators, we have

$$[e^{-tB/2n}e^{-tA/n}e^{-tB/2n}]^n - e^{-t(A+B)}$$
$$= \exp\left(-t(A+B) - n^{-2}\frac{t^2}{24}[2A+B,[A,B]] - O_p(n^{-3})\right)$$
$$= e^{-t(A+B)} - n^{-2}\frac{t^2}{24}\int_0^t e^{-(t-s)(A+B)}[2A+B,[A,B]]e^{-s(A+B)}ds + O_p(n^{-3}),$$

where $O_p(n^{-3})$ is an operator with norm of $O(n^{-3})$. In our case where $A = -\frac{1}{2}\partial_x^2$, $B = \frac{1}{2}x^2$, we can show $R(t,x,y)$ is just the integral kernel of the operator

$$-\frac{t^2}{24}\int_0^t e^{-(t-s)H}[2H_0+V,[H_0,V]]e^{-sH}ds,$$

which *does* make sense, though H_0 and V are unbounded operators. We have $[2H_0+V,[H_0,V]] = -4H_0+2V = -4H+6V$.

4. Idea of proof

Put $K(\tau) = e^{-\tau B/2}e^{-\tau A}e^{-\tau B/2}$. Note that $0 \le K(\tau) \le 1$. Then we need to estimate the difference between $K(t/n)^n$ and e^{-tH}. The general technique of proof is: (i) to establish an appropriate version of Chernoff's theorem ([8]):

$$\begin{aligned}&[(1+\tau^{-1}(1-K(\tau))^{-1} - (1+H)^{-1}] \to 0, \;\; \tau \downarrow 0 \\ &\qquad\qquad \Longrightarrow \quad [K(t/n)^n - e^{-tH}] \to 0, \;\; n \to \infty\end{aligned}$$

and/or (ii) to do telescoping:

$$e^{-tH} - K(t/n)^n = \sum_{k=1}^{n} e^{-(k-1)tH/n}(e^{-tH/n} - K(t/n))K(t/n)^{n-k}$$

to estimate each summand on the right. The former method (i) seems to be more efficient than the latter (ii).

In fact, to prove Theorem 1, we use the former method, establishing the following norm version of Chernoff's theorem with error bounds. The case without error bounds was noted by Neidhardt–Zagrebnov [33].

Lemma 4.1 (Ichinose-Tamura [23]).

I. *Let $\{F(t)\}_{t\ge 0}$ be a family of selfadjoint operators with $0 \le F(t) \le 1$, and $H \ge 0$ a selfadjoint operator in a Hilbert space $\mathcal{H}$. Define $S_t := t^{-1}(1-F(t))$. Then*

(a) *For $0 < \alpha \le 1$, $\|(1+S_t)^{-1} - (1+H)^{-1}\| = O(t^\alpha), \quad t \downarrow 0$*

implies

(b) *For every fixed $\delta > 0$,*

$$\|F(t/n)^n - e^{-tH}\| = \delta^{-2}t^{-1+\alpha}e^{\delta t}O(n^{-\alpha}), \;\; n \to \infty, \;\; t > 0.$$

Therefore for $\alpha = 1$ this convergence is uniform on each compact interval $[0, L]$ in the closed half-line $[0, \infty)$.

II. *Moreover, in case $H \ge \eta I$ for some constant $\eta > 0$, if for every $\varepsilon > 0$ there exists $\delta(\varepsilon) > 0$ such that $F(t) \le 1 - \delta(\varepsilon)$ for all $t \ge \varepsilon$, then*

$$\|F(t/n)^n - e^{-tH}\| = (1+2/\eta)^2 t^{-1+\alpha}O(n^{-\alpha}), \quad n \to \infty, \;\; t > 0.$$

Therefore for $\alpha = 1$ this convergence is uniform on the whole closed half-line $[0, \infty)$.

Condition II is satisfied, e.g., for $F(\tau) = e^{-\tau B/2}e^{-\tau A}e^{-\tau B/2}$. For the proof, we refer to [23].

For the proof of Theorems 2.2–2.5 we employ the latter method (ii), and further, for Theorems 2.2–2.4, make a crucial use of Agmon's kernel theorem:

Lemma 4.2 (Agmon's kernel theorem [1]). *Let $T : L^2(\mathbf{R}^d) \to L^2(\mathbf{R}^d)$ be a bounded operator with ranges of T and its adjoint T^* satisfying $R[T]$, $R[T^*] \subset H^m(\mathbf{R}^d)$,*

$m > d$. If T is an integral operator with integral kernel $T(x,y)$ being a bounded continuous function in $\mathbf{R}^d \times \mathbf{R}^d$ such that

$$(Tf)(x) = \int T(x,y)f(y)\,dy, \qquad f \in L^2,$$

then

$$|T(x,y)| \leq C(\|T\|_m + \|T^*\|_m)^{\frac{d}{m}} \|T\|^{1-\frac{d}{m}},$$

where $\|T\|_m := \|T\|_{\mathcal{L}(L^2 \to H^m)}$ is the operator norm of T as an operator of $L^2(\mathbf{R}^d)$ into the Sobolev space $H^m(\mathbf{R}^d)$.

Indeed, we estimate the $\mathcal{L}(L^2 \to H^m)$-operator norm of the difference

$$T = [e^{-tV/2n}e^{t\Delta/n}e^{-tV/2n}]^n - e^{t(-\Delta+V)}.$$

5. Concluding remarks

We have so far considered the case where the *operator sum* $H = A + B$ of two nonnegative selfadjoint operators A and B is selfadjoint. However, otherwise, the exponential product formula in norm does not in general hold for the *form sum* $H = A + B$ of two selfadjoint operators $A \geq 0$, $B \geq 0$, even if it is essentially selfadjoint on $D[A] \cap D[B]$ (see [43]). Nevertheless, there is some case where it holds:

Theorem 5.1 (Ichinose-Neidhardt-Zagrebnov 2004 [16]). *Let $H = A \dot{+} B$ be the form sum of A and B. If $D[H^\alpha] \subseteq D[A^\alpha] \cap D[B^\alpha]$ for some $\frac{1}{2} < \alpha < 1$, and $D[A^{\frac{1}{2}}] \subseteq D[B^{\frac{1}{2}}]$, then*

$$\|[e^{-tB/2n}e^{-tA/n}e^{-tB/2n}]^n - e^{-tH}\| = O(n^{-(2\alpha-1)}), \tag{5.1}$$

$$\|[e^{-tA/n}e^{-tB/n}]^n - e^{-tH}\| = O(n^{-(2\alpha-1)}), \tag{5.2}$$

locally uniformly in $[0,\infty)$.

This error bound in (5.1)/(5.2) is also optimal. For this we refer to [43]. The condition for the domains of A and B is not symmetric. It is an open question whether one may improve it so as to become symmetric with respect to A and B.

Finally, as we should like to mention, there is a very nice Feynman path integral formula which represents the Schrödinger semigroup, called the *Feynman–Kac formula*

$$\begin{aligned}(e^{-tH}f)(x) &= (e^{-t(-\Delta+V)}f)(x)\\ &= \int_{B \in C([0,\infty) \to \mathbf{R}^d), B(0)=x} \exp[-\int_0^t V(B(s))ds] f(B(t))d\mu(B),\end{aligned}$$

where $\mu(\cdot)$ is the Wiener measure on the path space $C([0,\infty) \to \mathbf{R}^d)$ (e.g., ([39]). We may use this formula to get whatever results, in fact, a lot of them. This is a big advantage! But disadvantage is that it is only restricted to the Schrödinger operator or Laplacian. For instance, if we think of the semigroup for the relativistic Schrödinger operator $H = \sqrt{-\Delta + m^2} + V(x)$, we have to establish another Feynman–Kac formula (cf. [28]).

Indeed, the Feynman–Kac formula is one of the realizations of Feynman path integral as a *true integral* on a path space. However, as Nelson [36] noted, the exponential/Trotter product formula also can give a meaning to the Feynman path integral as a *time-sliced approximation* by finite-dimensional integrals (cf. [15]). What it has advantage at is that we may apply it to the sum $H = A + B$ of any two selfadjoint operators A, B bounded from below.

Acknowledgements

One of the authors (T.I.) would like to thank Professor Vadim Adamyan for kind and warm hospitality during his stay in Odessa for the Mark Krein Centenary Conference, April 2007. This work (of T.I. and H.T.) was partially supported by Grant-in-Aid for Exploratory Research No. 17654030 and by Grant-in-Aid for Scientific Research (B) No. 18340049, Japan Society for the Promotion of Sciences.

References

[1] S. Agmon, *On kernels, eigenvalues, and eigenfunctions of operators related to elliptic problems.* Comm. Pure Appl. Math. **18** (1965), 627–663.

[2] Y. Azuma and T. Ichinose, *Note on norm and pointwise convergence of exponential products and their integral kernels for the harmonic oscillator.* to appear in Integral Equations Operator Theory (published on line first Nov. 12, 2007).

[3] V. Cachia, H. Neidhardt and V.A. Zagrebnov, *Accretive perturbations and error estimates for the Trotter product formula.* Integral Equations Operator Theory **39** (2001), 396–412.

[4] V. Cachia, H. Neidhardt and V.A. Zagrebnov, *Comments on the Trotter product formula error-bound estimates for nonself-adjoint semigroups.* Integral Equations Operator Theory **42** (2002), 425–448.

[5] V. Cachia and V.A. Zagrebnov, *Operator-norm convergence of the Trotter product formula for sectorial generators.* Lett. Math. Phys. **50** (1999), 203–211.

[6] V. Cachia and V.A. Zagrebnov, *Trotter product formula for nonself-adjoint Gibbs semigroups.* J. London Math. Soc. (2) **64** (2001), 436–444.

[7] V. Cachia and V.A. Zagrebnov, *Operator-norm convergence of the Trotter product formula for holomorphic semigroups.* J. Operator Theory **46** (2001), 199–213.

[8] P.R. Chernoff, *Product Formulas, Nonlinear Semigroups, and Addition of Unbounded Operators.* Memoirs Amer. Math. Soc. **140**, 1974.

[9] B.O. Dia and M. Schatzman, *An estimate on the Kac transfer operator*, J. Functional Analysis **145** (1997), 108–135.

[10] A. Doumeki, T. Ichinose and H. Tamura, *Error bounds on exponential product formulas for Schrödinger operators.* J. Math. Soc. Japan **50** (1998), 359–377.

[11] P. Exner and T. Ichinose, *A product formula related to quantum Zeno dynamics.* Ann. H. Poincaré **6** (2005), 195–215.

[12] I.C. Gohberg and M.G. Krein, Introduction to the theory of linear non-selfadjoint operators (in Hilbert space). Translations of Mathematical Monographs, Vol. 18, Amer. Math. Soc., Providence, R.I. 1969; its Russian edition, Izdat. "Nauka", Moscow 1965.

[13] B. Helffer, *Around the transfer operator and the Trotter–Kato formula.* Operator Theory: Advances and Appl. **79** (1995), 161–174.

[14] F. Hiai, *Trace norm convergence of exponential product formula.* Lett. Math. Phys. **33** (1995), 147–158.

[15] T. Ichinose, *Time-sliced approximation to path integral and Lie–Trotter–Kato product formula.* In: "A Garden of Quanta", Essays in Honor of Hiroshi Ezawa for his seventieth birthday, World Scientific 2003, pp. 77–93.

[16] T. Ichinose, H. Neidhardt and V.A. Zagrebnov, *Trotter–Kato product formula and fractional powers of self-adjoint generators.* J. Functional Analysis **207** (2004), 33–57; *Operator norm convergence of Trotter–Kato product formula,* Ukrainian Mathematics Congress – 2001 (Ukrainian), 100–106, Natsional. Akad. Nauk Ukraini, Inst. Mat., Kiev, 2002.

[17] T. Ichinose and S. Takanobu, *Estimate of the difference between the Kac operator and the Schrödinger semigroup.* Commun. Math. Phys. **186** (1997), 167–197.

[18] T. Ichinose and S. Takanobu, *The norm estimate of the difference between the Kac operator and the Schrödinger semigroup: A unified approach to the nonrelativistic and relativistic cases.* Nagoya Math. J. **149** (1998), 51–81.

[19] T. Ichinose and S. Takanobu, *The norm estimate of the difference between the Kac operator and the Schrödinger semigroup II: The general case including the relativistic case.* Electronic Journal of Probability **5** (2000), Paper 5, pages 1–47; URL http://www.math.washington.edu/ ejpecp/

[20] T. Ichinose and H. Tamura, *Error estimate in operator norm for Trotter–Kato product formula.* Integral Equations Operator Theory **27** (1997), 195–207.

[21] T. Ichinose and H. Tamura, *Error bound in trace norm for Trotter–Kato product formula of Gibbs semigroups.* Asymptotic Analysis **17** (1998), 239–266.

[22] T. Ichinose and H. Tamura, *Error estimate in operator norm of exponential product formulas for propagators of parabolic evolution equations.* Osaka J. Math. **35** (1998), 751–770.

[23] T. Ichinose and H. Tamura, *The norm convergence of the Trotter–Kato product formula with error bound.* Commun. Math. Phys. **217** (2001), 489–502; *Erratum*, ibid. **254** (2005), 255.

[24] T. Ichinose and H. Tamura, *Note on the norm convergence of the unitary Trotter product formula.* Lett. Math. Phys. **70** (2004), 65–81.

[25] T. Ichinose and H. Tamura, *Sharp error bound on norm convergence of exponential product formula and approximation to kernels of Schrödinger semigroups.* Comm. Partial Differential Equations **29** (2004), 1905–1918.

[26] T. Ichinose and H. Tamura, *Exponential product approximation to integral kernel of Schrödinger semigroup and to heat kernel of Dirichlet Laplacian.* J. Reine Angew. Math. **592** (2006), 157–188.

[27] T. Ichinose, H. Tamura, Hiroshi Tamura and V.A. Zagrebnov, *Note on the paper "The norm convergence of the Trotter–Kato product formula with error bound" by Ichinose and Tamura.* Commun. Math. Phys. **221** (2001), 499–510.

[28] T. Ichinose and Hiroshi Tamura, *Imaginary-time path integral for a relativistic spinless particle in an electromagnetic field.* Commun. Math. Phys. **105**, 239–257(1986); For the final result of this see: T. Ichinose, *Some results on the relativistic Hamiltonian: Selfadjointness and imaginary-time path integral.* Differential Equations and Mathematical Physics, 102–116, International Press, Boston 1995.

[29] T. Kato, *Trotter's product formula for an arbitrary pair of self-adjoint contraction semigroups.* In: Topics in Functional Analysis (essays dedicated to M.G. Krein on the occasion of his 70th birthday), edited by I. Gohberg and M. Kac, Academic Press, New York 1978, 185–195.

[30] H. Neidhardt and V.A. Zagrebnov, *The Trotter–Kato product formula for Gibbs semigroups.* Commun. Math. Phys. **131** (1990), 333–346.

[31] H. Neidhardt and V.A. Zagrebnov, *On the Trotter product formula for Gibbs semigroups.* Ann. Physik (7) **47** (1990), 183–191.

[32] H. Neidhardt and V.A. Zagrebnov, *On error estimates for the Trotter–Kato product formula.* Lett. Math. Phys. **44** (1998), 169–186.

[33] H. Neidhardt and V.A. Zagrebnov, *Trotter–Kato product formula and operator-norm convergence.* Commun. Math. Phys. **205** (1999), 129–159.

[34] H. Neidhardt and V.A. Zagrebnov, *Trotter–Kato product formula and symmetrically normed ideals.* J. Functional Analysis **167** (1999), 113–147.

[35] H. Neidhardt and V.A. Zagrebnov, *Fractional powers of self-adjoint operators and Trotter–Kato product formula.* Integral Equations Operator Theory **35** (1999), 209–231.

[36] E. Nelson, *Feynman integrals and the Schrödinger equation.* J. Math. Phys. **5** (1964), 332–343.

[37] Dzh.L. Rogava, *Error bounds for Trotter-type formulas for self-adjoint operators.* Functional Analysis and Its Applications **27** (1993), 217–219.

[38] R. Schatten, *Norm ideals of completely continuous operators.* Springer, Berlin 1960.

[39] B. Simon, *Functional Integration and Quantum Physics.* Academic Press, London 1979.

[40] M. Suzuki, *On the convergence of exponential operators – the Zassenhaus formula, BCH formula and systematic approximations.* Commun. Math. Phys. **57** (1977), 193–200.

[41] S. Takanobu, *On the estimate of the integral kernel for the Trotter product formula for Schrödinger operators.* Ann. Probab. **25** (1997), 1895–1952.

[42] S. Takanobu, *On the trace norm estimate of the Trotter product formula for Schrödinger operators.* Osaka J. Math. **35** (1998), 659–682.

[43] Hiroshi Tamura, *A remark on operator-norm convergence of Trotter–Kato product formula.* Integral Equations Operator Theory **37** (2000), 350–356.

[44] H.F. Trotter, *On the product of semigroups of operators.* Proc. Amer. Math. Soc. **10** (1959), 545–551.

[45] S. Varadarajan, *Lie Groups, Lie Algebras, and Their Representations.* Springer, Berlin–Heidelberg–New York–Tokyo 1974, 1984.

[46] V.A. Zagrebnov, *The Trotter–Lie product formula for Gibbs semigroups.* J. Math. Phys. **29** (1988), 888–891.

[47] V.A. Zagrebnov, *Topics in the Theory of Gibbs Semigroups.* Leuven Notes in Mathematical and Theoretical Physics, Vol. 10, Leuven University Press 2003.

Takashi Ichinose
Department of Mathematics
Kanazawa University
920–1192 Kanazawa, Japan
e-mail: `ichinose@kenroku.kanazawa-u.ac.jp`

Hideo Tamura
Department of Mathematics
Okayama University
700–8530 Okayama, Japan
e-mail: `tamura@math.okayama-u.ac.jp`

Operator Theory:
Advances and Applications, Vol. 190, 329–334

Generalized Selfadjoint Operators

Ivan Ya. Ivasiuk

Abstract. An essential problem in mathematical physics is to introduce and investigate operators perturbed by singular perturbations. Such operators are usually introduced with the aid of the theory of selfadjoint extensions of Hermitian operators. Berezansky and Brasche have proposed another viewpoint (see [3]) from which such objects are constructed by using operators in a Hilbert space chain (rigging).

My report is concerned with operators that act from a positive space into a negative space of some Hilbert rigging. We investigate the generalized selfadjointness of such operators.

Mathematics Subject Classification (2000). Primary 47A70, Secondary 47B25.

Keywords. Selfadjoint operators, generalized selfadjoint operators, Hilbert space rigging.

1. Basic definitions

Let us consider a complex Hilbert space with rigging

$$H_- \supset H_0 \supset H_+ \tag{1.1}$$

defined in the usual way ([4], Ch. 14, §1), $H_- = (H_+)'$. Let $\mathbb{I} : H_- \to H_+$, $\mathbb{J} : H_- \to H_0$, $J : H_0 \to H_+$ be standard isometric operators connected with (1.1) and such that the equality $\mathbb{I} = J\mathbb{J}$ holds.

Let $A : H_+ \to H_-$ be some linear operator with domain $D(A)$ dense in H_+. For A, it is easy to define the adjoint operator $A^+ : H_+ \to H_-$. So, let $\psi \in H_+$ be such that the functional $\varphi \to (A\varphi, \psi)_{H_0} \in \mathbb{C}$, defined on $D(A)$, is continuous and therefore has the representation $(A\varphi, \psi)_{H_0} = (\varphi, \psi^+)_{H_0}, \psi^+ \in H_-$. Such ψ form $D(A^+)$ of the operator A^+ and $A^+\psi := \psi^+$. If $H_+ = H_0$, then we have the usual definition of an adjoint operator.

If $A : H_+ \to H_-$ is continuous, then $A^+ : H_+ \to H_-$ exists, is bounded, and

$$(A\varphi, \psi)_{H_0} = (\varphi, A^+\psi)_{H_0}, \qquad \varphi, \psi \in H_0. \tag{1.2}$$

For any A, the operator A^+ is closed. If $A \subset B$, then $A^+ \supset B^+$.

Definition 1.1. A is generalized Hermitian if

$$(A\varphi, \psi)_{H_0} = (\varphi, A\psi)_{H_0}, \qquad \varphi, \psi \in D(A). \tag{1.3}$$

Indeed, (1.3) means that $A \subset A^+$ and the last inclusion is an alternative definition of the hermicity of A. It follows from the inclusion $A \subset A^+$ that A is closable and $(\widetilde{A})^+ = A^+$.

Definition 1.2. $A : H_+ \to H_-$ is called generalized selfadjoint if $A^+ = A$ and generalized essentially selfadjoint if $A^+ = \widetilde{A}$.

Remark 1.3. Consider a Schrödinger operator in the space $L_2(\mathbb{R}^d), d = 1, 2, \ldots$ with potential q generated by the mapping

$$L := -\Delta + q;\ L : C_0^\infty(R^d) \to L^2(R^d). \tag{1.4}$$

When q is real-valued and $q \in L^2_{\text{loc}}(R^d)$, the mapping (1.4) is well defined and determines a Hermitian operator in $L^2(R^d)$. But if q is a generalized function, the operator defined by (1.4) does not act in $L^2(R^d)$. In article [3], the following point of view on the construction was proposed.

We construct the operator $B : H_+ \to H_-$ corresponding to $L = -\Delta$. Then we construct the "selfadjoint" operator $C : H_+ \to H_-$ corresponding to $q(x)$; for a generalized $q(x)$, this is possible to do because H_- consists of generalized functions. Finally, we put $A := \widetilde{B} + C : H_+ \to H_-$.

2. Sufficient conditions for generalized selfadjointness

We will give some results represented by Yu.M. Berezansky and J. Brasche in [3].

Proposition 2.1. *The generalized selfadjointness and the generalized essentially selfadjointness of an operator A are equivalent to the same classical properties of the operator $\mathbb{I}A$ as an operator on H_+. Also, there holds the equality*

$$\mathbb{I}A^+ = (\mathbb{I}A)^*, \tag{2.1}$$

where the $*$ denotes the usual adjointness in H_+ .

Proposition 2.2. *Let the restriction of a generalized Hermitian operator A: $H_+ \to H_-$, $D(A) = H_+$, on H_0 be bounded (i.e., there exists $c > 0$ such that for every $\phi \in H_+$, $A\phi \in H_0$ and $\|A\phi\|_{H_0} \leq c\|\phi\|_{H_0}$). Then A is bounded and generalized selfadjoint.*

Proposition 2.3. *Let the restriction of A: $H_+ \to H_-$ on the space H_0 be essentially selfadjoint and strongly positive in the following sense: there exists $\varepsilon > 0$ such that*

$$(A\varphi, \varphi)_{H_0} \geq \varepsilon \|\varphi\|^2_{H_+}, \qquad \varphi \in D(A). \tag{2.2}$$

Then this operator is generalized essentially selfadjoint.

The next results were obtained by me.

Proposition 2.4. *Let for $A : H_+ \to H_-$ its action $A : H_0 \to H_0$, $D(A) \subset H_+$ on the space H_0 be essentially selfadjoint and*

$$\exists \varepsilon > 0, \exists b \in \mathbb{R} : \forall \varphi \in D(A)\, (A\varphi, \varphi)_{H_0} \geq \varepsilon ||\varphi||^2_{H_+} + b||\varphi||^2_{H_0}. \qquad (2.3)$$

Then A is generalized essentially selfadjoint.

This proposition shows that condition (2.2) can be weakened. This can be obtained from the following statement.

Theorem 2.5. *Let an operator $A\colon H_+ \to H_-$ be generalized selfadjoint, $D(A) \subset H_+$, and an operator B be Hermitian in H_0, $H_+ \subset D(B)$ and bounded as an operator from H_+ into H_-. Then the operator $A + B\colon H_+ \to H_-$ is generalized selfadjoint.*

The next result is an analogue of Relich-Kato's theorem.

Theorem 2.6. *Let $A\colon H_+ \to H_-$ be a generalized selfadjoint operator and $B\colon H_+ \to H_-$ be a Hermitian operator, $D(B) \supseteq D(A)$. If $\forall f \in D(A)$ there holds the inequality $\|Bf\|_{H_-} \leq p\,\|Af\|_{H_-} + q\,\|f\|_{H_+}$, $q \geq 0, p \in [0, 1)$, then $A+B$ also is generalized selfadjoint.*

3. Generalized essentially selfadjointness of the $i\frac{d}{dt}$ operator in weight Hilbert riggings

In their work, Yu.M. Berezansky and J. Brasche considered an example of an operator which is “good” (i.e., in particular, selfadjoint) an one acting from H_0 into H_0, but is not selfadjoint with respect to Hilbert rigging H_0 in the general sense ([3], Example 3.4). But it is easy to see that its construction is not correct (as was noticed by Yu.M. Berezansky).

To move in the right way, it is necessary to use one of M.G. Krein's results obtained in [6] (see, also, [1], Ch. 1, Theorem 1.2). In our case, it is stated as follows:

Suppose that in the space H_+ we have a continuous operator A with the norm $\|A\|_+$ and which is Hermitian in H_0, i.e., $(Au, v)_0 = (u, Av)_0$, $u, v \in H_+$. Then A is continuous on the space H_0 and $\|A\|_0 \leq \|A\|_+$.

So, let us consider, for example, some Hermitian operator $T : H_+ \to H_-$ with domain $D(T) = H_+$ which is continuous on H_+ and Hermitian in H_0. From the Krein's result and Proposition 2.2 we obtain that $T : H_+ \to H_-$ is selfadjoint. Then we have to look for an example in the class of operators unbounded and not determined on the whole H_+.

In view of this, it will be shown that such an example cannot be constructed for Hilbert weight riggings of spaces $L^2([0, 1])$ and $L^2(\mathbb{R})$ for the first order differential operator either. Moreover, it will be shown that in these cases the generalized selfadjointness is equivalent to the usual one. We will follow our article [5] and we will give sketches of the proofs.

3.1. Case of $L^2([0,1])$

Let us consider the rigging

$$L^2([0,1],p^{-1}dt) \supset L^2([0,1],dt) \supset L^2([0,1],pdt), \tag{3.1}$$

where $L^2([0,1],pdt)$ is the L^2 space with weight p such that

$$p \in C^1(0,1];\ p(t) \geq 1,\ t \in [0,1];\ p(t) \to +\infty, t \to 0. \tag{3.2}$$

Isometric operators, which are connected with (3.1), have representations: $(\mathbb{I}g)(t) = p^{-1}(t)g(t)$, $(\mathbb{J}g)(t) = p^{-\frac{1}{2}}(t)g(t)$, $(Jf)(t) = p^{-\frac{1}{2}}(t)f(t)$, where $g \in H_-, f \in H_0$.

Let us consider the operator $A\colon H_0 \to H_0$, $(Af)(t) = i\frac{df}{dt}$ with domain $D(A)$ dense in H_0 such that $D(A) = \{x \in AC[0,1] \mid x' \in L^2(0,1), x(0) = x(1)\}$. Here $AC[0,1]$ is the set of absolutely continuous functions on the segment $[0,1]$.

Remark 3.1. To consider the operator A as that acting in H_0, it is sufficient to determine it on $D(A)$. But, since in general $D(A) \not\subset H_+$, then, to consider the operator $A\colon H_+ \to H_-$, it is necessary to constrict its domain on H_+.

So, in what follows, the operator $A \upharpoonright D_+(A)$ will be denoted by A, where $D_+(A) = D(A) \cap H_+$.

Theorem 3.2. *Let p satisfy conditions* (3.2). *The operator $A\colon H_0 \to H_0$ with domain $D_+(A)$ is selfadjoint iff*

$$\int_0^1 p(t)\,dt < \infty. \tag{3.3}$$

Proof. Outline of the proof.

Necessity. Let us suppose the opposite that

$$\int_0^1 p(t)\,dt = \infty. \tag{3.4}$$

Then $D_+(A) \subset \{x \in AC[0,1] \mid x' \in L^2(0,1), x(0) = x(1) = 0\}$. So, A is not selfadjoint. We obtain a contradiction.

Sufficiency. It is easy to see that under condition (3.3), $D(A) \subset H_+$ and $D_+(A) = D(A)$. It is well known that the operator $A\colon D(A) \to H_0$ is selfadjoint. □

Let us consider an operator $A\colon H_+ \to H_-$ with domain $D_+(A)$ which is dense in H_+. According to Proposition 2.1, the generalized selfadjointness of this operator is equivalent to the selfadjointness of $\mathbb{I}A\colon H_+ \to H_+$. Then the generalized selfadjointness is equivalent to the usual selfadjointness of the operator $S := \mathbb{J}AJ =$

$J^{-1}\mathbb{I}AJ\colon H_0 \to H_0$, where $D(S) = \{x \in H_0 \big| Jx \in D_+(A)\}$. So, we will further investigate the operator $S = JAJ\colon L^2[0,1] \to L^2[0,1]$ with domain:

$$D(S) = \{x \in L^2[0,1] \mid p^{-\frac{1}{2}}x \in AC[0,1], \\ (p^{-\frac{1}{2}}x)' \in L^2(0,1), (p^{-\frac{1}{2}}x)(0) = (p^{-\frac{1}{2}}x)(1)\},$$

where $(Jx)(t) = p^{-\frac{1}{2}}(t)x(t)$.

Theorem 3.3. *Let p satisfy condition* (3.2). *The operator $A\colon H_+ \to H_-$ with domain $D_+(A)$ is generalized essentially selfadjoint iff p satisfies condition* (3.3).

Proof. Outline of the proof.

Necessity. Let us suppose the opposite that (3.4) holds. Then $D(S) = \{x \in L^2[0,1] \mid p^{-\frac{1}{2}}x \in AC[0,1], (p^{-\frac{1}{2}}x)' \in L^2(0,1), (p^{-\frac{1}{2}}x)(0) = (p^{-\frac{1}{2}}x)(1) = 0\}$. It is easy to show that
$D(S^*) \supseteq \{y \in L^2[0,1] \mid p^{-\frac{1}{2}}(t)y(t) \in AC[0,1], (p^{-\frac{1}{2}}(t)y(t))' \in L^2[0,1]\}$. It follows $\operatorname{Ker}(S^* \pm i) \neq \{0\}$. So, S is not essentially selfadjoint, and we obtain a contradiction.

Sufficiency. Let the weight p be such that (3.3) holds. Then $\exists x \in D(S)$ such that $(p^{-\frac{1}{2}}x)(0) = (p^{-\frac{1}{2}}x)(1) = c \neq 0$. So, for any $y \in D(S^*)$ there holds $(p^{-\frac{1}{2}}y)(0) = (p^{-\frac{1}{2}}y)(1)$. Then $\operatorname{Ker}(S^* \pm i) = \{0\}$ and S is essentially selfadjoint. □

It follows from the above-proved, that the operator $A\colon H_+ \to H_-$ with domain $D_+(A)$ is generalized essentially selfadjoint iff it is selfadjoint in the normal sense as an operator from H_0 into H_0.

3.2. Case of $L^2(\mathbb{R})$

Let us consider the rigging

$$L^2(\mathbb{R}, p^{-1}dt) \supset L^2(\mathbb{R}, dt) \supset L^2(\mathbb{R}, pdt), \tag{3.5}$$

where $L^2(\mathbb{R}, pdt)$ is the space with weight p such that $p \in C^\infty(\mathbb{R})$; $p(t) \geq 1$, $t \in \mathbb{R}$.

Consider an operator $A\colon H_+ \to H_-$ generated by the mapping $(Af)(t) = i\frac{df}{dt}$ and with domain $D(A) = C_0^\infty(\mathbb{R})$ ($C_0^\infty(\mathbb{R})$ is the space of finite and infinitely differentiable function on $\mathbb{R}$). It is dense in H_0 and H_+. As is well known, the operator $A\colon H_0 \to H_0$ with domain $D(A)$ is essentially selfadjoint.

Instead of investigating the generalized essentially selfadjointness of $A\colon H_+ \to H_-$, we will examine the essentially selfadjointness of $S := \mathbb{J}AJ = J^{-1}\mathbb{I}AJ\colon H_0 \to H_0$, where $D(S) = \{x \in H_0 \big| Jx \in D(A)\}$. Because $p^{-\frac{1}{2}} \in C^\infty(\mathbb{R})$ and $\forall t \in \mathbb{R}\ p^{-\frac{1}{2}}(t) > 0$, $D(S) = D(A) = C_0^\infty(\mathbb{R})$.

Theorem 3.4. *The operator S is generalized essentially selfadjoint in $L_2(\mathbb{R})$.*

Proof. Outline of the proof.

It is necessary to show that $\mathrm{Ker}(S^* \pm i) = \{0\}$. Let $y \in \mathrm{Ker}(S^* \pm i)$. Then y satisfies the equation

$$-(p^{-1}(t)y(t))' + ((p^{-\frac{1}{2}}(t))' p^{-\frac{1}{2}}(t) \mp 1)y(t) = 0,$$

whose solutions are $y_\pm(t) = c_\pm p^{\frac{1}{2}}(t) \exp\{\pm \int\limits_0^t p(s)ds\}$. But $y_\pm(t) \overline{\in} L_2(\mathbb{R})$. □

So, the operator $A\colon H_+ \to H_-$ is generalized essentially selfadjoint in the sense of rigging (3.5) with domain $D(A) = C_0^\infty(\mathbb{R})$ and, at the same time, $A\colon H_0 \to H_0$ is essentially selfadjoint in the usual sense on $D(A)$.

References

[1] Ju.M. Berezanskii, *Expansions in Eigenfunctions of Selfadjoint Operators.* Amer. Math. Soc., Providence, RI, 1968 (Russian edition: Naukova Dumka, Kiev, 1965).

[2] Yu.M. Berezansky, *Bilinear forms and Hilbert riggings.* Spectral Analysis of Differential Operators, Institute of Mathematics, Ukr. Acad. Sci., Kiev, (1980), 83–106.

[3] Yu.M. Berezansky and J. Brasche, *Generalized selfadgoint operators and their singular perturbations.* Methods Funct. Anal. Topol. **8** (2002), no. 4, 1–14.

[4] Yu.M. Berezansky, Z.G. Sheftel, G.F. Us, *Functional analysis.* Vols 1,2. Birkhäuser Verlag, Basel-Boston-Berlin, 1996 (Russian edition: Vyshcha shkola, Kiev, 1990).

[5] Ivan Ya. Ivasiuk, *Generalized selfadjointness of differentiation operator on weight Hilbert space.* Methods Funct. Anal. Topol. **13** (2007), no. 4, 333–337.

[6] M.G. Krein, *About linear quite continuous operators in functional spaces with two norms.* Zbirnyk prac' Inst. Math. Akad. Nauk. Ukrain. RSR (1947), no. 9, 104–129.

Ivan Ya. Ivasiuk
Kyiv National Taras Shevchenko University
Mechanics and Mathematics Faculty
Department of Mathematical Analysis
64 Volodymyrs'ka St.
01033 Kyiv, Ukraine
e-mail: `vanobsb@gmail.com`

Operator Theory:
Advances and Applications, Vol. 190, 335–340

Orthogonal Decomposition of Functions Subharmonic in the Unit Disc

Armen M. Jerbashian

To the memory of Mark Grigoryevich Krein

Abstract. The article gives a universal orthogonal decomposition for all subharmonic functions which are weighted square summable over the unit disc of the complex plane.

Mathematics Subject Classification (2000). Primary 32A35; Secondary 31A05.

Keywords. Classes of subharmonic functions, orthogonal decomposition, Green-type potentials.

1. Introduction

The present article is a continuation of a series of M.M. Djrbashian's investigations that have resulted in his factorization theory for all functions meromorphic in the unit disc of the complex plane [4, 5, 6, 7]. M.G. Krein considered M.M. Djrbashian's complex analysis results important for the development of the operator theory. He expressed this opinion orally on several occasions. Some applications of M.M. Djrbashian's results [4], along with the author's theory for the half-plane [11], are given in [8, 9, 11].

More precisely, the present article continues the author's work [10] which generalizes the early results of M.M. Djrbashian [2, 3] devoted to the analysis of some spaces and classes of regular functions defined by means of the Riemann–Liouville fractional integration. Namely, [2, 3] contain an investigation of the spaces A^p_α, $\alpha > -1$, $p \geq 1$ (or initially $H_p(\alpha)$), of functions $f(z)$ holomorphic in $|z| < 1$ which are defined by the condition

$$\iint_{|\zeta|<1} (1-|\zeta|)^\alpha |f(\zeta)|^p d\sigma(\zeta) = \sup_{0<r<1} \int_0^1 (1-t)^\alpha t dt \int_0^{2\pi} |f(tre^{i\vartheta})|^p d\vartheta < +\infty,$$

where $\sigma(\zeta)$ is Lebesgue's area measure, and Nevanlinna's classes of functions $f(z)$ meromorphic in $|z|<1$ (see [1], Section 216) which are defined by the condition

$$\int_0^1 (1-r)^\alpha T(r,f)dr = \sup_{0<r<1} \int_0^1 (1-t)^\alpha T(rt,f)dt < +\infty, \quad \alpha>-1, \tag{1.1}$$

where $T(r,f)$ is Nevanlinna's growth characteristic.

By the well-known equilibrium relation between Nevanlinna's growth and decrease characteristics, for functions $f(z)$ holomorphic in $|z|<1$ (1.1) is equivalent to the condition

$$\iint_{|z|<1} |\log|f(z)||(1-|z|)^\alpha d\sigma(z) < +\infty. \tag{1.2}$$

The present article deals with some classes of functions $u(z)$ subharmonic in $|z|<1$ (which, in particular, can be $\log|f(z)|$ with $f(z)$ holomorphic in $|z|<1$), the *squares* of which are summable *with some general measures.* As a result, the union of the considered classes coincides with the set of all functions subharmonic in $|z|<1$, and the considered representations become factorizations for all functions holomorphic in $|z|<1$.

2. Classes of square summable subharmonic functions

Everywhere below, we assume that $\omega(x)\in\Omega_{N^2}$, i.e., $\omega(x)$ is a continuously differentiable in $[0,1)$, strictly decreasing, real function, such that $\omega(0)=1$, $\omega(1)=\omega(1-0)=0$ and $|\omega'(x)|$ is non-increasing in $[0,1)$. Note that Ω_{N^2} is a subset of the class Ω_N of [10].

Further, we define N^2_ω as the set of functions $u(z)$ subharmonic in $|z|<1$, such that $u(z)$ belong to the Lebesgue space L^2_ω considered in [10], i.e.,

$$\|u\|^2_{L^2_\omega} = \frac{1}{2\pi}\iint_{|z|<1} \big[u(z)\big]^2 d\mu_\omega(z) < +\infty, \tag{2.1}$$

where $d\mu_\omega(re^{i\vartheta}) = -d\vartheta d\omega(r^2)$.

Proposition 2.1. *The union of the classes N^2_ω over all $\omega(x)\in\Omega_{N^2}$ coincides with the set of all functions subharmonic in $|z|<1$.*

Proof. Let $u(z)$ be any function subharmonic in $|z|<1$. Then one can see that

$$\varphi(r) = \frac{1}{2\pi}\int_0^{2\pi} \big[u(re^{i\vartheta})\big]^2 d\vartheta$$

is a continuous function in $[0,1)$. Besides, by setting

$$\omega(x) = M_\varphi \int_x^1 \frac{dt}{1+\left(\mathrm{V}_0^{\sqrt{t}}\right)^+\varphi}, \quad M_\varphi = \left(\int_0^1 \frac{dt}{1+\left(\mathrm{V}_0^{\sqrt{t}}\right)^+\varphi}\right)^{-1},$$

where $\left(\bigvee_0^{\sqrt{t}}\right)^+\varphi$ stands for a positive variation of $\varphi(r)$ in $[0,\sqrt{t}]$, one can verify that $\omega(x)\in\Omega_{N^2}$ and

$$\varphi(r)\left|\frac{d}{dr}\omega(r^2)\right| = \frac{2M_\varphi r\varphi(r)}{1+\left(\bigvee_0^r\right)^+\varphi} \le 2M_\varphi < +\infty.$$

Hence, $u(z)\in N^2_\omega$ for the chosen $\omega(x)$. □

Now, suppose that $d_0\in(0,1)$ is some fixed number and introduce the following Green potential formed by ordinary Blaschke factors and a bounded, nonnegative Borel measure $\nu(\zeta)$ in $|\zeta|\le d_0$:

$$G_0(z)=\iint_{|\zeta|<d_0}\log|b(z,\zeta)|\,d\nu(\zeta),\quad b(\lambda,\zeta)=\frac{\zeta-\lambda}{1-\lambda\overline{\zeta}}\frac{|\zeta|}{\zeta}.$$

Then, obviously, $G_0(z)$ belongs to N^2_ω. Hence, forming $G_0(z)$ by the Riesz associated measure of a function $u(z)\in N^2_\omega$ we conclude that also the subharmonic function

$$u_0(z)=u(z)-G_0(z)$$

belongs to N^2_ω. So, further we shall assume that the Riesz associated measure of a function $u(z)\in N^2_\omega$ is such that

$$\inf\left\{|\zeta|:\zeta\in\operatorname{supp}\nu\right\}\ge d_0,$$

where $d_0\in(0,1)$ is a fixed number, and this assumption will not affect the generality of our argument.

Further, one can see that $L^2_\omega\subset L^1_\omega$ for any $\omega(t)\in\Omega_N$, and, consequently, the inclusion $u(z)\in N^2_\omega$ implies $u(z)\in L^1_\omega$. Hence, by Theorem 4.3 of [10], the Riesz associated measure of $u(z)$ satisfies the density condition

$$\iint_{|\zeta|<1}\left(\int_{|\zeta|^2}^1\omega(t)dt\right)d\nu(\zeta)<+\infty, \tag{2.2}$$

and the following Riesz-type representation is true:

$$u(z)=G(z)+U(z),\quad |z|<1, \tag{2.3}$$

where

$$G(z)=\iint_{|\zeta|<1}\log|b_\omega(z,\zeta)|d\nu(\zeta)$$

is a Green-type potential formed by the Blaschke-type factors $b_\omega(z,\zeta)$ of [10]. This potential is convergent in $|z|<1$ in virtue of (2.2), and the integral

$$\begin{aligned}U(z)&=\frac{1}{\pi}\iint_{|\zeta|<1}u(\zeta)\operatorname{Re}\left\{C_\omega(z\overline{\zeta})\right\}d\mu_\omega(\zeta)-u(0)\\&=\frac{1}{\pi}\iint_{|\zeta|<1}\left[u(\zeta)-\frac{u(0)}{2}\right]\operatorname{Re}\left\{C_\omega(z\overline{\zeta})\right\}d\mu_\omega(\zeta),\end{aligned}$$

where

$$C_\omega(z) = \sum_{k=0}^{\infty} \frac{z}{\Delta_k}, \quad \Delta_k = -\int_0^1 t^k d\omega(t),$$

is M.M. Djrbashian's Cauchy-type kernel, is uniformly convergent in $|z| < 1$, where it represents a harmonic function.

Representation (2.3) is inherent in a class of functions which is wider than N^2_ω. Thus, the inclusion $u(z) \in N^2_\omega$ has to imply some additional statements.

Theorem 2.2. *The following statements are true:*

1°. *Both summands $G(z)$ and $U(z)$ in the right-hand side of representation* (2.3) *of a function $u(z) \in N^2_\omega$ $(\omega \in \Omega_{N^2})$ are of N^2_ω.*

2°. *The operator*

$$Q_\omega u(z) = \frac{1}{\pi} \iint_{|\zeta|<1} \left[u(\zeta) - \frac{u(0)}{2}\right] \operatorname{Re}\left\{C_\omega(z\overline{\zeta})\right\} d\mu_\omega(\zeta), \quad |z| < 1,$$

is identical on the set of harmonic functions of N^2_ω and it paps Green-type potentials $G(z) \in N^2_\omega$ to identical zero.

3°. *If $U(z)$ is any harmonic function of N^2_ω, then in L^2_ω the function $U(z) - U(0)$ is orthogonal to any Green-type potential $G(z) \in N^2_\omega$.*

Proof. 1°. By Theorem 3.1 of [10], $U(z) = \operatorname{Re} f(z)$ where

$$f(z) = \frac{1}{\pi} \iint_{|\zeta|<1} u(\zeta) C_\omega(z\overline{\zeta})\, d\mu_\omega(\zeta) - u(0)$$

is a holomorphic function in $|z| < 1$, such that $f(z) \in L^2_\omega$, i.e., $f(z)$ belongs to the Hilbert space A^2_ω of [10]. Thus, $U(z)$ belongs to the harmonic space A^2_ω, i.e., $U \in N^2_\omega$. Hence, $G(z) \in N^2_\omega$.

2°. One can see that representation (2.3) can be written in the form

$$u(z) = G(z) + Q_\omega u(z).$$

For any harmonic function of N^2_ω, the same representation is true without the summand $G_\omega(z)$, i.e., Q_ω is identical operator on the set of harmonic functions from N^2_ω. Consequently, by applying Q_ω to the above representation, we get

$$Q_\omega u(z) = Q_\omega G(z) + Q^2_\omega u(z) = Q_\omega G(z) + Q_\omega u(z).$$

Thus, $Q_\omega G(z) \equiv 0$.

3°. Denote $U_0(z) = U(z) - U(0)$ for brevity. Then $U_0(z) = \operatorname{Re} F(z)$ where

$$F(z) = \frac{1}{\pi} \iint_{|\zeta|<1} U_0(\zeta) C_\omega(z\overline{\zeta}) d\mu_\omega(\zeta)$$

is a holomorphic function in $|z| < 1$, such that $F(0) = 0$ and

$$\|F(z)\|^2_{A^2_\omega} = \sum_{k=1}^{\infty} |a_k|^2 \Delta_k < +\infty,$$

where $\{a_k\}_1^\infty$ are the coefficients of the Taylor expansion for $F(z)$ in $|z| < 1$. It is obvious that for any $r \in (0,1)$

$$F_r(z) = \frac{1}{\pi} \iint_{|\zeta|<r} U_0(\zeta) C_\omega(z\overline{\zeta}) d\mu_\omega(\zeta)$$

is a function of A^2_ω, which possesses the Taylor series expansion

$$F_r(z) = \sum_{k=1}^{\infty} \frac{a_k}{\Delta_k} \Big(- \int_0^{r^2} t^k d\omega(t) \Big) z^k, \quad |z| < 1.$$

Besides, one can see that

$$\big\| F(z) - F_r(z) \big\|_{A^2_\omega} = \sum_{k=1}^{\infty} \frac{|a_k|^2}{\Delta_k} \Big(\int_0^{r^2} t^k d\omega(t) \Big)^2 \leq \sum_{k=1}^{\infty} |a_k|^2 \Big| \int_0^{r^2} t^k d\omega(t) \Big| \to 0$$

as $r \to 1-0$. Consequently, setting

$$U_0(z,r) = \text{Re } F_r(z) = \frac{1}{\pi} \iint_{|\zeta|<r} U_0(\zeta) \text{Re } \{C_\omega(z\overline{\zeta})\} d\mu_\omega(\zeta)$$

we conclude that $\|U_0(z) - U_0(z,r)\|_{L^2_\omega} \to 0$ as $r \to 1-0$, and hence the following equalities are true for the inner product of $U_0(z)$ and $G(z)$ in L^2_ω:

$$\begin{aligned}
\big(U_0(z), G(z)\big)_\omega &= \lim_{r\to 1-0} \big(U_0(z,r), G(z)\big)_\omega \\
&= \lim_{r\to 1-0} \frac{1}{2\pi} \iint_{|z|<1} \Big(\frac{1}{\pi} \iint_{|\zeta|<r} U_0(\zeta) \text{Re } \{C_\omega(z\overline{\zeta})\} d\mu_\omega(\zeta) \Big) G(z) d\mu_\omega(z) \\
&= \lim_{r\to 1-0} \frac{1}{2\pi} \iint_{|\zeta|<r} \Big(\frac{1}{\pi} \iint_{|z|<1} G(z) \text{Re } \{C_\omega(\zeta\overline{z})\} d\mu_\omega(z) \Big) U_0(\zeta) d\mu_\omega(\zeta) \\
&= \big(0, U_0(z)\big)_\omega + \frac{1}{2} G(0) U_0(0) = 0.
\end{aligned}$$

□

In addition, the following statement is true for the L^2_ω-norms of the Green-type potentials, which we give without proof.

Theorem 2.3. *If $\nu(\zeta)$ is a nonnegative Borel measure in $|\zeta| < 1$ which satisfies (2.2) with some $\omega(x) \in \Omega_{N^2}$, then the Green type potential $G_\omega(z)$ with the Riesz measure $\nu(\zeta)$ belongs to N^2_ω if and only if there exists finite*

$$\|G\|^2_{L^2_\omega} = \lim_{\rho\to 1-0} \iint_{|\zeta|<1} \iint_{|\zeta'|<1} \big(\log|b_\omega(z,\zeta)|, \log|b_\omega(z,\zeta')| \big)_{\omega_\rho} d\nu(\zeta) d\nu(\zeta'), \tag{2.4}$$

where $\omega_\rho(x) = \omega(x)\chi_{[0,\rho]}(x)$ and $\chi_{[0,\rho]}(x)$ is the characteristic function of $[0,\rho]$.

Remark 2.4. In virtue of Theorems 2.2 and 2.3, the class N^2_ω coincides with the set of all those functions $u(z)$ subharmonic in $|z| < 1$ which are representable in the form (2.3), where $U(z)$ is a harmonic function of N^2_ω and $G(z)$ is a convergent Green-type potential with finite L^2_ω-norm. Note that with the use of some formulas from [10], the value of the inner product in (2.4) can be exactly calculated. This results in a density condition for the Riesz measure $d\nu(\zeta)d\nu(\zeta')$, which includes the arguments of ζ and ζ'.

References

[1] R. Nevanlinna, *Eindeutige Analytische Funktionen.* Springer, Berlin, 1936.

[2] M.M. Djrbashian, *On Canonical Representation of Functions Meromorphic in the Unit Disc.* DAN of Armenia **3** (1945), 3–9.

[3] M.M. Djrbashian, *On the Representability Problem of Analytic Functions.* Soobshch. Inst. Math. and Mech. AN Armenia **2** (1948), 3–40.

[4] M.M. Djrbashian, *Integral Transforms and Representations of Functions in the Complex Domain.* Moscow, Nauka, 1966.

[5] M.M. Djrbashian, *A Generalized Riemann–Liouville Operator and Some of its Applications.* Math. USSR Izv. **2** (1968), 1027–1065.

[6] M.M. Djrbashian, *Theory of Factorization of Functions Meromorphic in the Unit Disc.* Math. USSR Sbornik **8** (1969), 493–591 .

[7] M.M. Djrbashian, *Theory of Factorization and Boundary Properties of Functions Meromorphic in the Disc.* In: Proceedings of the ICM, Vancouver, B.C., 1974, **2**, 197–202, USA 1975.

[8] M.S. Livšic, *Linear Discrete Systems and Their Connection With the Theory of Factorization of Meromorphic Functions of M.M. Djrbashian.* USSR Ac. of Sci. Dokladi, **219** (1974), 793–796.

[9] G.M. Gubreev, A.M. Jerbashian, *Functions of Generalized Bounded Type in Spectral Theory of Non-Weak Contractions.* Journal of Operator Theory **26** (1991), 155–190.

[10] A.M. Jerbashian, *On the Theory of Weighted Classes of Area Integrable Regular Functions,* Complex Variables **50** (2005), 155–183.

[11] A.M. Jerbashian, *Functions of α-Bounded Type in the Half-Plane.* Springer, Advances in Complex Analysis and Applications, 2005.

Armen M. Jerbashian
Institute of Mathematics
National Academy of Sciences of Armenia
24-b Marshal Baghramian Avenue
375019 Yerevan, Armenia
e-mail: armen_jerbashian@yahoo.com

Operator Theory:
Advances and Applications, Vol. 190, 341–359

Asymptotics of Toeplitz Matrices with Symbols in Some Generalized Krein Algebras

Alexei Yu. Karlovich

To the memory of Mark Krein (1907–1989)

Abstract. Let $\alpha, \beta \in (0,1)$ and

$$K^{\alpha,\beta} := \left\{ a \in L^\infty(\mathbb{T}) : \sum_{k=1}^{\infty} |\widehat{a}(-k)|^2 k^{2\alpha} < \infty, \sum_{k=1}^{\infty} |\widehat{a}(k)|^2 k^{2\beta} < \infty \right\}.$$

Mark Krein proved in 1966 that $K^{1/2,1/2}$ forms a Banach algebra. He also observed that this algebra is important in the asymptotic theory of finite Toeplitz matrices. Ten years later, Harold Widom extended earlier results of Gabor Szegő for scalar symbols and established the asymptotic trace formula

$$\operatorname{trace} f(T_n(a)) = (n+1)G_f(a) + E_f(a) + o(1) \quad \text{as } n \to \infty$$

for finite Toeplitz matrices $T_n(a)$ with matrix symbols $a \in K^{1/2,1/2}_{N\times N}$. We show that if $\alpha + \beta \geq 1$ and $a \in K^{\alpha,\beta}_{N\times N}$, then the Szegő-Widom asymptotic trace formula holds with $o(1)$ replaced by $o(n^{1-\alpha-\beta})$.

Mathematics Subject Classification (2000). Primary 47B35; Secondary 15A15, 47B10.

Keywords. Toeplitz matrix, generalized Krein algebra, Szegő-Widom limit theorem, Wiener-Hopf factorization.

1. Introduction and the main result

For $1 \leq p \leq \infty$, let $L^p := L^p(\mathbb{T})$ and $H^p := H^p(\mathbb{T})$ be the standard Lebesgue and Hardy spaces on the unit circle $\mathbb{T}$, respectively. Denote by $\{\widehat{a}(k)\}_{k\in\mathbb{Z}}$ the sequence of the Fourier coefficients of a function $a \in L^1(\mathbb{T})$,

$$\widehat{a}(k) = \frac{1}{2\pi} \int_0^{2\pi} a(e^{i\theta}) e^{-ik\theta} d\theta \quad (k \in \mathbb{Z}).$$

This work is partially supported by the grant FCT/FEDER/POCTI/MAT/59972/2004.

For $\alpha, \beta \in (0,1)$, put

$$K^{\alpha,0} := \left\{ a \in L^\infty(\mathbb{T}) : \sum_{k=1}^{\infty} |\widehat{a}(-k)|^2 k^{2\alpha} < \infty, \right\},$$

$$K^{0,\beta} := \left\{ a \in L^\infty(\mathbb{T}) : \sum_{k=1}^{\infty} |\widehat{a}(k)|^2 k^{2\beta} < \infty \right\},$$

$$K^{\alpha,\beta} := K^{\alpha,0} \cap K^{0,\beta}.$$

It was Mark Krein [18] who first discovered that $K^{1/2,1/2}$ forms a Banach algebra under pointwise multiplication and the norm

$$\|a\|_{1/2,1/2} := \|a\|_{L^\infty} + \left(\sum_{k=-\infty}^{\infty} |\widehat{a}(k)|^2 (|k|+1) \right)^{1/2}.$$

By the same method, one can show that if $\alpha, \beta \in [1/2, 1)$, then $K^{\alpha,0}$ and $K^{0,\beta}$ are Banach algebras under pointwise multiplication and the norms

$$\|a\|_{\alpha,0} := \|a\|_{L^\infty} + \left(\sum_{k=0}^{\infty} |\widehat{a}(-k)|^2 (k+1)^{2\alpha} \right)^{1/2},$$

$$\|a\|_{0,\beta} := \|a\|_{L^\infty} + \left(\sum_{k=0}^{\infty} |\widehat{a}(k)|^2 (k+1)^{2\beta} \right)^{1/2},$$

respectively. Further, if $\max\{\alpha, \beta\} \geq 1/2$, then $K^{\alpha,\beta}$ is a Banach algebra under pointwise multiplication and the norm

$$\|a\|_{\alpha,\beta} := \|a\|_{L^\infty} + \left(\sum_{k=0}^{\infty} |\widehat{a}(-k)|^2 (k+1)^{2\alpha} \right)^{1/2} + \left(\sum_{k=0}^{\infty} |\widehat{a}(k)|^2 (k+1)^{2\beta} \right)^{1/2}$$

(see [4, Chap. 4] and also [6, Sections 10.9–10.11] and [2, Theorem 1.3]). In these sources even more general algebras are considered. The algebra $K^{1/2,1/2}$ is referred to as the *Krein algebra*. The algebras $K^{\alpha,0}$, $K^{0,\beta}$, and $K^{\alpha,\beta}$ will be called *generalized Krein algebras*.

Suppose $N \in \mathbb{N}$. For a Banach space X, let X_N and $X_{N\times N}$ be the spaces of vectors and matrices with entries in X, respectively. Let I be the identity operator, P be the Riesz projection of L^2 onto H^2, $Q := I - P$, and define I, P, and Q on L^2_N elementwise. For $a \in L^\infty_{N\times N}$ and $t \in \mathbb{T}$, put $\widetilde{a}(t) := a(1/t)$ and $(Ja)(t) := t^{-1}\widetilde{a}(t)$. Define *Toeplitz operators*

$$T(a) := PaP|\operatorname{Im} P, \quad T(\widetilde{a}) := JQaQJ|\operatorname{Im} P$$

and *Hankel operators*

$$H(a) := PaQJ|\operatorname{Im} P, \quad H(\widetilde{a}) := JQaP|\operatorname{Im} P.$$

The function a is called the *symbol* of $T(a)$, $T(\widetilde{a})$, $H(a)$, $H(\widetilde{a})$. We are interested in the asymptotic behavior of *finite block Toeplitz matrices*

$$T_n(a) := \left(\widehat{a}(j-k)\right)_{j,k=0}^{n}$$

generated by (the Fourier coefficients of) the symbol a as $n \to \infty$. It should be noted that asymptotics of Toeplitz matrices was one of the topics of Mark Krein's interests. In particular, he proved [18] that $K^{1/2,1/2}$ is an optimal smoothness class for the validity of the strong Szegő limit theorem for scalar positive symbols (independently this result was obtained by Devinatz [8]; for an extension of this result to matrix positive definite symbols, see Böttcher and Silbermann [3]). Many results about asymptotic properties of $T_n(a)$ as $n \to \infty$ are contained in the books by Grenander and Szegő [11], Böttcher and Silbermann [4, 5, 6], Hagen, Roch, and Silbermann [12], Simon [24], and Böttcher and Grudsky [1].

Let $\operatorname{sp} A$ denote the spectrum of an operator A. If f is an analytic function in an open neighborhood of $\operatorname{sp} A$, then we will simply say that f is analytic on $\operatorname{sp} A$. We assume that the reader is familiar with basics of trace class operators and their operator determinants (see Gohberg and Krein [10, Chap. 3 and 4] or Section 3). If A is a trace class operator, then $\operatorname{trace} A$ denotes the *trace* of A and $\det(I - A)$ denotes the *operator determinant* of $I - A$.

The following result was proved by Widom [26, Theorem 6.2] (see also [6, Section 10.90]). It extends earlier results by Szegő (see [11]) and now it is usually called the Szegő-Widom asymptotic trace formula.

Theorem 1.1 (Widom). *Let $N \geq 1$. If a belongs to $K^{1/2,1/2}_{N\times N}$ and f is analytic on $\operatorname{sp} T(a) \cup \operatorname{sp} T(\widetilde{a})$, then*

$$\operatorname{trace} f(T_n(a)) = (n+1)G_f(a) + E_f(a) + o(1) \quad \text{as } n \to \infty, \tag{1.1}$$

where

$$G_f(a) := \frac{1}{2\pi}\int_0^{2\pi} (\operatorname{trace} f(a))(e^{i\theta})d\theta,$$

$$E_f(a) := \frac{1}{2\pi i}\int_{\partial\Omega} f(\lambda)\frac{d}{d\lambda}\log\det T[a-\lambda]T[(a-\lambda)^{-1}]d\lambda,$$

and Ω is any bounded open set containing $\operatorname{sp} T(a) \cup \operatorname{sp} T(\widetilde{a})$ on the closure of which f is analytic.

Our main result is the following refinement of Theorem 1.1, which gives a higher-order asymptotic trace formula.

Theorem 1.2. *Let $N \geq 1$ and $\alpha, \beta \in (0,1)$. Suppose that $\alpha + \beta \geq 1$. If $a \in K^{\alpha,\beta}_{N\times N}$ and f is analytic on $\operatorname{sp} T(a) \cup \operatorname{sp} T(\widetilde{a})$, then (1.1) is true with $o(1)$ replaced by $o(n^{1-\alpha-\beta})$.*

Notice that higher-order asymptotic trace formulas are known for other classes of symbols: see [25] for $W \cap K^{\alpha,\alpha}$ with $\alpha > 1/2$ (here W stands for the Wiener algebra of functions with absolutely convergent Fourier series), [14] for weighted Wiener algebras, [15] for Hölder-Zygmund spaces, [16] for generalized Hölder spaces. All these classes consist of continuous functions only. More precisely, they are decomposing algebras of continuous functions in the sense of Budyanu and Gohberg. An invertible matrix function in such an algebra admits a Wiener-Hopf factorization within the algebra. The proofs of [14, 15, 16] are based on a combination of this observation and an approach of Böttcher and Silbermann [3] (see also [4, Sections 6.15–6.22] and [6, Sections 10.34–10.40]) to higher-order asymptotic formulas of Toeplitz determinants with Widom's original proof of Theorem 1.1 (see [26] and [6, Section 10.90]). As far as we know, Vasil'ev, Maximenko, and Simonenko have never published a proof of the result stated in the short note [25], however, their result can be proved by the same method.

Generalized Krein algebras $K^{\alpha,\beta}$ may contain discontinuous functions. To study them we need a more advanced factorization theory in decomposing algebras of L^∞ functions developed by Heinig and Silbermann [13]. We present main results of this theory in Section 2 and then apply them to $K^{\alpha,\beta}$ with $\alpha+\beta \geq 1$ and $\max\{\alpha,\beta\} > 1/2$. Under these assumptions, if both Toeplitz operators $T(a)$ and $T(\widetilde{a})$ are invertible, then a admits simultaneously canonical right and left Wiener-Hopf factorizations $a = u_-u_+ = v_+v_-$ in $K^{\alpha,\beta}_{N\times N}$. The factors and their inverses in these factorizations are stable under small perturbations of a in the norm of $K^{\alpha,\beta}_{N\times N}$. We will use this fact in Section 4 for factorizations of $a-\lambda$, where λ belongs to a compact neighborhood Σ of the boundary of a set Ω containing $\operatorname{sp} T(a) \cup \operatorname{sp} T(\widetilde{a})$.

Section 3 contains some preliminaries on trace class operators and their determinants. Further we formulate the Borodin-Okounkov formula under weakened smoothness assumptions. This is an exact formula which relates determinants of finite Toeplitz matrices $\det T_n(a)$ and operator determinants of $I - Q_nH(b)H(\widetilde{c})Q_n$, where $Q_nH(b)H(\widetilde{c})Q_n$ are truncations of the product of Hankel operators $H(b)$ and $H(\widetilde{c})$ with $b := v_-u_+^{-1}$ and $c := u_-^{-1}v_+$. Here $Q_n := I - P_n$ and P_n is the finite section projection.

If $a - \lambda \in K^{\alpha,\beta}_{N\times N}$, then we can effectively estimate the speed of convergence of the trace class norm of $I - Q_nH[b(\lambda)]H[\widetilde{c(\lambda)}]Q_n$ to zero as $n\to\infty$ uniformly in $\lambda\in\Sigma$. This speed is $o(n^{1-\alpha-\beta})$. Combining this estimate with the Borodin-Okounkov formula for $a-\lambda$ and then applying Widom's "differentiate-multiply-integrate" arguments with respect to $\lambda \in \Sigma$, we prove Theorem 1.2 in Section 4.

2. Wiener-Hopf factorization and generalized Krein algebras

2.1. Wiener-Hopf factorization in decomposing algebras

For a unital algebra A, let $\mathcal{G}A$ denote the its group of invertible elements.

Mark Krein [17] was the first to understand the Banach algebraic background of Wiener-Hopf factorization and to present the method in a crystal-clear manner. Gohberg and Krein [9] proved that $a \in \mathcal{G}W_{N\times N}$ admits a Wiener-Hopf factorization. Later Budyanu and Gohberg developed an abstract factorization theory in decomposing algebras of *continuous* functions. Their results are contained in

[7, Chap. 2]. Heinig and Silbermann [13] extended the theory of Budyanu and Gohberg to the case of decomposing algebras which may contain *discontinuous* functions. The following definitions and results are taken from [13] (see also [4, Chap. 5]).

Let A be a Banach algebra of complex-valued functions on the unit circle $\mathbb{T}$ under a Banach algebra norm $\|\cdot\|_A$. The algebra A is said to be *decomposing* if it possesses the following properties:

(a) A is continuously embedded in L^∞;
(b) A contains all Laurent polynomials;
(c) $PA \subset A$ and $QA \subset A$.

Using the closed graph theorem it is easy to deduce from (a)–(c) that P and Q are bounded on A and that PA and QA are closed subalgebras of A. For $k \in \mathbb{Z}$ and $t \in \mathbb{T}$, put $\chi_k(t) := t^k$. Given a decomposing algebra A put

$$A_+ = PA, \quad \overset{\circ}{A}_- = QA, \quad \overset{\circ}{A}_+ = \chi_1 A_+, \quad A_- = \chi_1 \overset{\circ}{A}_- .$$

Let A be a decomposing algebra. A matrix function $a \in A_{N\times N}$ is said to *admit a right* (resp. *left*) *Wiener-Hopf factorization in* $A_{N\times N}$ if it can be represented in the form

$$a = a_- d a_+ \quad (\text{resp. } a = a_+ d a_-),$$

where $a_\pm \in \mathcal{G}(A_\pm)_{N\times N}$ and

$$d = \operatorname{diag}(\chi_{\kappa_1}, \dots, \chi_{\kappa_N}), \quad \kappa_i \in \mathbb{Z}, \quad \kappa_1 \le \kappa_2 \le \dots \le \kappa_N.$$

The integers κ_i are usually called the *right* (resp. *left*) *partial indices* of a; they can be shown to be uniquely determined by a. If $\kappa_1 = \dots = \kappa_N = 0$, then the Wiener-Hopf factorization is said to be *canonical.* A decomposing algebra A is said to have the *factorization property* if every matrix function in $\mathcal{G}A_{N\times N}$ admits a right Wiener-Hopf factorization in $A_{N\times N}$.

Let $\mathcal{R}$ be the restriction to the unit circle $\mathbb{T}$ of the set of all rational functions defined on the whole plane $\mathbb{C}$ and having no poles on $\mathbb{T}$.

Theorem 2.1. *Let A be a decomposing algebra. If at least one of the sets*

$$(\mathcal{R} \cap \overset{\circ}{A}_-) + A_+ \quad \text{or} \quad \overset{\circ}{A}_- + (\mathcal{R} \cap A_+)$$

is dense in A, then A has the factorization property.

2.2. Stability of factors and their inverses under small perturbations

Let A be a Banach algebra equipped with a norm $\|\cdot\|_A$. We will always consider an admissible norm $\|\cdot\|_{A_{N\times N}}$ in $A_{N\times N}$. Recall that a Banach algebra norm is said to be admissible (see [6, Section 1.29]) if there exist positive constants m and M such that

$$m \max_{1\le i,j\le N} \|a_{ij}\|_A \le \|a\|_{A_{N\times N}} \le M \max_{1\le i,j\le N} \|a_{ij}\|_A$$

for every matrix $a = (a_{ij})_{i,j=1}^N \in A_{N\times N}$.

The following result can be extracted from a stability theorem for factors and their inverses in the Wiener-Hopf factorization in decomposing algebras given in

[20, Theorem 6.15]. There it was assumed, in addition, that a decomposing algebra is continuously embedded in the set of all continuous functions. However, the result is also true for decomposing algebras in the sense of Heinig and Silbermann adopted in this paper.

Theorem 2.2. *Let A be a decomposing algebra and $N \geq 1$. Suppose $a, c \in A_{N\times N}$ both admit canonical right (resp. left) Wiener-Hopf factorizations in $A_{N\times N}$. Then for any $\varepsilon > 0$ there exists a $\delta > 0$ such that if*

$$\|a - c\|_{A_{N\times N}} < \delta,$$

then for every canonical right (resp. left) Wiener-Hopf factorization $a = a_-^{(r)} a_+^{(r)}$ (resp. $a = a_+^{(l)} a_-^{(l)}$) one can choose a canonical right (resp. left) Wiener-Hopf factorization $c = c_-^{(r)} c_+^{(r)}$ (resp. $c = c_+^{(l)} c_-^{(l)}$) such that

$$\|a_\pm^{(r)} - c_\pm^{(r)}\|_{A_{N\times N}} < \varepsilon, \quad \|[a_\pm^{(r)}]^{-1} - [c_\pm^{(r)}]^{-1}\|_{A_{N\times N}} < \varepsilon$$

$$(\textit{resp.} \quad \|a_\pm^{(l)} - c_\pm^{(l)}\|_{A_{N\times N}} < \varepsilon, \quad \|[a_\pm^{(l)}]^{-1} - [c_\pm^{(l)}]^{-1}\|_{A_{N\times N}} < \varepsilon\ ).$$

2.3. Invertibility in generalized Krein algebras

For $1 \leq p \leq \infty$, let $\overline{H^p} := \{a \in L^p : \overline{a} \in H^p\}$ and let $C := C(\mathbb{T})$ denote the set of all continuous functions on $\mathbb{T}$. If $\alpha, \beta \geq 1/2$, then in view of [2, Lemma 6.2],

$$K^{\alpha,0} \subset C + H^\infty, \quad K^{0,\beta} \subset C + \overline{H^\infty}. \tag{2.1}$$

Hence, if $\alpha, \beta \in (0,1)$ and $\alpha + \beta \geq 1$, then

$$K^{\alpha,\beta} \subset (C + H^\infty) \cup (C + \overline{H^\infty}). \tag{2.2}$$

The following result was proved by Krein [18] for $\alpha = \beta = 1/2$.

Theorem 2.3 (see [2, Theorem 1.4]**).** *Let $\alpha, \beta \in (0,1)$.*

(a) *Suppose $\alpha \geq 1/2$ and K is either $K^{\alpha,0}$ or $K^{\alpha,1-\alpha}$. If $a \in K$, then*

$$a \in \mathcal{G}K \Longleftrightarrow a \in \mathcal{G}(C + H^\infty).$$

(b) *Suppose $\beta \geq 1/2$ and K is either $K^{0,\beta}$ or $K^{1-\beta,\beta}$. If $a \in K$, then*

$$a \in \mathcal{G}K \Longleftrightarrow a \in \mathcal{G}(C + \overline{H^\infty}).$$

Corollary 2.4. *Let $\alpha, \beta \in (0,1)$.*

(a) *Suppose $\alpha \geq 1 - \beta \geq 1/2$. If $a \in K^{\alpha,\beta}$, then*

$$a \in \mathcal{G}K^{\alpha,\beta} \Longleftrightarrow a \in \mathcal{G}(C + H^\infty).$$

(b) *Suppose $\beta \geq 1 - \alpha \geq 1/2$. If $a \in K^{\alpha,\beta}$, then*

$$a \in \mathcal{G}K^{\alpha,\beta} \Longleftrightarrow a \in \mathcal{G}(C + \overline{H^\infty}).$$

(c) *Suppose $\alpha \geq \beta \geq 1/2$ or $\beta \geq \alpha \geq 1/2$. If $a \in K^{\alpha,\beta}$, then*

$$a \in \mathcal{G}K^{\alpha,\beta} \Longleftrightarrow a \in \mathcal{G}\big((C + H^\infty) \cap (C + \overline{H^\infty})\big).$$

Proof. (a) Let $a \in K^{\alpha,\beta} = K^{1-\beta,\beta} \cap K^{\alpha,0}$. By Theorem 2.3(a),

$$a \in \mathcal{G}K^{1-\beta,\beta} \Longleftrightarrow a \in \mathcal{G}(C+H^\infty), \quad a \in \mathcal{G}K^{\alpha,0} \Longleftrightarrow a \in \mathcal{G}(C+H^\infty).$$

Thus $a \in \mathcal{G}K^{\alpha,\beta} \Longleftrightarrow a \in \mathcal{G}(C+H^\infty)$. Part (a) is proved. Part (b) follows form Theorem 2.3(b) in the same fashion.

(c) Let $a \in K^{\alpha,\beta} = K^{\alpha,0} \cap K^{0,\beta}$. From Theorem 2.3 it follows that

$$a \in \mathcal{G}K^{\alpha,0} \Longleftrightarrow \mathcal{G}(C+H^\infty), \quad a \in \mathcal{G}K^{0,\beta} \Longleftrightarrow \mathcal{G}(C+\overline{H^\infty}).$$

Hence $a \in \mathcal{G}K^{\alpha,\beta} = G(K^{\alpha,0} \cap K^{0,\beta}) \Longleftrightarrow a \in \mathcal{G}\big((C+H^\infty) \cap (C+\overline{H^\infty})\big)$. Part (c) is proved. □

2.4. Wiener-Hopf factorization in generalized Krein algebras

Lemma 2.5. *Let $\alpha, \beta \in (0,1)$ and $\max\{\alpha,\beta\} > 1/2$. Then $K^{\alpha,\beta}$ is a decomposing algebra with the factorization property.*

Proof. The statement is proved by analogy with [2, Lemma 7.7]. By [2, Lemma 6.1], the projections P and Q are bounded on $K^{\alpha,\beta}$. Hence $K^{\alpha,\beta}$ is a decomposing algebra. Assume that $\beta > 1/2$. Taking into account that

$$K^{\alpha,\beta} = L^\infty \cap (QB_2^\alpha + PB_2^\beta),$$

where B_2^α and B_2^β are Besov spaces, from [22, Sections 3.5.1 and 3.5.5] one can deduce that $\mathcal{R} \cap PK^{\alpha,\beta}$ is dense in $PK^{\alpha,\beta}$. Analogously, if $\alpha > 1/2$, then $\mathcal{R} \cap QK^{\alpha,\beta}$ is dense in $QK^{\alpha,\beta}$. Theorem 2.1 gives the factorization property of $K^{\alpha,\beta}$. □

Theorem 2.6. *Let $N \geq 1$, $\alpha, \beta \in (0,1)$, $\alpha + \beta \geq 1$, and $\max\{\alpha,\beta\} > 1/2$. If $a \in K^{\alpha,\beta}_{N\times N}$ and both $T(a)$ and $T(\widetilde{a})$ are invertible on H^2_N, then a is invertible in $K^{\alpha,\beta}_{N\times N}$ and admits canonical right and left Wiener-Hopf factorizations in $K^{\alpha,\beta}_{N\times N}$.*

Proof. Once one has at hands Corollary 2.4 and Lemma 2.5, the proof is developed as in [2, Theorem 1.7(a)]. For the convenience of the reader we give a complete proof here.

Suppose $\alpha = \max\{\alpha,\beta\}$. It is clear that for every $\beta \in (0,1)$ one has $\beta \geq 1/2$ or $1-\beta \geq 1/2$. Thus $\alpha \geq \beta \geq 1/2$ $(\geq 1-\beta)$ or $\alpha \geq 1-\beta \geq 1/2$ $(\geq \beta)$. In the first case from (2.1) it follows that

$$K^{\alpha,\beta}_{N\times N} \subset (C+H^\infty)_{N\times N} \cap (C+\overline{H^\infty})_{N\times N}.$$

Since $T(\widetilde{a})$ and $T(a)$ are invertible, from [6, Theorem 2.94(a)] we deduce that $\det a$ and $\det \widetilde{a}$ belong to $\mathcal{G}(C+H^\infty)$. Hence, $\det a$ belongs to $\mathcal{G}\big((C+H^\infty) \cap (C+\overline{H^\infty})\big)$. By Corollary 2.4(c), $\det a \in \mathcal{G}K^{\alpha,\beta}$. Then, in view of [19, Chap. 1, Theorem 1.1], $a \in \mathcal{G}K^{\alpha,\beta}_{N\times N}$.

The case $\alpha \geq 1-\beta \geq 1/2$ is treated with the help of Corollary 2.4(a). Then $a \in \mathcal{G}K^{\alpha,\beta}_{N\times N}$. Analogously, if $\beta = \max\{\alpha,\beta\}$, then by using Corollary 2.4(b) or (c), one cans how that $a \in \mathcal{G}K^{\alpha,\beta}_{N\times N}$.

By Simonenko's factorization theorem (see, e.g., [7, Chap. 7, Theorem 3.2] or [20, Theorem 3.14]), if $T(a)$ is invertible on H^2_N, then a admits a canonical right generalized factorization in L^2_N, that is, there exist functions a_-, a_+ such that $a = a_-a_+$ and $a_-^{\pm 1} \in (\overline{H^2})_{N\times N}$, $a_+^{\pm 1} \in (H^2)_{N\times N}$. On the other hand, in view of Lemma 2.5, a admits a right Wiener-Hopf factorization in $K^{\alpha,\beta}_{N\times N}$, that is, there exist functions $u_\pm \in \mathcal{G}(K^{\alpha,\beta}_\pm)_{N\times N}$ such that $a = u_-du_+$ and d is a diagonal term of the form $d = \operatorname{diag}(\chi_{\kappa_1}, \dots, \chi_{\kappa_N})$. It is clear that $u_-^{\pm 1} \in (\overline{H^2})_{N\times N}$ and $u_+^{\pm 1} \in (H^2)_{N\times N}$. Thus $a = u_-du_+$ is a right generalized factorization of a in L^2_N. It is well known that the set of partial indices of such a factorization is unique (see, e.g., [20, Corollary 2.1]). Thus d is the identity matrix and $a = u_-u_+$.

Since $T(\widetilde{a})$ is invertible on H^2_N, from [6, Proposition 7.19(b)] it follows that $T(a^{-1})$ is also invertible on H^2_N. By what has just been proved, there exist functions $f_\pm \in \mathcal{G}(K^{\alpha,\beta}_\pm)_{N\times N}$ such that $a^{-1} = f_-f_+$. Put $v_\pm := f_\mp^{-1}$. Then $a = v_+v_-$ and $v_\pm \in \mathcal{G}(K^{\alpha,\beta}_\pm)_{N\times N}$. □

3. The Borodin-Okounkov formula

3.1. Trace class operators, Hilbert-Schmidt operators, and operator determinants

In this subsection we collect necessary facts from general operator theory in Hilbert spaces (see [10, Chap. 3–4]).

Let $\mathcal{H}$ and $\mathcal{K}$ be separable Hilbert spaces. For a bounded linear operator $A : \mathcal{H} \to \mathcal{K}$ and $n \in \mathbb{Z}_+$, we define

$$s_n(A) := \inf\{\|A - F\| : \dim F \le n\}.$$

For $1 \le p < \infty$, the collection $\mathcal{C}_p(\mathcal{H}, \mathcal{K})$ of all bounded linear operators $A : \mathcal{H} \to \mathcal{K}$ satisfying

$$\|A\|_{\mathcal{C}_p(\mathcal{H},\mathcal{K})} := \left(\sum_{n=0}^{\infty} s_n^p(A)\right)^{1/p} < \infty$$

is called the p-Schatten-von Neumann class. If $p = 1$, then $\mathcal{C}_1(\mathcal{H}, \mathcal{K})$ is called the *trace class* and if $p = 2$, then $\mathcal{C}_2(\mathcal{H}, \mathcal{K})$ is called the class of Hilbert-Schmidt operators. We will simply write $\mathcal{C}_p(\mathcal{H})$ instead of $\mathcal{C}_p(\mathcal{H}, \mathcal{H})$.

One can show that, for every $A \in \mathcal{C}_1(\mathcal{H})$ and for every orthonormal basis $\{\varphi_j\}_{j=0}^\infty$ of $\mathcal{H}$, the series $\sum_{j=0}^\infty \langle A\varphi_j, \varphi_j\rangle_{\mathcal{H}}$ converges absolutely and that its sum does not depend on the particular choice of $\{\varphi_j\}_{j=0}^\infty$. This sum is denoted by $\operatorname{trace} A$ and is referred to as the *trace* of A. It is well known that

$$|\operatorname{trace} A| \le \|A\|_{\mathcal{C}_1(\mathcal{H})}$$

for all $A \in \mathcal{C}_1(\mathcal{H})$.

The Hilbert-Schmidt norm of an operator $A \in \mathcal{C}_2(\mathcal{H}, \mathcal{K})$ can be expressed in the form

$$\|A\|_{\mathcal{C}_2(\mathcal{H},\mathcal{K})} = \left(\sum_{j,k}^{\infty} |\langle A\varphi_j, \psi_k\rangle_{\mathcal{K}}|^2\right)^{1/2},$$

where $\{\varphi_j\}_{j=0}^\infty$ and $\{\psi_k\}_{k=0}^\infty$ are orthonormal bases of $\mathcal{H}$ and $\mathcal{K}$, respectively.

We will need the following version of the Hölder inequality. If $B \in \mathcal{C}_2(\mathcal{H},\mathcal{K})$ and $A \in \mathcal{C}_2(\mathcal{K},\mathcal{H})$, then $AB \in \mathcal{C}_1(\mathcal{H})$ and

$$\|AB\|_{\mathcal{C}_1(\mathcal{H})} \le \|A\|_{\mathcal{C}_2(\mathcal{K},\mathcal{H})}\|B\|_{\mathcal{C}_2(\mathcal{H},\mathcal{K})}. \tag{3.1}$$

Let A be a bounded linear operator on $\mathcal{H}$ of the form $I+K$ with $K \in \mathcal{C}_1(\mathcal{H})$. If $\{\lambda_j(K)\}_{j\ge 0}$ denotes the sequence of the nonzero eigenvalues of K (counted up to algebraic multiplicity), then $\sum_{j=0}^{\infty}|\lambda_j(K)| < \infty$. Therefore the (possibly infinite) product $\prod\limits_{j\ge 0}(1+\lambda_j(K))$ is absolutely convergent. The *operator determinant* of A is defined by

$$\det A = \det(I+K) = \prod_{j\ge 0}(1+\lambda_j(K)).$$

In the case where the spectrum of K consists only of 0 we put $\det(I+K) = 1$.

Lemma 3.1. *If $A \in \mathcal{C}_1(\mathcal{H})$ and $\|A\|_{\mathcal{C}_1(\mathcal{H})} < 1$, then $|\log\det(I-A)| \le 2\|A\|_{\mathcal{C}_1(\mathcal{H})}$.*

Proof. Since $A \in \mathcal{C}_1(\mathcal{H})$, by formula (1.16) of [10, Chap. IV],

$$\log\det(I-A) = \operatorname{trace}\log(I-A). \tag{3.2}$$

On the other hand,

$$\log(I-A) = -\sum_{j=1}^{\infty}\frac{1}{j}A^j. \tag{3.3}$$

From (3.2), (3.3), and $|\operatorname{trace} A| \le \|A\|_{\mathcal{C}_1(\mathcal{H})}$ we get

$$|\log\det(I-A)| \le \left|\operatorname{trace}\left[\sum_{j=1}^{\infty}\frac{1}{j}A^j\right]\right| \le \sum_{j=1}^{\infty}|\operatorname{trace} A^j| \le \sum_{j=1}^{\infty}\|A^j\|_{\mathcal{C}_1(\mathcal{H})}. \tag{3.4}$$

By Hölder's inequality,

$$\|A^j\|_{\mathcal{C}_1(\mathcal{H})} \le \|A\|_{\mathcal{C}_1(\mathcal{H})}\|A^{j-1}\|_{\mathcal{C}_\infty(\mathcal{H})} \le \|A\|_{\mathcal{C}_1(\mathcal{H})}\|A\|^{j-1}_{\mathcal{C}_\infty(\mathcal{H})}.$$

Taking into account that $\|A\|_{\mathcal{C}_\infty(\mathcal{H})} \le \|A\|_{\mathcal{C}_1(\mathcal{H})}$, we get $\|A^j\|_{\mathcal{C}_1(\mathcal{H})} \le \|A\|^j_{\mathcal{C}_1(\mathcal{H})}$. Hence, (3.4) yields

$$|\log\det(I-A)| \le \sum_{j=1}^{\infty}\|A\|^j_{\mathcal{C}_1(\mathcal{H})} = \frac{\|A\|_{\mathcal{C}_1(\mathcal{H})}}{1-\|A\|_{\mathcal{C}_1(\mathcal{H})}} \le 2\|A\|_{\mathcal{C}_1(\mathcal{H})}$$

because $\|A\|_{\mathcal{C}_1(\mathcal{H})} < 1$. □

3.2. The Borodin-Okounkov formula under weakened hypotheses

For $a \in L^\infty_{N\times N}$ and $n \in \mathbb{Z}_+$, define the operators

$$P_n : \sum_{k=0}^{\infty}\widehat{a}(k)\chi_k \mapsto \sum_{k=0}^{n}\widehat{a}(k)\chi_k, \quad Q_n := I - P_n.$$

The operator $P_nT(a)P_n : P_nH^2_N \to P_nH^2_N$ may be identified with the finite block Toeplitz matrix $T_n(a) = (\widehat{a}(j-k))^n_{j,k=0}$.

In June 1999, Its and Deift raised the question whether there is a general formula that expresses the determinant of the Toeplitz matrix $T_n(a)$ as the operator determinant of an operator $I - K$ where K acts on $\ell_2\{n+1, n+2, \dots\}$. Borodin and Okunkov showed in 2000 that such a formula exists (however, it was known even much earlier. In 1979, Geronimo and Case used it to prove the strong Szegő limit theorem). Further, in 2000, several different proofs of it were found by Basor and Widom and by Böttcher. We refer to the books by Simon [24], Böttcher and Grudsky [1, Section 2.8], Böttcher and Silbermann [6, Section 10.40] for the exact references, proofs, and historical remarks on this beautiful piece of mathematics. Below we formulate the Borodin-Okounkov formula in a form suggested by Widom under assumptions on the symbol a of $T_n(a)$ which are slightly weaker than in [6, Section 10.40].

Theorem 3.2. *Suppose $a \in (C + H^\infty)_{N\times N} \cup (C + \overline{H^\infty})_{N\times N}$ satisfies the following hypothesis:*

(i) *there are two factorizations $a = u_- u_+ = v_+ v_-$, where $u_-, v_- \in \mathcal{G}(\overline{H^\infty})_{N\times N}$ and $u_+, v_+ \in \mathcal{G}(H^\infty)_{N\times N}$;*

Put $b := v_- u_+^{-1}$ and $c := u_-^{-1} v_+$. Suppose that

(ii) $H(b)H(\widetilde{c}) \in \mathcal{C}_1(H^2_N)$.

Then the constants

$$G(a) := \lim_{r\to 1-0} \exp\left(\frac{1}{2\pi}\int_0^{2\pi} \log\det a_r(e^{i\theta})d\theta\right), \tag{3.5}$$

where

$$a_r(e^{i\theta}) := \sum_{n=-\infty}^{\infty} \widehat{a}(n) r^{|n|} e^{in\theta},$$

and

$$E(a) := \frac{1}{\det T(b)T(c)}$$

are well defined, are not equal to zero, and the Borodin-Okounkov formula

$$\det T_n(a) = G(a)^{n+1} E(a) \det\left(I - Q_n H(b) H(\widetilde{c}) Q_n\right) \tag{3.6}$$

holds for every $n \in \mathbb{N}$. If, in addition,

(iii) $H(a)H(\widetilde{a}^{-1}) \in \mathcal{C}_1(H^2_N)$,

then $E(a) = \det T(a)T(a^{-1})$.

Proof. From (i) and [6, Proposition 2.14] it follows that the operators $T(a)$ and $T(a^{-1})$ are invertible on H^2_N. If $a \in (C + H^\infty)_{N\times N}$, then from [6, Proposition 10.6(a)] we deduce that the limit in (3.5) exists, is finite and nonzero. Hence the constant $G(a)$ is well defined. The proof of [6, Proposition 10.6(a)] equally works also for the case $a \in (C + \overline{H^\infty})_{N\times N}$. Therefore, $G(a)$ is well defined in this case as well.

From [6, Proposition 2.14] it follows also that the operators $T(b)$, $T(c)$ are invertible and $T(b)T(c) = I - H(b)H(\widetilde{c})$. From (ii) and the above equality we get that $\det T(b)T(c)$ makes sense. Since $T(b)T(c)$ is invertible, $\det T(b)T(c) \neq 0$ and therefore the constant $E(a)$ is well defined.

The Borodin-Okounkov formula (3.6) is proved in [6, Section 10.40] under the assumption that $a \in K^{1/2,1/2}_{N\times N}$ admits right and left canonical Wiener-Hopf factorizations in $K^{1/2,1/2}_{N\times N}$. The two proofs given in [6, Section 10.40] work equally under weaker hypotheses (i)–(ii).

Applying [6, Proposition 2.14], we get

$$I - H(b)H(\widetilde{c}) = T(b)T(c) = T(v_-)T(u_+^{-1})T(u_-^{-1})T(v_+)$$

and

$$\begin{aligned} T(v_+)\big(I - H(b)H(\widetilde{c})\big)T^{-1}(v_+) &= I - T(v_+)H(b)H(\widetilde{c})T^{-1}(v_+) \\ &= T(v_+)T(v_-)T(u_+^{-1})T(u_-^{-1}) \\ &= T^{-1}(a^{-1})T^{-1}(a). \end{aligned}$$

From these equalities and [10, Chap. IV, Section 1.6] it follows that

$$\begin{aligned} \det T(b)T(c) &= \det\big(I - H(b)H(\widetilde{c})\big) \\ &= \det\big(I - T(v_+)H(b)H(\widetilde{c})T^{-1}(v_+)\big) \\ &= \det T^{-1}(a^{-1})T^{-1}(a). \end{aligned} \tag{3.7}$$

From (iii) and $T(a)T(a^{-1}) = I - H(a)H(\widetilde{a}^{-1})$ it follows that $\det T(a)T(a^{-1})$ makes sense. By [10, Chap. IV, Section 1.7],

$$\det T^{-1}(a^{-1})T^{-1}(a) \cdot \det T(a)T(a^{-1}) = \det T^{-1}(a^{-1})T^{-1}(a)T(a)T(a^{-1}) = 1.$$

Hence

$$\det T^{-1}(a^{-1})T^{-1}(a) = \frac{1}{\det T(a)T(a^{-1})}. \tag{3.8}$$

Combining (3.7) and (3.8), we arrive at $E(a) = \det T(a)T(a^{-1})$. □

4. Proof of the main result

4.1. Hilbert-Schmidt norms of truncations of Hankel operators

Let $\gamma \in \mathbb{R}$. By ℓ_2^γ we denote the Hilbert space of all sequences $\{\varphi_k\}_{k=0}^\infty$ such that

$$\sum_{k=0}^{\infty} |\varphi_k|^2 (k+1)^{2\gamma} < \infty.$$

Clearly, the sequence $\{e_k/(k+1)^\gamma\}_{k=0}^\infty$, where $(e_k)_j = \delta_{kj}$ and δ_{kj} is the Kronecker delta, is an orthonormal basis of ℓ_2^γ. If $\gamma = 0$, we will simply write ℓ_2 instead of ℓ_2^0.

In this subsection we will estimate Hilbert-Schmidt norms of truncations of Hankel operators acting between ℓ_2 and ℓ_2^γ by the rules

$$H(a) : \{\varphi_j\}_{j=0}^\infty \mapsto \Big\{\sum_{j=0}^\infty \widehat{a}(i+j+1)\varphi_j\Big\}_{i=0}^\infty,$$
$$H(\widetilde{a}) : \{\varphi_j\}_{j=0}^\infty \mapsto \Big\{\sum_{j=0}^\infty \widehat{a}(-i-j-1)\varphi_j\Big\}_{i=0}^\infty.$$

Notice that one can identify Hankel operators acting on H^2 and on ℓ^2. For $\varphi = \{\varphi_j\}_{j=0}^\infty$ and $n \in \mathbb{Z}_+$, define

$$(Q_n\varphi)_j = \begin{cases} \varphi_j & \text{if } j \geq n+1, \\ 0 & \text{otherwise.} \end{cases}$$

For $a \in K^{\alpha,\beta}$ and $n \in \mathbb{N}$, put

$$r_n^-(a) := \left(\sum_{k=n+1}^\infty |\widehat{a}(-k)|^2(k+1)^{2\alpha}\right)^{1/2},$$
$$r_n^+(a) := \left(\sum_{k=n+1}^\infty |\widehat{a}(k)|^2(k+1)^{2\beta}\right)^{1/2}.$$

Lemma 4.1. *Let $-1/2 < \gamma < 1/2$ and $\alpha, \beta \in (0,1)$. Suppose $b, c \in K^{\alpha,\beta}$.*

(a) *If $\alpha \geq \gamma + 1/2$, then there exists a positive constant $M(\alpha,\gamma)$ depending only on α and γ such that for all sufficiently large n,*

$$\|H(\widetilde{c})Q_n\|_{C_2(\ell_2,\ell_2^\gamma)} \leq M(\alpha,\gamma)\frac{r_{n+1}^-(c)}{n^{\alpha-\gamma-1/2}}. \tag{4.1}$$

(b) *If $\beta \geq -\gamma+1/2$, then there exists a positive constant $M(\beta,\gamma)$ depending only on β and γ such that for all sufficiently large n,*

$$\|Q_nH(b)\|_{C_2(\ell_2^\gamma,\ell_2)} \leq M(\beta,\gamma)\frac{r_{n+1}^+(b)}{n^{\beta+\gamma-1/2}}. \tag{4.2}$$

Proof. (a) It is easy to see that

$$(H(\widetilde{c})Q_ne_j)_k = \begin{cases} \widehat{c}(-k-j-1) & \text{if } j \geq n+1,\ k \in \mathbb{Z}_+, \\ 0 & \text{otherwise.} \end{cases}$$

Then

$$
\begin{aligned}
\|H(\widetilde{c})Q_n\|^2_{\mathcal{C}_2(\ell_2,\ell_2^\gamma)} &= \sum_{j,k=0}^{\infty} \left| \left\langle H(\widetilde{c})Q_n e_j, \frac{e_k}{(k+1)^\gamma} \right\rangle_{\ell_2^\gamma} \right|^2 \\
&= \sum_{j,k=0}^{\infty} |(H(\widetilde{c})Q_n e_j)_k|^2 (k+1)^{2\gamma} \\
&= \sum_{k=0}^{\infty} \sum_{j=n+1}^{\infty} |\widehat{c}(-k-j-1)|^2 (k+1)^{2\gamma} \\
&= \sum_{k=n+2}^{\infty} |\widehat{c}(-k)|^2 \sum_{j=1}^{k-n-1} j^{2\gamma}.
\end{aligned} \tag{4.3}
$$

If $-1/2 < \gamma < 1/2$, then

$$
\sum_{j=1}^{k-n-1} j^{2\gamma} \le \frac{(k-n)^{1+2\gamma}}{1+2\gamma}. \tag{4.4}
$$

From (4.3) and (4.4) it follows that

$$
\begin{aligned}
\|H(\widetilde{c})Q_n\|^2_{\mathcal{C}_2(\ell_2,\ell_2^\gamma)} &\le \frac{1}{1+2\gamma} \sum_{k=n+2}^{\infty} |\widehat{c}(-k)|^2 (k-n)^{1+2\gamma} \\
&\le \frac{1}{1+2\gamma} \sum_{k=n+2}^{\infty} |\widehat{c}(-k)|^2 (k+1)^{2\alpha} \frac{(k-n)^{1+2\gamma}}{k^{2\alpha}} \\
&\le \frac{1}{1+2\gamma} \left(\sup_{k\ge n+2} \frac{(k-n)^{1+2\gamma}}{k^{2\alpha}} \right) \left[r_{n+1}^-(c) \right]^2.
\end{aligned} \tag{4.5}
$$

If $1+2\gamma = 2\alpha$, then

$$
\sup_{k\ge n+2} \left(\frac{k-n}{k} \right)^{2\alpha} \le 1. \tag{4.6}
$$

Combining (4.5) and (4.6), we arrive at (4.1) with $M(\alpha,\gamma) = (2\alpha)^{-1/2}$.

If $\alpha > \gamma + 1/2$, then put

$$
A := \frac{2\gamma+1}{2\alpha - 2\gamma - 1} > 0.
$$

Let $n \ge 2/A$. Then

$$
x_n := (A+1)n = \frac{2\alpha n}{2\alpha - 2\gamma - 1} \in [n+2, \infty).
$$

It is not difficult to show that the function

$$
f_n(x) := (x-n)^{1+2\gamma} x^{-2\alpha}, \quad x \in [n+2, \infty)
$$

attains its absolute maximum at x_n. Thus

$$\sup_{k\geq n+2}\frac{(k-n)^{1+2\gamma}}{k^{2\alpha}}\leq f_n(x_n)=\frac{A^{1+2\gamma}}{(A+1)^{2\alpha}}n^{1+2\gamma-2\alpha}. \tag{4.7}$$

Combining (4.5) and (4.7), we arrive at (4.1) with

$$M(\alpha,\gamma):=(1+2\gamma)^{-1/2}A^{1/2+\gamma}(A+1)^{-\alpha}$$

for all $n\geq 2/A$. Part (a) is proved. The proof of part (b) is analogous. □

4.2. Trace class norms of truncations of products of two Hankel operators

The following fact is well known (see, e.g., [6, Section 10.12] and also [23], [21, Chap. 6]).

Lemma 4.2. *Let $N\geq 1$, $\alpha,\beta\in(0,1)$, and $\alpha+\beta\geq 1$. If $b,c\in K^{\alpha,\beta}_{N\times N}$, then $H(b)H(\widetilde{c})\in\mathcal{C}_1(H^2_N)$.*

We will also need a quantitative version of the above result for truncations of the product $H(b)H(\widetilde{c})$.

For $a\in K^{\alpha,\beta}_{N\times N}$ and $n\in\mathbb{N}$, put

$$R_n^-(a):=\left(\sum_{k=n+1}^{\infty}\|\widehat{a}(-k)\|^2_{\mathbb{C}_{N\times N}}(k+1)^{2\alpha}\right)^{1/2},$$

$$R_n^+(a):=\left(\sum_{k=n+1}^{\infty}\|\widehat{a}(k)\|^2_{\mathbb{C}_{N\times N}}(k+1)^{2\beta}\right)^{1/2}.$$

Lemma 4.3. *Let $N\geq 1$, $\alpha,\beta\in(0,1)$, and $\alpha+\beta\geq 1$. Then there exists a constant $L=L_{\alpha,\beta,N}$ depending only on N and α,β such that for every $b,c\in K^{\alpha,\beta}_{N\times N}$ and all sufficiently large n,*

$$\|Q_nH(b)H(\widetilde{c})Q_n\|_{\mathcal{C}_1(H^2_N)}\leq\frac{L}{n^{\alpha+\beta-1}}R^+_{n+1}(b)R^-_{n+1}(c). \tag{4.8}$$

Proof. Put $\gamma:=1/2-\beta$. Then $\gamma\in(-1/2,1/2)$ and $\alpha\geq\gamma+1/2$ because $\alpha+\beta\geq 1$. Let b_{ij} and c_{ij}, where $i,j\in\{1,\dots,N\}$, be the entries of $b,c\in K^{\alpha,\beta}_{N\times N}$, respectively. By Lemma 4.1, there exist positive constants $M(\alpha,\gamma)$ and $M(\beta,\gamma)$ depending only on α,β, and γ such that for all sufficiently large n and all $i,j\in\{1,\dots,N\}$,

$$\|Q_nH(b_{ij})\|_{\mathcal{C}_2(\ell_2^\gamma,\ell_2)}\leq M(\beta,\gamma)\frac{r^+_{n+1}(b_{ij})}{n^{\beta+\gamma-1/2}}, \tag{4.9}$$

$$\|H(\widetilde{c_{ij}})Q_n\|_{\mathcal{C}_2(\ell_2,\ell_2^\gamma)}\leq M(\alpha,\gamma)\frac{r^-_{n+1}(c_{ij})}{n^{\alpha-\gamma-1/2}}. \tag{4.10}$$

From Hölder's inequality (3.1) and (4.9)–(4.10) it follows that

$$\begin{aligned} \|Q_nH(b_{ij})H(\widetilde{c_{ij}})Q_n\|_{\mathcal{C}_1(\ell_2)} &\le \|Q_nH(b_{ij})\|_{\mathcal{C}_2(\ell_2^\gamma,\ell_2)}\|H(\widetilde{c_{ij}})Q_n\|_{\mathcal{C}_2(\ell_2,\ell_2^\gamma)} \\ &\le M(\alpha,\gamma)M(\beta,\gamma)\frac{r_{n+1}^+(b_{ij})r_{n+1}^-(c_{ij})}{n^{\alpha+\beta-1}} \end{aligned} \tag{4.11}$$

for all $i,j\in\{1,\dots,N\}$ and all large n.

It is not difficult to verify that there exist positive constants A_N and B_N depending only on the dimension N such that

$$\|Q_nH(b)H(\widetilde{c})Q_n\|_{\mathcal{C}_1(H_N^2)} \le A_N \max_{1\le i,j\le N}\|Q_nH(b_{ij})H(\widetilde{c_{ij}})Q_n\|_{\mathcal{C}_1(\ell_2)} \tag{4.12}$$

and

$$\max_{1\le i,j\le N} r_{n+1}^+(b_{ij}) \le B_N R_{n+1}^+(b), \qquad \max_{1\le i,j\le N} r_{n+1}^-(c_{ij}) \le B_N R_{n+1}^-(c) \tag{4.13}$$

for all sufficiently large n and all $b,c\in K_{N\times N}^{\alpha,\beta}$. Combining (4.11)–(4.13), we arrive at (4.8) with $L=L_{\alpha,\beta,N}:=A_NB_N^2M(\alpha,\gamma)M(\beta,\gamma)$. □

4.3. Tails of the norms of functions in generalized Krein algebras

Lemma 4.4. *Let $N\ge 1$, $\alpha,\beta\in(0,1)$, and $\max\{\alpha,\beta\}\ge 1/2$. Suppose Σ is a compact set in the complex plane. If $a:\Sigma\to K_{N\times N}^{\alpha,\beta}$ is a continuous function, then*

$$\lim_{n\to\infty}\sup_{\lambda\in\Sigma} R_n^-(a(\lambda))=0, \qquad \lim_{n\to\infty}\sup_{\lambda\in\Sigma} R_n^+(a(\lambda))=0. \tag{4.14}$$

Proof. This statement is proved by analogy with [14, Proposition 2.3] and [16, Lemma 6.2]. Let us prove the first equality in (4.14). Assume the contrary. Then there exist a constant $C>0$ and a sequence $\{n_k\}_{k=1}^\infty$ such that

$$\lim_{k\to\infty}\sup_{\lambda\in\Sigma} R_{n_k}^-(a(\lambda))\ge C.$$

Hence there are a number $k_0\in\mathbb{N}$ and a sequence $\{\lambda_k\}_{k=k_0}^\infty$ such that for all $k\ge k_0$,

$$R_{n_k}^-(a(\lambda_k))\ge\frac{C}{2}>0. \tag{4.15}$$

Since $\{\lambda_k\}_{k=k_0}^\infty$ is bounded, there is its convergent subsequence $\{\lambda_{k_j}\}_{j=1}^\infty$. Let λ_0 be the limit of this subsequence. Clearly, $\lambda_0\in\Sigma$ because Σ is closed. Since the function $a:\Sigma\to K_{N\times N}^{\alpha,\beta}$ is continuous at λ_0, for every $\varepsilon\in(0,C/2)$, there exists a $\Delta>0$ such that $|\lambda-\lambda_0|<\Delta,\lambda\in\Sigma$ implies $\|a(\lambda)-a(\lambda_0)\|_{K_{N\times N}^{\alpha,\beta}}<\varepsilon$. Because $\lambda_{k_j}\to\lambda_0$ as $j\to\infty$, for that Δ there is a number $J\in\mathbb{N}$ such that $|\lambda_{k_j}-\lambda_0|<\Delta$ for all $j\ge J$, and thus

$$\|a(\lambda_{k_j})-a(\lambda_0)\|_{K_{N\times N}^{\alpha,\beta}}<\varepsilon \quad \text{for all} \quad j\ge J. \tag{4.16}$$

On the other hand, (4.15) implies that

$$R_{n_{k_j}}^-(a(\lambda_{k_j}))\ge\frac{C}{2}>0 \quad \text{for all} \quad j\ge J. \tag{4.17}$$

By the Minkowski inequality,

$$\begin{aligned} R^-_{n_{k_j}}(a(\lambda_{k_j})) &\le R^-_{n_{k_j}}(a(\lambda_0)) + R^-_{n_{k_j}}(a(\lambda_{k_j}) - a(\lambda_0)) \\ &\le R^-_{n_{k_j}}(a(\lambda_0)) + \|a(\lambda_{k_j}) - a(\lambda_0)\|_{K^{\alpha,\beta}_{N\times N}}. \end{aligned} \tag{4.18}$$

From (4.16)–(4.18) we get for all $j \ge J$,

$$R^-_{n_{k_j}}(a(\lambda_0)) \ge \frac{C}{2} - \varepsilon > 0.$$

Therefore,

$$\sum_{k=0}^{\infty} \big\|[a(\lambda_0)]\hat{\ }(-k)\big\|^2_{\mathbb{C}_{N\times N}}(k+1)^{2\alpha} = +\infty$$

and this contradicts the fact that $a(\lambda_0) \in K^{\alpha,\beta}_{N\times N}$. Hence, the first equality in (4.14) is proved. The second equality in (4.14) can be proved by analogy. □

4.4. Proof of Theorem 1.2

Proof. The proof is developed similarly to the proofs of [14, Theorem 1.5], [15, Theorem 1.4], [16, Theorem 2.2] with some modifications. For the convenience of the reader, we provide some details.

Without loss of generality, we can suppose that $\max\{\alpha,\beta\} > 1/2$ (otherwise $\max\{\alpha,\beta\} \le 1/2$ and $\alpha+\beta \ge 1$ imply that $\alpha = \beta = 1/2$ and this is exactly the case of Theorem 1.1).

Let Ω be a bounded open set containing the set $\operatorname{sp} T(a) \cup \operatorname{sp} T(\widetilde{a})$ on the closure of which f is analytic. From (2.2) and [6, Theorem 7.20] it follows that Ω contains the spectrum (eigenvalues) of $T_n(a)$ for all sufficiently large n. Further, Corollary 2.4 and Theorem [6, Theorem 2.94] imply that the spectrum of a in $K^{\alpha,\beta}_{N\times N}$ is contained in Ω. Hence $f(a) \in K^{\alpha,\beta}_{N\times N}$ and $f(T_n(a))$ is well defined whenever f is analytic on $\operatorname{sp} T(a) \cup \operatorname{sp} T(\widetilde{a})$.

One can choose a closed neighborhood Σ of its boundary $\partial\Omega$ such that f is analytic on Σ and $\Sigma \cap (\operatorname{sp} T(a) \cup \operatorname{sp} T(\widetilde{a})) = \emptyset$. If $\lambda \in \Sigma$, then $T(a) - \lambda I = T[a-\lambda]$ and $T(\widetilde{a}) - \lambda I = T[(a-\lambda)\tilde{\ }]$ are invertible on H^2_N. From Theorem 2.6 it follows that $(a-\lambda)^{-1} \in K^{\alpha,\beta}_{N\times N}$ and that $a-\lambda$ admits canonical right and left Wiener-Hopf factorizations

$$a - \lambda = u_-(\lambda)u_+(\lambda) = v_+(\lambda)v_-(\lambda)$$

in the algebra $K^{\alpha,\beta}_{N\times N}$. Since $a - \lambda : \Sigma \to K^{\alpha,\beta}_{N\times N}$ is a continuous function with respect to λ, from Lemma 2.5 and Theorem 2.2 we that these canonical factorizations can be chosen so that the functions

$$u_-^{\pm1}, v_-^{\pm1} : \Sigma \to (K^{\alpha,\beta} \cap \overline{H^\infty})_{N\times N}, \quad u_+^{\pm1}, v_+^{\pm1} : \Sigma \to (K^{\alpha,\beta} \cap H^\infty)_{N\times N}$$

are continuous with respect to $\lambda \in \Sigma$.

Put $b(\lambda) := v_-(\lambda)u_+^{-1}(\lambda)$ and $c(\lambda) := u_-^{-1}(\lambda)v_+(\lambda)$. By Lemma 4.2,

$$H[b(\lambda)]H[\widetilde{c(\lambda)}] \in \mathcal{C}_1(H^2_N) \tag{4.19}$$

for all $\lambda \in \Sigma$. On the other hand, since $(a-\lambda)^{-1} \in K^{\alpha,\beta}_{N\times N}$, we also have $(\widetilde{a}-\lambda)^{-1} = [(a-\lambda)^{\sim}]^{-1} \in K^{\alpha,\beta}_{N\times N}$. Then from Lemma 4.2 it follows that

$$H[a-\lambda]H[(\widetilde{a}-\lambda)^{-1}] \in \mathcal{C}_1(H^2_N) \tag{4.20}$$

for all $\lambda \in \Sigma$. From (2.2), (4.19)–(4.20), and Theorem 3.2 we conclude that

$$\begin{aligned}\det T_n(a-\lambda) =& G(a-\lambda)^{n+1} \det T[a-\lambda]T[(a-\lambda)^{-1}] \\ & \times \det\left(I - Q_n H[b(\lambda)]H[\widetilde{c(\lambda)}]Q_n\right)\end{aligned} \tag{4.21}$$

for all $\lambda \in \Sigma$ and all $n \in \mathbb{N}$.

From Lemmas 4.3 and 4.4 it follows that there exists a number $n_0 \in \mathbb{N}$ such that

$$\left\|Q_n H[b(\lambda)]H[\widetilde{c(\lambda)}]Q_n\right\|_{\mathcal{C}_1(H^2_N)} \le \frac{L\left(\sup\limits_{\lambda\in\Sigma} R^+_n(b(\lambda))\right)\left(\sup\limits_{\lambda\in\Sigma} R^-_n(c(\lambda))\right)}{n^{\alpha+\beta-1}} < 1 \tag{4.22}$$

for all $\lambda \in \Sigma$ and all $n \ge n_0$. Here L is a positive constant depending only only on α, β an N. Combining (4.22) and Lemma 3.1, we arrive at the estimate

$$\begin{aligned}&\left|\log\det\left(I - Q_n H[b(\lambda)]H[\widetilde{c(\lambda)}]Q_n\right)\right| \\ &\le \frac{2L}{n^{\alpha+\beta-1}}\left(\sup_{\lambda\in\Sigma} R^+_n(b(\lambda))\right)\left(\sup_{\lambda\in\Sigma} R^-_n(c(\lambda))\right)\end{aligned} \tag{4.23}$$

for all $\lambda \in \Sigma$ and all $n \ge n_0$.

From (4.21), (4.23), and Lemma 4.4 we conclude that

$$\begin{aligned}\log\det T_n(a-\lambda) =& (n+1)\log G(a-\lambda) \\ & + \log\det T[a-\lambda]T[(a-\lambda)^{-1}] + o(n^{1-\alpha-\beta})\end{aligned} \tag{4.24}$$

as $n \to \infty$ uniformly with respect to $\lambda \in \Sigma$. Hence we can differentiate both sides with respect to λ, multiply by $f(\lambda)$, and integrate over $\partial\Omega$. The rest of the proof is the repetition of Widom's arguments [26, Theorem 6.2] (see also [6, Section 10.90] and [5, Theorem 5.6]) with $o(1)$ replaced by $o(n^{1-\alpha-\beta})$. □

Remark 4.5. Formula (4.24) for $\lambda = 0$ and $a \in K^{\alpha,\alpha}_{N\times N}$ with $\alpha > 1/2$ was obtained by Silbermann [23] by using methods of [3] (see also [4, Sections 6.15–6.23] and [6, Sections 10.34–10.37]). On the other hand, for $\alpha+\beta = 1$, formula (4.24) with $\lambda = 0$ was proved by Böttcher and Silbermann [4, Theorem 6.11] and [6, Theorem 10.30].

References

[1] A. Böttcher and S.M. Grudsky, *Spectral Properties of Banded Toeplitz Operators.* SIAM, Philadelphia, PA, 2005.

[2] A. Böttcher, A.Yu. Karlovich, and B. Silbermann, *Generalized Krein algebras and asymptotics of Toeplitz determinants.* Methods Funct. Anal. Topology **13** (2007), 236–261.

[3] A. Böttcher and B. Silbermann, *Notes on the asymptotic behavior of block Toeplitz matrices and determinants.* Math. Nachr. **98** (1980), 183–210.

[4] A. Böttcher and B. Silbermann, *Invertibility and Asymptotics of Toeplitz Matrices.* Akademie-Verlag, Berlin, 1983.

[5] A. Böttcher and B. Silbermann, *Introduction to Large Truncated Toeplitz Matrices.* Springer-Verlag, New York, 1999.

[6] A. Böttcher and B. Silbermann, *Analysis of Toeplitz Operators.* 2nd Edition. Springer, Berlin, 2006.

[7] K.F. Clancey and I. Gohberg, *Factorization of Matrix Functions and Singular Integral Operators.* Birkhäuser, Basel, 1981.

[8] A. Devinatz, *The strong Szegő limit theorem.* Illinois J. Math. **11** (1967), 160–175.

[9] I. Gohberg and M.G. Krein, *Systems of integral equations on a half-line with kernels depending upon the difference of arguments.* Amer. Math. Soc. Transl. (2) **14** (1960), 217–287.

[10] I. Gohberg and M.G. Krein, *Introduction to the Theory of Linear Nonselfadjoint Operators.* American Mathematical Society, Providence, RI, 1969.

[11] U. Grenander and G. Szegő, *Toeplitz Forms and Their Applications.* University of California Press, Berkeley, Los Angeles, 1958.

[12] R. Hagen, S. Roch, and B. Silbermann, *C^*-Algebras and Numerical Analysis.* Marcel Dekker, Inc., New York, 2001.

[13] G. Heinig and B. Silbermann, *Factorization of matrix functions in algebras of bounded functions.* In: "Spectral Theory of Linear Operators and Related Topics (Timişoara/Herculane, 1983)". Operator Theory: Advances and Applications **14** (1984), 157–177.

[14] A.Yu. Karlovich *Asymptotics of determinants and traces of Toeplitz matrices with symbols in weighted Wiener algebras.* Z. Anal. Anwend. **26** (2007), 43–56.

[15] A.Yu. Karlovich, *Higher order asymptotic formulas for traces of Toeplitz matrices with symbols in Hölder-Zygmund spaces.* In: "Recent Advances in Matrix and Operator Theory". Operator Theory: Advances and Applications **179** (2007), 185–196.

[16] A.Yu. Karlovich, *Higher order asymptotic formulas for Toeplitz matrices with symbols in generalized Hölder spaces.* In: "Operator Algebras, Operator Theory and Applications". Operator Theory: Advances and Applications **181** (2008), 207–228.

[17] M.G. Krein, *Integral equations on a half-line with kernel depending upon the difference of the arguments.* Amer. Math. Soc. Transl. (2) **22** (1962), 163–288.

[18] M.G. Krein, *Certain new Banach algebras and theorems of the type of the Wiener-Lévy theorems for series and Fourier integrals.* Amer. Math. Soc. Transl. (2) **93** (1970), 177–199.

[19] N.Ya. Krupnik, *Banach Algebras with Symbol and Singular Integral Operators.* Birkhäuser, Basel, 1987.

[20] G.S. Litvinchuk and I.M. Spitkovsky, *Factorization of Measurable Matrix Functions.* Birkhäuser, Basel, 1987.

[21] V.V. Peller, *Hankel Operators and Their Applications.* Springer, Berlin, 2003.

[22] H.-J. Schmeisser and H. Triebel, *Topics in Fourier Analysis and Function Spaces.* John Wiley & Sons, Chichester, 1987.

[23] B. Silbermann, *Some remarks on the asymptotic behavior of Toeplitz determinants.* Appl. Analysis **11** (1981), 185–197.

[24] B. Simon, *Orthogonal Polynomials on the Unit Circle. Part* 1. AMS, Providence, RI, 2005.

[25] V.A. Vasil'ev, E.A. Maksimenko, and I.B. Simonenko, *One Szegő-Widom limit theorem.* Doklady Math. **68** (2003), 361–362.

[26] H. Widom, *Asymptotic behavior of block Toeplitz matrices and determinants. II.* Advances in Math. **21** (1976), 1–29.

Alexei Yu. Karlovich
Departamento de Matemática
Faculdade de Ciências e Tecnologia
Universidade Nova de Lisboa
Quinta da Torre
2829–516 Caparica, Portugal
e-mail: oyk@fct.unl.pt

Operator Theory:
Advances and Applications, Vol. 190, 361–373

Generators of Random Processes in Ultrametric Spaces and Their Spectra

Witold Karwowski

Abstract. The $L^2(\mathbb{S})$ space of square integrable functions on an ultrametric space $\mathbb{S}$ has rather specific structure. As a consequence in a natural way there appear in $L^2(\mathbb{S})$ the operators of which unitary counterparts in $L^2(\mathbb{R}^n)$ would be difficult to construct. Such class of self-adjoint operators emerge from theory of random processes on ultrametric spaces. In this paper we collect known material on spectral properties of the generators of random processes on $\mathbb{S}_B$ an ultrametric space of sequences. (The set of p-adic numbers is a subset of $\mathbb{S}_B$.) Then we discuss structure of the eigenspaces of the generators.

Mathematics Subject Classification (2000). Primary 60J35; Secondary 47B25, 60J75.

Keywords. p-adics, nonarchimedean, random process, semigroup generator.

1. Introduction

It is a well-known fact that many systems of very diverse nature develop hierarchical structures. One can easily give examples ranging from social structure, government administration or descriptive sciences like taxonomy to complicated physical phenomena like spin glasses or turbulent cascades. Going in other direction one finds hierarchical structures in neural networks and cognitive systems [9]. Some phenomena (like for instance relaxations in spin glasses [15] or development of a turbulent cascades [14]) can be described by random processes in ultrametric spaces.This requires a more precise mathematical formulation. A hierarchical structure is conveniently visualized as a tree graph, and for its mathematical description one needs a suitable labelling system. A model example is the correspondence between the set of p-adic numbers $\mathbb{Q}_p$ [10] and the tree with p branches emerging from every branching point. Parts of the branches (edges) between two consecutive branching levels are of equal lengths. The ratio of the lengths of the edges between consecutive branching levels is constant and equals p.

There is a one to one correspondence between ends of the tree and p-adic numbers. The distance between ends of two branches is equal to the p-adic distance between corresponding p-adic numbers.

A natural generalization of the p-adic model would be to allow for the trees with number of edges that depends on branching point and that does not have to be a prime number. Such trees can be labelled by the spaces $\mathbb{S}_B$ of numerical sequences introduced and studied in [2]. When constructing a probabilistic model for time evolution of a hierarchical system one represents the system by appropriate tree and then study the random processes on the corresponding space $\mathbb{S}_B$. According to [2] $\mathbb{S}_B$ is a complete separable metric space with nonarchimedean metric. The p-adic spaces $\mathbb{Q}_p$ are special cases of $\mathbb{S}_B$. Note that $\mathbb{Q}_p$ has the algebraic structure of a field while it is not even possible to define addition in $\mathbb{S}_B$ in general.

There are different methods to construct random processes on p-adics. Evans [6] based his method on p-adic Fourier transform. Vladimirov [16],[17] constructed the α-stable processes representing their generators by fractional differential operator D^α. Kochubei [11],[12], [13] investigated strong stochastic differential equations on $\mathbb{Q}_p$. Aldous and Evans [5] and also Kaneko [7] constructed the processes in terms of Dirichlet forms. The method developed in [1] is independent of the algebraic structure of $\mathbb{Q}_p$ and hence carry over to $\mathbb{S}_B$. The random processes on $\mathbb{S}_B$ are constructed and studied in [2]. These constructions are based on solving the Chapman-Kolmogorov equations and then obtaining the transition functions. The coefficients in the Chapman-Kolmogorov equations are determined by distances between the initial and target states and the structure of the state space.

This fact is of consequence if the symmetry properties are concerned. The transition functions of a random process on $\mathbb{Q}_p$ is invariant under p-adic translations. Since there is no addition in $\mathbb{S}_B$ a similar statement for $\mathbb{S}_B$ cannot even be formulated in general.

Another class of processes on $\mathbb{Q}_p$ has been constructed in [8]. The general method was the same as in [1] and [2] but the coefficients in the Chapman-Kolmogorov equations were determined by the p-adic distance between initial and target states and in addition by the target states explicitly.
It is well known that $\mathbb{Q}_p$ can be made a measurable space. Roughly, the measure is defined as follows. One defines a set function α on the p-adic balls so that it is equal to the ball radius. The function α can be extended to a Haar measure invariant under p-adic translation.

Actually similar procedure works for $\mathbb{S}_B$. One defines a set function α to be equal 1 on an arbitrarily chosen ball. Then the structure of $\mathbb{S}_B$ and requirement of additivity determines the values of α on all balls in $\mathbb{S}_B$. Further one shows that α extends to a Borel measure on $\mathbb{S}_B$.

Let $P_t(x, A)$, $x \in \mathbb{S}_B$, $A \subset \mathbb{S}_B$ be a transition function. By standard procedure $P_t(x, A)$ extends to a strongly continuous Markovian semigroup $T_t, t > 0$ in

$L^2(\mathbb{S}_B)$. By the Stone theorem

$$T_t = e^{-Ht}, \tag{1.1}$$

where H is a nonnegative self-adjoint operator in $L^2(\mathbb{S}_B)$.

Spectral analysis of the generators H appears in probabilistic papers [1],[2] as a side product rather than the main point of investigations. Thus not being of much interest for probabilistic it stays unknown to the analysts. The aim of this note is to put together the results scattered in several publications and thus expose them to the analytically oriented readers. We also add a couple of new results on the structure of the eigenspaces. To make the main points readable we find it necessary to give basic facts about the space $\mathbb{S}_B$ (Section 2) and the probabilistic context (Section 3). The generators are discussed in Section 4.

2. The space $\mathbb{S}_B$

In this section we introduce the spaces $\mathbb{S}_B$ and list their basic properties. The presentation follows [2]. Put $\mathbb{Z}$, $\mathbb{N}$ and $\mathbb{N}_0$ for the set of integers, positive integers and non-negative integers respectively. For any $k \in \mathbb{Z}$ let S_k be the family of all sequences $\{\alpha_i\}_{i \le k}$ such that $\alpha_i \in \mathbb{N}_0$ and $\alpha_i = 0$ for all $i \le N$ where $N \le k$. Put

$$S = \bigcup_{k \in \mathbb{Z}} S_k.$$

Let $\alpha_{k+1} \in \mathbb{N}_0$. Then the product $\{\alpha_i\}_{i \le k} \times \{\alpha_{k+1}\}$ can be identified with $\{\alpha_i\}_{i \le k+1}$,

$$\{\alpha_i\}_{i \le k} \times \{\alpha_{k+1}\} = \{\alpha_i\}_{i \le k+1}. \tag{2.1}$$

To simplify notations we put $\{\alpha\}_k := \{\alpha_i\}_{i \le k}$.

If $\alpha_i = 0$ for all $i \le k$ then we write $\{\alpha\}_k = \{0\}_k$.

Definition 2.1. Let $B_{\{\alpha\}_k}$ be a function defined on S with values in $\mathbb{N} \setminus \{1\}$.

1. We say that $\{\alpha\}_{k+1}$ is the B-product of $\{\alpha\}_k$ and $\{\alpha_{k+1}\}$ iff

$$\{\alpha\}_{k+1} = \{\alpha\}_k \times \{\alpha_{k+1}\} \tag{2.2}$$

and

$$0 \le \alpha_{k+1} \le B_{\{\alpha\}_k} - 1. \tag{2.3}$$

2. We say that $\{\alpha\}_{k+l}$, $l \in \mathbb{N}$ is the B-product of $\{\alpha\}_k$ and the ordered l-tuple $\{\alpha_{k+1}, \dots, \alpha_{k+l}\}$

$$\{\alpha\}_{k+l} = \{\alpha\}_k \times \{\alpha_{k+1}, \dots, \alpha_{k+l}\}$$

iff

$$\{\alpha\}_{k+l} = (\dots((\{\alpha\}_k \times \{\alpha_{k+1}\}) \times \{\alpha_{k+2}\}) \times \dots \times \{\alpha_{k+l}\}) \tag{2.4}$$

where all products are B-products in the sense of 1. We then write

$$\{\alpha\}_{k+l} = \{\alpha\}_k \times \{\alpha_{k+1}\} \times \dots \times \{\alpha_{k+l}\}. \tag{2.5}$$

□

Remark 2.2. Whenever we write a formula like the right side of (2.5) we always mean the B-product. □

Definition 2.3. Given a function $B_{\{\alpha\}_k}$, we define $S_B \subset S$ by

1. $\{0\}_k \in S_B$ for all $k \in \mathbb{Z}$,
2. $\{\alpha\}_{k+1} \in S_B$ iff $\{\alpha\}_{k+1} = \{\alpha\}_k \times \{\alpha_{k+1}\}$, where $\{\alpha\}_k \in S_B$.

□

Proposition 2.4. *$\{\alpha\}_k \in S_B$ iff there is $l \in \mathbb{N}$ such that*

$$\{\alpha\}_k = \{0\}_{k-l} \times \{\alpha_{k-l+1}\} \times \cdots \times \{\alpha_k\}.$$

□

Definition 2.5. We say that a sequence $\{\alpha_i\}_{i\in\mathbb{Z}}$ belongs to the set $\mathbb{S}_B$ iff $\{\alpha\}_k \in S_B$ for all $k \in \mathbb{Z}$. □

To simplify notations we write

$$\alpha := \{\alpha_i\}_{i\in\mathbb{Z}}. \tag{2.6}$$

Let $q > 1$. For any pair $\alpha, \beta \in \mathbb{S}_B$ we define

$$\begin{aligned} \rho_q(\alpha,\alpha) &= 0 \\ \rho_q(\alpha,\beta) &= q^{-i_0}, \end{aligned} \tag{2.7}$$

where i_0 is such that $\alpha_{i_0} \neq \beta_{i_0}$ and $\alpha_i = \beta_i$ if $i < i_0$.

It is easy to verify that the following proposition holds:

Proposition 2.6. *ρ_q is a metric on $\mathbb{S}_B$ satisfying the non-Archimedean triangle inequality*

$$\rho_q(\alpha,\beta) \leq \max\{\rho_q(\alpha,\gamma), \rho_q(\gamma,\beta)\}. \tag{2.8}$$

□

It is clear that for any $q, q' > 1$ the metrics ρ_q and $\rho_{q'}$ are equivalent. Thus we fix a number $q > 1$ throughout the paper and drop the subscript q. Set

$$\mathbb{S}_B^k := \{\alpha \in \mathbb{S}_B; \alpha_i = 0 \ for \ i \geq k\}, \tag{2.9}$$

and

$$\mathbb{S}_{B,0} := \bigcup_{k\in\mathbb{Z}} \mathbb{S}_B^k. \tag{2.10}$$

Proposition 2.7. *$\mathbb{S}_B$ equipped with the metric ρ is a complete metric space and $\mathbb{S}_{B,0}$ is a dense subset of it.* □

Given $\alpha \in \mathbb{S}_B$ and $N \in \mathbb{Z}$ the set

$$K(\alpha, q^N) = \{\beta \in \mathbb{S}_B; \rho(\alpha,\beta) \leq q^N\} \tag{2.11}$$

will be called a ball of radius q^N centered at α. As consequences of (2.8) we have

1) If $\beta \in K(\alpha, q^N)$ then $K(\beta, q^N) = K(\alpha, q^N)$.
2) $K(\alpha, q^N), K(\beta, q^N)$ are either disjoint or identical, for any $\alpha, \beta \in \mathbb{S}_B$.
3) If $\alpha \in \mathbb{S}_B$ and $\alpha_i = 0$ for $i < -N$, then $K(\alpha, q^N) = K(0, q^N)$.

It follows from (2.11) that $K(\alpha, q^N)$ is uniquely defined by $\{\alpha\}_{-(N+1)}$ and thus we can identify

$$\{\alpha\}_{-(N+1)} = K(\alpha, q^N). \tag{2.12}$$

With this notation we have

4) $$K(\alpha, q^{N+1}) = \{\alpha\}_{-(N+2)} = \bigcup_{\gamma} \{\alpha\}_{-(N+2)} \times \{\gamma\}. \tag{2.13}$$

where the union is taken over γ satisfying: $0 \leq \gamma \leq B_{\{\alpha\}_{-(N+2)}} - 1$. Thus the ball $K(\alpha, q^{(N+1)})$ is the union of $B_{\{\alpha\}_{-(N+2)}}$ disjoint balls of radius q^N. Let $M \in \mathbb{Z}$ be given. Then according to 3) for any $\beta \in \mathbb{S}_B$ there is $N > M$ such that $\beta \in K(0, q^N)$. Thus

$$\mathbb{S}_B = \bigcup_{N>M} K(0, q^N). \tag{2.14}$$

Put $\mathcal{K}(M)$ for the family of all disjoint balls of radius q^M. It follows that $\mathcal{K}(M)$ is countable. Thus

$$\mathcal{K}(M) = \{K_i^M\}_{i \in \mathbb{N}}, \tag{2.15}$$

where K_i^M is a ball of radius q^M and $K_i^M \cap K_j^M = \emptyset$ iff $i \neq j$. As a consequence of (2.13), (2.14), we have

$$\mathbb{S}_B = \bigcup_{i \in \mathbb{N}} K_i^M. \tag{2.16}$$

We also have

Proposition 2.8. *A ball in $\mathbb{S}_B$ is both open and compact.* □

Looking back at the discussion leading to definition of the spaces $\mathbb{S}_B$ one realizes that $\mathbb{S}_B$ is uniquely determined by choice of the function $B_{\{\alpha\}_k}$. We shall now specify two classes of $\mathbb{S}_B$.

Let $j \in \mathbb{N}_0, l \in \mathbb{N}$. Set

$$S_j^0 = \{\{\alpha\}_j \in S_j; \alpha_i = 0, i \leq 0\} \tag{2.17}$$

$$S^l = \bigcup_{j=0}^{l-1} S_j^0. \tag{2.18}$$

Put $k = nl + s$, where $n \in \mathbb{Z}, s = 0, \ldots, l-1$. Then

$$L\{\alpha\}_k = \{\ldots, 0_{-1}, 0_0, \alpha_{nl+1}, \ldots, \alpha_k\}_s \tag{2.19}$$

defines a map of S onto S^l. Let $B^l_{\{\alpha\}_j}$, $\{\alpha\}^j \in S^l$ stand for a function defined on S^l with its values in $\mathbb{N} \backslash \{1\}$. We define the function $B_{\{\alpha\}_k}$ on S by

$$B_{\{\alpha\}_k} = B^l_{L\{\alpha\}_k}. \tag{2.20}$$

If $B_{\{\alpha\}}{}^k$ is defined by (2.20) then we say that the corresponding $\mathbb{S}_B$ belongs to the $\mathcal{L}_l$ class. It will be convenient to have following equivalence relation in S. We say that $\{\alpha\}_i, \{\beta\}_j \in S$ are L-equivalent if $i = j$ modulo l and $L\{\alpha\}_i = L\{\beta\}_j$.

Note that S^l is a countable infinite set. Hence the set of L-equivalence classes in S is also countable infinite. If $B\{\alpha\}_k$ is defined by (2.20) then the number of L-equivalence classes in $\mathbb{S}_B$ is finite and equal to the number of the B-products

$$\{0\}_0 \times \{\gamma_1\} \times \cdots \times \{\gamma_s\}\}, \quad s = 1, \ldots, l.$$

The important examples of $\mathbb{S}_B$'s are obtained by further specification.
Take $l = 1$. Then $S^l = S^1$ is a one element set $S^1 = \{\{0\}_0\}$. Define $B_{\{0\}_0} = p$, where p is a prime number.Then we have $B_{\{\alpha\}_k} \equiv p$. The corresponding space $\mathbb{S}_B$ will be denoted by $\mathbb{S}_{Bp}$. Let $\alpha \in \mathbb{S}_{Bp}$. Then $\alpha_k \in \{0, 1, \ldots, p-1\}$. If N is such that $\alpha_k = 0$ for $k < N$, then

$$a = \sum_{i=N}^{\infty} \alpha_i p^i \tag{2.21}$$

is the Hensel representation for a p-adic number. Conversely, let $a \in \mathbb{Q}_p$ be given by (2.21) then defining $\alpha_k = 0$ for $k < N$ we obtain $\alpha \in \mathbb{S}_{Bp}$. Thus we defined a one to one correspondence between $\mathbb{S}_{Bp}$ and $\mathbb{Q}_p$. The set $\mathbb{Q}_p$ is a field. Ignoring its algebraical structure we say only that it is equipped with a norm which defines the following metric. Let $a, b \in \mathbb{Q}_p$ be given by the Hensel representation (2.21) then

$$\rho_{(p)}(a, a) = 0 \qquad \text{and} \qquad \rho_{(p)}(a, b) = p^{-i_0},$$

where i_0 is such that $\alpha_{i_0} \neq \beta_{i_0}$ and $\alpha_i = \beta_i$ for $i < i_0$. Put $q = p$ in (2.7). If $\alpha, \beta \in \mathbb{S}_{Bp}$ correspond to $a, b \in \mathbb{Q}_p$ respectively then

$$\rho_p(\alpha, \beta) = \rho_{(p)}(a, b). \tag{2.22}$$

3. Stochastic processes on $\mathbb{S}_B$

In this section we outline the construction of a class of stochastic processes on $\mathbb{S}_B$. The main step in this direction will be a construction of the processes on $\mathcal{K}(M)$. Let $\alpha^i \in K_i^M$, $i \in \mathbb{N}$. Then according to (2.13)

$$K_i^M = K(\alpha^i, q^M) = \{\alpha^i\}_{-(M+1)}. \tag{3.1}$$

Put $P_{\{\alpha^i\}_{-(M+1)}\{\alpha^j\}_{-(M+1)}}(t), t \in \mathbb{R}_+$ for the transition probability from K_i^M to K_j^M in time t. Whenever possible we shall use the simplified notation

$$P_{ij}^{M,M} := P_{\{\alpha^i\}_{-(M+1)}\{\alpha^j\}_{-(M+1)}}(t). \tag{3.2}$$

Thus the forward and backward Chapman-Kolmogorov equations read:

$$\dot{P}_{ij}^{M,M}(t) = -\tilde{a}_j P_{ij}^{M,M}(t) + \sum_{\substack{l=1\\ l\neq j}}^{\infty} \tilde{u}_{lj} P_{il}^{M,M}, \tag{3.3a}$$

$$\dot{P}_{ij}^{M,M}(t) = -\tilde{a}_j P_{ij}^{M,M}(t) + \sum_{\substack{l=1\\ l\neq j}}^{\infty} \tilde{u}_{il} P_{lj}^{M,M}, \tag{3.3b}$$

$i, j \in \mathbb{N}$. We impose the initial condition

$$P_{ij}^{M,M}(0) = \delta_{ij}. \tag{3.4}$$

The coefficients $\tilde{u}_{lj}$ and $\tilde{a}_j$ will be defined as follows. Let $a(N)$, $N \in \mathbb{Z}$ be a sequence such that

$$a(N) \geq a(N+1) \tag{3.5a}$$

and

$$\lim_{N\to\infty} a(N) = 0, \tag{3.5b}$$

Put

$$U(N+1) = a(N) - a(N+1). \tag{3.6}$$

Let $\rho(\alpha^l, \alpha^j) = q^{N+m}$, $m \in \mathbb{N}$. Define

$$B(\alpha^j, m, M) \equiv (B_{\{\alpha^j\}_{-(M+m+1)}} - 1)B_{\{\alpha^j\}_{-(M+m)}} \cdots B_{\{\alpha^j\}_{-(M+2)}}. \tag{3.7}$$

Set

$$u(\alpha^j, m, M) := B^{-1}(\alpha^j, m, M)U(M+m) \tag{3.8}$$

and define

$$\tilde{u}_{lj} = u(\alpha^j, m, M). \tag{3.9}$$

To underline the fact that the elementary balls have radius q^M we write

$$\tilde{a}_j = \tilde{a}_j(M). \tag{3.10}$$

Lemma 3.1. *If* $\tilde{a}_j(M) = \sum_{l=1}^{\infty} \tilde{u}_{jl}$ *then*

$$\tilde{a}_j(M) = a(M). \tag{3.11}$$

The solution of the Chapman-Kolmogorov equations satisfying the initial conditions (3.4) reads

$$P_{ii}^{M,M} = \sum_{n=0}^{\infty} \left(B_{\{\alpha^i\}_{-(M+n+2)}} \cdots B_{\{\alpha^i\}_{-(M+2)}}\right)^{-1} \left(B_{\{\alpha^i\}_{-(M+n+2)}} - 1\right)$$
$$\exp\left\{-\left(a(M+n) + u(\alpha^i, 1, M+n)\right)t\right\}. \tag{3.12}$$

If $\rho(\alpha^i, \alpha^j) = q^{M+k}$, $k \in \mathbb{N}$ then

$$P_{ij}^{M,M}(t) = B^{-1}(\alpha^j, k, M) B^{-1}_{\{\alpha^i\}_{-(M+k+1)}} \left(B_{\{\alpha^i\}_{-(M+k+1)}} - 1\right)$$
$$\Big[\sum_{n=0}^{\infty} \left(B_{\{\alpha^i\}_{-(M+k+n+2)}} \cdots B_{\{\alpha^i\}_{-(M+k+2)}}\right)^{-1} \left(B_{\{\alpha^i\}_{-(M+k+n+2)}} - 1\right)$$
$$\exp\left\{-\left(a(M+k+n) + u(\alpha^i, 1, M+k+n)\right)t\right\}$$
$$- \exp\left\{-\left(a(M+k-1) + u(\alpha^i, 1, M+k-1)\right)t\right\}\Big]. \tag{3.13}$$

Thus we have found the transition functions for the processes on $\mathcal{K}(M)$. To obtain the transition functions for the processes on $\mathbb{S}_B$ we proceed as follows. Let $M, N \in \mathbb{Z}$ and $M \leq N$. Then $\{\beta\}_{-(N+1)}$ is an union of the balls of radius q^M i.e.,

$$\{\beta\}_{-(N+1)} = \bigcup_{\gamma} \{\beta\}_{-(N+1)} \times \{\gamma_{-N}\} \times \cdots \{\gamma_{-(M+1)}\}, \tag{3.14}$$

where the union runs over all B-products of $\{\beta\}_{-(N+1)}$ and the $(N-M)$-tuples γ. Set

$$P_{\{\alpha\}_{-(M+1)}\{\beta\}_{-(N+1)}}(t) = \sum_{\gamma} P_{\{\alpha\}_{-(M+1)}\{\beta\}_{-(N+1)} \times \{\gamma_{-N}\} \times \cdots \times \{\gamma_{-(M+1)}\}}(t). \tag{3.15}$$

One shows that (3.15) does not depend on M provided $M \leq N$.
Since $\alpha = \cap_{M \leq N} \{\alpha\}_{-(M+1)}$ we define

$$P(\alpha, \{\beta\}_{-(N+1)}, t) = P_{\{\alpha\}_{-(M+1)}\{\beta\}_{-(N+1)}}(t). \tag{3.16}$$

It is further shown that $P(\alpha, \{\beta\}_{-(N+1)}, t)$ extends to a transition function for a process on $\mathbb{S}_B$.

4. The Markovian semigroup and its generator

The next step of our discussion will be to define a Borel measure μ on $\mathbb{S}_B$ and the Markovian semigroup T_t, $t > 0$ of selfadjoint operators in $L^2(\mathbb{S}_B, \mu)$ determined by $P(\alpha, \{\beta\}_{-(N+1)}, t)$.

Recall that $\mathcal{K}(M)$ defined by (2.15) is the family of all disjoint balls of radius q^M. Then

$$\mathcal{K} := \bigcup_{M \in \mathbb{Z}} \mathcal{K}(M)$$

is the family of all balls in $\mathbb{S}_B$. We define a set function μ on $\mathcal{K}$ as follows

$$\mu(\{0\}_{-1}) = 1, \tag{4.1}$$

and

$$\mu(\{\alpha\}_{-(M+1)}) = B_{\{\alpha\}_{-(M+1)}} \mu(\{\alpha\}_{-M}) \tag{4.2}$$

for all $\alpha \in \mathbb{S}_B$ and $M \in \mathbb{Z}$. It follows from (4.2) that the numbers $\mu(\{\alpha\}_{-(M+1) \times \{\gamma\}})$, $0 \leq \gamma \leq B_{\{\alpha\}_{-(M+1)}} - 1$ are equal. By standard arguments μ can be extended to a Borel measure on $\mathbb{S}_B$. It is shown that the transition function $P(\alpha, \{\beta\}_{-(k+1)}, t)$ determines a μ-symmetric Markovian integral kernel which in turn extends uniquely to a self adjoint Markovian semigroup $T_t, t > 0$ in $L^2(\mathbb{S}_B, \mu)$. Moreover we have

Proposition 4.1. *The Markovian semigroup $T_t, t > 0$ in $L^2(\mathbb{S}_B, \mu)$ defined by $P(\alpha, \{\beta\}_{-(k+1)}, t)$ is strongly continuous.*

To summarize $T_t, t > 0$ is a strongly continuous self-adjoint contraction semigroup and hence it has the representation

$$T_t = e^{-Ht}, \qquad t \geq 0$$

where H is a non-negative self-adjoint operator. Let $f \in L^2(\mathbb{S}_B, \mu)$. Then by definition

$$(Hf)(\eta) = \lim_{t \downarrow 0} t^{-1} \left[f(\eta) - (T_t f)(\eta) \right] = \lim_{t \downarrow 0} t^{-1} \left[f(\eta) - \int_{\mathbb{S}_B} f(\xi) P(\eta, d\xi, t) \right]$$

whenever the strong limit exists. It is shown that the characteristic function of every ball $\chi_{\{\alpha\}_{-(M+1)}}$ belongs to the domain of H. We have the explicit formula which is valid for all $\alpha \in \mathbb{S}_B$ and $M \in \mathbb{Z}$

$$H\chi_{\{\alpha\}_{-(M+1)}}(\eta) = \begin{cases} a(M) \; if \; \eta \in \{\alpha\}_{-(M+1)}, \\ -B^{-1}(\alpha, k, M)\left(a(M+k-1) - a(M+k)\right) \\ \text{if } \rho(\eta, \alpha) = q^{M+k}. \end{cases} \tag{4.3}$$

Put D_0 for the linear hull spanned by the characteristic functions of all balls in $\mathbb{S}_B$. Then we have

Corollary 4.2. *D_0 is a core for H.*

The spectral properties of H are described by

Theorem 4.3. *Let $-H$ denote as above the generator of the strongly continuous semigroup T_t with the kernel defined by* (3.16)*. Then*

(a) *For any $M \in \mathbb{Z}$ such that $a(M) > 0$ and $\alpha \in \mathbb{S}_B$ there corresponds an eigenvalue $h_{M,\alpha}$ of H given by*

$$h_{M,\alpha} = \left(B_{\{\alpha\}_{-(M+2)}} - 1 \right)^{-1} \left(B_{\{\alpha\}_{-(M+2)}} a(M) - a(M+1) \right) \tag{4.4}$$

and a $B_{\{\alpha\}_{-(M+2)}} - 1$-dimensional eigenspace spanned by vectors of the form

$$e_{M,\alpha} = \sum_{\gamma=0}^{s} b_\gamma \chi_{\{\alpha\}_{-(M+2)} \times \{\gamma\}} \tag{4.5}$$

where

$$\sum_{\gamma=0}^{s} b_\gamma = 0 \text{ and } s = B_{\{\alpha\}_{-(M+2)}} - 1. \tag{4.6}$$

If $a(M) = 0$ then $\chi_{\{\alpha\}_{-(M+1)}}$ is an eigenvector of H to the eigenvalue 0.

(b) *The linear hull spanned by the vectors $e_{M,\alpha}$, $M \in \mathbb{Z}$, $\alpha \in \mathbb{S}_B$ is dense in $L^2(\mathbb{S}_B, \mu)$.*

Proposition 4.4. *Let $\mathbb{S}_B \in \mathcal{L}_l$, $\alpha, \beta \in \mathbb{S}_B$ and $M, N \in \mathbb{Z}$ be such that $a(M), a(N) > 0$. If $\{\alpha\}_{-(M+2)}$ and $\{\beta\}_{-(N+2)}$ are L-equivalent then $h_{M,\alpha}$ and $h_{N,\beta}$ have the same multiplicity. If in addition $M = N$ then $h_{M,\alpha} = h_{M,\beta}$.*

Proof. If $\{\alpha\}_{-(M+2)}$ and $\{\beta\}_{-(M+2)}$ are L-equivalent then

$$B_{\{\alpha\}_{-(M+2)}} = B_{L\{\alpha\}_{-(M+2)}} = B_{L\{\beta\}_{-(M+2)}} = B_{\{\beta\}_{-(M+2)}}.$$

By Theorem 4.3 the eigenspaces of $h_{M,\alpha}$ and $h_{M,\beta}$ are of equal dimensions. If in addition $M = N$ then the equality $h_{M,\alpha} = h_{N,\beta}$ follows from (4.4). □

The processes and their generators discussed so far have been constructed from the solutions of the Chapman-Kolmogorov equations with coefficients $\widetilde{u}_{lj}$ defined by (3.8), (3.9) and thus depending only on the distance $\rho(\alpha^l, \alpha^j)$ and the structure of $\mathbb{S}_B$. However the procedure we have been using can be extended to the models with $\widetilde{u}_{lj}$ depending also on the target state explicitly. Since the formulation allowing for the full generality of $\mathbb{S}_B$ would involve a very heavy notation we shall limit our considerations to the processes on $\mathbb{Q}_p$. Such processes have been studied in [8]. The generators and their spectral properties can be found in [3]. Here we present the results of [3] modified according to its generalization done in [4].

Let $\mathcal{R}$ be a σ-finite Borel measure on $\mathbb{Q}_p$. If $\rho(\alpha^l, \alpha^j) = p^{M+m}$ then we define

$$\widetilde{u}_{jl} = \mathcal{R}(\{\alpha^l\}_{-(M+1)})u(M+m). \tag{4.7}$$

Put

$$W_j^M = -\sum_{m=1}^{\infty} u(M+m)\mathcal{R}(\{\alpha^j\}_{-(M+1)} \setminus \{\alpha^j\}_{-(M+m+1)}). \tag{4.8}$$

Then we have

Lemma 4.5. *If* $\widetilde{a}_j(M) = \sum_{l=1}^{\infty} \widetilde{u}_{jl}$ *then*

$$\widetilde{a}_j(M) = -W_j^M \tag{4.9}$$

Let the coefficients $\widetilde{u}_{jl}$, $\widetilde{a}_j(M)$ in the Chapman-Kolmogorov equations be defined by (4.7) and (4.9) respectively. The solutions will be given in terms of following expressions

$$-\mathcal{W}_N^{\alpha} = \sum_{k=N}^{\infty} (u(k) - u(k+1))\mathcal{R}(\{\alpha\}_{-(k+1)}) \tag{4.10}$$

We have to distinguish two cases:

1. If $\mathcal{R}(\mathbb{Q}_p)$ is finite then we assume $\mathcal{R}(\mathbb{Q}_p) = 1$. This assumption is not significant if the corresponding stochastic process is concerned. Then the solutions read

$$P(\alpha, \{\alpha\}_{-(M+1)}, t) = \mathcal{R}(\{\alpha\}_{-(M+1)}) \left\{1 + \sum_{m=0}^{\infty} (\mathcal{R}^{-1}(\{\alpha\}_{-(M+m+1)}) - \mathcal{R}^{-1}(\{\alpha\}_{-(M+m+2)})) \exp\{t\mathcal{W}_{M+m+1}^{\alpha}\}\right\}. \tag{4.11}$$

If $\rho(\alpha, \beta) = p^{M+k}$ then

$$P(\alpha, \{\beta\}_{-(M+1)}, t) = \mathcal{R}(\{\beta\}_{-(M+1)}) \left\{1 + \sum_{m=0}^{\infty} (\mathcal{R}^{-1}(\{\beta\}_{-(M+k+m+1)}) - \mathcal{R}^{-1}(\{\beta\}_{-(M+k+m+2)})) \exp\{t\mathcal{W}_{M+k+m+1}^{\beta}\}\right\} - \mathcal{R}(\{\beta\}_{-(M+1)})\mathcal{R}^{-1}(\{\beta\}_{-(M+k+m+1)}) \exp\{t\mathcal{W}_{M+k}^{\beta}\}. \tag{4.12}$$

2. If $\mathcal{R}(\mathbb{Q}_p) = \infty$ then we have

$$P(\alpha, \{\alpha\}_{-(M+1)}, t) = \mathcal{R}(\{\alpha\}_{-(M+1)}) \sum_{m=0}^{\infty} \left(\mathcal{R}^{-1}(\{\alpha\}_{-(M+m+1)}) - \mathcal{R}^{-1}(\{\alpha\}_{-(M+m+2)})\right) \exp\{t\mathcal{W}^{\alpha}_{M+m+1}\}. \quad (4.13)$$

If $\rho(\alpha, \beta) = p^{M+k}$ then

$$P(\alpha, \{\beta\}_{-(M+1)}, t) = \mathcal{R}(\{\beta\}_{-(M+1)}) \left\{\sum_{m=0}^{\infty} \left(\mathcal{R}^{-1}(\{\beta\}_{-(M+k+m+1)}) - \mathcal{R}^{-1}(\{\beta\}_{-(M+k+m+2)})\right) \exp\{t\mathcal{W}^{\beta}_{M+k+m+1}\}\right\} \quad (4.14)$$

It is further found that $P(\alpha, \{\beta\}_{-(M+1)}, t)$ defines a strongly continuous semigroup $T_t, t \geq 0$ of self-adjoint operators in $L^2(\mathbb{Q}_p, \mathcal{R})$. Thus the semigroup can be represented by

$$T_t = e^{-Ht}, \qquad t \geq 0. \quad (4.15)$$

The generator H is given by

$$H\chi_{\{\alpha\}_{-(M+1)}}(\eta) = \begin{cases} -W^{\alpha}_{M} \ if\ \eta \in \{\alpha\}_{-(M+1)}, \\ -\mathcal{R}(\{\alpha\}_{-(M+1)})u(M+j) \\ \text{if } \rho(\eta, \alpha) = p^{M+j}. \end{cases} \quad (4.16)$$

In analogy to Theorem 4.3 we have

Theorem 4.6. *Let $-H$ stands for generator of the Markovian semigroup $T_t, t \geq 0$ in $L^2(\mathbb{Q}_p, \mathcal{R})$ with the integral kernel $P(\eta, A, t)$ defined by (4.13), (4.14). Then*

1. *For any $\alpha \in \mathbb{Q}_p, M \in \mathbb{Z}$ there corresponds an eigenvalue $h_{M,\alpha}$ of H given by*

$$h_{M,\alpha} = \mathcal{W}^{\alpha}_{M}. \quad (4.17)$$

To this eigenvalue there corresponds a $p-1$-dimensional eigenspace spanned by the vectors of the form

$$e_{M,\alpha} = \sum_{\gamma=0}^{p-1} b_\gamma \chi_{\{\ldots,\alpha_{-(M+2)},\alpha_{-(M+1)},\gamma\}}, \quad (4.18)$$

where $\sum_{\gamma=0}^{p-1} b_\gamma \mathcal{R}(\{\alpha\}_{-(M+1)} \times \gamma\}) = 0$.

2. *Denote by $e^s_{M,\alpha}$, $s = 1, 2, \ldots, p-1$ the orthonormalized eigenvectors corresponding to $h_{M,\alpha}$. If $\mathcal{R}(\mathbb{Q}_p) = \infty$ then the orthonormal system $\{e^s_{M,\alpha}\}, \alpha \in \mathbb{Q}_p, M \in \mathbb{Z}, s = 1, \ldots, p-1$ is a basis for $L^2(\mathbb{Q}_p, \mathcal{R})$. If $\mathcal{R}(\mathbb{Q}_p) = 1$ then the above vectors together with the constant function 1 form a basis for $L^2(\mathbb{Q}_p, \mathcal{R})$.*

Proposition 4.7. *Let $\alpha, \beta \in \mathbb{Q}_p$, $M \in \mathbb{Z}$, $l \in \mathbb{N}$ and $\rho(\alpha, \beta) = p^M$. Then*

$$h_{N,\alpha} = h_{N,\beta} \tag{4.19}$$

for $N \geq M$. If in addition

$$\mathcal{R}(\{\alpha\}_{-(M-l+1)}) = \mathcal{R}(\{\beta\}_{-(M-l+1)}) \tag{4.20}$$

for $l = 1, 2, \ldots, l_0$ then (4.19) holds for $N \geq M - l_0$.

Proof. We have to prove that $\mathcal{W}_N^\alpha = \mathcal{W}_N^\beta$. Since $\rho(\alpha, \beta) = p^M$ we have $\{\alpha\}_{-(N+1)} = \{\beta\}_{-(N+1)}$ if $N \geq M$. Thus

$$\mathcal{R}(\{\alpha\}_{-(k+1)}) = \mathcal{R}(\{\beta\}_{-(k+1)}), \qquad k \geq N \geq M$$

and consquently $\mathcal{W}_N^\alpha = \mathcal{W}_N^\beta$ by (4.8). If in addition (4.20) is assumed then (4.19) holds for $N \geq M - l_0$. □

References

[1] S. Albeverio, W. Karwowski, *A random walk on p-adics: generator and its spectrum.* Stochastic processes and their Applications **53** (1994), 1–22.

[2] S. Albeverio, W. Karwowski, *Jump processes on leaves of multibranching trees* (to be published).

[3] S. Albeverio, W. Karwowski, X. Zhao, *Asymptotic and spectral results for random walks on p-adics.* Stochastic Processes and their Applications **83** (1999), 39–59.

[4] S. Albeverio, X.L. Zhao, *A remark on nonsymmetric stochastic processes on p-adics.* Stoch. Anal. Appl. **20** (2002), 243–261.

[5] D. Aldous, S. Evans, *Dirichlet forms on totally disconnected spaces and bipartite Markov chains.* J. Theor. Prob. **12** (1999), 839–857.

[6] S.N. Evans, *Local properties of Levy processes on totally disconnected groups.* J. Theoret. Prob. **2** (1989), 209–259.

[7] H. Kaneko, *A class of spatially inhomogeneous Dirichlet spaces on the p-adic number field.* Stochastic Processes and their Applications **88** (2000), 161–174.

[8] W. Karwowski, R. Vilela Mendes, *Hierarchical structures and asymmetric processes on p-adics and adeles.* J. Math. Phys. **35** (1994), 4637–4650.

[9] A. Khrennikov, *p-adic discrete dynamical systems and their applications in physics and cognitive science.* Russian Journal of Mathematical Physics **11** (2004), 45–70.

[10] N. Koblitz, *p-adic numbers, p-adic analysis and zeta functions.* 2nd ed. Springer, New York 1984.

[11] A.N. Kochubei, *Parabolic equations over the field of p-adic numbers.* Math. USSR Izviestiya **39** (1992), 1263–1280.

[12] A.N. Kochubei, *Stochastic Integrals and stochastic differential equations over the field of p-adic numbers.* Potential Anal. **6** (1997), 105–1025.

[13] A.N. Kochubei, *Pseudo-differential Equation and Stochastics over non-Archimdean Fields.* Mongr. Textbooks Pure Appl. Math. 244 Marcel Dekker, New York, 2001.

[14] R. Lima, R. Vilela Mendes, *Stochastic processes for the turbulent cascade.* Physics Revue E **53** (1996), 3536–540.

[15] R. Ramal, G. Touluse, M.A. Virasoro, *Ultrametricity for Physicists*. Rev. Mod. Phys. **58** (1986), 765–788.

[16] V. Vladimirov, *Generalized functions over the field of p-adic numbers*. Russian Math. Surveys. **43** (1988), 19–64.

[17] V. Vladimirov, I. Volovich, E. Zelnov, *p-adic numbers in mathematical physics*. World Scientific, Singapore 1993.

[18] K. Yasuda, *Semistable processes on local fields*. Tohoku Math. Journal **58** (2006), 419–431.

Witold Karwowski
Institute of Physics,
Opole University
48 Oleska Str
45-052 Opole, Poland
e-mail: `witoldkarwowski@go2.pl`

Operator Theory:
Advances and Applications, Vol. 190, 375–385

On Pseudo-Hermitian Operators with Generalized $\mathcal{C}$-symmetries

S. Kuzhel

Abstract. The concept of $\mathcal{C}$-symmetries for pseudo-Hermitian Hamiltonians is studied in the Krein space framework. A generalization of $\mathcal{C}$-symmetries is suggested.

Mathematics Subject Classification (2000). Primary 47A55; Secondary 81Q05, 81Q15.

Keywords. Krein spaces, J-self-adjoint operators, $\mathcal{PT}$-symmetric quantum mechanics, $\mathcal{C}$-symmetries, pseudo-Hermitian Hamiltonians.

1. Introduction

The employing of non-Hermitian operators for the description of experimentally observable data goes back to the early days of quantum mechanics [14, 18, 30]. A steady interest to non-Hermitian Hamiltonians became enormous after it has been discovered that complex Hamiltonians possessing so-called $\mathcal{PT}$-symmetry (the product of parity and time reversal) can have a real spectrum (like self-adjoint operators) [9, 15, 26, 34]. The results obtained gave rise to a consistent complex extension of the standard quantum mechanics (see the review paper [8] and the references therein).

One of key moments in the $\mathcal{PT}$-symmetric quantum theory is the description of a previously unnoticed symmetry (hidden symmetry) for a given $\mathcal{PT}$-symmetric Hamiltonian A that is represented by a linear operator $\mathcal{C}$. The properties of $\mathcal{C}$ are nearly identical to those of the charge conjugation operator in quantum field theory and the existence of $\mathcal{C}$ provides an inner product whose associated norm is positive definite and the dynamics generated by A is then governed by a unitary time evolution. However, the operator $\mathcal{C}$ depends on the choice of A and its finding is a nontrivial problem [7, 10, 12].

The concept of $\mathcal{PT}$-symmetry can be placed in a more general mathematical context known as *pseudo-Hermiticity.* A linear densely defined operator A acting in a Hilbert space $\mathfrak{H}$ is pseudo-Hermitian if there is an invertible bounded self-adjoint operator $\eta : \mathfrak{H} \to \mathfrak{H}$ such that

$$A^*\eta = \eta A, \tag{1.1}$$

where the sign $*$ stands for the adjoint of the corresponding operator. Including the concept of $\mathcal{PT}$-symmetry in a pseudo-Hermitian framework enables one to make more clear basic constructions of $\mathcal{PT}$-symmetric quantum mechanics and to achieve a lot of nontrivial physical results [1, 6, 17, 19, 27, 28].

The related notion of quasi-Hermiticity and its physical implications were discussed in detail in [16, 31].

Using Langer's observation [22] that a Hilbert space $\mathfrak{H}$ with the indefinite metric $[f, g]_\eta = (\eta f, g)$ $(0 \in \rho(\eta))$ is a Krein space, one can reduce the investigation of pseudo-Hermitian operators to the study of self-adjoint operators in a Krein space [2, 3, 25, 29, 32]. The present paper continues such trend of investigations and its aim is to analyze pseudo-Hermitian operators with $\mathcal{C}$-symmetries in the Krein space setting. The special attention will be paid to the generalization of the concept of $\mathcal{C}$ operators.

The existence of a $\mathcal{C}$-symmetry for a pseudo-Hermitian operator A means that A has a maximal dual pair $\{\mathfrak{L}_+, \mathfrak{L}_-\}$ of invariant subspaces of $\mathfrak{H}$ [23, 24].

The paper is organized as follows. Section 2 contains all necessary Krein space results in the form convenient for our presentation. Their proofs and detailed analysis can be found in [5]. Section 3 deals with the study of $\mathcal{C}$-symmetries by the Krein space methods. A more physical presentation of the subject can be found in [7]–[13, 19, 29, 32]. Section 4 contains some generalization of the concept of $\mathcal{C}$-symmetry where operators $\mathcal{C}$ are supposed to be unbounded in $\mathfrak{H}$. The case of unbounded metric operator η has been recently studied in [20]. Examples of $\mathcal{C}$-symmetries and generalized $\mathcal{C}$-symmetries are presented in Section 5.

For the sake of simplicity, we restrict ourselves to the case where a self-adjoint operator η is also unitary. For such a self-adjoint and unitary operator η the notation J (i.e., $\eta \equiv J$) will be used. Note that the requirement of unitarity of η is not restrictive because the general case of a self-adjoint operator η is reduced to the case above if, instead of the original scalar product $(\cdot, \cdot)$ in $\mathfrak{H}$, ones consider another (equivalent to it) scalar product $(|\eta|\cdot, \cdot)$, where $|\eta| = \sqrt{\eta^2}$ is the modulus of η.

2. Elements of the Krein's spaces theory

Let $\mathfrak{H}$ be a Hilbert space with scalar product $(\cdot, \cdot)$ and let J be a fundamental symmetry in $\mathfrak{H}$ (i.e., $J = J^*$ and $J^2 = I$). The corresponding orthoprojectors $P_+ = 1/2(I + J)$, $P_- = 1/2(I - J)$ determine the fundamental decomposition of $\mathfrak{H}$

$$\mathfrak{H} = \mathfrak{H}_+ \oplus \mathfrak{H}_-, \qquad \mathfrak{H}_- = P_-\mathfrak{H}, \quad \mathfrak{H}_+ = P_+\mathfrak{H}. \tag{2.2}$$

The space $\mathfrak{H}$ with an indefinite scalar product (indefinite metric)

$$[x, y] := (Jx, y), \qquad \forall x, y \in \mathfrak{H} \tag{2.3}$$

is called *a Krein space* if $\dim \mathfrak{H}_+ = \dim \mathfrak{H}_- = \infty$.

A (closed) subspace $\mathfrak{L} \subset \mathfrak{H}$ is called *nonnegative, positive, uniformly positive* if, respectively, $[x,x] \geq 0$, $[x,x] > 0$, $[x,x] \geq \alpha\|x\|^2$ for all $x \in \mathfrak{L} \setminus \{0\}$. Nonpositive, negative, and uniformly negative subspaces are introduced similarly. The subspaces $\mathfrak{H}_+$ and $\mathfrak{H}_-$ in (2.2) are maximal uniformly positive and maximal uniformly negative, respectively.

Let $\mathfrak{L}_+$ be a maximal positive subspace (i.e., the closed set $\mathfrak{L}_+$ does not belong as a subspace to any positive subspace). Its J-orthogonal complement

$$\mathfrak{L}_- = \mathfrak{L}_+^{[\perp]} = \{x \in \mathfrak{H} \mid [x,y] = 0, \ \forall y \in \mathfrak{L}_+\}$$

is a maximal negative subspace of $\mathfrak{H}$ and the direct J-orthogonal sum

$$\mathfrak{H}' = \mathfrak{L}_+[\dot{+}]\mathfrak{L}_- \tag{2.4}$$

is a dense set in $\mathfrak{H}$. (Here the brackets $[\cdot]$ means the orthogonality with respect to the indefinite metric.) The linear set $\mathfrak{H}'$ coincides with $\mathfrak{H}$ if and only if $\mathfrak{L}_+$ is a maximal uniformly positive subspace. In that case $\mathfrak{L}_- = \mathfrak{L}_+^{[\perp]}$ is a maximal uniformly negative subspace.

The subspaces $\mathfrak{L}_+$ and $\mathfrak{L}_-$ in (2.4) can be decomposed as follows:

$$\mathfrak{L}_+ = (I+K)\mathfrak{H}_+ \qquad \mathfrak{L}_- = (I+Q)\mathfrak{H}_-,$$

where $K : \mathfrak{H}_+ \to \mathfrak{H}_-$ is a contraction and ($Q = K^* : \mathfrak{H}_- \to \mathfrak{H}_+$) coincides with the adjoint of K.

The self-adjoint operator $T = KP_+ + K^*P_-$ acting in $\mathfrak{H}$ is called *an operator of transition* from the fundamental decomposition (2.2) to (2.4). Obviously,

$$\mathfrak{L}_+ = (I+T)\mathfrak{H}_+, \qquad \mathfrak{L}_- = (I+T)\mathfrak{H}_-. \tag{2.5}$$

The collection of operators of transition admits a simple 'external' description. Namely, a self-adjoint operator T in $\mathfrak{H}$ is an operator of transition if and only if

$$\|Tx\| < \|x\| \quad (\forall x \neq 0) \qquad \text{and} \qquad JT = -TJ. \tag{2.6}$$

The important particular case of (2.4), where $\mathfrak{H}'$ coincides with $\mathfrak{H}$ (i.e., $\mathfrak{H} = \mathfrak{L}_+[\dot{+}]\mathfrak{L}_-$) corresponds to the more strong condition $\|T\| < 1$ in (2.6).

Let $P_{\mathfrak{L}_\pm} : \mathfrak{H}' \to \mathfrak{L}_\pm$ be the projectors onto $\mathfrak{L}_\pm$ with respect to decomposition (2.4). Repeating step by step the proof of Proposition 9.1 in [21] where the case $\mathfrak{H} = \mathfrak{L}_+[\dot{+}]\mathfrak{L}_-$ has been considered, one gets

$$P_{\mathfrak{L}_-} = (I-T)^{-1}(P_- - TP_+), \qquad P_{\mathfrak{L}_+} = (I-T)^{-1}(P_+ - TP_-), \tag{2.7}$$

where T is the operator of transition from (2.2) to (2.4).

3. The condition of $\mathcal{C}$-symmetry

A linear densely defined operator A acting in a Krein space $(\mathfrak{H}, [\cdot,\cdot])$ is called *J-self-adjoint* if its adjoint A^* satisfies the condition $A^*J = JA$. Obviously, J-self-adjoint operators are pseudo-Hermitian ones in the sense of (1.1).

Since a J-self-adjoint operator A is self-adjoint with respect to the indefinite metric (2.3), one can attempt to develop a consistent quantum theory for J-self-adjoint Hamiltonians with real spectrum. However, in this case, we encounter the difficulty of dealing with the indefinite metric $[\cdot,\cdot]$. Since the norm of states carries a probabilistic interpretation in the standard quantum theory, the presence of an indefinite metric immediately raises problems of interpretation. One of the natural ways to overcome this problem consists in the construction of a hidden symmetry of A that is represented by the linear operator $\mathcal{C}$. This symmetry operator $\mathcal{C}$ guarantees that the pseudo-Hermitian Hamiltonian A can be used to define a unitary theory of quantum mechanics [8, 11].

By analogy with [8] the definition of $\mathcal{C}$ can be formalized as follows.

Definition 3.1. A J-self-adjoint operator A has the property of $\mathcal{C}$-symmetry if there exists a bounded linear operator $\mathcal{C}$ in $\mathfrak{H}$ such that: (i) $\mathcal{C}^2 = I$; (ii) $J\mathcal{C} > 0$; (iii) $A\mathcal{C} = \mathcal{C}A$.

The next simple statement clarifies the structure of pseudo-Hermitian operators with $\mathcal{C}$-symmetries.

Proposition 3.2. *Let A be a J-self-adjoint operator. Then A has the property of $\mathcal{C}$-symmetry if and only if A admits the decomposition*

$$A = A_+[\dot{+}]A_-, \qquad A_+ = A \upharpoonright \mathfrak{L}_+, \quad A_- = A \upharpoonright \mathfrak{L}_- \tag{3.8}$$

with respect to a certain choice of J-orthogonal decomposition of $\mathfrak{H}$

$$\mathfrak{H} = \mathfrak{L}_+[\dot{+}]\mathfrak{L}_-, \quad \mathfrak{L}_- = \mathfrak{L}_+^{[\perp]}, \tag{3.9}$$

where $\mathfrak{L}_+$ is a maximal uniformly positive subspace of the Krein space $(\mathfrak{H}, [\cdot,\cdot])$.

Proof. Let A admit the decomposition (3.8) with respect to (3.9). Denote $\mathcal{C} = P_{\mathfrak{L}_+} - P_{\mathfrak{L}_-}$, where $P_{\mathfrak{L}_\pm}$ are projectors onto $\mathfrak{L}_\pm$ according to (3.9). Obviously, the bounded linear operator $\mathcal{C}$ satisfies $\mathcal{C}^2 = I$ and $\mathcal{C}A = A\mathcal{C}$. Furthermore, by virtue of the second relation in (2.6) and (2.7),

$$\mathcal{C} = P_{\mathfrak{L}_+} - P_{\mathfrak{L}_-} = (I-T)^{-1}(I+T)(P_+ - P_-) = J(I-T)(I+T)^{-1}, \tag{3.10}$$

where T is the operator of transition from (2.2) to (3.9). Hence $J\mathcal{C} = (I-T)(I+T)^{-1} > 0$ (since $\|T\| < 1$). Thus A has $\mathcal{C}$-symmetry.

Conversely, assume that A has $\mathcal{C}$-symmetry and denote $\mathfrak{L}_+ = (I+\mathcal{C})\mathfrak{H}$ and $\mathfrak{L}_- = (I-\mathcal{C})\mathfrak{H}$. Since $\mathcal{C}^2 = I$, one gets $\mathfrak{H} = \mathfrak{L}_+\dot{+}\mathfrak{L}_-$ and

$$\mathcal{C}f_- = -f_-, \quad \mathcal{C}f_+ = f_+, \qquad \forall f_\pm \in \mathfrak{L}_\pm. \tag{3.11}$$

Therefore,

$$[f_+, f_+] = [\mathcal{C}f_+, f_+] = (J\mathcal{C}f_+, f_+) > 0, \ [f_-, f_-] = -[\mathcal{C}f_-, f_-] = -(J\mathcal{C}f_-, f_-) < 0.$$

Thus $\mathfrak{L}_+$ ($\mathfrak{L}_-$) is a positive (negative) linear set of $\mathfrak{H}$.

The property of $J\mathcal{C}$ to be self-adjoint in $\mathfrak{H}$ implies that $\mathcal{C}^*J = J\mathcal{C}$, i.e., $\mathcal{C}$ is J-self-adjoint. In that case

$$[f_+, f_-] = [\mathcal{C}f_+, f_-] = [f_+, \mathcal{C}f_-] = -[f_+, f_-]$$

and hence, $[f_+, f_-] = 0$.

Summing the results established above, one concludes that the operator $\mathcal{C}$ determines an J-orthogonal decomposition (3.9) of $\mathfrak{H}$, where $\mathfrak{L}_+$ and $\mathfrak{L}_-$ are positive and negative linear subspaces of $\mathfrak{H}$. Such type of decomposition is possible only in the case where $\mathfrak{L}_+$ is a maximal uniformly positive subspace of $\mathfrak{H}$ and $\mathfrak{L}_- = \mathfrak{L}_+^{[\perp]}$ [5].

To complete the proof it suffices to observe that (3.8) follows from the relations $A\mathcal{C} = \mathcal{C}A$ and $\mathfrak{L}_\pm = (I \pm \mathcal{C})\mathfrak{H}$. □

Corollary 3.3. *A J-self-adjoint operator A has the property of $\mathcal{C}$-symmetry if and only if $H = \sqrt{J\mathcal{C}}A(\sqrt{J\mathcal{C}})^{-1}$ is a self-adjoint operator in $\mathfrak{H}$.*

Proof. Denote for brevity $F = J\mathcal{C}$. It follows from the conditions (i), (ii) of the definition of $\mathcal{C}$-symmetry that F is a bounded uniformly positive operator in $\mathfrak{H}$.

If a J-self-adjoint operator A has the property of $\mathcal{C}$-symmetry, then

$$(\sqrt{F}Ax, \sqrt{F}y) = [\mathcal{C}Ax, y] = [A\mathcal{C}x, y] = [\mathcal{C}x, Ay] = (\sqrt{F}x, \sqrt{F}Ay).$$

This means that $H = \sqrt{F}A(\sqrt{F})^{-1}$ is a self-adjoint operator in $\mathfrak{H}$ with respect to the initial product $(\cdot,\cdot)$.

Conversely, if $H = \sqrt{F}A(\sqrt{F})^{-1}$ is self-adjoint, then

$$[\mathcal{C}Ax, y] = (H\sqrt{F}x, \sqrt{F}y) = (\sqrt{F}x, H\sqrt{F}y) = [\mathcal{C}x, Ay] = [A\mathcal{C}x, y]$$

for any $x, y \in \mathfrak{H}$. Therefore, $\mathcal{C}A = A\mathcal{C}$ and A has $\mathcal{C}$-symmetry. □

It follows from the proof that the scalar product $(x, y)_\mathcal{C} := [\mathcal{C}x, y]$ determined by $\mathcal{C}$ is equivalent to the initial scalar product $(\cdot,\cdot)$ in $\mathfrak{H}$. Thus the existence of a $\mathcal{C}$-symmetry for a J-self-adjoint operator A ensures unitarity of the dynamics generated by A in the norm $\|\cdot\|^2_\mathcal{C} = (\cdot,\cdot)_\mathcal{C}$ equivalent to the initial one.

In contrast to Proposition 3.2, Corollary 3.3 does not emphasize the property of A to be diagonalizable into two operator parts in $\mathfrak{H}$.

Denote

$$U = \frac{1}{2}[(I + \mathcal{C})P_+ + (I - \mathcal{C})P_-]. \tag{3.12}$$

It is clear that $\mathcal{C}U = UJ$. This gives $U : \mathfrak{H}_+ \to \mathfrak{L}_+$ and $U : \mathfrak{H}_- \to \mathfrak{L}_-$, where $\mathfrak{H}_\pm$ are subspaces of the fundamental decomposition (2.2) and $\mathfrak{L}_\pm = (I \pm C)\mathfrak{H}$ are reducing subspaces for A in Proposition 3.2. Furthermore, it follows from (3.12) that the operators $U_\pm = U \restriction \mathfrak{H}_\pm$ determine bounded invertible mappings of $\mathfrak{H}_\pm$ onto $\mathfrak{L}_\pm$ (since the subspaces $\mathfrak{H}_+$ and $\mathfrak{L}_+$ are maximal uniformly positive and $\mathfrak{H}_-$ and $\mathfrak{L}_-$ are maximal uniformly negative).

By virtue of Proposition 3.2, the transformation U decomposes an J-self-adjoint operator A with $\mathcal{C}$ symmetry into the 2×2-block form

$$U^{-1}AU = \begin{pmatrix} U_{+}^{-1}AU_{+} & 0 \\ 0 & U_{-}^{-1}AU_{-} \end{pmatrix}$$

with respect to the fundamental decomposition (2.2). For this reason the mapping determined by U can be considered as a generalization of the Foldy-Wouthuysen transformation well-known in quantum mechanics (see, e.g., [33]).

4. Generalized $\mathcal{C}$-symmetry

The concept of $\mathcal{C}$-symmetry can be weakened as follows.

Definition 4.1. A J-self-adjoint operator A has the property of generalized $\mathcal{C}$-symmetry if there exists a linear densely defined operator $\mathcal{C}$ in $\mathfrak{H}$ such that:

(i) $\mathcal{C}^2 = I$;
(ii) the operator $J\mathcal{C}$ is positive self-adjoint in $\mathfrak{H}$;
(iii) $\mathcal{D}(A) \subset \mathcal{D}(\mathcal{C})$ and $A\mathcal{C} = \mathcal{C}A$.

The main difference with Definition 3.1 is that the operator $\mathcal{C}$ is not assumed to be bounded.

Proposition 4.2 (cf. Proposition 3.2**).** *A J-self-adjoint operator A has the property of generalized $\mathcal{C}$-symmetry if and only if A admits the decomposition*

$$A = A_{+}[\dot{+}]A_{-}, \qquad A_{+} = A \upharpoonright \mathfrak{L}_{+}, \quad A_{-} = A \upharpoonright \mathfrak{L}_{-} \tag{4.13}$$

with respect to a certain choice of J-orthogonal sum

$$\mathfrak{H} \supset \mathfrak{H}' = \mathfrak{L}_{+}[\dot{+}]\mathfrak{L}_{-} \qquad \mathfrak{L}_{-} = \mathfrak{L}_{+}^{[\perp]}, \tag{4.14}$$

where $\mathfrak{L}_{+}$ is a maximal positive subspace of the Krein space $(\mathfrak{H}, [\cdot,\cdot])$.

Proof. If $\mathfrak{L}_{+}$ satisfies the condition of Proposition 4.2, then $\mathfrak{L}_{+}[\dot{+}]\mathfrak{L}_{-}$ is a dense set in $\mathfrak{H}$ [5]. Denote $\mathcal{C} = P_{\mathfrak{L}_{+}} - P_{\mathfrak{L}_{-}}$, where $P_{\mathfrak{L}_{\pm}}$ are projectors in $\mathfrak{L}_{+}[\dot{+}]\mathfrak{L}_{-}$ onto $\mathfrak{L}_{\pm}$. It follows from the definition of $\mathcal{C}$ and (4.13) that $\mathcal{C}^2 = I$, $\mathcal{C}A = A\mathcal{C}$, and $\mathcal{D}(\mathcal{C}) = \mathfrak{L}_{+}[\dot{+}]\mathfrak{L}_{-}$.

The operator $\mathcal{C}$ is also defined by (3.10), where T is the operator of transition from (2.2) to $\mathfrak{L}_{+}[\dot{+}]\mathfrak{L}_{-}$. Since T satisfies (2.6), the operator $J\mathcal{C} = (I-T)(I+T)^{-1}$ is positive self-adjoint in $\mathfrak{H}$. Thus A has a generalized $\mathcal{C}$-symmetry.

Conversely, assume that A has generalized $\mathcal{C}$-symmetry and denote

$$T = (I - F)(I + F)^{-1}, \qquad F = J\mathcal{C}. \tag{4.15}$$

Since F is positive self-adjoint, the operator T satisfies $\|Tx\| < \|x\|$ $(x \neq 0)$ and

$$J\mathcal{C} = F = (I - T)(I + T)^{-1}. \tag{4.16}$$

The conditions (i) and (ii) of Definition 4.1 imply $JF = \mathcal{C} = F^{-1}J$. Combining this relation with (4.15) one gets $JT = -TJ$. So, the operator T satisfies the conditions of (2.6). This means that T is the operator of transition from (2.2) to the direct J-orthogonal sum (2.4) (or (4.14)), where a maximal positive subspace $\mathfrak{L}_+$ has the form (2.5) and $\mathfrak{L}_- = \mathfrak{L}_+^{[\perp]}$.

Since the projectors $P_{\mathfrak{L}_\pm}$ are defined by (2.7), relation (4.16) implies (cf. (3.10)) $P_{\mathfrak{L}_+} - P_{\mathfrak{L}_-} = J(I-T)(I+T)^{-1} = \mathcal{C}$. Hence, $\mathfrak{L}_+ = (I+\mathcal{C})\mathcal{D}(\mathcal{C})$ and $\mathfrak{L}_- = (I-\mathcal{C})\mathcal{D}(\mathcal{C})$. In that case, the decomposition (4.13) immediately follows from the relation $A\mathcal{C} = \mathcal{C}A$. □

Corollary 4.3. *If a J-self-adjoint operator A possesses a generalized $\mathcal{C}$-symmetry given by an operator $\mathcal{C}$ in the sense of Definition* 4.1, *then its adjoint $\mathcal{C}^*$ provides the property of a generalized $\mathcal{C}$-symmetry for A^*.*

Proof. Since A is J-self-adjoint, the relation $AJ = JA^*$ holds. Therefore, if A is decomposed with respect to (4.14) in the sense of Proposition 4.2, then A^* has the similar decomposition with respect to the dense subset $J\mathfrak{L}_+[\dot{+}]J\mathfrak{L}_-$ of $\mathfrak{H}$, where $J\mathfrak{L}_+$ is a maximal positive subspace of the Krein space $(\mathfrak{H}, [\cdot,\cdot])$. Therefore, A^* has a generalized $\mathcal{C}$-symmetry.

The J-orthogonal sum (4.14) is uniquely determined by the operator $\mathcal{C} = J(I-T)(I+T)^{-1}$, where T is the operator of transition from (2.2) to $\mathfrak{L}_+[\dot{+}]\mathfrak{L}_-$. It follows from (2.5) and (2.6) that $T' = -T$ is the operator of transition from (2.2) to $J\mathfrak{L}_+[\dot{+}]J\mathfrak{L}_-$. According to the proof of Proposition 4.2 and (2.6), the operator

$$\mathcal{C}' = J(I-T')(I+T')^{-1} = J(I+T)(I-T)^{-1} = (I-T)(I+T)^{-1}J = \mathcal{C}^*$$

provides the property of generalized $\mathcal{C}$-symmetry for A^*. □

Corollary 4.4. *If a J-self-adjoint operator A has a generalized $\mathcal{C}$-symmetry, then $\mathbb{C}\setminus\mathbb{R}$ belongs to the continuous spectrum of A (i.e., $\sigma_c(A) \supset \mathbb{C}\setminus\mathbb{R}$).*

Proof. Assume that a J-self-adjoint operator A with generalized $\mathcal{C}$-symmetry has a non-real eigenvalue $z \in \mathbb{C}\setminus\mathbb{R}$. By Proposition 4.2, at least one of the operators $A_\pm$ in (4.13) have the eigenvalue z. However, this is impossible because $A_\pm$ are symmetric in the pre-Hilbert spaces $\mathfrak{L}_\pm$ with scalar products $[\cdot,\cdot]$ and $-[\cdot,\cdot]$, respectively. Therefore, $\sigma_p(A)\cap(\mathbb{C}\setminus\mathbb{R}) = \emptyset$. Further $\sigma_r(A)\cap(\mathbb{C}\setminus\mathbb{R}) = \emptyset$ since A^* has a generalized $\mathcal{C}$-symmetry.

In view of (4.13) and (4.14), $\mathcal{R}(A-zI) \subseteq \mathfrak{L}_+[\dot{+}]\mathfrak{L}_- = \mathfrak{H}' \neq \mathfrak{H}$ for any non-real z. Hence, $\sigma_c(A) \supset \mathbb{C}\setminus\mathbb{R}$. □

Denote by $\mathfrak{H}_\mathcal{C}$ the completion of $\mathcal{D}(\mathcal{C})$ with respect to the positive sesquilinear form

$$(f,g)_\mathcal{C} := [\mathcal{C}f, g] = (J\mathcal{C}f, g), \qquad \forall f,g \in \mathcal{D}(\mathcal{C}).$$

In contrast to the case of $\mathcal{C}$-symmetry (see Section 3), the norm $\|\cdot\|_\mathcal{C}^2 = (\cdot,\cdot)_\mathcal{C}$ is not equivalent to the initial one.

If a J-self-adjoint operator A has a generalized $\mathcal{C}$-symmetry, then

$$(Af,g)_\mathcal{C} = [\mathcal{C}Af, g] = [A\mathcal{C}f, g] = [\mathcal{C}f, Ag] = (f, Ag)_\mathcal{C}, \quad \forall f,g \in \mathcal{D}(A).$$

Hence, A is a symmetric operator in $\mathfrak{H}_\mathcal{C}$.

5. Schrödinger operator with $\mathcal{PT}$-symmetric zero-range potentials

5.1. An example of $\mathcal{C}$-symmetry

Let $\mathfrak{H} = L_2(\mathbb{R})$ and let $J = \mathcal{P}$, where $\mathcal{P}f(x) = f(-x)$ is the space parity operator in $L_2(\mathbb{R})$. In that case, the fundamental decomposition (2.2) of the Krein space $(L_2(\mathbb{R}), [\cdot,\cdot])$ takes the form

$$L_2(\mathbb{R}) = L_2^{\text{even}} \oplus L_2^{\text{odd}}, \tag{5.17}$$

where $\mathfrak{H}_+ = L_2^{\text{even}}$ and $\mathfrak{H}_- = L_2^{\text{odd}}$ are subspaces of even and odd functions in $L_2(\mathbb{R})$.

Consider the one-dimensional Schrödinger operator with singular zero-range potential

$$-\frac{d^2}{dx^2} + V_\gamma, \qquad V_\gamma = i\gamma[< \delta', \cdot > \delta + < \delta, \cdot > \delta'], \quad \gamma \geq 0 \tag{5.18}$$

where δ and δ' are, respectively, the Dirac δ-function and its derivative (with support at 0).

It is easy to verify that $\mathcal{PT}[-\frac{d^2}{dx^2} + V_\gamma] = [-\frac{d^2}{dx^2} + V_\gamma]\mathcal{PT}$, where $\mathcal{T}$ is the complex conjugation operator $\mathcal{T}f(x) = \overline{f(x)}$. Thus the expression (5.18) is $\mathcal{PT}$-symmetric [8, 10].

The operator realization A_γ of $-d^2/dx^2 + V_\gamma$ in $L_2(\mathbb{R})$ is defined as

$$A_\gamma = A_{\text{reg}} \restriction \mathcal{D}(A_\gamma), \quad \mathcal{D}(A_\gamma) = \{f \in W_2^2(\mathbb{R}\backslash\{0\}) \ : \ A_{\text{reg}} f \in L_2(\mathbb{R})\}, \tag{5.19}$$

where the regularization of $-d^2/dx^2 + V_\gamma$ onto $W_2^2(\mathbb{R}\backslash\{0\})$ takes the form

$$A_{\text{reg}} = -\frac{d^2}{dx^2} + i\gamma[< \delta'_{\text{ex}}, \cdot > \delta + < \delta_{\text{ex}}, \cdot > \delta'].$$

Here $-d^2/dx^2$ acts on $W_2^2(\mathbb{R}\backslash\{0\})$ in the distributional sense and

$$< \delta_{\text{ex}}, f > = \frac{f(+0) + f(-0)}{2}, \qquad < \delta'_{\text{ex}}, f > = -\frac{f'(+0) + f'(-0)}{2}$$

for all $f(x) \in W_2^2(\mathbb{R}\backslash\{0\})$.

The operator A_γ defined by (5.19) is a $\mathcal{P}$-self-adjoint operator in the Krein space $L_2(\mathbb{R})$.

According to [4], A_γ has $\mathcal{C}$-symmetry for all $\gamma \neq 2$. The corresponding operator $\mathcal{C} \equiv \mathcal{C}_\gamma$ takes the form

$$\mathcal{C}_\gamma = \alpha_\gamma \mathcal{P} + i\beta_\gamma \mathcal{R}, \tag{5.20}$$

where $\alpha_\gamma = \frac{\gamma^2+4}{|\gamma^2-4|}$ and $\beta_\gamma = \frac{4\gamma}{|\gamma^2-4|}$ are 'hyperbolic coordinates' $(\alpha_\gamma^2 - \beta_\gamma^2 = 1)$ and $\mathcal{R}f(x) = (\text{sign } x)f(x)$.

The operator A_γ is reduced by the decomposition

$$L_2(\mathbb{R}) = \mathfrak{L}_+^\gamma [\dot{+}] \mathfrak{L}_-^\gamma, \qquad \mathfrak{L}_+^\gamma = (I + \mathcal{C}_\gamma)L_2(\mathbb{R}), \quad \mathfrak{L}_-^\gamma = (I - \mathcal{C}_\gamma)L_2(\mathbb{R}). \tag{5.21}$$

Here $\mathcal{C}_\gamma = \mathcal{P}(I - T_\gamma)(I + T_\gamma)^{-1}$, where the operator T_γ

$$T_\gamma = i\frac{2}{\gamma}\mathcal{RP} \quad (\gamma > 2); \qquad T_\gamma = i\frac{\gamma}{2}\mathcal{RP} \quad (\gamma < 2) \tag{5.22}$$

is the operator of transition from the fundamental decomposition (5.17) to (5.21).

Let us assume $\gamma > 2$ (the case $\gamma < 2$ is completely similar). By (2.5), (5.17), and (5.22)

$$\mathfrak{L}_+^\gamma = \{f_{\text{even}} + i\frac{2}{\gamma}\mathcal{R}f_{\text{even}} \ : \ f_{\text{even}} \in L_2^{\text{even}}\}, \quad \mathfrak{L}_-^\gamma = \{f_{\text{odd}} - i\frac{2}{\gamma}\mathcal{R}f_{\text{odd}} \ : \ f_{\text{odd}} \in L_2^{\text{odd}}\}.$$

If $\gamma \to 2$, then the invariant subspaces $\mathfrak{L}_+^\gamma$ and $\mathfrak{L}_-^\gamma$ for A_γ 'tend' to each other and for $\gamma = 2$ they coincide with the hyper-maximal neutral subspace

$$\mathfrak{L}^2 = \{f_{\text{even}} + i\mathcal{R}f_{\text{even}} \ : \ f_{\text{even}} \in L_2^{\text{even}}\} = \{f_{\text{odd}} - i\mathcal{R}f_{\text{odd}} \ : \ f_{\text{odd}} \in L_2^{\text{odd}}\}.$$

The spectrum of A_2 coincides with $\mathbb{C}$ and any point $z \in \mathbb{C}\backslash\mathbb{R}_+$ is an eigenvalue of A_2 [4]. Therefore, the $\mathcal{P}$-self-adjoint operator A_2 has no $\mathcal{C}$-symmetry as well as generalized $\mathcal{C}$-symmetry (see Corollary 4.4).

The obtained result is in accordance with the 'physical' concept of $\mathcal{PT}$-symmetry. Indeed, it is easy to verify that the $\mathcal{PT}$-symmetry of (5.18) is unbroken for $\gamma \neq 2$ and broken for $\gamma = 2$ in the sense of [7] – [12]. According to the general concepts of the theory [10, 11], the existence of a hidden $\mathcal{C}$-symmetry is an intrinsic property of unbroken $\mathcal{PT}$-symmetry.

5.2. An example of generalized $\mathcal{C}$-symmetry

Let us consider a Hilbert space $\mathfrak{H} = \bigoplus_1^\infty L_2(\mathbb{R})$ with elements $\mathfrak{f} = \{f_1, f_2, \ldots\}$, where $f_i \in L_2(\mathbb{R})$ and the scalar product $(\cdot,\cdot)_{\mathfrak{H}}$ is defined by the formula

$$(\mathfrak{f}, \mathfrak{g})_{\mathfrak{H}} = \sum_{i=1}^{\infty}(f_i, g_i)_{L_2(\mathbb{R})}.$$

The operator $J\mathfrak{f} = \{\mathcal{P}f_1, \mathcal{P}f_2, \ldots\}$ is a fundamental symmetry in $\mathfrak{H}$ (i.e., $J = J^*$ and $J^2 = I$) and $(\mathfrak{H}, [\cdot,\cdot])$ endowed by the indefinite metric (2.3) is a Krein space.

The operator

$$A_{\vec{\gamma}}\mathfrak{f} = \{A_{\gamma_1}f_1, A_{\gamma_1}f_2, \ldots\}, \qquad \vec{\gamma} = \{\gamma_i\}, \quad \gamma_i \geq 0,$$

where A_{γ_i} are defined by (5.19), is J-self-adjoint in $\mathfrak{H}$. If 2 is not a limit point for the set $\vec{\gamma}$ (i.e., 2 does not belong to the closure of $\vec{\gamma}$), the operator $A_{\vec{\gamma}}$ has $\mathcal{C}$-symmetry with $\mathcal{C} = \bigoplus_1^\infty \mathcal{C}_{\gamma_i}$ where $\mathcal{C}_{\gamma_i}$ are given by (5.20).

Let us assume that $2 \neq \gamma_i$ ($i \in \mathbb{N}$) and there exists a subsequence γ_j of $\vec{\gamma}$ such that $\gamma_j \to 2$. In that case the operator $T = \bigoplus_1^\infty T_{\gamma_i}$ with T_{γ_i} determined by (5.22) satisfies the conditions (2.6) and $\|T\| = 1$. Therefore, T is the operator of transition from the fundamental decomposition of $(\mathfrak{H}, [\cdot,\cdot])$ to the J-orthogonal sum (2.4) that is dense in $\mathfrak{H}$. Since $\|T\| = 1$ the subspaces $\mathfrak{L}_\pm$ in (2.4) are determined by the unbounded operator

$$\mathcal{C} = \bigoplus_1^\infty \mathcal{C}_{\gamma_i} = J(I - T)(I + T)^{-1}.$$

Thus the operator $A_{\vec{\gamma}}$ has a generalized $\mathcal{C}$-symmetry.

Acknowledgments

The author thanks Dr. Uwe Günther for valuable discussions. The supports by DFG 436 UKR 113/88/0-1 and DFFD of Ukraine 14.01/003 research projects are gratefully acknowledged.

References

[1] Z. Ahmed, *$\mathcal{P}$, $\mathcal{T}$, $\mathcal{PT}$, and $\mathcal{CPT}$-invariance of Hermitian Hamiltonians.* Phys. Lett. A. **310** (2003), 139–142.

[2] S. Albeverio, S.M. Fei, and P. Kurasov, *Point interactions: $\mathcal{PT}$-Hermiticity and reality of the spectrum.* Lett. Math. Phys. **59** (2002), 227–242.

[3] S. Albeverio and S. Kuzhel, *Pseudo-Hermiticity and theory of singular perturbations.* Lett. Math. Phys. **67** (2004), 223–238.

[4] S. Albeverio and S. Kuzhel, *One-dimensional Schrödinger operators with $\mathcal{P}$-symmetric zero-range potentials.* J. Phys. A. **38** (2005) 4975-4988.

[5] T.Ya. Azizov and I.S. Iokhvidov *Linear Operators in Spaces with Indefinite Metric.* Wiley, Chichester, 1989.

[6] A. Batal and A. Mostafazadeh, *Physical aspects of pseudo-Hermitian and $\mathcal{PT}$-symmetric quantum mechanics.* J. Phys. A. **37** (2004), 11645–11679.

[7] C.M. Bender, *Calculating the $\mathcal{C}$ operator in $\mathcal{PT}$-quantum mechanics.* Czechoslovak J. Phys. **54** (2004), 1027–1038.

[8] C.M. Bender, *Making sense of non-Hermitian Hamiltonians.* Rep. Prog. Phys. **70** (2007), 947–1018.

[9] C.M. Bender and S. Boettcher, *Real spectra in non-Hermitian Hamiltonians having $\mathcal{PT}$-symmetry.* Phys. Rev. Lett. **80** (1998), 5243–5246.

[10] C.M. Bender, D.C. Brody, and H.F. Jones, *Complex extension of quantum mechanics.* Phys. Rev. Lett. **89** (2002), 401–405.

[11] C.M. Bender, D.C. Brody, and H.F. Jones, *Must a Hamiltonian be Hermitian?* Amer. J. Phys., **71** (2003), 1095–1102.

[12] C.M. Bender and H.F. Jones, *Semiclassical calculation of the $\mathcal{C}$ operator in $\mathcal{PT}$-quantum mechanics.* Phys. Lett. A. **328** (2004), 102–109.

[13] C.M. Bender and B. Tan, *Calculation of the hidden symmetry operator for a $\mathcal{PT}$-symmetric square well.* J. Phys. A. **39** (2006), 1945–1953.

[14] P.A.M. Dirac, *The physical interpretation of quantum mechanics.* Proc. Roy. Soc. London A. **180** (1942), 1-40.

[15] P. Dorey, C. Dunning, and R. Tateo, *Spectral equivalence, Bethe ansatz, and reality properties in $\mathcal{PT}$-symmetric quantum mechanics.* J. Phys. A: Math. Gen. **34** (2001), 5679–5704.

[16] H.B. Geyer, W.D. Heiss, and F.G. Scholtz, *Non-Hermitian Hamiltonians, metric, other observables and physical implications.* ArXiv:0710.5593v1 [quant-ph] 30 Oct. 2007.

[17] U. Gunther, F. Stefani, and M. Znojil, *MHD α^2-dynamo, Squire equation and $\mathcal{PT}$-symmetric interpolation between square well and harmonic oscillator.* J. Math. Phys. **46** (2005), 063504–063526.

[18] S. Gupta, *Theory of longitudinal photons in quantum electrodynamics.* Proc. Phys. Soc. **63** (1950), 681–691.

[19] H.F. Jones, *On pseudo-Hermitian Hamiltonians and their Hermitian counterparts.* J. Phys. A. **38** (2005) 1741–1746.

[20] R. Kretschmer and L. Szymanowski, *Quasi-Hermiticity in infinite-dimensional Hilbert spaces.* Phys. Lett. A. **325** (2004) 112–122.

[21] A. Kuzhel and S. Kuzhel, *Regular Extensions of Hermitian Operators.* VSP, Utrecht, 1998.

[22] H. Langer, *Zur Spektraltheorie $\mathcal{J}$-selbstadjungierter Operatoren.* Math. Ann. **146** (1962), 60–85.

[23] H. Langer, *Maximal dual pairs of invariant subspaces of J-self-adjoint operators.* Mat. Zametki **7** (1970), 443–447. (Russian)

[24] H. Langer, *Invariant subspaces for a class of operators in spaces with indefinite metric.* J. Funct. Anal. **19** (1975), 232–241.

[25] H. Langer and C. Tretter, *A Krein space approach to $\mathcal{PT}$-symmetry.* Czechoslovak J. Phys. **54** (2004), 1113–1120.

[26] G. Lévai and M. Znojil, *Systematic search for $\mathcal{PT}$-symmetric potentials with real energy spectra.* J. Phys. A. **33** (2000), 7165–7180.

[27] A. Mostafazadeh, *Pseudo-Hermiticity versus $\mathcal{PT}$-symmetry: the necessary condition for the reality of the spectrum of a non-Hermitian Hamiltonian.* J. Math. Phys. **43** (2002), 205–214.

[28] A. Mostafazadeh, *Pseudo-Hermiticity and generalized $\mathcal{PT}$- and $\mathcal{CPT}$-symmetries.* J. Math. Phys. **44** (2003), 979–989.

[29] A. Mostafazadeh, *Krein-space formulation of $\mathcal{PT}$-symmetry, $\mathcal{CPT}$-inner products, and pseudo-Hermiticity.* J. Math. Phys. **44** (2003), 979–989.

[30] W. Pauli, *On Dirac's new method of field quantization.* Rev. Mod. Phys. **15** (1943), 175–207.

[31] F.G. Scholtz, H.B. Geyer, and F.J.W. Hahne, *Quasi-Hermitian operators in quantum mechanics and the variational principle.* Ann. Phys. (N.Y.) **213** (1992).

[32] T. Tanaka, *General aspects of $\mathcal{PT}$-symmetric and $\mathcal{P}$-self-adjoint quantum theory in a Krein space.* J. Phys. A. **39** (2006), 14175–14203.

[33] B. Thaller, *The Dirac Equation.* Springer-Verlag, Berlin 1992.

[34] M. Znojil, *$\mathcal{PT}$-symmetric harmonic oscillators.* Phys. Lett. A. **259** (1999), 220–223.

S. Kuzhel
Institute of Mathematics
National Academy of Sciences of Ukraine
3 Tereshchenkivska Str
01601 Kiev, Ukraine
e-mail: kuzhel@imath.kiev.ua

Operator Theory:
Advances and Applications, Vol. 190, 387–395

Nonisospectral Flows on Self-adjoint, Unitary and Normal Semi-infinite Block Jacobi Matrices

Oleksii Mokhonko

Abstract. The article gives an overview of recent results in the theory of difference-differential lattices generated by various forms of the Lax equation $\dot{\boldsymbol{J}}(t) = \Phi(\boldsymbol{J}(t), t) + [\boldsymbol{J}(t), A(\boldsymbol{J}(t), t)]$ of the following type. It is required that $\boldsymbol{J}(t) : \mathbf{l}_2 \to \mathbf{l}_2$ be able to be mapped into a self-adjoint, or unitary, or normal operator $L(t)$ of multiplication by an independent variable in separable Hilbert space $L^2(\mathbb{C}, d\rho(\cdot, t))$. Here $d\rho$ is a probability measure with infinite compact support defined on the Borel σ-algebra $\mathfrak{B}(\mathbb{C})$ (spectral measure of $L(t)$).

The article presents an algorithm that solves such lattices via the Inverse Spectral Problem. For the case of unitary $\boldsymbol{J}(t)$, three applications (in terms of the Verblunsky coefficients) are presented.

The results of the article are closely related to the classical theory of Jacobi matrices (Toda lattice, etc.) and the OPUC theory (Schur flows, etc.).

Mathematics Subject Classification (2000). Primary 47A70; Secondary 46A11.
Keywords. Jacobi field, spectral measure, orthogonal polynomial, differential lattice, inverse spectral problem.

1. Introduction

This article gives a brief sketch of the recently-born theory of difference-differential lattices generated by various forms of the Lax equation $\dot{\boldsymbol{J}}(t) = \Phi(\boldsymbol{J}(t), t) + [\boldsymbol{J}(t), A(\boldsymbol{J}(t), t)]$ of the following type. It is required that $\boldsymbol{J}(t) : \mathbf{l}_2 \to \mathbf{l}_2$ be able to be mapped into a self-adjoint, or unitary, or normal operator $L(t)$ of multiplication by an independent variable in separable Hilbert space $L^2(\mathbb{C}, d\rho(\cdot, t))$. The probability measure $d\rho$ has an infinite compact support and is defined on the Borel σ-algebra $\mathfrak{B}(\mathbb{C})$. The operator L is bounded. The Lax equation is treated in the weak (coordinate-wise) sense. These restrictions define a class of difference-differential lattices that can be integrated using the method presented here.

The results of this article are based on the classical theory of Jacobi matrices (see Mark Krein's article [1]) and are inspired by recent advances in the spectral theory of block unitary and normal matrices (see [2, 3]). The basic ideas for the self-adjoint case can be found in [4]. A complete proof for the unitary case is presented in [5]. The normal case is analyzed in [6].

The most simple (but not so trivial) case of a self-adjoint L (this case will be referred to as the "self-adjoint case") was extensively studied during last few decades. A classical lattice of this type is the well-known Toda lattice

$$\begin{aligned} \dot{a}_n &= \tfrac{1}{2} a_n (b_{n+1} - b_n) \\ \dot{b}_n &= a_n^2 - a_{n-1}^2 \end{aligned} \tag{1.1}$$

Here, $n = 0, 1, \ldots; t \in [0, \infty)$, and $a_{-1} = 0$ is the boundary condition. The corresponding Lax equation has the form

$$\dot{L}(t) = [L, A] = LA - AL. \tag{1.2}$$

Here

$$A = \begin{pmatrix} 0 & -a_0(t) & & & & \cdot \\ a_0(t) & 0 & -a_1(t) & & & \cdot \\ & a_1(t) & 0 & -a_2(t) & & \cdot \\ & & a_2(t) & 0 & -a_3(t) & \cdot \\ \cdot & \cdot & \cdot & \cdot & \cdot & \cdot \end{pmatrix}$$

The initial boundary value problem is stated as follows: *find the solution* $(a(t), b(t))$ *subject to to the given initial data* $(a(0), b(0))$ *in the class*

$$K[0,T] = \{a_n(t), b_n(t) \in C^1_{[0,T] \to \mathbb{R}} \mid \forall t \in [0,T],$$

$$a_n(t) > 0, \sup_{n \in \mathbb{N}_0} \max_{t \in [0,T]} \max(a_n(t), b_n(t)) < +\infty\}.$$

The solution can be found in [7]. The main idea is to connect the Toda lattice with the measure transformation

$$d\rho(\lambda, t) = C(t) \cdot e^{\lambda t} d\rho(\lambda, 0),$$

where $\lambda \in \mathbb{R},\ t \in [0, \infty)$, and C is a normalizing factor.

In [8], one can find examples of the mapping + multiplication measure transformations for the case where the corresponding chains of differential equations can efficiently be found and integrated. Article [8] deals with the following more general Lax equation:

$$\dot{L}(t) = \Phi(L(t), t) + [L(t), \langle \Phi(L(t), t) D_{L(t)} + \frac{1}{2} \Psi(L(t), t) \rangle]$$

Here $D_L = \frac{\partial}{\partial \lambda}$. For the matrix $A = (a_{jk})_{j,k=0}^{\infty}$, the matrix $\langle A \rangle$ is built in the following way: $\langle A \rangle_{jk} = a_{jk}$ if $j > k$, $\langle A \rangle_{jk} = 0$ if $j = k$, and $\langle A \rangle_{jk} = -a_{kj}$ if $j < k$.

Analogous results for the case of a unitary multiplication operator L can be found in the OPUC theory. In [9], one can find the so-called Schur flow. The following theorem can be found in that article:

Theorem 1.1. *Let* $\alpha_n(t)$, $n \in \mathbb{Z}_+$, *be a sequence of complex-valued functions,* $|\alpha_n(t)| < 1$ *for* $t \geqslant 0$ *and* $\alpha_{-1} = -1$. *The following three statements are equivalent:*

1. α_n *solve the Schur flow equations*
$$\alpha_n'(t) = (1 - |\alpha_n(t)|^2)(\alpha_{n+1}(t) - \alpha_{n-1}(t)), \qquad t > 0;$$
2. *The CMV matrices* $\mathcal{C}(t)$ *satisfy the Lax equation*
$$\mathcal{C}'(t) = [B, \mathcal{C}];$$
(B *is a matrix, whose exact form we do not give here. Instead, we will later represent a more abstract construction from which this form results as a very simple technical side effect.)*
3. *Due to Verblunsky's theorem, there exists a one-to-one correspondence between the coefficients* $\alpha_n(t)$ *and the nontrivial probability measures on the unitary circle* $\mathbb{T}$. *The third statement of the theorem says that the measure* $d\mu(\zeta, t)$, *having* $\alpha_n(t)$ *as its Verblunsky coefficients, satisfies the relation*
$$d\mu(\zeta, t) = C(t)e^{t(\zeta+\zeta^{-1})}d\mu(\zeta, 0),$$
where C *is a normalizing factor.*

All these cases are quite similar: they use a common spectral technique.

Now, let us note that on $\mathbb{R}$, the operator of multiplication by an independent variable L is self-adjoint due to the equality $z = \bar{z}, z \in \mathbb{R}$). In $L^2(\mathbb{T}, d\rho)$, it is unitary because $\bar{z} = z^{-1} \quad \forall z \in \mathbb{T} = \{z \in \mathbb{C} : |z| = 1\}$. And in $L^2(\mathbb{C}, d\rho)$, it is normal because $z \cdot \bar{z} = |z|^2 = \bar{z} \cdot z$. No other cases exist. This observation gives us an idea to unite all three cases (difference-differential lattices generated by the Lax equation for a self-adjoint, unitary and normal L) into one theory with a single spectral approach.

A breakthrough in this direction became possible due to articles [2], [3]. The most significant result of these articles is the way of building the orthonormal basis in $L^2(\mathbb{T}, d\rho)$ and $L^2(\mathbb{C}, d\rho)$. In this basis, the operator $I^{-1}LI$ has a convenient block structure. The mapping I is built in such a way that the matrices of L and $I^{-1}LI$ coincide. This observation furnishes an opportunity to efficiently build difference-differential lattices starting from an abstract measure transformation.

The united theory still has practical applications. The corresponding lattices now have matrix-valued coefficients and arguments. The theory can be represented in a coordinate-wise form if necessary.

2. The unitary case: theory and examples

Consider a three-diagonal block Jacobi matrix

$$L(t) = \begin{pmatrix} b_0(t) & c_0(t) & & & \cdot \\ a_0(t) & b_1(t) & c_1(t) & & \cdot \\ & a_1(t) & b_2(t) & c_2(t) & \cdot \\ \cdot & \cdot & \cdot & \cdot & \cdot \end{pmatrix} \tag{2.1}$$

in the space

$$\mathbf{l}_2 = \mathcal{H}_0 \oplus \mathcal{H}_1 \oplus \mathcal{H}_2 \oplus \cdots, \quad \mathcal{H}_0 = \mathbb{C}, \quad \mathcal{H}_n = \mathbb{C}^2, \qquad n \geq 1. \tag{2.2}$$

$\mathbf{l}_2$ is a Hilbert space with natural scalar product: for $f, g \in \mathbf{l}_2$ with coordinates in the standard orthonormal basis

$$e_0 = \left(1, \begin{pmatrix}0\\0\end{pmatrix}, \begin{pmatrix}0\\0\end{pmatrix}, \dots\right),$$

$$e_{n,1} = \left(0, \begin{pmatrix}0\\0\end{pmatrix}, \dots, \begin{pmatrix}0\\0\end{pmatrix}, \underset{n}{\begin{pmatrix}1\\0\end{pmatrix}}, \begin{pmatrix}0\\0\end{pmatrix} \dots\right),$$

$$e_{n,2} = \left(0, \begin{pmatrix}0\\0\end{pmatrix}, \dots, \begin{pmatrix}0\\0\end{pmatrix}, \underset{n}{\begin{pmatrix}0\\1\end{pmatrix}}, \begin{pmatrix}0\\0\end{pmatrix} \dots\right),$$

the norm and the scalar product are defined as

$$\|f\|_{\mathbf{l}_2}^2 = \sum_{n=0}^{\infty} \|f_n\|_{\mathcal{H}_n}^2, \quad (f, g)_{\mathbf{l}_2} = \sum_{n=0}^{\infty} (f_n, g_n)_{\mathcal{H}_n}$$

Here,

$$\|f_n\|_{\mathcal{H}_n}^2 = \|f_n\|_{\mathbb{C}^2}^2 = |f_{n,1}|^2 + |f_{n,2}|^2,$$

$$(f_n, g_n)_{\mathcal{H}_n} = (f_n, g_n)_{\mathbb{C}^2} = f_{n,1} \cdot \overline{g_{n,1}} + f_{n,2} \cdot \overline{g_{n,2}}.$$

Assume that in the standard orthonormal basis, L has the following form:

$$L(t) = \left(\begin{array}{c|cc|cc|cc}
b_{0;22} & c_{0;21} & c_{0;22} & & & & \\
\hline
a_{0;12} & b_{1;11} & b_{1;12} & 0 & 0 & & \\
0 & b_{1;21} & b_{1;22} & c_{1;21} & c_{1;22} & & \\
\hline
 & a_{1;11} & a_{1;12} & b_{2;11} & b_{2;12} & 0 & 0 \\
 & 0 & 0 & b_{2;21} & b_{2;22} & c_{2;21} & c_{2;22} \\
\hline
 & & & a_{2;11} & a_{2;12} & b_{3;11} & b_{3;12} \\
 & & & 0 & 0 & b_{3;21} & b_{3;22}
\end{array}\right) \tag{2.3}$$

All the entries are assumed to be continuously differentiable functions on the interval $[0, T]$. Consider the following polynomials in λ :

$$\Phi(\lambda, t) = \sum_{j=0}^{l} \varphi_j(t)\lambda^j, \quad \varphi_j(t) \in C^1_{[0,\infty)\to\mathbb{R}}, \lambda \in \mathbb{T}, \tag{2.4}$$

$$\Psi(\lambda, t) = \sum_{j=0}^{m} \psi_j(t)\lambda^j, \quad \psi_j(t) \in C^1_{[0,\infty)\to\mathbb{C}}, \lambda \in \mathbb{T}. \tag{2.5}$$

Denote by D the differential operator $\frac{\partial}{\partial\lambda}$ and consider the following operators:

$$\Omega = \Phi(L(t), t) \circ D,$$

$$\widehat{\Omega} = \Phi(L^*(t), t) \circ D,$$

$$\Psi = \Psi(L(t), t), \quad \Xi = -\Omega - \widehat{\Omega}^* - \Psi.$$

$$I = \left(\begin{array}{c|cc|cc} \frac{1}{2}\Xi_{0,1;0,1} & \Xi_{0,1;1,1} & \Xi_{0,1;1,2} & \Xi_{0,1;2,1} & \Xi_{0,1;2,2} \\ \hline 0 & \frac{1}{2}\Xi_{1,1;1,1} & \Xi_{1,1;1,2} & \Xi_{1,1;2,1} & \Xi_{1,1;2,2} \\ 0 & 0 & \frac{1}{2}\Xi_{1,2;1,2} & \Xi_{1,2;2,1} & \Xi_{1,2;2,2} \\ \hline 0 & 0 & 0 & \frac{1}{2}\Xi_{2,1;2,1} & \Xi_{2,1;2,2} \\ 0 & 0 & 0 & 0 & \frac{1}{2}\Xi_{2,2;2,2} \end{array}\right) \qquad (2.6)$$

Consider the following differential equation:

$$\frac{d}{dt}L(t) = \Phi(L(t),t) + [L(t), \Omega + I + \frac{1}{2}\Psi]. \qquad (2.7)$$

Here $[A, B] = AB - BA$. This Lax equation can be rewritten as the lattice in coordinate-wise form by using the matrix-variables a_n, b_n, c_n. The Cauchy problem for the differential equation (2.7) can be stated as follows.

Suppose we have a bounded unitary block Jacobi matrix L_0 with entries $a_{n;11} > 0$, $c_{n,22} > 0$. Find $L(t)$, $t \in [0;T]$, with continuously differentiable entries such that $L(t)$ is a solution of (2.7) for $t \in [0,T]$ where T depends only on the initial condition

$$L(0) = L_0 \qquad (2.8)$$

and functions Φ, Ψ (see (2.4), (2.5)).

Theorem 2.1. *A solution of the Cauchy problem (2.7), (2.8) exists and can be found in the following way.*

Let $\rho(\cdot,0)$ be the spectral measure of the Jacobi matrix L_0. Denote $M =$ $\operatorname{supp}\rho(\cdot,0) \subset \mathbb{T}$. Consider the Cauchy problem

$$\frac{d\lambda(t)}{dt} = \Phi(\lambda(t),t), \quad \lambda(0) = \mu, \qquad \mu \in M, \quad t \geq 0. \qquad (2.9)$$

From the standard theory of differential equations, it is well known that one can choose $T > 0$ such that for every $\mu \in M$, there exists a unique solution $\lambda(\cdot,\mu)$ of the Cauchy problem (2.9) defined on the interval $[0,T]$.

For every fixed $t \in [0,T]$, consider the mapping

$$\begin{array}{rcl} \omega_t : M & \longrightarrow & \mathbb{T} \\ \mu & \longmapsto & \lambda(t,\mu) \end{array} \qquad (2.10)$$

and construct the measure q

$$\tilde{\rho}(\Delta,t) = \rho(\omega_t^{-1}(\Delta),0), \qquad \Delta \in \mathfrak{B}(\mathbb{T}). \qquad (2.11)$$

Here $\omega_t^{-1}(\Delta)$ is the full preimage of the set Δ under the mapping ω_t. Since $\lambda(t,\mu)$ is continuous, we obtain that $\tilde{\rho}(\cdot,t) \in \mathfrak{M}$. Here $\mathfrak{M}$ is the set of all non-zero finite measures on $\mathfrak{B}(\mathbb{T})$ with infinite compact support.

Let us consider the following partial differential equation:

$$\frac{\partial s(\lambda,t)}{\partial\lambda}\Phi(\lambda,t) + \frac{\partial s(\lambda,t)}{\partial t} = \Psi(\lambda,t)s(\lambda,t), \qquad (2.12)$$

$$s(\lambda,0) = 1, \qquad \lambda \in \mathbb{T}, \quad t \geq 0.$$

Let $s(\lambda, t)$ be its nonnegative solution. Let us build a new measure

$$\rho(\Delta, t) = \int_{\Delta} s(\lambda, t) d\tilde{\rho}(\lambda, t), \qquad \Delta \in \mathfrak{B}(\mathbb{T}). \tag{2.13}$$

It is obvious that $\rho(\cdot, t) \in \mathfrak{M}$. Thus, the functions Φ and Ψ define, through equations (2.9), (2.12) and equalities (2.11), (2.13), some measure transformation of type "mapping + multiplication":

$$\mathfrak{M} \ni \rho(\cdot, 0) \mapsto \rho(\cdot, t) \in \mathfrak{M}, \qquad t \in [0, T]. \tag{2.14}$$

The last step is to reconstruct $L(t)$ from its spectral measure $\rho(\cdot, t), t \in [0, T]$ by solving the Inverse Spectral Problem. This can be done as follows.

Consider the following family of functions:

$$1, \quad z, \quad \bar{z} = \frac{1}{z}, \quad z^2, \quad \bar{z}^2 = \frac{1}{z^2}, \quad \dots \tag{2.15}$$

Let us build the orthonormal basis

$$P_0(z, t), P_{1,1}(z, t), P_{1,2}(z, t), P_{2,1}(z, t), P_{2,2}(z, t), \dots$$

in the space $L^2(\mathbb{T}, d\rho(\cdot, t))$ (by using the standard Shmidt orthogonalization procedure). $L(t)$ is the operator of multiplication by an independent variable in the space $L^2(\mathbb{T}, d\rho(\cdot, t))$. Thus, the solution can be represented as follows:

$$L_{j,\alpha;k,\beta}(t) = \int_{\mathbb{T}} \lambda P_{k,\beta}(\lambda, t) \overline{P_{j,\alpha}(\lambda, t)} d\rho(\lambda, t).$$

Theorem 2.2. *Let the coefficients $\varphi_j(t), \psi_j(t)$ in (2.4), (2.5) be analytical functions in the neighbourhood of the interval $[0; T]$. Then the solution of the Cauchy problem (2.7), (2.8) is unique.*

Corollary 2.3. *Let $\Phi(\lambda, t) \equiv 0, \Psi(\lambda, t) = \lambda + \frac{1}{\lambda}$. Then we obtain the above-mentioned Leonid Golinskii's case (see [9]):*

Proof. At $\Phi(\lambda, t) \equiv 0$ we have $\Omega = \Phi(L(t), t) \circ D(t) = \mathbb{O}$, $\widehat{\Omega} = \Phi(L^*(t), t) \circ D(t) = \mathbb{O}$, $\Psi(L(t), t) = L + L^*$, $\Xi = -\Omega - \widehat{\Omega}^* - \Psi = -\Psi = -L - L^*$. Substitute this into (2.7):

$$\frac{d}{dt} L(t) = \Phi(L(t), t) + [L(t), \Omega + I + \frac{1}{2}\Psi] = [L, B],$$

where $B = I + \frac{1}{2}\Psi = \frac{(L+L^*)_- - (L+L^*)_+}{2}$. □

Corollary 2.4. *Let $\Phi(\lambda, t) \equiv 0, \Psi(\lambda, t) = \lambda$. Then we obtain a two-dimensional analog of the Toda lattice for the unitary case.*

Proof. At $\Phi(\lambda, t) \equiv 0$ we have $\Omega = \Phi(L(t), t) \circ D(t) = \mathbb{O}$, $\widehat{\Omega} = \Phi(L^*(t), t) \circ D(t) = \mathbb{O}$, $\Psi(L(t), t) = L(t)$, $\Xi = -\Omega - \widehat{\Omega}^* - \Psi = -\Psi = -L$.

Substitute this into (2.7):

$$\frac{d}{dt} L(t) = \Phi(L(t), t) + [L(t), \Omega + I + \frac{1}{2}\Psi] = [L, A],$$

where

$$A = I + \frac{1}{2}\Psi = \frac{1}{2}\left(\begin{array}{c|cc|cc} 0 & -c_{0;10} & -c_{0;11} & 0 & 0 \\ \hline a_{0;01} & 0 & -b_{1;01} & 0 & 0 \\ 0 & b_{1;10} & 0 & -c_{1;10} & -c_{1;11} \\ \hline 0 & a_{1;00} & a_{1;01} & 0 & -b_{2;01} \\ 0 & 0 & 0 & b_{2;10} & 0 \end{array}\right)$$

Note that A is not uniquely determined: the differential equation does not change if we replace A with $A+T$ where T is an arbitrary operator commuting with L.

Now, if we rewrite A and L in terms of the Verblunsky coefficients,

$$L = \mathcal{C}(\{\alpha_n\}) = \left(\begin{array}{c|cc|cc} \bar\alpha_0 & \bar\alpha_1\rho_0 & \rho_0\rho_1 & & \\ \hline \rho_0 & -\bar\alpha_1\alpha_0 & -\alpha_0\rho_1 & 0 & 0 \\ 0 & \bar\alpha_2\rho_1 & -\bar\alpha_2\alpha_1 & \bar\alpha_3\rho_2 & \rho_2\rho_3 \\ \hline & \rho_1\rho_2 & -\alpha_1\rho_2 & -\bar\alpha_3\alpha_2 & -\alpha_2\rho_3 \\ & 0 & 0 & \bar\alpha_4\rho_3 & -\bar\alpha_4\alpha_3 \end{array}\right) \tag{2.16}$$

then we obtain the "Toda" flow for the unitary case:

$$\alpha_n'(t) = (|\alpha_n|^2 - 1)\alpha_{n-1}. \tag{2.17}$$

There are many ways to prove this. The simplest one is to slightly modify [9, Theorem 2]. Similarly, the next example is obtained. □

Corollary 2.5. *Let $\Phi(\lambda,t) \equiv 0, \Psi(\lambda,t) = \lambda^2$. Then we obtain the analog of the Kac-van Moerbeke lattice.*

Proof. Recall that the classical Kac-van Moerbeke lattice for a self-adjoint L has the following form:

$$\dot x_n(t) = x_n(x_{n+1} - x_{n-1}), \qquad n = 0,1,\ldots; \quad x_{-1} = 0. \tag{2.18}$$

In our case, the Lax equation has the same form as that in the previous example with $A = I + \frac{1}{2}\Psi$ where $\Psi = L^2$ and $\Xi = -L^2$. In terms of the Verblunsky coefficients, the Kac-van Moerbeke flow is:

$$\alpha_n'(t) = (1 - |\alpha_n|^2)(\alpha_{n+1}\bar\alpha_n\alpha_{n-1} - \alpha_{n-2} + |\alpha_{n-1}|^2(\alpha_n + \alpha_{n-2})). \tag{2.19}$$

The theory presented above provides a possibility ofbuilding many other difference-differential flows. □

3. Some generalizations

3.1. Lattices generated by normal operators

For the case of a normal operator L of multiplication by an independent variable, a similar result holds. $L(t)$ has the following block structure:

$$L(t) = \begin{pmatrix} b_0(t) & c_0(t) & & & \cdot \\ a_0(t) & b_1(t) & c_1(t) & & \cdot \\ & a_1(t) & b_2(t) & c_2(t) & \cdot \\ \cdot & \cdot & \cdot & \cdot & \cdot \end{pmatrix}; \tag{3.1}$$

$$a_n = \underbrace{\begin{bmatrix} a_{n;0,0} & a_{n;0,1} & a_{n;0,2} & \cdots & a_{n;0,n-1} & a_{n;0,n} \\ 0 & a_{n;1,1} & a_{n;1,2} & \cdots & a_{n;1,n-1} & a_{n;1,n} \\ 0 & 0 & a_{n;2,2} & \cdots & a_{n;2,n-1} & a_{n;2,n} \\ \vdots & \vdots & \vdots & \ddots & \vdots & \vdots \\ 0 & 0 & 0 & \cdots & a_{n;n-1,n-1} & a_{n;n-1,n} \\ 0 & 0 & 0 & \cdots & 0 & a_{n;n,n} \\ 0 & 0 & 0 & \cdots & 0 & 0 \end{bmatrix}}_{n+1} \left.\vphantom{\begin{bmatrix} a\\a\\a\\ \vdots\\a\\a\\a \end{bmatrix}}\right\} n+2,$$

$$c_n = \underbrace{\begin{bmatrix} c_{n;0,0} & c_{n;0,1} & 0 & \cdots & 0 & 0 \\ c_{n;1,0} & c_{n;1,1} & c_{n;1,2} & \cdots & 0 & 0 \\ c_{n;2,0} & c_{n;2,1} & c_{n;2,2} & \cdots & 0 & 0 \\ \vdots & \vdots & \vdots & \ddots & \vdots & \vdots \\ c_{n;n-1,0} & c_{n;n-1,1} & c_{n;n-1,2} & \cdots & c_{n;n-1,n} & 0 \\ c_{n;n,0} & c_{n;n,1} & c_{n;n,2} & \cdots & c_{n;n,n} & c_{n;n,n+1} \end{bmatrix}}_{n+2} \left.\vphantom{\begin{bmatrix} a\\a\\a\\ \vdots\\a\\a \end{bmatrix}}\right\} n+1,$$

$$a_{n;0,0} > 0, \quad a_{n;1,1} > 0, \dots, a_{n;n,n} > 0;$$

$$c_{n;0,1} > 0, \quad c_{n;1,2} > 0, \dots, c_{n;n,n+1} > 0, \qquad n \in \mathbb{N}_0.$$

The Lax equation is the same as that for the unitary case:

$$\frac{d}{dt}L(t) = \Phi(L(t),t) + [L(t), \Omega + I + \frac{1}{2}\Psi].$$

It is necessary to note that, for the time being, the author cannot prove the uniqueness theorem for the Cauchy problem in the normal case. Normal operators can have quite a complex spectrum: there can be interior points. This is the main source of difficulties in this case.

3.2. Difference-differential lattices on Fock spaces

There exists *Projective Spectral Theorem* (see [10]), which builds the unitary isomorphism between the Fock space $\mathcal{F}(H) = \bigoplus_{n=0}^{\infty} \mathcal{F}_n(H)$ and $L^2(H_-, d\rho)$. Instead of $L(t)$, we have here a commutative Jacobi field $(L(\varphi))_{\varphi\in H}$. Here H is a Hilbert space.

It is well known that $\dim \mathcal{F}_n(H) = \binom{n+\dim H-1}{n}$.

If $\dim H = 1$ $(H = \mathbb{R})$, we obtain the *self-adjoint case*, because

$$(\dim \mathcal{F}_n(H))_{n=0}^{\infty} = (1,1,1,\dots).$$

If $\dim H = 2$ $(H = \mathbb{R}^2)$, we obtain the *normal case*, because

$$(\dim \mathcal{F}_n(H))_{n=0}^{\infty} = (1,2,3,4,\dots).$$

Completely new results can be obtained for higher dimensions: for $H = \mathbb{R}^3$, we have

$$(\dim \mathcal{F}_n(H))_{n=0}^{\infty} = (1,3,6,10,15,\dots).$$

And for $H = \mathbb{R}^4$, we have

$$(\dim \mathcal{F}_n(H))_{n=0}^{\infty} = (1, 4, 10, 20, 35, \ldots).$$

The corresponding investigations are to be carried out in the nearest future. For details, see [11].

References

[1] M. Krein, *Infinite J-matrices and matrix moment problem.* Reports Ac. Sci. USSR **69** (1949), no. 2, 125–128 (in Russian).

[2] Yurij M. Berezansky and Mykola E. Dudkin, *The direct and inverse spectral problems for the block Jacobi type unitary matrices.* Methods of Functional Analysis and Topology **11** (2005), no. 4, 327–345.

[3] Yurij M. Berezansky and Mykola E. Dudkin, *The complex moment problem and direct and inverse spectral problems for the block Jacobi type bounded normal matrices.* Methods of Functional Analysis and Topology **12** (2006), no. 1, 1–31.

[4] Yu. Berezansky and M. Shmoish, *Nonisospectral flows on semi-infinite Jacobi matrices.* Nonlinear Math. Phys. **1** (1994), no. 2, 116–146.

[5] Oleksii Mokhonko, *Nonisospectral Flows on Semi-Infinite Unitary Block Jacobi Matrices.* Ukr. Math. J. **60** (2008), no. 4, 521–544.

[6] Yurij M. Berezansky and Oleksii Mokhonko, *Integration of some differential-difference nonlinear equations using the spectral theory of normal block Jacobi matrices.* Func. An. and Appl. **42** (2008), no. 1, 1–21 (in Russian).

[7] Yu.M. Berezansky, *The integration of semi-infinite Toda chain by means of inverse spectral problem.* Reports on Math. Phys. **24** (1986), 21–47.

[8] O. Mokhonko, *Some solvable classes of nonlinear nonisospectral difference equations.* Ukr. Math. J. **57** (2005), 356–365.

[9] Leonid Golinskii, *Schur flows and orthogonal polynomials on the unit circle.* Math. Coll. **197** (2006), no. 8, 41–62.

[10] Yu.M. Berezansky, *On the theory of commutative Jacobi fields.* Methods of functional analysis and topology **4** (1998), no. 1, 1–31.

[11] Yurij M. Berezansky, *Spectral Theory of Commutative Jacobi Fields: Direct and Inverse Problems.* Fields Institute Communications **25** (2000), 211–224.

Oleksii Mokhonko
Department of Mathematical Analysis
Mechanics and Mathematics Faculty
Kyiv National Taras Shevchenko University
01033 Kyiv, Ukraine
e-mail: `AlexeyMohonko@univ.kiev.ua`

Operator Theory:
Advances and Applications, Vol. 190, 397–417

Compact Extrema: A General Theory and Its Applications to Variational Functionals

I.V. Orlov

Abstract. The general properties of compact extrema and the conditions for compact extrema in terms of the compact derivatives in Hilbert space are considered. The compact-analytical properties of and analytical conditions for the compact extrema of variational functionals in Sobolev space W_2^1 are studied in detail.

Mathematics Subject Classification (2000). Primary 49K05, 49K20; Secondary 49K24.

Keywords. Variational functional, Sobolev space, Hilbert space, compact ellipsoid, compact extremum, compact derivative.

1. Introduction

The extremum problems for integral functionals play an important role in the modern nonlinear analysis and its applications. Diverse methods are employed to investigate these problems (see, e.g., [1]–[3]). One of the fundamental difficulties in these questions for the case of a Hilbert-type space, such as the Sobolev space W_2^1, is an essential deterioration of the analytical properties of integral functionals (see [4], [5]). Another aspect of the above problem is that the extrema of such functionals are sought, as a rule, on some compact sets. These cases exclude an application of the classical Fréchet conditions for local extrema.

Our investigations [13], [16], [18]–[20] (see also [14], [17], [21]) have showed that both the analytical properties and extrema of variations on Sobolev spaces of the W_2^1-type can be expressed in terms of subspaces of W_2^1, spanned by absolutely convex compact sets (with corresponding norms). This leads to the notions of K-continuity, K-differentiability, K-extrema, etc. Such an approach allows us to receive, for the K-extrema, analogs of both the general classical conditions for local extrema and the well-known conditions for the extrema of variational functionals on spaces of the C^1-type.

In the first two sections of this work, we consider the general properties of compact (or K-) extrema in Hilbert spaces and the conditions for K-extrema in terms of the K-derivatives. The following two sections deal with the K-analytical properties of and the analytical conditions for the K-extrema of the main variational functional in W_2^1.

2. Compact extremum in Hilbert spaces

In this and subsequent sections of the work, we pay our main attention to the compact (K-) extrema of functionals acting in Hilbert space. But the general notion of a K-extremum (as well as those of K-continuity and K-differentiability) can be introduced for an arbitrary complete locally convex space (LCS).

Definition 2.1. Let E be a complete LCS, $\Phi : E \to \mathbb{R}$. We say that the functional Φ has a *compact extremum* (or *K-extremum*) at a point $y \in E$ if for each absolutely convex (a.c.) compactum $C \subset E$, the restriction Φ to $(y + \operatorname{span} C)$ has a local extremum at y with respect to the norm $\|\cdot\|_C$ in span C generated by C.

Remark 2.2.

1) It follows from the well-known completeness criterion [6, I.1.6] that the spaces $(\operatorname{span} C, \|\cdot\|_C)$ are Banach spaces.
2) The correctness of the above Definition 2.1 follows from the compactness of $\overline{\operatorname{conv}}(C_1 \bigcup C_2)$ in the case of the compactness of both C_1 and C_2 [6, II.4.3]. So, Φ does not have to simultaneously possess a (strict) maximum at y with respect to $(y + \operatorname{span} C_1)$ and a minimum at y with respect to $(y + \operatorname{span} C_2)$.
3) One can propose a more elementary form of the definition of a K-extremum: Φ has a K-extremum at $y \in E$ if for each a.c. compactum $C \subset E$, there exists $\delta > 0$ such that an extremum of the restriction Φ to $(y + \operatorname{span} C)$ occurs on $(y + \delta \cdot C)$.
4) In view of the boundedness of compacta, every local extremum is a K-extremum as well. The converse of this statement is not true. Let us consider a simple example.

Example 1. Let B be a closed unit ball in a reflexive Banach space C. If we pass to the weak topology $\sigma(E, E^*)$ in E, then B is compact in E_σ. Since, in addition [7, 8.2.2], any compactum $C \subset E_\sigma$ is bounded in E, then $B \supset \delta \cdot C$ for a sufficiently small $\delta > 0$. Set

$$\Phi(x) = \begin{cases} \|x\|, & x \in B; \\ 1 - \|x\|, & x \notin B. \end{cases}$$

Then the restrictions $\Phi\big|_{\delta \cdot C}(x)$, $x \neq 0$, are positive for a sufficiently small $\delta > 0$, whence it follows that Φ has a strict K-minimum at zero in E_σ. However, Φ has no local extremum at zero in E_σ because each zero neighborhood in E_σ intersects both $B \backslash \{0\}$ and $E \backslash 2B$.

Note that in the above case, each K-extremum in E_σ is a local extremum in E and, therefore, the theory of K-extrema in E_σ is part of the theory of local extrema in E.

Let us pass to the case of Hilbert space. First of all, we state the fact that (nondegenerated) compact ellipsoids in Hilbert space are universal compacta that absorb all the other compacta. In what follows, H and H_i are complete infinite-dimensional separable Hilbert spaces.

Definition 2.3. A set $C_\varepsilon \subset H$, $\varepsilon = (\varepsilon_k)_{k=1}^\infty$, $\varepsilon_k > 0$, is called a (nondegenerated) *ellipsoid* in H if, for some choice of an orthonormal basis $(e_k)_{k=1}^\infty$ in H,

$$C_\varepsilon = \{x = \sum_{k=1}^\infty x_k e_k \in H \Big| \sum_{k=1}^\infty \frac{|x_k|^2}{\varepsilon_k^2} \le 1\}. \tag{2.1}$$

Remark 2.4. Note that the norm $\|\cdot\|_{C_\varepsilon}$, generated by C_ε in span C_ε, is obviously the Hilbert norm:

$$\|x\|_{C_\varepsilon}^2 = \sum_{k=1}^\infty \frac{|x_k|^2}{\varepsilon_k^2}; \quad \langle x, y\rangle_{C_\varepsilon} = \sum_{k=1}^\infty \frac{x_k \overline{y_k}}{\varepsilon_k^2}.$$

Let us mention the known test (see, e.g., [8], [9]).

Theorem 2.5. *The ellipsoid* (2.1) *is compact iff* $\varepsilon_k \to 0$.

The idea of proof of the following theorem belongs to Yu.V. Bogdansky.

Theorem 2.6. *A closed set $C \subset H$ is compact iff for an arbitrary choice of the orthonormal basis in H, there exists a compact ellipsoid $C_\varepsilon \supset C$.*

The proof is based upon two lemmas. The first one is the known statement on number series [10] complemented by an estimate of the series sum.

Lemma 2.7. *Let $\sum_{n=1}^\infty a_n$ be a convergent positive number series, $r_n = \sum_{k=n}^\infty a_k$, $n = 1, 2, \ldots$, $S = r_1$. Then the series $\sum_{n=1}^\infty (a_n/\sqrt{r_n})$ also converges; moreover,*

$$\sum_{n=1}^\infty \frac{a_n}{\sqrt{r_n}} \le \frac{a_1}{\sqrt{S}} + 2\sqrt{S}.$$

The second lemma is an analog of the known result on termwise integration in Lebesgue integrals [11, Ch. III]; the proof can be found in [22].

Lemma 2.8. *A convex set $C \subset l_2$ is compact iff*

$$\lim_{n\to\infty} \left(\sup_{x=(x_k)\in C} \sum_{k=n}^\infty |x_k|^2 \right) = 0.$$

Proof of Theorem 2.6. The proof of the necessity is obvious. Conversely, let C be compact in H. Let us choose an arbitrary orthonormal basis $(e_k)_{k=1}^{\infty}$ in H and set

$$\varepsilon_n^2 = \sqrt{\sup_{x\in C}\sum_{k=n}^{\infty}|x_k|^2} \quad (n=1,2,\ldots), \qquad x=\sum_{k=1}^{\infty}x_k e_k \in C, \qquad M_C=\sup_{x\in C}\|x\|.$$

According to Lemma 2.8, $\varepsilon_n \searrow 0$ as $n\to\infty$. Set $\varepsilon=(\varepsilon_k)_{k=1}^{\infty}$. Then for $x\in C$, according to Lemma 2.7,

$$\sum_{n=1}^{\infty}\frac{|x_n|^2}{\varepsilon_n^2} \le \sum_{n=1}^{\infty}\frac{|x_n|^2}{\sqrt{\sum\limits_{k=n}^{\infty}|x_k|^2}} \le \frac{|x_1|^2}{\sqrt{\sum\limits_{k=1}^{\infty}|x_k|^2}} + 2\sqrt{\sum_{k=1}^{\infty}|x_k|^2}$$

$$= \frac{|x_1|^2}{\|x\|} + 2\|x\| \le 3\|x\| \le 3M_C,$$

whence $C\subset 3M_C\cdot C_\varepsilon = C_{3M_C\cdot\varepsilon}$. □

Corollary 2.9. *For two arbitrary orthonormal bases in H, each compact ellipsoid C_ε relative to the first basis is contained in some compact ellipsoid $C_{\varepsilon'}$ relative to the second basis.*

Corollary 2.10. *Let $A: H_2\to H_1$ be some compact embedding of H_2 into H_1. Then the operator A transfers a unit ball of H_2 into some compact ellipsoid of H_1.*

In what follows, we denote by $\varepsilon_C(H)$ the set of all compact ellipsoids in H.

Remark 2.11.

1) The result of Theorem 2.6 extends, obviously, to the compact ellipsoids relative to the orthogonal bases $(e_k)_{k=1}^{\infty}$ in H that satisfy the condition
$$0<m\le\|e_k\|\le M<+\infty \quad (k=1,2,\ldots).$$
2) Condition (2.1) for the compact ellipsoid $C_\varepsilon\in\varepsilon_C(H)$ is related to the degree of convergence to zero of the Fourier coefficients for $x\in C_\varepsilon$. In addition, Corollary 2.10 allows us to choose a convenient basis in H. Let us demonstrate this by concrete examples.

Example 2. Let $H=L_2[0;2\pi]$, $e_k=e^{ikt}/\sqrt{2\pi}$ $(k\in\mathbb{Z})$. Then x_k are the Fourier coefficients of a function $x=x(t)$ relative to the normed exponential system in $L_2[0;2\pi]$. Since the degree of convergence of x_k to zero determines, as is well known [12, Vol. 2, Ch. XVIII], the smoothness class of the function, then Condition (2.1) is in fact the condition for the smoothness of $x(t)$.

E.g., it follows from $\varepsilon_k=O\left(\frac{1}{|k|^m}\right)$, $m\in\mathbb{N}$, that $x_k=o\left(\frac{1}{|k|^m}\right)$, whence $x^{(m-1)}\in BV[0;2\pi]$. It follows from $\varepsilon_k=O(q^{|k|})$, $0<q<1$, that $x_k=o(q^{|k|})$, whence $x(t)$ is analytical on $[0;2\pi]$.

Example 3. Let $H=\overset{\circ}{W}{}_2^1[0;2\pi]$, $e_k=e^{ikt}/\sqrt{2\pi(k^2+1)}$ $(k\in\mathbb{Z})$. Then x_k are the Fourier coefficients of a function $x=x(t)$ relative to the normed exponential system in $\overset{\circ}{W}{}^1_2[0;2\pi]$. Hence, in this case Condition (2.1) is the higher condition for the smoothness of $x(t)$.

For example, it follows from $\varepsilon_k = O\left(\frac{1}{|k|^m}\right)$, $m \in \mathbb{N}$, that $x_k = o\left(\frac{1}{|k|^{m+1}}\right)$, whence $x^{(m)} \in BV[0;2\pi]$. Analogously, it follows from $\varepsilon_k = O\left(\frac{1}{|k|^m}\right)$ in the space $H = \overset{\circ}{W}{}^p_2[0;2\pi]$, $p \in \mathbb{N}$, that $x_k = o\left(\frac{1}{|k|^{m+p}}\right)$, whence $x^{(m+p-1)} \in BV[0;2\pi]$.

Let us pass to consideration of the K-extrema of real functionals in Hilbert space $\Phi : H \to \mathbb{R}$. From Theorem 2.6 and Definition 2.1 it obviously follows

Theorem 2.12. *A functional Φ has a K-extremum at a point $y \in H$ iff, for each $C_\varepsilon \in \varepsilon_C(H)$, the restriction of Φ to $(y + \operatorname{span} C_\varepsilon)$ has a local extremum at y relative to the Hilbert norm $\|\cdot\|_{C_\varepsilon}$ generated by C_ε.*

Let us introduce, based on Corollary 2.10, the notion of local realization of a K-extremum and check the possibility of such realization.

Definition 2.13. Let Φ have a K-extremum at a point $y \in H_1$. If for some $C_\varepsilon \in \varepsilon_C(H)$ and some compact injective embedding $A : H_2 \to H_1$, the condition

$$\Im A \subset \operatorname{span} C_\varepsilon \tag{2.2}$$

is fulfilled, then we say that the operator A carries out *local realization in H_2* of the given K-extremum that corresponds to C_ε.

Remark 2.14. As was shown in [13, Thm. 3], it follows from (2.2) that A transfers a unit ball $B \subset H_2$ into some multiple of $C_\varepsilon : A(B) \subset \delta \cdot C_\varepsilon$. Whence and from Remark 2.2 (3) it follows that the composition $\Phi \circ A$ has a local extremum at the point $A^{-1}(y) \in H_2$.

Theorem 2.15. *If Φ has a K-extremum at a point $y \in H$, then for any $C_\varepsilon \in \varepsilon_C(H)$ there exists local realization of the given K-extremum in H that corresponds to C_ε.*

Proof. If $\varepsilon = (\varepsilon_k)_{k=1}^\infty$ and $(e_k)_{k=1}^\infty$ is the corresponding orthonormal basis in H, then it suffices to set

$$A(x) = \sum_{k=1}^{\infty} \varepsilon_k x_k e_k \quad \text{for} \quad x = \sum_{k=1}^{\infty} x_k e_k \in H. \qquad \square$$

Example 4.

1) It follows from Example 2 (p. 400) that the compact identical embedding $W_2^m[0;2\pi] \hookrightarrow L_2[0;2\pi]$, $m \in \mathbb{N}$, carries out local realization of the K-extrema in $L_2[0;2\pi]$ for $\varepsilon_k = O\left(\frac{1}{|k|^{m+1}}\right)$, $k \in \mathbb{Z}$.
2) It follows from Example 3 (p. 400) that the compact identical embedding $\overset{\circ}{W}{}_2^{m+p}[0;2\pi] \hookrightarrow \overset{\circ}{W}{}_2^{p}[0;2\pi]$; $m, p \in \mathbb{N}$, carries out local realization of the K-extrema in $\overset{\circ}{W}{}_2^{p}[0;2\pi]$ for $\varepsilon_k = O\left(\frac{1}{|k|^m}\right)$, $k \in \mathbb{Z}$.

3. Compact extremum conditions in terms of the compact derivatives

The general notions of K-continuity and K-differentiability are introduced for an arbitrary complete LCS, although the applications to the K-extrema theory will further be connected in preference to Hilbert spaces.

Definition 3.1. Let E be a complete LCS, $\Phi : E \to \mathbb{R}$. We say that the functional Φ is *compactly continuous* (*K-continuous*) [or *compactly differentiable* (*K-differentiable*), *twice K-differentiable*, etc.] at a point $y \in E$ if for each absolutely convex compactum $C \subset E$, the restriction of Φ to $(y + \text{span } C)$ is continuous (or, respectively, Fréchet differentiable, twice Fréchet differentiable, etc.) at y with respect to the norm $\|\cdot\|_C$ in span C, generated by C. The corresponding *compact derivatives* (*K-derivatives*) will be denoted by $\Phi'_K(y)$, $\Phi''_K(y)$, etc.

Remark 3.2.

1) Note, by analogy with Remark 2.2, that the correctness of the definition of K-derivatives follows from the compactness of $\overline{\text{conv}}(C_1 \bigcup C_2)$ together with the compactness of C_1 and C_2 in E.
2) For arbitrary a.c. compacta C, C_1, C_2 in E, the definitions of the first- and second-order K-derivatives can be written out explicitly as

$$\Phi(y+h) - \Phi(y) = \Phi'_K(y) \cdot h + o(\|h\|_C); \tag{3.1}$$

$$(\Phi'_K(y+h) - \Phi'_K(y)) \cdot k = \Phi''_K(y) \cdot (h,k) + o(\|h\|_{C_1} \cdot \|k\|_{C_2}). \tag{3.2}$$

3) Formally, the linear functional $\Phi'_K(y)$ in (3.1) and bilinear form $\Phi''_K(y)$ in (3.2) are only K-continuous. But in the case of Hilbert space, as is shown below, the K-continuity for linear and bilinear forms coincides with the usual continuity.

Theorem 3.3. *Let F be an arbitrary LCS. A linear operator $A : H \to F$ is continuous iff all restrictions $A : C_\varepsilon \to F$, $C_\varepsilon \in \varepsilon_C(H)$, are continuous with respect to the norms $\|\cdot\|_{C_\varepsilon}$.*

Proof. Denote by H_K the space H equipped with inductive topology generated by the subspaces (span $C_\varepsilon, \|\cdot\|_{C_\varepsilon}$), $C_\varepsilon \in \varepsilon_C(H)$. The continuous embedding of H_K into H is evident. Let us prove the continuity of the inverse embedding.

Let, on the contrary, there exist a zero neighborhood $U \subset H_K$ such that is not zero neighborhood in H. Let B_1 be a unit ball in H. Then, by assumption, for every $n = 1, 2, \ldots$ there exists an element $x_n \in \frac{1}{4^n} B_1 \backslash U$. Set $C = \overline{\text{abs}\,.\,\text{conv}_H \{2^n x_n\}_{n \in \mathbb{N}}}$. Since $\|2^n x_n\| \leq \frac{1}{2^n} \to 0$ as $n \to \infty$, then C is a compactum in H. Then, by Theorem 2.6, C is contained in some $C_\varepsilon \in \varepsilon_C(H)$. Since the embedding $H_{C_\varepsilon} \hookrightarrow H_K$ is continuous, then C_ε and therefore C is absorbed by the set U. However, the embeddings $2^n x_n \in C \backslash 2^n U$ imply $C \backslash 2^n U \neq \emptyset$, $n \in \mathbb{R}$, i.e., C is not absorbed by U, a contradiction.

Thus, H_K and H are isomorphic; this fact allows us to apply to the operator $A : H_K \to F$ the well-known theorem on inductive limit [II.6.1][6]. □

Corollary 3.4. *The K-derivatives $\Phi'_K(y)$, $\Phi''_K(y), \dots$ of a functional $\Phi : H \to \mathbb{R}$ are continuous forms on H.*

Note that in the case of nonlinear functionals in H, the K-continuity can be different from the usual one [14]. Let us pass to the K-extremum conditions in terms of the K-derivatives. The classical necessary conditions for local extrema in Banach space [15] lead, together with Definitions 2.1 and 3.1, to analogous conditions for K-extrema.

Theorem 3.5. *Let E be a complete LCS and $\Phi : E \to \mathbb{R}$ have a K-minimum (or a K-maximum) at a point $y \in E$. Then:*

1) *If Φ is K-differentiable at y, then $\Psi'_K(y) = 0$.*
2) *If Φ is twice K-differentiable at y, then $\Phi''_K(y) \geq 0$ (or, respectively, $\Phi''_K(y) \leq 0$).*

To receive the sufficient condition for the Hilbert case, let us introduce an auxiliary notion.

Definition 3.6. Let g be a bilinear continuous form on H. We say that the form g is *K-positive definite*: $g \gg 0 \,(\mathrm{mod}\, K)$ if for an arbitrary compact ellipsoid $C_\varepsilon \in \varepsilon_C(H)$, the restriction of g to span C_ε is positive definite with respect to the inner product $\langle \cdot, \cdot \rangle_{C_\varepsilon}$, generated by C_ε.

Applying the classical sufficient condition for local extrema in Hilbert space [15], together with Definitions 2.1, 3.1 and Theorem 3.5, we receive an analogous sufficient condition for K-extrema.

Theorem 3.7. *Let a functional $\Phi : H \to \mathbb{R}$ be twice K-differentiable at a point $y \in H$. If:*

1) $\Phi'_K(y) = 0$;
2) $\Phi''_K(y) \gg 0(\mathrm{mod}\, K)$ *(or* $\Phi''_K(y) \ll 0(\mathrm{mod}\, K)$*)*;

then Φ has a strict K-minimum (or, respectively, a strict K-maximum) at y.

Further, the case of a functional of two variables will be of special importance to us. First, we formulate an auxiliary statement about operator matrices.

Lemma 3.8. ([16], *Prop.* 3.1) *Let $B = (B_{ij} : H_j \to H_i)^2_{i,j=1}$ be a selfadjoint linear continuous operator in $H_1 \times H_2$ (so, $B_{11} = B^*_{11}$, $B_{22} = B^*_{22}$, $B_{12} = B^*_{21}$). The operator B is positive definite on $H_1 \times H_2$ ($B \gg 0$) iff:*

1) $B_{11} \gg 0$, $B_{22} \gg 0$;
2) $\triangle^2_1(B) := B_{11} - B_{12} \cdot B^{-1}_{22} \cdot B_{21} \gg 0$, $\triangle^1_2(B) := B_{22} - B_{21} \cdot B^{-1}_{11} \cdot B_{12} \gg 0$.

Application of Lemma 3.8 to the subspaces $C_\varepsilon \in \varepsilon_C(H)$ immediately leads to a sufficient condition for the K-positive definiteness of a symmetric bilinear form in $H_1 \times H_2$.

Theorem 3.9. *Let g be a symmetric bilinear continuous form on $H_1 \times H_2$, and B be a linear continuous operator associated with g and acting in*

$$H_1 \times H_2, \quad B = (B_{ij} : H_j \to H_i)^2_{i,j=1}.$$

If:

1) $B_{11} \gg 0(\mathrm{mod}\ K)$, $B_{22} \gg 0(\mathrm{mod}\ K)$;
2) $\triangle_1^2(B) \gg 0(\mathrm{mod}\ K)$, $\triangle_2^1(B) \gg 0(\mathrm{mod}\ K)$.

then the form g is K-positive definite on $H_1 \times H_2$: $g \gg 0(\mathrm{mod}\ K)$.

It is obvious that the reversal of all signs in the above conditions leads to the sufficient condition for $g \ll 0(\mathrm{mod}\ K)$. Finally, applying Theorem 3.9 to the K-Hessian $(\partial_{ijK}\Phi)_{i,j=1}^2$, together with Theorem 3.7, we obtain the sufficient condition for K-extrema in terms of the second-order partial K-derivatives of Φ.

Theorem 3.10. *Let a functional $\Phi : H_1 \times H_2 \to \mathbb{R}$ be twice K-differentiable at a point $(y_1, y_2) \in H_1 \times H_2$. Suppose that:*

1) $\partial_{1K}\Phi(y_1, y_2) = 0$, $\partial_{2K}\Phi(y_1, y_2) = 0$;
2) $\partial_{11K}\Phi(y_1, y_2) \gg 0(\mathrm{mod}\ K)$, $\partial_{22K}\Phi(y_1, y_2) \gg 0(\mathrm{mod}\ K)$;
3) $\triangle_{1K}^2\Phi(y_1, y_2) = (\partial_{11K}\Phi - \partial_{12K}\Phi \cdot \partial_{22K}^{-1}\Phi \cdot \partial_{21K}\Phi)\big|_{(y_1,y_2)} \gg 0(\mathrm{mod}\ K)$,
$\triangle_{2K}^1\Phi(y_1, y_2) = (\partial_{22K}\Phi - \partial_{21K}\Phi \cdot \partial_{11K}^{-1}\Phi \cdot \partial_{12K}\Phi)\big|_{(y_1,y_2)} \gg 0(\mathrm{mod}\ K)$.

Then Φ has a strict K-minimum at (y_1, y_2).

The reversal of all signs in conditions 1)–3) leads to sufficient conditions for a strict K-maximum of Φ. Also, we represent here for references the corresponding conditions for a local minimum (which we failed to find in literature).

Theorem 3.11. *Let a functional $\Phi : H_1 \times H_2 \to \mathbb{R}$ be twice Fréchet differentiable at a point $(y_1, y_2) \in H_1 \times H_2$. Suppose that:*

1) $\partial_1\Phi(y_1, y_2) = 0$, $\partial_2\Phi(y_1, y_2) = 0$;
2) $\partial_{11}\Phi(y_1, y_2) \gg 0$, $\partial_{22}\Phi(y_1, y_2) \gg 0$;
3) $\triangle_1^2\Phi(y_1, y_2) = (\partial_{11}\Phi - \partial_{12}\Phi \cdot \partial_{22}^{-1}\Phi \cdot \partial_{21}\Phi)\big|_{(y_1,y_2)} \gg 0$,
$\triangle_2^1\Phi(y_1, y_2) = (\partial_{22}\Phi - \partial_{21}\Phi \cdot \partial_{11}^{-1}\Phi \cdot \partial_{12}\Phi)\big|_{(y_1,y_2)} \gg 0$.

Then Φ has a strict local minimum at (y_1, y_2).

Note that any of inequalities 3) in Theorem 3.10-3.11 and, similarly, any of inequalities 2) in Lemma 3.8 and Theorem 3.9 can be omitted. Note also, in conclusion of this Section, that in work [17] both sufficient and necessary conditions for the K-extrema of functionals of n Hilbert variables in terms of the second-order partial K-derivatives can be found.

4. K-analytical properties of the basic variational functional in Sobolev space W_2^1

It was explained in the 60–70ies of the last century ([3], [4]) that the Euler–Lagrange functional

$$\Phi(y) = \int_a^b f(x, y, y')dx \tag{4.1}$$

in spaces with Hilbert integral metric possesses considerably worse differential properties than those in the Banach space of the C^k-type. An attentive analysis of the variational situation in Sobolev spaces of the W_2^1-type shows that the conditions similar to quadratic ones in y' for the integrand in (4.1) are necessary even for the well-definiteness of the functional (4.1). For example, the functional

$$\Phi(y) = \int_a^b (y')^6 dx = (\|y'\|_{L_6})^6$$

is only densely defined in $W_2^1([a;b], \mathbb{R})$, although $f \in C^\infty$.

Further, the basic notion is one of pseudoquadratic mapping.

Definition 4.1. Let X be a compact space with finite Borel measure, Y, Z and F be real Banach spaces, and $f : X \times Y \times Z \to F$ be a Borel function. The mapping f is said to be *pseudoquadratic in* $z \in Z$: $f \in K_2(z)$ if it can be represented in the form

$$f(x, y, z) = P(x, y, z) + Q(x, y, z) \cdot \|z\| + R(x, y, z) \cdot \|z\|^2, \tag{4.2}$$

where, for each compactum $C_Y \subset Y$, the Borel mappings P, Q and R are essentially bounded on $X \times C_Y \times Z$ in $x \in X$.

At first, let us obtain the well-definiteness condition for the functional (4.1) in W_2^1. In what follows, E is a Banach space, $f : [a;b] \times E \times E \to \mathbb{R}$, $u = f(x, y, z)$.

Theorem 4.2. *If $f \in K_2(z)$, then the variational functional* (4.1) *is well defined everywhere on $W_2^1([a;b], E)$.*

Proof. Let us fix $y(\cdot) \in W_2^1([a;b], E)$ and set $C_Y = y([a;b])$. Using the notation (4.2), we see that $f(x, y, y')$ is measurable and

$$\Phi(y) = \int_a^b f(x, y, y')dx = \int_a^b P dx + \int_a^b Q \cdot \|y'\| dx + \int_a^b R \cdot \|y'\|^2 dx,$$

where

$$|P(x, y, y')| \leq M_P, \ |Q(x, y, y')| \leq M_Q, \ |R(x, y, y')| \leq M_R \text{ a.e. on } [a;b]. \tag{4.3}$$

Moreover,

$$\left| \int_a^b P(x, y, y')dx \right| \leq M_P \cdot (b - a), \quad \left| \int_a^b Q(x, y, y') \cdot \|y'\| dx \right| \leq M_Q \cdot \int_a^b \|y'\| dx$$

$$\leq M_Q \cdot \sqrt{(b-a)} \cdot \|y\|_{W_2^1}, \quad \left| \int_a^b R(x, y, y') \cdot \|y'\|^2 dx \right| \leq M_R \cdot \|y\|^2_{W_2^1},$$

whence $|\Phi(y)| < \infty$ for all $y(\cdot) \in W_2^1([a;b], E)$. □

The K-continuity of the functional (4.1) in W_2^1 requires more severe constraints on f and the Hilbert space of the values of $y(\cdot)$.

Definition 4.3. Let, in the preceding notation, $f \in K_2(z)$ and be continuous in (y, z). The mapping f is said to be *Weierstrass pseudoquadratic in* $z \in Z$: $f \in WK_2(z)$ if the representation (4.2) can be chosen such that the mappings P, Q and R are uniformly continuous and bounded on $X \times C_Y \times Z$ for each compactum $C_Y \subset Y$.

Let us demonstrate, in proving the theorem on the K-continuity, the methods that are further used to prove the K-differentiability and the twice K-differentiability of the variational functional in W_2^1. Further, in the preceding notation, $E = H$ is a Hilbert space, $f : [a; b] \times H \times H \to \mathbb{R}$.

Theorem 4.4. *If* $f \in WK_2(z)$, *then the functional* (4.1) *is* K*-continuous everywhere on* $W_2^1([a; b], H)$.

Proof. Let us fix $y(\cdot) \in W_2^1([a; b], H)$ and set $C_Y = y([a; b])$. In the the notation (4.2), we obtain:

$$\begin{aligned}\Phi(y+h) - \Phi(y) &= \int_a^b [P(x, y+h, y'+h') - P(x, y, y')]dx \\ &+ \int_a^b [Q(x, y+h, y'+h') \cdot \|y'+h'\| - Q(x, y, y') \cdot \|y'\|]dx \\ &+ \int_a^b [R(x, y+h, y'+h') \cdot \|y'+h'\|^2 - R(x, y, y') \cdot \|y'\|^2]dx,\end{aligned} \tag{4.4}$$

where P, Q and R are uniformly continuous on $[a; b] \times C_Y \times H$ and the estimates (4.3) hold. For an arbitrary $\delta > 0$, set

$$e_\delta = (|h|^2 + |h'|^2 > \delta^2), \quad e^\delta = (|h| < \delta, \ |h'| < \delta); \tag{4.5}$$

in addition,

$$(\|h\|^2_{W_2^1} < \delta^3) \Rightarrow (me_\delta < \delta), \ \text{ and } \ [a; b] = e_\delta \cup e^\delta. \tag{4.6}$$

Let us fix $\varepsilon > 0$, $\varepsilon < 1$, and choose $\delta = \delta(\varepsilon) > 0$ such that

$$\int_{e_\delta} \|y'\|^2 dx < \varepsilon^2; \quad (x < e^\delta) \Rightarrow (|\Delta P| < \varepsilon, \ |\Delta Q| < \varepsilon, \ |\Delta R| < \varepsilon) \tag{4.7}$$

Now, suppose that

$$\|h\|^2_{W_2^1} < \delta^3. \tag{4.8}$$

Let us fix an a.c. compactum C_Δ in $W_2^1([a; b], H)$. By virtue of the Arzéla–Ascoli theorem, $\forall \delta > 0 \ \exists \eta > 0$:

$$(h \in \eta \cdot C_\Delta) \Leftrightarrow (\|h\|_{C_\Delta} \le \eta) \Rightarrow (|h(x)| < \delta, \ \forall x \in [a; b]). \tag{4.9}$$

Hence, in view of (4.7) and (4.3),

$$|\Delta P| < 2M_P + \varepsilon, \ |\Delta Q| < 2M_Q + \varepsilon, \ |\Delta R| < 2M_R + \varepsilon \tag{4.10}$$

as $\|h\|_{C_\Delta} \leq \eta$, $x \in [a;b]$. Whence, using the estimates (4.3), (4.5)–(4.6) and (4.10), we obtain:

$$\left|\int_a^b [P(x,y+h,y'+h') - P(x,y,y')]dx\right| \leq \int_{e^\delta} |\Delta P|dx + \int_{e_\delta} |\Delta P|dx$$

$$< \varepsilon \cdot (b-a) + (2M_P + \varepsilon) \cdot \varepsilon < (b - a + 2M_P + 1) \cdot \varepsilon =: C_1 \cdot \varepsilon. \tag{4.11}$$

Next, using the identity

$$Q(x,y+h,y'+h') \cdot \|y'+h'\| - Q(x,y,y') \cdot \|y'\|$$
$$= \Delta Q \cdot \|y'+h'\| + Q(x,y,y') \cdot (\|y'+h'\| - \|y'\|)$$

and the estimates (4.3) and (4.6)–(4.10), we obtain:

$$\left|\int_a^b [Q(x,y+h,y'+h') \cdot \|y'+h'\| - Q(x,y,y') \cdot \|y'\|]dx\right|$$

$$\leq \int_a^b |\Delta Q| \cdot \|y'+h'\|dx + \int_a^b |Q| \cdot \big|\|y'+h'\| - \|y'\|\big|dx$$

$$\leq \int_{e^\delta} |\Delta Q| \cdot \|y'+h'\|dx + \int_{e_\delta} |\Delta Q| \cdot \|y'+h'\|dx + \int_a^b |Q| \cdot \|h'\|dx$$

$$\leq \varepsilon \cdot \sqrt{(b-a)} \cdot (\|y\|_{W_2^1} + \|h\|_{W_2^1}) + (2M_Q + \varepsilon) \cdot \sqrt{(b-a)} \cdot (\varepsilon + \|h\|_{W_2^1})$$
$$+ M_Q \cdot \sqrt{(b-a)} \cdot \|h\|_{W_2^1}$$
$$< \sqrt{(b-a)} \cdot \left[\varepsilon(\|y\|_{W_2^1} + \delta^{3/2}) + (2M_Q + \varepsilon) \cdot (\varepsilon + \delta^{3/2}) + M_Q \cdot \delta^{3/2}\right]$$
$$< \varepsilon \cdot \sqrt{(b-a)} \cdot (5M_Q + \|y\|_{W_2^1} + 3) =: C_2 \cdot \varepsilon \tag{4.12}$$

Further, using the identity

$$R(x,y+h,y'+h') \cdot \|y'+h'\|^2 - R(x,y,y') \cdot \|y'\|^2$$
$$= \Delta R \cdot \|y'+h'\|^2 + R(x,y,y') \cdot (\|y'+h'\|^2 - \|y'\|^2)$$
$$= \Delta R \cdot \|y'+h'\|^2 + R(x,y,y') \cdot (2\langle y',h'\rangle + \|h'\|^2)$$

and the estimates (4.3), (4.7)–(4.8) and (4.10), we obtain:

$$\left|\int_a^b [R(x,y+h,y'+h') \cdot \|y'+h'\|^2 - R(x,y,y') \cdot \|y'\|^2]dx\right|$$

$$\leq \int_a^b |\Delta R| \cdot \|y'+h'\|^2 dx + \int_a^b |R| \cdot (2|\langle y',h'\rangle| + \|h'\|^2)dx$$

$$\leq \int_{e^\delta} |\Delta R| \cdot \|y' + h'\|^2 dx + \int_{e_\delta} |\Delta R| \cdot \|y' + h'\|^2 dx$$

$$+ 2 \int_a^b |R| \cdot |\langle y', h' \rangle| dx + \int_a^b |R| \cdot \|h'\|^2 dx$$

$$\leq 2\varepsilon \cdot (\|y\|^2_{W_2^1} + \|h\|^2_{W_2^1}) + (2M_R + \varepsilon) \cdot 2\Big(\int_{e_\delta} \|y'\|^2 dx + \|h\|^2_{W_2^1} \Big)$$

$$+ 2M_R \cdot \|y\|_{W_2^1} \cdot \|h\|_{W_2^1} + M_R \cdot \|h\|^2_{W_2^1}$$

$$< 2\varepsilon \cdot \big(\|y\|^2_{W_2^1} + \delta^3\big) + (4M_R + 2\varepsilon) \cdot (\varepsilon^2 + \delta^3) + 2M_R \cdot \|y\|_{W_2^1} \cdot \delta^{\frac{3}{2}} + M_R \cdot \delta^3$$

$$< \varepsilon \cdot \Big(2\|y\|^2_{W_2^1} + 2M_R \cdot \|y\|_{W_2^1} + 9M_R + 6\Big) =: C_3 \cdot \varepsilon. \tag{4.13}$$

Finally, it follows from (4.4) and (4.11)–(4.13) that

$$|\Phi(y + h) - \Phi(y)| < (C_1 + C_2 + C_3) \cdot \varepsilon$$

as $\|h\|_{W_2^1} < \delta^{\frac{3}{2}}$, $\delta = \delta(\varepsilon)$, $\|h\|_{C_\Delta} \leq \eta = \eta(\varepsilon)$. This means the K-continuity of Φ at any point $y(\cdot) \in W_2^1([a; b], H)$. □

Note that it was shown in [14] that there exist variational functionals that are K-continuous, but not continuous in the usual sense (see Example 6 below).

Definition 4.5. Let, in the preceding notation, f be from C^1 in (y, z) and $f \in WK_2(z)$. We say that f *belongs to the class* $W^1K_2(z)$ if the representation (4.2) can be chosen such that not only P, Q and R, but also the gradients $\nabla P := \nabla_{yz} P$, $\nabla Q := \nabla_{yz} Q$ and $\nabla R := \nabla_{yz} R$ are uniformly continuous and bounded on $X \times C_Y \times Z$ for each compactum $C_Y \subset Y$.

Theorem 4.6. *If* $f \in W^1K_2(z)$, *then the variational functional* (4.1) *is* K*-differentiable everywhere on* $W_2^1([a; b], H)$. *In addition,*

$$\Phi'_K(y)h = \int_a^b \Big[\frac{\partial f}{\partial y}(x, y, y')h + \frac{\partial f}{\partial z}(x, y, y')h'\Big] dx.$$

The proof of Theorem 4.6 uses the "δ-procedure" from the preceding proof; this procedure becomes rather complicated when passing to the class $W^1K_2(z)$. Also, the following strengthening of the property of absolute continuity of the Lebesgue integral is used.

Lemma 4.7. *Let* (X, μ) *be a finite measure space, and* C_Δ *be a compactum in* $L_1(X, \mu)$. *Then* $\forall \varepsilon > 0\ \exists \delta > 0$:

$$\Big(e_\delta \subset X,\ \mu(e_\delta) < \delta,\ h \in C_\Delta\Big) \Rightarrow \Big(\int_{e_\delta} |h(x)| d\mu \leq \varepsilon \cdot \int_X |h(x)| d\mu\Big).$$

Remark 4.8. Note that, without loss of generality, the representation (4.2) for $f \in W^1K_2(z)$ can be replaced by

$$f(x,y,z) = \widetilde{P}(x,y,z) + \widetilde{R}(x,y,z) \cdot \|z\|^2$$

with analogous conditions for $\widetilde{P}$ and $\widetilde{R}$. Indeed, consider this partition of unity in H: $1 = \varphi_1(z) + \varphi_2(z)$ where, for some $M_z > 0$, $\varepsilon > 0$,

$$\operatorname{supp}\varphi_1(z) \subset (\|z\| \le M_z),\ \operatorname{supp}\varphi_2(z) \subset (\|z\| \ge M_z + \varepsilon), \quad 0 \le \varphi_1 \le 1,\, 0 \le \varphi_2 \le 1$$

and φ_1, φ_1', φ_2, φ_2' are uniformly continuous and bounded on H. Set

$$\widetilde{P} = (P + Q \cdot \|z\| + R \cdot \|z\|^2) \cdot \varphi_1(z),\ \widetilde{R} = \left(\frac{P}{\|z\|^2} + \frac{Q}{\|z\|} + R \right) \cdot \varphi_2(z).$$

Direct calculations show that $\widetilde{P}$, $\nabla\widetilde{P}$, $\widetilde{R}$, $\nabla\widetilde{R}$ are also uniformly continuous and bounded on $[a;b] \times C_H \times H$ for each compactum $C_H \subset H$. The same is true for the classes $WK_2(z)$ (see above) and $W^2K_2(z)$ (see below).

Definition 4.9. Let, in the preceding notation, f be from C^2 in (y,z) and $f \in W^1K_2(z)$. We say that f *belongs to the class* $W^2K_2(z)$ if the representation (4.2) can be chosen such that not only P, ∇P, Q, ∇Q, R, ∇R, but also the Hessians $H(P) := H_{yz}(P)$, $H(Q) := H_{yz}(Q)$, $H(R) := H_{yz}(R)$ are uniformly continuous and bounded on $X \times C_Y \times Z$ for each compactum $C_Y \subset Y$.

Theorem 4.10. *If $f \in W^2K_2(z)$, then the variational functional* (4.1) *is twice K-differentiable everywhere on $W_2^1([a;b],H)$. In addition,*

$$\Phi_K''(y)(h,k) = \int_a^b \Big[\frac{\partial^2 f}{\partial y^2}(x,y,y')(h,k) + \frac{\partial^2 f}{\partial y \partial z}(x,y,y')((h',k) + (h,k'))$$

$$+ \frac{\partial^2 f}{\partial z^2}(x,y,y')(h',k')\Big]\, dx. \tag{4.14}$$

The proof is also based on the "δ-procedure", which becomes even more complicated when passing to the class $W^2K_2(z)$.

Remark 4.11.

1) In [5], it was first proved that the variational functional (4.1) in W_2^1 is not twice Fréchet differentiable at all, except for the pure quadratic case with respect to y'.
2) If $f \in WK_2(z)$, then the application of the Cezàro operator

$$(Tf)(z) = \frac{1}{z} \int_0^z f(x,y,s)ds$$

leads to a function from $W^1K_2(z)$ and the repeated application of the Cezàro operator leads to a function from $W^2K_2(z)$.

5. Analytical conditions for the K-extrema of variational functionals in Sobolev space W_2^1

The main aim of this section is to describe both necessary and sufficient conditions for the K-extrema of variational functionals in W_2^1 in terms of the usual strong derivatives of the integrand, like in the C^1-situation.

In the situation under consideration, it is convenient for us to complement the notion of a K-extremum, termed further a *strong K-extremum*, by the notion of a *weak K-extremum*; the latter occurs only on some concrete types of compact ellipsoids. Taking into account Remark 2.2 (3) and Theorem 2.12, let us give the following

Definition 5.1. We say that the variational functional (4.1) has a *strong K-extremum* at a point $y(\cdot) \in W_2^1([a;b], H)$ if to each compact ellipsoid C_ε in $W_2^1([a;b], H)$, there corresponds $\delta > 0$ such that an extremum of Φ occurs on $(y + \delta \cdot C_\varepsilon)$.

We say that Φ has a *weak K-extremum* at $y(\cdot) \in W_2^1([a;b], H)$ if to some fixed (nondegenerated) compact ellipsoid C_{ε_0} in $W_2^1([a;b], H)$, there corresponds $\delta_0 > 0$ such that an extremum of Φ occurs on $(y + \delta_0 \cdot C_{\varepsilon_0})$.

Note that Φ can simultaneously have both a weak K-maximum and a weak K-minimum that correspond to two different compact ellipsoids with zero intersection. Now, let us consider examples of strong and weak K-extrema.

Example 5. Define $\varphi(t) = t$ as $0 \le t \le 1 - \delta$, $\varphi(t) = 2 - t$ as $1 + \delta \le t \le +\infty$, $\varphi \nearrow$ on $[1 - \delta; 1]$, $\varphi \searrow$ on $[1; 1 + \delta]$, where δ is small enough, $\varphi \in C^2[0, +\infty)$. Set

$$\Phi(y) = \int_0^1 \varphi(y^2 + y'^2)dx, \quad y(\cdot) \in W_2^1([0;1], \mathbb{R}). \tag{5.1}$$

Direct estimates show [18] that:

1) Φ has a strong K-minimum at the point $y_0(t) \equiv 0$.
2) Φ has no local minimum at y_0.

Below, we shall also prove the existence of this K-minimum with the help of the sufficient conditions for a K-extremum.

Example 6. Define $\varphi(0) = 0$, $\varphi(t) > 0$ as $0 < t < 6\pi$, $\varphi(t) < 0$ and $\varphi(t) \searrow$ as $t > 6\pi$, $\varphi(t) = O^*(t^2)$ as $t \to +\infty$, $\varphi \in C^2[0, +\infty)$. Set

$$\Phi(y) = \int_{2\pi}^{4\pi} \varphi(|y' \ln y'|^{1-\delta})dx; \quad y(\cdot) \in W_2^1([2\pi; 4\pi], \mathbb{C}).$$

where $0 < \delta < 1$.

Direct calculations show [13] that:

1) Φ is well defined in $W_2^1([2\pi;4\pi],\mathbb{C})$.
2) For the orthonormal basis $\{e^{ikx}/\sqrt{2\pi(k^2+1)}\}_{k\in\mathbb{Z}}$, the following estimate of lower and upper bounds for the semiaxis lengths of the compact ellipsoid C_ε realizing a minimum of Φ holds:

$$\frac{\lambda_0\sqrt{k^2+1}}{k^2} \le \varepsilon_k \le \frac{4\sqrt{2\pi(k^2+1)}}{k^2} \qquad (0<\lambda_0<\sqrt{2\pi},\ |k|\ge 2).$$

Here, the lower estimates confirm the weak K-extremum of Φ at zero and the upper estimates confirm that the this extremum is not a strong K-extremum.

Let us pass to the description of the extremals for strong K-extrema. First, Theorem 3.5 (1) immediately implies

Theorem 5.2. *If $f\in W^1K_2(z)$ and the variational functional* (4.1) *has a strong K-extremum at a point $y(\cdot)\in W_2^1([a;b],H)$, then $\Phi'_K(y)=0$.*

Whence, by analogy with the derivation of the classical Euler–Lagrange equation, it is not difficult to obtain the equation for a K-extremal ([18], [19]).

Theorem 5.3. *Let $f\in W^2K_2(z)$, $y\in W_2^2([a;b],H)$, $y(a)=y(b)=0$. Then, for the variational functional* (4.1) *in $W_2^1([a;b],H)$, the equality $\Phi'_K(y)=0$ is equivalent to the generalized Euler–Lagrange equation*

$$\frac{\partial f}{\partial y}(x,y,y') - \frac{d}{dx}\left[\frac{\partial f}{\partial z}(x,y,y')\right] \overset{a.e.}{=} 0. \tag{5.2}$$

The last two results allow us to refer to the solutions of the equation (5.2) as *K-extremals* of the functional (4.1) in space $\overset{\circ}{W}{}_2^1([a;b],H)$. Further, let us consider an analog of the necessary Legendre extremum condition. The main difference between our approach and the classical one to proving the above-mentioned necessary and below sufficient conditions is the use of the density points instead of the interior points. Let us demonstrate this with the proof of Theorem 5.4.

Theorem 5.4. *Let $f\in W^2K_2(z)$, $y(\cdot)$ be a K-extremal of the functional* (4.1) *in $\overset{\circ}{W}{}_2^1([a;b],H)$ and the function $\frac{\partial^2 f}{\partial y\partial z}(x,y,y')$ be absolutely continuous on $[a;b]$. If the variational functional* (4.1) *has a strong K-minimum (or a K-maximum) at a point $y(\cdot)$, then*

$$\frac{\partial^2 f}{\partial z^2}(x,y,y')\ge 0 \ \left(\text{or, respectively, } \frac{\partial^2 f}{\partial z^2}(x,y,y')\le 0\right) \ \text{a.e. on } [a;b]. \tag{5.3}$$

Proof. Let us transform the expression (4.14). For this purpose, we use Theorems 4.6, 5.4, and integrate by parts the second summand on the right of (4.14):

$$\int_a^b \frac{\partial^2 f}{\partial y \partial z}(x,y,y')((h',k)+(h,k'))dx = \int_a^b \frac{\partial^2 f}{\partial y \partial z}(x,y,y')d(h(x),k(x))$$

$$= \frac{\partial^2 f}{\partial y \partial z}(x,y,y')(h(x),k(x))\Big|_a^b - \int_a^b \frac{d}{dx}\left(\frac{\partial^2 f}{\partial y \partial z}(x,y,y')\right)(h,k)dx$$

$$= (\text{under } h(a)=h(b)=0,\ k(a)=k(b)=0) = -\int_a^b \frac{d}{dx}\left(\frac{\partial^2 f}{\partial y \partial z}(x,y,y')\right)(h,k)dx.$$

Then (4.14) takes the form

$$\Phi''_K(y)(h,k) = \int_a^b \left(\left[-\frac{d}{dx}\left(\frac{\partial^2 f}{\partial y \partial z}(x,y,y')\right) + \frac{\partial^2 f}{\partial y^2}(x,y,y')\right](h,k) + \frac{\partial^2 f}{\partial z^2}(x,y,y')(h',k')\right)dx. \quad (5.4)$$

Now, let us assume that (5.3) is not fulfilled and that there exists $k_0^2 > 0$ such that the inequality

$$\frac{\partial^2 f}{\partial z^2}(x,y(x)\,y'(x))(h_0,h_0) \le -k_0^2 < 0 \quad (x \in A_0^1)$$

holds true for some $A_0^1 \subset [a;b]$, $mA_0^1 > 0$, and some fixed $h_0 \in H$.

Let us choose a set $A_0^2 \subset [a;b]$, $mA_0^1 + mA_0^2 > b-a$, such that the inequality

$$\left[-\frac{d}{dx}\left(\frac{\partial^2 f}{\partial y \partial z}(x,y,y')\right) + \frac{\partial^2 f}{\partial y^2}(x,y,y')\right](h_0,h_0) \le C_0^2 < \infty \quad (x \in A_0^2)$$

holds true.

Then the set $A_0 := A_0^1 \bigcap A_0^2$ is also of a positive measure. Now, let x_0 be an arbitrary density point of A_0. Choose a neighborhood $O_{\delta_0}(x_0)$ such that

$$\frac{\mu(A_0 \bigcap O_\delta(x_0))}{2\delta} > 1-\varepsilon_0 \quad (0 < \varepsilon_0 < 1)$$

as $\delta < \delta_0$. Define the function

$$h(x) = \begin{cases} \sqrt{\delta}\left(1+\frac{x-x_0}{\delta}\right), & \text{as} \quad x_0-\delta \le x \le x_0; \\ \sqrt{\delta}\left(1-\frac{x-x_0}{\delta}\right), & \text{as} \quad x_0 \le x \le x_0+\delta; \\ 0, & \text{otherwise.} \end{cases}$$

Then for $x \in (x_0-\delta; x_0+\delta)$ we obtain

$$h'^2(x) = \frac{1}{\delta}, \quad h^2(x) \le \delta. \quad (5.5)$$

Taking into account (5.4), we obtain for $\widetilde{h}(x) = h(x) \cdot h_0$:

$$\Phi''_K(y)(\widetilde{h},\widetilde{h}) = \int_a^b \left[-\frac{d}{dx}\left(\frac{\partial^2 f}{\partial y \partial z}(x,y,y')\right) + \frac{\partial^2 f}{\partial y^2}(x,y,y')\right](\widetilde{h}(x),\widetilde{h}(x))dx$$

$$+ \int_a^b \frac{\partial^2 f}{\partial z^2}(x,y,y')(\widetilde{h}'(x),\widetilde{h}'(x))dx$$

$$= \int_{x_0-\delta}^{x_0+\delta} \left[-\frac{d}{dx}\left(\frac{\partial^2 f}{\partial y \partial z}(x,y,y')\right) + \frac{\partial^2 f}{\partial y^2}(x,y,y')\right](h_0,h_0) \cdot h^2(x)dx$$

$$+ \int_{x_0-\delta}^{x_0+\delta} \frac{\partial^2 f}{\partial z^2}(x,y,y')(h_0,h_0) \cdot h'^2(x)dx.$$

Taking into account (5.5), we see from here that

$$\Phi''_K(y)(\widetilde{h},\widetilde{h}) \le C_0^2 \cdot \delta \cdot 2\delta + [(1-\varepsilon_0)\cdot 2\delta]\cdot\frac{1}{\delta}\cdot(-k^2) = 2C_0^2\cdot\delta^2 + 2(1-\varepsilon_0)\cdot(-k^2) < 0$$

for sufficiently small $\delta \in (0;\delta_0)$. It follows that $\Phi''_K(y)(t\widetilde{h},t\widetilde{h}) < 0$ for all $t \in \mathbb{R}\backslash\{0\}$. Since $\Phi'_K(y)(t\widetilde{h},t\widetilde{h}) = 0$, it immediately follows from the second-order Taylor formula for the line $\mathbb{R}\cdot\widetilde{h}$ that $\Phi(y+t\widetilde{h})-\Phi(y) < 0$ for sufficiently small t and, therefore, Φ does not realize a minimum on each compact ellipsoid $C_\varepsilon \subset W_2^1([a;b],H)$ for which $C_\varepsilon \bigcap\{\mathbb{R}\cdot\widetilde{h}\} \neq \{0\}$. Hence, Φ has no strong K-minimum at y, in contradiction with the hypothesis of Theorem. □

To obtain an analog of the Legendre–Jacobi sufficient condition for the K-extrema of variational functionals in W_2^1, we need to investigate preliminarily in $\overset{\circ}{W}{}_2^1([a;b],H)$ the quadratic functional

$$\widehat{\Phi}(h) = \int_a^b (P(h',h') + Q(h,h))\,dx, \tag{5.6}$$

where $P(x)$ and $Q(x)$ are bilinear continuous forms for every $x \in [a;b]$ and the mappings P and Q from $[a;b]$ into the space $(H,H)^* \cong (H,H)$ of the bilinear continuous forms over H are also continuous. Let us introduce the Jacobi condition for the functional (5.6).

Definition 5.5. Define a *Jacobi equation*

$$-\frac{d}{dx}(PU') + QU \overset{\text{a.e.}}{=} 0, \quad U(a) = 0, \quad U'(a) = I_H. \tag{5.7}$$

in the class of the mappings $U \in W_2^1([a;b],\mathcal{L}(H))$. We say that the functional (5.6) *satisfies the Jacobi condition* if, for each solution $U(x)$ of the equation (5.7), the operators $U(x)$ are continuously invertible as $a < x \le b$.

Remark 5.6. It can be obtained [Rem. 2.1][20] that the solutions of the equation (5.7) belong, in fact, to class C^1. In particular, $U'(a)$ is well defined.

Theorem 5.7. *Let $P(x) \gg 0$ for $a \le x \le b$ and the Jacobi condition from Definition 5.5 be fulfilled. Then the quadratic functional (5.6) is positive definite on $\overset{\circ}{W}{}_2^1([a;b], H)$.*

The proof can be found in [Thm. 2.1][20]. Now, let us formulate a K-analog of the sufficient Legendre–Jacobi extremum condition.

Theorem 5.8. *Let $f \in W^2K_2(z)$ and the assumptions of Theorem 5.4 be fulfilled. If for a K-extremal $y(\cdot) \in \overset{\circ}{W}{}_2^1([a;b], H)$,*

1) *$(\partial^2 f/\partial z^2)(x, y, y') \gg 0$ everywhere on $[a;b]$ (a strengthened Legendre condition);*
2) *for the quadratic functional $\Phi''_K(y)$ (see (4.14)) the Jacobi condition is fulfilled, i.e., for each solution $U(x)$ of the Jacobi equation*

$$-\frac{d}{dx}\left(\frac{\partial^2 f}{\partial z^2}(x,y,y')U'\right) + \left[-\frac{d}{dx}\left(\frac{\partial^2 f}{\partial y \partial z}(x,y,y')\right) + \frac{\partial^2 f}{\partial y^2}(x,y,y')\right]U \overset{a.e.}{=} 0$$

in the class $W_2^1([a;b], \mathcal{L}(H))$ with initial conditions $U(a) = 0$, $U'(a) = I_H$, the operator $U(x)$ is continuously invertible for all $a < x \le b$.

Then the Euler–Lagrange functional (4.1) has a strict K-minimum at $y(\cdot)$.

Proof. By virtue of Theorem 5.7, the quadratic functional (4.14) is positive definite on $\overset{\circ}{W}{}_2^1([a;b], H)$. Here, in view of (5.4),

$$P(x) = \frac{\partial^2 f}{\partial z^2}(x,y,y'), \quad Q(x) = -\frac{d}{dx}\left(\frac{\partial^2 f}{\partial y \partial z}(x,y,y')\right) + \frac{\partial^2 f}{\partial y^2}(x,y,y').$$

Hence, $\Phi''_K(y) \gg 0$ at the K-extremal $y(\cdot)$. Whence, by virtue of Theorem 3.7, Φ has a strong strict K-minimum at $y(\cdot)$. □

The Legendre–Jacobi condition for a K-maximum can be obtained, obviously, by changing the sign in condition 1) of Theorem 5.8. Let us give an example of application of Theorem 3.7.

Example 7. Consider once more the functional (5.1) from Example 5. In this case, the variational Euler–Lagrange equation (5.2) takes the form

$$\frac{d\varphi}{dt}(y - y'') - 2\frac{d^2\varphi}{dt^2}(y + y'')(y')^2 = 0. \tag{5.8}$$

Thus, the function $y(x) \equiv 0$ satisfies the equation (5.8), i.e., it is a K-extremal for the functional (5.1). Direct calculation shows that

$$\frac{\partial^2 f}{\partial y^2} = 4\frac{d^2\varphi}{dt^2}y^2 + 2\frac{d\varphi}{dt}, \quad \frac{\partial^2 f}{\partial y \partial z} = 4\frac{d^2\varphi}{dt^2}yz, \quad \frac{\partial^2 f}{\partial z^2} = 4\frac{d^2\varphi}{dt^2}z^2 + 2\frac{d\varphi}{dt}, \tag{5.9}$$

whence $(\partial^2 f/\partial y^2) = 2 > 0$ on the K-extremal, and the Jacobi equation takes, on the K-extremal $y(x) \equiv 0$, the form

$$U'' - U = 0; \quad U(0) = 0,\ U'(0) = 1.$$

The solution $U(x) = shx$ to this equation is not zero on $(0;1]$, i.e., it satisfies the Jacobi condition in the class $W_2^1([0;1],\mathbb{R})$. Hence, by Theorem 5.8, the functional (5.1) has a strong strict K-minimum at zero. As was noted above, this minimum is not local.

Let us give one more sufficient condition for a strong K-extremum of the variational functional (4.1) in W_2^1, based on Theorems 3.7 and 3.11.

Lemma 5.9. *Let $f \in W^2K_2(z)$ and $y(\cdot)$ be a K-extremal of the functional* (4.1) *in $\overset{\circ}{W}{}_2^1([a;b],H)$. If $f''(x,y,y') \gg 0$ (in (y,z)) for all $x \in [a;b]$, then the functional* (4.1) *has a strong strict K-minimum at $y(\cdot)$.*

Proof. According to the well-known positive definiteness test (see, e.g., [15]), for every $x \in [a;b]$ there exists $\alpha(x) > 0$ such that

$$\begin{aligned} f''(x,y,y')\cdot((h_1,h_2),(h_1,h_2)) &= \frac{\partial^2 f}{\partial y^2}(h_1,h_1) + 2\frac{\partial^2 f}{\partial y\partial z}(h_1,h_2) + \frac{\partial^2 f}{\partial z^2}(h_2,h_2) \\ &\geq \alpha(x)\cdot(\|h_1\|^2 + \|h_2\|^2) \end{aligned}$$

for all $(h_1,h_2) \in H^2$. Using the compactness of $[a;b]$, we can choose $\alpha > 0$ such that α does not depend on x. In particular,

$$\begin{aligned} &\frac{\partial^2 f}{\partial y^2}(h(x),h(x)) + 2\frac{\partial^2 f}{\partial y\partial z}(h(x),h'(x)) + \frac{\partial^2 f}{\partial z^2}(h'(x),h'(x)) \\ &\quad \geq \alpha(x)\cdot(\|h(x)\|^2 + \|h'(x)\|^2) \end{aligned}$$

for all $x \in [a;b]$ and $h(\cdot) \in \overset{\circ}{W}{}_2^1([a;b],H)$. It follows from here and (4.14) that

$$\Phi''_K(y)(h,h) \geq \alpha\cdot\left(\int_a^b \|h(x)\|^2dx + \int_a^b \|h'(x)\|^2dx\right) = \alpha\cdot\|h(x)\|^2_{W_2^1},$$

that is $\Phi''_K(y) \gg 0$. Whence, by Theorem 3.7, Φ has a strong strict K-minimum at $y(\cdot)$. □

Theorem 5.10. *Let, under the assumptions of Lemma* 5.9, *the inequalities*

1) $(\partial^2 f/\partial y^2)(x,y,y') \gg 0$, $(\partial^2 f/\partial z^2)(x,y,y') \gg 0$;
2) $\Delta_1^2 f(x,y,y') \gg 0$, $\Delta_2^1 f(x,y,y') \gg 0$;

be fulfilled on the K-extremal $y(\cdot)$ for all $x \in [a;b]$. Then the variational functional (4.1) *has a strong strict K-minimum at $y(\cdot)$.*

Proof. According to Theorem 3.11, the assumptions 1)–2) of Theorem 5.10 provide the positive definiteness of the form $f''(x,y,y')$ for all $x \in [a;b]$. It remains to apply Lemma 5.9. □

The reversal of all signs in inequalities 1)–2) of Theorem 5.10 leads, obviously, to conditions for a strong strict K-maximum.

Note also that assumptions 1)–2) of Theorem 5.10 can be applied to the variational functional (5.1) (see Examples 5–7 above). Here, in view of (5.9), on the extremal $y(x) \equiv 0$,

$$\frac{\partial^2 f}{\partial y^2} = 2, \quad \frac{\partial^2 f}{\partial y \partial z} = 0, \quad \frac{\partial^2 f}{\partial z^2} = 2,$$

and hence $\triangle_1^2 f = \triangle_2^1 f = 2 > 0$. This confirms again the existence of a strict K-minimum of Φ at zero.

Final Remarks

1) In the work [17], a nontrivial generalization of Theorem 5.10 to the case of variational functionals of several variables is obtained. The analogous necessary conditions for K-extrema are also considered in there.
2) The notion of local realization of a strong K-extremum (see Def. 2.13) can be extended to a weak K-extremum. In particular, the estimates (2) (Example 6) make possible local realization of the weak K-minimum from Example 6 in Sobolev space W_2^2.
3) In most cases, the results of this paper can be extended to mappings taking values in locally convex spaces from a rather extensive class of so-called semi-nuclear spaces (see [21]).

References

[1] M.Z. Zgurovsky, *Calculus of variationals in the Banach spaces.* Naukova Dumka, Kiev, 1999. (In Russian)

[2] A.V. Uglanov, V.S. Melnik, *Nonlinear analysis and control of infinite-dimensional systems.* Mathematical Transactions **191** (2000), no. 10, 105–118. (In Russian)

[3] B. Ricceri, *Integral functionals on Sobolev spaces having multiple local minima.* arXiv:math.OC/0402445, **1** (2004).

[4] M.M. Vainberg, *Variational method and method of monotone operators.* Nauka, Moskow, 1972. (In Russian)

[5] I.V. Skrypnik, *Nonlinear elliptic high order equations.* Naukova Dumka, Kiev, 1973. (In Russian)

[6] H.H. Shaefer, *Topological vector spaces.* MacMillan Comp., New York–London, 1966.

[7] R.E. Edwards, *Functional analysis. Theory and applications.* Holt, Rinehart and Winston, New York, 1965.

[8] V.A. Trenogin, B.M. Pisarevsky, T.S. Soboleva, *Problems and exercises in functional analysis.* Nauka, Moscow, 1984. (In Russian)

[9] Yu.V. Bogdansky, G.B. Podkolzin, Yu.A. Chapovsky, *Functional analysis. Collection of exercises.* Politechnika, Kyiv, 2005. (In Ukrainian)

[10] W. Rudin, *Basic mathematical analysis.* Mir, Moscow, 1975. (In Russian)

[11] Yu.M. Berezansky, Z.G. Sheftel, G.F. Us, *Functional Analysis.* Vol.1, Birkhäuser Verlag, Basel–Boston–Berlin, 1995.

[12] V.A. Zorich, *Mathematical analysis.* Vol. 2, Nauka, Moskow, 1984. (In Russian)

[13] I.V. Orlov, *Hilbert compacta, compact ellipsoids and compact extrema.* Contemporary mathematics. Fundamental directions, Vol. 29, 2008, 165–175. (In Russian)

[14] E.V. Bozhonok, *Example of K-continuous, but discontinuous variation functional in Sobolev space.* Dynamic Systems **22** (2007), 140–144. (In Russian)

[15] H. Cartan, *Calcul différentiel. Formes différentielles.* Hermann, Paris, 1967.

[16] I.V. Orlov, *Extreme Problems and Scales of the Operator Spaces*, North–Holland Math. Studies. Funct. Anal. & Appl., Elsevier, Amsterdam–Boston–... **197** (2004), 209–227.

[17] E.V. Bozhonok, *Sufficient and necessary functional extreme conditions in nuclear locally convex space (case of several variables).* Scientific Notes of Taurida National University, **18(57)** (2005), no. 1, 3–26. (In Russian)

[18] I.V. Orlov, *K-differentiability and K-extrema.* Ukrainian Mathematical Bulletin **3** (2006), no. 1, 97–115. (In Russian)

[19] I.V. Orlov, *Normal differentiability and functional extremum on locally convex spaces.* Cybernetics and System Analysis (2002), no. 4, 23–34. (In Russian)

[20] E.V. Bozhonok, I.V. Orlov, *Legendre and Jacobi conditions of compact extrema for variation functionals in Sobolev spaces.* Proceedings of Institute of Mathematics, NAS of Ukraine **3** (2006), no. 4, 282–293. (In Russian)

[21] E.V. Bozhonok, *Pseudoquadratic functionals in Sobolev space.* Integral Transforms and Special Functions, (to appear).

[22] A.Ya. Khelemsky, *Lessons in functional analysis.* MCNMO, Moscow, 2004. (In Russian)

I.V. Orlov
Taurida National University
4 V.Vernadsky Ave.
95007 Simferopol, Ukraine
e-mail: `old@crimea.edu`

Operator Theory:
Advances and Applications, Vol. 190, 419–428

On Extremal Problem for Algebraic Polynomials in Loading Spaces

B.P. Osilenker

Abstract. We consider discrete loading spaces under the inner product

$$\langle f,g\rangle = \int_{\mathbb{R}} f(x)g(x)d\mu(x) + \sum_{j=1}^{m} M_j f(x_j)g(x_j)$$
$$(M_j \geq 0,\ \ x_j \in \mathbb{R}\ \ (j=1,2,\dots,m)),$$

where μ be a finite positive Borel measure such that the moments are finite and support is an infinite set. In this spaces we study the problem of finding

$$\inf_{a_0,a_1,\dots,a_{N-r}} \Big\{ \langle \Pi_N^{(r)}, \Pi_N^{(r)} \rangle ; \Pi_N^{(r)}(x) = \sum_{k=0}^{N-r} a_k x^k + \sum_{k=N-r+1}^{N} a_k^0 x^k \Big\},$$

where $a_N^0 > 0, a_{N-1}^0, \dots, a_{N-r+1}^0$ are fixed real numbers. The extremal polynomials are also constructed.

Mathematics Subject Classification (2000). Primary: 41A10; Secondary: 33C45.

Keywords. Extremal problems, Orthogonal polynomials, Loading spaces.

1. Introduction

Let $\Phi(x)$ be a positive linear functional on the linear space $\mathbb{P}$ of polynomials with real coefficients. Define by

$$\hat{q}_n(x) = \hat{k}_n^{(n)} x^n + \hat{k}_{n-1}^{(n)} x^{n-1} + \cdots, \quad \hat{k}_n^{(n)} > 0 \qquad (n \in \mathbb{Z}_+)$$

polynomials orthonormal with respect to (w.r.t.) the linear functional Φ:

$$\Phi(\hat{q}_m, \hat{q}_n) = \delta_{m,n} (m, n \in \mathbb{Z}_+).$$

Define by $\Re_N^{(r)}$ a class of all polynomials $\Pi_N(x) \in \mathbb{P}$ of degree N with r-fixed coefficients:

$$\Re_N^{(r)} = \Big\{ \Pi_N^{(r)}, \Pi_N^{(r)}(x) = \sum_{j=N-r+1}^{N} a_j^0 x^j + \sum_{j=0}^{N-r} a_j x^j, a_N^0 > 0 \Big\},$$

where $a_N^0, a_{N-1}^0, \dots, a_{N-r+1}^0$ are fixed real numbers.

We expand polynomial $\Pi_N^{(r)}(x)$ by the system $\{\hat{q}_n\}(n \in \mathbb{Z}_+)$:

$$\Pi_N^{(r)}(x) = \sum_{j=0}^{N} \alpha_j^{(N)} \hat{q}_j(x).$$

As in [1], one gets the following statement.

Theorem 1.

i) *The following representation*

$$\inf_{\Pi_N^{(r)} \in \Re_N^{(r)}} \Phi\{|\Pi_N^{(r)}(x)|^2\} = \sum_{j=N-r+1}^{N} [\alpha_j^{(N),0}]^2$$

holds, where the coefficients $\alpha_s^{(N),0}(s = N-r+1, N-r+2, \ldots, N-1, N)$ *are the solutions of the system*

$$\sum_{j=s}^{N} \hat{k}_s^{(j)} \alpha_j^{(N)} = a_s^0 (s = N-r+1, N-r+2, \ldots, N-1, N);$$

ii) *extremal polynomials is*

$$\Pi_N^{(r),extr}(x) = \sum_{j=N-r+1}^{N} \alpha_j^{(N),0} \hat{q}_j(x).$$

We consider this extremal problem in a discrete loading space. On the space $\mathbb{P}$ we introduce a discrete loading inner product

$$\langle p, q \rangle = \int_{\mathbb{R}} p(x)q(x)d\mu(x) + \sum_{j=1}^{m} M_j p(x_j) q(x_j) \quad (p, q \in \mathbb{P}),$$

where μ be a finite positive Borel measure on $\mathbb{R}$ whose moments are finite and whose support is an infinite set, $M_j \geq 0,\ x_j \in \mathbb{R}\ (j = 1, 2, \ldots, m)$. In what follows we assume that support of measure μ without the discrete part.

Completion of $\mathbb{P}$ w.r.t. the norm $\|f\|^2 = \langle f, f \rangle$ we will call a discrete loading space S w.r.t. the inner product

$$\langle f, g \rangle = \int_{\mathbb{R}} f(x)g(x)d\mu(x) + \sum_{j=1}^{m} M_j f(x_j) g(x_j). \tag{1.1}$$

Let $q_n(x)(n \in \mathbb{Z}_+)$ be the polynomials orthonormal w.r.t. the inner product (1.1):

$$\langle q_m, q_n \rangle = \delta_{m,n}.$$

We will call their a discrete loading orthonormal polynomials.

Space S and polynomial system $\{q_n\}(n \in \mathbb{Z}_+)$ have attracted the interest of many researches. They play an important role in some problems of functional analysis, function theory and mathematical physics [2]–[6].

Polynomial system $\{q_n\}(n \in \mathbb{Z}_+)$ is used in some investigations of boundary problems with spectral parameter in the boundary conditions; differential equation with discontinuous coefficients; the loading integral equations.

Example. (*A discrete loading Jacobi polynomials*).
First, we remind some properties of the classical Jacobi polynomials $P_n^{(\alpha,\beta)}(x)$ orthogonal on interval (-1,1) w.r.t the Jacobi measure

$$d\mu_{\alpha,\beta}(x) = \frac{\Gamma(\alpha+\beta+2)}{2^{\alpha+\beta+1}\Gamma(\alpha+1)\Gamma(\beta+1)}(1-x)^\alpha(1+x)^\beta dx \qquad (\alpha,\beta > -1) \quad (1.2)$$

(see [7]).

By the Rodrigues formula

$$P_n^{(\alpha,\beta)}(x) = \frac{(-1)^n}{n!2^n}(1-x)^{-\alpha}(1+x)^{-\beta}[(1-x)^{n+\alpha}(1+x)^{n+\beta}]^{(n)}$$
$$(x \in (-1,1); n \in \mathbb{Z}_+).$$

One has

$$P_n^{(\alpha,\beta)}(x) = \frac{1}{n!2^n}\frac{\Gamma(2n+\alpha+\beta+1)}{\Gamma(n+\alpha+\beta+1)}x^n + \cdots .$$

For classical orthogonal Jacobi polynomials the values at the points ± 1

$$P_n^{(\alpha,\beta)}(1) = \frac{(\alpha+1)_n}{n!}, \quad P_n^{(\alpha,\beta)}(-1) = (-1)^n\frac{(\beta+1)_n}{n!},$$

where "shifted factorial" (Pochhammer symbol) is defined by

$$(a)_n = a(a+1)(a+2)\cdots(a+n-1) = \frac{\Gamma(n+a)}{\Gamma(a)} \quad (n = 1,2,\ldots), \ (a)_0 = 1.$$

The squared norm of classical Jacobi orthogonal polynomials is

$$||P_n^{(\alpha,\beta)}||^2_{d\mu_{\alpha,\beta}} := \int_{-1}^{1}[P_n^{(\alpha,\beta)}(x)]^2 d\mu_{\alpha,\beta}(x)$$
$$= \frac{1}{2n+\alpha+\beta+1}\frac{\Gamma(\alpha+\beta+2)}{\Gamma(\alpha+1)\Gamma(\beta+1)}\frac{\Gamma(n+\alpha+1)\Gamma(n+\beta+1)}{n!\Gamma(n+\alpha+\beta+1)}.$$

Let $\widehat{P}_n^{(\alpha,\beta)}(x)$ be polynomials orthonormal on the interval $(-1,1)$ w.r.t. the measure $d\mu_{\alpha,\beta}(x)$ (see (1.2)). Then

$$\widehat{P}_n^{(\alpha,\beta)}(x) = \sqrt{\frac{\Gamma(\alpha+1)\Gamma(\beta+1)}{\Gamma(\alpha+\beta+2)}}\frac{1}{2^n}\sqrt{\frac{2n+\alpha+\beta+1}{n!}}$$
$$\cdot\frac{\Gamma(2n+\alpha+\beta+1)}{\sqrt{\Gamma(n+\alpha+1)\Gamma(n+\beta+1)\Gamma(n+\alpha+\beta+1)}}x^n + \cdots . \qquad (1.3)$$

We consider the inner product

$$\langle f,g\rangle_{\alpha,\beta} = \int_{-1}^{1} f(x)g(x)d\mu_{\alpha,\beta}(x) + Lf(1)g(1) + Mf(-1)g(-1), \quad L,\, M \geq 0. \quad (1.4)$$

and define by $\widehat{P}_n^{\alpha,\beta;L,M}(x)(n \in \mathbb{Z}_+)$ polynomial system orthonormal w.r.t. the inner product (1.4). These polynomials were introduced by T. Koornwinder [8]. It is called the loading Jacobi polynomials (or generalized Jacobi polynomials). It should be noted that the loading Jacobi polynomials have some properties other than the classical Jacobi polynomials (behaviour at the points ± 1; linear differential operator, for which polynomials $\widehat{P}_n^{(\alpha,\beta;L,M)}(x)$ are eigenfunctions and so on; see [9]–[12]).

2. Extremal problem for algebraic polynomials in a discrete loading space

Let

$$p_n(x) = \sum_{i=0}^{n} l_i^{(n)} x^i, \quad l_n^{(n)} > 0 \ (n \in \mathbb{Z}_+) \tag{2.1}$$

be the polynomials orthonormal w.r.t. the measure μ:

$$(p_m, p_n) = \int_{\mathbb{R}} p_m(x) p_n(x) d\mu(x) = \delta_{m,n} \quad (m, n \in \mathbb{Z}_+) \tag{2.2}$$

and let

$$q_n(x) = \sum_{s=0}^{n} k_s^{(n)} x^s, \quad k_n^{(n)} > 0 \ (n \in \mathbb{Z}_+) \tag{2.3}$$

be the polynomials orthonormal w.r.t. the inner product (1.1).

We expand polynomial $q_n(x)$ by basis $\{p_n\}(n \in \mathbb{Z}_+)$:

$$q_n(x) = \sum_{s=0}^{n} a_s^{(n)} p_s(x). \tag{2.4}$$

Using (2.1)–(2.3) and comparing the coefficients at x^n in the relation (2.4), one obtains

$$a_s^{(n)} = \int_{\mathbb{R}} q_n(x) p_s(x) d\mu(x) \quad (s = 0, 1, \dots, n-1) \tag{2.5}$$

and

$$a_n^{(n)} = \frac{k_n^{(n)}}{l_n^{(n)}} = \int_{\mathbb{R}} q_n(x) p_n(x) d\mu(x) \quad (n \in \mathbb{Z}_+). \tag{2.6}$$

Substituting relations (2.5) and (2.6) in (2.4) ,one gets

$$q_n(x) = \frac{k_n^{(n)}}{l_n^{(n)}} p_n(x) - \sum_{j=1}^{m} M_j q_n(x_j) D_{n-1}(x_j, x), \tag{2.7}$$

where

$$D_n(t, x) = \sum_{i=0}^{n} p_i(t) p_i(x) \quad (t, x \in \mathbb{R}; n \in \mathbb{Z}_+)$$

is the Dirichlet kernel of the system $\{p_n\}$ $(n \in \mathbb{Z}_+)$.

Taking into account (2.7) at the points x_i $(i = 1, 2, \ldots, m)$, one has

$$[1 + M_i D_{n-1}(x_i, x_i)]q_n(x_i) + \sum_{j=1, j\neq i}^{m} M_j D_{n-1}(x_j, x_i) q_n(x_j) = \frac{k_n^{(n)}}{l_n^{(n)}} p_n(x_i) \qquad (2.8)$$

$$(i = 1, 2, \ldots, m).$$

Denote by

$$\mathbb{D}_{n-1} := \mathbb{D}_{n-1}\begin{pmatrix} x_1, & x_2, & \ldots, & x_m \\ x_1, & x_2, & \ldots, & x_m \end{pmatrix}$$

$$= \begin{pmatrix} D_{n-1}(x_1, x_1) & D_{n-1}(x_2, x_1) & \ldots & D_{n-1}(x_m, x_1) \\ D_{n-1}(x_1, x_2) & D_{n-1}(x_2, x_2) & \ldots & D_{n-1}(x_m, x_2) \\ \ldots & \ldots & \ldots & \ldots \\ D_{n-1}(x_1, x_m) & D_{n-1}(x_2, x_m) & \ldots & D_{n-1}(x_m, x_m) \end{pmatrix}.$$

Note that the $m \times m$ matrix $\mathbb{D}_{n-1}$ is symmetric and positive definite.

Similarly we put

$$P_n = (p_n(x_1), p_n(x_2), \ldots, p_n(x_m))^T, \ Q_n = (q_n(x_1), q_n(x_2), \ldots, q_n(x_m))^T,$$

where T is transposition, and

$$M = \operatorname{diag}(M_1, M_2, \ldots, M_m).$$

Then the system (2.8) one can rewrite in the form

$$\left[I + \mathbb{D}_{n-1}\begin{pmatrix} x_1, & x_2, & \ldots, & x_m \\ x_1. & x_2, & \ldots, & x_m \end{pmatrix} M\right] Q_n = \frac{k_n^{(n)}}{l_n^{(n)}} P_n, \qquad (2.9)$$

where I is the identity matrix of order m. Denote by

$$\Delta_{n-1} = |I + \mathbb{D}_{n-1}\begin{pmatrix} x_1, & x_2, & \ldots, & x_m \\ x_1, & x_2, & \ldots, & x_m \end{pmatrix} M|. \qquad (2.10)$$

It follows from (2.9) by Cramer's rule

$$q_n(x_j) = \frac{k_n^{(n)}}{l_n^{(n)}} \frac{\tilde{\Delta}_{n-1}^{(j)}}{\Delta_{n-1}}, \qquad (2.11)$$

where $\tilde{\Delta}_{n-1}^{(j)}$ is a determinant was obtained from Δ_{n-1} by substitution j-column by column matrix P_n.

Theorem 2. *For a loading orthonormal polynomials $q_n(x)$ the following representation holds*

$$q_n(x) = \sqrt{\frac{\Delta_{n-1}}{\Delta_n}} p_n(x) - \frac{1}{\sqrt{\Delta_{n-1}\Delta_n}} \sum_{s=1}^{n-1} M_s \tilde{\Delta}_{n-1}^{(s)} D_{n-1}(x_s, x). \qquad (2.12)$$

Proof. We expand polynomial $p_n(x)$ by basis $\{q_n\}$ $(n \in \mathbb{Z}_+)$:

$$p_n(x) = \sum_{j=0}^{n} b_j^{(n)} q_j(x). \qquad (2.13)$$

Then, as above

$$b_n^{(n)} = \frac{l_n^{(n)}}{k_n^{(n)}} = \langle p_n, q_n\rangle (n \in \mathbb{Z}_+) \tag{2.14}$$

and

$$p_n(x) = \frac{l_n^{(n)}}{k_n^{(n)}} q_n(x) + \sum_{j=0}^{n-1} b_j^{(n)} q_j(x).$$

Taking into account (2.6) and by definition of the inner product (1.1), one has

$$\frac{k_n^{(n)}}{l_n^{(n)}} = \int_{\mathbb{R}} q_n(x)p_n(x)d\mu(x) = \langle p_n, q_n\rangle - \sum_{j=1}^{m} M_j q_n(x_j)p_n(x_j).$$

It follows from (2.14) that

$$\frac{k_n^{(n)}}{l_n^{(n)}} = \frac{l_n^{(n)}}{k_n^{(n)}} - \sum_{j=1}^{m} M_j q_n(x_j)p_n(x_j).$$

By (2.11) one obtains

$$\frac{l_n^{(n)}}{k_n^{(n)}} = \frac{1}{\Delta_{n-1}} \frac{k_n^{(n)}}{l_n^{(n)}} \Big[\Delta_{n-1} + \sum_{j=1}^{m} M_j p_n(x_j)\tilde{\Delta}_{n-1}^{(j)}\Big]. \tag{2.15}$$

We show that

$$\Delta_{n-1} + \sum_{j=1}^{m} M_j p_n(x_j)\tilde{\Delta}_{n-1}^{(j)} = \Delta_n. \tag{2.16}$$

It is not difficult to see from (2.10) that

$$\Delta_n = |I + \mathbb{D}_{n-1}M + \mathbb{H}_{n-1}M|,$$

where

$$\mathbb{H}_{n-1} := \mathbb{H}_{n-1}\begin{pmatrix} x_1, & x_2, & \dots & , x_m \\ x_1, & x_2, & \dots & , x_m \end{pmatrix} = \begin{pmatrix} p_n^2(x_1) & p_n(x_1)p_n(x_2) & \dots & p_n(x_1)p_n(x_m) \\ p_n(x_1)p_n(x_2) & p_n^2(x_2) & \dots & p_n(x_2)p_n(x_m) \\ \dots & \dots & \dots & \dots \\ p_n(x_1)p_n(x_m) & p_n(x_2)p_n(x_m) & \dots & p_n^2(x_m) \end{pmatrix}.$$

Consequently, using properties of the determinant, one gets (2.16).

Taking into account (2.15), one obtains

$$\frac{l_n^{(n)}}{k_n^{(n)}} = \frac{\Delta_n}{\Delta_{n-1}} \frac{k_n^{(n)}}{l_n^{(n)}} (n \in \mathbb{Z}_+).$$

Then

$$k_n^{(n)} = \sqrt{\frac{\Delta_{n-1}}{\Delta_n}} l_n^{(n)} \quad (n \in \mathbb{Z}_+). \tag{2.17}$$

Finally, substituting expression (2.11) and (2.17) in (2.7), one obtains (2.12). Theorem 2 is completely proved. □

Corollary 1. *For the coefficients of a loading orthonormal polynomials $q_n(x)$ ($n \in \mathbb{Z}_+$) (see (2.3)) the following representation*

$$k_i^{(n)} = \sqrt{\frac{\Delta_{n-1}}{\Delta_n}} l_i^{(n)} - \frac{1}{\sqrt{\Delta_{n-1}\Delta_n}} \sum_{j=1}^{n-1} M_j \tilde{\Delta}_{n-1}^{(j)} \sum_{s=i}^{n-1} p_s(x_j) l_i^{(s)} \tag{2.18}$$

$$(i = 0, 1, \ldots, n-1)$$

holds, where $l_i^{(n)} (i = 0, 1, \ldots, n-1)$are the coefficients of polynomials $p_n(x)$ (see (2.1)).

Corollary 1 follows from (2.1), (2.3) and (2.12) by comparing the coefficients at x^i $(i = 0, 1, 2, \ldots, n-1)$. In particular, one obtains

$$k_{n-1}^{(n)} = \sqrt{\frac{\Delta_{n-1}}{\Delta_n}} l_{n-1}^{(n)} - \frac{1}{\sqrt{\Delta_{n-1}\Delta_n}} \Big[\sum_{j=1}^{n-1} M_j \tilde{\Delta}_{n-1}^{(j)} p_{n-1}(x_j)\Big] l_{n-1}^{(n-1)} \tag{2.19}$$

and

$$\frac{k_{n-1}^{(n)}}{k_n^{(n)}} = \frac{l_{n-1}^{(n)}}{l_n^{(n)}} - \frac{1}{\Delta_{n-1}} \Big[\sum_{j=1}^{n-1} M_j \tilde{\Delta}_{n-1}^{(j)} p_{n-1}(x_j)\Big] \frac{l_{n-1}^{(n-1)}}{l_n^{(n)}}. \tag{2.20}$$

Using Theorem 1 and (2.7), (2.19), one gets

Corollary 2.*For the monic polynomials*

$$\Pi_N^{(1)}(x) = x^N + \sum_{s=0}^{N-1} a_s^{(N-1)} x^s, \quad \Pi_N^{(2)}(x) = x^N - \sigma x^{N-1} + \sum_{s=0}^{N-2} b_s^{(N-2)} x^s$$

(σ is a fixed real number) the following assertions are valid:

i)
$$\begin{aligned} \kappa_N^{(1)} &:= \inf_{\Pi_N^{(1)} \in \Re_N^{(1)}} \langle \Pi_N^{(1)}, \Pi_N^{(1)} \rangle = \frac{\Delta_N}{\Delta_{N-1}} \frac{1}{(l_N^{(N)})^2}; \\ \Pi_N^{(1),\mathrm{extr}}(x) &= \sqrt{\frac{\Delta_N}{\Delta_{N-1}}} \frac{1}{l_N^{(N)}} q_N(x); \end{aligned} \tag{2.21}$$

ii) (*Zolotarev's problem in the metric of a discrete loading space S*):

$$\kappa_N^{(2)} := \inf_{\Pi_N^{(2)} \in \Re_N^{(2)}} \langle \Pi_N^{(2)}, \Pi_N^{(2)} \rangle = \frac{\Delta_N}{\Delta_{N-1}} \frac{1}{(l_N^{(N)})^2} + \frac{\Delta_{N-1}}{\Delta_{N-2}} \frac{1}{(l_{N-1}^{(N-1)})^2} \Big[\sigma + \frac{k_{N-1}^{(N)}}{k_N^{(N)}}\Big]^2$$

and

$$\Pi_N^{(2),\mathrm{extr}}(x) = \sqrt{\frac{\Delta_N}{\Delta_{N-1}}} \frac{1}{l_N^{(N)}} q_N(x) - \sqrt{\frac{\Delta_{N-1}}{\Delta_{N-2}}} \frac{1}{l_{N-1}^{(N-1)}} \Big[\sigma + \frac{k_{N-1}^{(N)}}{k_N^{(N)}}\Big] q_{N-1}(x).$$

Using relations (2.19) and (2.20) one can calculate $\kappa_N^{(1)}$, $\kappa_N^{(2)}$ and $\Pi_N^{(1),\mathrm{extr}}(x)$, $\Pi_N^{(2),\mathrm{extr}}(x)$.

Remark. *It should be noted that one can calculate the determinants using the following formula* [13]: *determinant*

$$\begin{vmatrix} a_{11}+1 & a_{12} & \dots & a_{1n} \\ a_{21} & a_{22}+1 & \dots & a_{2n} \\ \dots & \dots & \dots & \dots \\ a_{n1} & a_{n2} & \dots & a_{nn}+1 \end{vmatrix}$$

is equal to

$$1+\sum_{k=1}^{n} S_k,$$

where $S_k(k=1,2,\dots,S_n)$ *are sum of main minors of order* k *for*

$$\begin{vmatrix} a_{11} & a_{12} & \dots & a_{1n} \\ a_{21} & a_{22} & \dots & a_{2n} \\ \dots & \dots & \dots & \dots \\ a_{n1} & a_{n2} & \dots & a_{nn} \end{vmatrix}$$

Example. *A discrete loading Jacobi polynomials.* We put $x_1 = 1, x_2 = -1$. Then for Dirichlet kernel $D_n(t,x)$ we have the following representation (see [7]):

$$\begin{cases} D_{n-1}^{(\alpha,\beta)}(1,1) = \dfrac{\lambda^2}{2^{\alpha+\beta+1}\Gamma(\alpha+1)\Gamma(\alpha+2)} \dfrac{\Gamma(n+\alpha+1)\Gamma(n+\alpha+\beta+1)}{\Gamma(n)\Gamma(n+\beta)}; \\ D_{n-1}^{(\alpha,\beta)}(-1,1) = \dfrac{\lambda^2}{2^{\alpha+\beta+1}\Gamma(\beta+1)\Gamma(\beta+2)} \dfrac{\Gamma(n+\beta+1)\Gamma(n+\alpha+\beta+1)}{\Gamma(n)\Gamma(n+\alpha)}; \\ D_{n-1}^{(\alpha,\beta)}(1,-1) = D_{n-1}^{(\alpha,\beta)}(-1,1) = \dfrac{(-1)^{n-1}\lambda^2}{2^{\alpha+\beta+1}\Gamma(\alpha+1)\Gamma(\beta+1)} \dfrac{\Gamma(n+\alpha+\beta+1)}{\Gamma(n)}, \end{cases}$$

where

$$\lambda^2 = \frac{2^{\alpha+\beta+1}\Gamma(\alpha+1)\Gamma(\beta+1)}{\Gamma(\alpha+\beta+2)}.$$

Substituting last formulas in (2.10) for

$$\Delta_{n-1} = \begin{vmatrix} 1+D_{n-1}^{(\alpha,\beta)}(1,1) & D_{n-1}^{(\alpha,\beta)}(-1,1) \\ D_{n-1}^{(\alpha,\beta)}(1,-1) & 1+D_{n-1}^{(\alpha,\beta)}(-1,-1) \end{vmatrix}$$

we obtain

$$\begin{aligned} \Delta_{n-1} = \Delta_{n-1}^{\alpha,\beta;L,M} &= 1 + \frac{\Gamma(\beta+1)}{\Gamma(\alpha+2)\Gamma(\alpha+\beta+2)} \frac{\Gamma(n+\alpha+1)\Gamma(n+\alpha+\beta+1)}{\Gamma(n)\Gamma(n+\beta)} L \\ &+ \frac{\Gamma(\alpha+1)}{\Gamma(\beta+2)\Gamma(\alpha+\beta+2)} \frac{\Gamma(n+\beta+1)\Gamma(n+\alpha+\beta+1)}{\Gamma(n)\Gamma(n+\alpha)} M \\ &+ \frac{1}{(\alpha+1)(\beta+1)\Gamma^2(\alpha+\beta+2)} \\ &\times \frac{\Gamma(n+\alpha+\beta+1)\Gamma(n+\alpha+\beta+2)}{\Gamma(n-1)\Gamma(n)} LM. \end{aligned}$$

It follows from (1.3) and (2.17) that

$$\widehat{P}_n^{\alpha,\beta;L,M}(x) = \frac{1}{2^n}\sqrt{\frac{\Gamma(\alpha+1)\Gamma(\beta+1)(2n+\alpha+\beta+1)}{\Gamma(\alpha+\beta+2)\Gamma(n+1)}}$$
$$\times \frac{\Gamma(2n+\alpha+\beta+1)}{\sqrt{\Gamma(n+\alpha+1)\Gamma(n+\beta+1)\Gamma(n+\alpha+\beta+1)}}$$
$$\times \sqrt{\frac{\Delta_n^{\alpha,\beta;L,M}}{\Delta_{n-1}^{\alpha,\beta;L,M}}}\, x^n + \cdots .$$

By (2.21) we obtain the following solution of extremal problem for a discrete loading Jacobi space.

Theorem 3. *The following representation*

$$\inf_{\Pi_N^{(1)}\in\Re_N^{(1)}} \langle \Pi_N^{(1)}, \Pi_N^{(1)}\rangle_{\alpha,\beta} = \frac{\Gamma(\alpha+\beta+2)}{\Gamma(\alpha+1)\Gamma(\beta+1)}\frac{2^{2N}N!}{2N+\alpha+\beta+1}$$
$$\times \frac{\Gamma(N+\alpha+\beta+1)\Gamma(N+\alpha+1)\Gamma(N+\beta+1)}{\Gamma^2(2N+\alpha+\beta+1)}\frac{\Delta_N^{\alpha,\beta;L,M}}{\Delta_{N-1}^{\alpha,\beta;L,M}}$$

holds. In addition, equality in the last relation is realized on the following polynomial

$$\Pi_N^{(1),extr} = \sqrt{\frac{\Gamma(\alpha+\beta+2)}{\Gamma(\alpha+1)\Gamma(\beta+1)}\frac{2^N\sqrt{N!}}{\sqrt{2N+\alpha+\beta+1}}}$$
$$\times \frac{\sqrt{\Gamma(N+\alpha+1)\Gamma(N+\beta+1)\Gamma(N+\alpha+\beta+1)}}{\Gamma(2N+\alpha+\beta+1)}$$
$$\times \sqrt{\frac{\Delta_N^{\alpha,\beta;L,M}}{\Delta_{N-1}^{\alpha,\beta;L,M}}}\,\widehat{P}_N^{\alpha,\beta;L,M}(x),$$

where $\widehat{P}_n^{\alpha,\beta;L,M}(x)$ *is a polynomial system orthonormal w.r.t. the inner product* (1.4).

Acknowledgment

The research of the author was supported by the Russian Foundation for Basic Research (Grant 00-01-00286)

References

[1] B.P. Osilenker, *On extremal problem for algebraic polynomials in a symmetric discrete Gegenbauer–Sobolev space.* Matem Zametki **82** (2007), 411–425 (in Russian).

[2] F.V. Atkinson, *Discrete and continuous boundary problems.* Academic Press, 1964.

[3] J. Arvesu, F. Marcellan, R. Alvarez–Nodarse, *On a modification of the Jacobi linear functional:properties and zeros the corresponding orthogonal polynomials.* J. Comput. Appl. Math. **71** (2002), 127–158.

[4] R. Courant, D. Hilbert, *Methoden der mathematischen Physik.* Vol. 1. Berlin, 1931.

[5] A. Krall, *Hilbert space Boundary value problems and orthogonal polynomials.* Birkhäuser, 2002.

[6] F. Marcellan, M. Alfaro, M.L. Rezola, *Orthogonal polynomials: old and new directions.* J. Comput. Appl. Math. **48** (1993), 113–132.

[7] G. Szegö, *Orthogonal Polynomials.* 4rd Edition, Amer. Math. Soc., Providence, 1975.

[8] T.H. Koorwinder, *Orthogonal polynomials with the weight function* $M\delta(x+1)+N\delta(x-1)$. Canad. Math. Bull. **27(2)** (1984), 205–214.

[9] R. Koekoek, *Differential equations for symmetric generalized ultraspherical polynomials.* Trans. Amer. Math. Soc. **345** (1994), 47–72.

[10] J. Koekoek, R. Koekoek, *Differential equations for generalized Jacobi polynomials.* J. Comput. Appl. Math. **126** (2000), 1–31.

[11] L.L. Littlejohn, *The Krall polynomials: A new class of orthogonal polynomials.* Quaest. Math. **5** (1982), 255–265.

[12] B.P. Osilenker, *Generalized trace formula and asymptotics of the averaged Turan determinant for polynomials orthogonal with a discrete Sobolev inner product.* J. Approx. Theory **141** (2006), 70–97.

[13] F.R. Gantmacher, *Matrix Theory.* Moscow, Nauka, 1966. (in Russian).

B.P. Osilenker
Mathematical Department
Moscow State Civil Engineering University
26 Yaroslavskoe Shosse
129336 Moscow, Russia
e-mail: b_osilenker@mail.ru

Operator Theory:
Advances and Applications, Vol. 190, 429–450

On Coxeter Graph Related Configurations of Subspaces of a Hilbert Space

N.D. Popova, Yu.S. Samoĭlenko and A.V. Strelets

Abstract. For a Hilbert space $\mathcal{H}$, we study configurations of its subspaces related to Coxeter graphs $\mathbb{G}_{4,4}$, which are arbitrary trees such that two edges have type 4 and the rest are of type 3. We prove that such irreducible configurations exist only in a finite-dimensional $\mathcal{H}$, where the dimension of $\mathcal{H}$ does not exceed the number of vertices of the graph plus the number of vertices of the subgraph that lies between the edges of type 4. We give a description of all irreducible nonequivalent configurations; they are indexed with a continuous parameter. As an example, we study all irreducible configurations related to a graph that consists of three vertices and two type 4 edges.

Mathematics Subject Classification (2000). Primary 46C07; Secondary 46K10.

Keywords. Family of subspaces, Coxeter graph, Temperley–Lieb algebras, $*$-representation.

Introduction

One of important and fruitful directions of M.G. Krein's original mathematical work was a development of algebraic methods in functional analysis and operator theory.

The main object we study in this paper is a family $S = (\mathcal{H}; \mathcal{H}_1, \dots, \mathcal{H}_n)$ of subspaces $\mathcal{H}_1, \dots, \mathcal{H}_n$ of a Hilbert space $\mathcal{H}$. For any family of subspaces S there is a unique collection of orthogonal projections $P_k = P_{\mathcal{H}_k}$, which are orthogonal projections of $\mathcal{H}$ onto $\mathcal{H}_k$, $k = 1, \dots, n$.

A family $S = (\mathcal{H}; \mathcal{H}_1, \dots, \mathcal{H}_n)$ is *unitarily equivalent* to a family $\tilde{S} = (\tilde{\mathcal{H}}; \tilde{\mathcal{H}}_1, \dots, \tilde{\mathcal{H}}_n)$ of subspaces $\tilde{\mathcal{H}}_1, \dots, \tilde{\mathcal{H}}_n$ of a Hilbert space $\tilde{\mathcal{H}}$ if there exists a unitary operator U acting from $\mathcal{H}$ onto $\tilde{\mathcal{H}}$ such that U maps $\mathcal{H}_k$ onto $\tilde{\mathcal{H}}_k$, $k = 1, \dots, n$, that is, the corresponding collections of the orthogonal projections P_k and $\tilde{P}_k$, $k = 1, \dots, n$, are unitarily equivalent.

A family $S = (\mathcal{H}; \mathcal{H}_1, \dots, \mathcal{H}_n)$ is *irreducible* if the identity $[A, P_k] = 0$ satisfied for all $k = 1, \dots, n$, where $A \in \mathcal{B}(\mathcal{H})$, implies that $A = \lambda I$, where $\lambda \in \mathbb{C}$, I

is the identity operator on $\mathcal{H}$, that is, the collection of orthogonal projections $P_k, k = 1, \dots, n$, is irreducible.

Irreducible pairs of subspaces exist only in one- or two-dimensional Hilbert spaces $\mathcal{H}$. A list of such subspaces, up to unitary equivalence, is the following:

1. $\mathcal{H} = \mathbb{C}^1$,

$$(\mathbb{C}^1; 0, 0), \qquad (\mathbb{C}^1; \mathbb{C}^1, 0), \qquad (\mathbb{C}^1; 0, \mathbb{C}^1), \qquad (\mathbb{C}^1; \mathbb{C}^1, \mathbb{C}^1);$$

2. $\mathcal{H} = \mathbb{C}^2 = \langle\, e_1, e_2 \,\rangle$ ($\|e_k\| = 1$, $k = 1, 2$; $e_1 \perp e_2$),

$$(\mathbb{C}^2; \langle\, e_1 \,\rangle, \langle\, \cos\varphi e_1 + \sin\varphi e_2 \,\rangle),$$

($\varphi \in (0, \frac{\pi}{2})$ is the angle between the subspaces $\langle\, e_1 \,\rangle$ and $\langle\, \cos\varphi e_1 + \sin\varphi e_2 \,\rangle$). Denoting $P_1 = P_{\langle\, e_1 \,\rangle}$, $P_2 = P_{\langle\, \cos\varphi e_1 + \sin\varphi e_2 \,\rangle}$ we have

$$P_1 P_2 P_1 = \tau P_1,$$
$$P_2 P_1 P_2 = \tau P_2,$$

where $\tau = \cos^2\varphi \in (0, 1)$.

For any pair of subspaces $S = (\mathcal{H}; \mathcal{H}_1, \mathcal{H}_2)$ there is a spectral theorem, – the space $\mathcal{H}$ can be decomposed into a direct sum, or an integral, of irreducible pairs of subspaces that are equivalent to pairs of subspaces listed in 1 and 2 above, see [2].

The problem of describing irreducible n-tuples of subspaces,

$$S = (\mathcal{H}; \mathcal{H}_1, \dots, \mathcal{H}_n),$$

and a similar problem of representing n-tuples as a direct sum, or an integral, of irreducible ones for $n \geqslant 3$ is a $*$-wild problem. The problem of describing triples of subspaces, $S = (\mathcal{H}; \mathcal{H}_1, \mathcal{H}_2, \mathcal{H}_3)$, such that $\mathcal{H}_2 \perp \mathcal{H}_3$ is also $*$-wild [3, 4].

It is thus natural to study n-tuples of subspaces $\mathcal{H}_1, \dots, \mathcal{H}_n$, with a condition on the angles between the subspaces $\mathcal{H}_i, \mathcal{H}_j$ for $i, j = \overline{1, n}, i \neq j$.

Configurations of subspaces, $S_{\Gamma,\tau,\perp}$, related to a simple graph Γ with edges labelled with $\tau = \{\tau_{ij}\}$, where $\tau_{ij} \in (0; 1)$, are studied in [5, 9] and others. (By a simple graph we mean a finite non-oriented graph without multiple edges and loops.) In the configuration $S_{\Gamma,\tau,\perp}$, the number of subspaces coincides with the number of vertices in the graph, and if vertices i and j are not connected with an edge, the corresponding subspaces are assumed to be orthogonal,

$$P_i P_j = P_j P_i = 0,$$

and if there is an edge, then we fix an angle between $\mathcal{H}_i$ and $\mathcal{H}_j$, $\phi_{ij} \in (0, \frac{\pi}{2})$, ($\cos^2\phi_{ij} = \tau_{ij}$, where $\tau_{ij} \in (0, 1)$ is the number located above the edge connecting the vertices i and j), that is, we have

$$P_i P_j P_i = \tau_{ij} P_i \text{ and } P_j P_i P_j = \tau_{ij} P_j.$$

Since the subspaces corresponding to vertices in different connected components are orthogonal, we always assume that Γ is a connected graph.

The problem of describing the configurations $S_{\Gamma,\tau,\perp}$ can be reformulated in terms of $*$-representations of corresponding $*$-algebras $TL_{\Gamma,\tau,\perp}$ that are defined as follows:

$$\begin{aligned} TL_{\Gamma,\tau,\perp} = \mathbb{C}\langle p_1,\dots,p_n \mid & p_i^2 = p_i^* = p_i; \\ & p_i p_j = p_j p_i = 0, \text{ if the vertices } i,j \text{ are not connected with an edge;} \\ & p_i p_j p_i = \tau_{ij} p_i, p_j p_i p_j = \tau_{ij} p_j, \text{ otherwise } \rangle. \end{aligned}$$

Note that $TL_{\Gamma,\tau,\perp}$ are the quotient algebras of introduced in [1] generalized Temperley–Lieb algebras (which are, in their turn, quotient algebras of the Hecke algebras related to Γ).

The dimension and the growth of these algebras do not depend on the numbers τ_{ij}, see [8], namely, the algebra $TL_{\Gamma,\tau,\perp}$

- has a finite dimension equal to n^2+1 if Γ is a tree, where n is the number of the vertices;
- has linear growth if Γ has exactly one cycle;
- contains a free subalgebra with two generators if Γ has more than one cycle.

If we fix a graph Γ, then for each collection of the parameters τ_{ij} there is a finite number of irreducible unitarily nonequivalent configurations $S_{\Gamma,\tau,\perp}$ if and only if the graph Γ is a tree, see [7,9]. The dimension of the space $\mathcal{H}$ does not exceed the number of vertices, and the dimensions of the subspaces are not greater than 1.

If the graph Γ has one cycle, i.e., it is unicyclic, then there is an arrangement τ of numbers at its edges such that there exist infinitely many irreducible nonequivalent configurations $S_{\Gamma,\tau,\perp}$. The dimensions of the subspaces do not exceed 1, and the dimension of the whole space is not greater than the number of vertices in the graph [9].

A more general case is obtained by not fixing the angles between the subspaces but allowing them to assume values from a fixed finite set. Hence, we come to configurations $S_{\mathbb{G},g,\perp}$, where $\mathbb{G}$ is a Coxeter graph with the vertices corresponding to subspaces, and g is a collection of polynomials that define permissible angles between the subspaces.

By a *Coxeter graph* $\mathbb{G}$ we mean a finite non-oriented marked graph without multiple edges and with no loops. We will write $\mathbb{G} = (V, R)$, where $V = \{1,\dots,n\}$ is the set of vertices, R is the set of edges. The edge between i and j will be denoted by γ_{ij} (we take $\gamma_{ij} = \gamma_{ji}$). All edges of a Coxeter graph $\mathbb{G}$ can be subdivided into different types,

$$R = \bigsqcup_{s=3}^{\infty} R_s.$$

The corresponding edges will be called R_3-edges, R_4-edges, etc., or we will say that the edge has type 3, 4, etc.

Let g be a mapping that assigns a polynomial g_{ij} to each edge $\gamma_{ij} \in R_s$, $s = 2k + \sigma \geqslant 3$, $k \in \mathbb{N}$, $\sigma \in \{0,1\}$, such that $\deg g_{ij} \leqslant k-1$ and $g_{ij}(0) = 0$

if $\sigma = 0$,

$$g : R \to \mathbb{R}[x] : \gamma_{ij} \mapsto g_{ij}(x) = \sum_{m=1-\sigma}^{k-1} \tau_{ij}^{(m)} x^m \in \mathbb{R}[x].$$

The number of subspaces in a configuration $S_{\mathbb{G},g,\perp}$, as in the case of a simple graph, coincides with the number of vertices in the graph. If vertices i, j are not connected with an edge, the corresponding subspaces $\mathcal{H}_i$, $\mathcal{H}_j$ are considered to be orthogonal,

$$P_i P_j = P_j P_i = 0,$$

and if there is an edge of type $s = 2k+\sigma \geqslant 3$, $k \in \mathbb{N}$, $\sigma \in \{0,1\}$, then the following relations hold:

$$(P_i P_j)^k P_i^\sigma = g_{ij}(P_i P_j) P_i^\sigma \quad \text{and} \quad (P_j P_i)^k P_j^\sigma = g_{ij}(P_j P_i) P_j^\sigma. \tag{0.1}$$

It is convenient to reformulate the problem about the configurations $S_{\mathbb{G},g,\perp}$ in terms of finding $*$-representations of the corresponding $*$-algebras $TL_{\mathbb{G},g,\perp}$.

Definition 0.1. $TL_{\mathbb{G},g,\perp}$ is an associative $*$-algebra over $\mathbb{C}$, with an identity e, and is defined by generators and relations determined by the graph $\mathbb{G}$ and the mapping g,

$$\begin{aligned} TL_{\mathbb{G},g,\perp} = \mathbb{C}\langle p_1, \dots, p_n \,|\, p_i^2 = p_i^* = p_i;\quad & p_i p_j = p_j p_i = 0, \text{ if } \gamma_{ij} \notin R; \\ & (p_i p_j)^k p_i^\sigma = g_{ij}(p_i p_j) p_i^\sigma,\ (p_j p_i)^k p_j^\sigma = g_{ij}(p_j p_i) p_j^\sigma, \\ & \text{if } \gamma_{ij} \in R_s, \quad s = 2k+\sigma \geqslant 3, \sigma \in \{0,1\} \rangle. \end{aligned} \tag{0.2}$$

For an edge $\gamma_{ij} \in R_s$, $s = 2k+\sigma$, and a polynomial g_{ij}, we will also consider the polynomial $f_{ij}(x) = x^k - g_{ij}(x)$. Then relations (0.2) can be rewritten in the form $f_{ij}(p_i p_j) p_i^\sigma = f_{ij}(p_j p_i) p_j^\sigma = 0$.

If an edge γ_{ij} has type 3, then the corresponding polynomial is $g_{ij}(x) = \tau_{ij}$, and the relations satisfied by p_i and p_j will be

$$p_i p_j p_i = \tau_{ij} p_i, \quad p_j p_i p_j = \tau_{ij} p_j.$$

If an edge γ_{ij} has type 4, then $g_{ij}(x) = \tau_{ij} x$ and the relations satisfied by p_i and p_j will be

$$(p_i p_j)^2 = \tau_{ij} p_i p_j, \quad (p_j p_i)^2 = \tau_{ij} p_j p_i.$$

The dimension and the growth of the algebras $TL_{\mathbb{G},g,\perp}$ were studied in [6]. The dimension and the growth of the algebra $TL_{\mathbb{G},g,\perp}$ depend on the type of the graph $\mathbb{G}$, which can be a tree, an unicyclic graph, a graph with two or more cycles, and on the type of its edges, but do not depend on the way the polynomials g_{ij} are positioned at its edges. Let us give here brief formulations of the results. To this end, we introduce the following notations: $\hat{R}_r = \bigsqcup_{s=r}^{\infty} R_s$, and $s_{\mathbb{G}} \geqslant 3$ denotes the index s such that $R_{s_{\mathbb{G}}} \neq \varnothing$ and $R_s = \varnothing$ if $s > s_{\mathbb{G}}$.

Theorem 0.2. *Let $\mathbb{G}$ be a tree. Then we have the following.*

0) *If $|\hat{R}_4| = 0$, then* $\dim TL_{\mathbb{G},g,\perp} = |V|^2 + 1$.

1) *If* $|\hat{R}_4| = |R_{s_{\mathbb{G}}}| = 1$*, then* $\mathbb{G}$ *consists of two* R_3*-connected components* $\mathbb{G}_1$ *and* $\mathbb{G}_2$*, and*

$$\dim TL_{\mathbb{G},g,\perp} = \begin{cases} m|V|^2 + 1, & \text{if } s_{\mathbb{G}} = 2m+1, \\ (m-1)|V|^2 + |V_1|^2 + |V_2|^2 + 1, & \text{if } s_{\mathbb{G}} = 2m. \end{cases}$$

2) *If* $|\hat{R}_4| \geqslant 2$*, then* $\dim TL_{\mathbb{G},g,\perp} = \infty$*. In the case where* $|\hat{R}_4| = 2$*, the algebra has polynomial growth if* $|\hat{R}_6| = 0$*, and contains a free subalgebra with two generators if* $|\hat{R}_6| \geqslant 1$*.*
3) *If* $|\hat{R}_4| \geqslant 3$*, then* $TL_{\mathbb{G},g,\perp}$ *contains a free algebra with two generators.*

Theorem 0.3. *Let* $\mathbb{G}$ *be a connected Coxeter graph.*

0) *If* $\mathbb{G}$ *contains a cycle and* $|\hat{R}_4| = 0$*, then* $\dim TL_{\mathbb{G},g,\perp} = \infty$*. If also there is precisely one cycle, then the algebra has linear growth, and if* $\mathbb{G}$ *has two cycles, then the algebra has a free algebra with two generators.*
1) *If* $\mathbb{G}$ *contains at least one cycle and* $|\hat{R}_4| \geqslant 1$*, then* $TL_{\mathbb{G},g,\perp}$ *contains a free algebra with two generators.*

The authors in [7] studied $*$-representations of the algebras $TL_{\mathbb{G},g,\perp}$. They gave a description of all irreducible $*$-representations of the finite-dimensional algebras $TL_{\mathbb{G},g,\perp}$ on a Hilbert space. In particular, it was shown that, if an irreducible pair of projections P_1 and P_2 satisfies relations (0.1), then at least one of the roots $\lambda \in [0,1]$ of the polynomial $f_{12}(x)$ satisfies the relations

$$\begin{cases} P_1P_2P_1 = \lambda P_1 \text{ and } P_2P_1P_2 = \lambda P_2, & \lambda \neq 0, \\ P_1P_2 = P_2P_1 = 0, & \lambda = 0. \end{cases}$$

Hence, using roots of the polynomial $f_{12}(x)$ we can define admissible angles between the subspaces. There was also proved a theorem stating that *only graphs that are trees with no more than one edge of type* $s > 3$ *define algebras of finite Hilbert type, i.e., the algebras that have a finite number of irreducible* $*$*-representations on a Hilbert space for all values of the parameters.*

In this paper, we use $*$-representations of the algebra $TL_{\mathbb{G}_{4,4},g,\perp}$ to study configurations of the subspaces, $S_{\mathbb{G}_{4,4},g,\perp}$, that are related to the Coxeter graph $\mathbb{G}_{4,4}$, which is an arbitrary tree with exactly two edges of type 4 and others are of type 3. As Theorem 0.2 shows, this is the simplest case where the algebra is already not finite-dimensional but still has polynomial growth.

In Section 1, we give necessary definitions and introduce notations; we also introduce the notion of a proper $*$-representation of the algebra $TL_{\mathbb{G}_{4,4},g,\perp}$.

Section 2 gives a construction of a family of irreducible proper $*$-representations π_ν of the algebra $TL_{\mathbb{G}_{4,4},g,\perp}$. This family is parametrized with a parameter ν running over a subset $\Sigma_{\mathbb{G}_{4,4},g}$ of the interval $(0,1)$. It is shown that $*$-representations corresponding to different values of the parameter are not unitarily equivalent.

In Section 3, we show that any irreducible proper $*$-representation π of the algebra $TL_{\mathbb{G}_{4,4},g,\perp}$ is unitarily equivalent to one of the $*$-representations π_ν

constructed in Section 2, which finishes the description of all irreducible proper $*$-representations.

As an example, in Section 4, we study all irreducible $*$-representations of the algebra $TL_{\mathbb{G}_{4,4},g,\perp}$, where the graph $\mathbb{G}_{4,4}$ consists of three vertices and two edges of type 4.

1. Definitions and notations

A path of length m in a Coxeter graph $\mathbb{G}$,

$$l = l(i_0) = (i_0, i_1, \ldots, i_m), \qquad \gamma_{i_{k-1}, i_k} \in R,$$

will be called a *path without repetitions* if $i_k \neq i_j$ for $k, j = 0, \ldots, m$, $k \neq j$. The path $l = (i_0)$ is considered as a path of length 0 without repetitions, and it is convenient to consider the path $l = ()$ as "empty". For a path $l = (i_0, i_1, \ldots, i_m)$, define $l^* = (i_m, i_{m-1}, \ldots, i_0)$. A *union of paths* $l_1 = (i_0, \ldots, i_{k-1}, i_k)$ and $l_2 = (i_k, i_{k+1} \ldots, i_t)$ is defined to be the path $l_1 \cup l_2 = (i_0, \ldots, i_{k-1}, i_k, i_{k+1}, \ldots, i_t)$. To any path $l = (i_0, i_1, \ldots, i_m)$, we make correspond the product $\Pi_l = p_{i_0} \cdots p_{i_m}$ in the algebra, to the "empty" path, we set $\Pi_l = e$.

By $\mathbb{G}_{4,4}$, we will denote a Coxeter graph, which is a tree having two edges of type 4 and the rest of edges are of type 3. We index the vertices in such a way that the edges $\gamma_{0,1}$, $\gamma_{m-1,m}$ are of type 4 and the vertices 1, $m-1$ are connected with the path $(1, 2, \ldots, m-1)$,

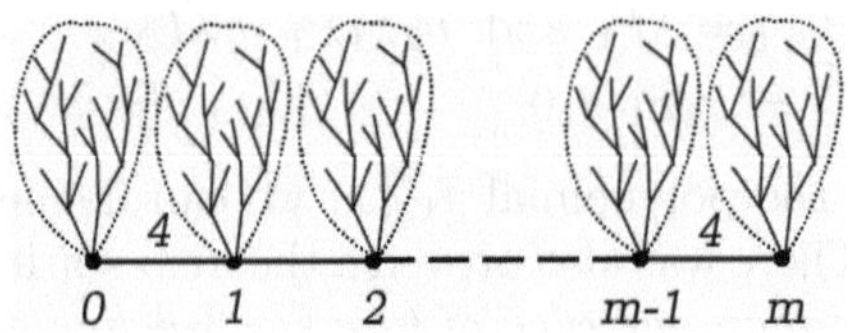

It is natural to split the graph into three parts,

$$V = V_0 \cup V_{in} \cup V_m,$$

namely, any two different vertices in each part are connected with a path consisting of type 3 edges.

Denote by $\hat{l}$ the path $(m, m-1, \ldots, 1, 0)$, and by $\mathcal{P}$ the set of all paths $l = (i_0, i_1, \ldots, 0)$ such that Π_l is a normal word that does not have $\Pi_{\hat{l}^* \cup \hat{l}}$ as a subword. For normal words, Groebner bases, the composition lemma, we refer to, e.g., [10]. For the algebra $TL_{\mathbb{G},g,\perp}$, normal words are precisely the words that do not contain, as subwords, the leading words of the defining relations (0.2) of the algebra $TL_{\mathbb{G},g,\perp}$, see [6]. That is, a normal word should not contain, as subwords, the following words:

$p_i^2, i \in V$;

$p_i p_j,\ p_j p_i$, if $\gamma_{ij} \notin R$;

$(p_i p_j)^k p_i^\sigma,\ (p_j p_i)^k p_j^\sigma$, if $\gamma_{ij} \in R_s$, $\quad s = 2k + \sigma \geqslant 3$, $\sigma \in \{0, 1\}$.

It is clear that the set $\mathcal{P}$ consists of two parts containing paths $l \in \mathcal{P}$ without and with repetitions, denoted by $\mathcal{S}$ and $\mathcal{L}'$, respectively. The set $\mathcal{L}'$, in its turn, can be slitted into two more parts containing the the paths l that do or do not have $\Pi_{\hat{l}}$ as a subword of Π_l, and are denoted by $\mathcal{L}$ and $\mathcal{L}_0$, respectively. We set $\hat{\mathcal{P}} = \mathcal{S} \cup \mathcal{L}$.

Proposition 1.1.

1) *For each path $l \in \mathcal{L}$ there exists a unique vertex $j \in \{1, \ldots, m-1\}$ and a unique collection of paths without repetitions,*

$$\begin{aligned} l_s &= (i_0, i_1, \ldots, j), \\ l_e &= (j, j-1, \ldots, 0), \\ \tilde{l} &= (j, j+1, \ldots, m), \end{aligned}$$

such that $l = l_s \cup \tilde{l} \cup \hat{l}$ and $\tilde{l}^* \cup l_e = \hat{l}$. The identity

$$\omega(l) = l_s \cup l_e$$

defines an injective mapping $\omega : \mathcal{L} \to \mathcal{S}$.

2) *There are naturally defined bijective mappings $\psi_* : V \to \mathcal{S}$ and $\varphi_* : V_{in} \to \mathcal{L}$. We also have $\omega(\varphi_*(j)) = \psi_*(j)$ for any $j \in V_{in}$.*

For any

$$l = (i_0, i_1, \ldots, 0) \in \hat{\mathcal{P}} \backslash \{\psi_*(0)\},$$

define a truncation operation, η, of the path l by

$$\eta(l) = (i_1, \ldots, 0).$$

For a $*$-representation π of the algebra $TL_{\mathbb{G}_{4,4,g,\perp}}$, denote $P_i = \pi(p_i)$, $i \in V$.

Definition 1.2. A $*$-representation π of the algebra $TL_{\mathbb{G}_{4,4,g,\perp}}$ will be called *proper* if none of the following relations is satisfied:

$$P_0 P_1 = 0, \tag{1.1}$$

$$P_{m-1} P_m = 0, \tag{1.2}$$

$$P_1 P_0 P_1 - \tau_{0,1} P_1 = 0, \tag{1.3}$$

$$P_{m-1} P_m P_{m-1} - \tau_{m-1,m} P_{m-1} = 0. \tag{1.4}$$

If some of relations (1.1)–(1.4) are satisfied, this would mean that the $*$-representation π is a lift of a certain $*$-representation of the algebra obtained by taking the quotient with respect to an ideal generated by a corresponding relation. All such quotient algebras are finite-dimensional, and their $*$-representations were studied before, see [9], [7]. Thus, a complete description of all irreducible $*$-representations of the algebra $TL_{\mathbb{G}_{4,4,g,\perp}}$ require a description of all irreducible proper $*$-representations.

For each $\nu \in \Sigma_{\mathbb{G}_{4,4,g}}$, we will construct an irreducible proper $*$-representation π_ν, where $\Sigma_{\mathbb{G}_{4,4,g}}$ is a subset of the interval $(0, 1)$. An exact definition of this set will be given later. Let us note that, in general, this set could be empty. This means that the algebra $TL_{\mathbb{G}_{4,4},g,\perp}$ does not have proper $*$-representations. On the other hand, for any Coxeter graph $\mathbb{G}_{4,4}$ there is g such that $\Sigma_{\mathbb{G}_{4,4},g}$ contains an entire interval [7]. It will be shown that $*$-representations π_{ν_1} and π_{ν_2} are not unitarily equivalent if $\nu_1 \neq \nu_2$. The fact that for every irreducible proper $*$-representation π there exists $\nu \in \Sigma_{\mathbb{G}_{4,4},g}$ such that π is unitarily equivalent to π_ν finishes the description of all irreducible proper $*$-representations.

2. The set $\Sigma_{\mathbb{G}_{4,4},g}$, proper $*$-representations π_ν, $\nu \in \Sigma_{\mathbb{G}_{4,4},g}$

Let L be a linear complex space obtained as a set of all formal linear combinations of paths taken from the set $\hat{\mathcal{P}}$.

For each $\nu \in (0,1)$, introduce a Hermitian sesquilinear form $B^{\nu}_{\mathbb{G}_{4,4},g}$ on the linear space L by defining it on the formal linear basis $\hat{\mathcal{P}}$ to be

$$\begin{aligned}
B^{\nu}_{\mathbb{G}_{4,4},g}(l,l) &= 1, \qquad l \in \hat{\mathcal{P}},\\
B^{\nu}_{\mathbb{G}_{4,4},g}(l,\eta(l)) &= B^{\nu}_{\mathbb{G}_{4,4},g}(\eta(l),l)\\
&= \begin{cases} \sqrt{\tau_{ij}}, & l \in \hat{\mathcal{P}}\backslash\{\psi_*(0),\psi_*(m),\varphi_*(m-1)\},\\ \sqrt{\nu\tau_{m-1,m}}, & l = \psi_*(m),\\ \sqrt{(1-\nu)\tau_{m-1,m}}, & l = \varphi_*(m-1),\end{cases}\\
B^{\nu}_{\mathbb{G}_{4,4},g}(l_1,l_2) &= 0, \qquad l_1,l_2 \in \hat{\mathcal{P}},\ l_1 \neq l_2,\ l_1 \neq \eta(l_2),\ l_2 \neq \eta(l_1),
\end{aligned}$$

where the paths l and $\eta(l)$ start at i and j, correspondingly.

Let $\Sigma_{\mathbb{G}_{4,4},g}$ be the set of all $\nu \in (0,1)$ such that the sesquilinear form $B^{\nu}_{\mathbb{G}_{4,4},g}$ is nonnegative definite. For $\nu \in \Sigma_{\mathbb{G}_{4,4},g}$, denote by $\mathcal{H}_\nu$ the Hilbert space obtained by equipping the linear space L/L_ν, where $L_\nu = \{x \,|\, B^{\nu}_{\mathbb{G}_{4,4},g}(x,x) = 0\}$, with the scalar product $\langle\, y_1 + L_\nu, y_2 + L_\nu\,\rangle_\nu = B^{\nu}_{\mathbb{G}_{4,4},g}(y_1,y_2)$. Let ρ_ν be the linear mapping

$$\rho_\nu : L \to \mathcal{H}_\nu : y \mapsto y + L_\nu.$$

By the definition of the form $B^{\nu}_{\mathbb{G}_{4,4},g}$, any $l \in \hat{\mathcal{P}}$ does not belong to L_ν. Hence, there is a bijection between the set $\rho_\nu(\hat{\mathcal{P}}) = \{l + L_\nu \,|\, l \in \hat{\mathcal{P}}\} \subset \mathcal{H}_\nu$ and the set $\hat{\mathcal{P}}$, where the former generates a linear space $\mathcal{H}_\nu$, however, if the form is not positive definite, the set $\rho_\nu(\hat{\mathcal{P}})$ is not a set of linearly independent vectors. Set

$$\begin{aligned}
\psi = \psi_\nu = \rho_\nu \circ \psi_* : \mathcal{S} \to \mathcal{H}_\nu,\\
\varphi = \varphi_\nu = \rho_\nu \circ \varphi_* : \mathcal{L} \to \mathcal{H}_\nu.
\end{aligned}$$

For an arbitrary vertex $i \in V_{in}$, define an operator $P_{\nu,i}$ to be the orthogonal projection onto the linear span of the pair of vectors $\psi(i), \varphi(i) \in \mathcal{H}_\nu$, and, for an arbitrary vertex $i \in V\backslash V_{in}$, the operator $P_{\nu,i}$ is defined to be an orthogonal projection onto the linear span of the vector $\psi(i) \in \mathcal{H}_\nu$.

Proposition 2.1. *For any $x \in \mathcal{H}_\nu$, we have the formula*

$$P_{\nu,i}x = \begin{cases} \langle x, \psi(i) \rangle_\nu \psi(i) + \langle x, \varphi(i) \rangle_\nu \varphi(i), & i \in V_{in}, \\ \langle x, \psi(i) \rangle_\nu \psi(i), & i \in V \backslash V_{in}. \end{cases}$$

Proof. It is sufficient to notice that $\langle P_{\nu,i}x, \psi(i) \rangle_\nu = \langle x, \psi(i) \rangle_\nu$ for any $i \in V$, and $\langle P_{\nu,i}x, \varphi(i) \rangle_\nu = \langle x, \varphi(i) \rangle_\nu$ for any $i \in V_{in}$. □

Lemma 2.2. *For each $\nu \in \Sigma_{\mathbb{G}_{4,4,g}}$, the mapping*

$$\pi_\nu : TL_{\mathbb{G}_{4,4,g},\perp} \to \mathcal{B}(\mathcal{H}_\nu) : p_i \mapsto P_{\nu,i}$$

is an irreducible proper $*$-representation.

Proof. For the sake of brevity, set $\varphi(i) = 0$ for $i \in V \backslash V_{in}$. Then, for any $i \in V$, $x \in \mathcal{H}_\nu$, we have

$$P_{\nu,i}x = \langle x, \psi(i) \rangle_\nu \psi(i) + \langle x, \varphi(i) \rangle_\nu \varphi(i).$$

Let us show that π_ν is a $*$-representation.

It is clear that for any $x \in \mathcal{H}_\nu$, $P_{\nu,i}^2 x = P_{\nu,i}x$, since $\langle \psi(i), \varphi(i) \rangle_\nu = 0$.

Now, if the vertices i and j are not connected with an edge, then for any vector $x \in \mathcal{H}_\nu$,

$$\begin{aligned} P_{\nu,i}P_{\nu,j}x &= P_{\nu,i}(\langle x, \psi(j) \rangle_\nu \psi(j) + \langle x, \varphi(j) \rangle_\nu \varphi(j)) \\ &= \langle x, \psi(j) \rangle_\nu (\langle \psi(j), \psi(i) \rangle_\nu \psi(i) + \langle \psi(j), \varphi(i) \rangle_\nu \varphi(i)) \\ &+ \langle x, \varphi(j) \rangle_\nu (\langle \varphi(j), \psi(i) \rangle_\nu \psi(i) + \langle \varphi(j), \varphi(i) \rangle_\nu \varphi(i)) = 0. \end{aligned}$$

Let now the vertices i and j be joined with a type 3 edge. Then we can assume that $\psi_*(i) = \eta(\psi_*(j))$. Moreover, we have that either $i, j \in V \backslash V_{in}$, then $\varphi(i) = \varphi(j) = 0$, or $i, j \in V_{in}$ implying that one of the following two identities is verified: if $j \in \{2, \ldots, m-1\}$ then $\varphi_*(j) = \eta(\varphi_*(i))$, otherwise $\varphi_*(i) = \eta(\varphi_*(j))$. Hence, for an arbitrary vector $x \in \mathcal{H}_\nu$, we have

$$\begin{aligned} P_{\nu,j}P_{\nu,i}P_{\nu,j}x &= P_{\nu,j}P_{\nu,i}(\langle x, \psi(j) \rangle_\nu \psi(j) + \langle x, \varphi(j) \rangle_\nu \varphi(j)) \\ &= \sqrt{\tau_{ij}} P_{\nu,j}(\langle x, \psi(j) \rangle_\nu \psi(i) + \langle x, \varphi(j) \rangle_\nu \varphi(i)) \\ &= \tau_{ij}(\langle x, \psi(j) \rangle_\nu \psi(j) + \langle x, \varphi(j) \rangle_\nu \varphi(j)) \\ &= \tau_{ij} P_{\nu,j}x. \end{aligned}$$

It remains to check the relations for the orthogonal projections corresponding to the vertices joined with edges of type 4. For an arbitrary $x \in \mathcal{H}_\nu$, we have

$$\begin{aligned} P_{\nu,0}P_{\nu,1}P_{\nu,0}x &= P_{\nu,0}P_{\nu,1}\langle x, \psi(0) \rangle_\nu \psi(0) \\ &= \sqrt{\tau_{0,1}} P_{\nu,0}\langle x, \psi(0) \rangle_\nu \psi(1) \\ &= \tau_{0,1}\langle x, \psi(0) \rangle_\nu \psi(0) \\ &= \tau_{0,1} P_{\nu,0}x, \end{aligned}$$

$$\begin{aligned}P_{\nu,m}P_{\nu,m-1}P_{\nu,m}x &= P_{\nu,m}P_{\nu,m-1}\langle x,\psi(m)\rangle_\nu\psi(m)\\ &= \sqrt{\tau_{m-1,m}}\langle x,\psi(m)\rangle_\nu P_{\nu,m}(\sqrt{\nu}\psi(m-1)+\sqrt{1-\nu}\varphi(m-1))\\ &= \tau_{m-1,m}\langle x,\psi(m)\rangle_\nu(\nu+(1-\nu))\psi(m)\\ &= \tau_{m-1,m}P_{\nu,m}x.\end{aligned}$$

This implies that

$$\begin{aligned}P_{\nu,0}P_{\nu,1}P_{\nu,0}P_{\nu,1} &= \tau_{0,1}P_{\nu,0}P_{\nu,1},\\ P_{\nu,1}P_{\nu,0}P_{\nu,1}P_{\nu,0} &= \tau_{0,1}P_{\nu,1}P_{\nu,0},\\ P_{\nu,m}P_{\nu,m-1}P_{\nu,m}P_{\nu,m-1} &= \tau_{m-1,m}P_{\nu,m}P_{\nu,m-1},\\ P_{\nu,m-1}P_{\nu,m}P_{\nu,m-1}P_{\nu,m} &= \tau_{m-1,m}P_{\nu,m-1}P_{\nu,m}.\end{aligned}$$

It is easy to see that $*$-representation π_ν is proper.

Let us show that the constructed $*$-representation π_ν is irreducible.

Assume that an operator $A \in \mathcal{B}(\mathcal{H}_\nu)$ commutes with all $P_{\nu,i}$, $i \in V$. Then, $A(\mathrm{Im}P_{\nu,i}) \subset \mathrm{Im}P_{\nu,i}$. Consequently, there exists a number $\lambda \in \mathbb{C}$ such that $A\psi(0) = \lambda\psi(0)$. Then $\lambda\sqrt{\tau_{0,1}}\psi(1) = \lambda P_{\nu,1}\psi(0) = P_{\nu,1}A\psi(0) = A\sqrt{\tau_{0,1}}\psi(1)$. In a similar way, it is easy to show that if $A\psi(i) = \lambda\psi(i)$ and the vertices j and i are joined with an edge, then $A\psi(j) = \lambda\psi(j)$.

Let us show that $A\varphi(m-1) = \lambda\varphi(m-1)$. It was proved before that $A\psi(m) = \lambda\psi(m)$. Then, on the one hand,

$$\begin{aligned}P_{\nu,m-1}A\psi(m) &= \lambda P_{\nu,m-1}\psi(m)\\ &= \lambda(\sqrt{\nu\tau_{m-1,m}}\psi(m-1)+\sqrt{(1-\nu)\tau_{m-1,m}}\varphi(m-1)).\end{aligned}$$

On the other hand, we have also shown that $A\psi(m-1) = \lambda\psi(m-1)$ and, consequently,

$$\begin{aligned}AP_{\nu,m-1}\psi(m) &= A(\sqrt{\nu\tau_{m-1,m}}\psi(m-1)+\sqrt{(1-\nu)\tau_{m-1,m}}\varphi(m-1))\\ &= \lambda\sqrt{\nu\tau_{m-1,m}}\psi(m-1)+\sqrt{(1-\nu)\tau_{m-1,m}}A\varphi(m-1).\end{aligned}$$

So, $A\varphi(m-1) = \lambda\varphi(m-1)$.

Now, for any vertices $i, j \in V_{in}$, which are connected with an edge, the identity $A\varphi(i) = \lambda\varphi(i)$ implies that $A\varphi(j) = \lambda\varphi(j)$. Indeed, $P_{\nu,j}A\varphi(i) = \lambda P_{\nu,j}\varphi(i) = \lambda\sqrt{\tau_{ij}}\varphi(j) = AP_{\nu,j}\varphi(i) = \sqrt{\tau_{ij}}A\varphi(j)$.

Thus, $A = \lambda I$ and, consequently, the $*$-representation π_ν is irreducible. □

Lemma 2.3. *The $*$-representations π_{ν_1} and π_{ν_2}, $\nu_1, \nu_2 \in \Sigma_{G_{4,4,g}}$, are unitarily equivalent if and only if $\nu_1 = \nu_2$.*

Proof. Let us consider the operator

$$W_\nu = U^\nu_{0,1}U^\nu_{1,2}\cdots U^\nu_{m-1,m}U^\nu_{m,m-1}\cdots U^\nu_{2,1}U^\nu_{1,0}, \qquad U^\nu_{i,j} = \frac{P_{\nu,i}P_{\nu,j}}{\sqrt{\tau_{i,j}}}.$$

It is easy to see that if the vertices $i, j \in V_{in}$ are joined with an edge, then

$$U^{\nu}_{i,j}(\mu_1\psi(j) + \mu_2\varphi(j)) = \mu_1\psi(i) + \mu_2\varphi(i).$$

and, consequently,

$$\begin{aligned}
W_\nu\psi(0) &= U^{\nu}_{0,1}U^{\nu}_{1,2}\dots U^{\nu}_{m-1,m}U^{\nu}_{m,m-1}\dots U^{\nu}_{2,1}\psi(1)\\
&= U^{\nu}_{0,1}U^{\nu}_{1,2}\dots U^{\nu}_{m-1,m}U^{\nu}_{m,m-1}\psi(m-1)\\
&= \sqrt{\nu}\,U^{\nu}_{0,1}U^{\nu}_{1,2}\dots U^{\nu}_{m-1,m}\psi(m)\\
&= \sqrt{\nu}\,U^{\nu}_{0,1}U^{\nu}_{1,2}\dots U^{\nu}_{m-2,m-1}(\sqrt{\nu}\psi(m-1) + \sqrt{1-\nu}\varphi(m-1))\\
&= \sqrt{\nu}\,U^{\nu}_{0,1}(\sqrt{\nu}\psi(1) + \sqrt{1-\nu}\varphi(1))\\
&= \nu\psi(0).
\end{aligned}$$

Assume that the $*$-representations π_{ν_1} and π_{ν_2} are unitarily equivalent, that is, there exists a unitary operator

$$V : \mathcal{H}_{\nu_1} \to \mathcal{H}_{\nu_2}$$

such that $V\pi_{\nu_1}(a) = \pi_{\nu_2}(a)V$ for any $a \in TL_{\mathbb{G}_{4,4,g},\perp}$. Then

$$W_{\nu_2}V\psi_{\nu_1}(0) = VW_{\nu_1}\psi_{\nu_1}(0) = \nu_1 V\psi_{\nu_1}(0),$$

so that $V\psi_{\nu_1}(0)$ is an eigenvector of the operator W_{ν_2} with an eigenvalue ν_1. Consequently, this vector belongs to $\mathrm{Im}P_{\nu_2,0}$, so that there is a number $\beta \in \mathbb{C}\setminus\{0\}$ such that $V\psi_{\nu_1}(0) = \beta\psi_{\nu_2}(0)$. But, in this case,

$$W_{\nu_2}V\psi_{\nu_1}(0) = \beta W_{\nu_2}\psi_{\nu_2}(0) = \beta\nu_2\psi_{\nu_2}(0) = \beta\nu_1\psi_{\nu_2}(0),$$

whence we get that $\nu_1 = \nu_2$.

We have thus shown that if $\nu_1, \nu_2 \in \Sigma_{\mathbb{G}_{4,4,g}}$ are different, the $*$-representations π_{ν_1} and π_{ν_2} are not unitarily equivalent. □

3. A description of all proper $*$-representations of the algebra $TL_{\mathbb{G}_{4,4,g},\perp}$

Let now π be some $*$-representation of the algebra $TL_{\mathbb{G}_{4,4,g},\perp}$ on a Hilbert space $\mathcal{H}$, $P_i = \pi(p_i)$, $i \in V$. Denote $w = \Pi_{\hat{l}^*\cup\hat{l}}$, $W = \pi(w)$. Note that $W = (\pi(\Pi_{\hat{l}}))^*\pi(\Pi_{\hat{l}})$, where $\pi(\Pi_{\hat{l}}) = P_mP_{m-1}\dots P_0$, thus W is a positive operator.

Proposition 3.1. *Let π be an irreducible $*$-representation, $P_0P_1 \neq 0$, and $P_{m-1}P_m \neq 0$. Then $W \neq 0$.*

Proof. Assume that $W = 0$. Since $P_0P_1 \neq 0$, there exists x_0 such that $P_0x_0 = x_0$, $P_1P_0x_0 \neq 0$.

Consider the linear span $\mathcal{H}'$ of a finite set of the vectors $\{\pi(\Pi_l)x_0\}_{l\in\mathcal{P}}$. It is easy to see that $\mathcal{H}'$ is invariant with respect to π and, consequently, it coincides with $\mathcal{H}$.

It is clear that $\pi(\Pi_{\hat{l}}) = 0$. Moreover, $\pi(\Pi_l) = 0$ for any $l \in \mathcal{P}$ such that Π_l contains $\Pi_{\hat{l}}$ as a subword. Let us show that $P_mP_{m-1}\pi(\Pi_l)x_0 = 0$ for any $l \in \mathcal{P}$.

Indeed, $P_{m-1}\pi(\Pi_l) \neq 0$ only if the initial point of l coincides with the vertex $m-1$ or is joined to it with an edge, and Π_l does not contain $\Pi_{\hat{l}}$ as a subword. There are three cases possible,

1) $l = (m-1, m-2, \dots, 0)$,
2) $l = (m-2, \dots, 0)$,
3) $l = (j, m-1, m-2, \dots, 0)$,

where j is any vertex distinct from $m-2$ and joined to the vertex $m-1$ with an edge of type 3. In the first two cases, $P_m P_{m-1}\pi(\Pi_l) = \pi(\Pi_{\hat{l}}) = 0$, and in the third one, $P_m P_{m-1}\pi(\Pi_l) = \tau_{j,m-1}\pi(\Pi_{\hat{l}}) = 0$. Hence, $P_m P_{m-1}\mathcal{H} = \{0\}$, and this contradicts to $P_m P_{m-1} \neq 0$. □

Since we are interested in proper $*$-representations of the algebra $TL_{\mathbb{G}_{4,4,g,\perp}}$, in the sequel we will consider only $*$-representations π such that $W \neq 0$.

Proposition 3.2. *Let π be an irreducible $*$-representation of the algebra $TL_{\mathbb{G}_{4,4,g,\perp}}$, such that $W \neq 0$. Then there exist $\xi \in (0,1]$ and $0 \neq x_0 \in \mathcal{H}$ such that $P_0 x_0 = x_0$ and $W x_0 = \xi x_0$.*

Proof. Consider the linear subspace $L = \operatorname{Im} W \neq \{0\}$ invariant with respect to the operator W. If some $x \in \mathcal{H}$ satisfies $W^2 x = 0$, then

$$0 = \langle W^2 x, x \rangle = \langle Wx, Wx \rangle = \|Wx\|^2,$$

that is, we have shown that $\operatorname{Im} W \cap \ker W = \{0\}$.

Denote by $\tilde{\pi}$ the $*$-representation of the generated by an element w commutative subalgebra of the algebra $TL_{\mathbb{G}_{4,4,g,\perp}}$ on the Hilbert space $\tilde{\mathcal{H}} = \overline{L}$, which is defined by

$$\tilde{\pi}(w) = \tilde{W} = W|_{\tilde{\mathcal{H}}}.$$

If $\tilde{\pi}$ is irreducible, then $\dim \tilde{\mathcal{H}} = 1$ and $\tilde{W} = \xi \in \mathbb{C}$. Moreover, since W is a nonzero positive operator with the norm being less or equal to one, we see that $\xi \in (0,1]$. Take $0 \neq x_0 \in \tilde{\mathcal{H}}$. Then $W x_0 = \xi x_0$, $P_0 x_0 = x_0$.

Assume that $\tilde{\pi}$ is reducible. Then there are nonzero subspaces $\tilde{\mathcal{H}}_1$ and $\tilde{\mathcal{H}}_2$ of the space $\tilde{\mathcal{H}}$ that are invariant with respect to the action of $\tilde{\pi}$ and such that $\tilde{\mathcal{H}} = \tilde{\mathcal{H}}_1 \oplus \tilde{\mathcal{H}}_2$. Since $\tilde{W} \neq 0$, there exist $0 \neq x \in \tilde{\mathcal{H}}_1 \cap L$ and $0 \neq y \in \tilde{\mathcal{H}}_2 \cap L$.

Let $\mathcal{H}_0$ be the closure of the linear span of the vectors $\{\pi(\Pi_l)x\}_{l \in \mathcal{N}}$, where $\mathcal{N}$ is a set of paths $l = (i_0, i_1, \dots, 0)$ such that Π_l is a normal word. It is clear that $\mathcal{H}_0$ is invariant with respect to π and, hence, $\mathcal{H} = \mathcal{H}_0$. Then, for any $l \in \mathcal{N}$,

$$\langle y, \pi(\Pi_l)x \rangle = \sum_{l' \in \mathcal{N}_0} \lambda_{l,l'} \langle y, \pi(\Pi_{l'})x \rangle,$$

where $\mathcal{N}_0 = \{l \in \mathcal{N} \mid i_0 = 0\}$.

For any $l' \in \mathcal{N}_0$, the monomial $\Pi_{l'}$ can be written in one of the following forms:

$$p_0, \quad p_0 p_1 p_0, \quad w^n,\ n \in \mathbb{N}.$$

To show that $\langle y, \pi(\Pi_{l'})x \rangle = 0$, it is sufficient to prove that $\pi(\Pi_{l'})x \in \tilde{\mathcal{H}}_1$. It is clear that $P_0 x = x \in \tilde{\mathcal{H}}_1$, $W^n x \in \tilde{\mathcal{H}}_1$. Now, $P_0 P_1 P_0 x = \tau_{0,1} x \in \tilde{\mathcal{H}}_1$, since $x \in \operatorname{Im} W$.

Hence, $\langle y, \pi(\Pi_l)x \rangle = 0$, that is, $y \in \mathcal{H}_0^{\perp} = \mathcal{H}^{\perp}$ and, consequently, $y = 0$, which contradicts to $y \neq 0$. □

Fix $x_0 \in \mathcal{H}$ and $\xi \in (0,1]$, which exist by the previous proposition. Everywhere in the sequel, we consider $\|x_0\| = 1$. Denote

$$\nu = \xi\Big(\prod_{i=1}^{m} \tau_{i-1,i}\Big)^{-1} > 0, \tag{3.1}$$

and introduce

$$\tau'_{i,j} = \begin{cases} \sqrt{\tau_{i,j}}, & \gamma_{i,j} \neq \gamma_{m-1,m}, \\ \sqrt{\nu\tau_{m-1,m}}, & \gamma_{i,j} = \gamma_{m-1,m}. \end{cases}$$

For a path $l = (i_0, i_1, \ldots, i_k)$, let us introduce

$$W_l = \begin{cases} P_{i_0}, & \text{if } k = 0, \\ \frac{P_{i_0}P_{i_1}}{\tau'_{i_0,i_1}} \cdot \frac{P_{i_1}P_{i_2}}{\tau'_{i_1,i_2}} \cdot \ldots \cdot \frac{P_{i_{k-1}}P_{i_k}}{\tau'_{i_{k-1},i_k}}, & \text{otherwise.} \end{cases}$$

Proposition 3.3. *The following identity holds:*

$$W_{\hat{l}^* \cup \hat{l}} x_0 = x_0.$$

Proof. We have $W_{\hat{l}^* \cup \hat{l}} x_0 = \big(\prod_{i=1}^{m} \tau'_{i-1,i}\big)^{-2} W x_0 = \big(\nu \prod_{i=1}^{m} \tau_{i-1,i}\big)^{-1} \xi x_0 = x_0$. □

Proposition 3.4. *The linear span $\mathcal{H}'$ of the vectors $\{W_l x_0\}_{l \in \hat{\mathcal{P}}}$ is invariant with respect to an irreducible $*$-representation π.*

Proof. For any l such that Π_l is a normal word, W_l coincides with $\pi(\Pi_l)$ up to a numeric coefficient. Let us show that, for any $a \in TL_{\mathbb{G}_{4,4},g,\perp}$ and any $l_0 \in \hat{\mathcal{P}}$, the following identities are verified:

$$\pi(a)W_{l_0}x_0 = \sum_{l \in \mathcal{N}} \lambda_l \pi(\Pi_l)x_0 = \sum_{l \in \mathcal{N}} \lambda'_l W_l x_0 = \sum_{l \in \mathcal{P}} \lambda''_l W_l x_0 = \sum_{l \in \hat{\mathcal{P}}} \lambda'''_l W_l x_0,$$

where $\mathcal{N}$ is a set of paths l that end in the vertex 0 and such that Π_l is a normal word. To prove the identity before the last one, it will suffice to note that any $l \in \mathcal{N}$ can be represented as

$$l = l' \cup \underbrace{(\hat{l}^* \cup \hat{l}) \cup \cdots \cup (\hat{l}^* \cup \hat{l})}_{k \text{ times}}, \quad k \geq 0,$$

where $l' \in \mathcal{P}$. To prove the last identity, we note that, for $l \in \mathcal{L}_0$,

$$W_l x_0 = W_{l'} W_{(0,1,0)} W_{\hat{l}^* \cup \hat{l}} x_0 = W_{l'} W_{\hat{l}^* \cup \hat{l}} x_0 = W_{l'} x_0,$$

where $l' \in \mathcal{S}$. □

Corollary 3.5. *If π is an irreducible $*$-representation of the algebra $TL_{\mathbb{G}_{4,4},g,\perp}$, then $\mathcal{H} = \mathcal{H}'$ and $\dim \mathcal{H} \leqslant |V| + |V_{in}|$.*

Proposition 3.6.

1) *For any* $l \in \mathcal{S}$,

$$\langle W_l x_0, W_l x_0 \rangle = 1.$$

2) *For any* $l \in \mathcal{L}$,

$$\langle W_l x_0, W_{\omega(l)} x_0 \rangle = 1,$$
$$\langle W_l x_0, W_l x_0 \rangle = \nu^{-1}.$$

Proof. 1) It is clear that $\langle W_{\psi_*(0)} x_0, W_{\psi_*(0)} x_0 \rangle = \langle x_0, x_0 \rangle = 1$. Now,

$$\begin{aligned}\langle W_{\psi_*(1)} x_0, W_{\psi_*(1)} x_0 \rangle &= \left\langle \frac{P_1P_0}{\tau'_{0,1}} x_0, \frac{P_1P_0}{\tau'_{0,1}} x_0 \right\rangle = \left\langle \frac{P_0P_1}{\tau'_{0,1}} \cdot \frac{P_1P_0}{\tau'_{0,1}} W_{\hat{l}^* \cup \hat{l}} x_0, x_0 \right\rangle \\ &= \left\langle \frac{P_0P_1}{\tau'_{0,1}} \cdot \frac{P_1P_0}{\tau'_{0,1}} \cdot \frac{P_0P_1}{\tau'_{0,1}} W_{\eta(\hat{l}^* \cup \hat{l})} x_0, x_0 \right\rangle \\ &= \left\langle \frac{P_0P_1}{\tau'_{0,1}} W_{\eta(\hat{l}^* \cup \hat{l})} x_0, x_0 \right\rangle = \langle W_{\hat{l}^* \cup \hat{l}} x_0, x_0 \rangle = 1,\end{aligned}$$

$$\langle W_{\psi_*(m)} x_0, W_{\psi_*(m)} x_0 \rangle = \langle W_{\hat{l}} x_0, W_{\hat{l}} x_0 \rangle = \langle W_{\hat{l}^* \cup \hat{l}} x_0, x_0 \rangle = 1.$$

Let $l \in \mathcal{S} \backslash \{\psi_*(0), \psi_*(1), \psi_*(m)\}$, and j, j' be the initial points of the paths l and $\eta(l)$, correspondingly. Assuming that $\langle W_{\eta(l)} x_0, W_{\eta(l)} x_0 \rangle = 1$ has been proved, we have

$$\begin{aligned}\langle W_l x_0, W_l x_0 \rangle &= \left\langle \frac{P_jP_{j'}}{\tau'_{j,j'}} W_{\eta(l)} x_0, \frac{P_jP_{j'}}{\tau'_{j,j'}} W_{\eta(l)} x_0 \right\rangle \\ &= \langle W_{\eta(l)} x_0, W_{\eta(l)} x_0 \rangle = 1.\end{aligned}$$

2) Let $l = l_s \cup \tilde{l} \cup \hat{l} \in \mathcal{L}$, $\omega(l) = l_s \cup l_e$, see Proposition 1.1. Then

$$\begin{aligned}\langle W_l x_0, W_{\omega(l)} x_0 \rangle &= \langle W_{l_s} W_{\tilde{l}} W_{\hat{l}} x_0, W_{l_s} W_{l_e} x_0 \rangle \\ &= \langle W_{\hat{l}} x_0, W_{\tilde{l}}^* W_{l_e} x_0 \rangle \\ &= \langle W_{\hat{l}} x_0, W_{\hat{l}} x_0 \rangle = 1.\end{aligned}$$

Now, consider $l = \varphi_*(m-1)$. Then $\eta(l) = \hat{l}$ and

$$\begin{aligned}\langle W_l x_0, W_l x_0 \rangle &= \left\langle \frac{P_{m-1}P_m}{\tau'_{m-1,m}} W_{\hat{l}} x_0, \frac{P_{m-1}P_m}{\tau'_{m-1,m}} W_{\hat{l}} x_0 \right\rangle \\ &= \left\langle \frac{P_mP_{m-1}}{\tau'_{m-1,m}} \cdot \frac{P_{m-1}P_m}{\tau'_{m-1,m}} \cdot \frac{P_mP_{m-1}}{\tau'_{m-1,m}} W_{\eta(\hat{l})} x_0, W_{\hat{l}} x_0 \right\rangle \\ &= \frac{1}{\nu} \left\langle \frac{P_mP_{m-1}}{\tau'_{m-1,m}} W_{\eta(\hat{l})} x_0, W_{\hat{l}} x_0 \right\rangle \\ &= \nu^{-1} \langle W_{\hat{l}} x_0, W_{\hat{l}} x_0 \rangle = \nu^{-1}.\end{aligned}$$

Let now $l \in \mathcal{L}\backslash\{\varphi_*(m-1)\}$, and j, j' be initial points of the paths l and $\eta(l)$, correspondingly. Assuming that $\langle W_{\eta(l)}x_0, W_{\eta(l)}x_0 \rangle = \nu^{-1}$, we have

$$\langle W_l x_0, W_l x_0 \rangle = \left\langle \frac{P_j P_{j'}}{\tau'_{j,j'}} W_{\eta(l)}x_0, \frac{P_j P_{j'}}{\tau'_{j,j'}} W_{\eta(l)}x_0 \right\rangle = \langle W_{\eta(l)}x_0, W_{\eta(l)}x_0 \rangle = \nu^{-1}.$$

This finishes the proof. □

Let us recall that π is an irreducible $*$-representation of the algebra $TL_{\mathbb{G}_{4,4,g,\perp}}$, and the number ν is given by the formula (3.1).

Proposition 3.7. *The parameter ν belongs to the interval* $(0,1]$. *Here, if* $\nu = 1$, *then* $P_{m-1}P_mP_{m-1} = \tau_{m-1,m}P_{m-1}$ *and* $\dim \mathcal{H} \leqslant |V|$.

Proof. It follows from (3.1) that $\nu > 0$.

Consider $l \in \mathcal{L}$ and calculate the norm of $(W_l - W_{\omega(l)})x_0$,

$$\|(W_l - W_{\omega(l)})x_0\|^2 = \langle W_l x_0 - W_{\omega(l)}x_0, W_l x_0 - W_{\omega(l)}x_0 \rangle = \nu^{-1} - 1 \geqslant 0.$$

Hence, $\nu \leqslant 1$.

If $\nu = 1$, then $(W_l - W_{\omega(l)})x_0 = 0$, that is, $W_l x_0 = W_{\omega(l)}x_0$. This shows that the linear span of $\{W_l x_0\}_{l\in\mathcal{S}}$ coincides with the linear span of $\{W_l x_0\}_{l\in\hat{\mathcal{P}}}$ and, consequently, $\dim\mathcal{H} \leqslant |V|$, see Proposition 3.4.

Let us show that in this case, $P_{m-1}P_mP_{m-1} - \tau_{m-1,m}P_{m-1} = 0$.

Indeed, $P_{m-1}W_l \neq 0$ only if the initial point of $l \in \mathcal{S}$ coincides with the vertex $m-1$ or is joined to it with an edge. There are four cases to consider,

1) $l = \psi_*(m-1) = (m-1, m-2, \ldots, 0)$,
2) $l = \psi_*(m-2) = (m-2, \ldots, 0)$,
3) $l = \hat{l} = \psi_*(m) = (m, m-1, m-2, \ldots, 0)$,
4) $l = \psi_*(j) = (j, m-1, m-2, \ldots, 0)$,

where j is any vertex, distinct from $m-2$, joined to the vertex $m-1$ with an edge of type 3. For the first case, we get

$$(P_{m-1}P_mP_{m-1} - \tau_{m-1,m}P_{m-1})W_l x_0 = \tau_{m-1,m}(W_{\varphi_*(m-1)} - W_{\psi_*(m-1)})x_0 = 0.$$

For the second case,

$$(P_{m-1}P_mP_{m-1} - \tau_{m-1,m}P_{m-1})W_l x_0 = \tau_{m-1,m}\tau'_{m-2,m-1}(W_{\varphi_*(m-1)} - W_{\psi_*(m-1)})x_0 = 0.$$

In the third case,

$$(P_{m-1}P_mP_{m-1} - \tau_{m-1,m}P_{m-1})W_{\hat{l}} x_0 = \frac{1}{\tau'_{m-1,m}} P_{m-1}P_m(P_{m-1}P_m - \tau_{m-1,m})W_{\psi_*(m-1)}x_0 = 0.$$

For the fourth case,

$$(P_{m-1}P_mP_{m-1} - \tau_{m-1,m}P_{m-1})W_l x_0 = \tau'_{m-1,j}(P_{m-1}P_mP_{m-1} - \tau_{m-1,m}P_{m-1})W_{\psi_*(m-1)}x_0 = 0.$$

Hence, we have shown that $(P_{m-1}P_mP_{m-1} - \tau_{m-1,m}P_{m-1})\mathcal{H} = \{0\}$. □

Corollary 3.8. *If π is a proper $*$-representation, then $\nu \in (0,1)$.*

Definition 3.9. For any $l \in \hat{\mathcal{P}}$, define

$$y_l = \begin{cases} W_l x_0, & l \in \mathcal{S}, \\ \sqrt{\frac{\nu}{1-\nu}}\big(W_l - W_{\omega(l)}\big)x_0, & l \in \mathcal{L}. \end{cases}$$

Recall that $\mathcal{H}'$ denotes the linear span of the vectors $\{W_l x_0\}_{l\in\hat{\mathcal{P}}}$.

Proposition 3.10. *The linear span of the vectors $\{y_l\}_{l\in\hat{\mathcal{P}}}$ coincides with $\mathcal{H}'$.*

Proposition 3.11. *The following identities are satisfied:*

1. *for any $l, l' \in \hat{\mathcal{P}}$ such that their initial points do not coincide and are not joined with an edge,*
$$\langle y_l, y_{l'} \rangle = 0;$$
2. (a) *for any $l \in \hat{\mathcal{P}}$,*
$$\langle y_l, y_l \rangle = 1;$$
   (b) *for any $l \in \mathcal{L}$,*
$$\langle y_l, y_{\omega(l)} \rangle = 0;$$
3. (a) *for any $l \in \hat{\mathcal{P}}\backslash\{\psi_*(0)\}$,*
$$\langle y_l, y_{\eta(l)} \rangle = \begin{cases} \sqrt{\nu}\sqrt{\tau_{m-1,m}}, & l = \psi_*(m), \\ \sqrt{1-\nu}\sqrt{\tau_{m-1,m}}, & l = \varphi_*(m-1), \\ \sqrt{\tau_{j,j'}}, & \textit{otherwise}, \end{cases}$$
   where j and j' are initial points of l and $\eta(l)$, correspondingly;
   (b) *for any $l \in \mathcal{L}\backslash\{\varphi_*(m-1)\}$,*
$$\langle y_{\eta(l)}, y_{\omega(l)} \rangle = 0, \qquad \langle y_l, y_{\omega(\eta(l))} \rangle = 0.$$

Proof. 1. In this case, the proof is clear.
2. Let $l \in \mathcal{S}$. Then $\langle y_l, y_l \rangle = \langle W_l x_0, W_l x_0 \rangle = 1$. Now, for any $l \in \mathcal{L}$,

$$\begin{aligned}\langle y_l, y_l \rangle &= \frac{\nu}{1-\nu}\langle W_l x_0 - W_{\omega(l)}x_0, W_l x_0 - W_{\omega(l)}x_0 \rangle \\ &= \frac{\nu}{1-\nu}\big(\nu^{-1} - 2 + 1\big) = 1.\end{aligned}$$

Hence, (a) is proved for $l \in \hat{\mathcal{P}}$. Let us now prove (b) for $l \in \mathcal{L}$,

$$\langle y_l, y_{\omega(l)} \rangle = \sqrt{\frac{\nu}{1-\nu}}\langle W_l x_0 - W_{\omega(l)}x_0, W_{\omega(l)}x_0 \rangle = 0.$$

3. Let $l \in \mathcal{S}\backslash\{\psi_*(0)\}$, and j, j' be initial points of the paths l and $\eta(l)$, respectively. Then

$$\begin{aligned}\langle y_l, y_{\eta(l)} \rangle &= \langle P_j W_l x_0, P_{j'} W_{\eta(l)} x_0 \rangle = \langle W_l x_0, P_j P_{j'} W_{\eta(l)} x_0 \rangle \\ &= \tau'_{j,j'}\langle W_l x_0, W_l x_0 \rangle = \tau'_{j,j'}.\end{aligned}$$

Before proving (a) for $l \in \mathcal{L}$, let us prove (b).

Let $l \in \mathcal{L}\backslash\{\varphi_*(m-1)\}$, and j, j' be initial points of the paths l and $\eta(l)$, respectively. Then

$$\begin{aligned}
\langle\, y_l, y_{\omega(\eta(l))} \,\rangle &= \sqrt{\frac{\nu}{1-\nu}}\langle\, W_l x_0 - y_{\omega(l)}, y_{\omega(\eta(l))} \,\rangle \\
&= \sqrt{\frac{\nu}{1-\nu}}\Big(\langle\, W_l x_0, y_{\omega(\eta(l))} \,\rangle - \tau'_{j,j'}\Big) \\
&= \sqrt{\frac{\nu}{1-\nu}}\Big(\Big\langle\, P_{j'}P_j \cdot \frac{P_j P_{j'}}{\tau'_{j,j'}} W_{\eta(l)} x_0, y_{\omega(\eta(l))} \,\Big\rangle - \tau'_{j,j'}\Big) \\
&= \tau'_{j,j'} \cdot \sqrt{\frac{\nu}{1-\nu}}\Big(\langle\, W_{\eta(l)} x_0, y_{\omega(\eta(l))} \,\rangle - 1\Big) = 0; \\
\langle\, y_{\eta(l)}, y_{\omega(l)} \,\rangle &= \sqrt{\frac{\nu}{1-\nu}}\langle\, W_{\eta(l)} x_0 - y_{\omega(\eta(l))}, y_{\omega(l)} \,\rangle \\
&= \sqrt{\frac{\nu}{1-\nu}}\Big(\langle\, W_{\eta(l)} x_0, y_{\omega(l)} \,\rangle - \tau'_{j,j'}\Big) \\
&= \sqrt{\frac{\nu}{1-\nu}}\Big(\langle\, P_j P_{j'} W_{\eta(l)} x_0, y_{\omega(l)} \,\rangle - \tau'_{j,j'}\Big) \\
&= \tau'_{j,j'} \cdot \sqrt{\frac{\nu}{1-\nu}}\Big(\langle\, W_l x_0, y_{\omega(l)} \,\rangle - 1\Big) = 0.
\end{aligned}$$

Now we prove (a) for $l \in \mathcal{L}$.

Consider $l = \varphi_*(m-1)$. Then $\eta(l) = \psi_*(m) = \hat{l}$, $\eta(\hat{l}) = \omega(l) = \psi_*(m-1)$, and

$$\begin{aligned}
\langle\, y_l, y_{\hat{l}} \,\rangle &= \sqrt{\frac{\nu}{1-\nu}}\langle\, W_l x_0 - y_{\omega(l)}, y_{\hat{l}} \,\rangle = \sqrt{\frac{\nu}{1-\nu}}\Big(\langle\, W_l x_0, W_{\hat{l}} x_0 \,\rangle - \tau'_{m-1,m}\Big) \\
&= \tau'_{m-1,m} \cdot \sqrt{\frac{\nu}{1-\nu}}\Big(\Big\langle\, \frac{P_{m-1}P_m}{\tau'_{m-1,m}} W_{\hat{l}} x_0, \frac{P_{m-1}P_m}{\tau'_{m-1,m}} W_{\hat{l}} x_0 \,\Big\rangle - 1\Big) \\
&= \tau'_{m-1,m}\sqrt{\frac{1-\nu}{\nu}} = \sqrt{(1-\nu)\tau_{m-1,m}}.
\end{aligned}$$

Let $l \in \mathcal{L}\backslash\{\varphi_*(m-1)\}$, and j, j' be initial points of the paths l and $\eta(l)$, correspondingly. Then

$$\begin{aligned}
\langle\, y_l, y_{\eta(l)} \,\rangle &= \sqrt{\frac{\nu}{1-\nu}}\langle\, y_l, W_{\eta(l)} x_0 - y_{\omega(\eta(l))} \,\rangle \\
&= \frac{\nu}{1-\nu}\langle\, W_l x_0 - y_{\omega(l)}, W_{\eta(l)} x_0 \,\rangle \\
&= \frac{\nu}{1-\nu}\langle\, W_l x_0 - y_{\omega(l)}, P_j P_{j'} W_{\eta(l)} x_0 \,\rangle \\
&= \tau'_{j,j'} \frac{\nu}{1-\nu}\langle\, W_l x_0 - y_{\omega(l)}, W_l x_0 \,\rangle = \tau'_{j,j'}. \qquad \square
\end{aligned}$$

This implies the following.

Lemma 3.12. *If π is a proper irreducible $*$-representation, then $\nu \in \Sigma_{\mathbb{G}_{4,4},g}$. Moreover, the unitary operator $V : \mathcal{H}_\nu \to \mathcal{H} : \rho_\nu(l) \mapsto y_l$, $l \in \hat{\mathcal{P}}$, defines a unitary equivalence between the $*$-representations π_ν and π.*

Proof. It is clear that for any $l, l' \in \hat{\mathcal{P}}$, we have $B^\nu_{\mathbb{G}_{4,4},g}(l, l') = \langle y_l, y_{l'} \rangle$. Hence, V is a unitary operator and the form $B^\nu_{\mathbb{G}_{4,4},g}(\cdot,\cdot)$ is nonnegative definite.

The operator V intertwines the $*$-representations π_ν and π. Indeed, for any $j \in V_{in}$ and any $l \in \hat{\mathcal{P}}$,

$$\begin{aligned} VP_{\nu,j}\rho_\nu(l) &= \langle \rho_\nu(l), \psi(j) \rangle_\nu V\psi(j) + \langle \rho_\nu(l), \varphi(j) \rangle_\nu V\varphi(j) \\ &= \langle \rho_\nu(l), \psi(j) \rangle_\nu y_{\psi_*(j)} + \langle \rho_\nu(l), \varphi(j) \rangle_\nu y_{\varphi_*(j)}, \\ P_j V \rho_\nu(l) = P_j y_l &= \langle y_l, y_{\psi_*(j)} \rangle y_{\psi_*(j)} + \langle y_l, y_{\varphi_*(j)} \rangle y_{\varphi_*(j)}. \end{aligned}$$

For any $i \in V \backslash V_{in}$ and any $l \in \hat{\mathcal{P}}$, we similarly have

$$\begin{aligned} VP_{\nu,i}\rho_\nu(l) &= \langle \rho_\nu(l), \psi(i) \rangle_\nu V\psi(i) = \langle \rho_\nu(l), \psi(i) \rangle_\nu y_{\psi_*(i)}, \\ P_i V \rho_\nu(l) &= P_i y_l = \langle y_l, y_{\psi_*(i)} \rangle y_{\psi_*(i)}. \end{aligned}$$ □

This proves the following theorem.

Theorem 3.13. *There is a one-to-one correspondence between the set $\Sigma_{\mathbb{G}_{4,4},g}$ and the set of classes of unitarily equivalent irreducible proper $*$-representations of the algebra $TL_{\mathbb{G}_{4,4},g,\perp}$.*

4. A description of all irreducible $*$-representations of the algebra $TL_{\mathbb{G}_{4,4},g,\perp}$ generated by three projections

As an example, we consider the case where the graph $\mathbb{G}_{4,4}$ consists of precisely three vertices and two edges of type 4, that is, $V = \{0, 1, 2\}$, $R = R_4 = \{\gamma_{01}, \gamma_{12}\}$, $g_{01}(x) = \tau_1 x$, $g_{12}(x) = \tau_2 x$, $\tau_j \in (0, 1)$, $j \in \{1, 2\}$. It is clear that $V_{in} = \{1\}$ and, consequently, the dimension of proper $*$-representations does not exceed 4.

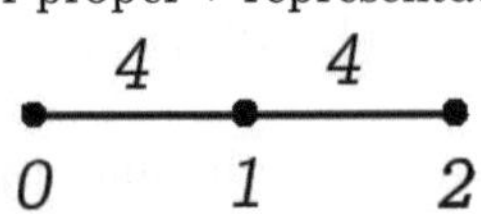

Theorem 4.1. *All irreducible $*$-representations of $TL_{\mathbb{G}_{4,4},g,\perp}$, up to unitary equivalence, are as follows:*

a) *four improper one-dimensional $*$-representations,*

$$\begin{aligned} P_0 = 0, \quad P_1 = 0, \quad P_2 = 0; \\ P_0 = 1, \quad P_1 = 0, \quad P_2 = 0; \\ P_0 = 0, \quad P_1 = 1, \quad P_2 = 0; \\ P_0 = 0, \quad P_1 = 0, \quad P_2 = 1; \end{aligned}$$

b) *two improper two-dimensional $*$-representations,*

$$P_0 = \begin{pmatrix} 1 & 0 \\ 0 & 0 \end{pmatrix}, \ P_1 = \begin{pmatrix} \tau_1 & \sqrt{\tau_1(1-\tau_1)} \\ \sqrt{\tau_1(1-\tau_1)} & 1-\tau_1 \end{pmatrix}, \ P_2 = \begin{pmatrix} 0 & 0 \\ 0 & 0 \end{pmatrix};$$

$$P_0 = \begin{pmatrix} 0 & 0 \\ 0 & 0 \end{pmatrix}, \ P_1 = \begin{pmatrix} \tau_2 & \sqrt{\tau_2(1-\tau_2)} \\ \sqrt{\tau_2(1-\tau_2)} & 1-\tau_2 \end{pmatrix}, \ P_2 = \begin{pmatrix} 1 & 0 \\ 0 & 0 \end{pmatrix};$$

in the case where $\tau_1 + \tau_2 = 1$, there is a third two-dimensional improper $*$-representation,

$$P_0 = \begin{pmatrix} 1 & 0 \\ 0 & 0 \end{pmatrix}, \ P_1 = \begin{pmatrix} \tau_1 & \sqrt{\tau_1\tau_2} \\ \sqrt{\tau_1\tau_2} & \tau_2 \end{pmatrix}, \ P_2 = \begin{pmatrix} 0 & 0 \\ 0 & 1 \end{pmatrix};$$

c) *one improper three-dimensional $*$-representation, if $\tau_1 + \tau_2 < 1$,*

$$P_0 = \begin{pmatrix} 1 & 0 & 0 \\ 0 & 0 & 0 \\ 0 & 0 & 0 \end{pmatrix}, \ P_1 = \begin{pmatrix} \tau_1 & \sqrt{c\tau_1} & \sqrt{\tau_1\tau_2} \\ \sqrt{c\tau_1} & c & \sqrt{c\tau_2} \\ \sqrt{\tau_1\tau_2} & \sqrt{c\tau_2} & \tau_2 \end{pmatrix}, \ P_2 = \begin{pmatrix} 0 & 0 & 0 \\ 0 & 0 & 0 \\ 0 & 0 & 1 \end{pmatrix},$$

where $c = 1 - \tau_1 - \tau_2$;
one proper three-dimensional $*$-representation π_{ν_0}, if $\tau_1 + \tau_2 > 1$,

$$P_0 = \begin{pmatrix} 1 & 0 & 0 \\ 0 & 0 & 0 \\ 0 & 0 & 0 \end{pmatrix}, \ P_1 = \begin{pmatrix} \tau_1 & r(c,\tau_1) & r(\tau_1,\tau_2) \\ r(c,\tau_1) & c & -r(c,\tau_2) \\ r(\tau_1,\tau_2) & -r(c,\tau_2) & \tau_2 \end{pmatrix}, \ P_2 = \begin{pmatrix} 0 & 0 & 0 \\ 0 & 0 & 0 \\ 0 & 0 & 1 \end{pmatrix},$$

where $c = 2 - \tau_1 - \tau_2$, $r(x,y) = \sqrt{(1-x)(1-y)}$,

$$\nu_0 = \frac{r^2(\tau_1,\tau_2)}{\tau_1\tau_2} = \left(\frac{1}{\tau_1} - 1\right)\left(\frac{1}{\tau_2} - 1\right) \in \Sigma_{\mathbb{G}_{4,4},g};$$

d) *a family of four-dimensional proper $*$-representations $\pi_\nu(\cdot)$, where $\nu \in (0,1)$, if $\tau_1 + \tau_2 \leqslant 1$, and $\nu \in (0,\nu_0)$, if $\tau_1 + \tau_2 > 1$,*

$$P_0 = \begin{pmatrix} 1 & 0 & 0 & 0 \\ 0 & 0 & 0 & 0 \\ 0 & 0 & 0 & 0 \\ 0 & 0 & 0 & 0 \end{pmatrix}, \ P_2 = \begin{pmatrix} 0 & 0 & 0 & 0 \\ 0 & 0 & 0 & 0 \\ 0 & 0 & 0 & 0 \\ 0 & 0 & 0 & 1 \end{pmatrix},$$

$$P_1 = \begin{pmatrix} \tau_1 & \sqrt{c\tau_1} & 0 & \sqrt{\nu\tau_1\tau_2} \\ \sqrt{c\tau_1} & c + \frac{\nu(1-\nu)\tau_2^2}{c} & -\frac{\sqrt{d(1-\nu)\nu\tau_2}}{c} & \frac{b\sqrt{\nu\tau_2}}{\sqrt{c}} \\ 0 & -\frac{\sqrt{d(1-\nu)\nu\tau_2}}{c} & \frac{d}{c} & \frac{\sqrt{d(1-\nu)\tau_2}}{\sqrt{c}} \\ \sqrt{\nu\tau_1\tau_2} & \frac{b\sqrt{\nu\tau_2}}{\sqrt{c}} & \frac{\sqrt{d(1-\nu)\tau_2}}{\sqrt{c}} & \tau_2 \end{pmatrix},$$

where $b = 1 - \tau_1 - \tau_2$, $c = b + (1-\nu)\tau_2$, $d = b + (1-\nu)\tau_1\tau_2$.

Proof. With respect to the formal linear basis

$$\hat{\mathcal{P}} = \{\psi(0) = (0),\, \psi(1) = (1,0),\, \psi(2) = (2,1,0),\, \varphi(1) = (1,2,1,0)\},$$

the matrix of the sesquilinear form $B^{\nu}_{\mathbb{G}_{4,4},g}$ will be

$$\begin{pmatrix} 1 & \sqrt{\tau_1} & 0 & 0 \\ \sqrt{\tau_1} & 1 & \sqrt{\nu\tau_2} & 0 \\ 0 & \sqrt{\nu\tau_2} & 1 & \sqrt{(1-\nu)\tau_2} \\ 0 & 0 & \sqrt{(1-\nu)\tau_2} & 1 \end{pmatrix}.$$

For the form to be positive definite, it is necessary and sufficient that

$$\begin{aligned} &1-\tau_1 > 0, \\ &1-\tau_1-\tau_2+(1-\nu)\tau_2 > 0, \\ &1-\tau_1-\tau_2+(1-\nu)\tau_1\tau_2 > 0. \end{aligned}$$

It is clear that the first inequality holds, since $\tau_1 \in (0,1)$. Moreover, if $\tau_1+\tau_2 \leqslant 1$, then the second and the third inequalities hold for any $\nu \in (0,1)$. Hence, if the condition $\tau_1+\tau_2 \leqslant 1$ is satisfied, then $\Sigma_{\mathbb{G}_{4,4},g} = (0,1)$ and $\dim \mathcal{H}_\nu = 4$ for any $\nu \in (0,1)$.

Let now $\tau_1+\tau_2 > 1$. It is clear that

$$1-\tau_1 > 1-\tau_1-\tau_2+(1-\nu)\tau_2 > 1-\tau_1-\tau_2+(1-\nu)\tau_1\tau_2.$$

Now, the third inequality can be rewritten as

$$\nu < \left(\frac{1}{\tau_1}-1\right)\left(\frac{1}{\tau_2}-1\right) = \nu_0.$$

This shows that $(0,\nu_0) \subset \Sigma_{\mathbb{G}_{4,4},g}$, and $\dim \mathcal{H}_\nu = 4$ for any $\nu \in (0,\nu_0)$.

It is also clear that $1-\tau_1-\tau_2+(1-\nu)\tau_1\tau_2 < 0$ for $\nu > \nu_0$. If $\nu = \nu_0$, we get

$$\begin{aligned} &1-\tau_1 > 0, \\ &1-\tau_1-\tau_2+(1-\nu_0)\tau_2 > 0, \\ &1-\tau_1-\tau_2+(1-\nu_0)\tau_1\tau_2 = 0. \end{aligned}$$

Hence, $\Sigma_{\mathbb{G}_{4,4},g} = (0,\nu_0]$ and $\dim \mathcal{H}_{\nu_0} = 3$.

Let $\nu \in \Sigma_{\mathbb{G}_{4,4},g}$ be such that $\dim \mathcal{H}_\nu = 4$. Then, using the system of vectors $\psi(0), \psi(2), \psi(1), \varphi(1)$ we get an orthonormal basis y_0, y_3, y_1, y_2 such that

$$\begin{aligned} \psi(0) &= y_0, \\ \psi(1) &= \sqrt{\tau_1}y_0 + \sqrt{c}y_1 + \sqrt{\nu\tau_2}y_3, \\ \varphi(1) &= -\frac{\sqrt{\nu(1-\nu)\tau_2}}{\sqrt{c}}y_1 + \frac{\sqrt{d}}{\sqrt{c}}y_2 + \sqrt{(1-\nu)\tau_2}y_3, \\ \psi(2) &= y_3, \end{aligned}$$

where $b = 1-\tau_1-\tau_2$, $c = b+(1-\nu)\tau_2$, $d = b+(1-\nu)\tau_1\tau_2$.

It is clear that $P_{\nu,0}$ and $P_{\nu,2}$ are orthogonal projections onto the subspaces generated by the vectors y_0 and y_3, correspondingly. Let us calculate $P_{\nu,1}$ on the basis vectors by the formula $P_{\nu,1}y_i = \langle\, y_i, \psi(1)\,\rangle\psi(1) + \langle\, y_i, \varphi(1)\,\rangle\varphi(1)$,

$$P_{\nu,1}y_0 = \tau_1 y_0 + \sqrt{c\tau_1}y_1 + \sqrt{\nu\tau_1\tau_2}y_3,$$

$$P_{\nu,1}y_1 = \sqrt{c\tau_1}y_0 + (c + \frac{\nu(1-\nu)\tau_2^2}{c})y_1 - \frac{\tau_2\sqrt{d(1-\nu)\nu}}{c}y_2 + \frac{b\sqrt{\nu\tau_2}}{\sqrt{c}}y_3,$$

$$P_{\nu,1}y_2 = -\frac{\tau_2\sqrt{d(1-\nu)\nu}}{c}y_1 + \frac{d}{c}y_2 + \frac{\sqrt{d(1-\nu)\tau_2}}{\sqrt{c}}y_3,$$

$$P_{\nu,1}y_3 = \sqrt{\nu\tau_1\tau_2}y_0 + \frac{b\sqrt{\nu\tau_2}}{\sqrt{c}}y_1 + \frac{\sqrt{d(1-\nu)\tau_2}}{\sqrt{c}}y_2 + \nu\tau_2 y_3.$$

Similar calculations give the tree-dimensional proper $*$-representation for

$$\tau_1 + \tau_2 > 1, \quad \nu = \nu_0.$$

By writing all irreducible $*$-representations for the quotient algebras

$$TL_{\mathbb{G}_{4,4,g,\perp}}/\langle\, p_1p_0p_1 - \tau_1p_1\,\rangle,$$
$$TL_{\mathbb{G}_{4,4,g,\perp}}/\langle\, p_1p_2p_1 - \tau_1p_1\,\rangle$$

and

$$TL_{\mathbb{G}_{4,4,g,\perp}}/\langle\, p_0p_1, p_1p_0, p_1p_2, p_2p_1\,\rangle,$$

we obtain all improper $*$-representations of $TL_{\mathbb{G}_{4,4,g,\perp}}$. □

References

[1] J.J. Graham, *Modular representations of Hecke algebras and related algebras.* Ph.D. thesis, University of Sydney, 1995.

[2] P.R. Halmos, *Two subspaces.* Trans. of the Amer. Math. Soc. **144** (1969), 381–389.

[3] S.A. Kruglyak and Yu.S. Samoilenko, *On unitary equivalence of collections of self-adjoint operators.* Funktsional. Anal. i Prilozhen. **14** (1980), no. 1, 60–62 (Russian).

[4] S.A. Kruglyak and Yu.S. Samoilenko, *On complexity of description of representations of *-algebras generated by idempotents.* Proc. Am. Math. Soc. **128** (2000), no. 6, 1655–1664.

[5] N.D. Popova, *On ∗-representations of one deformed quotient of affine Temperley-Lieb algebra.* Proc. of Institute of Math. of NAS of Ukraine **50** (2004), part 3, 1169–1171.

[6] N.D. Popova, Yu.S. Samoilenko, and O.V. Strilets, *On growth of deformations of algebras connected with Coxeter graphs.* Ukrain. Mat. Zh. **59** (2007), no. 6, 826–837 (Ukrainian).

[7] N.D. Popova, Yu.S. Samoilenko, and O.V. Strilets, *On ∗-representation of a class of algebras connected with Coxeter graphs.* Ukrain. Mat. Zh. **60** (2008), no. 4, 545–556 (Ukrainian).

[8] M. Vlasenko, *On the growth of an algebra generated by a system of projections with fixed angles.* Methods of Funct. Anal. and Topology **10** (2004), no. 1, 98–104.

[9] M.A. Vlasenko and N.D. Popova, *On configurations of subspaces of a Hilbert space with fixed angles between them.* Ukrain. Mat. Zh. **56** (2004), no. 5, 606–615 (Russian, English); translation in Ukr. Math. J. **56**, (2004), no. 5, 730–740.

[10] V.A. Ufnarovskii, *Combinatorial and asymptotic methods in algebra.* Itogi Nauki Tekh. Ser. Sovrem. Mat. Prilozh. Temat. Obz. **57** (1990), Moscow, VINITI, 5–177 (Russian).

N.D. Popova, Yu.S. Samoĭlenko and A.V. Strelets
Institute of Mathematics
National Academy of Science of Ukraine
3 Tereshchenkivs'ka St.
01601 Kyiv, Ukraine
e-mail: popova@imath.kiev.ua
yurii_sam@imath.kiev.ua
sav@imath.kiev.ua

Operator Theory:
Advances and Applications, Vol. 190, 451–470

Correlation Functions of Intrinsically Stationary Random Fields

Zoltán Sasvári

Dedicated to the Centenary of Mark Grigorievich Krein

Abstract. In the introduction we give a short historical survey on the theory of correlation functions of intrinsically stationary random fields. We then prove the existence of generalized correlation functions for intrinsically stationary fields on $\mathbb{R}^d$ as well as an integral representation for these functions. At the end of the paper we show that intrinsically stationary fields are related to unitary operators in Pontryagin spaces in a similar way as stationary fields are related to unitary operators in Hilbert spaces.

Mathematics Subject Classification (2000). Primary 60G10, 42A82; Secondary 43A35, 46C20, 86A32.

Keywords. Intrinsically stationary, generalized correlation, conditionally positive definite, k negative squares, Pontryagin space.

1. Introduction

Let $(\Omega, \mathcal{A}, P)$ be a probability space and denote by $L_2(\Omega, \mathcal{A}, P)$ the corresponding complex Hilbert space. Let further G be a commutative topological group, we will write the group operation as addition. A *second-order* (*complex*) *random field* Z on G is a mapping

$$Z : G \mapsto L_2(\Omega, \mathcal{A}, P).$$

If $L_2(\Omega, \mathcal{A}, P)$ is replaced by the real Hilbert space $L_2^r(\Omega, \mathcal{A}, P)$ generated by real-valued square integrable functions then Z is said to be a *second-order real random field.* In both cases, $(\cdot, \cdot)$ will denote the inner product in the corresponding Hilbert space:

$$(X, Y) = \int_\Omega X \cdot \overline{Y} \, \mathrm{d}P = \mathbb{E}(X \cdot \overline{Y}).$$

Here the symbol $\mathbb{E}$ stands for expectation. We will also use the notation $\|\cdot\|$ for the Hilbert space norm:

$$\|X\| = \sqrt{(X,X)}.$$

If $G = \mathbb{R}$ then the term *random field* is usually replaced by *random process*. A second-order field Z is called *continuous* if

$$\lim_{x\to x_0} \|Z(x) - Z(x_0)\| = 0$$

holds for all $x_0 \in G$. The field Z is said to be *second-order stationary*, or simply *stationary*[1], if $\mathbb{E}\,(Z(x))$ does not depend on x and $\mathbb{E}\,(Z(x)\cdot\overline{Z(y)}\,)$ is a function of $x-y$:

$$(Z(x), Z(y)) = C(x-y) \quad x, y \in G.$$

We call C the *correlation function* of Z. Correlation functions are *positive definite* in the sense that the matrix

$$(C(x_i - x_j))_{i,j=1}^n$$

is nonnegative definite for an arbitrary choice of n and $x_1, \ldots, x_n \in G$. Indeed,

$$\sum_{i,j=1}^n C(x_i - x_j)c_i\overline{c_j} = \sum_{i,j=1}^n (Z(x_i), Z(x_j))c_i\overline{c_j} = \left\|\sum_{j=1}^n Z(x_j)c_j\right\|^2 \geq 0$$

for all $c_j \in \mathbb{C}$.

The so-called *correlation theory* of stationary processes started with A.I. Khintchin's paper [9]. Using Bochner's theorem [2] on the integral representation of positive definite functions he proved that a continuous real-valued function C on $G = \mathbb{R}$ is the correlation function of a continuous real stationary process on $\mathbb{R}$ if and only if

$$C(t) = \int_{-\infty}^{\infty} \cos(tx)\, \mathrm{d}\mu(x), \quad t \in \mathbb{R}$$

where μ is a finite, nonnegative measure. Replacing here $\cos(tx)$ by e^{itx} we obtain the general form of correlation functions in the complex case.

Closely related results were obtained by J. von Neumann and I.J. Schoenberg [19] during the second half of the 1930's. They called a continuous mapping $f: \mathbb{R} \mapsto \mathcal{H}$, where $\mathcal{H}$ is a real Hilbert space, a *screw function* of $\mathcal{H}$ if the distance of $f(t)$ and $f(s)$ depends on $t-s$ only:

$$\|f(t) - f(s)\| = F(t-s), \quad t, s \in \mathbb{R}.$$

Since the space $\mathcal{H}$ is real this condition is equivalent to the independence of r of the inner product

$$(f(t+r) - f(r), f(s+r) - f(r)).$$

[1] In the literature the term *homogeneous field* is used as well.

Neumann and Schoenberg proved that the class of screw functions F is identical with the class of functions whose squares are of the form

$$F^2(t) = \int_0^\infty \frac{\sin^2 tx}{x^2}\, \mathrm{d}\mu(x)$$

where μ is a nonnegative measure on $[0, \infty)$ such that

$$\int_0^1 1\, \mathrm{d}\mu(x) < \infty, \quad \int_1^\infty \frac{1}{x^2}\, \mathrm{d}\mu(x) < \infty.$$

We remark that by well-known results[2], for an arbitrary mapping $f: G \mapsto \mathcal{H}$ where $\mathcal{H}$ is a real (complex) Hilbert space there exists a second-order real (complex, respectively) field Z on G such that $Z(t)$ is gaussian, $\mathbb{E}(Z(t)) = 0$ and

$$(f(t), f(s)) = (Z(t), Z(s)), \quad t, s \in G.$$

Using this probabilistic setting, von Neumann and Schoenberg obtained results about, as we now call them, *processes with stationary increments.* However, in [19], von Neumann and Schoenberg did not mention any relation to processes. This class of processes is the invention of A.N. Kolmogorov. In the papers [10] and [11], he considered the following problem. For each $t \in \mathbb{R}$ let L_t be an operator in a complex Hilbert space $\mathcal{H}$ such that

(i) $L_{s+t} = L_s L_t \quad (t, s \in \mathbb{R})$;

(ii) $L_t h = a_t + U_t h \quad (h \in \mathcal{H})$ where U_t is a unitary operator and $a_t \in \mathcal{H}$;

(iii) $\lim_{t \to s} \|L_t h - L_s h\| = 0 \quad (s \in \mathbb{R},\ h \in \mathcal{H})$.

Fix $h_0 \in \mathcal{H}$ and define $f: \mathbb{R} \mapsto \mathcal{H}$ by

$$f(t) = L_t h_0, \quad t \in \mathbb{R}.$$

If $a_t = 0$ $(t \in \mathbb{R})$ then

$$(f(t), f(s)) = (U_{t-s} h_0, h_0)$$

so that this special case corresponds to stationary processes. In the general case f has *stationary increments* in the sense that the function B defined by

$$B(t, s) = (f(t + r) - f(r), f(s + r) - f(r)), \quad t, s, r \in \mathbb{R}$$

does not depend on r. Kolmogorov proved, among others[3], that B admits an integral representation

$$B(t, s) = \int_{-\infty}^{\infty} (\mathrm{e}^{\mathrm{i}tx} - 1)(\mathrm{e}^{-\mathrm{i}sx} - 1)\, \mathrm{d}\mu(x) + cts$$

[2]See, e.g., Chapter II, §3 in the book [4] of J.L. Doob.

[3]For example he treats the special case

$$B(t, s) = c \cdot [\,|t|^\gamma + |s|^\gamma - |t - s|^\gamma], \quad 0 \le c,\ 0 \le \gamma \le 2.$$

Note that the case $\gamma = 1$ corresponds to the Wiener process.

where $c \in \mathbb{C}$ and μ is a nonnegative measure such that

$$\int_{|x|\leq 1} x^2 \,\mathrm{d}\mu(x) < \infty, \quad \int_{|x|\geq 1} 1 \,\mathrm{d}\mu(x) < \infty.$$

Motivated by problems posed by Kolmogorov[4], M.G. Krein ([12], [13]) obtained solutions of the extrapolation problem for stationary processes and for processes with stationary increments from a finite time interval[5].

The general theory of processes with stationary increments of order n was developed independently by A.M. Yaglom and M.S. Pinsker who published the first results in the joint paper [24]. Denote by Δ_s the backward difference operator defined by

$$\Delta_t f(s) = f(s) - f(s-t), \quad s, t \in \mathbb{R}$$

where f is an arbitrary function on $\mathbb{R}$ with values in a Hilbert space. For each nonnegative integer n we then have

$$\Delta_t^n f(s) = \sum_{j=0}^{n} (-1)^j \binom{n}{j} f(s - jt).$$

A second-order process Z on $\mathbb{R}$ is said to have *stationary increments of order n* if the following expectations do not depend on s:

$$\mathbb{E}\,\Delta_t^n Z(s) =: c^{(n)}(t), \quad \mathbb{E}\,\Delta_{t_1}^n Z(s+t) \cdot \overline{\Delta_{t_2}^n Z(s)} =: D^{(n)}(t; t_1, t_2).$$

Yaglom and Pinsker showed that for continuous processes there exist $c \in \mathbb{C}$ and a finite nonnegative measure on $\mathbb{R}$ such that

$$c^{(n)}(t) = ct^n$$

and

$$D^{(n)}(t; t_1, t_2) = \int_{-\infty}^{\infty} \mathrm{e}^{\mathrm{i}tx} \cdot (\mathrm{e}^{-\mathrm{i}t_1 x} - 1)^n \cdot (\mathrm{e}^{\mathrm{i}t_2 x} - 1)^n \cdot \frac{1 + x^{2n}}{x^{2n}} \,\mathrm{d}\mu(x).$$

Moreover, it can be shown that Z admits a spectral representation

$$Z(t) = \int_{-1}^{1} \mathrm{e}^{\mathrm{i}t\omega} - 1 - \mathrm{i}t\omega - \cdots - \frac{(\mathrm{i}t\omega)^{n-1}}{(n-1)!} \,\mathrm{d}Y(\omega) \qquad (1.1)$$

$$+ \int_{|x|\geq 1} \mathrm{e}^{\mathrm{i}t\omega} \,\mathrm{d}Y(\omega) + X_0 + X_1 t + \cdots + X_n t^n.$$

Here the X_j's are random variables and Y is a random orthogonal measure with some additional properties. We refer to Section 24.3 of Yaglom's book [23] for more details. K. Itô [8] and I.M. Gelfand [6] obtained analogous representations for the so-called *generalized processes* with stationary increments of order n.

The theory of (ordinary) processes with stationary increments of order n was further developed and applied to geostatistical problems by G. Matheron, see, e.g.,

[4]In [13] Krein notes a letter of Kolmogorov to him.

[5]Sections 11 and 12 of the (unfortunately) unpublished manuscript M.G. Krein–H. Langer [16] contains a survey of results about the extrapolation problem.

[17][6]. To formulate some of Matherons definitions denote by $\mathcal{M}_f = \mathcal{M}_f(G)$ the set of all finitely supported complex measures on G. Each element μ of $\mathcal{M}_f$ can be written as

$$\mu = \sum_{i=1}^{n} c_i \delta_{x_i}$$

where $c_i \in \mathbb{C}$ and δ_x denotes the one-point or Dirac measure concentrated at x. We define the measure $\tilde{\mu}$ by

$$\tilde{\mu} = \sum_{i=1}^{n} \overline{c_i} \delta_{-x_i}$$

and the convolution $\mu * g$ by

$$\mu * g(t) = \int g(t-x)\, \mathrm{d}\mu(x) = \sum_{i=1}^{n} c_i g(t - x_i), \quad t \in G$$

where g is an arbitrary function on G with values in a Hilbert space. If $\nu = \sum_{j=1}^{m} d_i \delta_{y_i} \in \mathcal{M}_f$, then

$$\mu * \nu = \sum_{i=1}^{n} \sum_{j=1}^{m} c_i d_j \delta_{x_i + y_j}.$$

Matheron called a second-order real (complex) process Z on $\mathbb{R}$ an *intrinsic random function of order k* or a *k-IRF*, if the process $t \mapsto \mu * Z(t)$ is stationary for all real (complex, respectively) measures $\mu \in \mathcal{M}_f(\mathbb{R})$ such that

$$\int x^l \, \mathrm{d}\mu(x) = 0, \quad l = 0, \dots, k.$$ [7]

Since $\Delta_t^{k+1} x^l = 0$ $(0 \le l \le k)$, a k-IRF has stationary increments of order $k+1$. On the other hand, the spectral representation (1.1) can be used to show that every continuous process with stationary increments of order $k+1$ is a k-IRF. To formulate a more general definition of Matheron, let $\mathcal{F}(G)$ denote the set of all complex-valued functions on G. For an arbitrary translation invariant linear subspace $F \subset \mathcal{F}(G)$ we write

$$F^{*\perp} = \{\mu \in \mathcal{M}_f : \mu * f = 0 \text{ for all } f \in F\}.$$

A second-order field Z on G is called *F-stationary* if the field $\mu * Z$ is stationary for all $\mu \in F^{*\perp}$. Matheron notes that the most important case is when F is finite dimensional. In this case we will say that Z is *intrinsically stationary.* We emphasize that this definition is only essentially but not exactly the same as Matherons. Moreover, he considered only the case $G = \mathbb{R}^d$.

[6] At present, technical reports and manuscripts of G. Matheron are available for free in PDF file format from the home page of the library of the Center of Geostatistics, Fontainebleau. An overview of geostatistical applications can be found in the book [3] by J.-P. Chilès and P. Delfiner.
[7] We remark that Matheron gave also another version of this definition in which a k-IRF is not defined on $\mathbb{R}$ but on a certain set of measures. However, we will not use this version in the present paper.

In the case of an IRF-k on $\mathbb{R}$ the subspace F is the set of all polynomials of degree at most k. In general, a finite-dimensional translation invariant linear space of continuous functions on $\mathbb{R}$ is spanned by *exponential polynomials*, i.e., by functions of the form $x \mapsto p(x)e^{cx}$ where $c \in \mathbb{C}$ and p is a polynomial (see Section 5.4 in [21]).

A complex-valued function K on G is called a *generalized correlation* of an F-stationary field Z if

$$(\mu * Z(x), \nu * Z(y)) = \mu * \tilde{\nu} * K(x-y)$$

holds for all $\mu, \nu \in F^{*\perp}$. Note that this equation implies that K is conditionally positive definite in the sense that the inequality

$$\mu * \tilde{\mu} * K(0) = \sum_{i,j=1}^{n} K(x_j - x_i) c_i \overline{c_j} \geq 0 \tag{1.2}$$

holds for all $\mu = \sum_{i=1}^{n} c_i \delta_{x_i} \in F^{*\perp}$. Matheron proved the existence of the generalized correlation function for continuous k-IRF's on $\mathbb{R}^d$ and for continuous intrinsic stationary fields on $\mathbb{R}$ and gave integral representations (see [17] and [18]).

In the present paper we prove the existence of intrinsic correlation functions for continuous intrinsically stationary fields on $\mathbb{R}^d$. Actually, the method of proof can be extended to arbitrary locally compact abelian groups with a denumerable basis of the topology but we will not consider this general setting here. We will give an integral representation for these functions which shows that they have a finite number of negative squares.

Recall that a hermitian function f on G is said to have k *negative squares*, where k is a nonnegative integer, if $f(-x) = \overline{f(x)}$ $(x \in G)$, and the hermitian matrix

$$A = (f(x_j - x_i))_{i,j=1}^{n}$$

has at most k negative eigenvalues (counted with their multiplicities) for any choice of n and $x_1, \ldots, x_n \in G$, and for some choice of n and $x_1, \ldots, x_n$ the matrix A has exactly k negative eigenvalues. We denote by $P_k(G)$ the set of all functions on G with k negative squares while $P_k^c(G)$ denotes the set of continuous functions $f \in P_k(G)$. It turns out that functions $f \in P_k(G)$, where G is an arbitrary commutative group, are *definitizable* in the following sense: There exists a finite-dimensional translation invariant linear space F of functions on G such that $\mu * \tilde{\mu} * f$ is positive definite whenever $\mu \in F^{*\perp}$ (see Section 5.5 in [21]). This fact shows an obvious analogy to intrinsically stationary fields, which we will explain more detailed in the second part of the paper.

Functions with a finite number of negative squares are the invention of M.G. Krein. In [14] he proved the definitizability of real-valued functions in $P_1(\mathbb{Z})$, where $\mathbb{Z}$ denotes the set of integers, and $P_1^c(\mathbb{R})$ and gave integral representations for these functions. The definitizablitity of functions in $P_k(\mathbb{Z})$ has been proved by Iohvidov [7]. In [15] Krein proved that for every function $f \in P_k^c(\mathbb{R})$ there exists a polynomial

Q of degree k such that the inequality[8]

$$\int_{-\infty}^{\infty}\int_{-\infty}^{\infty} f(x-y)Q\left(-\mathrm{i}\frac{\mathrm{d}}{\mathrm{d}y}\right)h(y)\,\overline{Q\left(-\mathrm{i}\frac{\mathrm{d}}{\mathrm{d}x}\right)h(x)}\,\mathrm{d}y\,\mathrm{d}x \geq 0$$

holds for every infinitely differentiable function h with compact support. Using this, he obtained the integral representation

$$f(x) = p(x) + \int_{-\infty}^{\infty} \frac{\mathrm{e}^{itx} - S(x,t)}{|Q_0(t)|^2}\,\mathrm{d}\mu(t) \tag{1.3}$$

where p is a hermitian solution of the differential equation

$$\overline{Q}\left(-\mathrm{i}\frac{\mathrm{d}}{\mathrm{d}x}\right)Q\left(-\mathrm{i}\frac{\mathrm{d}}{\mathrm{d}x}\right)p(x) = 0 \qquad \left(\overline{Q}(t) = \overline{Q(\bar{t})}\right)$$

Q_0 is a polynomial that obtains by deleting the non-real zeros of Q, S is a regularizing correction compensating for the real zeros of Q, and μ is a nonnegative measure satisfying

$$\int_{-\infty}^{\infty} \frac{1}{(1+t^2)^m}\,\mathrm{d}\mu(t) < \infty$$

where m denotes the degree of Q_0. The definitizability of a function $f \in P_k(G)$ where G is an arbitrary commutative group has first been proved in [20]. Chapters 5 and 6 of [21] contain an introduction to the theory of functions with negative squares and of definitizable functions. More historical remarks on these functions can be found in the Notes to Chapter 6.

2. Definition and some basic properties

In this section G denotes an arbitrary commutative Hausdorff topological group. We will use the notations $\mathcal{M}_f = \mathcal{M}_f(G)$, $\mathcal{F} = \mathcal{F}(G)$ and $F^{*\perp}$ introduced in the previous section but we repeat the basic definitions for the sake of clarity.

Definition 2.1. Let F be a translation invariant linear subspace of $\mathcal{F}$. A second-order random field Z on G is said to be *F-stationary* if the field $\mu * Z$ is stationary whenever $\mu \in F^{*\perp}$. If F is finite dimensional then any F-stationary field is called *intrinsically stationary.*[9]

Assume that Z is F-stationary and for an arbitrary $\mu \in F^{*\perp}$ denote by C_μ the correlation function of $\mu * Z$. We then have

$$(\nu * \mu * Z(x), \nu * \mu * Z(y)) = \nu * \tilde{\nu} * C_\mu(x-y) \tag{2.1}$$

for all $\nu \in \mathcal{M}_f$, $x, y \in G$. In particular,

$$(\mu * Z(x), \mu * Z(y)) = C_\mu(x-y), \quad \mu \in F^{*\perp}. \tag{2.2}$$

[8]This inequality is the integral version of (1.2).

[9]It would be possible to distinguish between weak and strong intrinsically stationarity as in the case of stationary fields but strong stationarity plays no role in the present paper.

From (2.1) we conclude that

$$\nu * \tilde{\nu} * C_\mu = \mu * \tilde{\mu} * C_\nu, \quad \mu, \nu \in F^{*\perp}. \tag{2.3}$$

Equation (2.3) plays a key role in Section 4 where for continuous intrinsically stationary fields on $\mathbb{R}^d$ we prove the existence of a complex-valued function K on G such that

$$C_\mu = \mu * \tilde{\mu} * K, \quad \mu \in F^{*\perp}.$$

This motivates the following definition.

Definition 2.2. Let Z be an F-stationary random field on G. A complex-valued function K on G is called a *generalized correlation* of Z if it is hermitian, i.e., $K(-x) = \overline{K(x)}$, and

$$(\mu * Z(x), \nu * Z(y)) = \mu * \tilde{\nu} * K(x - y)$$

holds for all $\mu, \nu \in F^{*\perp}$ and $x, y \in G$.

The generalized correlation function is not unique in general. Indeed, if K is a generalized correlation of Z then so is the function $L = K + P$ for an arbitrary function P satisfying

$$\mu * \tilde{\nu} * P = 0 \tag{2.4}$$

for all $\mu, \nu \in F^{*\perp}$. Conversely, if L and K are generalized correlation functions of Z then $P := L - K$ satisfies the equation (2.4).

Recall that a non-zero complex-valued function γ on G is called *multiplicative* if $\gamma(x + y) = \gamma(x)\gamma(y)$ holds for all $x, y \in G$. Note that for a multiplicative function γ and for a measure $\mu \in \mathcal{M}_f$ the equation $\mu * \gamma = 0$ holds if and only if $\int \gamma(-x)\, \mathrm{d}\mu(x) = 0$. Continuous multiplicative functions on $\mathbb{R}^d$ have the form $\gamma(x) = \mathrm{e}^{(x,c)}$ with some $c \in \mathbb{C}^d$ where $(\cdot,\cdot)$ denotes the Euclidean inner product.

The next theorem follows immediately from Theorem 5.4.10 in [21].

Theorem 2.3. *For every intrinsically stationary field Z there exist positive integers $k_1, \dots, k_n$ and multiplicative functions $\gamma_1, \dots, \gamma_n$ on G such that $\mu * Z$ is stationary for all measures μ of the form*

$$\mu = \mu_{11} * \cdots * \mu_{1k_1} * \cdots * \mu_{n1} * \cdots * \mu_{nk_n} \tag{2.5}$$

where $\mu_{ij} \in \mathcal{M}_f$ and $\mu_{ij} * \gamma_i = 0$ for $j = 1, \dots, k_i;\ i = 1, \dots, n$.

The set of all measures of the form (2.5) will be denoted by

$$\mathcal{M}_f(\gamma_1, k_1; \dots; \gamma_n, k_n).$$

If all functions γ_j are continuous then

$$\mathcal{M}_c(\gamma_1, k_1; \dots; \gamma_n, k_n)$$

denotes the set of all measures of the form (2.5) such that all factors μ_{ij} are compactly supported finite Borel measures.

Definition 2.4. A second-order field Z on G is called *stationarizable*[10] if there exist positive integers $k_1, \dots, k_n$ and multiplicative functions $\gamma_1, \dots, \gamma_n$ on G such that $\mu * Z$ is stationary for all measures

$$\mu \in \mathcal{M}_f(\gamma_1, k_1; \dots; \gamma_n, k_n).$$

We denote by

$$S(\gamma_1, k_1; \dots; \gamma_n, k_n)$$

the set of stationarizable fields Z on G for which $\mu * Z$ is stationary whenever $\mu \in \mathcal{M}_f(\gamma_1, k_1; \dots; \gamma_n, k_n)$. The set of continuous fields in $S(\gamma_1, k_1; \dots; \gamma_n, k_n)$ is denoted by $S^c(\gamma_1, k_1; \dots; \gamma_n, k_n)$. We say that a field $Z \in S(\gamma_1, k_1; \dots; \gamma_n, k_n)$ has a *singularity of order k_j at γ_j* if

$$Z \notin S(\gamma_1, k_1; \dots; \gamma_j, k_j - 1; \dots; \gamma_n, k_n).$$

Here we omit the term $\gamma_j, k_j - 1$ if $k_j - 1 = 0$ and agree that $S()$ consists of all stationary fields. When dealing with $G = \mathbb{R}^d$ and identifying the multiplicative function $\mathrm{e}^{\mathrm{i}(\cdot, z)}$ with $z \in \mathbb{C}^d$ we will also write

$$S(z_1, k_1; \dots; z_n, k_n) \quad \text{or} \quad \mathcal{M}_f(z_1, k_1; \dots; z_n, k_n)$$

and speak of singularity at $z_j \in \mathbb{C}^d$.

Lemma 2.5. *If the functions γ_j are continuous then*

$$M_f := \mathcal{M}_f(\gamma_1, k_1; \dots; \gamma_n, k_n)$$

is a dense subset of

$$M_c := \mathcal{M}_c(\gamma_1, k_1; \dots; \gamma_n, k_n)$$

with respect to the weak topology.

Proof. Let $\mu \in M_c$, $\mu \neq 0$. Since μ is the convolution product of finitely many compactly supported measures μ_{ij}, its support, which we denote by K, is compact. By continuity, the functions γ_j are bounded on K. In view of Proposition 3.5 in [1], there exist nets $\{\nu_{ij}^\alpha\}$ in $\mathcal{M}_f$ converging weakly to μ_{ij}. Let $x_0 \in K$ be arbitrary. Since

$$\lim_\alpha \nu_{ij}^\alpha * \gamma_j(x_0) = \lim_\alpha \int \gamma_j(x_0 - x)\, \mathrm{d}\nu_{ij}^\alpha(x) = \mu_{ij} * \gamma_j(x_0) = 0$$

the nets

$$\mu_{ij}^\alpha := \nu_{ij}^\alpha - \nu_{ij}^\alpha * \gamma_j(x_0) \cdot \delta_{x_0}$$

converge weakly to μ_{ij} for all i, j. Moreover, $\mu_{ij}^\alpha * \gamma_j = 0$. By Corollary 3.4 in [1], the convolution product of these measures converges to μ. □

[10] Note that the terminology *stationarizable* has been used by W.A. Gardner [5] in a different sense. In spite of this fact, we suggest this terminology because of its analogy with that of definitizable functions (see Chapter 6 in [21]).

Corollary 2.6. *If $Z \in S^c(\gamma_1, k_1; \dots; \gamma_n, k_n)$ where the γ_j's are continuous then $\mu * Z$ is stationary for all*

$$\mu \in M_c := \mathcal{M}_c(\gamma_1, k_1; \dots; \gamma_n, k_n).$$

Moreover, the analogue of (2.3) *holds:*

$$\nu * \tilde{\nu} * C_\mu = \mu * \tilde{\mu} * C_\nu, \quad \mu, \nu \in M_c. \tag{2.6}$$

3. Auxiliary results

We will need some technical results on the Fourier transform

$$\hat{\mu}(y) = \int_{\mathbb{R}^d} \mathrm{e}^{-\mathrm{i}(y,x)} \, \mathrm{d}\mu(x), \quad y \in \mathbb{R}^d$$

of compactly supported complex (Borel) measures μ. The first lemma is a special case of Lemma 6.4.3 in [21].

Lemma 3.1. *For every $y \in \mathbb{R}^d$ and every neighborhood V of y there exist a compactly supported complex measure μ such that*

$$\hat{\mu}(y) = 0 \quad \text{and} \quad |\hat{\mu}(x)| \geq \frac{1}{2}, \quad x \notin V.$$

Proof. Let φ be a continuous positive definite function such that $0 \leq \varphi \leq 1$, $\varphi(0) = 1$ and $\operatorname{supp}(\varphi) \subset V - y$ and denote by ν the probability measure satisfying $\hat{\nu} = \varphi$. Choose a compact set $K \subset \mathbb{R}^d$ such that $\nu(K) \geq \frac{3}{4}$ and put

$$\sigma = \frac{1}{\nu(K)} \cdot \nu_K$$

where ν_K denotes the restriction of ν to the set K. We have

$$\sup_{x \in \mathbb{R}^d} |\varphi(x) - \hat{\sigma}(x)| \leq \|\nu - \sigma\| \leq \frac{1}{2}$$

and hence $|\hat{\sigma}(x)| \leq \frac{1}{2}$ for all $x \notin V - y$. Setting

$$\mathrm{d}\mu(t) := \mathrm{e}^{\mathrm{i}(t,y)} \, \mathrm{d}(\delta_0 - \sigma)(t)$$

we have $\hat{\mu}(x) = 1 - \hat{\sigma}(x - y)$. Thus, $\mu(y) = 0$ and

$$|\hat{\mu}(x)| \geq 1 - |\hat{\sigma}(x - y)| \geq \frac{1}{2}, \; x \notin V.$$ □

Lemma 3.2. *Let $y_1, \dots, y_m \in \mathbb{R}^d$ be mutually different, $k_1, \dots, k_m$ be positive integers and V be an open set containing $\{y_1, \dots, y_m\}$. Then there exist a measure*

$$\mu \in \mathcal{M}_c(y_1, k_1; \dots; y_m, k_m)$$

and a positive number δ such that

$$|\hat{\mu}(x)| \geq \delta, \quad x \notin V.$$

Proof. By (3.1) there exists a compactly supported complex measure μ_j such that $\hat{\mu}_j(y_j) = 0$ and $|\hat{\mu}_j(x)| \geq \frac{1}{2}$ for $x \notin V$. The measure $\mu = \mu_1^{k_1} * \cdots * \mu_m^{k_m}$ has the desired properties with $\delta = 2^{-(k_1+\cdots+k_m)}$. □

Lemma 3.3. *Let $y_1, \ldots, y_m \in \mathbb{R}^d$ be mutually different and $k_1, \ldots, k_m$ be positive integers. For each $x \in \mathbb{R}^d$ and for each $j = 1, \ldots, m$ there exist a neighborhood $V = V_j$ of y_j, a measure*

$$\mu = \mu_{x,j} \in \mathcal{M}_f(y_1, k_1; \ldots; y_m, k_m)$$

and $\delta = \delta_{x,j} > 0$ such that the inequality

$$|\hat{\mu}(y)| \geq \delta \cdot |(x, y - y_j)|^{k_j}, \quad y \in V$$

holds.

Proof. Without loss of generality assume that $j = 1$. Let $y_0 \in \mathbb{R}^d$ be such that $(y_0, y_i - y_1) \neq 0$ for all $i = 2, \ldots, n$ and choose a neighborhood $W = W_x$ of y_1 so that

$$|(x, y - y_1)| \leq \pi, \quad y \in W. \tag{3.1}$$

Setting $\mu_1 := \delta_0 - \mathrm{e}^{\mathrm{i}(x,y_1)}\delta_x$ and $\mu_i := \delta_0 - \mathrm{e}^{\mathrm{i}(y_0,y_i)}\delta_{y_0}$ $(i > 1)$ we have $\hat{\mu}_i(y_i) = 0$ for all $i = 1, \ldots, m$ and $\hat{\mu}_i(y_1) \neq 0$ if $i > 1$. Using the inequality

$$1 - \cos t \geq \frac{t^2}{\pi^2}, \quad -\pi \leq t \leq \pi$$

and (3.1) we obtain

$$\begin{aligned} |\hat{\mu}_1(y)| &= \left|1 - \mathrm{e}^{\mathrm{i}(x,y_1-y)}\right| = 2 \cdot [1 - \cos(x, y - y_1)]^{\frac{1}{2}} \\ &\geq \frac{2}{\pi} \cdot |(x, y - y_1)|, \quad y \in W. \end{aligned}$$

Now it is not hard to see that the measure $\mu = \mu_1^{k_1} * \cdots * \mu_m^{k_m}$ has the desired properties. □

Lemma 3.4. *Let $z_1, \ldots, z_n \in \mathbb{C}^d \backslash \mathbb{R}^d$ be mutually different and $k_1, \ldots, k_n$ be positive integers. Then there exist a measure*

$$\mu \in \mathcal{M}_f(z_1, k_1; \ldots; z_n, k_n)$$

such that

$$|\hat{\mu}(x)| \geq 1, \quad x \in \mathbb{R}^d.$$

Proof. Since $z_j \notin \mathbb{R}^d$ we can choose $x_j \in \mathbb{R}^d$ so that $|\mathrm{e}^{\mathrm{i}(x_j,z_j)}| \geq 2$. We write $\mu_j := \delta_0 - \mathrm{e}^{\mathrm{i}(x_j,z_j)}\delta_{x_j}$. Then $\mu_j * \mathrm{e}^{\mathrm{i}(\cdot,z_j)} = 0$ and

$$|\hat{\mu}_j(x)| = |1 - \mathrm{e}^{\mathrm{i}(x_j,z_j)}\mathrm{e}^{-\mathrm{i}(x_j,x)}| \geq 1, \quad x \in \mathbb{R}^d$$

and hence the measure $\mu = \mu_1^{k_1} * \cdots * \mu_n^{k_n}$ has the desired properties. □

Lemma 3.5. *Let $y_1, \dots, y_m \in \mathbb{R}^d$ be mutually different and $l_1, \dots, l_m$ be positive integers. A nonnegative Radon measure σ on $\mathbb{R}^d$ satisfies the condition*

$$\int_{\mathbb{R}^d} |\hat{\mu}(y)| \, \mathrm{d}\sigma(y) < \infty \tag{3.2}$$

for all $\mu \in \mathcal{M}_c(y_1, l_1; \dots; y_m, l_m)$ if and only if

$$\sigma(\mathbb{R}^d \setminus V) < \infty \tag{3.3}$$

for every open set V containing $\{y_1, \dots, y_m\}$ and

$$\int_{V_j} \|y - y_j\|_2^{l_j} \, \mathrm{d}\sigma(y) < \infty \tag{3.4}$$

for each j and every relative compact neighborhood V_j of y_j.

Proof. Assume that (3.2) holds. The validity of (3.3) follows immediately from (3.2). By Lemma 3.3, the function $y \mapsto |(x, y-y_j)|^{l_j}$ is for each $x \in \mathbb{R}^d$ μ-integrable in a certain neighborhood of y_j. Let $\{e_1, \dots, e_d\}$ be an orthonormal basis of $\mathbb{R}^d$. Then the function

$$y \mapsto \sum_{k=1}^{d} |(e_k, y - y_j)|^{l_j} = \|y - y_j\|_{l_j}^{l_j}$$

is μ-integrable in a neighborhood of y_j. Relation (3.4) follows now from the fact that all l_p-norms on $\mathbb{R}^d$ are equivalent.

Suppose that (3.3) and (3.4) hold and let $\mu \in \mathcal{M}_c(y_1, l_1; \dots; y_m, l_m)$. Since $\hat{\mu}$ is bounded it suffices to show its integrability in a neighborhood of y_j ($j = 1, \dots, m$). This follows from the fact that the inequality

$$|\hat{\nu}(y)| \le D \cdot \|y - y_j\|_2$$

where D is a suitable constant, holds in a neighborhood of y_j for any compactly supported measure ν such that $\hat{\nu}(y_j) = 0$ (note that $\hat{\nu}$ is continuously differentiable). □

Lemma 3.6. *Let $y_1, \dots, y_m \in \mathbb{R}^d$ be mutually different and $l_1, \dots, l_m$ be positive integers. Choose mutually disjoint bounded neighborhoods V_j of the y_j's and compactly supported continuous functions $d_j : \mathbb{R}^d \mapsto \mathbb{R}$ such that $d_j(y) = 1$ on V_j and $d_j = 0$ on V_k, $k \neq j$. Then the function*

$$P(x, y) = \sum_{j=1}^{m} \mathrm{e}^{\mathrm{i}(x, y_j)} \cdot Q_j(x, y), \quad x, y \in \mathbb{R}^d$$

where

$$Q_j(x, y) = d_j(y) \cdot \sum_{l=0}^{l_j} \frac{[\mathrm{i} \cdot (x, y - y_j)]^l}{l!}$$

has the following properties:

(i) *P is continuous and for each x the function $y \mapsto P(x,y)$ has compact support.*
(ii) *For each y and j the function $x \mapsto Q_j(x,y)$ is a Hermitian polynomial of degree at most l_j.*
(iii) *There exist a constant M such that*

$$|e^{i(x,y)} - P(x,y)| \leq M \cdot |(x, y - y_j)|^{l_j+1}, \quad y \in V_j, \ x \in \mathbb{R}^d.$$

Proof. Properties (i) and (ii) are trivial while (iii) follows from the fact that

$$P(x,y) = T_{k_j}(i(x,y), i(x,y_j)), \quad y \in V_j$$

where

$$T_k(z, z_0) = e^{z_0} \sum_{l=0}^{k} \frac{1}{l!} \cdot (z - z_0)^l$$

denotes the Taylor polynomial of order k at the point z_0 of the function $z \mapsto e^z$, $z \in \mathbb{C}$. □

4. Existence of generalized correlations

In this section we prove the existence of generalized correlation functions for stationarizable fields on $\mathbb{R}^d$ as well as an integral representation for these functions. The proof goes along the same lines as the proof of the integral representation for definitizable functions (cf. Theorem 6.4.7 in [21]).

Theorem 4.1. *Any random field $Z \in S^c(y_1, k_1; \ldots; y_n, k_n)$ on $\mathbb{R}^d$ has a generalized correlation.*

Proof. Without loss of generality assume that y_j is a singularity of Z of order k_j. We enumerate the singularities y_j in such a way that $y_j \in \mathbb{R}^d$ if $1 \leq j \leq m$, $y_j \in \mathbb{C}^d$ if $m < j \leq n$ and write

$$M = \mathcal{M}_c(y_1, k_1; \ldots; y_n, k_n).$$

By Corollary 2.6, the field $\mu * Z$ is stationary for an arbitrary $\mu \in M$. Let C_μ be the correlation function of $\mu * Z$ and denote by τ_μ its spectral measure, i.e.,

$$C_\mu(x) = \int_{\mathbb{R}^d} e^{i(x,y)} \, d\tau_\mu(y), \quad x \in \mathbb{R}^d.$$

From equation (2.6) we conclude that

$$|\hat{\nu}(y)|^2 \, d\tau_\mu(y) = |\hat{\mu}(y)|^2 \, d\tau_\nu(y), \quad \mu, \nu \in M. \tag{4.1}$$

By Lemmas 3.2 and 3.4, the family

$$O_\mu := \{y \in \mathbb{R}^d : \hat{\mu}(y) \neq 0\}, \quad \mu \in M$$

is an open covering of $\mathbb{R}^d \setminus \{y_1, \ldots, y_m\}$. On each set O_μ we define the Borel measure σ_μ by

$$d\sigma_\mu(y) = \frac{1}{|\hat{\mu}(y)|^2} \, d\tau_\mu(y).$$

In view of (4.1), the measures σ_μ satisfy the compatibility condition of Theorem 1.18 in [1]. Consequently, there exists a (nonnegative) Borel measure σ on $\mathbb{R}^d \setminus \{y_1, \ldots, y_m\}$ such that the restriction of σ to O_μ is equal to σ_μ. We extend σ to a measure on $\mathbb{R}^d$ by setting $\sigma(\{y_j\}) = 0$ $(j = 1, \ldots, m)$. It follows from the definition of σ_μ that

$$\int_{\mathbb{R}^d} |\hat{\mu}(y)|^2 \, \mathrm{d}\sigma(y) = \int_{O_\mu} |\hat{\mu}(y)|^2 \, \mathrm{d}\sigma(y) = \int_{O_\mu} 1 \, \mathrm{d}\tau_\mu(y) < \infty \tag{4.2}$$

for all $\mu \in M$.

Let now P be the function from Lemma 3.6 with $l_j = 2k_j - 1$. Define the function K by

$$K(x) = \int_{\mathbb{R}^d} \mathrm{e}^{\mathrm{i}(x,y)} - P(x,y) \, \mathrm{d}\sigma(y), \quad x \in \mathbb{R}^d.$$

That the integral exists follows from (4.2), from Lemma 3.5 with $l_j = 2k_j$ and from the properties of P.

We now show that

$$C_\mu = \mu * \tilde{\mu} * K, \quad \mu \in \mathcal{M}_f(y_1, k_1; \ldots; y_n, k_n).$$

By Theorem 5.4.10 in [21], $\mu * \tilde{\mu} * P(\cdot, y) = 0$, $y \in \mathbb{R}^d$. Using this we obtain

$$\mu * \tilde{\mu} * K(x) = \int_{\mathbb{R}^d} \mathrm{e}^{\mathrm{i}(x,y)} \cdot |\hat{\mu}(y)|^2 \, \mathrm{d}\sigma(y) = \int_{\mathbb{R}^d} \mathrm{e}^{\mathrm{i}(x,y)} \, \mathrm{d}\tau_\mu(y) = C_\mu(x)$$

completing the proof. □

Theorem 4.2. *Let $Z \in S^c(y_1, k_1; \ldots; y_n, k_n)$ be a random field where $y_j \in \mathbb{R}^d$ if $j \le m$ and $y_j \in \mathbb{C}^d \setminus \mathbb{R}^d$ if $j > m$. Then*

$$K(x) = \int_{\mathbb{R}^d} \mathrm{e}^{\mathrm{i}(x,y)} - P(x,y) \, \mathrm{d}\sigma(y), \quad x \in \mathbb{R}^d \tag{4.3}$$

is a generalized correlation function of Z, where:

(i) *P is the function from Lemma 3.6 with $l_j = 2k_j - 1$;*
(ii) *σ is a certain nonnegative Borel measure on $\mathbb{R}^d$ such that*
(iii) $\sigma(\{y_j\}) = 0 \quad (j = 1, \ldots, m)$;
(iv) *$\sigma(\mathbb{R}^d \setminus V) < \infty$ for every open set V containing $\{y_1, \ldots, y_m\}$;*
(v) *Each y_j has a neighborhood V_j such that*

$$\int_{V_j} \|y - y_j\|_2^{2k_j} \, \mathrm{d}\sigma(y) < \infty.$$

The measure σ is the only Borel measure on $\mathbb{R}^d$ satisfying (4.3) *and having the properties* (iii)–(v).

If

$$\int_{W_i} \|y - y_i\|_2^{2k_i - 2} \, \mathrm{d}\sigma(y) = \infty \tag{4.4}$$

for every neighborhood of y_i then Z has a singularity of order k_i at y_i.

Proof. The integral representation (4.3) and the properties (i)–(iii) have already been established in the proof of Theorem 4.1. Properties (iv) and (v) follow from (4.2) and Lemma 3.5.

Let μ be a Borel measure on $\mathbb{R}^d$ satisfying (iii)–(v) and such that

$$\int_{\mathbb{R}^d} e^{i(x,y)} - P(x,y)\, d\sigma(y) = \int_{\mathbb{R}^d} e^{i(x,y)} - P(x,y)\, d\nu(y), \quad x \in \mathbb{R}^d.$$

By the same argument as at the end of the proof of Theorem 4.1 we conclude that

$$\int_{\mathbb{R}^d} e^{i(x,y)} \cdot |\hat{\mu}(y)|^2\, d\sigma(y) = \int_{\mathbb{R}^d} e^{i(x,y)} \cdot |\hat{\mu}(y)|^2\, d\nu(y)$$

for all $\mu \in M_f = \mathcal{M}_f(y_1, k_1; \ldots; y_m, k_m)$. Thus,

$$|\hat{\mu}(y)|^2\, d\sigma(y) = |\hat{\mu}(y)|^2\, d\nu(y), \quad \mu \in M_f. \tag{4.5}$$

Since M_f is dense in $M_c = \mathcal{M}_f(y_1, k_1; \ldots; y_m, k_m)$ we see that the above equation holds for all $\mu \in M_c$. Applying Lemma 3.2 and (iii) we see that $\nu = \sigma$.

To prove the last statement assume that (4.4) holds for some i but the order of the singularity at y_i is less than k_i, i.e.,

$$Z \in S^c(y_1, k_1; \ldots; y_i, k_i - 1; \ldots; y_m, y_m).$$

Then, by what we have already proved, we obtain another integral representation of K of the form (4.3) with some measure ν and with $k_i - 1$ instead of k_i. Hence,

$$\int_{O_i} \|y - y_i\|_2^{2k_i - 2}\, d\nu(y) < \infty$$

for some neighborhood O_i of y_i and therefore $\sigma \neq \mu$. The function P corresponding to this representation still satisfies $\mu * \tilde{\mu} * P(\cdot, y) = 0$ for all $\mu \in M_f$ and hence equation (4.5) holds. We conclude again that σ and μ must be equal. This contradiction completes the proof. □

Remark 4.3. It follows from Theorem 5.4.10 in [21] that $\mu * \tilde{\nu} * P(\cdot, t) = 0,\ t \in \mathbb{R}^d$ if $\mu, \nu \in M_f = \mathcal{M}_f(y_1, k_1; \ldots; y_n, k_n)$. Using this, from (4.3) we see that

$$\mu * \tilde{\nu} * K(x - y) = \int_{\mathbb{R}^d} e^{i(x-y,t)} \cdot \hat{\mu}(t) \cdot \overline{\hat{\nu}(t)}\, d\sigma(t) \tag{4.6}$$

holds for all $\mu, \nu \in M_f$.

Theorem 4.4. *For every function K of the form (4.3) there exists a random field $Z \in S^c(y_1, k_1; \ldots; y_m, k_m)$ such that K is a generalized correlation function of Z.*

Proof. Without loss of generality we may assume that the order of singularity at y_i is equal to k_i, i.e., relation (4.4) holds. Let q be a nonnegative, continuous, bounded function on $\mathbb{R}^d$ such that $q(y) = \|y - y_j\|_2^{k_j}$ for all y in some neighborhood of y_j and $q(y) \geq \delta$ holds for some $\delta > 0$ and for all y not contained in the union of these neighborhoods. We define the nonnegative measure σ_q by $d\sigma_q(y) = q(y)^2\, d\sigma(y)$. In view of (4.2.v), this measure is finite. Multiplying q with a suitable constant we may assume that σ_q is a probability measure. Consider now the probability space $(\mathbb{R}^d, \mathcal{B}(\mathbb{R}^d), \sigma_q)$ and define $Y(x)$ by

$$Y(x) = [\mathrm{e}^{\mathrm{i}(x,\cdot)} - P(x,\cdot)] \cdot \frac{1}{q(\cdot)}, \quad x \in \mathbb{R}^d$$

where $\cdot$ indicates the variable $\omega \in \Omega = \mathbb{R}^d$ and P is the function from Lemma 3.5 with $l_j = k_j - 1$. By the properties of P and σ, the random variable $Y(x)$ is square integrable with respect to σ_q. It is easy to check that

$$\mu * Y(x) = \hat{\mu}(\cdot)Y(x), \quad \mu \in M = \mathcal{M}_f(y_1, k_1; \ldots; y_m, k_m).$$

Using this and (4.6) we obtain that

$$\begin{aligned}(\mu * Y(x), \nu * Y(y)) &= \int_\Omega \hat{\mu}(\omega)Y(x)\overline{\hat{\nu}(\omega)Y(y)}\,\mathrm{d}\sigma_q(\omega) \\ &= \int_\Omega \mathrm{e}^{\mathrm{i}(x-y,\omega)}\hat{\mu}(\omega)\overline{\hat{\nu}(\omega)}\,\mathrm{d}\sigma(\omega) = \mu * \tilde{\nu} * K(x-y)\end{aligned}$$

holds for all $\mu, \nu \in M$. As we mentioned in the introduction there exists a second-order random field Z on $\mathbb{R}^d$ such that $Z(x)$ is gaussian, $\mathbb{E}(Z(x)) = 0$ and

$$(Y(x), Y(y)) = (Z(x), Z(y)), \quad x, y \in \mathbb{R}^d$$

completing the proof. □

Remark 4.5. Using the notations of the preceding proof but setting

$$Y(x) = \mathrm{e}^{\mathrm{i}(x,\cdot)} \cdot \frac{1}{q(\cdot)}, \quad x \in \mathbb{R}^d$$

we obtain a random variable which is not necessarily square integrable. However, $\mu * Y$ is square integrable for each $\mu \in M$ and the equation

$$(\mu * Y(x), \nu * Y(y)) = \mu * \tilde{\nu} * K(x-y), \quad \mu, \nu \in M$$

still holds. If in the definition of a stationarizable field we drop the condition that Z be of second-order and only require that $\mu * Z$ is of second order for all $\mu \in M$, we obtain a wider class of random fields. Several results, e.g., the existence of the generalized correlation, could then be proved in the same way.

Theorem 4.6. *Let Z be as in Theorem 4.2. There exist ordinary continuous correlation functions C_n $(n = 1, 2, \ldots)$ and exponential polynomials*

$$Q_n(x) = \sum_{j=1}^{m} \mathrm{e}^{\mathrm{i}(x,y_j)} \cdot P_{j,n}(x), \quad x \in \mathbb{R}^d$$

where $P_{j,n}$ is a hermitian polynomial with $\deg(P_{j,n}) \le 2k_j - 1$, such that

$$K(x) = \lim_{n\to\infty} C_n(x) + Q_n(x), \quad x \in \mathbb{R}^d$$

is a generalized correlation function for Z and the convergence is uniform on compact sets.

Proof. Choose an arbitrary compact subset F of $\mathbb{R}^d$. Let K be as in (4.2) and for each n let $O_n \subset \mathbb{R}^d$ be a union of disjoint open balls with centers at the y_j's, the radii tending to zero as n tends to infinity. We also assume that the open balls with center y_j are contained in the set V_j of Lemma 3.6. In view of (iii) of Lemma 3.6, there exists a constant M_F such that

$$|\mathrm{e}^{\mathrm{i}(x,y)} - P(x,y)| \leq M_F \cdot \|y - y_j\|_2^{2k_j}, \quad y \in V_j,\ x \in F. \tag{4.7}$$

Using that the restriction of σ to $\mathbb{R}^d \setminus O_n$ is a finite measure we obtain

$$\begin{aligned} K(x) &= \int_{\mathbb{R}^d} \mathrm{e}^{\mathrm{i}(x,y)} - P(x,y)\,\mathrm{d}\sigma(y) \\ &= \int_{O_n} \mathrm{e}^{\mathrm{i}(x,y)} - P(x,y)\,\mathrm{d}\sigma(y) \\ &\quad + \int_{\mathbb{R}^d \setminus O_n} \mathrm{e}^{\mathrm{i}(x,y)}\,\mathrm{d}\sigma(y) + \int_{\mathbb{R}^d \setminus O_n} P(x,y)\,\mathrm{d}\sigma(y). \end{aligned}$$

Inequality (4.7) shows that the first integral on the right, as a function of x, tends to zero uniformly on compact sets. The second integral on the right is an ordinary correlation function while the third one is, by the definition of P, a hermitian polynomial of degree at most $2k_j - 1$. □

Corollary 4.7. *Continuous generalized correlation functions on $\mathbb{R}^d$ have a finite number of negative squares.*

Proof. Let C_n and Q_n be as in Corollary 4.6. Since $\deg(P_{j,n}) \leq 2k_j - 1$, it follows immediately from Lemma 3.11 in [22] that there exists an integer k such that the dimension of the linear space spanned by all translates of Q_n is less than or equal to k for all n. Thus, Q_n has at most k negative squares. Since C_n is positive definite, we conclude that $K = \lim_n (C_n + Q_n)$ has at most k negative squares, as well. The proof is finished by noting that any other generalized correlation function of the given field can be obtained by adding an exponential polynomial to K. □

5. Connection to Pontryagin spaces

In this section G denotes an arbitrary abelian group and $f \in P_k(G)$ is a function with k negative squares. We will construct intrinsically stationary fields having f as generalized correlation. These fields will be obtained as projections of trajectories of a unitary representation of G in a Pontryagin space. We refer to Appendices A and B in [21] for basic facts on Pontryagin spaces and on unitary operators in these spaces. Recall that a unitary representation (U_x) of G in a π_k-space Π_k is a mapping $x \mapsto U_x$ such that U_x is a unitary operator in Π_k and $U_{x+y} = U_x U_y$ $(x, y \in G)$.

The construction is now as follows. By Theorem 5.1.7 in [21], to the function f there corresponds a π_k-space $\Pi_k(f)$ with inner product $(\cdot,\cdot)$ such that elements of $\Pi_k(f)$ are complex-valued functions on G and the linear span of all translates of f is dense in $\Pi_k(f)$. Moreover, setting

$$(U_x g)(y) := g(y-x), \quad g \in \Pi_k(f),\ x, y \in G$$

we obtain a unitary representation of G satisfying

$$g(x) = (g, U_x f), \quad g \in \Pi_k(f),\ x \in G.$$

In view of Theorem B.7 in [21], there exists a k-dimensional, non-positive, (U_x)-invariant subspace $F \subset \Pi_k(f)$. Let H be the closed linear subspace of $\Pi_k(f)$ generated by all functions of the form $\mu * f$, $\mu \in F^{*\perp}$. By Theorem 5.5.1 in [21], the function $\mu * \tilde{\mu} * f$ is positive definite implying that H is nonnegative. Let $H_0 = H \cap H^{\perp}$ be the isotropic subspace of H. Both subspaces H and H_0 are (U_x)-invariant and, by the definition of H,

$$\mu * f = \sum_j c_j U_{x_j} f \in H \tag{5.1}$$

holds whenever $\mu = \sum_j c_j \delta_{x_j} \in F^{*\perp}$. Since H and H_0 are (U_x)-invariant, we conclude that

$$\tilde{U}_x(w + H_0) := U_{-x} w + H_0, \quad w \in H$$

defines a unitary representation $(\tilde{U}_x)$ of G in the Hilbert space $\tilde{H} := H/H_0$. Let now P be a linear projection of Π_k onto H. For each $x \in G$ we define $Y(x) \in H$ and $Z(x) \in \tilde{H}$ by

$$Y(x) = PU_{-x} f, \quad Z(x) = Y(x) + H_0.$$

Choosing μ as above and using (5.1) and the fact that $PU_{-x}h = U_{-x}h$ for all $h \in H$ and $x \in G$ we obtain

$$\begin{aligned} \mu * Y(x) &= \sum_j c_j Y(x - x_j) = \sum_j c_j PU_{-x+x_j} f \\ &= PU_{-x} \sum_j c_j U_{x_j} f = U_{-x} \sum_j c_j U_{x_j} f. \end{aligned}$$

Therefore

$$\mu * Z(x) = \tilde{U}_x h_\mu, \quad x \in G,\ \mu \in F^{*\perp}$$

where $h_\mu = \sum_j c_j U_{x_j} f + H_0$. This shows that Z can be identified with an intrinsically stationary field. If $\mu = \sum_j c_j \delta_{x_j}$ and $\nu = \sum_j d_j \delta_{y_j}$ are two measures in $F^{*\perp}$ then

$$\begin{aligned} (\mu * Z(x), \nu * Z(y)) &= (\tilde{U}_x h_\mu, \tilde{U}_y h_\nu) = (\sum_j c_j U_{x_j - x} f, \sum_j d_j U_{y_j - y} f) \\ &= \sum_{i,j} c_i \overline{d_j} f(x - y - x_i + y_j) = \mu * \tilde{\nu} * f(x-y). \end{aligned}$$

Thus, f is a generalized covariance for Z.

References

[1] C. Berg, J.P.R. Christensen, P. Ressel, *Harmonic Analysis on Semigroups. Theory of Positive Definite and Related Functions.* Berlin-Heidelberg-New York-Tokyo: Springer-Verlag 1984.

[2] S. Bochner, *Vorlesungen über Fouriersche Integrale.* Leipzig: Akademische Verlagsgesellschaft 1932.

[3] J.-P., Chilès, P. Delfiner, Geostatistics. Modelling Spatial Uncertainty. New York: John-Wiley 1999.

[4] J.L. Doob, *Stochastic processes.* New York: John Wiley & Sons, London: Chapman & Hall 1953.

[5] W.A. Gardner, *Stationarizable random processes.* IEEE Trans. Inform. Theory **24** (1978), no. 1, 8–22.

[6] I.M. Gelfand, *Generalized random processes.* (Russian) Doklady Akad. Nauk. SSSR **100** 5 (1955), 853–856.

[7] I.S. Iohvidov, *Unitary and selfadjoint operators in spaces with an indefinite metric.* (Russian) Dissertation, Odessa 1950.

[8] K. Itô, *Stationary random distributions.* Mem. Coll. Sci. Univ. Kyoto, Ser. A **28** (1954), no. 3, 209–223.

[9] A.I. Khinchin, *Korrelationstheorie der stationären stochastischen Prozesse.* Math. Ann. Vol. **109** (1934), 604–615.

[10] A.N. Kolmogorov, *Kurven im Hilbertschen Raum, die gegenüber einer einparametrigen Gruppe von Bewegungen invariant sind.* C. R. Acad. Sci. URSS, Vol. **XXVI** (1940), no. 1, 6–9.

[11] A.N. Kolmogorov, *Wienersche Spiralen und einige andere interessante Kurven im Hilbertschen Raum.* C. R. Acad. Sci. URSS, Vol. **XXVI** (1940), no. 2, 115–118.

[12] M.G. Krein, *On the problem of continuation of helical arcs in Hilbert space.* C.R. Acad. Sci. URSS, Vol. **XLV** (1944), no. 4, 139–142.

[13] M.G. Krein, *On a problem of extrapolation of A. N. Kolmogoroff.* C.R. Acad. Sci. URSS, Vol. **XLVI** (1945), no. 8, 306–309.

[14] M.G. Krein, *Helical lines in infinite-dimensional Lobachevski space and the Lorentz transformation.* (Russian) Usp. Mat. Nauk **3**(3) (1948), 158–160.

[15] M.G. Krein, *On the integral representation of a continuous Hermitian-indefinite function with a finite number of negative squares.* (Russian) Dokl. Akad. Nauk SSSR **125**(1) (1959), 31–34.

[16] M.G. Krein–H. Langer, *Continuation of Hermitian positive definite functions and related questions.* Unpublished manuscript.

[17] G. Matheron, *The intrinsic random functions and their applications.* Adv. Appl. Prob. **5** (1973), 439–468.

[18] G. Matheron, *Comment translater les catastrophes ou La structure des F.A.I. generales.* Manuscript **N-167**, Fontainebleau (1979), 1–36.

[19] J. von Neumann, I.J. Schoenberg, *Fourier integrals and metric geometry.* Trans. Amer. Math. Soc. **50** (1941), no. 2, 225–250.

[20] Z. Sasvári, *Indefinite functions on commutative groups.* Monatshefte für Math. **100** (1985), 223–238.

[21] Z. Sasvári, *Positive Definite and Definitizable Functions.* Akademie Verlag, Berlin, 1994.

[22] Z. Sasvári, *On the number of negative squares of certain functions.* Operator Theory. Advances and Applications, Vol. **106**, 337–353, Birkhäuser Verlag, Basel, 1998.

[23] A.M. Yaglom, *Correlation Theory of Stationary and Related Random Functions I.* Springer Verlag, Berlin, Heidelberg, New York, 1986.

[24] A.M. Yaglom, M.S. Pinsker, *Random processes with stationary increments of order n.* (Russian) Doklady Akad. Nauk. SSSR, **XC** 5 (1953), 731–734.

Acknowledgment

We would like to thank Christiane Weber for her contribution to the historical survey by finding numerous papers, not only those listed in the References, some of them not easily accessible.

Zoltán Sasvári
Technical University of Dresden
Department of Mathematics
Mommsenstr. 13
01062 Dresden, Germany
e-mail: `Zoltan.Sasvari@math.tu-dresden.de`

Operator Theory:
Advances and Applications, Vol. 190, 471–490

Inverse Problem for Conservative Curved Systems

Alexey Tikhonov

Abstract. Conservative curved systems over multiply connected domains are introduced and relationships of such systems with related notions (functional model, characteristic function, and transfer function) are studied. In contrast to standard theory for the unit disk, characteristic functions and transfer functions are essentially different objects. We study possibility to recover the characteristic function for a given transfer function. As the result we obtain the procedure to construct the functional model for a given conservative curved system.

Mathematics Subject Classification (2000). Primary 47A48; Secondary 47A45, 47A56, 47A55.

Keywords. Conservative system, functional model, transfer function, characteristic function.

0. Introduction

In order to present the main result of the paper, we need to start with some background. On the whole, the well-developed and well-known theory of unitary colligations [1, 2] and related topics [3, 4, 5] can be represented by the following diagram (see details in the above-mentioned references).

$$\begin{array}{ccc} \text{Sys} & \xrightarrow{\mathcal{F}_{ts}} & \text{Tfn} \\ \mathcal{F}_{sm}\,\big\uparrow\big\downarrow\,\mathcal{F}_{ms} & & \mathcal{F}_{tc}\,\big\uparrow\big\downarrow\,\mathcal{F}_{ct} \\ \text{Mod} & \underset{\mathcal{F}_{mc}}{\overset{\mathcal{F}_{cm}}{\leftrightarrows}} & \text{Cfn} \end{array} \tag{Dgr}$$

The research for this article was supported by INTAS grant, project 05-1000008-7883.

Here Sys is the class of all simple unitary colligations (or, in terminology of systems theory, conservative controllable observable linear systems), that is, $\mathfrak{A} \in \mathrm{Sys}$ iff

$$\mathfrak{A} = \begin{pmatrix} T & N \\ M & L \end{pmatrix} \in \mathcal{L}(H \oplus \mathfrak{N}, H \oplus \mathfrak{M}), \quad \mathfrak{A}^*\mathfrak{A} = I, \ \mathfrak{A}\mathfrak{A}^* = I$$

and T is c.n.u. contraction [1], where $H, \mathfrak{M}, \mathfrak{N}$ are separable Hilbert spaces. Classes of transfer Tfn and characteristic Cfn functions coincide with the Schur class of analytic contractive-valued functions

$$S = \{\Theta \in H^\infty(\mathbb{D}, \mathcal{L}(\mathfrak{N}, \mathfrak{M})) : ||\Theta||_\infty \leq 1\}.$$

The class of functional models Mod is a class of all pairs $\Pi = (\pi_+, \pi_-)$ of operators $\pi_\pm \in \mathcal{L}(L^2(\mathbb{T}, \mathfrak{N}_\pm), \mathcal{H})$ such that

(i) $(\pi_\pm^* \pi_\pm) = I$;

(ii)$_1$ $(\pi_-^* \pi_+)z = z(\pi_-^* \pi_+)$; (ii)$_2$ $P_-(\pi_-^* \pi_+)P_+ = 0$;

(iii) $\operatorname{Ran} \pi_+ \vee \operatorname{Ran} \pi_- = \mathcal{H}$,

where $\mathfrak{N}_\pm, \mathcal{H}$ are separable Hilbert spaces, P_+ is the orthoprojection onto the Hardy space H^2, and $P_- = I - P_+$.

The transformations $\mathcal{F}_{ts}, \mathcal{F}_{cm}, \mathcal{F}_{ct}, \mathcal{F}_{tc}$ are defined by formulas

$$\Upsilon(z) = L + zM(I - zT)^{-1}N, \quad \Theta = \pi_-^* \pi_+, \quad \Upsilon = \Theta^\sim,$$

where $\Theta^\sim(z) := \Theta(\bar{z})^*$. The transformation $\mathfrak{A} = \mathcal{F}_{sm}(\Pi)$ is defined by formulas

$$\begin{aligned}
&\widehat{T} \in \mathcal{L}(\mathcal{K}_\Theta) && \widehat{T}f := \mathcal{U}f - \pi_+ \widehat{M}f; \\
&\widehat{M} \in \mathcal{L}(\mathcal{K}_\Theta, \mathfrak{N}_+) && \widehat{M}f := \frac{1}{2\pi i} \int_{\mathbb{T}} (\pi_+^* f)(z)\, dz; \\
&\widehat{N} \in \mathcal{L}(\mathfrak{N}_-, \mathcal{K}_\Theta) && \widehat{N}n := P_\Theta \pi_- n; \\
&\widehat{L} \in \mathcal{L}(\mathfrak{N}_-, \mathfrak{N}_+) && \widehat{L} = (\pi_-^* \pi_+)(0)^*,
\end{aligned}$$

where $f \in \mathcal{K}_\Theta := \operatorname{Ran} P_\Theta$, $P_\Theta := (I - \pi_+ P_+ \pi_+^*)(I - \pi_- P_- \pi_-^*)$, $n \in \mathfrak{N}_-$, and the unitary operator $\mathcal{U}$ with absolutely continuous spectrum is uniquely determined by conditions $\mathcal{U}\pi_\pm = \pi_\pm z$.

The transformation $\Pi = \mathcal{F}_{ms}(\mathfrak{A})$ can be obtained on the standard way of constructing of unitary dilation for the unitary colligation $\mathfrak{A}$ and mapping the dilation space to functional spaces with the aid of Fourier representations [1]. Note that the isometries $\pi_\pm$ are adjoint to that Fourier representations.

The construction of the transformation $\Pi = \mathcal{F}_{mc}(\Theta)$ can be found in [1, 4] (see the so-called Sz.-Nagy-Foiaş's transcription of functional model).

Remark 0.1. Note that colligations $\mathfrak{A}$ and $\mathcal{F}_{sm}\mathcal{F}_{ms}(\mathfrak{A})$ are unitarily equivalent. This is a reason to regard unitary equivalent colligation as equal in this theory. Similarly, models Π, $\mathcal{F}_{ms}\mathcal{F}_{sm}(\Pi)$, and $\mathcal{F}_{mc}\mathcal{F}_{cm}(\Pi)$ are unitarily equivalent and we regard them as equal. With this remark the above diagram becomes a commutative diagram.

Remark 0.2. Note also that any Schur class function Θ can be represented in the form of orthogonal sum $\Theta = \Theta_p \oplus \Theta_u$ of pure operator-valued function Θ_p and unitary constant Θ_u. The unitary constant Θ_u coincides with unitary part $L_u = L|\mathfrak{N}_u$ of the operator L from the corresponding unitary colligation, where $\mathfrak{N}_u := \{n \in \mathfrak{N} : ||Ln|| = ||n||\}$. Note that a unitary colligation $\mathfrak{A}$ can be recovered from the part L_u and the operators T, M, N. The latter observation explains some details of our further constructions.

Thus one may start from any vertex of the diagram and obtain all information relating to other vertexes. A starting point depends on object area of a researcher. For instance, one of approaches in mathematical physics is the study of an operator by means of its functional model which is unitary equivalent to the initial one [6]. On the other hand, from point of view of systems theory it is interesting to study a conservative system knowing only its transfer function and vice versa [3]. In turn, some researches restrict themselves to study only the vertex Mod and this approach gives a nice opportunity to connect operator theory and function theory in a very deep and fruitful manner [5]. In the latter case corresponding authors avoid to use the word "model" and say about restricted shift. Thus the term "model" can be interpreted as some enough simple operator acting in functional space.

At first sight the functional model is a universal tool to study any contraction (and therefore any linear operator in Hilbert space), but on the very early stage of development it was discovered and noticed that the functional model is efficient and useful mainly for operators that are closed to unitary ones. For general operators the functional model is not very informative. For instance, it is possible to construct Sz.-Nagy-Foiaş functional model for the operator of multiplication by the independent variable in $L^2(0,1)$. Obviously this operator is a c.n.u. contraction, but its Sz.-Nagy-Foiaş functional model brings almost nothing in regard to its spectral analysis. The standard Spectral Theorem for self-adjoint operators is a lot more informative and natural tool for this operator.

We are not so arrogant to suggest a model that pretends to study efficiently every operator. Our intentions are:

1) to modify the definition of Mod with purpose of more convenient studying operators with continuous spectrum on a curve;
2) to develop corresponding links and reveal changes for other vertexes and arrows of the diagram (Dgr).

Thus our approach is model-centric and we will consider constructing of any transformation from the vertex Mod as a direct problem and towards Mod as an inverse one, respectively.

Let G_+ be a finite-connected domain of the complex plane $\mathbb{C}$ bounded by a rectifiable Carleson curve C, $G_- = \mathbb{C} \setminus \operatorname{clos} G_+$ and $\infty \in G_-$. We shall consider pairs $\Pi = (\pi_+, \pi_-)$ of operators $\pi_\pm \in \mathcal{L}(L^2(C, \mathfrak{N}_\pm), \mathcal{H})$ such that

$$\begin{array}{lll} \text{(i)}_1\ (\pi_\pm^* \pi_\pm) z = z(\pi_\pm^* \pi_\pm); & \text{(i)}_2\ \pi_\pm^* \pi_\pm >> 0; & \\ \text{(ii)}_1\ (\pi_-^\dagger \pi_+) z = z(\pi_-^\dagger \pi_+); & \text{(ii)}_2\ P_-(\pi_-^\dagger \pi_+) P_+ = 0; & \text{(Mod)} \\ \text{(iii)}\ \ \operatorname{Ran} \pi_+ \vee \operatorname{Ran} \pi_- = \mathcal{H}, & & \end{array}$$

where $\mathfrak{N}_\pm, \mathcal{H}$ are separable Hilbert spaces; $A >> 0$ means that $\exists\, c > 0$ such that $\forall u\ (Au, u) \geq c(u, u)$; the (non-orthogonal) projections $P_\pm$ are uniquely determined by conditions $\operatorname{Ran} P_\pm = E^2(G_\pm, \mathfrak{N}_\pm)$ and $\operatorname{Ker} P_\pm = E^2(G_\mp, \mathfrak{N}_\pm)$ (since the curve C is a Carleson curve, the projections $P_\pm$ are bounded); the spaces $E^2(G_\pm, \mathfrak{N}_\pm)$ are Smirnov's spaces [7] of vector-valued functions with values in $\mathfrak{N}_\pm$; the operators $\pi_\pm^\dagger$ are adjoint to $\pi_\pm$ if we regard $\pi_\pm : L^2(C, \Xi_\pm) \to \mathcal{H}$ as operators acting from weighted L^2 spaces with operator-valued weights $\Xi_\pm = \pi_\pm^* \pi_\pm$. In this interpretation $\pi_\pm$ are isometries. For such pairs $\Pi = (\pi_+, \pi_-)$, we shall say that Π is a free functional model and write $\Pi \in \operatorname{Mod}$. A reader interested in motivation of this definition of functional model can found corresponding explanations and discussion in [8].

It can easily be shown (see, e.g., [8]) that there is an one-to-one correspondence (up to unitary equivalence of models) between models $\Pi \in \operatorname{Mod}$ and characteristic functions $\Theta = (\Theta^+, \Xi_+, \Xi_-) \in \operatorname{Cfn}$ from weighted Schur classes

$$\begin{aligned} S_\Xi := \ & \{ (\Theta^+, \Xi_+, \Xi_-) \ : \ \Theta^+ \in H^\infty(G_+, \mathcal{L}(\mathfrak{N}_+, \mathfrak{N}_-)), \\ & \forall\, \zeta \in C \ \forall\, n \in \mathfrak{N}_+ \quad \|\Theta^+(\zeta) n\|_{-,\zeta} \leq \|n\|_{+,\zeta} \}, \end{aligned} \tag{Cfn}$$

where $\mathfrak{N}_\pm$ are separable Hilbert spaces and $\Xi_\pm$ are operator-valued weights such that $\Xi_\pm, \Xi_\pm^{-1} \in L^\infty(C, \mathcal{L}(\mathfrak{N}_\pm))$, $\Xi_\pm(\zeta) \geq 0$, $\zeta \in C$, $\|n\|_{\pm,\zeta} := (\Xi_\pm(\zeta) n, n)^{1/2}$, $n \in \mathfrak{N}_\pm$. Note that again (see Remark 0.1) we regard unitary equivalent models as equal. The transformation $\Theta = \mathcal{F}_{cm}(\Pi)$ is defined by the formula

$$\Theta = (\pi_-^\dagger \pi_+, \pi_+^* \pi_+, \pi_-^* \pi_-) \in S_\Xi . \tag{MtoC}$$

The construction of the inverse transformation $\Pi = \mathcal{F}_{mc}(\Theta)$ can be found in [8].

We cannot give independent definition for extension of notion of conservative systems to this new context and have to make use of the above-introduced functional model. First we define the transformation $\widehat{\Sigma} = \mathcal{F}_{sm}(\Pi)$. We put

$$\widehat{\Sigma} = \mathcal{F}_{sm}(\Pi) := (\widehat{T}, \widehat{M}, \widehat{N}, \widehat{\Theta}_u, \widehat{\Xi}; \mathcal{K}_\Theta, \mathfrak{N}_+, \mathfrak{N}_-)$$

with

$$\begin{aligned} & \widehat{T} \in \mathcal{L}(\mathcal{K}_\Theta), && \widehat{T} f := \mathcal{U} f - \pi_+ \widehat{M} f, \ \ f \in \mathcal{K}_\Theta ; \\ & \widehat{M} \in \mathcal{L}(\mathcal{K}_\Theta, \mathfrak{N}_+), && \widehat{M} f := \frac{1}{2\pi i} \int_C (\pi_+^\dagger f)(z)\, dz ; \\ & \widehat{N} \in \mathcal{L}(\mathfrak{N}_-, \mathcal{K}_\Theta), && \widehat{N} n := P_\Theta \pi_- n, \ \ n \in \mathfrak{N}_- ; \\ & \widehat{\Xi} := (\pi_+^* \pi_+, \ \pi_-^* \pi_-) ; \end{aligned} \tag{MtoS}$$

where $\mathcal{K}_\Theta := \operatorname{Ran} P_\Theta$, $P_\Theta := (I - \pi_+ P_+ \pi_+^\dagger)(I - \pi_- P_- \pi_-^\dagger)$; the normal operator $\mathcal{U}$, which spectrum is absolutely continuous and lies on C, is uniquely determined by conditions $\mathcal{U} \pi_\pm = \pi_\pm z$. Since, in general, we cannot define the operator $\widehat{L} = (\pi_-^* \pi_+)(0)^*$, we content ourselves with the "unitary part" $\widehat{\Theta}_u$ of the characteristic function (see Remark 0.2 and explanation of meaning of the unitary part of weighted Schur class function in Section 1). In the sequel, we shall refer $\widehat{\Sigma}$ as the model system and the operator $\widehat{T}$ as the (main) model operator.

A coupling of operators and Hilbert spaces $\Sigma = (T, M, N, \Theta_u, \Xi; H, \mathfrak{N}, \mathfrak{M})$ is called a *simple conservative curved system* if there exists a functional model Π with $\mathfrak{N}_+ = \mathfrak{N}$, $\mathfrak{N}_- = \mathfrak{M}$, and an invertible operator $X \in \mathcal{L}(H, \mathcal{K}_\Theta)$ such that

$$\Sigma = (T, M, N, \Theta_u, \Xi; H, \mathfrak{N}, \mathfrak{M}) \overset{X}{\sim} \widehat{\Sigma} = \mathcal{F}_{sm}(\Pi), \tag{Sys}$$

where we write $\Sigma_1 \overset{X}{\sim} \Sigma_2$ if

$$XT_1 = T_2X, \quad M_1 = M_2X, \quad N_1X = N_2, \quad \Theta_{1u} = \Theta_{2u}, \quad \Xi_1 = \Xi_2 .$$

In our theory we shall regard similar systems $\Sigma_1 \overset{X}{\sim} \Sigma_2$ as equal.

Passing on to transfer functions, again at first we define the transformation $\Upsilon = \mathcal{F}_{ts}(\Sigma)$,

$$\Upsilon = (\Upsilon(z), \Theta_u, \Xi) \text{ , where } \Upsilon(z) := M(T - z)^{-1}N . \tag{StoT}$$

Υ is called the transfer function of a curved conservative system Σ. Within functional model the transformation $\mathcal{F}_{tc} := \mathcal{F}_{ts} \circ \mathcal{F}_{sm} \circ \mathcal{F}_{mc}$ can be computed as

$$\Upsilon(z) = (\mathcal{F}_{tc}(\Theta))(z) = \begin{cases} \Theta_+^-(z) - \Theta^+(z)^{-1}, & z \in G_+ \cap \rho(T); \\ -\Theta_-^-(z), & z \in G_-, \end{cases} \tag{CtoT}$$

where the operator-valued functions $\Theta_\pm^-(z)$ are defined by the formulas

$$\Theta_\pm^-(z)\, n := (P_\pm \Theta^- n)(z), \quad z \in G_\pm, \quad n \in \mathfrak{N}_- ;$$
$$\Theta^-(\zeta) := (\pi_+^\dagger \pi_-)(\zeta) = \Xi_+(\zeta)^{-1}\Theta^+(\zeta)^*\Xi_-(\zeta), \quad \zeta \in C .$$

Remark 0.3. Note that, in terminology of [9], the relationship (CtoT) means that the transfer function Υ "corresponds to the function" Θ^+. It should also be noted that the class of models considered by D. Yakubovich in [9] is a particular case of our class Mod. In fact, he studied models with Ξ-inner (in our terminology) function Θ for which $\Xi_+ = I$ and $\Xi_- = (\Theta^+\Theta^{+*})^{-1}$, though he did not introduce the weights $\Xi_\pm$ explicitly. From the point of view of operators, this means that the absolutely continuous spectral component of the main operator T is trivial.

Thus we arrive at the following diagram

$$\begin{array}{ccc} \text{Mod} & \underset{\mathcal{F}_{mc}}{\overset{\mathcal{F}_{cm}}{\rightleftarrows}} & \text{Cfn} \\ {\scriptstyle\mathcal{F}_{sm}}\downarrow & & \downarrow{\scriptstyle\mathcal{F}_{tc}} \\ \text{Sys} & \underset{\mathcal{F}_{ts}}{\longrightarrow} & \text{Tfn} \end{array} \tag{Dgr}$$

The main problem which arises here is to invert arrows $\mathcal{F}_{tc}$, $\mathcal{F}_{ts}$, and $\mathcal{F}_{sm}$. It is sufficient to invert only the arrow $\mathcal{F}_{tc}$, that is, to recover characteristic function for a given transfer function. We emphasize that, in contrast to standard theory, now characteristic functions and transfer functions are essentially different objects and therefore our main problem is to solve the equations (CtoT). Note also that actually the projections $P_\pm$ are singular integral operators and one can regard (CtoT) as a system of (non-linear) singular integral equations with unknown Θ^+

and given $\Upsilon, \Xi_\pm$. The main result of the paper is the uniqueness of a solution of the equations (CtoT), which is expressed in the following theorem.

Theorem A. *If* $\Theta_1, \Theta_2 \in \mathrm{Cfn}$ *and* $\mathcal{F}_{tc}(\Theta_1) = \mathcal{F}_{tc}(\Theta_2) \in \mathcal{N}(G_+ \cup G_-, \mathcal{L}(\mathfrak{N}_-, \mathfrak{N}_+))$, *then* $\Theta_1 = \Theta_2$.

Here $\mathcal{N}(G_+ \cup G_-, \mathcal{L}(\mathfrak{N}_-, \mathfrak{N}_+))$ is the Nevanlinna class of operator-valued functions that admit representation of the form $\Upsilon(z) = 1/\delta(z)\,\Omega(z)$, where $\delta \in H^\infty(G_+ \cup G_-)$ and $\Omega \in H^\infty(G_+ \cup G_-, \mathcal{L}(\mathfrak{N}_-, \mathfrak{N}_+))$.

Analyzing the proof of Theorem A, one can obtain a procedure of recovering the characteristic function for a given transfer function. Then we are able to construct functional model system for a given conservative curved system.

We also consider a problem how to determine whether a conservative curved system $\Sigma = (T, M, N, \Theta_u, \Xi; H, \mathfrak{N}, \mathfrak{M})$ is similar to the system $\widehat{\Sigma} = (\mathcal{F}_{sm} \circ \mathcal{F}_{mc})(\Theta)$ for a given characteristic function Θ. Here we extend results from [9] to our more general context. In [9] D.Yakubovich calls this problem "to establish that two 2-systems are dual with respect to the function Θ ". Note that namely using that procedure he obtained generalized Sz.-Nagy-Foiaş functional model for generators of systems with delay [9, 10]. We refer a reader to [9] for this and many other interesting examples of conservative curved systems. In fact, we take a next step in this direction. Combining all the above, it is possible to give intrinsic description of simple conservative curved systems with known weights $\Xi_\pm$, but not with known whole characteristic function $\Theta = (\Theta^+, \Xi_+, \Xi_-)$ as it was assumed in [9].

It should also be noted that, in the case when the domain G_+ is simply connected and the weights $\Xi_\pm$ are scalar, i.e., $\Xi_\pm = \delta_\pm I$, the problem to describe all conservative curved systems was studied in [11]. Therein it was established that any simple conservative curved system is similar to a system of the form

$$(\varphi(T_0),\ M_0\psi_+(T_0),\ \psi_-(T_0)N_0), \quad \text{where} \quad \mathfrak{A}_0 = \begin{pmatrix} T_0 & N_0 \\ M_0 & L_0 \end{pmatrix}$$

is a simple unitary colligation, $\varphi : \mathbb{D} \to G_+$ is a conformal map, $\psi_+ = \sqrt{\varphi'}/(\eta_+ \circ \varphi)$, $\psi_- = \sqrt{\varphi'}\,(\eta_- \circ \varphi)$, and $\eta_\pm, \eta_\pm^{-1} \in H^\infty(G_+)$ are scalar outer functions such that $|\eta_\pm|^2 = \delta_\pm$. The form of this representation give us justification and additional reason to use the specifier *curved conservative* for our class of systems.

The latter result (with $\Xi_+ = |\varphi'|I$ and $\Xi_- = \frac{1}{|\varphi'|}I$) plays important role in [12] where it was shown that any trace class perturbation of normal operator with continuous spectrum lying on a smooth curve can be represented as a special form of perturbation of conservative curved system. This allowed us to apply all power of machinery of Sz.-Nagy-Foiaş-Naboko's functional model [13, 14] to solve the problem of duality of spectral components [12].

The paper is organized as follows. In Section 1 we summarize (without proofs) relevant material to categories Mod, Sys, Cfn, Tfn. We introduce morphisms (which are determined by transformations Φ_η^X) in these categories and show that $\mathcal{F}_{mc}$, $\mathcal{F}_{cm}$, $\mathcal{F}_{tc}$, $\mathcal{F}_{ts}$, and $\mathcal{F}_{sm}$ are covariant functors. We use the language of categories and functors for short and because of its more systematic character (anyway we need some notation for our classes of objects and their transformations). Note that by means of the functors $\mathcal{F}_{mc}$, $\mathcal{F}_{cm}$, $\mathcal{F}_{tc}$, $\mathcal{F}_{ts}$, and $\mathcal{F}_{sm}$ we can translate relationships from some category into language of parallel category. For instance, it is possible to give a complete description of the spectrum of a model operator $\widehat{T}$ in terms of its characteristic function. Another example is the existence of the one-to-one correspondence between regular factorizations of a characteristic function and invariant subspaces of the operator $\widehat{T}$ [8]. The transformations Φ_η^X are useful when we want to transfer an object to another object of more simple form, e.g., we can pass to a circular domain or get rid of weights on some connected component of the boundary. For simply connected domains, in such a way we can reduce our problems to the case of the unit circle and $\Xi_\pm \equiv I$ (see, e.g., [11, 21]). Besides, in this section we explain and clarify some subtle points of the theory.

In Section 2 our main results in regard to inverse problems (see the above discussion) are stated and proved. We use notation and basic facts from Section 1 and some results from theory of operators over multiply connected domains and related topological facts [15, 16, 17, 18, 19, 20].

The author is grateful to D. Yakubovich for interesting and useful information and J. Ball for initial stimulating questions.

1. Categories and functors

In this Section we survey (without proofs) basic facts relating to the categories of *functional models* Mod, *characteristic functions* Cfn, *conservative curved systems* Sys, *transfer functions* Tfn, and corresponding functors.

1.1. Category Cfn

Objects of the category Cfn are defined by the condition (Cfn) from the Introduction

$$\mathrm{Ob(Cfn)} := \{\Theta : \Theta \text{ satisfies the condition (Cfn)}\},$$

i.e., $\Theta = (\Theta^+, \Xi_+, \Xi_-) \in \mathrm{Ob(Cfn)}$ are weighted Schur class functions. For short we shall often use the notation $\Theta \in \mathrm{Cfn}$.

To define morphisms in the category Cfn, we need to introduce an important class of transformations

$$\Phi_\eta^{\mathrm{Cfn}} : \mathrm{Ob(Cfn)} \to \mathrm{Ob(Cfn)},$$

which are defined by the formula

$$\Theta_2 = \Phi_\eta^{\mathrm{Cfn}}(\Theta_1) := (\eta_-^{-1}(\Theta_1^+ \circ \varphi^{-1})\eta_+,\ \eta_+^*(\Xi_{1+} \circ \varphi^{-1})\eta_+,\ \eta_-^*(\Xi_{1-} \circ \varphi^{-1})\eta_-).$$

These transformations are parameterized by class CM_η, where $\eta = (\varphi, \eta_+, \eta_-) \in CM_\eta$ iff φ is a conformal mapping of the domain G_{1+} onto G_{2+} and $\eta_\pm, \eta_\pm^{-1} \in H^\infty(G_{2+}, \mathcal{L}(\mathfrak{N}_\pm))$. It is clear that

$$\Phi_{\mathrm{id}}^{\mathrm{Cfn}} = \mathrm{id}_{\mathrm{Cfn}}, \qquad \Phi_{\eta_{32}}^{\mathrm{Cfn}} \circ \Phi_{\eta_{21}}^{\mathrm{Cfn}} = \Phi_{\eta_{32}\circ\eta_{21}}^{\mathrm{Cfn}}, \qquad (\Phi_\eta^{\mathrm{Cfn}})$$

where $\eta_{32} \circ \eta_{21} := (\varphi_{32} \circ \varphi_{21},\ \eta_{3+} \cdot (\eta_{2+} \circ \varphi_{32}^{-1}),\ \eta_{3-} \cdot (\eta_{2-} \circ \varphi_{32}^{-1}))$.

We define morphisms in Cfn by the following rule. Let $\Theta_1, \Theta_2 \in$ Cfn. We shall say that a triple $m_{\Theta_1\Theta_2} = (\Theta_1, \Theta_2, \eta)$ is a morphism in the category Cfn if $\Theta_2 = \Phi_\eta^{\mathrm{Cfn}}(\Theta_1)$, $\eta \in CM_\eta$. In other words,

$$\mathrm{Mor}(\Theta_1, \Theta_2) := \{m_{\Theta_1\Theta_2} : \exists \eta \in CM_\eta \quad \Theta_2 = \Phi_\eta^{\mathrm{Cfn}}(\Theta_1)\}$$

The basic property of morphisms $m_{\Theta_1\Theta_3} = m_{\Theta_2\Theta_3} \circ m_{\Theta_1\Theta_2}$ easily follows from (Φ_η).

Remark 1.1. The above procedure is a pattern for definition of morphisms in other our categories and in the sequel we shall restrict ourselves only to definition of corresponding transformations $\Phi_\eta^X : \mathrm{Ob}(X) \to \mathrm{Ob}(X)$ and verification of their basic property $\Phi_{\eta_{32}}^X \circ \Phi_{\eta_{21}}^X = \Phi_{\eta_{32}\circ\eta_{21}}^X$, where $X \in \{\mathrm{Mod}, \mathrm{Cfn}, \mathrm{Sys}, \mathrm{Tfn}\}$.

In the Introduction we mentioned that any Schur class function can be decomposed into the orthogonal sum of its pure and unitary parts. We have no such a decomposition for an arbitrary *weighted* Schur class function, but using the transformations Φ_η^{Cfn} one can consider an analogue of it.

Let $\chi_\pm$ be outer operator-valued functions such that $\chi_\pm^* \chi_\pm = \Xi_\pm$. If the domain G_+ is multiply connected, these functions can be many-valued and character of their multivalence is completely determined by some unitary representation α of the fundamental group of domain G_+. Recall [15, 19] that any such function can be lift to the universal cover space $\mathbb{D}$ of the domain G_+ and there is one-to-one correspondence between many-valued functions on G_+ and their character-automorphic one-valued liftings on $\mathbb{D}$. Note that we shall use the specifier *character-automorphic* for many-valued functions on G_+ too.

A characteristic function $\Theta \in$ Cfn is called Ξ-pure if the (possibly many-valued character-automorphic) function $\Theta_0^+ = \chi_- \Theta^+ \chi_+^{-1}$ satisfies the condition

$$\forall z \in G_+ \; \forall n \in \mathfrak{N}_+, \; n \neq 0 \qquad \|\Theta_0^+(z)n\|_{\mathfrak{N}_-} < \|n\|_{\mathfrak{N}_+}.$$

$\Theta \in$ Cfn is called a Ξ-unitary constant if $\Theta^- = (\Theta^+)^{-1} \in H^\infty(G_+, \mathcal{L}(\mathfrak{N}_-, \mathfrak{N}_+))$.

It can be shown that for any $\Theta \in$ Cfn there exist $\eta_\pm$, $\eta_\pm^{-1} \in H^\infty(G_+, \mathcal{L}(\mathfrak{N}_\pm))$ such that the characteristic function $\Phi_\eta^{\mathrm{Cfn}}(\Theta)$ can be represented in the form $\Phi_\eta^{\mathrm{Cfn}}(\Theta) = (\Theta_p^+ \oplus \Theta_u^+, \Xi_{p+} \oplus \Xi_{u+}, \Xi_{p-} \oplus \Xi_{u-})$, where Θ_p is Ξ-pure and Θ_u is a Ξ-unitary constant. If $\Phi_{\eta'}^{\mathrm{Cfn}}(\Theta) = (\Theta_p^{+\prime} \oplus \Theta_u^{+\prime}, \Xi_{p+}' \oplus \Xi_{u+}', \Xi_{p-}' \oplus \Xi_{u-}')$ is another such representation with some $\eta_\pm'$, then the functions $\psi_\pm = \eta_\pm^{-1}\eta_\pm'$ admit the decompositions $\psi_\pm = \psi_\pm^p \oplus \psi_\pm^u : \mathfrak{N}_\pm^{p\,\prime} \oplus \mathfrak{N}_\pm^{u\,\prime} \to \mathfrak{N}_\pm^p \oplus \mathfrak{N}_\pm^u$. If we take into account the latter remark, we see that pure-unitary decompositions for a characteristic function are consistent and therefore we can consider different characteristic functions with equal Ξ-pure (Ξ-unitary) "parts". In this sense we shall consider Ξ-unitary part Θ_u of an arbitrary characteristic function Θ.

1.2. Category Mod

$$\mathrm{Ob}(\mathrm{Mod}) := \{\Pi : \Pi \text{ satisfies the condition (Mod)}\}$$

and we regard two models $\Pi_1, \Pi_2 \in \mathrm{Ob}(\mathrm{Mod})$ as equal if there exists an unitary operator $U : \mathcal{H}_1 \to \mathcal{H}_2$ such that $\pi_{2\pm} = U\pi_{1\pm}$.

Transformations Φ_η^{Mod} are defined by the formula

$$\Pi_2 = \Phi_\eta^{\mathrm{Mod}}(\Pi_1) := (\pi_{1+}C_\varphi\eta_+,\ \pi_{1-}C_\varphi\eta_-)\,,$$

where the operator $C_\varphi : L^2(C_2, \mathfrak{N}) \to L^2(C_1, \mathfrak{N})$,

$$(C_\varphi f(\cdot))(\zeta) := \sqrt{\varphi'(\zeta)}f(\varphi(\zeta))\,, \quad \zeta \in C_2\,, \quad f \in L^2(C_2, \mathfrak{N})$$

is unitary. It is easily checked that

$$\Phi_{\mathrm{id}}^{\mathrm{Mod}} = \mathrm{id}_{\mathrm{Mod}}\,, \qquad \Phi_{\eta_{32}}^{\mathrm{Mod}} \circ \Phi_{\eta_{21}}^{\mathrm{Mod}} = \Phi_{\eta_{32}\circ\eta_{21}}^{\mathrm{Mod}}\,. \qquad (\Phi_\eta^{\mathrm{Mod}})$$

1.3. Category Sys

$$\mathrm{Ob(Sys)} := \{\Sigma : \Sigma \sim \widehat{\Sigma},\ \widehat{\Sigma} \in \mathrm{Ran}(\mathcal{F}_{sm})\}$$

and we regard two systems $\Sigma_1, \Sigma_2 \in \mathrm{Ob(Sys)}$ as equal if $\Sigma_1 \sim \Sigma_2$. Hence any conservative curved system $\Sigma \in \mathrm{Sys}$ is similar to $\widehat{\Sigma} = \mathcal{F}_{sm}(\Pi)$ for some $\Pi \in \mathrm{Mod}$ and Ob(Sys) is a quotient set with the similarity as relation of equivalence.

Note that the main operator $\widehat{T}$ can be computed on the following way (see the Introduction for the definitions of objects using below). Let $\mathcal{D}_+ := \mathrm{Ran}\ \pi_+P_+\pi_+^\dagger$ and $\mathcal{H}_+ := \mathrm{Ran}\,(I - \pi_-P_-\pi_-^\dagger)$. It is easily shown that

$$\mathcal{D}_+ \subset \mathcal{H}_+ \subset \mathcal{H} \qquad \text{and} \qquad \mathcal{U}\mathcal{D}_+ \subset \mathcal{D}_+\,,\ \mathcal{U}\mathcal{H}_+ \subset \mathcal{H}_+\,.$$

Moreover,

$$\forall z \in G_- \quad (\mathcal{U} - z)^{-1}\mathcal{D}_+ \subset \mathcal{D}_+\,, \quad (\mathcal{U} - z)^{-1}\mathcal{H}_+ \subset \mathcal{H}_+\,.$$

This means that the operator $\mathcal{U}$ is a normal dilation of the operator $\widehat{T} = P_\Theta\mathcal{U}|\mathcal{K}_\Theta$. Straightforward computations lead to the formula $\widehat{T}f = \mathcal{U}f - \pi_+\widehat{M}f$, where $\widehat{M} : \mathcal{K}_\Theta \to L^2(C, \mathfrak{N}_+)$. Taking into account that $\mathrm{Ker}\,P_+ = E^2(G_-)$ and $\infty \in G_-$, we get $\widehat{M} : \mathcal{K}_\Theta \to \mathfrak{N}_+$ and the explicit formula for $\widehat{M}$ is

$$\widehat{M}f = (\pi_+^\dagger\mathcal{U}f)(\infty) = \frac{1}{2\pi i}\int_C (\pi_+^\dagger f)(z)\,dz\,.$$

Using the dual model (see the ending of this Section) we can define $\widehat{N}$ by

$$(\widehat{N}n, g) = (n, \widehat{M}_*g)\,, \quad g \in \mathcal{K}_{*\Theta}, \quad n \in \mathfrak{N}_-$$

and therefore $\widehat{N}n = P_\Theta\pi_-n\,,\ n \in \mathfrak{N}_-$.

Remark 1.2. Note that it is possible to consider the functional model under more weak assumption, when the projection P_+ is any projection (including the *orthoprojection*) onto $E^2(G_+, \mathfrak{N}_+)$. But casting away the property $\mathrm{Ker}\,P_+ = E^2(G_-, \mathfrak{N}_+)$ we lose the most part of computing power of our functional model. Note that we intensively make use of these computing abilities, especially in expressions for the resolvent of $\widehat{T}$ in both the domains G_+ and G_-.

Remark 1.3. Note that we have only $\mathcal{H}_+ = \mathcal{K}_\Theta \dot{+} \mathcal{D}_+$ while in the case of the unit disk and $\Xi_\pm \equiv I$ we have $\mathcal{H}_+ = \mathcal{K}_\Theta \oplus \mathcal{D}_+$. Thus we can also consider the subspace $\mathcal{K}_0 = \mathcal{H}_+ \ominus \mathcal{D}_+$. Let $\widehat{T}_0 = P_0\mathcal{U}|\mathcal{K}_0$, where P_0 is the orthoprojection onto $\mathcal{K}_0$. Recall that in the unit disk case the inclusions $\mathcal{U}\mathcal{D}_+ \subset \mathcal{D}_+$, $\mathcal{U}\mathcal{H}_+ \subset \mathcal{H}_+$ with some unitary in $\mathcal{H}$ operator $\mathcal{U}$ provide an alternative characterization for the class of all contractions in a Hilbert space [1, 4]. Therefore in our case the operator $\widehat{T}_0$ is an analogue of contraction. We claim that the operators $\widehat{T}$ and $\widehat{T}_0$ are similar. This fact follows from the observation that operators T_2 and T_3 are similar if they are entries in the triangular representations of an operator T

$$T = \begin{pmatrix} T_1 & * \\ 0 & T_2 \end{pmatrix} \quad \text{and} \quad T = \begin{pmatrix} T_1 & * \\ 0 & T_3 \end{pmatrix},$$

where $H = H_1 \dot{+} H_2 = H_1 \dot{+} H_3$ and the subspace H_1 is invariant under the operator T.

Transformations Φ_η^{Sys} are defined by the formula

$$\Sigma_2 = \Phi_\eta^{\mathrm{Sys}}(\Sigma_1) := (T_2, M_2, N_2, \Theta_{2u}, \Xi_2),$$

where

$$\begin{aligned} T_2 &= \varphi(T_1), \\ M_2 f &= -\frac{1}{2\pi i}\int_{C_1} \sqrt{\varphi'(\zeta)}\, \eta_+(\varphi(\zeta))^{-1}[M_1(T_1 - \cdot)^{-1} f]_-(\zeta)\, d\zeta, \\ N_2^* g &= -\frac{1}{2\pi i}\int_{C_{1*}} \sqrt{\bar{\varphi}'(\bar{\zeta})}\, \eta_-(\varphi(\bar{\zeta}))^*\, [N_1^*(T_1^* - \cdot)^{-1} g]_-(\zeta)\, d\zeta, \\ \Xi_{2\pm} &= \eta_\pm^*(\Xi_{1\pm} \circ \varphi^{-1})\, \eta_\pm \end{aligned}$$

and it can be checked that

$$\Phi_{\mathrm{id}}^{\mathrm{Sys}} = \mathrm{id}_{\mathrm{Sys}}, \qquad \Phi_{\eta_{32}}^{\mathrm{Sys}} \circ \Phi_{\eta_{21}}^{\mathrm{Sys}} = \Phi_{\eta_{32}\circ\eta_{21}}^{\mathrm{Sys}}. \qquad (\Phi_\eta^{\mathrm{Sys}})$$

1.4. Functors $\mathcal{F}_{cm}$, $\mathcal{F}_{mc}$, $\mathcal{F}_{sm}$

The corresponding transformations is already defined on Ob(Mod) and Ob(Cfn). It just needs to note that $\mathcal{F}_{cm}$ is well defined, i.e., if models Π_1 and Π_2 are unitarily equivalent, then $\mathcal{F}_{cm}(\Pi_1) = \mathcal{F}_{cm}(\Pi_2)$. To verify that $\mathcal{F}_{cm}$, $\mathcal{F}_{mc}$, $\mathcal{F}_{sm}$ are functors we must show that they are consistent with compositions of morphisms. This easily follows from $(\Phi_\eta^{\mathrm{Mod}})$, $(\Phi_\eta^{\mathrm{Cfn}})$, $(\Phi_\eta^{\mathrm{Sys}})$, and the identities $\mathcal{F}_{cm} \circ \Phi_\eta^{\mathrm{Mod}} = \Phi_\eta^{\mathrm{Cfn}} \circ \mathcal{F}_{cm}$, $\mathcal{F}_{mc} \circ \Phi_\eta^{\mathrm{Cfn}} = \Phi_\eta^{\mathrm{Mod}} \circ \mathcal{F}_{mc}$ and $\mathcal{F}_{sm} \circ \Phi_\eta^{\mathrm{Mod}} = \Phi_\eta^{\mathrm{Sys}} \circ \mathcal{F}_{sm}$, which, in turn, can be verified by straightforward computations (see also [11] for the latter identity).

Remark 1.4. Note that it is possible to obtain $(\Phi_\eta^{\mathrm{Cfn}})$ and $(\Phi_\eta^{\mathrm{Sys}})$ from $(\Phi_\eta^{\mathrm{Mod}})$, $\mathcal{F}_{cm} \circ \Phi_\eta^{\mathrm{Mod}} = \Phi_\eta^{\mathrm{Cfn}} \circ \mathcal{F}_{cm}$, and $\mathcal{F}_{sm} \circ \Phi_\eta^{\mathrm{Mod}} = \Phi_\eta^{\mathrm{Sys}} \circ \mathcal{F}_{sm}$.

Remark 1.5. Actually, models $(\mathcal{F}_{mc} \circ \Phi_\eta^{\mathrm{Cfn}})(\Theta)$ and $(\Phi_\eta^{\mathrm{Mod}} \circ \mathcal{F}_{mc})(\Theta)$ are unitarily equivalent. Systems $(\mathcal{F}_{sm} \circ \Phi_\eta^{\mathrm{Mod}})(\Pi)$ and $(\Phi_\eta^{\mathrm{Sys}} \circ \mathcal{F}_{sm})(\Pi)$ are similar. This is one of the reasons to use quotient spaces for Ob(Mod) and Ob(Sys). Another reason is the following similarities from [8] : $\mathcal{F}_{sm}(\Pi_1) \cdot \mathcal{F}_{sm}(\Pi_2) \sim \mathcal{F}_{sm}(\Pi_1 \cdot \Pi_2)$ and $\Sigma_1 \cdot (\Sigma_2 \cdot \Sigma_3) \sim (\Sigma_1 \cdot \Sigma_2) \cdot \Sigma_3$. Note that the observation from Remark 1.3 explains all the above similarities. For yet another reason see Remark 1.7.

1.5. Category Tfn **and functors** $\mathcal{F}_{tc}$, $\mathcal{F}_{ts}$

Again, coming back to the Introduction, we have

$$\mathrm{Ob(Tfn)} := \{\Upsilon : \Upsilon \in \mathrm{Ran}(\mathcal{F}_{ts}) = \mathrm{Ran}(\mathcal{F}_{tc})\} .$$

We will not define transformations Φ_η^{Tfn} explicitly and put

$$(\Phi_\eta^{\mathrm{Tfn}} \circ \mathcal{F}_{tc})(\Theta) := (\mathcal{F}_{tc} \circ \Phi_\eta^{\mathrm{Cfn}})(\Theta) , \qquad \Theta \in \mathrm{Cfn} .$$

The transformations Φ_η^{Tfn} are well defined (at least for $\varphi' \in \cup_{p>1} L^p(C_1)$) since one can show that $\mathcal{F}_{tc}(\Theta_1) = \mathcal{F}_{tc}(\Theta_2) \implies \mathcal{F}_{tc}(\Phi_\eta^{\mathrm{Cfn}}(\Theta_1)) = \mathcal{F}_{tc}(\Phi_\eta^{\mathrm{Cfn}}(\Theta_2))$. Similarly to Remark 1.4, it is easily checked that

$$\Phi_{\mathrm{id}}^{\mathrm{Tfn}} = \mathrm{id}_{\mathrm{Tfn}} , \qquad \Phi_{\eta_{32}}^{\mathrm{Tfn}} \circ \Phi_{\eta_{21}}^{\mathrm{Tfn}} = \Phi_{\eta_{32} \circ \eta_{21}}^{\mathrm{Tfn}} , \qquad (\Phi_\eta^{\mathrm{Tfn}})$$

Therefore the transformations $\mathcal{F}_{tc}$ and $\mathcal{F}_{ts}$ are functors.

The most simple case when we can present explicit formulaes for the transformation Φ_η^{Tfn} is boundedness of $\Upsilon(z)$, $z \in G_-$ (recall that $\Upsilon(z)$ is bounded whenever the boundary of G_+ is a simple closed $C^{2+\varepsilon}$ curve and the weights $\Xi_\pm$ are scalar and $C^{1+\varepsilon}$-smooth [21]). The desired formula looks like

$$\Upsilon_{2-}(z) = (\Phi_\eta^{\mathrm{Tfn}}(\Upsilon_1))(z) = (P_-[\eta_+(\zeta)^{-1}\Upsilon_{1-}(\varphi^{-1}(\zeta))\eta_-(\zeta)])(z) , \quad z \in G_- ,$$

where $\Upsilon_1(\cdot)_\pm(\zeta)$ are the angular boundary values (in strong operator convergence) of $\Upsilon_1(z)$ from the domains $G_\pm$. In the general case, $\Upsilon(z) \notin H^\infty(G_-)$, but it belongs to the class of strong H^2-functions (see, e.g., [5, 9]), that is, $\forall n \in \mathfrak{N}_-$ $\Upsilon(z)n \in E^2(G_-, \mathfrak{N}_+)$. In this case we have to decompose the above transformation into composition of two transformations of the form $F_{\eta,\varphi}(u)(z) := [P_-(\eta(\zeta)(u \circ \varphi^{-1})(\zeta))](z)$. Taking into account that P_- is actually a singular integral operator, we see that the transformation $F_{\eta,\varphi}$ is defined for $\varphi' \in \cup_{p>1} L^p(C_1)$ and $u \in \cap_{q>1} L^q(C_1, \mathfrak{N})$. So that

$$\begin{aligned} \Upsilon_2(z) = \Lambda_2^\sim(z) , \qquad & \Lambda_2(\bar z)\, m = F_{\eta_-^\sim, z}((\Lambda_1^\sim(\cdot)\, m)_-)(\bar z) , \\ & \Lambda_1(z)\, n = F_{\eta_+^{-1}, \varphi}((\Upsilon_1(\cdot)\, n)_-)(z) , \qquad z \in G_{2-} . \end{aligned}$$

Recall that $\Lambda^\sim(z) = \Lambda(\bar z)^*$.

Remark 1.6. If $\eta(\zeta) \equiv 1$, G_- is simply-connected, $\infty \in G_+$, and φ is normed at the infinity, the transformation $F_{\eta,\varphi}$ is the well-known Faber transformation [22, 23].

To calculate $\Upsilon_2(z)$ for $z \in G_{2+}$, we need some additional assumptions. It suffices to admit existence of boundary values of operator-valued function $\Theta_1^+(z)^{-1}$ a.e. on C_1. In this case $\Upsilon_2(z)$ is the analytic continuation of

$$\Upsilon_2(\varphi(\zeta))_+ = \Upsilon_2(\varphi(\zeta))_- + \eta_+^{-1}(\varphi(\zeta))\,(\Upsilon_1(\zeta)_+ - \Upsilon_1(\zeta)_-)\,\eta_-(\varphi(\zeta)) , \quad \zeta \in C_1$$

into the domain G_{2+} .

In terms of the transformations Φ_η^{Tfn} one can define Ξ-pure transfer functions. Under the assumption $\rho(T) \cap G_+ \neq \emptyset$ we shall call a transfer function $\Upsilon \in \mathrm{Ob}(\mathrm{Tfn})$ Ξ-pure if $\forall \eta \in CM_\eta\ \forall n \in \mathfrak{N}_- \quad \Phi_\eta^{\mathrm{Tfn}}(\Upsilon)(z)\, n \equiv 0 \Rightarrow n = 0$. This definition agrees with the corresponding definition for characteristic functions and therefore there exists a decomposition of transfer function $\Upsilon = \Upsilon_p \oplus 0$ in the same sense as in Subsection 1.1.

1.6. Duality

In the language of categories and functors duality means existence of the transformation $\mathcal{F}_{*X} : \mathrm{Ob}(X) \to \mathrm{Ob}(X)$ such that $\mathcal{F}_{*X} \circ \Phi_\eta^X = \Phi_{\eta_{-*}}^X \circ \mathcal{F}_{*X}$ and $\mathcal{F}_{*X}^2 = \mathrm{id}_X$, where $\eta_{-*} = (\varphi^\sim, \eta_-^{\sim-1}, \eta_+^{\sim-1})$. This transformation can be extended to morphisms and therefore $\mathcal{F}_{*X}$ is a contravariant functor in the category X. For $X \in \{\mathrm{Mod}, \mathrm{Cfn}, \mathrm{Sys}, \mathrm{Tfn}\}$, the following formulas define the corresponding functors

$$\Theta_* = \mathcal{F}_{*c}(\Theta) := (\Theta^{+\sim}, \Xi_*), \quad \Xi_{*\pm} = \Xi_\mp^{\sim-1};$$

$$\Pi_* = \mathcal{F}_{*m}(\Pi) \ : \ (f, \pi_{*\mp} v)_{\mathcal{H}} = < \pi_\pm^\dagger f, v >_C, \ f \in \mathcal{H}, \ v \in L^2(\bar{C}, \mathfrak{N}_\mp);$$

$$\Sigma_* = \mathcal{F}_{*s}(\Sigma) := (T^*, N^*, M^*, \Theta_u^\sim, \Xi_*);$$

$$\Upsilon_* = \mathcal{F}_{*t}(\Upsilon) := (\Upsilon^\sim, \Theta_u^\sim, \Xi_*),$$

where $< u, v >_C := \frac{1}{2\pi i} \int_C (u(z), v(\bar{z}))_{\mathfrak{N}}\, dz$, $u \in L^2(C, \mathfrak{N})$, $v \in L^2(\bar{C}, \mathfrak{N})$.
For $\mathcal{F}_{cm}$, $\mathcal{F}_{sm}$, $\mathcal{F}_{tc}$, the identities $\mathcal{F}_{*Y} \mathcal{F}_{YX} = \mathcal{F}_{YX} \mathcal{F}_{*X}$ can easily be checked.

Remark 1.7. In fact, we have $\mathcal{F}_{*s}(\mathcal{F}_{sm}(\Pi)) \sim \mathcal{F}_{sm}(\mathcal{F}_{*m}(\Pi))$ and this is an additional reason to use a quotient space for systems.

2. Inverse problem

Our aim is to construct the functional model for a given conservative curved system $\Sigma = (T, M, N, \Xi)$.

2.1. Uniqueness theorem

We divide the proof of the main uniqueness Theorem A into parts, which we arrange as separate assertions. We start with following elementary lemma.

Lemma 2.1. *Suppose* $L \in \mathcal{L}(H_1, H_2)$, $L^{-1} \in \mathcal{L}(H_2, H_1)$, *and* $||L|| \leq 1$. *Let* $L = |L|U$ *be the polar decomposition of* L. *Then* $|L| = \psi(B^*B)$, *where*

$$B = L^{-1} - L^* \quad \text{and} \quad \psi(z) = \frac{1}{2}\sqrt{2 + z - \sqrt{z^2 + 4z}}$$

Besides, $U^*|\,\mathrm{Ran}(I - |L|^2) = B(|L|^{-1} - |L|)^{-1}|\,\mathrm{Ran}(I - |L|^2)$.

Proposition 2.2. *Suppose* $\Theta \in \mathrm{Cfn}$ *and* $\Theta^+(\zeta)^{-1}$ *possesses boundary values a.e. on* C. *Then the operator-valued function* $\Delta^+(\zeta) := (I - \Theta^+(\zeta)\Theta^-(\zeta))^{1/2}$ *can be expressed in terms of the transfer function* $\Upsilon(z)$.

Proof. The identity $\Theta^-(\zeta) - \Theta^+(\zeta)^{-1} = \Upsilon(\zeta)_+ - \Upsilon(\zeta)_-$, $\zeta \in C$ easily follows from (CtoT). Then, taking into account that $\Theta^-(\zeta)$ is adjoint to $\Theta^+(\zeta) : \mathfrak{N}_{+,\zeta} \to \mathfrak{N}_{-,\zeta}$, it remains to make use of the above Lemma. □

For $\Theta \in \mathrm{Cfn}$, we introduce the Hilbert space

$$\mathcal{H}_{NF} := L^2(C, \Xi_+) \oplus \operatorname{clos} \Delta^+ L^2(C, \Xi_-), \tag{NF}$$

which is endowed with the inner product $(f_\pi, g_\pi)_{L^2(C,\Xi_+)} + (f_\tau, g_\tau)_{L^2(C,\Xi_-)}$. Let $\Pi = \mathcal{F}_{mc}(\Theta)$ and $\tau_+ := ((\Delta^+)^{-1}(\pi_-^\dagger - \Theta^+ \pi_+^\dagger))^\dagger$. Then the pair $(\pi_+^\dagger f, \tau_+^\dagger f)$ defines a unitary operator $W_{NF} : \mathcal{H} \to \mathcal{H}_{NF}$. The inverse operator can easily be calculated $W_{NF}^{-1}(f_\pi, f_\tau) = \pi_+ f_\pi + \tau_+ f_\tau$. It is clear that $W_{NF}\mathcal{U} = zW_{NF}$. We define the subspaces

$$\mathcal{K}_{\Theta NF} := W_{NF}\mathcal{K}_\Theta, \quad \mathcal{H}_{NF+} := \mathcal{K}_{\Theta NF} \dot{+} (E^2(G_+, \mathfrak{N}_+) \oplus \{0\}).$$

The subspace $\mathcal{H}_{NF+}$ is invariant under the operator of multiplication by the independent variable z. Therefore $z|\mathcal{H}_{NF+}$ is a subnormal operator with the normal spectrum on the curve C. By [15], there exists a unique generalized Wold-Kolmogorov decomposition of the space $\mathcal{H}_{NF+} = \mathcal{H}_{NF+}^{pur} \oplus \mathcal{H}_{NF+}^{nor}$ with respect to the operator z. Note that $\mathcal{H}_{NF+}^{pur} = W_{NF}\pi_- E^2(G_+, \mathfrak{N}_-)$. Now we are ready to state and prove the following assertions.

Proposition 2.3. *Suppose $\Theta_1, \Theta_2 \in \mathrm{Cfn}$; $\Theta_1^+(\zeta)^{-1}, \Theta_2^+(\zeta)^{-1}$ possesses boundary values a.e. on C, and $\mathcal{F}_{tc}(\Theta_1) = \mathcal{F}_{tc}(\Theta_2)$. Then $\mathcal{K}_{1\Theta NF} = \mathcal{K}_{2\Theta NF}$.*

Proof. It can easily be calculated

$$(\widehat{T} - z)^{-1} f = (\mathcal{U} - z)^{-1}(f - \pi_+ n), \quad n = \begin{cases} \Theta^+(z)^{-1}(\pi_-^\dagger f)(z) & , z \in G_+ \cap \rho(\widehat{T}) \\ (\pi_+^\dagger f)(z) & , z \in G_- \end{cases},$$

where $f \in \mathcal{K}_\Theta$. We will consider family of vectors

$$r_{nz} := (T - z)^{-1} N n, \quad n \in \mathfrak{N}_-, \quad z \in \rho(T). \tag{RK}$$

For model systems we have

$$\hat{r}_{nz} = (\widehat{T} - z)^{-1} \widehat{N} n = \begin{cases} -P_\Theta \pi_+ \frac{\Theta^+(z)^{-1} n}{\zeta - z}, & z \in G_+ \cap \rho(\widehat{T}) \\ P_\Theta \pi_- \frac{n}{\zeta - z}, & z \in G_- \end{cases}.$$

In the space $\mathcal{K}_{\Theta NF}$ this family looks like (see [21])

$$\pi_+^\dagger \hat{r}_{nz} = \frac{(\Upsilon n)(\zeta)_- - \Upsilon(z)\, n}{\zeta - z}, \qquad \tau_+^\dagger \hat{r}_{nz} = \frac{-\Delta^+(\zeta)\, n}{\zeta - z}.$$

Since $\Upsilon_1 = \mathcal{F}_{tc}(\Theta_1) = \mathcal{F}_{tc}(\Theta_2) = \Upsilon_2$, and using Proposition 2.2, we get $W_{1NF}\hat{r}_{1nz} = W_{2NF}\hat{r}_{2nz}$. To complete the proof it remains to recall that $\bigvee_{z \in \rho(\widehat{T})} \hat{r}_{nz} = \mathcal{K}_\Theta$ if $G_+ \cap \rho(\widehat{T}) \neq \emptyset$. The condition $G_+ \cap \rho(\widehat{T}) \neq \emptyset$ follows from existence of inverse operators $\Theta^+(z)^{-1}$ for some $z \in G_+$. □

Remark 2.1. Recall [8] that we call a curved conservative system Σ simple if

$$\rho(T) \cap G_+ \neq \emptyset \qquad \text{and} \qquad \bigcap_{z\in\rho(T)} \operatorname{Ker} M(T-z)^{-1} = \{0\}\,.$$

Note that if for a model system $\rho(\widehat{T}) \cap G_+ \neq \emptyset$, then $\widehat{\Sigma}$ is simple and this is equivalent to the completeness $\bigvee_{z\in\rho(\widehat{T})} \hat{r}_{nz} = \mathcal{K}_\Theta$.

Proposition 2.4. *Let* $\Theta_1 = (\Theta_1^+, \Xi)$ *and* $\Theta_2 = (\Theta_2^+, \Xi)$. *Let* $\chi_\pm$ *be outer (possible character-automorphic) operator-valued functions such that* $\chi_\pm^* \chi_\pm = \Xi_\pm$. *Suppose* $\mathcal{K}_{1\Theta NF} = \mathcal{K}_{2\Theta NF}$. *Then there exists an unitary operator* $U \in \mathcal{L}(\mathfrak{N}_-)$ *such that* $\chi_- \Theta_2^+ \chi_+^{-1} = U \chi_- \Theta_1^+ \chi_+^{-1}$.

Proof. Since $W_{NF}\,\pi_+ = I$, we have $W_{1NF}\pi_{1+} = W_{2NF}\pi_{2+}$. On the other hand, on account of the uniqueness of the generalized Wold-Kolmogorov decomposition of the space $\mathcal{H}_{NF+} = \mathcal{K}_{1\Theta NF} \dot{+} (E^2(G_+, \mathfrak{N}_+) \oplus \{0\}) = \mathcal{K}_{2\Theta NF} \dot{+} (E^2(G_+, \mathfrak{N}_+) \oplus \{0\})$, we get for the pure part $\mathcal{H}_{NF+}^{pur} = W_{1NF}\pi_{1-}E^2(G_+, \mathfrak{N}_-) = W_{2NF}\pi_{2-}E^2(G_+, \mathfrak{N}_-)$. Let α be the unitary representation of the fundamental group of domain G_+, which is associated with operator-valued function χ_-. Then we have

$$W_{1NF}\pi_{1-}\chi_-^{-1} E_\alpha^2(G_+, \mathfrak{N}_-) = W_{2NF}\pi_{2-}\chi_-^{-1} E_\alpha^2(G_+, \mathfrak{N}_-)\,,$$

where $E_\alpha^2(G_+, \mathfrak{N}_-) = \chi_- E^2(G_+, \mathfrak{N}_-)$ is the subspace of character-automorphic functions corresponding to the representation α. In view that $W_{kNF}\pi_{k-}\chi_-^{-1}$, $k = 1, 2$ are isometries, by the generalized Beurling theorem [15], there exists an unitary operator $U \in \mathcal{L}(\mathfrak{N}_-)$ such that $W_{1NF}\pi_{1-}\chi_-^{-1} = W_{2NF}\pi_{2-}\chi_-^{-1} U$ and therefore we have

$$\begin{aligned}
\chi_- \Theta_2^+ \chi_+^{-1} &= \chi_- \pi_{2-}^\dagger \pi_{2+} \chi_+^{-1} = (\pi_{2-}\chi_-^{-1})^* \pi_{2+}\chi_+^{-1} \\
&= (W_{2NF}^{-1} W_{1NF}\pi_{1-}\chi_-^{-1} U^{-1})^* W_{2NF}^{-1} W_{1NF}\pi_{1+}\chi_+^{-1} \\
&= U(\pi_{1-}\chi_-^{-1})^* \pi_{1+}\chi_+^{-1} = U\chi_- \pi_{1-}^\dagger \pi_{1+}\chi_+^{-1} = U\chi_- \Theta_1^+ \chi_+^{-1}\,. \qquad \square
\end{aligned}$$

Proposition 2.5. *Suppose* $\Theta_1, \Theta_2 \in \mathrm{Cfn}$; $\Theta_1^+(\zeta)^{-1}, \Theta_2^+(\zeta)^{-1}$ *possesses boundary values a.e. on* C, *and* $\mathcal{F}_{tc}(\Theta_1) = \mathcal{F}_{tc}(\Theta_2)$. *Then* $\Theta_1 = \Theta_2$.

Proof. First we note that

$$\mathcal{F}_{tc}(\Phi_\eta^{\mathrm{Cfn}}(\Theta_1)) = \Phi_\eta^{\mathrm{Tfn}}(\mathcal{F}_{tc}(\Theta_1)) = \Phi_\eta^{\mathrm{Tfn}}(\mathcal{F}_{tc}(\Theta_2)) = \mathcal{F}_{tc}(\Phi_\eta^{\mathrm{Cfn}}(\Theta_2))$$

and, since we can take the parameter η such that $\Phi_\eta^{\mathrm{Cfn}}(\Theta_1) = \Theta_{1p} \oplus \Theta_{1u}$, without loss of generality one can assume that the transfer function $\Upsilon = \mathcal{F}_{ct}(\Theta_1) = \mathcal{F}_{ct}(\Theta_2)$ is Ξ-pure.

By Propositions 2.3 and 2.4, we have $\chi_-\Theta_2^+\chi_+^{-1} = U\chi_-\Theta_1^+\chi_+^{-1}$. We can rewrite this identity in the form $\chi_-^{-1}U\chi_- = \Theta_1^+\Theta_2^{+-1}$. Hence $\chi_-^{-1}U\chi_-$ is one-valued and therefore the operator U commutes with the W^*-algebra $\mathcal{W}^*(\alpha)$ generated by the unitary representation α. By the theorem of Bungart [17], the analytic vector bundles corresponding to the operator-valued character-automorphic function χ_- is trivial with respect to the group of invertible operators $G(\alpha)$ of the algebra $\mathcal{W}^*(\alpha)$. That means there exists a character-automorphic function $\chi(z)$ with values in $G(\alpha)$ such that this function trivializes the analytic vector bundle. In particular, $U\chi = \chi U$. It is also clear that $\eta_- = \chi_-^{-1}\chi$ is one-valued and $\eta_-, \eta_-^{-1} \in H^\infty(G_+, \mathfrak{N}_-)$.

Let $\eta = (z, I, \eta_-)$, $\Theta_{1\eta} = \Phi_\eta^{\mathrm{Cfn}}(\Theta_1)$, and $\Theta_{2\eta} = \Phi_\eta^{\mathrm{Cfn}}(\Theta_2)$. Then (see the beginning of this proof) the transfer function $\Upsilon_\eta = \mathcal{F}_{tc}(\Theta_{1\eta}) = \mathcal{F}_{tc}(\Theta_{2\eta})$ corresponds to both the characteristic functions $\Theta_{1\eta}$ and $\Theta_{2\eta}$. We have

$$\begin{aligned} U\Theta_{1\eta}^+ &= U\eta_-^{-1}\Theta_1^+ = U\chi^{-1}\chi_-\Theta_1^+\chi_+^{-1}\chi_+ \\ &= \chi^{-1}(U\chi_-\Theta_1^+\chi_+^{-1})\chi_+ = \chi^{-1}\chi_-\Theta_2^+\chi_+^{-1}\chi_+ = \eta_-^{-1}\Theta_2^+ = \Theta_{2\eta}^+ . \end{aligned}$$

Since $\Xi_{1\eta-} = \Xi_{2\eta-} = \chi^*\chi$ and $U\chi^*\chi = \chi^*\chi U$, we get

$$\begin{aligned} \Theta_{1\eta}^- U^* &= \Xi_+^{-1}\Theta_{1\eta}^{+*}\chi^*\chi U^* = \Xi_+^{-1}(U\Theta_{1\eta}^+)^*\chi^*\chi \\ &= \Xi_+^{-1}\Theta_{2\eta}^{+*}\chi^*\chi = \Theta_{2\eta}^- . \end{aligned}$$

Then, taking into account (CtoT), we have $\Upsilon_{2\eta}(z) = \Upsilon_{1\eta}(z)U$. Since $\Upsilon_{2\eta}(z) = \Upsilon_{1\eta}(z)$, it follows that $\Upsilon_{2\eta}(z)(U - I) = 0$ and it remains to recall that $\Upsilon_{2\eta}$ is Ξ-pure. Hence $U = I$ and therefore $\Theta_1^+ = \Theta_2^+$. □

Remark 2.2. In the above proof we made use of the fact of triviality of any analytic vector bundles over multiply connected domains [18, 17, 15]. Note that this assertion is equivalent to similarity of the bundle shift $z|E_\alpha^2(G_+, \mathfrak{N})$ to the trivial shift $z|E^2(G_+, \mathfrak{N})$ [15] (see also the scalar case in [16]). Generally we prefer to deal systematically with the trivial shift $z|E^2(G_+, \mathfrak{N})$ only and hide possible multiple-valuedness using the technique of operator-valued weights $\Xi_\pm$ as adequate replacement of the traditional uniformization technique or analytic vector bundles. Our axiomatics for category Mod is a typical example of this approach.

Note that it is possible to show that $\mathcal{F}_{ct}(\Theta) \in \mathcal{N}(G_+ \cup G_-, \mathcal{L}(\mathfrak{N}_-, \mathfrak{N}_+)) \implies \Theta_\pm^\mp \in \mathcal{N}(G_\pm, \mathcal{L}(\mathfrak{N}_-, \mathfrak{N}_+)) \implies (\Theta^+)^{-1} \in \mathcal{N}(G_+, \mathcal{L}(\mathfrak{N}_-, \mathfrak{N}_+))$ and therefore existence of boundary values of $(\Theta^+)^{-1}$ a.e. on C. Thus we arrive at the following important theorem.

Theorem A. *If* $\Theta_1, \Theta_2 \in \mathrm{Cfn}$ *and* $\mathcal{F}_{ct}(\Theta_1) = \mathcal{F}_{ct}(\Theta_2) \in \mathcal{N}(G_+ \cup G_-, \mathcal{L}(\mathfrak{N}_-, \mathfrak{N}_+))$, *then* $\Theta_1 = \Theta_2$.

2.2. Procedure of recovery

Analyzing the proof of Theorem A, we can obtain a recovery procedure of the characteristic function $\Theta = (\Theta^+, \Xi_+, \Xi_-)$ for a given transfer function $\Upsilon = (\Upsilon(z), \Theta_u, \Xi)$. Below we present such a procedure, which is effective at least in the case when $\dim \mathfrak{N}_+ = \dim \mathfrak{N}_- < \infty$. We keep on the assumption that $\Theta^+(\zeta)^{-1}$ possesses boundary values a.e. on C.

Let α be the unitary representation of the fundamental group of domain G_+, which is associated with operator-valued function χ_- defined by the factorization $\chi_-^*\chi_- = \Xi_-$ and $E_\alpha^2(G_+, \mathfrak{N}_-) = \chi_- E^2(G_+, \mathfrak{N}_-)$. If the domain G_+ has n boundary components, there exists a scalar inner function ϑ on G_+ with precisely

n zeroes (see, e.g., [20, 24]). Define the unitary operator $\vartheta = \vartheta(z)$ on the space $L^2(C, \mathfrak{N}_-)$ and the wandering subspace $\mathfrak{M}_\alpha = E^2_\alpha(G_+, \mathfrak{N}_-) \ominus \vartheta E^2_\alpha(G_+, \mathfrak{N}_-)$, for which we have

$$L^2(C, \mathfrak{N}_-) = \bigoplus_{k=-\infty}^{+\infty} \vartheta^k \mathfrak{M}_\alpha \,, \quad E^2_\alpha(G_+, \mathfrak{N}_-) = \bigoplus_{k=0}^{+\infty} \vartheta^k \mathfrak{M}_\alpha \,.$$

Let $\mathfrak{M}_0 = E^2(G_+, \mathfrak{N}_-) \ominus \vartheta E^2(G_+, \mathfrak{N}_-)$. Taking into account the vector representation of scalar Hardy spaces for multiply connected domains [20], it can easily be shown that $\dim \mathfrak{M}_0 = n \cdot \dim \mathfrak{N}_-$. Since $\mathfrak{M}_\alpha = (\chi^*_-)^{-1}\mathfrak{M}_0$, we get $\dim \mathfrak{M}_\alpha = n \cdot \dim \mathfrak{N}_-$. Now we take an arbitrary unitary operator $V_\alpha \in \mathcal{L}(\mathfrak{N}^n_-, \mathfrak{M}_\alpha)$ and define a wave operator $W_\alpha : L^2(C, \mathfrak{N}_-) \to L^2(\mathbb{T}, \mathfrak{N}^n_-)$ by

$$W_\alpha f := \sum_{k=-\infty}^{+\infty} \mathcal{Z}^k V_\alpha^{-1} P_{\mathfrak{M}_\alpha} \vartheta^{-k} f \,, \quad f \in L^2(C, \mathfrak{N}_-) \,,$$

where $P_{\mathfrak{M}_\alpha}$ is the orthoprojection onto $\mathfrak{M}_\alpha$ and $\mathcal{Z}$ is the operator of multiplication by the independent variable in $L^2(\mathbb{T}, \mathfrak{N}^n_-)$. The operator W_α is unitary, for which we have $W_\alpha E^2_\alpha(G_+, \mathfrak{N}_-) = H^2(\mathbb{D}, \mathfrak{N}^n_-)$ and $W_\alpha \vartheta = \mathcal{Z} W_\alpha$.

In the same way as in the proofs of Proposition 2.3 and Proposition 2.4, we have $\mathcal{H}_{NF+} = \mathcal{K}_{\ominus NF} \dot{+} (E^2(G_+, \mathfrak{N}_+) \oplus \{0\}) \subset \mathcal{H}_{NF}$, where $\mathcal{K}_{\ominus NF} = \bigvee_{z \in \rho(\widehat{T})} W_{NF} \hat{r}_{nz}$. Consider the unitary operator $\vartheta_{NF} = \vartheta(z)$, where z is the operator of multiplication by the independent variable in $\mathcal{H}_{NF}$. Since the subspace $\mathcal{H}_{NF+}$ is invariant under the unitary operator ϑ_{NF}, the operator $\vartheta_{NF} | \mathcal{H}_{NF+}$ is an isometry with the wandering subspace $\mathfrak{M} = \mathcal{H}_{NF+} \ominus \vartheta_{NF} \mathcal{H}_{NF+}$. For an arbitrary unitary operator $V \in \mathcal{L}(\mathfrak{N}^n_-, \mathfrak{M})$, define a wave operator $W : \mathcal{H}_{NF} \to L^2(\mathbb{T}, \mathfrak{N}^n_-)$ by

$$Wf := \sum_{k=-\infty}^{+\infty} \mathcal{Z}^k V^{-1} P_{\mathfrak{M}} \vartheta^{-k}_{NF} f \,, \qquad f \in \mathcal{H}_{NF} \,,$$

where $P_{\mathfrak{M}}$ is the orthoprojection onto $\mathfrak{M}$. The operator W^* is an isometry with $\operatorname{Ran} W^* = \operatorname{Ran} W_{NF} \pi_-$. Besides, we have

$$W^*(H^2(\mathbb{D}, \mathfrak{N}^n_-)) = \mathcal{H}^{\mathrm{pur}}_{NF+} = W_{NF}\pi_- E^2(G_+, \mathfrak{N}_-) = W_{NF}\pi_- \chi_-^{-1} E^2_\alpha(G_+, \mathfrak{N}_-)$$

and $W \vartheta_{NF} = \mathcal{Z} W$. Hence we see that

$$W^*_\alpha H^2(\mathbb{D}, \mathfrak{N}^n_-) = E^2_\alpha(G_+, \mathfrak{N}_-) = \chi_- \pi_-^\dagger W^*_{NF} W^*(H^2(\mathbb{D}, \mathfrak{N}^n_-))$$

and $[W_\alpha \chi_- \pi_-^\dagger W^*_{NF} W^*]\mathcal{Z} = \mathcal{Z}[W_\alpha \chi_- \pi_-^\dagger W^*_{NF} W^*]$. By Beurling theorem, the unitary operator $W_\alpha \chi_- \pi_-^\dagger W^*_{NF} W^*$ is an operator of multiplication by unitary constant, that is, there exists a unitary operator $V_n : \mathfrak{N}^n_- \to \mathfrak{N}^n_-$ such that $W_{NF}\pi_- \chi_-^{-1} = W^* V_n W_\alpha$. Then we have

$$W^* V_n W_\alpha z = W_{NF}\pi_- \chi_-^{-1} z = z W_{NF} \pi_- \chi_-^{-1} = z W^* V_n W_\alpha \,.$$

Thus, the linear equation

$$W^* V_n W_\alpha z | \mathfrak{M}_\alpha = z W^* V_n W_\alpha | \mathfrak{M}_\alpha \tag{V$_n$}$$

have at least one unitary solution V_n.

Let $\widetilde{V}_n$ be another solution of (V$_\mathrm{n}$). The identity (V$_\mathrm{n}$) can easily be extended to the whole space $L^2(\mathbb{T}, \mathfrak{N}_-^n)$. Since we have $W^* V_n W_\alpha(E_\alpha^2(G_+, \mathfrak{N}_-)) = W^* \widetilde{V}_n W_\alpha(E_\alpha^2(G_+, \mathfrak{N}_-))$, by the generalized Beurling theorem [15], there exists an unitary operator $U \in \mathcal{L}(\mathfrak{N}_-)$ such that $W_{NF} \pi_- \chi_-^{-1} = W^* V_n W_\alpha = W^* \widetilde{V}_n W_\alpha U$. Further, since $\pi_-^\dagger = \chi_-^{-1} W_\alpha^* V_n^* W W_{NF}$ and $\pi_+ = W_{NF}^*$, we get

$$\Theta^+ = \pi_-^\dagger \pi_+ = \chi_-^{-1} W_\alpha^* V_n^* W = \chi_-^{-1} U^* W_\alpha^* \widetilde{V}_n^* W = \chi_-^{-1} U^* \chi_- \widetilde{\Theta}^+ ,$$

where $\widetilde{\Theta}^+ = \chi_-^{-1} W_\alpha^* \widetilde{V}_n^* W$. By Proposition 2.5, the unitary operator U is uniquely determined by the condition $\Upsilon = \mathcal{F}_{tc}(\chi_-^{-1} U^* \chi_- \widetilde{\Theta}^+)$. Since for the adjoint operator we have $\Theta^- = (\chi_+^* \chi_+)^{-1} \Theta^{+*} (\chi_-^* \chi_-)$, the identity (CtoT) can be rewritten in the form

$$\begin{aligned} {}_+(z) - \widetilde{\Theta}^+(z)^{-1} \chi_-^{-1} U \chi_- &= \Upsilon(z) , \quad z \in G_+ \\ -[(\chi_+^* \chi_+)^{-1} \widetilde{\Theta}^{+*} (\chi_-^* U \chi_-)]_-(z) &= \Upsilon(z) , \quad z \in G_- \end{aligned} . \tag{U}$$

Hence we have obtained linear equations (U) with respect to unknown operator U. These equations have a unique unitary solution.

Thus to recover characteristic function Θ from a given transfer function Υ we need to fulfill the following steps:

- find the inner ϑ function with precisely n zeroes;
- find the outer functions $\chi_\pm$ from the factorizations $\chi_\pm^* \chi_\pm = \Xi_\pm$;
- construct the wave operator W_α;
- construct the space $\mathcal{H}_{NF+}$ and the wave operator W;
- find any unitary solution $\widetilde{V}_n$ of the linear equation (V$_\mathrm{n}$);
- find any unitary solution U of the linear equations (U).

As result we obtain $\Theta^+ = \chi_-^{-1} U^* W_\alpha^* \widetilde{V}_n^* W$.

Remark 2.3. If $\operatorname{clos} \Delta^+(\zeta) \mathfrak{N}_- = \mathfrak{N}_-$ a.e. on C (that is, the absolutely continuous spectrum of the operator T is rich enough), we do not need the above procedure of recovery and the characteristic function Θ^+ can be easily restored from the relation

$$\Theta_+^-(\zeta) - \Theta^+(\zeta)^{-1} = \Upsilon_+(\zeta) - \Upsilon_-(\zeta) , \; \zeta \in C$$

using the last formula from Lemma 2.1.

2.3. Constructing functional model for a given system

We pass to the problem to construct the functional model for a given simple conservative curved system $\Sigma = (T, M, N, \Theta_u, \Xi)$. So, we can consider its transfer function Υ. If $\Upsilon(z) \in \mathcal{N}(G_+ \cup G_-, \mathcal{L}(\mathfrak{N}_-, \mathfrak{N}_+))$, then, by the above procedure, we are able to restore the corresponding characteristic function Θ.

Remark 2.4. Note that for $\Upsilon(z) \in \mathcal{N}(G_+ \cup G_-, \mathcal{L}(\mathfrak{N}_-, \mathfrak{N}_+))$ it suffices to assume $T - U \in \mathfrak{S}_1$, $U^* U = U U^*$, $\sigma(U) \subset C$, $\sigma_c(T) \subset C$, $M, N \in \mathfrak{S}_2$, and $C^{1+\varepsilon}$ smooth of the curve C (see [12, 25]).

Knowing the characteristic function Θ, one can easily to build the corresponding functional model. We can merely take $\mathcal{H} = \mathcal{H}_{NF}$, $\pi_+ = (I,0)^T \in \mathcal{L}(L^2(C,\Xi_+),\mathcal{H}_{NF})$ and $\pi_- = (\Theta^-,\Delta^+)^T \in \mathcal{L}(L^2(C,\Xi_-),\mathcal{H}_{NF})$.

Let $X \in \mathcal{L}(H,\mathcal{K}_\Theta)$ be an invertible operator, which realizes similarity of the system Σ to the model system $\widehat{\Sigma}$. Then it can be calculated that

$$\begin{aligned}(Xf)_\pi &= -[M(T-\zeta)^{-1}f]_-\,,\\ (Xf)_\tau &= (\Delta^+)^{-1}\Theta^+([M(T-\zeta)^{-1}f]_+ - [M(T-\zeta)^{-1}f]_-)\,,\end{aligned} \tag{Sim}$$

where $f \in H$ and $[M(T-\zeta)^{-1}f]_\pm$ are the boundary values of $M(T-z)^{-1}f$ from the domains $G_\pm$, respectively.

2.4. Test (T,M,N) to be a system

In conclusion we consider the following problem. Let $\Upsilon(z) = M(T-z)^{-1}N$ and $\Upsilon = \mathcal{F}_{tc}(\Theta)$ for some characteristic function Θ. Note that the transfer function $\Upsilon(z)$ can admit other realizations in the form $\Upsilon(z) = M(T-z)^{-1}N$ and the triplet (T,M,N) have not to be a curved conservative system. That is, we want to know how to determine whether $\Sigma = (T,M,N,\Theta_u,\Xi;H,\mathfrak{N},\mathfrak{M})$ is similar to the system $\widehat{\Sigma} = (\mathcal{F}_{sm}\circ\mathcal{F}_{mc})(\Theta)$. The following assertions answer this question.

Proposition 2.6. *Suppose a triplet (T,M,N) is simple and there exists $\Theta \in \mathrm{Cfn}$ such that $\Theta^+(z)^{-1}$ possesses boundary values a.e. on C and $\mathcal{F}_{ct}(\Theta)(z) = \Upsilon(z) = M(T-z)^{-1}N$. Suppose that $\forall f \in H$ there exist $[M(T-\zeta)^{-1}f]_\pm$ a.e. on the curve C and an operator $X : H \to \mathcal{K}_\Theta$ is defined by* (Sim). *Then*

$$\Sigma = (T,M,N,\Theta_u,\Xi) \in \mathrm{Sys}$$

if and only if

$$\|X\| < \infty \quad \textit{and} \quad \|X^{-1}\| < \infty\,.$$

Proof. By straightforward computations it can be shown that $Xr_{nz} = \hat{r}_{nz}$. It remains to recall that $\vee_{z\in\rho(T)} r_{nz} = H$ and $\vee_{z\in\rho(\widehat{T})} \hat{r}_{nz} = \mathcal{K}_\Theta$. □

More symmetric variant of this assertion can be formulated if we take into account the dual model.

Proposition 2.7. *Under the hypotheses of Proposition* 2.6, *let an operator X_* : $H \to \mathcal{K}_{*\Theta}$ be defined by formulas those are dual to* (Sim). *Then*

$$\Sigma = (T,M,N,\Theta_u,\Xi) \in \mathrm{Sys}$$

if and only if

$$\|X\| < \infty \quad \textit{and} \quad \|X_*\| < \infty\,.$$

Proof. Additionally to the identity $Xr_{nz} = \hat{r}_{nz}$ we need to make use of the identities $X_* r_{*nz} = \hat{r}_{*nz}$ and $(Xr_{nz}, X_* r_{*mw})_{\mathcal{H}} = (r_{nz}, r_{*mw})_H$, which can be again obtained by straightforward computations. Considering X and X_* as operators acting into $\mathcal{H}$, we get $X_*^* X = I$ and therefore $\operatorname{Ker} X = \{0\}$ and $\operatorname{clos}\operatorname{Ran} X = \operatorname{Ran} X$. □

References

[1] Szökefalvi-Nagy B., Foiaş C., *Harmonic analysis of operators on Hilbert space.* North-Holland, Amsterdam-London, 1970.

[2] Brodskiy M.S., *Unitary operator nodes and their characteristic functions*, Uspehi Mat. Nauk **33** (1978), no. 4, 141–168.

[3] Arov D.Z., *Passive linear dynamic systems*, Sibirsk. Math. Zh., **20** (1979), no. 2, 211-228.

[4] Nikolski N.K., Vasyunin V.I., *Elements of spectral theory in terms of the free functional model. Part I: Basic constructions*, Holomorphic spaces (eds. Sh. Axler, J. McCarthy, D. Sarason), MSRI Publications **33** (1998), 211–302.

[5] Nikolski N.K., *Operators, functions, and systems: an easy reading.* **1** Hardy, Hankel, and Toeplitz. **2** Model operators and systems, Math. Surveys and Monographs, 92, 93, AMS, Providence, RI, 2002.

[6] B. Pavlov, *Spectral Analysis of a Dissipative Schrödinger operator in terms of a functional model* in the book : "Partial Differential Equations", ed. by M. Shubin in the series Encycl. Math. Sciences, **65**, Springer, 1995, 87–153.

[7] Duren P.L., *Theory of H^p spaces*, Pure Appl. Math., **38**, Academic Press, New York–London, 1970.

[8] Tikhonov A.S., *On connection between factorizations of weighted Schur function and invariant subspaces.* Operator Theory: Adv. and Appl. **174** (2007), 205–246.

[9] Yakubovich D.V., *Linearly similar model of Sz.-Nagy–Foias type in a domain.* Algebra i Analiz **15** (2003), no. 2, 180–227.

[10] Verduyn Lunel S.M., Yakubovich D.V., *A functional model approach to linear neutral functional differential equations*, Integr. Equ. Oper. Theory **27** (1997), 347–378.

[11] Tikhonov A.S., *Free functional model related to simply-connected domains.* Operator Theory: Adv. and Appl. **154** (2004), 405–415.

[12] Tikhonov A.S., *Functional model and duality of spectral components for operators with continuous spectrum on a curve.* Algebra i Analiz **14** (2002), no. 4, 158–195.

[13] Naboko S.N., *Functional model for perturbation theory and its applications to scattering theory.* Trudy Mat. Inst. Steklov **147** (1980), 86–114.

[14] Makarov N.G., Vasyunin V.I., *A model for noncontraction and stability of the continuous spectrum.* Lect. Notes in Math. **864** (1981), 365–412.

[15] Abrahamse M.B., Douglas R.G., *A class of subnormal operators related to multiply connected domains*, Adv. in Math. **19** (1976), 106–148.

[16] Sarason D.E., *The H^p spaces of an annulus.* Mem. Amer. Math. Soc. **56** (1956).

[17] Bungart L., *On analytic fiber bundles.* 1, Topology **7** (1968), 55–68.

[18] Grauert H., *Analytische Faserungen über holomorph vollständigen Räumen.* Math. Ann. **135** (1958), 263–273.

[19] Pavlov B.S., Fedorov S.I. *Group of shifts and harmonic analysis on Riemann surface of genus one*, Algebra i Analiz **1** (1989), no. 2, 132–168.

[20] Fedorov S.I., *On harmonic analysis in multiply connected domain and character-automorphic Hardy spaces.* Algebra i Analiz **9** (1997), no. 2, 192–239.

[21] Tikhonov A.S., *Transfer functions for "curved" conservative systems.* Operator Theory: Adv. and Appl. **153** (2004), 255–264.

[22] Suetin P.K., *Series of Faber polynomials.* Nauka, Moscow, 1984.

[23] Gaier D., *Lectures on approximation in the complex domain.* Birkhäuser, Basel-Boston, 1980.

[24] Ball J.A., *Operators of class C_{00} over multiply connected domains*, Michigan Math. J. **25** (1978), 183–195.

[25] Tikhonov A.S., *Boundary values of operator-valued functions and trace class perturbations.* Rom. J. of Pure and Appl. Math. **47** (2002), no. 5–6, 761–767.

Alexey Tikhonov
Mathematical department
Taurida National University
4 Yaltinskaya St.
95007 Simferopol
Crimea, Ukraine
e-mail: tikhonov@club.cris.net

MIX
Papier aus verantwortungsvollen Quellen
Paper from responsible sources
FSC® C105338

If you have any concerns about our products,
you can contact us on
ProductSafety@springernature.com

In case Publisher is established outside the EU,
the EU authorized representative is:
Springer Nature Customer Service Center GmbH
Europaplatz 3, 69115 Heidelberg, Germany

Printed by Libri Plureos GmbH
in Hamburg, Germany